ADVANCES IN MOLECULAR GENETICS OF PLANT-MICROBE INTERACTIONS

Current Plant Science and Biotechnology in Agriculture

VOLUME 14

Aims and Scope
The book series is intended for readers ranging from advanced students to senior research scientists and corporate directors interested in acquiring in-depth, state-of-the-art knowledge about research findings and techniques related to plant science and biotechnology. While the subject matter will relate more particularly to agricultural applications, timely topics in basic science and biotechnology will also be explored. Some volumes will report progress in rapidly advancing disciplines through proceedings of symposia and workshops while others will detail fundamental information of an enduring nature that will be referenced repeatedly.

The titles published in this series are listed at the end of this volume.

Advances in Molecular Genetics of Plant-Microbe Interactions

Vol. 2

Proceedings of the 6th International Symposium on Molecular Plant-Microbe Interactions, Seattle, Washington, U.S.A., July 1992

edited by

EUGENE W. NESTER

University of Washington, Seattle, Washington, U.S.A.

and

DESH PAL S. VERMA

The Ohio State University, Columbus, Ohio, U.S.A.

SPRINGER-SCIENCE+BUSINESS MEDIA, B.V.

Library of Congress Cataloging-in-Publication Data

Advances in molecular genetics of plant-microbe interactions / edited by Eugene W. Nester and Desh Pal S. Verma.
p. cm. -- (Current plant science and biotechnology in agriculture ; v. 14)
Papers presented at the Sixth International Symposium on the Molecular Genetics of Plant-Microbe Interactions, held in Seattle, Washington, July 1992.
"Volume 2."
Includes bibliographical references and index.
ISBN 978-90-481-4221-7 ISBN 978-94-017-0651-3 (eBook)
DOI 10.1007/978-94-017-0651-3
1. Plant-microbe relationships--Congresses. 2. Plant molecular genetics--Congresses. 3. Plant molecular biology--Congresses.
I. Nester, Eugene W. II. Verma, D. P. S. (Desh Pal S.), 1944- .
III. Series: Current plant science and biotechnology in agriculture ; 14.
QR351.A38 1993
581.5'2482--dc20 92-35081

ISBN 978-90-481-4221-7

Printed on acid-free paper

PREFACE

Research on the interaction between plants and microbes continues to attract increasing attention, both within the field as well as in the scientific community at large. Many of the major scientific journals have recently reviewed various aspects of the field. Several papers dealing with plant-microbe interactions have been featured on the covers of scientific publications in the past several months, and the lay press have recently presented feature articles of this field. An additional sign of the interest in this field is that the International Society of Molecular Plant-Microbe Interactions has almost 500 members.

This book is a collection of the papers that were given at the Sixth International Symposium on the Molecular Genetics of Plant-Microbe Interactions which was held in Seattle, Washington in July, 1992. Approximately 650 scientists attended and approximately 50 lectures covering the topics of *Agrobacterium*-plant interactions, *Rhizobium*-plant interactions, bacteria-plant interactions, fungal-plant interactions and new aspects of biotechnology were presented. In addition, many sessions were devoted to the plant response to the microbe. Over 400 posters were presented of which the authors of 20 were selected to give an oral presentation. These papers are included in this volume as well. The symposium also included speakers whose research interests are not directly related to plant-microbe interactions but who are at the cutting edge of research areas that impact on the theme of the symposium. These individuals kindly agreed to summarize their talks and their papers are also included.

This volume should prove useful in providing significant advances that have been made over the past two years in this rapidly advancing field. Because the pace of research in this area is so rapid, it may serve as a useful reference work only until the next meeting in 1994.

The Symposium held in Seattle would not have been possible without the generous support of several federal agencies, as well as private corporations and foundations. I am also grateful to Ms. Jane Halsey for her excellent secretarial work during the preparation of this book.

Seattle, Washington
August 1992

Eugene W. Nester, Ph.D.

We gratefully acknowledge the generous grant support provided by the following agencies:

- National Science Foundation Grant IBN 9207289
- United States Department of Agriculture Grant 92-37303-7616
- United States Department of Energy Grant DE-FG06-92ER20084

Corporate, institutional, and organizational support for the symposium was provided by the following sponsors:

- Agritope, Inc.
- Calgene
- CIBA-GEIGY AG Biotechnology
- CIBA-GEIGY BASLE
- Dekalb Plant Genetics
- International Society for Plant Molecular Biology
- Monsanto
- Pioneer Hi-Bred
- Plant Biology Division, The Samuel Roberts Noble Foundation, Inc.
- Rhône-Poulenc
- The American Phytopathological Society
- University of Washington, Graduate School

TABLE OF CONTENTS

Section 3 / *Rhizobium*-Plant Interactions: *Rhizobium* Side

Section 4 / Bacterial-Plant Interactions: Bacterial Side

Section 6 / *Rhizobium*-Plant Interactions: Plant Response

Section 7 / Bacteria-Plant Interactions: Plant Response

Section 8 / Fungal Plant Interactions: Plant Response

Section 1 / Keynote Address

RESEARCH ON PLANT-MICROBE INTERACTIONS: MAKING IT RELEVANT

LUIS SEQUEIRA
Department of Plant Pathology
University of Wisconsin
Madison, Wisconsin, 53706
U.S.A.

ABSTRACT. Significant advances in our understanding of host-microbe interactions have been made in the past 20 yr as a result of the application of the techniques of molecular biology and genetic engineering. In terms of potential applications to agriculture, however, the field has remained perennially promising. Relatively few products are planned for release and corporate interest continues to be limited to a few crops important to Western countries. Biotechnology for Third World countries, where such programs do not exist but severe shortages of food are the rule, is not developing at an adequate pace. I propose the creation of an International Science Foundation, which would focus on biotechnology and the improvement of crops that are important to agricultural development in the Third World. This research must be appropriate to small scale agriculture and must take into account practical and ethical issues related to the limited resources of developing countries.

1. Introduction

The Organizing Committee suggested that the subject of this keynote address should be a discussion of the changing scene in research on plant-microbe interactions. I am sure that they had in mind the fact that I have been in this field for nearly 40 yr. and might have: a) a perspective on how the field has changed and b) a vision as to what the future holds for our area of plant science. Thus, my purpose this evening is to trace the historical events that have brought us to today's exciting and rather lofty place in the field of molecular biology. From this vantage point, the future certainly looks bright in terms of our rapidly increasing understanding as to what transpires in the interaction between plants and microbes. However, I hope to impress upon you this evening my view that research in our field must become more relevant to the needs of a world full of social unrest, ravaged by widespread starvation, unchecked population growth, increasing pollution, and decreasing natural resources.

Let me indicate at the outset that I do not wish to become embroiled in arguments about the relative merits of fundamental versus applied research. It is clear in my mind that the need is for good research, whether fundamental or applied. My plea is for a more efficient transfer of technology so that research findings can be translated more rapidly into better and more productive crops. I believe that the impressive progress that has been made in fundamental areas of our science has not been matched by research on and development of useful products. We have remained a perennially promising field. It is perplexing to contrast the impressive number of research papers

E. W. Nester and D. P. S. Verma (eds.),
Advances in Molecular Genetics of Plant-Microbe Interactions, 3–14.

being published this year in our area of research, as compared with 20 yr ago, with the fact that many private companies have abandoned plant biotechnology because they see a difficult and uncertain future for many of its potential products. In addition, capital for investment in plant biotechnology has dried up, with the result that very few new companies are being formed. Perhaps total investment in plant biotechnology has not declined, but the rate of investment certainly has. Something is amiss with plant biotechnology, and, inevitably, the losers will be the Third World countries that desperately need the products of this industry. A second problem deals with the relevance to the needs of developing countries of: a) the products that are being developed by the very large companies that remain, and b) the basic plant science programs of the USDA and academic institutions. The need to become more relevant, therefore, is the message that I hope to convey to you this evening. The body of this talk is an attempt to support this message, but, first, a bit of history.

2. A Century of Waiting: From Form to Function and Beyond

The field of host-parasite interactions began as a cohesive area of research in the plant sciences soon after Pasteur and Koch established the role of microbes in animal diseases. Interest centered early on three different areas of plant-microbe interactions: a) the delicate relationships between plant cells and obligate parasites, such as rusts, as contrasted with the macerating effects of the more primitive soft-rotting fungi; b) the nature of plant tumors, as exemplified by the crown gall disease, and c) the role of bacteria in nitrogen fixation by legume root nodules.

In the field of plant pathology, the German scientist Anton deBary and his students, including Ward in England, Woronin in Russia, Brefeld in Germany, and Millardet in France, described the morphological changes that occurred in plant cells upon infection. For example, deBary described in great detail the development of haustoria, finger-like projections that penetrate individual plant cells from the intercellular hyphae of rusts, but it was Ward who early in this century openly discussed the possibility of exchange of information across the haustorial wall [11]. deBary also suggested that fungal "ferments" might be responsible for the tissue maceration that is typical of soft-rot diseases, but it was Ward who first demonstrated that the active principle in fungal extracts was heat-labile and must be an enzyme. Thus began in 1910 or thereabouts the era of physiological studies of host-parasite interactions. The emphasis was biochemical but still highly descriptive, focusing mainly on the study of the production and characteristics of cell wall-degrading enzymes secreted by plant pathogens. This type of research continued unabated for 50 yr and resulted in the description of a large number of pathogen enzymes by countless investigators. Unfortunately, none of this work resolved the fundamental issue of the role of these enzymes in virulence.

At the same time that Ward and his successors (Brown in England, Jones in the U.S.) were concerned with the role of enzymes in tissue maceration, Erwin F. Smith had begun investigations on the crown gall bacterium, *Agrobacterium tumefaciens*, which unlike the soft rot pathogens, causes hypertrophy and disorganized growth of host cells. The tumorigenic, semi-symbiotic relationship of host and pathogen intrigued Smith, but in spite of his detailed descriptions of cellular responses of host tissues, he could not determine how the bacterium was able to affect the host. He mistakenly described the bacterium as being inside the host cells and it remained for Riker to show, much to the dismay of Smith, that the bacterium grew outside of the plant cells [17]. The action at a distance of *A. tumefaciens* and the further demonstration by Braun a few years later that sterile secondary tumors were formed in infected plants attracted scores of scientists to study this system for they saw a possible relationship to animal cancer. As a result, during the following half century, investigators produced more "off the wall" explanations for the tumor-inducing properties of *A. tumefaciens* than for any other plant disease in the history of plant pathology. A lot of bad science was involved in the long litany of theories that were advanced to explain the autonomous development of secondary tumors, but the process of transformation could not have been divined without the tools of molecular biology, which did not become available until the late 1970's. A good example of the fact that you cannot run until you can learn to walk!

From the middle of the last century on, there was considerable interest in the peculiar outgrowths or nodules that occurred in legumes, but it was Woronin who observed in 1866 that the nodule tissue was filled with rod-shaped bodies that resembled bacteria. The causal relationship of these bacteria to nodule formation was not recognized at that time, but by 1887 Ward had shown that crushed nodules could be used to inoculate leguminous plants grown in sterilized soil. Soon thereafter, Beijerinck managed to isolate and grow the nodule-inducing bacterium in pure culture. The connection between nitrogen content and the presence of nodules had been established in a series of remarkable experiments by Hellriegel and Willfarth in 1885-1886 [3]. An important fact soon became clear: the bacteria could not fix nitrogen by themselves, but fixation resulted in some way from the symbiotic association of the bacterium with the plant. The unraveling of this symbiosis became the subject of a great deal of descriptive research during the greater part of the next half century. In particular, the advent of the electron microscope after the Second World War allowed extensive and careful scrutiny of the process of nodulation, from the earliest stages of shepherd's crook formation on root hairs to the envelopment of bacteroids within the interior of cortical cells. The nature of the specificity of *Rhizobium* and of other root-associated nitrogen fixing bacteria to their plant hosts was the subject of countless research projects that, without the tools of modern genetics and molecular biology, could not resolve any of the critical issues.

While biologists made little progress on the nature of the process of nodulation, biochemists, on the other hand, were very successful in unraveling the process of nitrogen fixation and in establishing the role of both plant and symbiont at each step [3]. The necessary factors were delineated: nitrogenase, reducing power, ATP, and absence of oxygen. In addition, the details of the path of nitrogen from N_2 to organic forms were worked out in great detail.

3. The Advent of Molecular Biology

This very brief historical account illustrates how three different areas of research in plant-microbe interactions had reached a common end point by the end of the 1960's: the chemistry and biochemistry of some of the components were known, but the biology of the systems remained intractable. In particular, the nature of the interactions could not be elucidated. How did the host know when and how to respond to infection? How did the pathogen or symbiont induce the host to respond? What was the nature of specificity of the interaction? These and many other fundamental questions vexed investigators then and even now, for we still do not have many of the answers. However, the important difference is that we now know how to get there. Molecular genetics has given us the tools to do it. The technological revolution that started in the 1970's with the development of methods for digesting, cloning, and sequencing DNAs has continued to widen our understanding of plant-microbe interactions to a degree that would have been unthinkable 20 yr ago.

My own research interests in the sixties and seventies mirrored what was then the state of the art: the description of changes that occurred in the host in response to infection. In retrospect, however, this was a very inexact science. For instance, one could not distinguish changes that were related from those that were unrelated to the resistant response in plants. One could not determine which of the many products excreted by pathogens or present at their surfaces were important in pathogenesis. While we were busily searching for answers to these questions, disease physiology began to wane as a traditional area of research. Molecular genetics replaced plant physiology as a primary area of interest in most departments of botany, plant pathology, and microbiology [14].

There were many reasons for this shift in approach. The primary one was the perception that after a century of descriptive morphological and physiological approaches there were still no answers to many basic questions about host-parasite interactions. Modern genetic approaches, on the other hand, offered the opportunity to obtain unambiguous answers. Support for physiological research from government and private foundations began to dwindle. Colleges and universities began to hire molecular geneticists, rather than traditional plant physiologists, plant pathologists, or microbiologists, because this new breed of scientists was deemed to be in a position to garner research funds and to initiate programs that were attractive to graduate students. These changes were not only inevitable but essential to the field of plant-microbe interactions as, indeed, they were to all of biology.

Among the most significant advances was the development of methods to obtain single-site mutants of microbes for which systems for gene transfer (conjugation, transformation, transduction, etc.) existed [14]. The availability of transposable elements was crucial. Mutants affected in a specific factor, i.e. nodulation ability, could be selected and the gene of importance could be identified since it was already tagged by the presence of the transposon. Wild-type equivalents could be searched for in a genomic library, the DNA fragments could be cloned in *E. coli* and the limits of the gene could be identified. Eventually, the nucleotide sequence of the gene could be determined and the promoter regions could be delineated. By the early 1980's scientists were able to begin to ask questions as to the expression of genes that were important in plant-microbe interactions and as to the function of these genes. The power of this approach lies in its directness and simplicity.

4. Genetic Engineering of Plants: The Promised Land

An equally revolutionary advancement in the 1980s was the development of methods for plant transformation and for the enhancement of expression of introduced, foreign genes in plants. It is significant that it was the study of a lowly bacterium, *Agrobacterium tumefaciens*, that led to the first practical method for plant transformation. The *Agrobacterium* system remained the mainstay of genetic engineering of plants until the recent advent of biolistics, the high-energy particle propulsion systems used to introduce DNA in plants. These methods are more convenient and have wider applicability than more conventional methods. It should be pointed out that equal in importance to the development of systems for plant transformation was the use of the 35S promoter from cauliflower mosaic virus, which allowed the construction of vectors for high level expression of introduced genes in plants. Again, it was the detailed knowledge of the genome of a plant pathogen, a virus this time, that allowed important advances in genetic engineering of plants.

By the early 1980's, the possibilities of improving crop plants via genetic engineering seemed limitless and optimism was rampant. Genetic engineering firms sprang up almost weekly, venture capital was readily available, and investors were promised untold riches derived from the sale of seed of improved cultivars of the major crop plants. Also, it was thought that advances in basic research would lead rapidly to the identification of genes for resistance in plants, to the transfer of the ability to fix nitrogen from legumes to corn and other crops, to pest resistant plants that would forever eliminate the use of insecticides, etc. Nirvana was at hand!

5. Back to Reality

This great optimism was short lived, however. In spite of overall increases in investment in biotechnology, the last few years have seen the demise of a large number of firms primarily involved in plant genetic engineering. Some of these failures are understandable, given the limited financial support and lack of business experience of many would-be entrepreneurs among our colleagues. Of course, too much financing often can be as much of a problem as too little. More disappointing is the fact that some large companies (ARCO, ENICHEM, SHELL) have given up completely, while others (MONSANTO, GRACE) have substantially reduced their efforts in the plant biotechnology area. Most companies are spending less on R&D today than they were in 1988. This has occurred at about the same time that the first products of genetic engineering are being tested at a reasonably large experimental scale. The reasons for the sudden lack of interest among investors are varied, but most have to do with a regulatory environment that for a long time seemed to be increasingly oppressive and confusing. Although the regulatory environment in 1992 has changed significantly for the better, there has not been a concomitant improvement in the investment climate. In addition, the cold economic fact is that growers are not likely to pay the exceedingly high costs of research and development for products that may have only a relatively modest impact on production of the major crops in this country. All the figures I have seen indicate that it would take at least a 10% increase in yield per acre of corn, soybeans, or cotton to induce farmers to pay a substantially higher price for improved seed, better inoculants, or more effective biocontrol agents. Although there are possible trade-offs (i.e. less money spent on chemicals), there is less elasticity in marketing new products today than in the past.

The demise of some biotechnology companies is a reflection of the fact that there is as yet not a single, effective product of plant biotechnology in the market after 10 or more years of continued effort. The failure to deliver improved cultivars, however, does not affect growers in the developed nations to any large extent. Most of the major crops that may have been improved by biotechnology, such as corn, soybeans, cotton, etc. are being harvested in record amounts and surpluses have clogged the markets and depressed prices for many years.

We seem to have forgotten the crops that are important to developing countries, where biotechnology programs do not exist but severe shortages of food are the rule. There is an important moral issue here and our lack of concern for the scientific support that these countries need will come back to haunt us. The difficulties that we have today in coping with hordes of boat people from Vietnam or Haiti, who are fleeing starvation and are arriving at our coasts in ever increasing numbers, is but the beginning: a minuscule search party for the huge, unstoppable army of impoverished, hungry invaders that is upon us. The best way to keep this depressing vision of the future from becoming reality is to strengthen the agricultural economy of these developing countries. Let me impress upon you the fact that inefficient agriculture is not a problem restricted to developing countries: the demise of the Soviet Union, a former superpower, can be traced to its inability to maintain an adequate food supply.

No one can deny, however, that the greatest needs for the products of biotechnology are in developing countries [4,15]. The enormity of the task ahead was illustrated well by M. S. Swaminathan who pointed out that to provide a balanced diet for the 8 billion human beings that will populate the world in the year 2020, "we have to duplicate what has been achieved on the production front during the past 12,000 years" [1] Yet, developing countries have neither the resources nor the intellectual infrastructure to cope with development of their own biotechnology programs. Help is on the way, but on a limited scale. For example, at the annual AAAS meeting earlier this year, Robert Fraley of Monsanto predicted that insect and disease resistant rice would be commercially introduced into some developing nations "by the middle to end of this decade" [13]. Similarly, after Roger Beachey and scientists at Monsanto discovered five years ago that expressing the coat protein gene of a virus in a transformed plant would lead to resistance against viral diseases, resistance has been introduced into tomato, alfalfa, tobacco, melons, potato, and even sweet potato, yams, and cassava [13]. These are all crops of importance in developing countries. In addition, genetic engineering has provided some cotton cultivars with the ability to express *Bacillus thuringiensis* toxins, thereby providing selective control against caterpillars that damage cotton everywhere. As a result of these developments, Ann S. Moffat concluded in a recent article in SCIENCE that biotechnology is finally "reaching beyond the high-tech West"[13].

Another encouraging sign is the increasing emphasis on plant molecular biology at international research centers. For example, the United Nations (UNIDO) is helping India with the establishment of a biotechnology center in New Delhi. Similarly, the Consultative Group on International Agricultural Research (a consortium of U.S. charitable foundations and international development organizations) is providing increased funding for biotechnological approaches in international centers dealing with major crops, such as potato (in Peru), cassava (in Colombia), and rice (in the Philippines) [15]. In particular, the Rockefeller Foundation is providing broad base support for rice biotechnology worldwide. This is encouraging, of course, but my feeling is that the present effort is small in proportion to the enormity of the task ahead.

I am fully cognizant of the fact that the primary problem in developing countries is unchecked population growth. Given that photosynthesis is a finite process energetically and that the total acreage of arable land is decreasing, one cannot pretend that, in the face of unchecked population growth, improved agriculture alone holds the answer to the betterment of humankind. Birth control is desperately needed worldwide, but this is a problem that must be addressed by world health organizations. We, as agricultural scientists, can do very little about it. Rather, we must dedicate our efforts to feed those who are already here and to attempt to provide for the many more millions of children that demographers tell us will be here before the turn of the century. All that one can hope is to buy some time until, perhaps by the year 2000, more practical methods of contraception will be available and common sense will finally prevail over politics in the solution of population control problems worldwide.

6. The International Science Foundation: An Idea Whose Time Has Come

It is evident that the products of plant biotechnology could help alleviate the problems of quality and quantity of food in developing countries. The dilemma is that these needs are increasing exponentially while there is now decreasing interest in corporate America in the products of plant biotechnology and, particularly, in those destined for markets outside of the U.S. and Europe. Furthermore, many of the products that are being developed are not pertinent to the needs of the Third World. To compound the problem, basic research at academic institutions in this country is focused on crops and products that are of little relevance to the needs of developing countries. Therefore, how can we fulfill our responsibilities to the coming millions of deprived children whose future may be one of misery and starvation?

I propose that one solution is the establishment by Western countries of an International Science Foundation (ISF), a functional equivalent of the National Science Foundation. While its main mission would be to support basic and applied research of international significance, the primary objective should be to improve the way research is structured and administered in the Third World. Although its mission should cover science and technology in general, the ISF could deal initially with biotechnology and all related sciences that are important to agricultural development in the Third World.

It was about 20 yr ago that I first discussed the possibility of an International Science Foundation with Dr. John Niederhauser, formerly on the staff of the Rockefeller Foundation and a leader in the development of national programs to improve potato production in Latin America, Asia, and Africa. I believe that there is a lot of merit in his idea. Economics rule the world of research at corporations and universities alike. With only a relatively modest budget, ISF could direct research along certain lines, both basic and applied, that deal with immediate agricultural problems that can be resolved by biotechnology. To my mind, the Foundation should establish the program areas, but the research would be investigator-driven, as it is now at NSF. Experience shows that one cannot direct research from above; investigators have to tell research administrators what is important and what is irrelevant. Program areas could be focused on important agricultural problems of the Third World, and grants could be awarded on a competitive basis to scientists who are willing to tackle those problems from fundamental as well as applied standpoints. Grants could be made available to scientists working for corporations, government agencies, and academic institutions. The only important requirement would be that the research projects establish proper collaborative agreements with researchers in developing countries. The emphasis should be on facilitating the return home of researchers from developing countries who have been trained in the United States, Europe, and Australia. Such programs already exist and are being financed by the Rockefeller Foundation, the International Foundation for Science, and others, but the number of scientists being supported is very small in relation to the needs of developing countries [16].

I believe that most scientists involved in plant-oriented research would be willing to dedicate at least part of their research efforts to help resolve crop production problems in developing nations. All that is needed is adequate funding and a strong administrative structure. Most of us want to find some application for the results of our research on host-parasite interactions. The emphasis should be on continuous collaborative efforts rather than on short visits by Western scientists to carry out demonstrations or collect data. We have an outstanding example of good international collaboration in our colleague, Roger Beachey, who has made great efforts to extend the benefits of his work on resistance to plant viruses to Third World countries, where these diseases decimate many of the important food crops [12].

7. Plant Biotechnology for the Third World: Practical and Ethical Issues

In recommending that more research in plant biotechnology be directed to developing countries, I am aware that there are practical as well as ethical issues that have to be considered. The practical issues are of two different types: a) the appropriateness of biotechnology to small scale agriculture, and b) the lack of resources, both structural and intellectual, in much of the developing world. Let me address both of these issues.

7.1 BIOTECHNOLOGY IN SMALL SCALE AGRICULTURE

This first issue has to do with the vulnerability of the small scale farmers, people who live on the edge of survival. As practiced today, biotechnology is oriented to the needs of large scale agriculture; thus, the gap between rich and poor farmers is expected to increase. The new technological advances may have a negative impact on the small-scale farmers if, for example, the cost of the improved seed is beyond their means. In a recent book, "Appropriate Biotechnology in Small Scale Agriculture", Joske Bunders and Jacqueline Broerse [2] have addressed the problem of how to reorient research and development in the West for the benefit of the poor farmers of the developing countries. Let me attempt to summarize some of their major findings:

a) The range of technologies that are appropriate in a given small scale farming situation is limited because of the prevailing economic, political, and infrastructural environment. The technologies deemed to be most useful, in order of complexity, involve: fermentation, microbial biocontrol agents, plant cell and tissue culture, enzyme technology, embryo transfer, protoplast fusion, monoclonal antibodies, and recombinant DNA. It will not have escaped most of you that rDNA technologies are listed last, although the improvement of microbes, either for controlled fermentations or for use in preventing frost or pathogen/pest damage, can be achieved most easily by rDNA techniques that are now commonplace.

b) Although small-scale agriculture requires immediate, appropriate solutions to many problems, the accelerating pace of biotechnology does not allow sufficient time to anticipate any possible negative consequences. One has to be mindful of cultivars of wheat and rice, products of the Green Revolution of 30 years ago, that did wonders for increasing yields of these crops in developing countries, but required large amounts of chemical fertilizer. Thus these cultivars effectively made farmers more dependent on the West, promoted intensive farming that displaced many small farmers, and resulted in reduced cultivation of the more traditional crops of the region. In these days of high energy costs, it is clear that small farmers are not well served by high input agriculture.

c) Governments, research institutes, donor organizations and other providers of support must incorporate the research priorities of small farmers into their biotechnology policies. Unfortunately, those who make the decisions (scientists, industrialists, and civil servants) are usually the least familiar with the specific problems and needs of the small-scale farmers. There must be an interactive "bottom-up approach" that establishes priorities of biotechnological research and incorporates these priorities into research agenda. The success of any innovation is dependent on coalition building, that is, on the development of a network of backers who are willing to provide resources (funds, materials, space, time, etc.).

7.2 LACK OF RESOURCES IN DEVELOPING COUNTRIES

The Third World has a huge store of genetic information in the form of tremendous diversity of plants and animals but lacks the infrastructure to exploit these resources. To be able to exploit this genetic information before it disappears, Third World countries have to reshape their governmental and academic institutions so that they are dedicated to creative research. As I see it, Third World leaders are worried about foreign debt and social services, but have no conception of the importance of science and technology as a prime mover of the area's economy. No one has made this point more clearly than Daniel Goldstein in a series of articles in the Journal of Agricultural Ethics in 1989 [7,8]. To quote Goldstein, "The isolation of useful genes cannot be made without strong schools of molecular plant biology, plant physiology, phytochemistry, and physiological ecology. The development of strong capability in modern botanical sciences is an absolute strategic priority for the Third World". And further, "The international organizations should help to make science understood by the politicians and administrators of the peripheral countries and to introduce new standards of behavior and management in their scientific establishments".

Clearly, a case must be made for the development of biotechnological capabilities in developing countries. One cannot limit the contributions of these countries to the mere testing of the products of the West. Some think that the Third World should concentrate on applied research, but this is a role that most scientists in these countries reject. The scientific community in developed nations insists that the plant genetic information in the developing countries belongs to the world at large. Yet, once that information travels to the West it is packaged in the form of patented products or improved seed that are for sale at prices that farmers in Third World countries often cannot afford. This is an impossible, unjust situation that can only be resolved by developing strong, biotechnological centers in the Third World that can focus activities on the needs of the region. Yet, the degree of official neglect, political corruption, and lack of perception as to what science and technology can do, make the creation of these centers almost an impossibility in many countries. The existence of an institution like ISF, however, could provide the impetus for creative research that is dependent upon intimate collaboration between scientists from developed and underdeveloped nations.

One can predict that genetic engineering will have an impact in the agriculture of Third World countries but only as a powerful adjunct to conventional plant improvement practices. The new technology offers the greatest advantage when it deals with crops that are difficult to manipulate by conventional means. In addition, it provides the ability to transfer genes into plants from any source in the whole of the biological world, thus circumventing the natural breeding barriers. While there is nothing inherently dangerous in this process, the choice of crops to be improved and of the genes to be transferred raises ethical issues that must be considered. These have been widely publicized, but it may be useful to give a few examples at this juncture.

7.3 SOME ETHICAL ISSUES

The most commonly expressed concern is that the same economically powerful corporations that control the pesticide and chemical weed killer markets also own the companies that provide the improved seed. It is no secret that the large agricultural chemical companies have moved massively into the seed sector. There are large profits to be made from the sale of improved seed and these companies are the logical conduit for the delivery of the products of genetic engineering to the public. In his book, "New Hope or False Promise? Biotechnology and Third World Agriculture", Henk Hobbelink makes the point that some 15 international corporations now own most of the international seed trade [10]. Although his book is couched in extreme terms, Hobbelink does bring out important problems related to the direction of biotechnological research that is under the control of multinational agricultural corporations. He points out that the double-barrelled control of both seed and chemicals by a few, large corporations creates a rather untenable dependency of the small farmer on products of the West. The same corporation provides not only the seed but also the chemicals that make it possible for the farmer to use the seed.

For example, several years ago scientists at the MONSANTO CORPORATION discovered that resistance to the widely used herbicide, ROUNDUP or glyphosate, could be obtained by deliberately overexpressing the enzyme, 5-enolpyruvylshikimate 3-phosphate (EPSP) synthase in plants [5]. A petunia EPSP synthase and a transit peptide were engineered to be expressed under the 35S promoter. The transgenic plants had a remarkable degree of tolerance to glyphosate. Herbicide tolerant plants, therefore, may allow the use of ROUNDUP as a postemergence herbicide. Because weeds are extremely difficult to control in the tropics, it is in underdeveloped countries that this improved seed may find its widest applications. This achievement is overshadowed, however, by the notion that the improved seed requires the use of ROUNDUP, and that both are products of the MONSANTO CORPORATION.

A second ethical concern is the fact that biotechnology has caused a transfer to the developed nations of products that are traditional in Third World countries. One example is sugar production [10]. It is estimated that over 50 million workers in the tropics derive their sustenance from the harvesting, milling and processing of sugarcane. You are all aware that sugar prices have been depressed for years due to overproduction. This world overproduction has been exacerbated by government support of the sugarbeet industry in the United States as well as in Europe. In addition, sugar-exporting nations are now being affected by a double whammy. First was the increased use of high fructose corn syrup, as a result of genetic improvements of bacteria that produce the enzyme glucose isomerase in high yields. This opened the door for the production of high fructose corn syrup, which is now used widely by the soft drink industry. Second was the introduction of artificial sweeteners, such as ASPARTAME (produced by SEARLE, now part of MONSANTO) which already have a market of over one billion dollars in the U.S. alone. Thus an important product of the tropics, or its equivalent, is now being produced in factories in the West, a fact that has had dramatic negative consequences on the economies of countries that are largely dependent on sugar exports. You may argue that these are inevitable technological advances, but can we really justify biotechnology focused on products, such as sugar, which are already produced in excess of world needs, without consideration of the social consequences?

A third ethical problem concerns the possibility that improved cultivars will lead to genetic uniformity over large areas, thus creating the possibility of widespread epidemics. For example, Hobbelink [10] points out that the UNILEVER CORPORATION, which controls about one third of the world market for vegetable oils, has developed techniques for micropropagation of the best clones of the African oil palm, the main crop and source of export revenues for Malaysia and several other countries. It is estimated that these few clones can increase world vegetable oil production by about 30%. Thus, UNILEVER is likely to increase its domination of the world's vegetable oil market, with negative effects on countries that depend on vegetable oil exports to maintain their economies. For example, the Philippines (coconut oil) and Senegal (peanut oil). More potentially damaging, however, is the high degree of uniformity that will be created in a crop that heretofore had been planted from seed.

These are but a few examples of technological developments that may affect the Third World negatively because these countries have not been brought into the total picture. Unlike prior revolutions in agriculture, the biotechnological revolution is very much in the hands of the private sector. Companies exist to generate profits for their stockholders and one cannot blame them for seizing opportunities to dominate the large markets of western countries. The technological know-how of the private sector, however, is of vital importance to developing nations where commercial opportunities are less rewarding than in other parts of the world. As Robert Goodman points out, dialogue on ways to resolve this dilemma is critically needed [9].

To conclude on a positive note, one must remember that the biotechnological revolution is knowledge-intensive and that most of this basic knowledge is generated at public institutions. Thus, the public sector can define the goals and priorities of biotechnology research, to make certain that the needs of impoverished nations are taken into account. Academia in this country has an opportunity to reorient its biotechnology research and development efforts in the direction of the most needy in the world. We cannot shy away from this responsibility to the developing nations: the truth is that we cannot survive unless they do.

8. Acknowledgments

My sincere appreciation to Dr. Robert M. Goodman and Dr. Caitilyn Allen for their help in the preparation of this article.

9. Literature Cited

1. Bilger, B. 1992. Dinner for eight billion. Earthwatch, March/April, 1992. pp. 13-15.
2. Bunders, J. F. G., and J. E. W. Broerse, eds. 1991. Appropriate Biotechnology in Small Scale Agriculture. Univ. Arizona Press, Tucson. 160 pp.
3. Burris, R. H. 1988. 100 years of discoveries in biological N_2 fixation. pp. 21-30 in Proc. 7th. Intl. Congr. on Nitrogen Fixation, H. Bothe, F. J. deBruijn, and W. E. Newton, eds. Gustav Fischer, Stuttgart. 878 pp.
4. Farrington, J. (ed.). 1989. Agricultural Biotechnology: Prospects for the Third World. Overseas Development Institute, London.
5. Gasser, C. S. and R. T. Fraley. 1989. Genetically engineering plants for crop improvement. Science 244:1293-1299.
6. Gibbons, A. 1990. Biotechnology takes root in the Third World. Science 248:962-963.
7. Goldstein, D. J. 1989. Ethical and political problems in Third World biotechnology. J. Agr. Ethics 2:5-36.
8. Goldstein, D. J. 1989. A biotechnological agenda for the Third World. J. Agr. Ethics 2:37-51.
9. Goodman, R. M. 1986. New technology and its role in enhancing global food production. Federation Proc. 45:2432-2437.
10. Hobbelink, H. 1987. Biotechnology: New Hope or False Promise? Intl. Coalition for Development Action, Brussels. 72 pp.
11. Keitt, G. W. 1959. History of plant pathology. pp. 61-97, in Plant Pathology, an Advanced Treatise, Vol.1, J. G. Horsfall and A. E. Dimond, eds. Academic Press, N. Y.
12. Moffet, A. S. 1992. Putting the moves on plant viruses. Science 255:291.
13. Moffet, A. S. 1992. Biotechnology reaches beyond the high-tech West. Science 255:919.
14. Sequeira, L. 1988. On becoming a plant pathologist: the changing scene. Annu. Rev. Phytopathol. 26:1-13.
15. Swaminathan, M. S. Biotechnology research and Third World agriculture. Science 218:467-472.
16. Toenniessen, G. 1992. Building plant science research capacity in developing countries. Plant Cell 4:5-6.
17. Williams, P. H. and M. Marosy (eds.) 1986. With One Foot in the Furrow, Kendall/Hunt, Dubuque, Iowa. 471 pp.

CELL-CELL INTERACTIONS DURING POLLINATION IN BRASSICA

J. B. NASRALLAH and M. E. NASRALLAH
Section of Plant Biology
Division of Biological Sciences
Cornell University
Ithaca, New York 14853

ABSTRACT. Self-incompatibility is a highly specific, genetically controlled recognition phenomenon that operates during the plant reproductive phase. This phenomenon is based on the ability of the female reproductive structure, the pistil, to distinguish between self (genetically related) and non-self (genetically unrelated) pollen. Recent developments in the molecular and genetic analysis of the self-incompatibility system of *Brassica* are described, and possible analogies to host-pathogen interactions are discussed.

Introduction

The male gametophyte (pollen and pollen tube) is a unique and highly specialized plant cell system whose function in fertilization is dependent on its ability to undergo directed growth from the surface of the pistil, through the transmitting tissue of the style and ovary. Pollen tubes extend via tip growth from the pollen grain body. The result of tip growth is a tubular cell that has a uniform diameter but can be several centimeters long in certain species such as maize and lilies. The fact that a pollen tube has to traverse and literally invade the somatic tissues of the pistil has created opportunity for the evolution of incompatibility reactions between the growing pollen tubes and cells of the pistil. Elaborate genetic systems have evolved that exploit such reactions and allow plants to avoid self-fertilization and the concomitant deleterious effects of inbreeding. In many flowering plants including crucifers, the genetic barrier to self-pollination, known as self-incompatibility, is under the control of a single multiallelic Mendelian locus, the S locus. Self-incompatibility (SI) involves highly specific interactions triggered by the activity of identical S "alleles" in pollen and pistil with subsequent inhibition of pollen-tube growth.

It has not escaped the attention of biologists that parallels may exist between self/non-self recognition phenomena that operate in pollen-pistil interactions and host-pathogen interactions that conform to the gene-for-gene hypothesis (for a review, see Hodgkin et al. 1988). It is the purpose of this paper to summarize current knowledge of the SI system of the crucifer *Brassica*. It is hoped that a consideration of the manner in which plant cells communicate with other plant cells during the reproductive phase of the plant life cycle might provide a useful backdrop for the study of plant-pathogen interactions.

E. W. Nester and D. P. S. Verma (eds.),
Advances in Molecular Genetics of Plant-Microbe Interactions, 15–21.

The Genetic and Cellular Basis of Self-Incompatibility in Brassica

The study of SI in crucifers began in earnest in the early 1950s with Bateman's analysis of its inheritance in *Iberis amara* (Bateman, 1954). His model of single multiallelic locus control, sporophytic determination of pollen reaction, with codominant and/or dominant allelic interactions, is widely accepted (for review, see Nasrallah and Nasrallah, 1989). Briefly, an incompatible response occurs when stigma and pollen express the same "*S* allele". Estimates or counts of the number of alleles at the *S* locus are: 22 in *Iberis*, 34 in *Raphanus* and 60 in *Brassica oleracea*. At the cytological level, SI in crucifers is manifested by the arrest of pollen and pollen-tube development at the stigma surface. In a compatible pollination, pollen hydrates, germinates, and the emerging tube grows into the wall of the papillar cells that line the surface of the stigma. In contrast, in an incompatible pollination, pollen fails to germinate at the stigmatic surface and the emerging pollen tube fails to invade the papillar cell wall.

Molecular Genetic Analysis of Self-Incompatibility

Molecular genetic analyses carried out in our laboratory have demonstrated so far that two transcribed genes map to the genetically defined *S* locus. The analysis of restriction fragment length polymorphisms in F_2 populations segregating for different SI genotypes has demonstrated that these two genes are not genetically separable from each other nor from SI phenotype (Stein et al. 1991), and thus behave as though only a single locus existed. We have shown by pulsed-field gel electrophoresis (Boyes and Nasrallah, submitted) that the two genes are physically linked and are separated by ~200 kilobases of DNA. To accomodate the new data, we have adopted the term "*S* haplotype" to replace the classical "*S* allele" designation for genetic constitution at the *S*-locus region.

The two *S*-locus linked genes are related members of the *S* superfamily of genes (Nasrallah et al. 1988) which also includes genes (designated *S*-*L*ocus *R*elated or *SLR* genes) not linked to the *S* locus (reviewed in Nasrallah et al. 1991). One of the two genes is the *S*-*L*ocus *G*lycoprotein (*SLG*) gene which encodes a secreted glycoprotein. The other gene is the *S*-locus *R*eceptor *K*inase (*SRK*) gene which is predicted to encode a single-pass transmembrane protein kinase the extracellular domain of which shares extensive sequence similarity with *SLG* (Stein et al. 1991). Our analysis of *SLG*/*SRK* gene pairs isolated from several different *S* haplotypes in two different species of *Brassica* has demonstrated that *SLG* and the extracellular *S* domain of *SRK* share approximately 90% amino-acid sequence identity within a haplotype, and only 67%-80% sequence identity between haplotypes (Stein et al. 1991). These observations suggest that the two genes have evolved in concert, perhaps by a mechanism involving gene conversion. Homogenizing gene conversion events have been postulated to explain the coevolution of genes in several systems (e.g. Gally and Edelman, 1992).

The structure of the *SRK* gene is similar to that predicted from the sequence of a recently described vegetatively expressed maize root cDNA clone, *ZmPK1* (Walker and Zhang, 1990). While both *SRK* and *ZmPK1* predict proteins similar to the growth factor receptor tyrosine kinases in animals (Yarden and Ullrich, 1988), the sequence of their predicted catalytic domains is more similar to that of the catalytic domains of protein serine/threonine kinases. In addition to the genotype-specific polymorphisms and the demonstration of genetic linkage of the *SLG* and *SRK* genes to the *S* locus --shown for *SLG* and *SRK* in our laboratory and for *SLG* by K. Hinata's laboratory in Japan (Hinata and Nishio, 1978; Nishio and Hinata, 1982), several lines of evidence implicate these genes in the SI response:

1) *SLG* and *SRK* genes derived from different *S* haplotypes exhibit a high degree of sequence polymorphism (Nasrallah et al. 1987; Stein et al. 1991) as expected for genes involved in specific pollen recognition. An example of *S*-associated polymorphism is provided by the comparison of the S_2 and S_6 haplotypes in which the SLG_2/SRK_2 gene pair showed only 70% amino-acid sequence identity to the SLG_6/SRK_6 gene pair (Stein et al. 1991). In contrast, members of the *S* gene family that are not linked to the *S* locus are highly conserved even in different *Brassica* species and in different cruciferous genera.

2) *SLG* and *SRK* are expressed in pistils and anthers, consistent with models of SI in which both pollen and stigma bear determinants of recognition derived from the *S* locus. The *SLG* gene, which has been characterized in detail, is expressed exclusively in reproductive structures as demonstrated by genetic ablation experiments in which transgenic plants expressing a chimeric gene consisting of the *SLG* promoter fused to the subunit A protein of diptheria toxin (Thorsness et al. 1991). In the pistil, *SLG* expression is developmentally regulated in correlation with the onset of SI (Nasrallah et al. 1988). The stigmas of immature buds are self-compatible and contain only very low levels of the *SLG* gene product. Stigmas acquire the capacity to discriminate between self-pollen (i.e. pollen expressing an *S* haplotype identical to that expressed in the stigma) and cross-pollen in buds at one day before flower opening, and these stigmas contain maximal levels of *SLG* RNA and SLG glycoprotein. Furthermore, *SLG* is expressed predominantly in the surface papillar cells of the stigma (Nasrallah et al. 1988; Sato et al. 1991) and its glycoprotein product accumulates to high levels in the walls of these cells (Kandasamy et al. 1989). This localization is consistent with cytological observations that demonstrate the arrest of pollen/pollen-tube development at the stigma surface. In anthers, the *SLG* gene is expressed sporophytically in the tapetum and gametophytically in the developing microspores (Sato et al. 1991). Sporophytic expression is predicted from the sporophytic control of SI phenotype in *Brassica* pollen.

3) The introduction of a cloned *SLG* gene into a self-incompatible strain of *B. oleracea* can result in the modification of SI phenotype (Toriyama et al. 1991). In these experiments, a self-compatible transgenic phenotype was generated which was associated with the down-regulation of the resident *SLG* gene, in a manner similar to the recently observed but poorly understood phenomenon of "sense" suppression (e.g. Napoli et al. 1990). This down-regulation was reflected in a drastic reduction in the amount of *SLG* glycoprotein product but did not affect the *SRK*, *SLR1* and *SLR2* genes (unpublished).

The Genetic Breakdown of Self-Incompatibility

One approach to understanding the function of *S*-locus linked genes is to analyze mutations that affect SI. Of these, the most easily identifiable mutants are self-compatible (SC) mutants in which self-pollen grains escape from inhibition and produce pollen tubes, thus resulting in the successful fertilization of ovules and seed production. In the two predominantly self-incompatible species *B. oleracea* and *B. campestris*, variants which are perfectly self-fertile have been observed. In the literature, these SC variants have been attributed to mutations in "modifier" genes that are unlinked to the *S* locus (Nasrallah, 1974; Thompson and Taylor, 1971; Hinata et al. 1983). The molecular basis of these mutations is not known except for a variant *B. oleracea* var. *capitata* (cabbage) strain in which self-compatibility was associated with a reduction in *S*-locus specific (SLG) antigen in the stigma (Nasrallah, 1974). More recently, we identified a mutation in *B. campestris* that also resulted in drastically reduced levels of stigma SLG and in the loss of the incompatibility response in the stigma but not in pollen (Nasrallah et al. 1992). Genetic analysis of this strain has identified a locus unlinked to the *S* locus, designated *SCF1* that is required for the normal developmental induction of SI and regulates *in trans* *S*-locus function. The SC strain carries a single recessive mutation at the *SCF1* locus which down-regulates at the RNA level the *SLG* gene (linked to the *S* locus) as well as the *SLR1* and *SLR2* genes (unlinked to the *S* locus). However, the expression of the *S*-locus linked *SRK* gene is unaffected in *scf1scf1* homozygotes. The *SLG, SLR1* and *SLR2* genes encode secreted glycoproteins that are present only at very low levels in young immature stigmas prior to the onset of self-incompatibility and are abundantly expressed in mature self-incompatible stigmas. The simultaneous down-regulation of these genes and breakdown of SI in *scf1scf1* homozygotes supports the involvement of these *S* gene family members in the pollen-stigma interaction of SI. We postulate that the mutation disrupts a regulatory gene that encodes a positive *trans*-acting factor required for high-level stigmatic transcription of the *SLG, SLR1* and *SLR2* genes.

A Model of S-Gene Action

The structure of the *SRK* gene suggests that the mechanism of SI in *Brassica* is different from the RNase-based mechanism proposed for SI in the Solanaceae (McClure et al. 1989). Rather, pollen-pistil interactions in the Brassicaceae appear to be similar to the animal signalling systems that are mediated by ligand-activated receptor tyrosine kinases (Ullrich and Schlessinger, 1990; Yarden and Ullrich, 1988). Ligand binding to the extracellular domain of these receptors results in receptor oligomerization and the subsequent activation of the kinase domain. The ligand-binding domain acts as a negative regulator of the catalytic activity, and the ability to phosphorylate is essential to receptor signalling, since mutations that destroy kinase activity also destroy receptor function. Activation of the receptor and inter-receptor transphosphorylation is followed by the phosphorylation of specific target substrates, resulting in a phosphorylation cascade that ultimately leads to a cellular response.

We propose that the SRK protein kinase is activated by contact between a papillar cell and self-pollen. By phosphorylating intracellular substrates, the SRK protein would couple the initial molecular recognition events at the papillar cell-pollen interface to the signal transduction chain that leads to pollen rejection. The apparent co-evolution of *SLG* and *SRK* implies a functional interaction between their products (Stein et al. 1991). A possible model of SI is that the *SRK* and *SLG* gene products act in concert, perhaps as components of a ligand-binding complex or as unassociated competitors for this binding. The allelic specificity of the SI response might involve the allele-specific homophilic binding of the SLG and SRK proteins. Alternatively "self" recognition may result from the allele-specific binding of an as yet unidentified ligand. More generally, a signalling mechanism based on the interplay of transmembrane receptors and secreted forms of the extracellular "recognition" domain of the receptor may be a consequence of the presence of the plant cell wall. The secreted glycoproteins, being 50-60 kilodaltons (kd) in size and therefore freely diffusible within the wall [the cell wall constitutes no barrier to molecules under 80 kd (Philip Low, personal communication)] , would bridge the distance between the external milieu and the membrane. This process would allow membrane-based signalling in plants, the occurrence of which had been thought problematic on account of the cell wall.

This model of SI allows specific predictions relating to the operation of the *Brassica* SI system. One prediction is that the activity of both *SLG* and *SRK* and the production of adequate levels of their functional protein products are required. A second prediction is that a high degree of sequence identity must be maintained between the extracellular domain of the *SRK* gene and the *SLG* gene (i.e. the *SLG* and *SRK* genes must be "matched"). Thus, mutations that affect the expression or function of either or both genes, and mutations or recombination events that lead to a "mismatch" between the two genes would result in the breakdown of SI. The correlation, discussed in the previous section, between a reduction in the SLG glycoprotein product and the breakdown of SI in *B. campestris scf1scf1* plants and in the transgene-induced *B. oleracea* self-compatible plants demonstrates a functional requirement for *SLG* in the operation of SI, and is consistent with the proposed model. In addition, the observation that the expression of the *SRK* gene is unaffected in these self-compatible strains is consistent with the prediction that the down-regulation of just one of the two *S*-locus genes is sufficient to disrupt self-pollen recognition. Regulatory *cis-* or *trans-*acting mutations that down-regulate the *SRK* gene and structural mutations that inactivate its protein product are therefore expected to lead to the breakdown of SI.

Similarities to Host-Pathogen Interactions?

Several parallels have been drawn between pollen-pistil interactions and host-pathogen interactions (Hodgkin et al. 1988). Similarities have been noted between the growth of a fungal pathogen and a pollen tube: both are characterized by tip growth and both require the activity of degrading enzymes that allow the extending cell to breach the plant cell wall. In addition, both types of interactions can involve a rapid response: the inhibition of self-pollen development in crucifers is manifested within minutes of pollen contact at the surface of the stigma, and the pathogen-induced transcription of pathogenesis-related protein genes has been observed within minutes of pathogen attack in some systems (Somssich et al. 1986). A common visible response in both incompatible pollinations and incompatible host-pathogen interactions is the deposition of callose at the site of the cellular interaction -- at the surface of stigma cells in SI, and of the plant cell in host-pathogen interactions. Despite these similar cytological manifestations of incompatible interactions in the two systems, the induced responses exhibit significant differences. In SI, cell necrosis which is characteristic of host-pathogen interactions does not occur. In addition, the phenomenon of cross-protection by which an incompatible host-pathogen interaction can affect the outcome of a compatible interaction is not a characteristic of pollen-pistil interactions, in which the independent nature of the interaction between an individual pollen grain and the papillar cell is the rule. Finally, while phytoalexins have been shown to inhibit pollen germination and pollen-tube growth, and low molecular weight inhibitors of pollen germination have been detected in self-pollinated stigmas of *Brassica* (reviewed in Hodgkin et al. 1988), such molecules have not been characterized nor have they been demonstrated to play a role in pollination. In fact, the induction of general inhibitory compounds such as occurs in host-pathogen response would be counterproductive in the case of pollen-pistil interactions since subsequent pollination with compatible pollen would be inhibited as well.

Perhaps the most significant parallels between SI, host-pathogen interactions, and biological recognition phenomena in general, lie in the genetic control of specificity in these interactions. Both SI and host-pathogen interactions are described as being based on one gene-to-one gene interactions and are characterized by a high degree of polymorphism. In the SI system of *Brassica*, one locus determines the specificity of the interaction, but several unlinked loci are required for the operation of the primary SI locus and affect either the expression of the S-locus genes or steps downstream of the initial recognition events. A similar situation might occur in host-pathogen interactions where an incompatible interaction may require the operation of several genes in addition to the primary resistance locus. A further parallel emerges from a consideration of the structure of the *Brassica* S locus and of certain disease resistance loci, and the emerging evidence that these loci have a complex organization (Pryor, 1987). As stated by Dangl (1992) in his comparison of host-pathogen interactions and immune recognition by the major histocompatibility complex, the concept of "multiple polymorphic loci encoded in a complex, and giving rise to one concerted function of recognition" may prove to be the most useful model by which to evaluate various specific recognition systems.

REFERENCES

Bateman, A.J. (1955) 'Self-incompatibility systems in angiosperms. III. Cruciferae', Heredity 9, 52-68.

Dangl, J.L. (1992) 'The major histocompatibility complex a la carte: are there analogies to plant disease resistance genes on the menu?', Plant Journal 2, 3-11.

Gally, J.A. and Edelman, G.M. (1992) 'Evidence for gene conversion in genes for cell-adhesion molecules', Proc. Natl. Acad. Sci. USA 89, 3276-3279.

Hinata, K. and Nishio, T. (1978) 'Stigma proteins in self-incompatible *Brassica campestris* L. and self-incompatible relatives, with special reference to S-allele specificity', Jpn. J. Genet. 53, 27-33.

Hinata , K., Okasaki, K. and Nishio, T. (1983) 'Gene analysis of self-incompatibility in *Brassica campestris* var. Yellow Sarson (a case of recessive epistatic modifier)', In Proceedings 6th International Rapeseed Conference, vol. I, Paris. pp. 354-359.

Hodgkin, T., Lyon, G.D. and Dickinson, H.G. (1988) 'Recognition in flowering plants: A comparison of the *Brassica* self-incompatibility system and plant-pathogen interactions', New Phytol. 110, 557-569.

Kandasamy, M.K., Paolillo, D.J., Faraday, C.D., Nasrallah, J.B. and Nasrallah, M.E. (1989) 'The S-locus specific glycoproteins of *Brassica* accumulate in the cell wall of developing stigma papillae', Dev. Biol. 134, 462-472.

McClure, B.A., Haring, V., Ebert, P.R., Anderson, M.A., Simpson, R.J. *et al* (1989) 'Style self-incompatibility gene products of *Nicotiana alata* are ribonuclease',. Nature 342, 955-957.

Napoli, C., Lemieux, C. and Jorgensen, R. (1990) 'Introduction of a chimeric chalcone synthase gene into Petunia results in reversible co-suppression of homologous genes *in trans*', Plant Cell 2, 279-289.

Nasrallah, J.B., Kao, T.H., Chen, C.H., Goldberg, M.L. and Nasrallah, M.E. (1987)

'Amino-acid sequence of glycoproteins encoded by three alleles of the *S* locus of *Brassica oleracea*', Nature 326, 617-619.

Nasrallah, J.B. and Nasrallah, M.E. (1989) 'The molecular genetics of self-incompatibility in *Brassica*', Annu. Rev. Genet. 23, 121-139.

Nasrallah, J.B., Nishio, T. and Nasrallah, M.E. (1991) 'The self-incompatibility genes of *Brassica:* expression and use in genetic ablation of floral tissues', Annu. Rev. Plant Physiol. Plant Mol. Biol. 42, 393-422.

Nasrallah, J.B., Yu, S.M. and Nasrallah, M.E. (1988) 'Self-incompatibility genes of *Brassica:* expression, isolation and structure', Proc. Natl. Acad. Sci. USA 85, 5551-5555.

Nasrallah, M.E. (1974) ' Genetic control of quantitative variation in self-incompatibility proteins detected by immunodiffusion', Genetics 76, 45-50.

Nasrallah, M.E., Kandasamy, M.K. and Nasrallah, J.B. (1992) 'A genetically defined *trans*-acting locus regulated *S*-locus function in *Brassica*', Plant Journal (in press).

Nishio, T. and Hinata, K. (1977) 'Analysis of S-specific proteins in stigmas of *Brassica oleracea* L. by isoelectric focusing', Heredity 38, 391-396.

Nishio, T. and Hinata, K. (1982) 'Comparative studies on S-glycoproteins purified from different S-genotypes in self-incompatible *Brassica* species. I. Purification and chemical properties', Genetics 100, 641-647.

Pryor, T. (1987) 'The origin and structure of fungal disease resistance genes in plants', T.I.G. 3, 157-161.

Sato, T., Thorsness, M.K., Kandasamy, M.K., Nishio, T., Hirai, M., Nasrallah, J.B. and Nasrallah, M.E. (1991) 'Activity of an S locus gene promoter in pistils and anthers of transgenic *Brassica*', Plant Cell 3, 867-876.

Somssich, I.E., Schmelzer, W., Bollman, J. and Hahlbrook, K. (1986) 'Rapid activation by fungal elicitor of genes encoding pathogenesis-related proteins in cultured parsley cells', Proc. Natl. acad. Sci. USA 83, 2427-2430.

Stein, J.C., Howlett, B., Boyes, D.C., Nasrallah, M.E. and Nasrallah, J.B. (1991) 'Molecular cloning of a putative receptor protein kinase gene encoded at the self-incomaptibility locus of *Brassica oleracea*', Proc. Natl. Acad. Sci. USA 88, 8816-8820.

Thompson, K.F. and Taylor, J.P. (1971) 'Self-incompatibility in kale' Heredity 27, 459-471.

Thorsness, M.K., Kandasamy, M.K., Nasrallah, M.E. and Nasrallah, J.B. (1991) 'A *Brassica* S-locus gene promoter targets gene expression and cell death to the pistil and pollen of transgenic *Nicotiana*', Dev. Biol. 143, 173-184.

Toriyama, K., Stein, J.C., Nasrallah, M.E. and Nasrallah, J.B. (1991) 'Transformation of *Brassica oleracea* with an S-locus gene from *B. campestris* changes the self-incompatibility phenotype', Theor. Appl. Genet. 81, 769-776.

Ullrich, A. and Schlessinger, J. (1990) 'Signal transduction by receptors with tyrosine kinase activity', Cell 61, 203-212.

Walker, J. and Zhang, R. (1990) 'Relationship of a putative receptor kinase from maize to the *S*-locus glycoproteins of *Brassica*', Nature 345, 743-746

Yarden, Y. and Ullrich, A. (1988) 'Growth factor receptor tyrosine kinases', Annu. Rev. Biochem. 57, 443-478.

OVERCOMING PEST AND PATHOGEN ADAPTABILITY USING INSIGHTS FROM EVOLUTIONARY BIOLOGY

P. KAREIVA, J. WINTERER, and B. KLEPETKA
Department of Zoology
University of Washington
Seattle, WA 98195
USA

ABSTRACT. Biotechnology is often sold as a cure-all for our pest problems. However, if we have learned anything from the "rise and fall" of chemical insecticides, it should be that the evolution of pest species can frustrate mankind's best laid technological solutions. Already, there is evidence that cultivars genetically engineered to be resistant to herbivores or pathogens may be useful for only brief time periods as a result of pest evolution. Fortunately, insights from evolutionary biology point towards strategies for deploying transgenic plants that could enhance their durability by impeding the evolution of virulence among those species that attack the genetically engineered cultivars. Importantly, the translation of this theoretical insight into practical action will require information best supplied by molecular genetics -- information on the molecular basis of virulence versus avirulence, information on mutation rates and the likelihood of cross-resistance, and quantification of gene flow. The future of biotechnology will be as bright as advertised only if evolutionary biologists and molecular biologists join forces in planning the design of agroecosystems.

1. The Hyperbole of Biotechnology

Both scientists (Davis 1991) and the popular media (Weintraub 1992) have told us that the wedding of molecular biology and biotechnology is "revolutionary". Claims about the extraordinary benefits to be spawned by this technological wedding include visions of a future world in which pesticides are no longer needed, and agroecosystems are so well-designed that high yields can be sustained on what are now marginal lands. However, contrary to such positive advertisements for biotechnology, environmentalists have cautioned that genetic engineering may do little more than enrich short-term profits for select corporations at the expense of ecological sustainability (Goldburg et al 1990).

Much of debate regarding the promises versus drawbacks of biotechnology is simply calculated overstatement for a specific goal -- exaggeration in order to encourage investment and funding, or excessive caution in order to encourage the

E. W. Nester and D. P. S. Verma (eds.),
Advances in Molecular Genetics of Plant-Microbe Interactions, 23–33.

development of a regulatory framework. There are, however, major scientific problems that must be solved before biotechnology can live up to its promise. We focus here on the challenge posed by the evolutionary adaptability of pests and pathogens. We begin by looking at the "pesticide revolution" of the 1940's and 1950's as a cautionary tale. Then we turn to specific ideas from evolutionary biology that represent a toolbox for avoiding the mistakes of the pesticide era. In addition to sketchily reviewing a wide variety of models from evolutionary biology, we focus in particular on the issue of "pyramids of resistance genes" as a technological solution to pest and pathogen evolution. Finally, we emphasize the information that molecular biology might offer to population biologists who are seeking strategies for maintaining durable resistance in newly engineered cultivars.

2. Lessons form the "Pesticide Revolution"

2.1 ENTOMOLOGISTS AT ONE TIME FELT LIKE WIZARDS WITH ENORMOUS TECHNOLOGICAL POWER, NOT UNLIKE MODERN-DAY MOLECULAR BIOLOGISTS

In 1947, the President of the American Society of Economic Entomologists heralded the advent of pesticides with the words, "*the entomologist has become a wizard in the eyes of the uninitiated -- and indeed some of the achievements seem little short of magic*" (Lyle, 1947). Such language is much like what we now hear being said on behalf of genetic engineers and their ability to insert into cultivars a wide variety of recombinant genes (e.g., bt endotoxins, enhanced starch production, modified oil production, virus coat proteins, resistance to herbicides, etc.). A second similarity behind the pesticide and the genetic engineering "revolutions" is the way both new technologies have dominated entire scientific disciplines, often to the exclusion of long-standing academic traditions. In particular, whereas economic entomologists traditionally emphasized the natural history of pest insects, during their honeymoon with pesticides these researchers roundly abandoned ecological studies in favor of endless pesticide efficacy tests. This trend can be documented by categorizing articles published in the leading American journal for applied entomology, *The Journal of Economic Entomology*. When one reads this journal from the late 1940's, the dominance of pesticide research astounding -- -- for example, in the 1946 volume over three-quarters of all the articles appearing in the *Journal of Economic Entomology* dealt with pesticide tests, leaving only a handful of articles to be concerned with understanding the biology of pests (this 1946 figure had reversed itself by 1990, when less than one-fifth of the articles were concerned with pesticide tests and most research dealt with pest ecology). It was as though there was a belief that chemicals would allow us to kill or suppress pests, without having to know much of anything about the demography and natural history of the pests themselves. Moreover, not only did entomologists fail to anticipate the environmental hazards of their new technology, they never expected the evolution of resistance to occur on the scale it has. Indeed, returning to our categorization of journal articles, whereas approximately one-quarter to one-third of the research published in *Journal of Economic Entomology* currently addresses the problem of pesticide resistance, in 1946 it is impossible to find a single paper in the journal addressing the resistance problem. Apparently, for all their technical achievements, post World War II entomologists were blind to the resistance dilemma, which now occupies much of their research. In summary, the response of entomology to the advent of pesticides shows that scientists are not especially prescient about the shortcomings of new technologies that sweep through their disciplines.

2.2 RESISTANCE AGAINST PESTICIDES AND FUNGICIDES EXPLODED SHORTLY AFTER THE WIDESPREAD USE OF THESE NEW CHEMICALS

When pesticides were first introduced they commonly killed anywhere between 80 and 100% of their target populations (Georghiou 1986). Because of this success, they exerted an enormous selective pressure on pest populations -- and where pesticide use was widespread and frequent, this selective pressure inexorably favored resistant variants. The result should not have suprised any evolutionary biologist; hundreds of pest species evolved resistance against one or more chemical insecticides over the timespan of a few decades (see figure 2 in Georghiou 1986). In some cases, a pest insect would evolve resistance against first one chemical, then another, then another, and another, until our chemical arsenal was virtually exhausted. For example populations of the Colorada Potato Beetle in Long Island evolved resistance against: DDT in 1952, Dieldrin in 1957, Endrin in 1958, Carbaryl in 1963, Azinphosmethyl in 1964, Monocrotophos in 1973, Phosmet in 1973, Phorate in 1974, Disulfoton in 1974, Carbofuran in 1976, Oxamyl in 1978, Fenvalerate in 1981, and so on (Georghiou 1986). The combination of resistance and environmental regulations have turned pesticides from a miracle technology into a sad failure in the eyes of many environmentalists (and no better than a modest success in the eyes of pesticides' greatest advocates).

2.3 LESSONS LEARNED ABOUT RESISTANCE MANAGEMENT

Population biologists have examined data on the rate and dosage of pesticide applications in relation to the onset of resistance. From such analyses and simple genetic models it has become clear that by reducing the frequency of application, resistance to pesticides can be forestalled (Goerghiou and Taylor 1986; May and Dobson 1986). Of course, it may be that the reduction in pesticide applications which is required to delay the evolution of resistance leaves a crop effectively "unprotected". Consequently, more sophisticated approaches to "resistance management" have also been explored. One approach that is firmly grounded in evolutionary theory is the idea of alternating in time different insecticides so that a pest is never presented with the same evolutionary pressure, year after year. Two strategic principles have emerged from empirical and theoretical studies of pesticide resistance (Tabashnik 1986; Gould 1991):

(i) resistance can be delayed by reducing the average selective pressure favoring resistant compared to susceptible variants

(ii) resistance can be delayed if the selective regime acting on the pest is heterogeneous and presents the pests with a "moving target".

As we will discuss below, both of these principles could and should be applied as we plan for the deployment of anti-pest or anti-pathogen cultivars engineered to be protected against pests and pathogens. Their actual implementation is, however, dependent on additional key features of pest biology -- primarily the interplay of pest dispersal and pest population growth, the likelihood of cross-resistance (a mutation conferring resistance to one pesticide also conferring complete or partial resistance to alternative pesticides), and the "cost of resistance" as measured by comparing in the absence of pesticides the fitness of susceptible pests relative to resistant pests.

3. Resistance Management in the Context of Biotechnology

3.1 DATA ALREADY INDICATE THAT GENETICALLY ENGINEERED PLANT DEFENSES ARE VULNERABLE TO PEST EVOLUTIONARY RESPONSES

Molecular biologists have inserted genes that code for viral coat proteins and *Bacillus thuringiensis* (or herafter bt) endotoxins into several different cultivars. Crops transgenic for bt endotoxin production are especially close to commercial production, having been field tested for tobacco, canola, tomato and cotton. However, evolutionary biologists familiar with the history of pesticides, have pointed out that massive deployment of resistant transgenic crops may simply select for pests that are immune to the plant's defenses (Gould 1988a,b). This is not just theoretical speculation. Direct applications of *Bacillus thuringiensis* has prompted the evolution of resistance in at least one natural pest population (diamondback moths) after only fifteen applications, which corresponded to only a three-year timespan. Moreover, in laboratory experiments resistance to bt endotoxins has evolved in tobacco budworm, Indian meal moth, and almond moths -- sometimes after only one generation of selection (McGaughey 1985; McGaughey and Beeman 1988; Sims and Stone 1991). Indeed, there is reason to believe that plants engineered to contain anti-pest or anti-pathogen properties may face a briefer future than pesticides. This gloomy scenario arises simply because when a transgenic plant containing an anti-herbivore toxin is deployed, the selective pressure acting on the pest is present as long as the plant is alive, whereas with pesticides the selective pressure is present only when the chemical is sprayed. In other words, certain types of engineered resistance traits will impose a selective regime analagous to spraying insecticide every day; and we know from past experience that such relentless selection can lead to rapid counter-evolution on the part of pests.

3.2 MODELS PROVIDE SOME GUIDANCE ON HOW TO MAXIMIZE THE DURABILITY OF PLANT DEFENSES

In anticipation of the problems posed by pest and pathogen evolution upon deployment of transgenic plants, Fred Gould (1988a,b and 1991) has championed the use of simple genetic models as a way of examining different strategies for resistance management. Among his most pragmatic findings is a clear case for seeking resistance genes that are either induced defenses or are expressed only in particularly valuable tissues (in terms of contribution to yield). The idea behind this strategy is simple. Most herbivores and pathogens reduce yield only when they attck certain portions of plants during certain time periods and at sufficiently high levels. Low or even moderate pest pressure can often be tolerated without any noticeable yield losses. If plants could be engineered so that anti-pest toxins were expressed only where and when needed, then there would be adequate protection with a minimum selection pressure. For example, if a recombinant gene for the production of bt endotoxin were expressed only in cotton bolls as opposed to throughout the cotton plant, then selection for resistance among bollworm populations could be reduced sufficiently that the durability of cotton resistance is doubled or even tripled (Gould 1988a)

Another strategy for enhancing the durability of resistance is the planting of some "susceptible" plants amongst the resistant cultivar; models indicate that by planting as little as 10% of a crop with susceptible variants, the durability of the resistant variant can be prolonged two to tenfold (depending on the genetic details of virulence, see Gould 1991, figure 8). A final intriguing lesson from Gould's models is the realization that the presence of natural enemies could dramatically alter the rate at which pests evolved virulence in the face of resistant plants; thus consideration of biotechnology solutions need to include attention to the role of natural enemies in the agroecosystem. In some cases the presence of natural enemies may mean that a resistance trait which merely slowed pest development could provide the crop with the "edge" it needs to escape severe damage (Gould 1988a,b); in other circumstances, the presence of natural enemies could accelerate the rate at which pests are likely to adapt to a plant's defenses (Gould et al 1991; Johnson and Gould 1992). Unfortunately, biotechnology often entails such a narrow perspective that broader pictures of a crop in a complex ecosystem are neglected.

As we gain sophistication at manipulating plant genomes, possibilities for interaction between molecular biology and ecology will expand enormously. In particular, instead of engineering plants with simple "toxic" defenses, ideas from evolutionary biology may inspire tactics that include plants that tolerate as opposed to kill pests. The advantage of resistance based on "tolerance" is that it does not select so severely for pest evolution, and in the long run would be much more durable than the current emphasis on bt endotoxins.

3.3 "PYRAMIDING" MULTIPLE RESISTANCE FACTORS AS A TECHNICAL SOLUTION TO THE ADAPTABILITY OF PESTS AND PATHOGENS

Perhaps the most appealing technical solution to the problem of resistance is "pyramiding" or stacking together in one germplasm multiple resistance factors. Pyramiding resistance factors may deter the counter-evolution of pathogen or pest virulence by two major mechanisms (Mundt 1991):

(i) a low probability that a pest or pathogen can simultaneosuly mutate at several different loci in a way that overcomes multiple resistance factors

(ii) an intrinsic constraint on pest or pathogen design such that any collection of mutations that simultaneously conferred resistance to several resistance factors would also entail substantial reductions in pest performance.

We do not know to what extent the above mechanisms actually apply to real pest or pathogen species (Mundt 1991). Indeed, we generally lack experimental and observational data pertaining to the efficacy of pyramiding resistance genes (Pedersen and Leath 1988). However, one can (and many have -- see Gould 1986; Wilhoit 1992) use simulation models to investigate the merits of pyramiding genes for resistance into a single germplasm. Our own simulations describing the evolution of virulence in rice blast fungus indicate that pyramided resistance in rice should generally break down if the selection pressure against avirulent fungus is on the order of .95 or higher. Perhaps more importantly, when contrasted to rotating resistance genes or planting mixtures of resistant and susceptible plants, the pyramiding strategy compares favorably only when selection coefficients are quite low (less than .4; see Winterer et al, *in press*). This does not mean that the pyramid approach is doomed. For example, one way of extending the durability of pyramided resistance under severe selection pressures is to ensure a heterogeneous environment in which dispersal is sufficently high that vast numbers of adapted pathogen propagules are lost by dispersal, and large numbers of avirulent propagules arrive through immigration (Winterer et al, *in press*). In fact, if one examines a strategy that includes cultivar rotations, so that germplasm with pyramids of resistance factors are rotated with susceptible plantings or mixes of susceptible and resistant plantings in a spatial mosaic, then the likelihood of a pathogen evolving virulence is extremely low. In general, the value of pyramiding resistance genes depends on pest or pathogen dispersal, mutation rates, and selection coefficents -- parameters about which we are amazingly ignorant. Of course, if cross-resistance exists and if there is minimal cost to virulence, then it would be hard to imagine any setting under which it would make sense to combine resistance genes in one "super genotype" as a technological soultion to pest evolution. Our enthusiasm for stacking up one resistance gene after another in transgenic germplasm needs to be tempered by the realization that this approach will be durable for only certain ecological settings. Largescale homogeneous plantings of super-resistant cultivars that have been engineered with resistance pyramids are prone to suffering the same fate as pesticides, especially if the pest or pathogen populations exhibit dispersal regimes that favor local adaptation (low dispersal and site fidelity).

4. Molecular biology can provide key information that is missing from our analyses of resistance management.

Although we have the theoretical sophistication to devise optimal strategies for resistance managemnt, too often we lack basic biological information. For example, when considering the construction of genomes with bt endotxin protection, the durability of resistance provided by multiple toxins depends on the mechanism by which herbivores acquire virulence (i.e., overide the toxins). Molecular studies have begun to document the mechanisms of herbivore adaptation to bt endotoxins (Ferre et al, 1991; McGaughey 1985; McGaughey and Beeman 1988; Van Rie 1990), and the answer is encouraging -- apparently adaptation to each toxin involves a specific gut receptor which is lost or modified such that the mutant herbivore does not take up the toxin and hence does not die (Lambert and Peferoen 1992). Mutations that lead to resistance at one receptor type rarely confer resistance at the receptor sites associated with different bt endotoxins. If this mechanistic portrait of the molecular details of bt endotxin resistance proves to be general, then the dangers of cross-resistance should be minimal. One caveat is that one study has hinted that tobacco hornworms resistant to one bt endotoxin are also readily developing resistance to other toxins (Gould 1991).

In addition to uncovering crucial information about the mechanisms of virulence and avirulence in pests or pathogens, molecular approaches can also help us quantify those parameters that evolutionary theory indicates are central to resistance management. For instance, using molecular markers one can obtain estimates of dispersal and gene flow that were previously impossible at a large scale (Kareiva et al 1991; Lindow et al 1988). Using genetic markers it is also possible to contrast replication rates of virulent versus avirulent pathogens, which can then be converted into estimates of all-important selection coefficients. Finally, now that we can routinely sequence DNA, it is practical to measure mutation rates in different genes, and thereby determine within an order of magnitude the likelihood of mutations arrising that convert avirulent pests into virulent pests. In short, tools from molecular biology offer evolutionary theorists the best hope for parameterizing their models of resistance managment.

5. Biotechnology will live up to its promise only if there is better communication between molecular biologists and evolutionary biologists.

Most researchers at the forefront of biotechnology are ignorant of population and evolutionary biology; conversely, most evolutionary biologists rarely communicate with those directly involved with genetic engineering (except as critics). Unless there is more cooperation and communication between these distinct scientific disciplines, progress will be slow in terms of resistance management. We need to identify mechanisms of virulence, and estimate mutation, selection and dispersal rates if there is to be any hope of "outsmarting" the evolutionary process. Even with perfect knowledge, it is not clear that the evolution of pests and pathogens can be arrested.

A wedding of biotechnology and population biology promises much more than simply prescriptions for resistance management. Studies of natural plant-pathogen or plant-herbivore systems may uncover fundamentally novel resistance factors (Kennedy and Barbour 1992). One of the advantages of recombinant technology is that classical constraints on breeding for resistance are relaxed, such that traits which were once unavailable, may become practical in the near future.

Finally, the greatest risk of any technology is smugness on the part of its advocates -- and biotechnology is no different in this respect. Constructive exchange between ecologists and molecular biologists would do much to counter the smugness of biotechnologists with respect to control of the natural world, and the smugness of ecologists with respect to their confidence in "knowing what is best". The fate of the widely advertised "biotechnology revolution in agriculture" depends on our ability to integrate insights from evolutionary biology and ecology into our design and deployment of trangenic crops.

6. References

Davis, B. (1991) The Genetic Revolution, Johns Hopkins University Press, Baltimore, 295 pp.

Ferre, J. et al (1991) Resistance to the *Bacillus thuringiensis* bioinsecticide in a field population of *Plutella xylostella* is due to a change in a midgut membrane receptor. PNAS 88: 5119-5123.

Georghiu, G. (1986) `The magnitude of the resistance problem' in Pesticide Resistance: Strategies and Tactics for Management, National Research Council, National Academy Press, Washington, D.C., pp. 14-43.

Georghiou, G. and C. Taylor (1986) `Factors influencing the evolution of resistance' in Pesticide Resistance: Strategies and Tactics for Management, National Research Council, National Academy Press, Washington, D.C., pp. 157-169.

Goldburg, D., et al. (1990) Biotechnology's Bitter Harvest, Biotechnology Working Group, Washington, D.C., 73 pp.

Gould, F. (1986) Simulation models for predicting the durability of insect-resistant germplasm: a deterministic two-locus diploid model. Environm. Entom. 15: 1-10.

Gould, F. (1988a) Evolutionary biology and genetically engineered crops. Bioscience 38: 26-33.

Gould, F. (1988b) Genetic engineering, integrated pest management and the evolution of pests. Trends in Ecology and Evolution 3: 515-518.

Gould, F. (1991) The evolutionary potential of crop pests. American Scientist 79: 496-507.

Gould, F. et al. (1991) Effects of natural enemies on the rate of herbivore adaptation to resistant host plants. Entomol. Expt. Applic. 58: 1-14.

Johnson, M. and F. Gould. (1992) Interaction of genetically engineered host plant resistance and natural enemies of *Heliothus virescens* in tobacco. Environ. Entomol. 21: 586-597.

Kareiva, P. et al (1991) `Using models to integrate data from field trials and estimate risks of gene escape and gene spread', in D. MacKenzie and S. Henry (eds.), Biological Monitoring of Genetically Engineered Plants and Microbes, ARI Press, Washington, D.C., pp. 31-44.

Kennedy, G. and J. Barbour (1992) `Resistance variation in natural and managed systems', in R. Fritz and E. Simms (eds.), Plant Resistance to Herbivores and Pathogens, University of Chicago press, Chicago, pp. 13-41.

Lambert, B. and M. Peferoen (1992) Insecticidal promise of *Bacillus thuringiensis*. Bioscience 42: 112-121.

Lindow, S. et al., (1988) Aerial dispersal and epiphytic survival of *Pseudomonas syringae* during a pretest for the release of genetically engineered strains into the environment. Envt. Microbiology 54: 15557-1563.

Lyle, C. (1947) Achievements and possibilities in pest eradication. J. Econ. Entomology 40: 1-8.

May, R. and A. Dobson (1986) `Population dynamics and the rate of evolution of pesticide resistance' in Pesticide Resistance: Stratgeies and Tactics for Management, National Research Council, National Academy Press, Washington, D.C., pp. 170-193.

McGaughey, W. (1985) Insect resistance to the biological insecticide *Bacillus thuringiensis*. Science 229:193-194.

McGaughey, W. and R. Beeman (1988) Resistance to *Bacillus thuringiensis* in colonies of Indian meal moth and almond moth. J. Econ. Entomol. 81: 28-33.

Mundt, C. (1991) Probability of mutation to multiple virulence and durability of resistance gene pyramids: further comments. Phytopathology 81:240-242.

Pedersen, W. and S. Laeth (1988) Pyramiding major genes for resistance to maintain residual effects. Ann. Rev. Phytopathology 26: 369-378.

Sims, S. and T. Stone (1991) Genetic basis of tobacco budworm resistance to an engineered *Pseudomonas fluorerscens* expressing the delta-endotoxin of *Bacillus thuringiensis*. J. Inv. Path. 57: 206-210.

Tabashnik, B. (1986) `Computer simulation as a tool for pesticide resistance management' in Pesticide Resistance: Strategies and Tactics for Management, National Research Counil, National Academy Press, Washington, D.C., pp. 194-206.

Van Rie, J. et al. (1990) Mechanism of insect resistance to the microbialinsecticide *Bacillus thuringiensis*. Science 247: 72-74.

Weintraub, P. (1992) The coming of the high-tech harvest. Audobon 94: 92-103.

Wilhout, L. (1992) `Evolution of herbivore virulence to plant resistance: influence of variety mixtures', in R. Fritz and E. Simms, (eds.), Plant Resistance to Herbivores and Pathogens, University of Chicago Press, Chicago, pp. 91-119.

Winterer, J. et al. (in press) `Crop intensification and increased pest problems: are there strategies for minimizing the vulnerability of rice to pest epidemics ?' in P. Pingali (ed.), Proceedings of 1992 International Rice Research Conference, International Rice Research Institute, Manila, Phillipines.

Wolfe, M. (1985) The current status and prospects of multiline cultivars and variety mixtures for disease resistance. Ann. Rev. Phytopath. 23: 251-273.

Section 2 / *Agrobacterium*-Plant Interactions

THE VIRULENCE SYSTEM OF *AGROBACTERIUM TUMEFACIENS*

Alice Beijersbergen and Paul J.J. Hooykaas
Institute of Molecular Plant Sciences
Clusius Laboratory
Leiden University
Wassenaarseweg 64
2333 AL Leiden, The Netherlands

1. Introduction

1.1. VIRULENCE OF *AGROBACTERIUM*

The gram-negative soil bacterium *Agrobacterium tumefaciens* causes the plant disease crown gall. This disease is characterized by the formation of tumors or crown galls at wound sites of infected dicotyledonous plants (for recent reviews see Kado, 1991; Winans, 1992; Zambryski, 1992; Hooykaas and Schilperoort, 1992). During tumor induction *Agrobacterium* attaches to the plant and transfers part of its tumor inducing (Ti) plasmid (Fig.1) to plant cells at wound sites. The transferred or T-DNA, which is surrounded by 24 basepair (bp) imperfect direct repeats or border repeats, becomes integrated in the plant cell nuclear DNA. Upon expression of the genes located on the T-DNA, proteins are produced involved in the production of the plant hormones auxin (IAA) and cytokinin (isopentenyl-AMP). These hormones cause the tumorous phenotype, characterized by the ability of the cells to proliferate unlimited and autonomously in the absence of added phytohormones. The T-DNA also encodes synthases catalyzing the formation of opines. These opines are mostly built from an amino acid and a sugar. Based on the kind of opines produced, *Agrobacterium* strains are classified as octopine, nopaline, succinamopine and leucinopine strains. The opines formed in the tumors are metabolized by the agrobacteria which induced tumor formation, but not by most other soil organisms. Thus, by genetic manipulation of plant cells *Agrobacterium* creates a favorable niche for itself, a process also called "genetic colonization".

Besides the T-region the approximately 200 kilobasepairs (kb) large Ti plasmid contains another region involved in tumorigenesis, the virulence or Vir region (Fig.1). The Vir region determines the system

E. W. Nester and D. P. S. Verma (eds.),
Advances in Molecular Genetics of Plant-Microbe Interactions, 37–49.

for T-DNA transfer to plant cells. Besides, both the Vir and T-region contain genes influencing the plant host range of tumor formation. While strains containing a wide host range (WHR) Ti plasmid cause tumors on a large variety of plants, limited host range (LHR) strains, which are often isolated from *Vitis*, are restricted to few plant species (Panagopoulos and Psallidas, 1973; see 2.2. "The VirA protein"). Outside the Vir and T-regions the Ti plasmid may have loci involved in uptake and catabolism of opines, Ti plasmid transfer, phage AP1 exclusion, sensitivity to agrocin 84 and replication and incompatibility (Fig.1). Certain opines induce not only the opine catabolism genes but also the transfer (*tra*) genes responsible for conjugative transfer of the Ti plasmid. The inducers for the *tra* genes of octopine, nopaline and leucinopine Ti plasmids are octopine, agrocinopine A and agrocinopine C respectively.

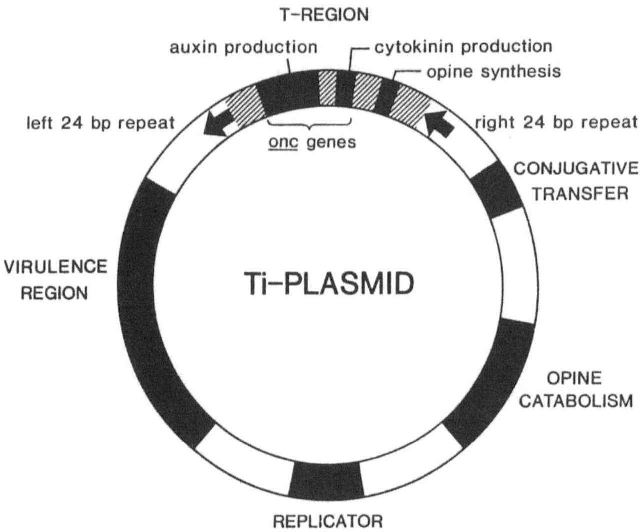

Fig.1. Genetic map of an octopine Ti plasmid.

1.2. CHROMOSOMAL VIRULENCE GENES

Chromosomally located loci involved in attachment of *Agrobacterium* to the plant cell include *chvA*, *chvB* and *pscA* (*exoC*). Mutants in these loci are avirulent on many plant species (Douglas *et al*, 1985). The *chvA* gene codes for an innermembrane protein, which is homologous to export proteins like HlyB and Mdr (Cangelosi *et al*, 1989). Because in *chvA* mutants ß1,2-glucan is formed but not exported (Cangelosi *et al*, 1989; Inon de Iannino and Ugalde, 1989), it is concluded that the ChvA protein is important for transport of ß1,2-glucan to the periplasm. In *chvB* mutants ß1,2-glucan is not produced. The *chvB* gene codes for an innermembrane protein, which probably is directly involved in ß1,2-glucan synthesis (Zorreguieta *et al*, 1988). Also the *exoC* mutants do not produce ß1,2-glucan. Expression of the *vir*-genes is affected in several chromosomal mutants (for *chvE* see 2.2. "The VirA protein"). Host range functions may also be determined by chromosomal genes such as *chvC* (Hooykaas *et al*, 1984).

1.3. VIRULENCE GENES OF THE TI PLASMID

The Vir region, which has a size of about 40 kb, is located adjacent to the left border repeat of the T-DNA (Fig.2). The octopine Ti plasmid Vir region consists of eight operons, *virA* to *virH*. Four of these operons are essential for virulence (*virA*, *virB*, *virD* and *virG*). While mutations in one of these operons abolish tumor formation on all plant species, mutations in the other loci (*virC*, *virE*, *virF* and *virH*) lead to a restriction in plant host range or to an attenuation of tumorigenicity. Contrary to the octopine Ti plasmid, the nopaline Ti plasmid lacks the *virF* gene but contains the *tzs* gene. The protein encoded by the latter gene is involved in the production of *trans*-zeatin, a cytokinin similar to the product of the *ipt*- or *cyt*-gene of the T-region. Recently, the *virH* operon was reported to encode two cytochrome P450-like enzymes, which may play a role in the detoxification of bacteriostatic and bacteriocidal compounds present in plant wound sap (Kanemoto *et al*, 1989).

1.4. T-DNA PROCESSING

In the presence of plant phenolic compounds, which are excreted from wound sites, the regulatory proteins VirA and VirG activate the other *vir* genes (see 2. "VirA and VirG: a two-component regulatory system"). Following expression of these genes, nicks are introduced into the bottom strands of the border repeats. Hereby, the proteins VirD1 and VirD2, encoded by the *virD* operon, act as an endonuclease. The precise site of the single-stranded DNA breakage was localized between the third and fourth base of the 24 bp repeats (Yanofsky *et al*, 1986; Wang *et al*, 1987). The VirD1 protein has topoisomerase activity catalyzing the conversion of supercoiled DNA to relaxed DNA (Ghai and Das, 1989). Another *vir*-encoded protein, the VirC1 protein, binds specifically to the overdrive or enhancer sequence located adjacent to the right-border repeat, thereby enhancing T-strand formation. This interaction might promote direction of the VirD endonuclease to the border sequence (Toro *et al*, 1988). Single-stranded T-DNA copies or T-strands are generated probably by replacement synthesis of the bottom strand (Fig.2). The protein VirD2 remains covalently attached to the 5'end of the T-strand and protects it from 5' to 3' exonucleolytic degradation (for review Zambryski, 1992). The VirD2 protein is thought to act as a pilot protein in the T-DNA transfer process because it has nuclear localization signals (Herrera-Estrella *et al*, 1990) and can direct a ß-glucuronidase reporter protein to the nucleus of plant cells (Howard *et al*, 1992). Another Vir protein, the VirE2 protein, can bind cooperatively to single-stranded DNA *in vitro*, thereby creating a thin and extended protein-DNA complex (Zambryski, 1992). Covering of the T-strand with VirE2 proteins was shown to protect the DNA against degradation by endonucleases *in vitro* (Sen *et al*, 1989).

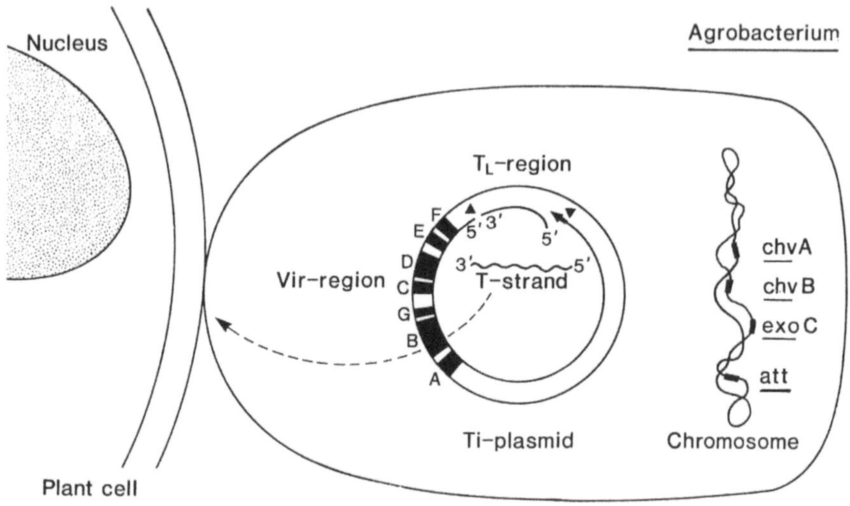

Fig.2. T-strand formation in *Agrobacterium.*

2. VirA and VirG: a two-component regulatory system

VirA and VirG together mediate the induction of the other *vir* operons in the presence of plant phenolic compounds such as acetosyringone (AS) (Fig.3). Sequence analysis revealed striking homology of VirA and VirG to the proteins of two-component regulatory systems like EnvZ/OmpR, PhoR/PhoB of *E.coli* and NtrB/NtrC of *K.pneumoniae* (for review Winans, 1992; Hooykaas and Schilperoort, 1992). The VirA protein is thought to be a sensor protein acting as a receptor for plant phenolic compounds like AS. The VirG protein propbably acts as a transcriptional activator of the other *vir* operons by binding to the promoter regions of these operons. Recently, it was shown that, like other sensor proteins, VirA has autokinase activity *in vitro* (Jin *et al*, 1990a; Huang *et al*, 1990). The conserved amino acid histidine 474 probably is phosphorylated, because changing this residue to glutamine resulted in loss of VirA autophosphorylation and inhibition of tumor induction. The phosphate group can directly be transferred from the phosphorylated VirA molecule to the VirG protein, whereby the VirG protein probably becomes activated (Jin *et al*, 1990b). In VirG the conserved aspartate residue at position 52 probably is phosphorylated.

2.1. CONDITIONS OF VIRULENCE GENE INDUCTION

The *vir* genes are induced by compounds present in plant exudates like AS and α-hydroxy-AS. The structural requirements of the inducing phenolic compounds were studied. Results showed that guiacol, a derivative of phenol with a methoxylgroup in the *ortho*-position, is the simplest inducer. Substitutions in the *para*-position lead to either an enhancement or a reduction of *vir*-induction (Spencer and Towers, 1988; Melchers *et al*, 1989a). It has been reported that the level of inducing phenolics may be a limiting factor in the efficiency of tumor formation.

Besides on the presence of an inducer the induction of the virulence genes is dependent on the pH of the medium, the growth phase of the cells, the growth temperature and the presence of certain monosaccharides. For octopine Ti plasmids induction is optimal at pH 5.3, a temperature below 30°C and a high sugar concentration in the medium. The slight difference with the optimal inducing pH of 5.8 for nopaline Ti plasmids is probably mainly due to differences in the VirA proteins (Turk *et al*, 1991).

2.2. THE VIRA PROTEIN : A SENSOR FOR PHENOLICS

The 91.6 kilodalton (kDa) large VirA protein is localized in the inner membrane of *Agrobacterium* (Leroux *et al*, 1987, Melchers, *et al*, 1987). Fusions between different N-terminal parts of VirA and PhoA showed that VirA consists of two transmembrane domains with a periplasmic domain at the N-terminus and a cytoplasmic domain at the C-terminus. Studying the properties of chimeric VirA/Tar receptor proteins we found to our surprise that the sensor domain for phenolic compounds is not located in the periplasmic region of VirA (Melchers *et al*, 1989b). The periplasmic domain has a role, however, in the detection of external pH conditions. A region including the second transmembrane domain of VirA was shown to be an essential element of VirA and probably contains the sensor function for phenolics.

Several sugars were reported to act synergistically with AS in *vir* gene induction (Shimoda *et al*, 1990) For this sugar-mediated induction the product of the chromosomal *chvE* gene is required (Huang *et al*, 1990, Cangelosi *et al*, 1990). The ChvE protein shows homology to periplasmically located sugar-binding proteins involved in active transport and chemotaxis. Sugars that were most efficient were the acidic monosaccharides D-galacturonic acid and D-glucuronic acid, which are monomers of major plant cell wall polysaccharides (Ankenbauer and Nester, 1990). The periplasmic domain of the VirA protein is required for sugar-mediated *vir*-induction. This suggests that after binding of the monosaccharide to the ChvE protein, the complex formed interacts with the periplasmic domain of VirA and thereby leads to a further activation of the VirA protein.

Contrary to earlier findings we showed that the *vir* genes of the LHR octopine Ti plasmid pTiAG57 can be induced by AS. Although the response was slower and reduced as compared to that of the WHR octopine Ti plasmid *vir* genes, by change of environmental conditions, i.e. high concentration of AS, a reduction of inoculum size and a decreased growth temperature, considerable *vir*-induction was observed (Turk *et al*, 1991).

Upon induction by AS the WHR *virA* promoter is also induced, albeit to a lower level than the promoters of the other *vir* operons. Recent work in our group showed that introduction of the WHR *virA* gene promoter in front of the LHR *virA* gene leads to *vir* gene expression nearly at WHR level in a LHR Ti strain and an extension of the host range of tumor induction (Turk *et al*, 1992). Thus, the non-inducibility of the promoter of the LHR *virA* gene is partially responsible for the LHR phenomenon. Because the LHR *virA* gene is induced by AS and this induction is enhanced by addition of D-glucose, the domains involved in binding of AS and ChvE are conserved between the LHR and WHR VirA receptor proteins (Turk *et al*, 1992).

2.3. THE VIRG PROTEIN : TRANSCRIPTIONAL ACTIVATOR OF *VIR* GENES

The VirG protein is a cytoplasmic protein, but is also partially associated with the *Agrobacterium* membrane (Rodenburg, 1992). The *virG* gene has two potential translational start sites : a TTG and an ATG startcodon. The TTG codon, surrounded by a Shine-Dalgarno sequence is the one used (Aoyama, 1989, Pazour and Das, 1990). Conserved regions in the 5'-noncoding region of all *vir* genes have been identified by DNA sequence analysis. A 12 bp large consensus sequence, called *vir*-box, was identified. This may be a *cis*-acting regulatory sequence to which the VirG protein binds. In analogy to other regulators of two-component regulatory systems direct binding of the VirG protein to the *vir*-boxes was demonstrated *in vitro* (Tamamoto *et al*, 1990). The C-terminal domain of the VirG protein is required and sufficient for DNA-binding (Powell and Kado, 1990; Roitsch *et al*, 1990). It is thought that by binding of the VirG protein to the *vir*-boxes the *vir* genes become expressed. The phosphorylated VirG protein might, compared to the unphosphorylated form, have a higher affinity for the *vir*-boxes. Also, it is possible that the phosphorylated VirG interacts more effectively with RNA polymerase, thereby increasing the efficiency of RNA polymerase to recognize the *vir* promoters or to convert the closed into an open DNA complex. The *vir*-boxes have imperfect dyad symmetry. Taken together with the cooperative binding of VirG, this suggests that the VirG protein binds as a dimer or a multimer (Tamamoto *et al*, 1990). The *virG* gene is expressed from two promoters. Promoter P1 is inducible by AS in strains containing VirA and VirG, but also by phosphate starvation in a VirA-VirG independent way. The second promoter, P2, is located 50 bp downstream from the first and is inducible independent of Ti plasmid genes by various forms of environmental stress like acidic media and, albeit less, by DNA damaging compounds and heavy metals (Winans, 1992). The P2 promoter resembles the *E.coli* heat shock promoters. Indeed, recently it was shown that the *Agrobacterium tumefaciens* heat shock regulatory system is similar to that of *E.coli* and that *Agrobacterium* has a σ^{32}-like sigma factor (Mantis and Winans, 1992). However, P2 regulation is at least partially independent of the heat shock and S.O.S. responses.
In our group random mutations were introduced into the *virG* gene of octopine Ti plasmid pTi15955 (Rodenburg, 1992). One of the mutants, in which aminoacid residue asparagine 54 had been changed into an aspartate residue, showed constitutive *vir* gene expression independent

of the presence of AS and the *virA* gene. Another mutant, in which amino acid residue isoleucine 77 had been converted into valine, showed supersensitivity to AS (i.e. a high level of induction at low concentrations of AS). Also, "loss-of-function" *virG* mutants were isolated. The structure of the N-terminal part of the VirG protein was predicted based on the crystal structure of the homologous *E.coli* CheY protein. It was shown that the VirG protein has an acidic pocket, which is similar to that at the phosphorylation site of CheY (Rodenburg, 1992). The isolated mutants together with the aid of the three-dimensional structural model will be used to further elucidate the function of the different domains of VirG.

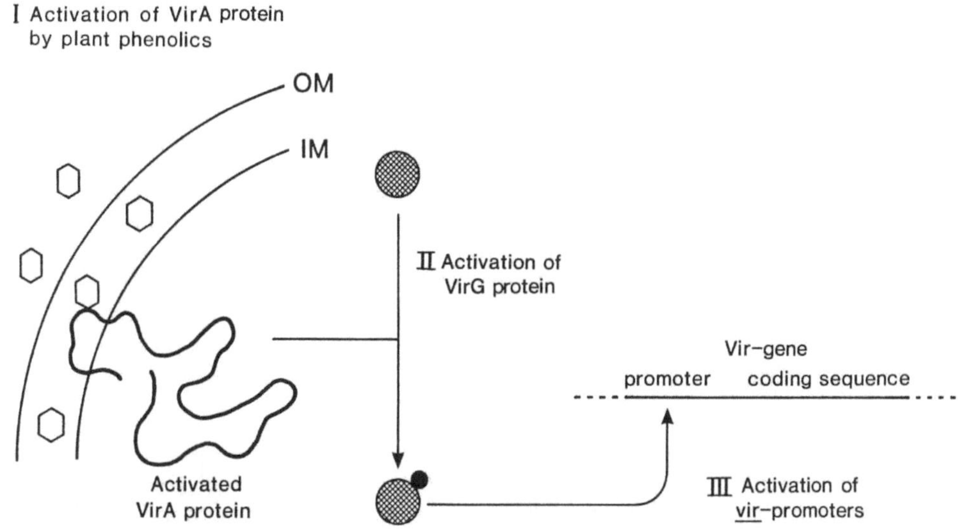

Fig.3. Activation of the *vir* genes via VirA and VirG.

3. T-DNA transfer and similarities to bacterial conjugation

3.1. THE VIRB PROTEINS : COMPONENTS OF A MEMBRANE PORE?

The *virB* operons of several Ti plasmids were sequenced (Thompson *et al*, 1988; Ward *et al*, 1988; Kuldau *et al*, 1990; Shirasu *et al*, 1990). This largest *vir* operon encompasses 9600 bp and has 11 open reading frames (ORFs) (Fig.4). Analysis of the predicted proteins showed that they nearly all contain hydrophobic stretches and mostly have signal sequences. Some localization studies were done using VirB antibodies or creating fusions between VirB proteins and the enzyme alkaline phosphatase (Zambryski, 1992; our unpublished results). We found that several VirB proteins containing hydrophobic stretches gave enzymatically active PhoA fusions, which means that these proteins probably are located in the inner membrane with a part protruding into the periplasm. To our knowledge, we showed for the first time that the small *virB7* gene indeed codes for a protein, since an enzymatically active fusion with PhoA was found. Regarding the presence of a signal sequence and the homology of VirB7 with several outer membrane lysis proteins (Shirasu *et al*, 1990), we suppose that the 6kDa VirB7 protein is located in the outer membrane. The VirB11 protein is an adenosine triphosphatase with autophosphorylating activity and is a cytoplasmic protein associated with the inner membrane (Christie *et al*, 1989). VirB11 is homologous to ORF1 of the *comG* operon of *Bacillus subtilis*, a protein required for DNA uptake (Albano *et al*, 1989). Also, the ATP-binding domain of VirB11 shows sequence similarity to that of PulC, an innermembrane protein of the pullulanase operon of *Klebsiella pneumoniae*, necessary for export of pullulanase (d'Enfert *et al*, 1989). ATP hydrolysis by VirB11 could be the energy source for T-complex (i.e. T-DNA coated with proteins) transfer across the membrane. The VirB10 protein was shown to be anchored in the innermembrane and to exist as a native oligomer or an aggregate with other membrane proteins (Ward *et al*, 1990). Based on the requirement of the *virB* operon for tumor formation and the probable location of most of the VirB proteins in the agrobacterial membrane, it is hypothesized that the VirB proteins, possibly together with other (Vir?) proteins, constitute a membrane pore through which the T-complex is transported.

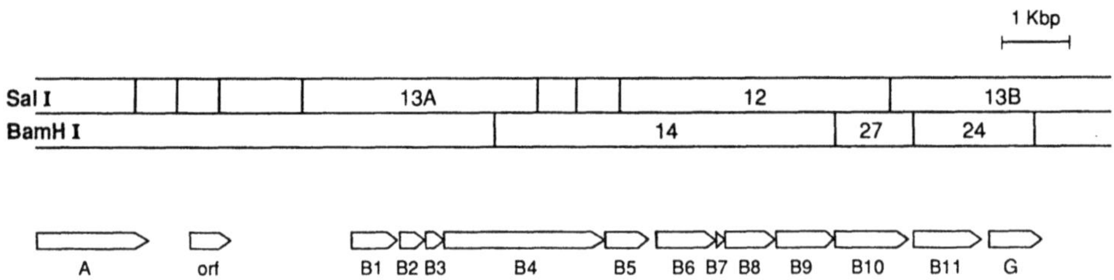

Fig.4. Structure of the *virB* operon as predicted by nucleotide sequence analysis.

3.2. CONJUGATIVE TRANSFER BY THE VIRULENCE SYSTEM

Similarities between T-DNA transfer from *Agrobacterium* to plants and conjugative transfer of plasmid DNA between bacteria have been reported. During both processes single-stranded breaks are introduced in the DNA that is transferred. While in case of T-DNA transfer VirD proteins cause a nick in the bottom strands of the border repeats (see 1.4."T-DNA processing"), in bacterial conjugation *mob* region encoded proteins make a similar break at the origin of transfer (*oriT*). Homology was found between the TraG protein from plasmids RP4 and R751, and the VirD4 protein of *Agrobacterium* (Ziegelin *et al*, 1991).

Furthermore, also nick sites and nicking enzymes in both processes show similarity (Pansegrau and Lanka, 1991). Part of the border sequences of both octopine and nopaline Ti plasmids share a nucleotide sequence stretch around the *oriT* of plasmids RK2, R751, R64 and pTF-FC2. The VirD2 protein has a sequence motif common with TraI of RP4 and R751, NikB of R64, Rlx of *S.aureus* plasmids and MobA of *Thiobacillus ferrooxidans* pTF-FC2 plasmid. In bacterial conjugation plasmid DNA probably is transferred through a channel encoded by the transfer (*tra*) region with help of a pilus structure for initial contact between the donor and acceptor cell. VirB proteins possibly form a comparable membrane pore for T-complex transport.

The first evidence for partial functional analogy between T-DNA transfer and bacterial conjugation was given in plant transformation experiments (Buchanan-Wollaston *et al*, 1987). Plasmid RSF1010, which lacks a T-region border repeat, was transferred to plant cells from *Agrobacterium* strain LBA4404 containing a Vir helper plasmid. This suggested that nicking of the *oriT* by Mob proteins can replace border repeat nicking by the VirD proteins in the T-DNA transfer process. In addition to this Ward *et al.*(1991) showed that an RSF1010-derived plasmid present in an octopine *Agrobacterium* strain greatly suppressed tumor formation. Overexpression of the *virB9*, *virB10* and *virB11* genes restored tumorigenicity. Thus, the T-complex and RSF1010 probably use the same transport machinery.

In our lab we studied whether Vir proteins can substitute for Tra proteins in the conjugal transfer of an *incQ* plasmid between agrobacteria. The wide host range *incQ* plasmid pKT230 is a derivative of RSF1010 and, like its parent, is nonconjugative but can be mobilized by certain conjugative plasmids. It encodes mobilization proteins that are active at its own *oriT*, but needs a conjugative plasmid such as RP4 (*incP*) for other transfer functions. As expected, in our experiments LBA4404, which contains the *tra*$^-$ Ti plasmid pAL4404, did not mobilize pKT230 during incubations on standard media. However, when this strain was incubated on minimal medium that contained the *vir*-inducer AS, mobilization of the *incQ* plasmid did occur. The presence of the Vir region was essential for transfer of pKT230 because a strain missing also the Vir region did not mobilize the *incQ* plasmid to recipients (Beijersbergen *et al*, 1992). These experiments indicate that the Vir system can mediate conjugation between agrobacteria. To analyze the involvement of the different *vir* loci, we introduced mutations in the Vir region of a helper plasmid. Mutations in the operons *virA* and *virG* abolished conjugative transfer. This was expected, because the regulatory VirA and VirG proteins mediate the induction of the *vir* regulon by AS. The *virE* mutant mobilized the *incQ* plasmid normally. Possibly the *incQ* plasmid itself (or the chromosome) provides a similar single-stranded DNA-binding protein or such a function is perhaps not necessary in this conjugative system. The plant-inducible host-range loci *virH* and *virC* were not required for mobilization of pKT230. The *virB* operon, however, was necessary, which suggests that VirB proteins function in the T-DNA transfer process like certain Tra proteins function in conjugation. Of the *virD* operon, at least VirD4, which is localized in the innermembrane (Okamoto *et al*, 1991) and homologous to TraG of RP4, was necessary for *incQ* mobilization. Therefore, we postulate that the VirB proteins together with the VirD4 protein mediate the transport of the T-complex through the membrane. Thus, the trans-kingdom transfer of DNA from *Agrobacterium* to plants may occur like the process of bacterial conjugation.

4. The host-range locus virF

The *virF* mutants of *Agrobacterium* show a strongly attenuated virulence on *Nicotiana glauca* (Melchers *et al*, 1990). Such mutants can be complemented for tumor formation by coinfection with a non-oncogenic helper strain such as LBA4404 (Otten *et al*, 1984, Otten *et al*, 1985). The helper strain must express the *virF* gene for complementation, but also the essential *vir* operons *virA*, *virB*, *virG* and *virD* (Melchers *et al*, 1990). The absence of the *virF* locus from nopaline strains causes its weak virulence on *N.glauca*. Jarchow et al. showed that *virF* is inhibitory in agro-infection on *Zea mays* (Jarchow *et al*, 1991). They propose that T-DNA transfer is affected because a potential maize pant cell receptor cannot accept a T-complex containing VirF. Thus, the presence of the *virF* gene can be either stimulatory or inhibitory for T-DNA transfer depending on the plant species involved. With help of promoter-probe vectors we directly showed that the *virF* gene is indeed part of the *vir* regulon. Nucleotide sequencing predicted one ORF for a hydrophylic 22 kDa protein (Melchers *et al*, 1990). By immunoblotting, using antibodies against in *E.coli* overproduced VirF protein, VirF was localized to the cytoplasm of agrobacterial cells, although some of the protein was also found in the membrane fraction (Regensburg-Tuïnk and Hooykaas, unpublished). In order to test the hypothesis that VirF is transported to plant cells during tumorigenesis and is functional in promoting T-DNA transfer from strains mutated in *virF*, we constructed *N.glauca* plants transgenic for *virF* (Regensburg-Tuïnk and Hooykaas, 1992). When such transgenic plants were infected with a *virF* mutant strain or with a nopaline strain, tumors were formed comparable to those found after infection with wild-type *virF*-containing octopine strains. Thus, the VirF protein mediates T-DNA transfer from *Agrobacterium* to the plant even when it is present in the plant cell itself. Because the complete T-complex transport system is required for complementation by coinfection, we postulate that the VirF protein (and probably also the VirE2 protein) uses the same membrane pore for crossing the membrane as the T-complex. In further research we will look at the localization of VirF in transgenic plants to get a more clear idea on its function in T-DNA transfer.

REFERENCES

Albano, M., Breitling, R., and Dubnau, D. A. (1989) Nucleotide sequence and genetic organization of the *Bacillus subtilis comG* operon. J.Bacteriol. *171*, 5386- 5404

Ankenbauer, R. G., and Nester, E. W. (1990) Sugar-mediated induction of *Agrobacterium tumefaciens* virulence genes: structural specificity and activities of monosaccharides. J.Bacteriol. *172*, 6442-6446

Aoyama, T., Hirayama, T., Tamamoto, S., and Oka, A. (1989) Putative start codon TTG for the regulatory protein VirG of the hairy-root-inducing plasmid pRiA4. Gene *78*, 173- 178

Beijersbergen, A., Den Dulk-Ras, A., Schilperoort, R. A., and Hooykaas, P. J. J. (1992) Conjugative transfer by the virulence system of *Agrobacterium tumefaciens*. Science *561*, 1324-1327

Buchanan-Wollaston, V., Passiatore, J. E., and Cannon, F. (1987) The *mob* and *oriT* mobilization functions of a bacterial plasmid promote its transfer to plants. Nature *328*, 172-175

Cangelosi, G. A., Martinetti, G., Leigh, J. A., Lee, C. C., Theines, C., and Nester, E. W. (1989) Role of *Agrobacterium tumefaciens* ChvA protein in export of ß-1,2 glucan. J.Bacteriol. *171*, 1609-1615

Cangelosi, G. A., Ankenbauer, R. G., and Nester, E. W. (1990) Sugars induce the *Agrobacterium tumefaciens* virulence genes via a periplasmic binding protein and a transmembrane signal protein. Proc.Natl.Acad.Sci.USA *87*, 6708-6712

Christie, P. J., Ward, J. E., Gordon, M. P., and Nester, E. W. (1989) A gene required for transfer of T-DNA to plants encodes an ATPase with autophosphorylating activity. Proc.Natl.Acad.Sci.USA *86*, 9677-9681

d'Enfert, C., Reyss, I., Wandersman, C., and Pugsley, A. (1989) Protein secretion by gram-negative bacteria. J. Biol.Chem. *264*, 17462-17468

Douglas, C. J., Staneloni, R. J., Rubin, R. A., and Nester, E. W. (1985) Identification and genetic analysis of an *Agrobacterium tumefaciens* chromosomal virulence region. J.Bacteriol. *161*, 850-860

Ghai, J., and Das, A. (1989) The *virD* operon of *Agrobacterium tumefaciens* Ti plasmid encodes a DNA- relaxing enzyme. Proc.Natl.Acad.Sci.USA *86*, 3109-3113

Herrera-Estrella, A., Van Montagu, M., and Wang, K. (1990) A bacterial peptide acting as a plant nuclear targeting signal: the amino-terminal portion of *Agrobacterium* VirD2 protein directs a ß-galactosidase fusion protein into tobacco nuclei. Proc.Natl.Acad.Sci.USA *87*, 9534- 9537

Hooykaas, P. J. J., Hofker, M., Den Dulk-Ras, H., and Schilperoort, R. A. (1984) A comparison of virulence determinants in an octopine Ti plasmid, a nopaline Ti plasmid, and an Ri plasmid by complementation analysis of *Agrobacterium tumefaciens* mutants. Plasmid *11*, 195-205

Hooykaas, P. J. J., and Schilperoort, R. A. (1992) *Agrobacterium* and plant genetic engineering. Plant Mol. Biol. *19*, 15-38

Howard, E. A., Zupan, J. R., Citovsky, V., and Zambryski, P. C. (1992) The VirD2 protein of *Agrobacterium tumefaciens* contains a C-terminal bipartite nuclear localization signal: implications for nuclear uptake of DNA in plant cells. Cell *68*, 109-118

Huang, M. - L. W., Cangelosi, G. A., Halperin, W., and Nester, E. W. (1990) A chromosomal *Agrobacterium tumefaciens* gene required for effective plant signal transduction. J.Bacteriol. *172*, 1814-1822

Huang, Y., Morel, P., Powell, B., and Kado, C. I. (1990) VirA, a coregulator of Ti-specified virulence genes, is phosphorylated *in vitro*. J.Bacteriol. *172*, 1142-1144

Inon de Iannino, N., and Ugalde, R. A. (1989) Biochemical characterization of avirulent *Agrobacterium tumefaciens chvA* mutants : synthesis and excretion of ß-(1,2) glucan. J.Bacteriol. *171*, 2842-2849

Jarchow, E., Grimsley, N. H., and Hohn, B. (1991) *virF*, the host-range-determining virulence gene of *Agrobacterium tumefaciens*, affects T-DNA transfer to *Zea mays*. Proc.Natl.Acad.Sci.USA *88*, 10426-10430

Jin, S., Roitsch, T., Ankenbauer, R. G., Gordon, M. P., and Nester, E. W. (1990 a) The VirA protein of *Agrobacterium tumefaciens* is autophosphorylated and is essential for *vir* gene regulation. J.Bacteriol. *172*, 525-530

Jin, S., Prusti, R. K., Roitsch, T., Ankenbauer, R. G., and Nester, E. W. (1990 b) Phosphorylation of the VirG protein of *Agrobacterium tumefaciens* by the autophosphorylated VirA protein : essential role in biological activity of VirG. J.Bacteriol. *172*, 4945-4950

Kado, C. I. (1991) Molecular mechanisms of crown gall tumorigenesis. Crit.Rev. in Plant.Sci. *10*, 1-32

Kanemoto, R. H., Powell, A. T., Akiyoshi, D. E., Regier, D. A., Kerstetter, R. A., Nester, E. W., Hawes, M. C., and Gordon, M. P. (1989) Nucleotide sequence and analysis of the plant-inducible locus *pinF* from *Agrobacterium tumefaciens*. J.Bacteriol. *171*, 2506-2512

Kuldau, G. A., De Vos, G., Owen, J., McCaffrey, G., and Zambryski, P. (1990) The *virB* operon of *Agrobacterium tumefaciens* pTiC58 encodes 11 open reading frames. Mol. Gen.Genet. *221*, 256-266

Leroux, B., Yanofsky, M. F., Winans, S. C., Ward, J. E., Ziegler, S. F., and Nester, E. W. (1987) Characterization of the *vir*A locus of *Agrobacterium tumefaciens*: a transcriptional regulator and host range determinant. EMBO J. *6*, 849-856

Mantis, N. J., and Winans, S. C. (1992) Characterization of the *Agrobacterium tumefaciens* heat shock response : evidence for a σ32-like sigma factor. J.Bacteriol. *174*, 991-997

Melchers, L. S., Thompson, D. V., Idler, K. B., Neuteboom, S. T. C., De Maagd, R. A., Schilperoort, R. A., and Hooykaas, P. J. J. (1987) Molecular characterization of the virulence gene *vir*A of the *Agrobacterium tumefaciens* octopine Ti plasmid. Plant Mol.Biol. *9*, 635- 645

Melchers, L. S., Regensburg-Tuïnk, A. J. G., Schilperoort, R. A., and Hooykaas, P. J. J. (1989 a) Specificity of signal molecules on the activation of *Agrobacterium* virulence gene expression. Mol.Microbiol. *3*, 969-977

Melchers, L. S., Regensburg-Tuïnk, A. J. G., Bourret, R. B., Sedee, N. J. A., Schilperoort, R. A., and Hooykaas, P. J. J. (1989 b) Membrane topology and functional analysis of the sensory protein VirA of *Agrobacterium tumefaciens*. EMBO J. *8*, 1919-1925

Melchers, L. S., Maroney, M. J., Den Dulk-Ras, A., Thompson, D. V., Van Vuuren, H. A. J., Schilperoort, R. A., and Hooykaas, P. J. J. (1990) Octopine and nopaline strains of *Agrobacterium tumefaciens* differ in virulence; molecular characterization of the *virF*-locus. Plant Mol. Biol. *14*, 249-259

Okamoto, S., Toyoda-Yamamoto, A., Ito, K., Takebe, I., and Machida, Y. (1991) Localization and orientation of the VirD4 protein of *Agrobacterium tumefaciens* in the cell membrane. Mol.Gen.Genet. *228*, 24-32

Otten, L. A. B. M., DeGreve, H., Leemans, J., Hain, R., Hooykaas, P. J. J., and Schell, J. (1984) Restoration of virulence of *vir* region mutants of *Agrobacterium tumefaciens* strain B6S3 by coinfection with normal and mutant *Agrobacterium* strains. Mol.Gen.Genet. *195*, 159- 163

Otten, L. A. B. M., Piotrowiak, G., Hooykaas, P. J. J., Dubois, M., Szegedi, E., and Schell, J. (1985) Identification of an *Agrobacterium tumefaciens* pTiB6S3 *vir* region fragment that enhances the virulence of pTiC58. Mol.Gen.Genet. *199*, 189-193

Panagopoulos, C. G., and Psallidas, P. G. (1973) Characteristics of Greek isolates of *Agrobacterium tumefaciens* (Smith and Townsend) Conn. J.Appl.Bact. *36*, 233-240

Pansegrau, W., and Lanka, E. (1991) Common sequence motifs in the DNA relaxases and nick regions from a variety of DNA transfer systems. Nucleic Acids Res. *19*, 3455

Pazour, G. J., and Das, A. (1990) *virG*, an *Agrobacterium tumefaciens* transcriptional activator, initiates translation at a UUG codon and is a sequence-specific DNA- binding protein. J.Bacteriol. *172*, 1241-1249

Powell, B. S., and Kado, C. I. (1990) Specific binding of VirG to the *vir* box requires a C-terminal domain and exhibits a minimum concentration treshold. Mol.Microbiol. *4*, 2159-2166

Regensburg-Tuïnk, A.J.G. and Hooykaas, P.J.J. (1992) Expression of the bacterial virulence gene *virF* in transgenic *Nicotiana glauca* plants converts this non-host into a host for nopaline strains of *Agrobacterium tumefaciens* (submitted)

Rodenburg, K. (1992) PhD thesis "Signal transduction in *Agrobacterium tumefaciens*: molecular and structural study into the function of the VirG protein", Leiden University, Netherlands

Roitsch, T., Wang, H., Jin, S., and Nester, E. W. (1990) Mutational analysis of the VirG protein, a transcriptional activator of *Agrobacterium tumefaciens* virulence genes. J.Bacteriol. *172*, 6054-6060

Sen, P., Pazour, G. J., Anderson, D., and Das, A. (1989) Cooperative binding of *Agrobacterium tumefaciens* VirE2 protein to single-stranded DNA. J.Bacteriol. *171*, 2573- 2580

Shimoda, N., Toyoda-Yamamoto, A., Nagamine, J., Usami, S., Katayama, M., Sakagami, Y., and Machida, Y. (1990) Control of expression of *Agrobacterium vir* genes by synergistic actions of phenolic signal molecules and monosaccharides. Proc.Natl.Acad.Sci USA *87*, 6684-6688

Shirasu, K., Morel, P., and Kado, C. I. (1990) Characterization of the *virB* operon of an *Agrobacterium tumefaciens* Ti plasmid: nucleotide sequence and protein analysis. Mol.Microbiol. *4*, 1153-1163

Spencer, P. A., and Towers, G. H. N. (1988) Specificity of signal compounds detected by *Agrobacterium tumefaciens*. Phytochemistry *27*, 2781-2785

Tamamoto, S., Aoyama, T., Takanami, M., and Oka, A. (1990) Binding of the regulatory protein VirG to the phased signal sequences upstream from virulence genes on the hairy-root-inducing plasmid. J.Mol.Biol. *215*, 537-547

Thompson, D. V., Melchers, L. S., Idler, K. B., Schilperoort, R. A., and Hooykaas, P. J. J. (1988) Analysis of the complete nucleotide sequence of the *Agrobacterium tumefaciens virB* operon. Nucleic Acids Res. *16*, 4621-4636

Toro, N., Datta, A., Yanofsky, M., and Nester, E. W. (1988) Role of the overdrive sequence in T-DNA border cleavage in *Agrobacterium*. Proc.Natl.Acad.Sci.USA *85*, 8558-8562

Turk, S. C. H. J., Melchers, L. S., Den Dulk-Ras, H., Regensburg-Tuïnk, A. J. G., and Hooykaas, P. J. J. (1991) Environmental conditions differentially affect *vir* gene induction in diffent *Agrobacterium* strains. Role of the VirA sensor protein. Plant Mol.Biol. *16*, 1051-1059

Turk, S. C. H. J., Nester, E. W. and Hooykaas, P. J. J. (1992) The *virA* promoter is a host range determinant in *Agrobacterium tumefaciens* (submitted)

Wang, K., Stachel, S. E., Timmerman, B., Van Montagu, M., and Zambryski, P. C. (1987) Site-specific nick in the T- DNA border sequence as a result of *Agrobacterium vir* gene expression. Science *235*, 587-591

Ward, J. E., Akiyoshi, D. E., Regier, D., Datta, A., Gordon, M. P., and Nester, E. W. (1988) Characterization of the *virB* operon from an *Agrobacterium tumefaciens* Ti plasmid. J.Biol.Chem. *263*, 5804-5814

Ward, J. E., Dale, E. M., Nester, E. W., and Binns, A. N. (1990) Identification of a VirB10 protein aggregate in the inner membrane of *Agrobacterium tumefaciens*. J. Bacteriol. *172*, 5200-5210

Ward, J. E., Dale, E. M., and Binns, A. N. (1991) Activity of the *Agrobacterium* T-DNA transfer machinery is affected by *virB* gene products. Proc.Natl.Acad.Sci.USA *88*, 9350- 9354

Winans, S. C. (1992) Two-way chemical signaling in *Agrobacterium* -plant interactions. Microbiol. Reviews *56*, 12-31

Yanofsky, M. F., Porter, S. G., Young, C., Albright, L. M., Gordon, M. P., and Nester, E. W. (1986) The *vir*D operon of *Agrobacterium tumefaciens* encodes a site-specific endonuclease. Cell *47*, 471-477

Zambryski, P. C. (1992) Chronicles from the *Agrobacterium* - plant cell DNA transfer story. Annu. Rev. Plant Physiol. Plant Mol. Biol. *43*, 465- 490

Ziegelin, G., Pansegrau, W., Strack, B., Balzer, D., Kröger, M., Kruft, V., and Lanka, E. (1991) Nucleotide sequence and organization of genes flanking the transfer origin of promiscuous plasmid RP4. DNA Sequence *1*, 303-327

Zorreguieta, A., Geremia, R. A., Cavaignac, S., Cangelosi, G. A., Nester, E. W., and Ugalde, R. A. (1988) Identification of the product of an *Agrobacterium tumefaciens* chromosomal virulence gene. Mol. Plant- Microbe Interactions *1*, 121-127

MOLECULAR AND CHEMICAL ANALYSIS OF SIGNAL PERCEPTION BY AGROBACTERIUM

Andrew N. Binns[1], Rolf D. Joerger[1], Lois M. Banta[1], Kyunghee Lee[2], and David G. Lynn[2]. 1. Plant Science Institute, Department of Biology, University of Pennsylvania, Philadelphia, PA.19104; 2. Department. of Chemistry, University of Chicago, Chicago, IL 60637

ABSTRACT Signal perception in the Agrobacterium:plant cell interaction leading to *vir* gene activation involves at least two gene products, ChvE and VirA. The working model is that the periplasmically localized ChvE protein, upon recognizing certain sugars, interacts with the VirA protein found in the inner membrane, increasing its sensitivity to phenolics such as acetosyringone. Using site specific mutagenesis and recently described phenolic derivatives that are inhibitors of *vir* gene induction we are testing the hypothesis that VirA is the phenolic binding protein. We have found that mutations in VirA that lie outside the predicted ChvE binding domain drastically alter the sensitivity of the system to phenolics, most likely by affecting the transduction of the ChvE mediated information. In addition we have utilized radiolabelled α-bromoacetophenones that are known to be specific, irreversible inhibitors of *vir* induction, as affinity labels to identify possible phenolic binding proteins.

Introduction

Perception of several different plant-produced signals by *Agrobacterium tumefaciens* is necessary for the activation of the virulence (*vir*) genes of the Ti plasmid involved in the processing and transfer of DNA from the bacterium into the plant cell (Winans 1992). These signals, produced at wound sites in most dicotyledonous plants, include substituted phenolics, various sugars and elevated [H^+]. In general, phenolics activate *vir* gene expression at low pH (5.5) , while the sensitivity of the system to the phenolics is vastly increased in the presence of certain sugars. Genetic studies have shown that the VirA and VirG proteins encoded by the Ti plasmid and the ChvE protein, encoded by the Agrobacterium chromosome, play a central role in transducing these signals (Stachel and Zambryski 1986; Huang et al. 1990). VirA spans the inner membrane, having two transmembrane domains, a large periplasmic domain and a large C-terminal cytoplasmic domain (Winans et al. 1989; Melchers et al. 1989). It shows significant sequence similarity to the "sensor" class of proteins involved in the "two-component" regulatory systems seen in a wide variety of bacteria (Stock et al. 1990). The VirA protein is capable of autophosphorylation (Jin et al. 1990b) and can transfer this phosphate to the cytoplasmic VirG protein (Jin et al. 1990a), which is homologous to the activator class of proteins in the "two component" system (Stock et al. 1990). Phosphorylated VirG appears

E. W. Nester and D. P. S. Verma (eds.),
Advances in Molecular Genetics of Plant-Microbe Interactions, 51–61.

to act as a transcriptional regulator that can bind to *vir* promoters and activate transcription of the various *vir* genes (Jin et al. 1990c; Pazour and Das 1990).

The mechanism of signal perception in this system is poorly understood. While genetic analysis has shown that VirA is required for the activation to proceed, no physical evidence has been presented to demonstrate that it interacts directly with any of the signals. In fact, the presence of sugar is sensed by the chromosomally encoded *chvE* gene product (Huang et al. 1990; Cangelosi et al. 1990; Shimoda et al. 1989); ChvE is predicted to be localized to the periplasm and thought to interact with VirA, altering the ability of VirA to respond to phenolics. Earlier analysis of phenolic derivatives that activate *vir* gene expression led us to present a model concerning the mechanism by which phenolics could activate a receptor protein (Hess et al. 1991). In this model, the phenolics are proposed to act as vinylogous carboxylic acids capable of transferring a proton from an acidic residue at one position on the receptor to a basic residue on another part of the protein. Several lines of evidence support this model. Phenolic derivatives that can act as acids are capable of *vir* induction, whereas similar molecules that are not readily deprotonated at the phenolic hydroxyl group are not inducers (Hess et al. 1991). Additionally, the α-haloketone derivative of acetosyringone (bromo-acetosyringone, ASBr), was predicted to bind covalently to acidic residues of the receptor protein. We found ASBr to be a specific, irreversible inhibitor of *vir* induction, suggesting that a specific, covalent modification of the phenolic receptor disrupted transcriptional activation of the *vir* genes (Hess et al 1991). Finally, analysis of the available sequences (Leroux et al. 1987; Melchers et al. 1987; Morel et al. 1989) has shown that there are two conserved glutamic acid residues near the second transmembrane domain of VirA, a region proposed to be important in signal recognition based on molecular genetic studies (Melchers et al. 1989). Here we describe the initial tests of our model of phenolic signal perception in Agrobacterium. First, we have begun to analyze the role of the conserved glutamate residues of VirA in phenolic signal transduction. Second, we have initiated affinity labelling studies utilizing radiolabelled versions of the α-haloketone type *vir* inhibitors.

Methods and Materials

1. Bacterial strains and plasmids.

Agrobacterium tumefaciens strain A348 is strain A136 (C58 cured of its Ti plasmid) containing the wild type octopine plasmid, pTiA6 (Garfinkel et al. 1981); strain 358mx is strain A136 containing pTi358, a version of pTiA6 containing Tn*3*HoHo1 in the *virE* locus (Stachel and Nester 1986). Strains 348-3 and 358-3 are strain A136 carrying pTiA6 ΔvirA and pTi358 ΔvirA respectively (Lee et al. 1992). Plasmid pVRA8 was constructed by cloning the 4.7 kbp Kpn fragment containing the wild type *virA* gene from pTiA6 into pUCD2 (Close et al. 1984) and pVRA10 is the same construct except that the sequences encoding residues 62-242 of the VirA periplasmic domain were deleted (Lee et al. 1992). pVRA11 is pUCD2 containing *virA* in which residues 242-257 have been deleted using oligonucleotide directed mutagenesis as described (Ward et al. 1990) using 5' CAAGAAAGATACGTGCCATCGCTGCAGGGCATCTCCGC3' as the mutagenic primer. pVRA15 is pUCD2 containing *virA* with residue 254 converted from glutamate to glutamine and residue 255 converted from glutamate to leucine using 5'

TGCGCTCCGCAGCTGTACATTTTT 3' as the mutagenic primer. These changes in the *virA* sequence were confirmed by sequence analysis.

2. Vir induction and virulence assays

Analysis of *vir* induction in strains carrying pTi358 or its derivatives was via assay of ß-galactosidase activity as described previously (Hess et al. 1991). AB induction medium (Ward et al. 1990) contained glycerol as a non-fermentable carbon source and either arabinose (inducing sugar) or sucrose (non-inducing sugar) at 10mM. In the induction assays, the Agrobacterium strains were cultured in induction medium for 24 hours at 30°C prior to assay.

Virulence assays were performed by co-cultivation of *Nicotiana tabacum* L cv. Havana 425 leaf explants and Agrobacterium strains for 48 hr on hormone free MS medium (Murashige and Skoog 1962) containing the indicated concentration of AS, followed by transfer to hormone free MS medium containing cefotaxime and vancomycin (Sigma Co.), 500 and 100 μg/ml respectively, to inhibit bacterial growth. After three weeks incubation in the dark at 25°C the number of leaf pieces with tumors was determined. At least 20 leaf explants were used at each AS concentration.

3. Affiinity labelling

Preparation of AV^{125}IBr and its utilization in affinity labelling studies has been described previously (Lee et al. 1992).

Results

1. VirA glutamate residues affect signal perception

The results of Melchers et al. (1989) demonstrated that a large portion of the periplasmic domain of VirA could be deleted without destroying its ability to respond to phenolics, whereas replacement of the second transmembrane domain rendered the protein inactive. These results suggested that sequences in or around the second transmembrane domain may be important in signal perception. Analysis of the amino acid sequence of this region indicated that the VirA protein of three different strains contained glutamates at or near residues 246 and 255 (of the pTiA6 VirA) (Fig 1). In order to test the hypothesis that acidic residues on the putative phenolic receptor were important, we decided to mutagenize this region. The first mutation tested was a deletion of the region in VirA from amino acid residues 242 to 257 (pVRA11), thus eliminating both conserved glutamates as well as two other glutamates in the region (243, 254; see Fig 1). For comparison, we constructed the same periplasmic deletion (Δ 62-242; pVRA10) as described by Melchers et al. (1989) and Cangelosi et al. (1990). These mutations, as well as the wild type version of *virA* were cloned into pUCD2 and electroporated into two different strains, 358-3 and 348-3, in which the wild type *virA* had been deleted. Strain 358-3, derived from 358mx (Stachel and

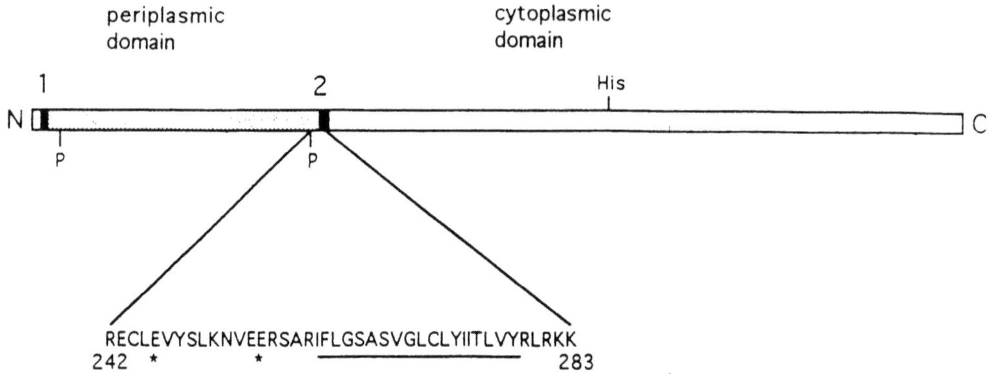

Figure 1. Schematic view of VirA. The two membrane-spanning domains are indicated by the numbers 1 and 2 above solid boxes. The region containing amino acid residues 242 through 283 is shown in an expanded view with the amino acid residues of the second membrane-spanning domain underlined. Conserved glutamic acid residues (246, 255) are marked by asterisks. His indicates the residue that is the site of autophosphorylation. P indicates Pst1 restriction site

Nester 1986) was chosen because it carries a transcriptional fusion of Tn*3*HoHo1 to *virE*; thus, the reporter enzyme ß-galactosidase is under the control of the *vir* inducing system. Strain 348-3 is derived from A348 and is wild type for all *vir* and T-DNA functions except *virA*. This strain was used in virulence (tumor formation) assays.

Analysis of *vir* induction by acetosyringone (AS) in the various mutants was carried out in either the presence or absence of arabinose, a sugar known to stimulate *vir* induction through the ChvE protein (Shimoda et al. 1990; Cangelosi et al. 1990). As previously reported by Melchers et al. (1989) and Cangelosi et al. (1990) deletion of most of the periplasmic domain (pVRA10) resulted in a loss of sensitivity to AS, compared to wild type, in the presence of arabinose (Fig 2A) but an increase in sensitivity in the absence of this sugar (Fig 2B). Interestingly, pVRA11, carring the small periplasmic deletion of residues 242-257, still allowed 358-3 to respond to AS in a dose dependent fashion, but was less sensitive to the phenolics than either the wild type or the large periplasmic deletion in either the presence or absence of arabinose. In addition to their effects on the sensitivity of 358-3 to the phenolics, the mutations also had substantial effects on the maximal levels of *vir* induction. We have consistently observed the result seen in Fig 2B: when tested in induction medium lacking arabinose (but containing sucrose) wild type *virA* in 358-3 results in maximal levels of ß-galactosidase activity 2-3 fold lower than the maximal activity of either of the mutants. This phenomenon is under further investigation.

The results described above indicated that the deletion of amino acids 242-257 from

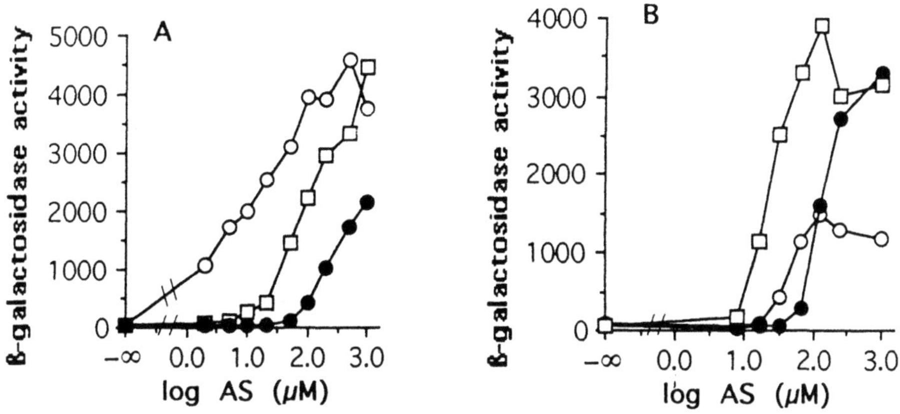

Figure 2. *vir* induction, assayed as ß-galactosidase activity, in 358-3 carrying pVRA8 (O), pVRA10 (□) and pVRA11 (●) in response to AS in the presence of arabinose (A) or sucrose (B).

the VirA protein drastically affected its ability to respond to the inducing sugar. While this suggests that these amino acids play a role in signal transduction, this deletion would also result in the placement of the putative ChvE binding domain of VirA (aa 225-239; Cangelosi et al. 1990) quite near the inner membrane, a situation that could negatively affect ChvE/VirA interactions. To further test the involvement of the glutamate residues and also lessen the possibility of steric hindrance of potential ChvE/VirA interactions, we used site-directed mutagenesis to change the glutamate residues at positions 254 and 255 to glutamine and leucine, respectively. This version of *virA* was cloned into pUCD2, resulting in pVRA15, and electroporated into 358-3 and 348-3. Analysis of *vir* induction in 358-3 (pVRA15) showed that this form of VirA has virtually the same effect on phenolic sensitivity as does *virA* Δ242-257 (pVRA11). In the presence of arabinose the wild type *virA* (pVRA8) is approximately 100 fold more sensitive to AS then the mutated version lacking glutamates 254 and 255 (Fig 3A). In the absence of arabinose, 358-3 (pVRA15) is slightly less sensitive to AS then 358-3 (pVRA8) (Fig 3B). Finally, 358-3 (pVRA15) showed a 2-3 fold higher level of maximal induction in comparison to the wild type strain when tested in the absence of arabinose. These results indicate that glutamates 254 and/or 255 of VirA are important in transducing the sugar/ChvE signalling that leads to increased phenolic sensitivity and in suppressing maximal *vir* induction in the absence of the inducing sugars.

Because the various *vir A* mutations resulted in altered responses to AS in terms of *vir* induction, we sought to determine whether they exhibited altered virulence phenotypes.

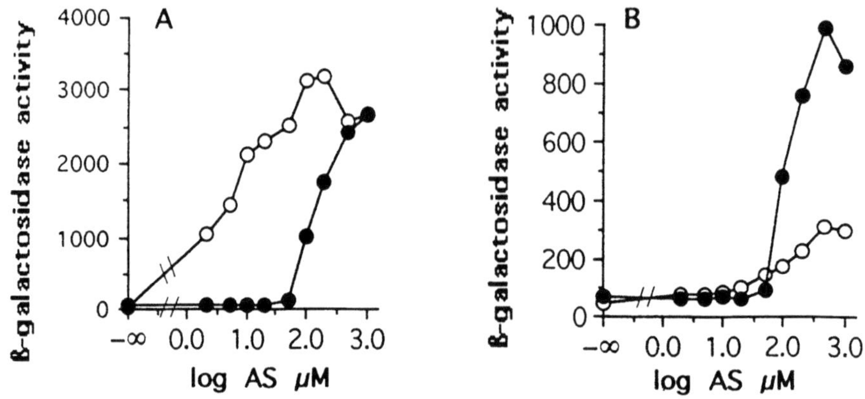

Figure 3. *vir* induction, assayed as ß-galactosidase activity, in 358-3 carrying pVRA8 (O), or pVRA15 (●) in response to AS in the presence of arabinose (A) or sucrose (B).

In these experiments tobacco leaf explants were prepared, co-cultivated with the indicated strains for 48 hr on medium containing the indicated concentration of AS, as described in Materials and Methods, and subsequently scored for tumor formation. Under these conditions strain 348-3 containing wild type *virA* was virulent, but could be stimulated to produce more tumors by increasing concentrations of AS (Fig 4). Strains 348-3 carrying pVRA11 or pVRA15 induced no tumors on the leaf explants except when co-cultivated at the highest AS levels (500μM). 348-3 containing pVRA10 gave a response intermediate between the wild type and the other mutants. These results indicate that the mutants' order

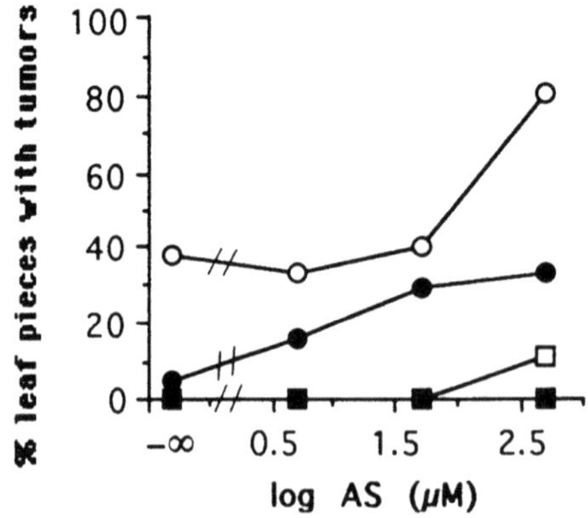

Figure 4. Tumor intitiation on tobacco leaf explants by strain 348-3 carrying pVRA8 (O), pVRA10 (●), pVRA11 (□) and pVRA15 (■) after co-cultivation in the presence of various doses of AS.

of responsiveness to AS in terms of virulence is similar to their relative sensitivities to AS in the *vir* induction assays carried out in the presence of arabinose. Addition of arabinose to the AS containing co-cultivation medium had no effect on tumor formation by either wild type or *virA* mutants (data not shown).

2. Affinity labelling of phenolic binding proteins

The fact that each of the various mutant forms of *virA* described above still responded to acetosyringone suggests that at least those regions mutated do not represent the sole phenolic binding (perception) sites. An alternate strategy to identify the phenolic binding site has been developed in which the specific and irreversible inhibitor of *vir* induction, ASBr (Hess et al. 1991), is utilized in affinity labelling studies. In these studies, the formal replacement of the 5' methoxy group on the phenolic ring of AS with iodine resulted in the formation of 5' iodo acetovanillone (AVI), which had equivalent *vir* inducing activity as AS (Lee et al. 1992). The brominated derivative of AVI (AVIBr) had the same inhibitory activities as ASBr (Lee et al 1992). For these reasons $AV^{125}IBr$ was chosen as a radiolabelled inhibitor for affinity labelling studies.

Analysis of either intact cells of Agrobacterium, or protein extracts prepared from such cells, exposed to the $AV^{125}IBr$ indicated that no proteins in the molecular weight range of VirA were labelled (Lee et al. 1992). However, two non-abundant, smaller proteins (10 and 21 kDa) were labelled. Interestingly, these proteins were not labelled when the labelling reactions were carried out on cells or extracts pre-incubated with non-radioactive ASBr, but were labelled when bromoacetophenone (APBr) was used as competitor. These competiton results correspond to studies of the inhibitory activities of the competitors: ASBr is an inhibitor of *vir* induction whereas APBr is not. However, the labelling experiments were all executed at pH 7.2, conditions in which the *vir* induction system, and particularly VirA, may be less active (Winans 1992). Therefore, intact cells of *A. tumefaciens* 358mx were exposed to $AV^{125}IBr$ in induction medium at pH5.5, and extracts were analyzed by SDS-PAGE. An audioradiogram of a typical gel (Fig. 5) shows that no radiolabelled proteins are visible in the molecular weight range of VirA whereas both p21 and p10 are distinctly labelled, essentially repeating the labelling pattern observed at pH 7.2 (Lee et al 1992). Western analysis of these extracts do show that VirA is present, and even when VirA is overexpressed no labelling by $AV^{125}IBr$ is detected (Lee et al. 1992).

Discussion

The experiments described here concern tests of a model that predicted acidic residues of the phenolic receptor would be critical in the perception and activity of phenolics such as AS. In the first series of experiments, site specific mutagenesis was used to construct versions of VirA that were altered at two conserved glutamate residues near the second transmembrane domain. If these residues are critical in phenolic perception, then their elimination should result in insensitivity to AS. Analysis of Agrobacterium strains carrying such mutations (pVRA11, pVRA15) showed that *vir*

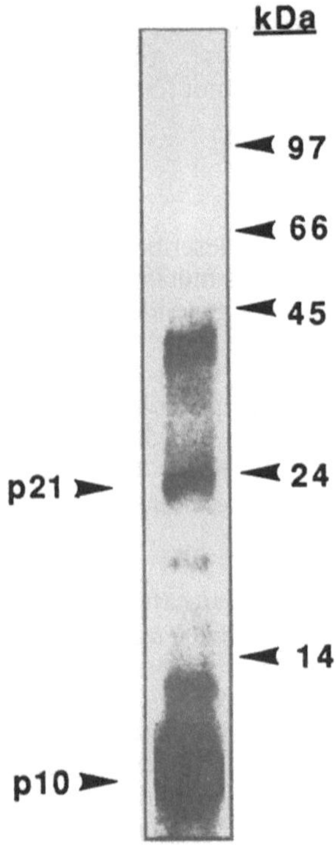

Figure 5. In vivo affinity labelling. *A. tumefaciens* strain 358mx was grown overnight, pelleted and resuspended in 40μl of AB induction medium (pH5.5) containing 1μM AV^{125}IBr. After a 30 minute incubation at 28°C the cells were pelleted and resuspended in 10μl of sucrose solution (0.75M sucrose in 0.55M Tris and 0.1M EDTA) and diluted with 10μl of TES (0.05Tris (pH8.0), 0.1M NaCl and 0.01 M EDTA). The crude extract resulting from a brief sonication was precipitated with acetone and electrophoresed in SDS-PAGE (10% acrylamide) and the gel used to expose X-ray film.

induction was still responsive to AS in a dose dependent fashion. However, these strains were 100 fold less sensitive to AS then wild type when "inducing" sugars such as arabinose were included in the induction medium. In the absence of inducing sugars the mutants consistently exhibited two phenotypes. First, they were slightly less responsive to AS. Second, in contrast to the wild type strains, they could be induced to maximal levels seen in the presence of inducing sugars. Finally, virulence assays demonstrated that the order of responsiveness to AS in terms of tumor formation corresponded to the responsiveness to AS for *vir* induction in the presence of inducing sugars such as arabinose.

These results suggest that the glutamate residues at positions 254 and/or 255 are involved in transducing signals of inducing sugars, thereby affecting the sensitivity of the *vir* inducing system to phenolics, but are not involved directly in sensing AS. Such a result is supported by the recent finding that a version of VirA missing the entire periplasmic domain and both transmembrane domains can still respond to AS in a dose dependent fashion (Winans 1992). The actual sites of ChvE/VirA interaction have not been identified, although VirA does contain a sequence (at residues 225-239) homologous to a region in an *E. coli* transmembrane protein, TRG, that has been implicated in interaction with sugar-binding proteins affecting chemotaxis (Cangelosi et al 1990). The glutamates at residues

254 and/or 255 could be directly involved in the interaction of VirA with ChvE or necessary to transduce the information resulting from that interaction. An alternate possibility is that more than one phenolic binding site exists and that the glutamates are involved in the activity of one of these sites.

A second test of the model concerning phenolic perception was that α-haloketones such as ASBr would interact with acidic residues in a covalent fashion. Earlier studies showed that ASBr is a specific and irreversible inhibitor of *vir* induction. The studies reported here and elsewhere (Lee et al 1992) utilized an iodinated version of this molecule, $AV^{125}IBr$, as an affinity labelling reagent. Interestingly, VirA was not labelled in these experiments, even when it was overexpressed (Lee et al 1992). In contrast, several other non-abundant, low molecular weight proteins were labelled when exposed to $AV^{125}I$. Two of these, p10 and p21, are of particular interest because prior exposure to non-radioactive ASBr will protect them from being labelled by $AV^{125}IBr$, whereas prior exposure to APBr does not provide such protection (Lee et al 1992). These results parallel the effects of ASBr and APBr on *vir* induction, the former being a specific inhibitor whereas the latter is not. The role that p10 and p21 may play in *vir* induction is currently under investigation.

Acknowledgements We gratefully ackowledge support form the NSF Chemistry of Life Processes Grant (CHE 9020872) to ANB and DGL

References

Cangelosi, G. A., Ankenbauer, R. G., and Nester, E. W. (1990) Sugars induce the *Agrobacterium* virulence genes through a periplasmic binding protein and a transmembrane signal protein. Proc. Natl. Acad. Sci. USA 87:6708-6712

Close, T. J., Zaitlin, D., and Kado, C. I. (1984) Design and development of amplifiable broad-host-range vectors: analysis of the *vir* region of *Agrobacterium tumefaciens* plasmid pTiC58. Plasmid 12:111-118

Garfinkel, D. J., Simpson, R. B., Ream, L. W., White, F. F., Gordon, M. P., and Nester, E. W. (1981) Genetic analysis of crown gall: a fine structure map of the T-DNA by site directed mutagenesis. Cell 27:143-153

Hess, K. M., Dudley, M. W., Lynn, D. G., Joerger, R. D., and Binns, A. N. (1991) Mechanism of phenolic activation of *Agrobacterium* virulence genes: development of a specific inhibitor of bacterial sensor/response systems. Proc. Natl. Acad. Sci. USA 88:7854-7858

Huang, M.-L. W., Cangelosi, G. A., Halperin, W., and Nester, E. W. (1990) A chromosomal *Agrobacterium tumefaciens* gene required for effective plant signal transduction. J. Bacteriol. 172:1814-1822

Jin, S., Prusti, R. K., Roitsch, T., Ankenbauer, R. G., Nester, E. W., (1990a) Phosphorylation of the VirG protein of *Agrobacterium tumefaciens* by the autophosphorylated VirA protein: essential role in biological activity of VirG. J. Bacteriol. 172:4945-4950

Jin, S., Roitsch, T., Ankenbauer, R. G., Gordon, M. P., and Nester, E. W. (1990b) The VirA protein of *Agrobacterium tumefaciens* is autophosphorylated and essential for *vir* gene regulation. J. Bacteriol. 172:525-530

Jin , S., Roitsch, T., Christie, P. J., Nester, E.W. (1990c) The regulatory VirG protein specifically binds to a cis-acting regulatory sequence involved in transcriptional activation of *Agrobacterium tumefaciens* virulence genes. J. Bacteriol. 172: 531-537

Lee, K., Hess, K. M., Dudley, M. W., Lynn, D. G., Joerger, R. D., and Binns, A. N. (1992) Mechanism of phenolic activation of *Agrobacterium* virulence genes: identification of binding phenolic proteins. Proc. Natl. Acad. Sci. USA in press

Leroux, B., Yanofsky, M. F., Winans, S. C., Ward, J. E., Zeigler, S. F., and Nester, E. W. (1987) Characterization of the *virA* locus of *Agrobacterium tumefaciens*: a transcriptional regulator and host range determinant. EMBO J. 6:849-856

Melchers, L. S., Regensburg-Tuink, J. G., Bourret, R. B., Sedee, N. J. A., Schilperoort, R. A., Hooykaas, P. J. J. (1989) Membrane topology and functional analysis of the sensory protein VirA of *Agrobacterium tumefaciens*. EMBO J. 8:1919-1925

Melchers, L. S., Thompson, D. V., Idler, K. B., Neuteboom, T. C., deMaagd, R. A., Schilperoort, R. A., and Hooykaas, P. J. J. (1987) Molecular characterization of the virulence gene *virA* of the *Agrobacterium tumefaciens* Ti plasmid. Plant Mol. Biol. 9:635-645

Morel, P., Powell, B. S., Rogowsky, P. M., and Kado, C. I. (1989) Characterization of the *virA* virulence gene of the nopaline plasmid, pTiC58 of *Agrobacterium tumefaciens*. Mol. Microbiol. 3:1237-1246

Murashige, T. and Skoog, F. (1962) A revised medium for rapid growth and biosassay with tobacco tissue culture. Physiol Plant. 15:473-496

Pazour, G. J., and Das, A. (1990) *vir G*, an *Agrobacterium tumefaciens* transcriptional activator, initiates translation at a UUG codon, and is a sequence specific DNA binding protein. J. Bacteriol. 172:1241-1249

Shimoda, N., Toyoda-Yamamoto, A., Nagime, J., Usami, S., Katayama, M., Sakagami, Y., and Machida, Y. 1990. Control of expression of *Agrobacterium vir* genes by synergistic actions of phenolic signal molecules and monosaccharides. Proc. Natl. Acad. Sci. USA 87:6684-6688

Stachel, S. E., and Nester, E. W. (1986) The genetic and transcriptional organization of the A6 Ti plasmid of *Agrobacterium tumefaciens*. EMBO J.. 5:1445-1454

Stachel, S. E., and Zambryski, P.C. (1986) *virA* and *virG* control the plant-induced activation of the T-DNA transfer process of *Agrobacterium tumefaciens*. Cell 46:325-333

Stock, J. B., Stock, A. M., Mottonen, J. M. (1990) Signal transduction in bacteria. Nature 344:395-400

Ward, J. E. Jr., Dale, E. M., Nester, E. W. and Binns, A. N. (1990) Identification of a VirB10 protein aggregate in the inner memnbrane of *Agrobacterium tumefaciens*. J. Bacteriol. 172:5200-5210

Winans, S. C. (1992) Two-way chemical signalling in *Agrobacterium*-plant interactions. Microbiol. Rev. 56:12-31

Winans, S. C., Kerstetter, R. A., Ward, J. E., and Nester, E. W. (1989) A protein required for transcriptional regulation of *Agrobacterium* virulence genes spans the cytoplasmic membrane. J. Bacteriol. 171:1616-1622

A. TUMEFACIENS T-DNA TRANSPORT: ROLES FOR VirB, VirD2 AND VirE2

B.G. McLean, Y. Thorstenson, V. Citovsky, J.R. Zupan, E. Greene, and P.C. Zambryski
Department of Plant Biology, University of California-Berkeley
Berkeley, CA 94720

ABSTRACT. The transfer of the T-DNA from *Agrobacterium* to the plant cell is a complicated process. Several of the steps involved have been the focus of intense study, while other steps remain essentially black boxes. Here we present an overview of this specialized interaction between bacteria and plant, focusing on aspects of the transfer process that have been primary subjects for recent research in our laboratory.

Agrobacterium tumefaciens is a gram-negative soil bacterium that causes neoplasm or crown gall on many dicotyledonous plants by genetically transforming the plant cells [reviewed recently in (7; 32)]. The *Agrobacterium*-plant cell interaction constitutes the only known natural example of DNA transfer between kingdoms. In *Agrobacterium*, three genetic components are required for plant cell transformation. One component is found on the *Agrobacterium* chromosome, the chromosomal virulence (*chv*) genes. The products of the *chv* genes are involved in physical interaction between the bacterial cell and the plant cell (10; 33). The two other genetic components are found on the large (~200 kbp) tumor-inducing or Ti plasmid located in the bacterial cell (7; 32). The first component is the T-DNA which is copied from a region of the Ti plasmid and eventually is transferred to the plant cell. T-DNA is delimited by two imperfect 25 bp repeats at its ends, the T-DNA borders. Any DNA between these borders can be transferred to the plant cell. Transfer of the T-DNA element is polar and initiates from the right border (7; 32). Consequently the right border is required for T-DNA transfer while deletion or manipulation of the left border has little effect. Further, the sequence context surrounding the T-DNA borders greatly affects their activity. On the native Ti plasmid, sequences surrounding the right border increase polar DNA transfer while sequences surrounding the left border decrease transfer (7; 32). While T-DNA is the mobile element, unlike classical transposable elements, it does not encode the products which mediate its transfer. These latter products are provided in *trans*.

Research into the Agrobacterium-plant cell interaction shows that the some of the underlying mechanisms involved in DNA transfer represent modification or adaptations of common prokaryotic processes: (1) a two component signal transduction system and (2) conjugative DNA transfer from a donor to a recipient cell (26). The transformation of the plant cell by *Agrobacterium* can be divided into several basic steps: (1) induction of virulence (*vir*) gene expression, (2) production of the T-strand, the single-stranded T-DNA copy that will be transferred, (3) formation of the T-complex, the T-strand coated with proteins, (4) exit of the T-complex from *Agrobacterium*, (5) T-complex entrance into the plant cell, (6) nuclear uptake of the T-complex, and (7) integration of the

E. W. Nester and D. P. S. Verma (eds.),
Advances in Molecular Genetics of Plant-Microbe Interactions, 63–71.

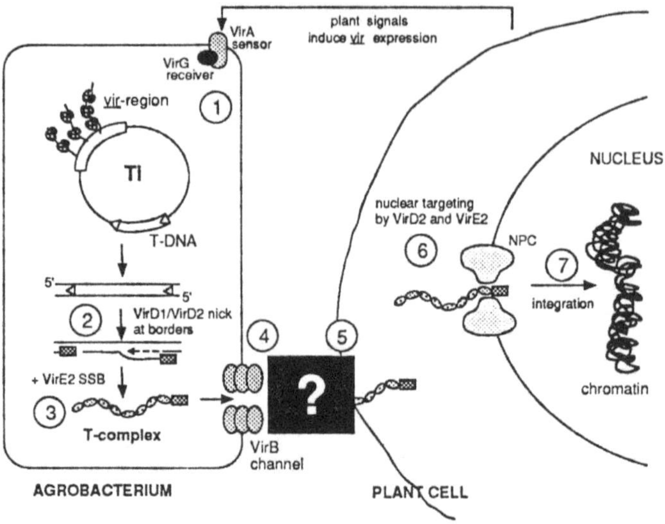

Figure 1. Basic steps in the transformation of plant cells by *Agrobacterium*.
The numbers in the diagram correspond to the seven steps listed in the text.
▩ - VirD2; ⬭ - VirE2; NPC - nuclear pore complex.

T-DNA into the plant genome (illustrated in figure 1). This article will focus on those steps that our laboratory has contributed to recently (steps 3, 4, and 6); relevant background is briefly presented (steps 1 and 2).

1. Induction of *vir* expression

The majority of the trans-acting products for T-DNA transfer are provided by the second genetic component on the Ti plasmid, the *vir* region. The *vir* region is an approximately 35 kbp stretch of DNA located outside the T-DNA and composed of seven complementation groups: *vir*A, *vir*B, *vir*C, *vir*D, *vir*E, *vir*G, *vir*H (18; 25).

The *vir*A and *vir*G loci constitute a two component positive regulatory system that activates *vir* gene expression in response to plant-derived inducing molecules (7; 31; 32). The sensor protein, VirA, spans the inner membrane of the bacterium and responds either directly or indirectly to inducer molecules such as phenolics and monosaccharides released by wounded plant cells. VirA then activates VirG, the regulator protein, to cause transcription of the *vir* loci. VirG contains a DNA binding region which binds a specific sequence, the *vir* box, in the promoter regions upstream of the *vir* genes. The activation of VirG by VirA presumably involves protein phosphorylation since VirA appears to be a protein kinase capable of autophosphorylation and phosphotransfer to VirG (15; 16; 17). The role of phosphorylation in the activation of transcription by VirG has not yet been determined.

2. Production of the T-strand

Induction of *vir* gene expression leads to the production of single-stranded (ss) DNA molecules called T-strands that are homologous to the bottom strand of the T-DNA (7; 32). The synthesis of the T-strand starts with ss endonucleolytic cleavages between the third and fourth base pairs of both 25 bp border repeats. The nicks are used as initiation and termination sites for displacement of a linear ss copy of the T-DNA region. Since the nopaline-type Ti plasmid contains a single T-DNA region, a single T-strand is produced; the more complex octopine Ti plasmid contains three T-DNA regions bracketed by four T-DNA border sequences, resulting in six possible T-strands. In a population of vir-induced *Agrobacteria*, approximately one T-strand is detected per cell.

Two *vir* specific products, VirD1 and VirD2 are required for both nicking and T-strand synthesis (7; 32). The functions of VirD1 and VirD2 have been assigned using overproduced proteins from E. coli. VirD1 extracts exhibit topoisomerase activity; although it can relax nonspecific DNA substrate, VirD1 shows affinity for the 25 bp repeat at the T-DNA borders (13). VirD2 extracts show site-specific nicking at the 25 bp repeats (13). These results suggest that VirD1 may relax the Ti plasmid in the region containing the 25 bp repeats and allow VirD2 to nick the DNA at a specific region in these repeats. While VirD1 and VirD2 are the only *vir*-specific products required for generation of the T-strand, the products of virC locus have been shown to enhance T-DNA border nicking in *Agrobacterium*.(28) VirC1 enhances T-strand production in a heterologous E. coli system when VirD1 and VirD2 are limiting but has no effect when VirD1 and VirD2 are produced in high amounts (8).

3. Formation of the T-complex

Following its production, the T-strand must traverse the bacterial cell membrane, the bacterial cell wall, the plant cell wall, and the plant cell and nuclear membranes before integrating into the plant genome. For the process of DNA transfer to be successful, the T-strand must avoid degradation by nucleases. Consequently, as a protective and possibly a guidance mechanism, the T-strand probably travels as a DNA-protein complex. At least two *vir*-encoded proteins, VirE2 and VirD2, are likely components of this T-complex.

Since the T-strand is ssDNA, a ssDNA binding protein (SSB) would be an obvious component of the T-complex; a vir-induced protein which functions as an SSB is VirE2 (7; 32). *In vitro* experiments show that VirE2 preferentially binds ssDNA without sequence specificity. This binding is rapid and highly cooperative. VirE2 also protects ssDNA from endonucleases and both 3' and 5' exonucleases. The stoichiometry of VirE2 production in *Agrobacterium* suggests that it can cover the entire length of the T-strand. VirE2 is the most abundant protein in *vir*-induced Agrobacterium cells, having an estimated production of 1000 molecules per cell. Since one VirE2 molecule is proposed to cover 30 nucleotides, a 20 kb nopaline-type T-strand requires 600 VirE2 molecules. Thus, more than enough VirE2 is present in the *Agrobacterium* cell to provide the T-strand with a protective coating from nucleases.

Electron microscopy of VirE2-ssDNA complexes suggests that VirE2 may serve another function in addition to nuclease protection (5). Dramatic changes occur in the conformation of ssDNA following incubation with VirE2. Free ssDNA appears as folded, collapsed molecules, but VirE2-bound ssDNA complexes appear unfolded and extended, forming exceptionally thin protein-ssDNA complexes. Thus, as proposed for chaperonin-mediated movement of proteins through organelle membranes (23), T-strands may be modified by VirE2 to an unfolded and more transferable form, capable of transfer across several membranes.

While VirE2 binds the T-strand along its entire length, the transfer of the T-strand is a polar process. Another protein, then, may provide a piloting function to directionally target the complex out of the bacterial cell and into the plant cell. The VirD2 protein may play this role. VirD2 is tightly linked to the 5' leading end of the T-strand (7; 32). This attachment is resistant to heat and strong denaturants and may be a covalent linkage.

4. Exit of the T-complex from *Agrobacterium*

Although VirE2 and VirD2 may supply protection and direction for T-strand transfer, the actual movement of the T-complex requires other proteins. For example, after formation of the complex, it must pass through the bacterial membrane. Thus a proposed function for some *vir* proteins is the modification of the bacterial membrane to create a transmembrane pore which would enable movement of the DNA out of the cell. The proteins of the *vir*B operon, an operon essential for DNA transfer, are likely candidates for formation of this transmembrane channel [reviewed extensively in (32)]. Amino acid sequence analyses, cell fractionation studies, TnPhoA translational fusions, and immuno-electron microscopy have allowed predictions of subcellular locations and possible functions of the VirB proteins.

VirB1. Cell fractionation indicates that VirB1 is a soluble protein in the periplasm (27). This result agrees with the prediction from amino acid sequence analysis that VirB1 contains a signal sequence for transport across the inner membrane.

VirB2 and VirB3. VirB2 and VirB3 have not been assayed by cell fractionation but, based on translational fusions with TnPhoA, are thought to be exposed to the periplasm (27).

VirB4. VirB4 has been localized to the total membrane fraction (11) and further pinpointed to the inner membrane (27). VirB4 exhibits a similar distribution to that of VirB11 (see below), suggesting VirB4 and VirB11 interact with one another. Although VirB4 does not have a predicted membrane spanning region, its conformation may allow association with the membrane.

VirB5. Cell fractionation indicates VirB5 is a soluble protein present in the cytoplasm (27). This localization of VirB5 is inconsistent with a predicted periplasmic location based on experiments using a VirB5::TnPhoA translational fusion, and with amino acid sequence analysis indicating a signal sequence for transport across the inner membrane (27). The predicted association of VirB5 with the membrane may be lost during cell disruption and consequently cytoplasmic location of VirB5 during cell fractionation may be incorrect. Alternatively, VirB5 may be found generally in the cytoplasm but have the capability of being transported to the periplasm, perhaps acting as a shuttle protein for the T-complex.

VirB8 and VirB9. VirB8 and B9 show similar distributions in cell fractions, being present in significant amounts in both the inner and outer membranes but most abundant in the inner membrane (27). As suggested for VirB4 and VirB11, VirB8 and VirB9 may also interact. The inner membrane location of VirB8 is also corroborated by use of a VirB8::TnPhoA translational fusion and transmission electron microscopy showing immunogold labeling of inner membranes(27).

VirB10. Consistent with predictions from amino acid sequence analysis (27), VirB10 has been shown to be an inner membrane protein by membrane fractionation; it forms large aggregates *in vivo*, by cross-linking either to itself or to other proteins (27; 30). We detected VirB10 in the inner membrane and in an equal amount in the outer membrane. Differences seen in its distribution may be due differences in the two methods used for cell fractionation, i.e. Ward *et al.* used lysozyme during their procedure; the detection of VirB10 in the inner and outer membranes following fractionation without lysozyme suggests VirB10 is associated with the outer membrane fraction when the peptidoglycan layer is intact.

VirB11. VirB11 has also been localized to the inner membrane by cell fractionation (27). Since there are no obvious membrane-spanning alpha helical regions in the predicted amino acid sequence for VirB11, the protein may either fold in a manner that creates a membrane-compatible domain or tightly associate with another integral membrane protein.

The functions of the *vir*B proteins are not well understood. VirB11 has been shown to have ATPase activity (4). Other proteins with homology to VirB11 confer competence for DNA uptake in *Bacillus subtilis* and pullulanase export in *Klebsiella oxytoca* (1; 21). All these processes probably share a requirement for energy to drive translocation of macromolecules across the membrane. Several other *vir*B proteins may be functionally related to proteins involved in general bacterial conjugation. For example, the *vir*B operon shows extensive sequence to the RP4 Tra2 operon involved in the conjugative transfer of incompatibility group P (IncP) plasmids in *E. coli* (19). Further, the *vir*B operon can substitute for transfer (*tra*) proteins to allow conjugative transfer of a conjugative plasmid between *Agrobacterium* cells (2). The *vir*B operon also can enable transfer of a conjugative plasmid into a plant cell, presumably by replacing the transfer functions of the plasmid (3; 29). Still, present studies cannot rule out the possibility that *vir*B proteins serve an alternate function for T-DNA transport. For example, rather than forming transmembrane channels, the *vir*B proteins could be an integral part of the T-complex, coating the complex with hydrophobic proteins to allow movement through bacterial and plant cell membranes. However, immuno-electron microscopy (27) showing the presence of *vir*B proteins in the membrane supports the speculation that these proteins may be components of transmembrane channels.

5. Nuclear uptake of the T-complex

The major unsolved aspects of the T-DNA transfer process are the plant cell surface receptors and the intracellular components involved in uptake of the T-complex from the bacterium to and through the plant cell cytoplasm. However, the targeting and transport of the T-complex into the plant cell nucleus are now under active investigation. Nuclear localizing sequences (NLS) have been identified in several animal proteins that function in the nucleus; specific receptor proteins recognize the amino acid sequences of the NLS and mediate protein transport into the nucleus. By analogy, the transport of the T-complex into the plant cell nucleus may be mediated by receptor proteins which recognize NLSs in the proteins of the T-complex. Examination of the two proteins associated with the T-strand, VirD2 and VirE2, show that both proteins likely play a role in nuclear localization of the T-strand (32).

The possible role of VirD2 in movement of T-complex in plants was indirectly suggested by the observation that *Agrobacterium* strains containing only the amino terminal half of VirD2 could not elicit tumors on infected plants (32). This half of the protein is sufficient for nicking the T-borders to generate T-strands; thus, the carboxy terminal half of VirD2 seemed to play an important role after T-strand formation. Furthermore, the tight linkage of VirD2 to the 5' leading end of the T-strand suggested a function as a pilot protein possibly to target the complex to the nucleus. Several experiments support that VirD2 directs nuclear uptake of the T-strand. First, *in vitro* nuclear localization assays using fusions of VirD2 to reporter proteins show that nopaline VirD2 is capable of nuclear localization. We found that VirD2 contains a bipartite NLS signal at the carboxy terminus (14), although another study using a different reporter protein detected an NLS in the amino terminus of VirD2 (12). Second, the precise removal of the carboxy terminal bipartite NLS signal from an octopine VirD2 produced an *Agrobacterium* strain greatly reduced in tumorigenicity; on tobacco stems, this strain formed tumors approximately 15% of wild type while on potato tuber discs, the effect was less pronounced, forming tumors at approximately 60% of wild type (24).

That an *Agrobacterium* strain containing a deletion of the strong carboxy terminal NLS was not completely avirulent suggests that another weaker NLS was used for nuclear uptake. This alternate NLS could be present in the amino terminus of VirD2; however, an *Agrobacterium* strain with both the amino terminal and carboxy terminal NLSs inactivated showed the same level of virulence as the strain in which only the carboxy terminal NLS was inactivated (22). This result suggests that the presence or absence of the amino terminal NLS-like sequence had no affect on virulence, and indicates that possibly another protein associated with the T-complex may function in nuclear uptake.

The rationale for another protein to aid in nuclear uptake is suggested by the size of the complex that must cross the nuclear membrane. The lone VirD2 protein would have to mediate localization of the extremely long and large T-complex that is approximately 60 times longer than the dimensions of a nuclear pore (32). Assuming a 20 kb T-strand with about 600 molecules of VirE2 attached, the T-complex would have an estimated mass of 50×10^6 daltons, approximately 1000 times greater than that of the single VirD2 pilot protein.

A proposed model for the active transport of molecules through the nuclear pore indirectly implies that a protein coating the T-strand may facilitate nuclear uptake (9). The central channel of the nuclear pore contains a plug or "transporter" consisting of two coaxial rings on the cytoplasmic and nuclear faces of the pore complex. The two aligned rings of the transporter are proposed to form a double iris. When a molecule docks at the central pore, the first iris opens to admit the molecule; then, the second iris opens and as the molecule moves further into the transporter, the first iris closes. Thus, a protein coating the T-strand may prevent the closing of the first iris after VirD2 has moved into the transporter and keep the iris open for continued translocation of the T-complex. A fibrous component of the nuclear complex also has been implicated as a mediator for nuclear import of molecules; these fibers pass through the pore complex and extend into the cytoplasm and the nucleoplasm. Perhaps nuclear import also involves association of the transported molecule with the fibers to direct the molecule to and through the nuclear pore. A protein coating the T-strand, then, may enable the extremely long T-complex to associate with these fibers and facilitate targeting to the nuclear pore.

Thus, the immense size of the T-complex suggests that a single nuclear targeting protein might not be sufficient. Our recent results provide evidence that VirE2 coating the T-strand facilitates nuclear uptake (6). Analyses of both nopaline and octopine type VirE2 protein sequences indicates two stretches of DNA similar to the bipartite NLS of VirD2 . *In vitro* nuclear localization assays show that VirE2 mediates nuclear localization. Complete localization of VirE2 to the nucleus is dependent on the presence of the two bipartite NLSs found in the amino acid sequence. If only one NLS is present in VirE2, partial nuclear uptake of the protein is observed, and a significant amount of VirE2 stays in the cytoplasm. Since VirE2 requires two bipartite NLSs while VirD2 requires only one bipartite NLS, the VirE2 NLS is weaker than that of VirD2 in directing complete uptake of the T-strand into the nucleus.

Experiments utilizing VirE2 transgenic plants also imply an *in planta* role for VirE2, possibly in nuclear localization of the T-strand (6). *vir*E2 mutants of *Agrobacterium* either do not cause tumors or show an attenuated tumorigenic phenotype on different host plants. If VirE2 serves a function *in planta*, transgenic plants should complement *vir*E2 mutant *Agrobacterium* strains. Indeed, wild-type tobacco plants did not form tumors when infected with *vir*E2 mutant strains but transgenic plants expressing VirE2 formed large wild-type tumors when infected. Further, complementation depended on the presence of both VirE2 NLSs. When transgenic plants expressed VirE2 containing one NLS, complementation of the *vir*E2 mutant *Agrobacterium* was blocked or greatly reduced. The observation that transgenic plants expressing an unrelated SSB did not complement tumor formation suggests that nuclear localization is at least one of the *in planta* functions of VirE2.

A model, then, for nuclear localization of the T-complex is that the VirD2 and VirE2 proteins work together to facilitate nuclear uptake (32). The VirD2 NLS targets the T-complex to the nuclear pore in a polar manner while the VirE2 NLSs allow the uninterrupted transport of this extremely long complex into the nucleus. The presence of VirD2 as a pilot protein ensures that the molecule is transported directionally.

Conclusions

The transfer of T-DNA from *Agrobacterium* to the plant cell consists of many complex steps. Some of these steps have been studied extensively while others have been recalcitrant to detailed analysis. The early steps of the T-DNA transfer process, the generation of T-strand and export of T-complex from the bacteria, resemble bacterial conjugation. In both T-DNA transport and bacterial conjugation, the transferred DNA is single-stranded and is polar with a nicking enzyme attached to the 5' end; single-stranded DNA binding proteins also are involved. The later steps of the transfer process, movement across plant cell and nuclear membranes and integration into the plant genome, have not been characterized to the extent that comparisons with other processes such as viral infection or illegitimate recombination are easily made. However, the transport of the T-complex through the nuclear membrane may be somewhat analogous to the translocation of premessenger ribonucleoprotein particles (RNP) through the nuclear pore to the cytoplasm (20). For movement through the pore, the RNP is oriented in a specific manner with the 5' end of the RNA leading. The normally bent structure of the RNP is gradually straightened and extended to allow transport. Further, the identification of nuclear localization signals in VirD2 and VirE2 indicates that these eukaryotic signal sequences have been maintained in the prokaryotic plasmid, suggesting that *Agrobacterium* has developed a unique combination of prokaryotic and eukaryotic tools to allow the bacteria to interact with two kingdoms. Although extensive research has uncovered the finer details of some of the steps involved in this unique interaction of bacteria and plant, other steps defy analysis and, at present, remain black boxes. How does the T-complex move out of the bacterial cell and enter the plant cell? What are the plant components involved in the movement of the T-complex out of the bacterium and across the plant membrane? How does the T-complex move through the plant cytoplasm, by an active or passive mechanism? What plant proteins are involved in nuclear uptake of the protein-DNA complex? How does the T-DNA stably integrate into the plant DNA? Thus, while the transfer of DNA from *Agrobacterium tumefaciens* to plants has been widely exploited for plant molecular biology and genetic engineering, a number of fascinating questions remain unanswered.

References

1. Albano, M., Breitling, R., and Dubnau, D.A. (1989) 'Nucleotide sequence and genetic organization of the *Bacillus subtilis com*G operon', J. Bacteriol., 171, 5386-404.
2. Beijersbergen, A., Dulk-Ras, A.D., Schilperoort, R.A., and Hooykaas, P.J.J. (1992) 'Conjugative transfer by the virulence system of *Agrobacterium tumefaciens*', Science, 256, 1324-1327.
3. Buchanan-Wollaston, V., Passiatore, J., and Cannon, F. (1987) 'The mobilization functions (mob and ori-T) of a bacterial plasmid promote its transfer to plants', Nature, 328, 172-75.
4. Christie, P.J., Ward, J.E., Gordon, M.P., and Nester, E.W. (1989) 'A gene required for transfer of T-DNA to plants encodes an ATPase with autophosphorylating activity', Proc. Natl. Acad. Sci. USA, 86, 9677-81.
5. Citovsky, V., Wong, M.L., and Zambryski, P. (1989) 'Cooperative interaction of *Agrobacterium* VirE2 protein with single-stranded DNA: implications for the T-DNA transfer process', Proc. Natl. Acad. Sci. USA, 86, 1193-7.
6. Citovsky, V., Zupan, J., Warnick, D., and Zambryski, P. (1992) 'Nuclear localization of *Agrobacterium* VirE2 protein in plant cells', Science, 256, 1802-1805.
7. Citovsky, V.C., McLean, B.G., Greene, E., Howard, E., Kuldau, G., Thorstenson, Y., Zupan, J., and Zambryski, P. (1991) '*Agrobacterium*-plant cell interaction: induction of *vir* genes and T-DNA transfer', in D. P. S. Verma (eds.), Molecular Signals in Plant-Microbe Communications, CRC Press Inc., pp. 169-199.
8. De Vos, G., and Zambryski, P. (1989) 'Expression of *Agrobacterium* nopaline specific VirD1, VirD2 and VirC1 proteins and their requirement for T-strand production in *E. coli*', Molec. Plant-Microbe Inter., 2, 42-52.
9. Dingwall, C. (1990) 'Plugging the nuclear pore', Nature, 346, 512-514.
10. Douglas, C.J., Staneloni, R.J., Rubin, R.A., and Nester, E.W. (1985) 'Identification and genetic analysis of an *Agrobacterium tumefaciens* chromosomal virulence region', J. Bacteriol., 161, 850-60.
11. Engstrom, P., Zambryski, P., Van Montagu, M., and Stachel, S. (1987) 'Characterization of *Agrobacterium tumefaciens* virulence proteins induced by the plant factor acetosyringone', J. Mol. Biol., 197, 635-45.
12. Herrera-Estrella, A., Van Montagu, M., and Wang, K. (1990) 'A bacterial peptide acting as a plant nuclear targeting signal: the amino-terminal portion of *Agrobacterium* VirD2 protein directs a beta-galactosidase fusion protein into tobacco nuclei', Proc. Natl. Acad. Sci. USA, 87, 9534-7.
13. Howard, E.A., Warnick, D., and Zambryski, P. Unpublished results.
14. Howard, E.A., Zupan, J.R., Citovsky, V., and Zambryski, P.C. (1992) 'The VirD2 protein of *A. tumefaciens* contains a C-terminal bipartite nuclear localization signal: implications for nuclear uptake of DNA in plant cells', Cell, 68, 109-118.
15. Huang, Y., Morel, P., Powell, B., and Kado, C.I. (1990) 'VirA, a coregulator of Ti-specified virulence genes, is phosphorylated *in vitro*', J. Bacteriol., 172, 1142-4.
16. Jin, S., Roitsch, T., Ankenbauer, R.G., Gordon, M.P., and Nester, E.W. (1990) 'The VirA protein of *Agrobacterium tumefaciens* is autophosphorylated and is essential for *vir* gene regulation', J. Bacteriol., 172, 525-30.
17. Jin, S.G., Prusti, R.K., Roitsch, T., Ankenbauer, R.G., and Nester, E.W. (1990) 'Phosphorylation of the VirG protein of *Agrobacterium tumefaciens* by the autophosphorylated VirA protein: essential role in biological activity of VirG', J. Bacteriol., 172, 4945-50.
18. Kanemoto, R.H., Powell, A.T., Akiyoshi, D.E., Regier, D.A., Kerstetter, R.A., Nester, E.W., Hawes, M.C., and Gordon, M.P. (1989) 'Nucleotide sequence and

analysis of the plant-inducible locus *pin*F from *Agrobacterium tumefaciens'*, J. Bacteriol., 171, 2506-12.
19. Lanka, E., and Wilkins, B. (1991) 'Sex, promiscuity, and the bacterial cell', New Biologist, 3, 1035-9.
20. Mehlin, H., Daneholt, B., and Skoglund, U. (1992) 'Translocation of a specific premessenger ribonucleoprotein partcle through the nuclear pore studied with electron microscope tomography', Cell, 69, 605-613.
21. Possot, O., d'Enfert, C., Reyss, I., and Pugsley, A.P. (1992) 'Pullulanase secretion in *Escherichia-coli* K-12 requires a cytoplasmic protein and a putative polytopic cytoplasmic membrane protein', Molec. Microbiol., 6, 95-105.
22. Ream, W. Personal communication.
23. Rothman, J.E., and Kornberg, R.D. (1986) 'An unfolding story of protein translocation', Nature, 322, 209-210.
24. Shurvinton, C.E., Hodges, L., and Ream, W. (1992) 'A nuclear localization signal in the *Agrobacterium tumefaciens* VirD2 endonuclease helps target T-DNA oncogenes to plant nuclei', submitted to Proc. Natl. Acad. Sci. USA.
25. Stachel, S.E., and Nester, E.W. (1986) 'The genetic and transcriptional organization of the vir region of the A6 Ti plasmid of *Agrobacterium tumefaciens'*, EMBO J., 5, 1445-54.
26. Stachel, S.E., and Zambryski, P.C. (1986) *'Agrobacterium tumefaciens* and the susceptible plant cell: a novel adaptation of extracellular recognition and DNA conjugation', Cell, 47, 155-7.
27. Thorstenson, Y., and Zambryski, P. Unpublished results.
28. Toro, N., Datta, A., Carmi, O.A., Young, C., Prusti, R.K., and Nester, E.W. (1989) 'The *Agrobacterium tumefaciens vir*C1 gene product binds to overdrive, a T-DNA transfer enhancer', J. Bacteriol., 171, 6845-9.
29. Ward, J.E., Dale, E.M., and Binns, A.N. (1991) 'Activity of the *Agrobacterium* T-DNA transfer machinery is affected by *vir*B gene products', Proc. Natl. Acad. Sci. USA, 88, 9350-9354.
30. Ward, J.E., Dale, E.M., Nester, E.W., and Binns, A.N. (1990) 'Identification of a *vir*B10 protein aggregate in the inner membrane of *Agrobacterium tumefaciens*', J. Bacteriol., 172, 5200-10.
31. Winans, S.C. (1992) 'Two-way chemical signaling in *Agrobacterium*-plant interactions', Microbiol. Rev., 56, 12-31.
32. Zambryski, P.C. (1992) 'Chronicles from the *Agrobacterium*-plant cell DNA transfer story', Annu. Rev. Plant Physiol. Plant Mol. Biol., 43, 465-490.
33. Zorreguieta, A., Geremia, R.A., Cavaignac, S., Cangelosi, G.A., Nester, E.W., and Ugalde, R.A. (1988) 'Identification of the product of an *Agrobacterium tumefaciens* chromosomal virulence gene', Molec. Plant Microbe Inter., 1, 121-127.

IDENTIFICATION OF GENETIC FACTORS CONTROLLING THE ABILITY OF *AGROBACTERIUM* TO TRANSFER DNA TO MAIZE.

MARGARET I. BOULTON*, DEANNA M. RAINERI$, JEFFREY W. DAVIES* & EUGENE W. NESTER$
* Dept of Virus Research, John Innes Institute, John Innes Centre for Plant Science Research, Colney Lane, Norwich NR4 7UH., U.K.
$ Dept. of Microbiology, University of Washington, Seattle, WA 98105, USA.

ABSTRACT. *Agrobacterium* strains containing nopaline-type Ti plasmids are able to transfer DNA to maize whereas strains containing octopine-type Ti plasmids cannot. In this study the genetic factors controlling the *Agrobacterium* strain-specificity of DNA transfer to maize have been investigated. "Agroinfection" of maize streak virus (MSV) DNA provided a "reporter" system for successful transfer. Mutation of the strain-specific *vir* loci showed that only the octopine-specific *vir*F locus affected DNA transfer. However, the increased efficiency of DNA transfer obtained using these mutants was insufficient to account for the strain specificity. Introduction of fragments of the octopine Ti plasmid into nopaline strains did not inhibit DNA transfer whereas the presence of a 10.6kb fragment of the nopaline Ti plasmid in octopine strains promoted efficient transfer. Efficient agroinfection was also obtained using an octopine *vir*A mutant strain which constitutively expresses the *vir*A gene; since the complementing fragment contained the *virA* locus these data suggest that the inability of octopine strains to promote DNA transfer is due to a lack of *vir* gene induction.

Introduction

Agrobacterium tumefaciens produces tumours on most dicotyledonous plants (dicots) and a limited number of monocotyledonous plants (monocots; De Cleene and Deley, 1976; De Cleene, 1985). The interaction between *Agrobacterium* and susceptible plant cells has been described in detail (see Zambryski *et al.*, 1989). Briefly, two regions of the tumour-inducing (Ti) plasmid are required for oncogenicity: the T-DNA which becomes integrated into the host chromosomal DNA, and the *vir* region which is not stably maintained in transformed plant cells but is essential for virulence. The T-DNA carries genes directing the synthesis of phytohormones as well as specific metabolites known as opines. *Agrobacterium* strains have been classified according to the type of opine which they specify; most Ti plasmids are classified as nopaline- or octopine-type. Several *vir* loci differ between the two strain types, *vir*H and *vir*F are specific to the octopine types, *tzs* to the nopaline strains.

Although *Agrobacterium* does not cause tumours on members of the Poaceae (cereals and grasses) the agroinfection technique has allowed the demonstration of viral DNA transfer to maize (Grimsley *et al.*, 1987), many cereals, and a number of forage grasses (Boulton *et al.*, 1989; Boulton and Davies, 1990; Dasgupta *et al.*, 1991). Agroinfection is a simple and sensitive

E. W. Nester and D. P. S. Verma (eds.),
Advances in Molecular Genetics of Plant-Microbe Interactions, 73–78.

assay for DNA transfer since the viral DNA sequences escape from the T-DNA and after replication and systemic spread their transfer is identified as the formation of disease symptoms.

Using the agroinfection technique it has been shown that DNA transfer to cereals is bacterial strain-specific (Boulton *et al.*, 1989; Marks *et al.*, 1989). Octopine strains were unable, or only weakly able to transfer DNA to maize whereas efficient transfer was seen with nopaline strains and mannopine and agropine strains of *A. rhizogenes* (Boulton *et al.*, 1989; Jarchow *et al.*, 1991). Although the (dicot) host range of a given strain of *Agrobacterium* is a specific character of that strain, most isolates induce crown galls on a wide range of plants although a few (termed the limited host range strains) exhibit high host specificity. However, opine type has only rarely been implicated in dicot host range (Byrne *et al.*, 1987; Melchers *et al.*, 1990).

Agroinfection studies have shown that the Ti plasmid determines the ability of *Agrobacterium* to transfer DNA to maize. Furthermore, as disarmed nopaline strains (deleted in the T-DNA region) are able to transfer MSV DNA the strain-specificity cannot be determined by the differences in phytohormone production (Boulton *et al.*, 1989).

In this study we have used the MSV agroinfection technique to determine whether the strain-specific transfer of DNA to maize is due to an inhibitor produced by the octopine Ti plasmid or an essential product encoded by the nopaline Ti plasmid.

Materials and Methods

BACTERIAL STRAINS AND PLASMIDS

The MSV-containing binary vectors are described in Boulton *et al.* (1989 or Raineri *et al.* (1992), pMSV-Ns was used for Tn3HoHo1 mutants, pMSV-Ns(G), pMSV-Ns-CGN or pMSV-Ns(S) were used for other strains, where necessary. The different vectors were required to facilitate antibiotic selection of recombinants in *Agrobacterium*, or because the original pBIN19 vector (pMSV-Ns) was incompatible with plasmids present within the *Agrobacterium* strains. Other plasmids used in this study and the culture conditions for *E.coli* and *Agrobacterium* have been described by Boulton *et al.* (1989) or Raineri *et al.* (1992).

Mutant Generation. Tn5 and Tn3HoHo1 mutants were made using standard procedures (Berg *et al.*, 1975; Stachel *et al.*, 1985). The *tzs* deletion mutant (*tzs*Δ) was generated by *Nru*I-*Mlu*I digestion of a 5.2kb *Sph*I fragment isolated from pMR19$_{mob}$ (Nunn *et al.*, 1990) followed by blunt-end ligation with the Tn903 kanamycin resistance cassette from pUC4K (Pharmacia Inc.). This resulted in a 291 base deletion (nucleotides 732-1021) of the *tzs* locus. Southern analysis was used to confirm the integrity of the mutants. The presence or absence of trans-zeatin in the culture fluid of *tzs*-complemented and tzs mutant strains, respectively, was confirmed using ELISA (Van Weemen and Schuurs, 1977).

Plasmid Constructions. A clone bank of pTiA6 (octopine strain; Knauf and Nester, 1982) was used for testing inhibition of DNA transfer. Plasmids containing fragments of the nopaline Ti plasmid, pTiC58, were as described by Raineri *et al.* (1992).

PLANT GROWTH AND INOCULATION

Growth of maize plants and preparation of inocula were as described by Boulton *et al.* (1989).

Results

THE OCTOPINE-SPECIFIC *VIR* LOCI DO NOT MARKEDLY INHIBIT DNA TRANSFER TO MAIZE.

The observed specificity for agroinfection is likely to reside within the *vir* region of the Ti plasmid, thus the regions of non-homology between octopine and nopaline strains were targeted for investigation.

Transposon insertion mutagenesis was used to disrupt the octopine-specific *virH* and *virF* loci. The agroinfection ability of the pTiA6 mutants was then compared with that of the wild type parental strain A348 (C58 chromosomal background, octopine pTiA6) and wild type C58.

Insertions in all five open reading frames (ORFs) of the *virH* region (*pinF1*, *pinF2* and ORFs 1, 2 and 3) resulted in MSV DNA transfer similar to that of the A348 control (Table 1).

TABLE 1. Effect of mutation of the octopine *virH* locus on agroinfection of maize.

Strain/mutation*	Agroinfection efficiency$ #Plants infected/# inoculated	%
A348/none	5/201	0-5
A348/Tn5 *virH*:ORF1:39	0/18	0
A348/Tn3HoHo1 *virH*:ORF2:212	0/24	0
A348/Tn3HoHo1 *virH*:ORF3:214	0/17	0
A348/Tn3HoHo1 *virH*:pinF1:219	1/18	5
" :231	0/18	0
A348/Tn3HoHo1 *virH*:pinF2:229	0/19	0
A348/pinF1/F2Δ	1/18	5
C58/none	327/340	95-100

* The location of the insertion within the ORF is denoted by :nucleotide number

$ Specific infectivities represent data from a single experiment, range of infectivities represent data from multiple experiments carried out over a period of several months.

In contrast, the *virF* mutants gave transfer frequencies ranging from 0%-20% (Fig. 1) compared with a maximum transfer efficiency of 5% (and a mean value of 2.5%) for A348 and a mean value of 98% for C58. It should be noted that the maximum transfer efficiency was obtained using mutant *virF*#37 in which the insertion maps outside of the *virF* region as reported by Melchers *et al.* (1990). Following inoculation of *virF* mutants disease symptoms were delayed and frequently were seen as discrete spots or streaks, rather than the almost complete leaf chlorosis seen after agroinfection with C58.

When the *virF* locus, carried on a plasmid having a copy number of five in *Agrobacterium*, was present in C58, mean agroinfection efficiency was 94%, showing that the *virF* locus is unable to suppress DNA transfer from a nopaline strain. In order to determine whether other regions of the octopine Ti plasmid were able to inhibit transfer from a nopaline strain, overlapping cosmid clones of pTiA6 were tested for their ability to inhibit DNA transfer mediated by pTiC58. Although initial experiments identified two inhibitory clones, these contained the origin of replication of the octopine Ti plasmid. Clones carrying identical Ti plasmid regions minus the origin failed to inhibit transfer.

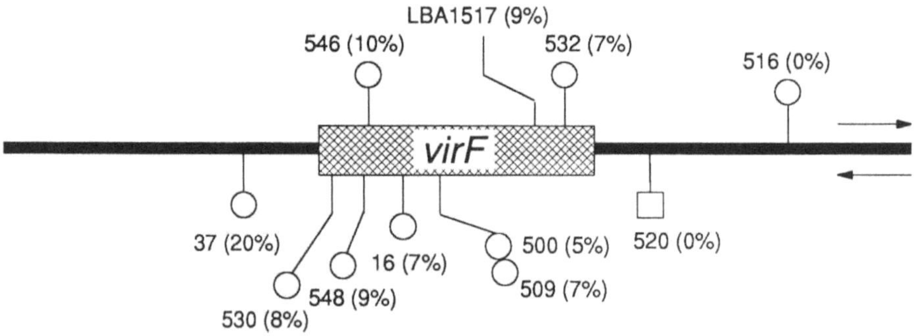

Figure 1. The locations of Tn3HoHo1 insertions in the *virF* region of pTiA6 and (in parentheses) their respective % agroinfection efficiency. The arrows indicate orientation of insertion. All mutants except #520 exhibit attenuated virulence on tomato, *Kalanchoe*, sunflower and pea. The cross-hatched area represents the *virF* region as reported by Melchers *et al.* (1990).

THE NOPALINE STRAIN-SPECIFIC TRANS ZEATIN SECRETION LOCUS (*TZS*) IS NOT REQUIRED FOR AGROINFECTION OF MAIZE

The *tzs* locus is specific to *A. rhizogenes* strains and nopaline strains of *A. tumefaciens*. Inactivation of this locus in the nopaline strain, T37, using Tn5 and deletion mutagenesis did not impair the ability of the mutant strains to transfer MSV DNA (Table 2).

TABLE 2. Effect of mutation of the nopaline *tzs* locus on agroinfection of maize.

Strain/mutation*	Agroinfection efficiency #Plants infected/# inoculated	%
T37/none	5/6	83
T37/Tn5 *tzs*:21	27/31	87
T37/Tn5 *tzs*:172	21/26	81
T37/Tn5 *tzs*:194	21/28	75
T37/*tzs* Δ	26/30	87

* The location of the insertion within the ORF is denoted by :nucleotide number

A clone (pUCD2614) containing the nopaline *vir* regulon plus some non-*vir* sequences was found to promote agroinfection from LBA4301 (an octopine strain cured of its Ti plasmid). The levels of infection were only slightly lower than those obtained with C58 (83% and 95%, respectively). Subcloning of the *vir* region showed that only the *Bam*HI fragment 2 of pTiC58 was able to complement DNA transfer when present in A348. This subclone contained the *virA* and *tzs* loci as well as a flanking non-*vir* region of about 5kb.

When plasmid pEB112 (Ankenbauer *et al.*, 1991) was introduced into A348, MSV DNA transfer efficiency was 81%. This plasmid contains the octopine $virA_{(AIS)}$ locus which mediates *vir* gene expression in the absence of inducing stimuli.

When considered together, the data presented in this paper strongly suggest that the nopaline *virA* locus is able to complement DNA transfer to maize from an octopine background, and that the inability of octopine strains to mediate transfer is due to a property of the octopine *virA* locus.

Discussion

In this study we have used the agroinfection technique to identify a region of the Ti plasmid which is responsible for the strain-specific transfer of DNA to maize, and have shown that the specificity determinant does not reside within the previously identified strain-specific loci *virF*, *virH*, or *tzs*.

The slight increase in agroinfection efficiency seen with some of the *virF* mutants agrees with the findings of Jarchow *et al.* (1991), although the authors reported a much higher range of infectivity (12-33% for LBA 1517) than was observed with the mutants used in this study. Our lower infection levels may reflect the different binary vector systems and/or maize cultivars used in the two studies, rather than the parental strains. That the *virF* locus does not play the major role in the transfer of DNA to maize is underlined by the ability to rapidly reach an end-point dilution of the *virF* mutant, but not the C58, inoculum. Furthermore, it is likely that the delay in onset of symptoms seen even with the concentrated mutant inocula reflects a low level of DNA transfer. The inability of the *virF* locus to suppress nopaline-mediated transfer in either nopaline or complemented octopine backgrounds (this study; Jarchow *et al.*, 1991), and the demonstration of *virF* activity in some *A. rhizogenes* strains, are also consistent with *virF* playing a minor role in host range determination.

We have shown that a 10.6kb subclone, containing the *virA* and *tzs* loci as well as an approximately 5kb non-*vir* region of the nopaline Ti plasmid, is able to complement an octopine plasmid for agroinfection of maize. Although the *tzs* locus has been implicated in host range determination in dicots (Fillatti *et al.*, 1987), our studies have shown that inactivation of this locus does not affect the ability of the mutant strain to transfer DNA to maize. Furthermore, since Jarchow *et al.* (1991) have shown that insertion of the *tzs* locus in an octopine strain did not result in increased transfer it is likely that the complementing factor resides, not within the *tzs*, but within the *virA* locus. Further evidence was provided for this conclusion with the observation that an octopine *virA* mutant strain, which is able to mediate *vir* gene expression in the absence of inducing agents, promoted high level transfer of DNA to maize. We therefore suggest that the strain-specific transfer of DNA to maize may be due to a lack of induction or an inhibition of virA activity in octopine strains. Future experiments will be designed to investigate inhibition of octopine *vir* gene induction by maize extracts and the domain of the *virA* locus which is responsible for the strain-specific interaction of *Agrobacterium* with cereals.

References

Ankenbauer, R.G., Best, E.A., Palanca, C.A. and Nester, E.W. (1991). Mutants of the *Agrobacterium tumefaciens virA* gene exhibiting acetosyringone-independent expression of the *vir* regulon. Mol. Plant. Microbe Int. **4**, 400-406.

Berg, D.E., Davies, J., Allet, B. and Rochaix, J.D. (1975). Transposition of R-factor genes to bacteriophage lamda. Proc. Natl. Acad. Sci., U.S.A. **72**, 3628.

Boulton, M.I. and Davies, J.W. (1990). Monopartite geminiviruses: markers for gene transfer to cereals. In: Aspects of Applied Biology, **24**, 79-86.

Boulton, M.I., Buccholz, W.G., Marks, M.S., Markham, P.G. and Davies, J.W. (1989). Specificity of *Agrobacterium*-mediated delivery of maize streak virus DNA to members of the Gramineae. Plant Mol. Biol. **12**, 31-40.

Byrne, M.C., McDonnell, R.E., Wright, M.S. and Carnes, M.G. (1987). Strain and cultivar specificity in the *Agrobacterium*-soybean interaction. Plant Cell, Tissue and Organ Culture **8**, 3-15.

Dasgupta, I., Hull, R., Eastop, S., Poggi-Pollini, C., Blakebrough, M., Boulton, M.I. and Davies, J.W. (1991). Rice tungro bacilliform virus DNA independently infects rice after *Agrobacterium* -mediated transfer. J. Gen. Virol. **72**, 1215-1221.

DeCleene, M. (1985). The susceptibility of monocotyledons to *Agrobacterium tumefaciens*. Phytopathologische Zeitschrift **113**, 81-89.

DeCleene, M. and DeLey, G. (1976). The host range of Crown Gall. Bot. Rev. **42**, 389-466.

Fillatti, J., Sellmer, J., McCown, B., Haissig, B. and Comai, L. (1987). *Agrobacterium*-mediated transformation and regeneration of *Populus*. Mol. Gen. Genet. **206**, 192-199.

Grimsley, N., Hohn, T., Davies, J.W. and Hohn, B. (1987). *Agrobacterium*-mediated delivery of infectious maize streak virus into maize plants. Nature **325**, 177-179.

Hooykaas, P.J., Hofker, M., Den Dulk-Ras, H. and Schilperoort, R. (1984). A comparison of virulence determinants in an octopine Ti plasmid, a nopaline Ti plasmid, and an Ri plasmid by complementation analysis of *Agrobacterium tumefaciens* mutants. Plasmid **11**, 195-205.

Jarchow, E., Grimsley, N.H. and Hohn, B. (1991). *virF*, the host range determining virulence gene of *Agrobacterium tumefaciens*, affects T-DNA transfer to *Zea mays*. Proc. Natl. Acad. Sci. U.S.A. **88**, 10426-10430.

Knauf, V., and Nester, E.W. (1982). Wide host range cloning vectors: a cosmid clone bank of an *Agrobacterium* Ti plasmid. Plasmid **8**, 45-54.

Melchers, L., Maroney, M., den Dulk-Ras, A., Thompson, D., van Vuuren, A., Schilperoort, R. and Hooykaas, P. (1990). Octopine and nopaline strains of *Agrobacterium tumefaciens* differ in virulence; molecular characterization of the *virF* locus. Plant Mol. Biol. **14**, 249-259.

Nunn, D.N., Bergman, S. and Lory, S. (1990). Products of three accessory genes, *pilB*, *pilC*, and *pilD*, are required for biogenesis of *Pseudomonas aeruginosa* pili. J. Bacteriol. **172**, 2911-2919.

Raineri, D.M., Boulton, M.I., Davies, J.W. and Nester, E.W. (1992). The plant signal receptor, *virA* is responsible for the strain specific transfer of DNA to maize by *Agrobacterium*. (submitted to Proc.Natl.Acad.Sci., U.S.A.).

Stachel, S., An, G., Flores, C. and Nester, E. W. (1985). A Tn3-*lacZ* transposon for the random generation of beta-galactosidase gene fusions: application to the analysis of gene expression in *Agrobacterium*. EMBO J. **4**, 891-898.

Van Weemen, B.K. and Schuurs, A.H.W.M.(1977). Immunoassay using antigen-enzyme conjugates. FEBS Letters **15**, 232-236.

Zambryski, P., Tempe, J. and Schell, J. (1989). Transfer and function of T-DNA genes from *Agrobacterium* Ti and Ri plasmids in plants. Cell **56**, 193-201.

MECHANISM OF T-DNA TRANSFER

BRUNO TINLAND[1], MICHAEL HALL[2], and BARBARA HOHN[1]
[1]Friedrich Miescher-Institut, P.O.Box 2543, CH-4002 Basel, Switzerland
[2]Department of Biochemistry, Biocenter, University of Basel, CH-4056 Basel, Switzerland

ABSTRACT

Upon cutting at the T-DNA borders the VirD2 protein of *Agrobacterium tumefaciens* is tightly (most probably covalently) attached to the 5' end of the processed T-DNA. Therefore, it was proposed to be transferred into the plant cell together with the T-DNA, and to perform several functions within the plant cell: it may protect the T-DNA against nucleases, it may target it into the plant nucleus and it (or a protein attached to it) may integrate it into the plant chromosome. Here we analyze the property of the VirD2 protein in nuclear targeting. We demonstrate that VirD2 protein contains two nuclear localization signals (NLS) which are functional in yeast as well as in plant cells. One of these signals is located in the N-terminal part of the protein and its sequence resembles that of a single-cluster type NLS, whereas the other is located in the C-terminal part and belongs to the bipartite type NLS. The involvement of these sequences in the entry of the T-DNA into the nucleus is discussed.

The T-DNA (transferred DNA) of *Agrobacterium tumefaciens* is the only non viral nucleic acid known to genetically transform a higher eukaryotic cell in nature. The T-DNA is a defined region of a large plasmid called Ti plasmid (Tumor inducing). The T-DNA is delimited by two almost perfect 25 bp repeat borders. In the presence of wounded plant cells, the T-DNA is liberated from the Ti plasmid and has been proposed to be transferred as a nucleoprotein complex to the plant cell, where it enters the nucleus and integrates into the genome. (For reviews see Ream, 1989; Kado, 1991; Zambryski, 1992; Hooykaas and Schilperoort, 1992).

The T-DNA mobilization to the plant cell is mediated by virulence (Vir) proteins which are encoded by a region of the Ti plasmid located outside the T-DNA and called *vir* region. This region is composed of six main operons, *vir*A, *vir*B, *vir*C, *vir*D, *vir*E, and *vir*G. *vir*A encodes an environmental sensor which is sensitive to small phenolic compounds emitted by wounded plants. Upon activation by VirA of a basal

E. W. Nester and D. P. S. Verma (eds.),
Advances in Molecular Genetics of Plant-Microbe Interactions, 79–84.

amount of VirG protein by phosphorylation, the latter then stimulates the transcription of all the other *vir* operons including itself (Leroux *et al.*, 1987 ; Winans *et al.*, 1987). A sequence specific endonuclease, composed of the VirD1 and the VirD2 proteins, recognizes the right and left borders. Upon cutting the bottom strand at the right border, VirD2 protein attaches (most likely covalently by a tyrosine) to the 5' end of the processed T-DNA (Young and Nester, 1988 ; Ward and Barnes, 1988 ; Howard *et al.*, 1989 ; Dürrenberger *et al.*, 1989), whereas a topoisomerase activity carried by the VirD1 protein may help in removing the corresponding strand, (T-strand) (Stachel *et al.*, 1986 ; Ghai and Das, 1989). It was shown that the VirE2 protein is a single-stranded DNA specific binding protein and it was proposed to cover the free single stranded T-DNA (Gietl *et al.*, 1987 ; Christie *et al.*, 1988). A nucleoprotein complex consisting of T-DNA-VirD2-VirE2 and possibly other proteins may be the active vector implicated in the transfer of the genetic information.

The biggest challenges to be met in the plant cell are nuclear entrance and efficient integration into the plant genome. This implies that plant cell proteins and/or bacterial proteins help the T-DNA in its journey to the plant cell. The entry of proteins into the nucleus is a selective process which requires the activation of the nuclear pore complex . This activation is mediated by a nuclear localization signal (NLS) carried either by the transported protein or by a helper protein (for review, see Garcia Bustos *et al.*, 1991 ; Silver, 1991 ; Goldfarb and Michaud, 1991). Two types of nuclear localization signals have been described. The first consists of a single cluster of positively charged amino-acids (Chelsky *et al.*, 1989). The second type is a bipartite signal in which two sequence elements made up of basic amino-acids are separated by about ten undefined amino-acids (Robbins *et al.*, 1991 ; Dingwall and Laskey, 1991). The recognition system involved in nuclear import is most likely universal (Wagner et Hall, submitted).

Because of this tight control, one could imagine that DNA delivered exogenously into the cytoplasm of a eukaryotic cell would not enter the nucleus efficiently. Indeed, microinjection of DNA into the cytoplasm of mammalian cells has yielded stable transformants at much lower frequency than microinjection directly into the nucleus (Capecchi, 1980). If this is interpreted as inability of the injected DNA to gain entrance into the nucleus, then specialized mechanisms have to be invoked to explain how the T-DNA complex finds its way into the plant nucleus.

The VirD2 protein is a good candidate for piloting the T-DNA complex into the plant nucleus for at least three reasons. First, VirD2 is the only protein which has been found tightly linked to the T-DNA in the bacterium. Second this link may persist in the plant cell since the right end of the T-DNA isolated from plant cell showed a particularly good conservation (Bakkeren *et al.*, 1989); this may be due to protection against nuclease by the bound protein. Third, the VirD2 protein contains two regions of homology with previously described nuclear localization sequences. One is located in the N-terminal part (monopartite type NLS) and the other in the C-terminal part of the VirD2 protein (bipartite type NLS). In addition, comparison of the amino-acid sequence of the VirD2 protein of three different *Agrobacterium* strains revealed that these sequences are particularly well conserved (Wang *et al.*, 1990).

Indeed, Herrera-Estrella *et al.*, 1990 showed that in a transgenic plant expressing the N-terminal 292 amino-acids of the VirD2 protein (70% of VirD2 protein) fused to β-galactosidase, the enzymatic activity is localized in the plant cell nucleus. In contrast, in plants expressing original β-galactosidase, the activity was found distributed throughout the entire cell. This result was contradicted by Howard *et al.*, 1992 who found that only the C-terminal NLS from VirD2 had nuclear targeting properties. These authors studied the transient production in plant protoplasts of different parts of VirD2 fused to β-glucuronidase. The enzymatic activity was localized in the corresponding area by its ability to convert X-glu (5-bromo-4-chloro-indolyl glucuronide) substrate into a blue derivative.

We tested the nuclear localization potential of VirD2 protein in stably transformed yeast cells. Different parts of VirD2 were fused to β-galactosidase. Each construct containing either one or both nuclear localization signals were expressed in yeast. Their products were localized by indirect immunofluorescence in the yeast nucleus, whereas in the absence of a nuclear localization signal, the β-galactosidase was identified all over the cell (Tinland *et al.*, 1992). The nuclear targeting function was precisely defined in yeast cells as two peptides of 11 and 20 amino-acids belonging to the N-terminal and to the C-terminal part of the VirD2 protein, respectively. Each of the peptides, tested individually, was then shown to target β-galactosidase into the nucleus of plant cell protoplasts (Fig.1).

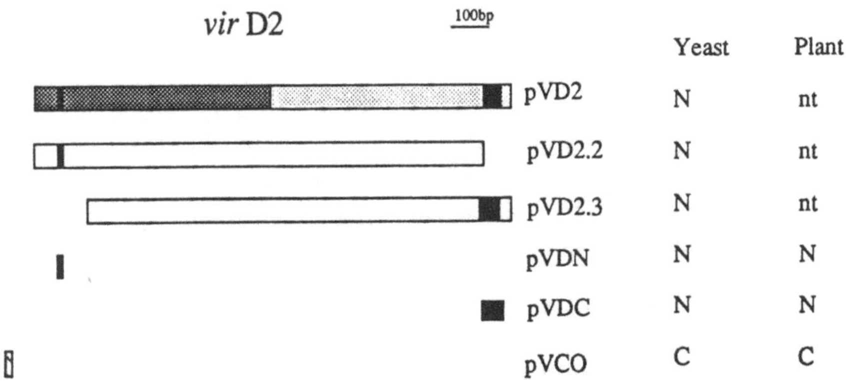

Figure 1. Nuclear targeting of VirD2 and derivatives in yeast and plant cells. The shadowed N-terminal part of VirD2 corresponds to the highly conserved part, the dotted C-terminal part to the less well conserved part ; black bars indicate the NLSs. The entire *vir*D2 gene, deletion derivatives lacking the N- or C-terminal NLS, the NLS sequences and a control sequence alone were fused to the *lacZ* gene at the 5' end of the latter and tested for the intracellular localization of the VirD2-β-galactosidase fusion derivatives. N, nuclear localization ; C, cytoplasmic localization ; nt, not tested. Data are from Tinland *et al.*, 1992.

These data clearly show that the VirD2 protein contains properties to target itself and others proteins to the nucleus. However, this does not automatically imply that the VirD2 protein also manages to import the linked T-DNA to the plant nucleus. Indirect evidence for the importance of at least the C-terminal NLS of VirD2 for full virulence of *Agrobacterium* has been reported. The virD2 gene of *Agrobacterium* has been precisely deleted in the DNA sequence corresponding to the C-terminal NLS and was tested for its ability to transfer DNA into the nucleus. *Agrobacterium* containing this deletion was shown to possess a drastically reduced transformation efficiency (Shurvinton *et al.*, in press; Rossi *et al.*, submitted).

ACKNOWLEDGEMENTS

We would like to thank Walt Ream and Luca Rossi for communicating data prior to publication, and EMBO for awarding a long term fellowship to Bruno Tinland.

REFERENCES

Bakkeren, G., Koukolíková-Nicola, Z., Grimsley, N. and Hohn, B. (1989) Recovery of *Agrobacterium tumefaciens* T-DNA molecules from whole plants early after transfer. Cell 57, 847-857

Capecchi, M.R. (1980) High efficiency transformation by direct microinjection of DNA into cultured mammalian cells. Cell 22, 479-488

Chelsky, D., Ralph, R. and Jonak, G. (1989) Sequence requirements for synthetic peptide-mediated translocation to the nucleus. Mol. Cell. Biol.9, 2487-2492

Christie, P.J., Ward, J.E., Winans, S.C. and Nester, E.W. (1988) The *Agrobacterium tumefaciens vir*E2 gene product is a single-stranded-DNA-binding protein that associates with T-DNA. J. Bacteriol. 170, 2659-2663

Dingwall, C. and Laskey, R.A. (1991) Nuclear targeting sequences- a consensus? TIBS 16, 478-481

Dürrenberger, F., Crameri, A., Hohn, B. and Koukolíková-Nicola, Z. (1989) Covalently bound VirD2 protein of *Agrobacterium tumefaciens* protects the T-DNA from exonucleolytic degradation. Proc. Natl. Acad. Sci. USA 86, 9154-9158

Garcia Bustos, J., Heitman, J. and Hall, M.N. (1991) nuclear protein localization. Biochem. Biophys. Acta Rev. 1071, 83-101

Ghai, J. and Das, A. (1989) The virD operon of *Agrobacterium tumefaciens* Ti plasmid encodes a DNA-relaxing enzyme. Proc. Natl. Acad. Sci. USA 86, 3109-3113

Gietl, C., Koukolíková-Nicola, Z. and Hohn, B. (1987) Mobilization of T-DNA from *Agrobacterium* to plant cells involves a protein that binds single-stranded DNA. Proc. Natl. Acad. Sci. USA 84, 9006-9010

Goldfarb, D. and Michaud, N. (1991) Pathways for the nuclear transport of protein. Trends Cell Biol. 1, 20-24

Herrera-Estrella, A. , Van Montagu, M. and Wang, K. (1990) A bacterial peptide acting as a plant nuclear targeting signal: the amino-terminal portion of *Agrobacterium* VirD2 protein directs a β-galactosidase fusion protein into tobacco nuclei. Proc. Natl. Acad. Sci. USA 87, 9534-9537

Hooykaas, P.J.J. and Schilperoort, R.A. (1992) *Agrobacterium* and plant genetic engineering. Plant Mol. Biol. 19, 15-38

Howard, E.A., Winsor, B., De Vos, G. and Zambryski, P. (1989) Activation of the T-DNA transfer process in *Agrobacterium* results in the generation of a T-strand-protein complex : tight association of virD2 with the 5' ends of T-strands. Proc. Natl. Acad. Sci. USA 86, 4017-4021

Howard, A.H., Zupan, J.R., Citovsky, V. and Zambryski, P.C. (1992) The VirD2 protein of *A. tumefaciens* contains a C-Terminal bipartite nuclear localization signal: implication for Nuclear uptake of DNA in plant cells. Cell 68, 109-118.

Kado, C.I. (1991) Molecular mechanisms of crown gall tumorigenesis. Crit. Rev. Plant Sci. 10, 1-32

Leroux, B., Yanofsky, M.F., Winans, S.C., Ward, J.E., Ziegler, S.F. and Nester, E.W. (1987) Characterization of the virA locus of *Agrobacterium tumefaciens*: a transcriptional regulator and host range determinant. EMBO J 6, 849-856

Ream, W. (1989) *Agrobacterium tumefaciens* and interkingdom genetic exchange. Annu. Rev. Phytopathol. 27, 583-618

Robbins, J., Dilworth, S.M., Laskey, R.A. and Dingwald, C. (1991) Two interdependent basic domains in nucleoplasmin nuclear targeting sequence: identification of a class of bipartite nuclear targeting sequence. Cell 64, 615-623

Rossi, L., Hohn, B. and Tinland, B. VirD2 protein carries a NLS important for transfer of T-DNA from bacteria to plants. Submitted

Shurvinton, C.E., Hodges, L., and Ream, W. A nuclear localization signal in the *Agrobacterium tumefaciens* VirD2 endonuclease is important for tumor formation. Proc. Natl. Acad. Sci. USA, in press

Silver, P.A. (1991) How proteins enter the nucleus. Cell 64, 489-497

Stachel, S.E., Timmerman, B. and Zambryski, P. (1986) Generation of single-stranded T-DNA molecules during the initial stages of T-DNA transfer from *Agrobacterium tumefaciens* to plant cells. Nature 322, 706-711

Tatzelt, J., Scholz, B., Fechteler, K., Jessberger, R. and Doerfler, W. (1992) Recombination between adenovirus type 12 DNA and a hamster preinsertion sequence in a cell-free system. J. Mol. Biol. 226, 117-126

Tinland, B., Koukolíková-Nicola, Z., Hall, M.N. and Hohn, B. (1992) The T-DNA linked VirD2 contains two distinct functional nuclear localization signals. Proc. Natl. Acad. Sci. USA. in press

Wang, K., Herrera-Estrella, A. and van Montagu, M. (1990) Overexpression of virD1 and virD2 genes in *Agrobacterium tumefaciens* Enhances T-complex formation and plant transformation. J. Bacteriol. 172, 4432-4440

Ward, E.R. and Barnes, W.M. (1988) VirD2 protein of *Agrobacterium tumefaciens* very tightly linked to the 5' end of T-strand DNA. Science 242, 927-930

Winans, S.C., Allenza, P., Stachel, S.E., Mc Bride, K.E. and Nester, E.W. (1987) Characterization of the virE operon of the *Agrobacterium* Ti plasmid pTiA6. Nucl Acids Res. 15, 825-837

Young, C. and Nester, E.W. (1988) Association of the VirD2 protein with the 5' end of T strands in *Agrobacterium tumefaciens*. J.Bacteriol. 170, 3367-3374

Zambryski, P.C. (1992) Chronicle from the *Agrobacterium*-plant cell DNA transfer story. Ann. Rev. Plant Physiol. Plant Mol. Biol. 43, 465-490

MOLECULAR INTERACTIONS BETWEEN *AGROBACTERIUM* AND PLANT CELLS

Y. MACHIDA, N. SHIMODA, A. YAMAMOTO-TOYODA, Y. TAKAHASHI, R. NISHIHAMA, S. AOKI, K. MATSUOKA*, K. NAKAMURA*, Y. YOSHIOKA, T. OHBA, R. T. OBATA
Department of Biology, School of Science, School of Agriculture*, Nagoya University, Chikusa-ku, Nagoya 464-01, Japan

ABSTRACT. The present paper describes two aspects of *Agrobacterium*-plant interaction. (1) Most *vir* genes are expressed in response to phenolic signal substances from plants. A group of aldoses such as glucose markedly enhance induction levels of *vir* gene expression by a mechanism involving action of products specified by *virA* and *chvE* genes. Our results of genetic analyses with base substitution mutations of these genes reveal that direct interaction between the periplasmic domain of VirA protein and ChvE protein is essential for enhancement by sugars. (2) We have developed a simple procedure to measure an efficiency of T-DNA transfer from *Agrobacterium* cells to plant nuclei. The results of experiments by this procedure indicate that products encoded by the *virB* locus and by the *virD4* gene are necessary for transfer of T-DNA. VirE2 protein is also required for efficient transfer of T-DNA, although it is not absolutely essential.

1. Introduction

Agrobacterium tumefaciens harboring the Ti plasmid incites crown gall tumors on a wide variety of dicotyledonous plants. Upon infection of plants, transferred DNA (T-DNA), a stretch which is flanked by 25 base pair direct repeats on Ti plasmid, is transferred by unknown mechanisms to plant cells and integrated into plant nuclear DNA. The T-DNA processing in *Agrobacterium* cells and its transfer require products encoded in the *vir* region. Mutations in this region abolish or lower the virulence such as tumorigenicity or transformation ability of *Agrobacterium*. The vir region of octopine Ti plasmid contains at least seven complementation loci (*virA, virB, virG, virC, virD, virE* and *virF*) (Stachel and Nester, 1986; Melchers et al. 1990). The expression of *virB, C, D* and *E* is positively regulated at the transcriptional level by plant signal molecules (Stachel and Nester, 1986; Stachel and Zambryski, 1986; Rogowsky et al. 1987). The regulatory genes *virA* and *virG* are expressed constitutively, the expression of virG increasing in the presence of the plant inducers (Stachel et al. 1986).

The plant signal is transduced into agrobacterial cells through functions of the VirA and VirG proteins which show similarities to the two-component regulatory system that is conserved in a variety of prokaryotes (Albright et al. 1989). VirA protein is thought to serve as a sensor or receptor to detect the signal of the inducers (Leroux et al. 1987; Melchers et al. 1987; Morel et al. 1989). It has been proposed that this protein spans the cytoplasmic membrane of *Agrobacterium* with approximately 270 amino acids, flanked by two hydrophobic transmembrane domains, and localized in the periplasmic space

E. W. Nester and D. P. S. Verma (eds.),
Advances in Molecular Genetics of Plant-Microbe Interactions, 85–96.

(Melchers et al. 1989; Winans et al. 1989). The signal detected by the VirA protein must be transduced to the VirG protein to activate the latter protein. The activated form is thought to act as a positive regulator for the transcription of other *vir* genes. This signal is likely transduced by a mechanism involving protein phosphorylation by VirA (Jin et al. 1990; Huang et al. 1990).

Plant signal molecules in tobacco have been identified as novel phenolics, acetosyringone (AS) and hydroxyl AS, which are exuded from wounded or actively growing cells (Stachel et al. 1985). Components of lignin or its precursors also act as signal molecules (Spencer and Towers 1989). The detection of AS by *Agrobacterium* cells is thought to occur in or near the second trans-membrane domain of VirA protein, because deletion mutants of the periplasmic domain of VirA still responded to AS to induce *vir* gene expression (Melchers et al. 1989). In addition to phenolics, monosaccharides including and 2-deoxy glucose, a non-metabolized sugar, have been shown to be markedly enhance AS-dependent expression of *vir* genes (Machida et al. 1989; Shimoda et al. 1990; Cangelosi et al. 1990). Enhancement by the sugars was eliminated by deleting the periplasmic portion of the VirA protein, but *vir* expression was induced by AS in this mutant. These results suggest that the periplasmic portion of the VirA protein is responsible for the effect of these sugars, which directly enhance a signalling process initiated by phenolic inducers. A mutant in another gene *chvE* also exhibited the same phenotype as the VirA mutant with the periplasmic deletion, indicating that a product encoded by the *chvE* gene is involved in the process of *vir* enhancement by sugars (Cangelosi et al. 1990). Since ChvE has the amino acid sequence highly homologous to that of the galactose/glucose binding protein of *A. radiobacter*, the model has been proposed that the sugar/ChvE complex interacts with the periplasmic domain of VirA to enhance cytoplasmic signal transduction (Cangelosi et al. 1990).

In the present paper, we have isolated a series of deletions and base substitution mutants of the coding region of the periplasmic domain of VirA protein. Some of base substitution mutants exhibited sugar-non-responsive phenotype. Furthermore, we have isolated *chvE* mutants which suppressed the non-responsive phenotype of one of the *virA* mutations. This genetic analysis indicates that the ChvE protein or the sugar/ChvE complex directly interacts with the periplasmic portion of the VirA sensor protein.

Following induction of *vir* gene expression, T-DNA processing reactions, which are mediated by VirD1 and VirD2 proteins, occur to generate transferable T-DNA molecules such as single-stranded and double-stranded T-DNA(Hohn et al. 1989; Zambryski 1992). VirD2 protein has been shown to bind to 5' regions of both single- and double-stranded T-DNA molecules (Kado 1991; Zambryski 1992). VirE2 protein has the ability to bind non-specifically to single-stranded DNA (Zambryski 1992), although the *virE* locus is not absolutely essential for the virulency (Stachel and Nester 1986). Such protein(s)/T-DNA complexes are thought to migrate to plant cells and to be targeted to the nuclei (Hohn 1989; Zambryski 1992). Recently, it have been shown that both VirD2 and VirE2 that are synthesized in plant cells are localized in plant nuclei, supporting the above idea (Herrera-Estrella et al. 1990; Howard et al. 1992; Citovsky et al. 1992). Since some of VirB proteins (Christie et al. 1989; Ward et al. 1990) and VirD4 (Okamoto et al. 1991) are anchored in the inner membrane, they are thought to promote movement of the protein(s)/T-DNA complexes through membranes. Functions of *virB* and *virD* loci which can encode a number of proteins, however, remain to be defined.

In the present paper, we report the transient gene expression system mediated by *Agrobacterium* to study the process of T-DNA transfer from *Agrobacterium* to plant cells. The results of experiments by this system indicated that *virB*, *virD4* and *virE2* are required for efficient transfer of T-DNA, although *virE2* is not absolutely essential.

2. Materials and Methods

2.1. BACTERIA

Agrobacterium tumefaciens C58C1Cm harboring pTiB6S3*tra*[c], EHA101 carrying pEHA101, A208 carrying pTiT37 and C58 carrying pTiC58 were described previously (Petit et al. 1978; Hood et al.1986; Sciaky et al. 1978). *A. tumefaciens* A348mx226, A348mx238, A348mx355 and A348mx341 harboring pTiA6 with insertions of transposon Tn*3*-HoHo1 in the *virA, virB, virD4* and *virE2*, respectively (Stachel and Nester 1986). The C58C1Cm strain was used for constructing the *virA* deletion mutant, the C58C1CmΔA/Blac strain, in which the internal region of the virA gene of pTiB6S3*tra*[c] was replaced with the *virB-lacZ* fusion. The C58C1CmΔA/Blac/ΔE strain which has a disruption mutation in the chromosomal *chvE* region was created by replacing with the *bla* gene by double homologous recombination. The *chvE* gene which was used for the disruption was cloned from *A. tumefaciens* C58C1Cm by using synthetic oligonucleotides, sequences of which were deduced from the amino acid sequence of GBP1 by Cornish et al (1990).

2.2. PLASMIDS

pCM110D having the *virD-lacZ* fusion has been described (Shimoda et al. 1990). Deletion mutations in the *virA* gene were generated by using suitable restriction endonucleases. Base substitution mutations were generated with synthetic oligonucleotides by the procedure as described by Ito et al (1991). All the mutations were confirmed by nucleotide sequencing. pIG221HM having the intron-containing GUS gene was previously described (Ohta et al. 1990). pAO416VG is constructed by inserting the DNA fragment covering the region from *virC* to *virG* of pEHA101 into mini Ri plasmid which was constructed by Nishiguchi et al (1987). Details of the procedure will be sent on request.

2.3. ENZYMES

Enzymes used for plasmid constructions were purchased from Toyobo Biochemicals (Kyoto, Japan), Takara Biochemicals (Osaka, Japan) and New England BioLabs (USA).

2.4. INCUBATION OF AGROBACTERIUM AND ASSAY OF ß-GALACTOSIDASE ACTIVITY

Cells of *A. tumefaciens* that carry Ti plasmid and other plasmid(s) were grown at 26 °C in L-broth. The bacterial cells at log phase were washed with 10 mM $MgSO_4$, and then resuspended at 2 x 10^8 cells/ml in MSPS medium (Murashige and Skoog medium supplemented with 62.5 mM sodium phosphate and 3% sucrose, pH 5.25) that contained 1 µM or 10 µM acetosyringone and various concentrations of sugars to be examined. The cells were incubated at 26 °C with shaking for 18 hours, a time sufficient for maximum induction. ß-galactosidase activity was assayed by the method of Usami et al (1988).

2.5. CO-CULTURE OF BY-2 CELLS WITH *AGROBACTERIUM*

BY-2 cells were co-cultured with *Agrobacterium tumefaciens* as described by An (1985). Stable transformants were selected for resistance to 200 µg/ml kanamycin. *Agrobacterium* cells were cultured until the absorbance at 600 nm was 1.5 and the culture was concentrated ten-fold in experiments with *vir* mutants. 100 µl of this culture was mixed with 4 ml of a suspension culture of BY-2 cells and the mixture was incubated with LS medium with 0.2 mg/l 2,4-D at 26°C in the dark for various periods.

2.6. ASSAY OF GUS ACTIVITY

Fluorometric assay for ß-glucuronidase (GUS) activity was performed by the procedures of Jefferson et al. (1986). BY-2 cells were collected various periods after co-culturing and lysed by sonication (three 30-sec pulses). After removal of cell debris by centrifugation, the supernatants were recovered and used for determination of protein concentration and GUS activity by a fluorometric assay.

3. Results

3.1. A GROUP OF MONOSACCHARIDES GREATELY ENHANCE ACETOSYRINGONE-DEPENDENT *VIR* GENE EXPRESSION

As shown in Fig. 1, aldoses were classified according to the level of activity to enhance AS-dependent expression of *vir* genes. There are clear correlations between activity and stereochemical structures of aldoses (Shimoda et al. 1990). Only aldoses that have the equatorial hydroxyl group at C-3 were effective. Aldoses which have the equatorial hydroxyl groups at C-2 as well as C-3 showed the highest activity for enhancement. Aldoses which have the different hydroxyl configuration at C-3 were inactive. Ketoses and disaccharides such as sucrose and lactose were not effective. These results indicate that stereochemical structures of sugars are important for enhancement of *vir* induction. In particular, the C-3 stereochemical structure is crucial. Acidic monosaccharides such as D-galacturonic acid were the most effective among aldoses we tested (Fig. 2). It is interesting to note that most of the effective sugars are

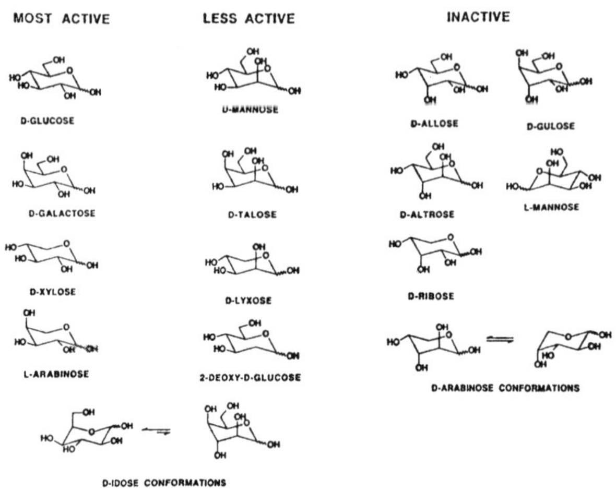

Fig. 1. Correlation between structures and activities of aldoses for sugar enhancement of *vir* induction. D-idose exhibited moderate activity for sugar enhancement. Since it can exist in two different chair conformations and both may be present at equilibrium, one of the conformations may have activity.

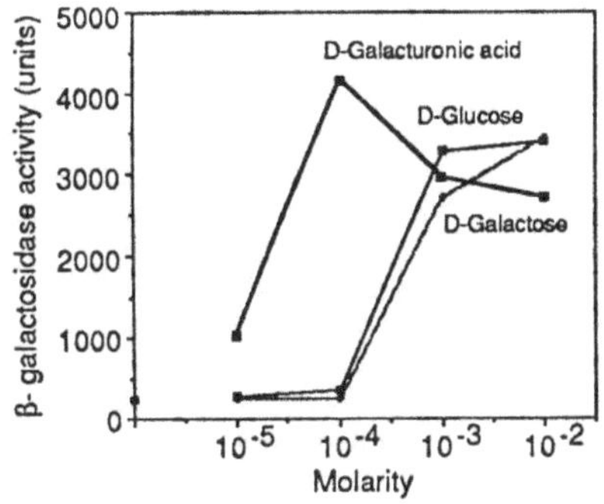

Fig. 2. Comparison of enhancing activity of an acidic sugar to that of neutral sugars. *A. tumefaciens* C58C1Cm cells carrying pTiB6S3^{trac} and pCM110D (*virD-lacZ*) were incubated in MSPS medium containing 10 µM AS and the indicated concentrations of sugars as described previously (Shimoda et al. 1990).

precursors of the major components of the cell wall polysaccharides of higher plants, i.e. cellulose, xyloglucans or arabinogalacturonans.

3.2. THE PERIPLASMIC DOMAIN OF VIRA PROTEIN IS RESPONSIBLE FOR SUGAR ENHANCEMENT OF *VIR* INDUCTION

In order to determine sites in the periplasmic domain of VirA protein responsible for sugar enhancement, we constructed *virA* gene mutants with various deletions in the region encoding the periplasmic domain as depicted in Fig. 3. For examining effects of these *virA* mutations on induction of *vir* expression, the *virA* gene on pTiB6S3 in *Agrobacterium* was disrupted by replacing with the *virB-lacZ* fusion gene, in which synthesis of ß-galactosidase is directed by the inducible *virB* promoter (We designated this *Agrobacterium* strain as ΔA/Blac). Plasmids carrying the mutant *virA* genes constructed above were introduced into ΔA/Blac agrobacterial cells and induction of *virB-lacZ* expression by AS was examined in the presence or the absence of glucose. As shown in Fig. 4, activity of ß-galactosidase was induced by AS in all the periplasmic mutants at various levels, but they were no longer enhanced by glucose. Even when the region between the *Pma*CI and the *Dra*II sites encoding only two amino acid residues was deleted (DM2), enhancement by glucose was not observed. Therefore, the large region of the periplasmic domain corresponding to the *Pst*I-*Dra*II DNA segment (see Fig. 3) is involved in enhancement by sugars. Deletion mutants, DM178 and DM2, exhibited the additional phenotype: the levels of vir gene expression induced by AS without glucose increased by

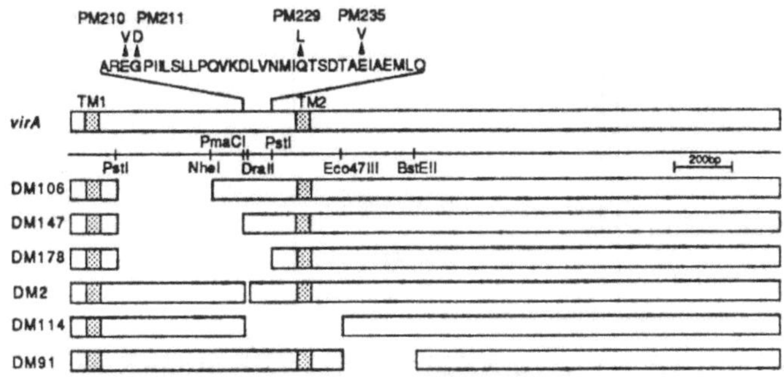

Fig. 3. Schematic representation of structures of *virA* mutants. Deletion and base substitution mutants are designated as DM and PM, respectively. The numbers in the mutant names of DM and PM refer to the number of amino acids deleted and positions of substituted residues, respectively. Dotted boxes marked by TM represent hydrophobic regions, which are thought to be transmembrane domains.

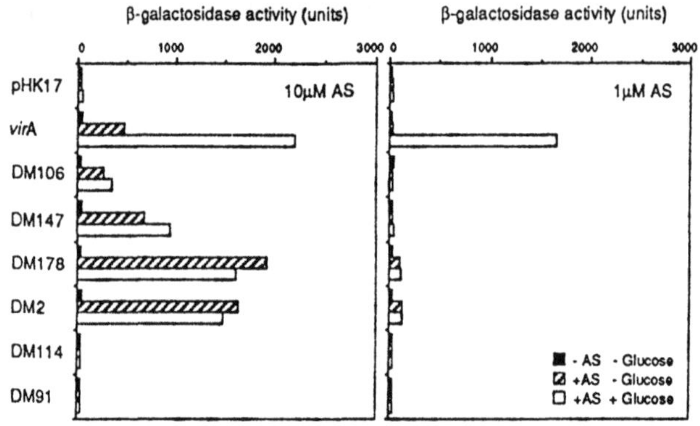

Fig. 4. Effects of glucose on *virB* gene expression in the deletion mutants of *virA*. *A. tumefaciens* strain C58C1CmΔA/Blac carrying the indicated DM plasmid with the *virA* mutation was incubated in AS-containing medium supplemented with or without 10 mM glucose. ß-galactosidase units represent the average of three independent experimental determinations.

several-fold. The *virA* mutant with the deletion near TM2 may have been more sensitive to AS than a wild type.

To precisely determine the region responsible for sugar enhancement, we then introduced base substitutions into the region between the *Pma*CI site and the *Pst*I site near the TM2-coding region (the 2nd *Pst*I site) (Fig. 3), which caused the non-conservative amino acid changes. Effects of these base substitutions on *vir* induction were examined in ΔA/Blac agrobacterial cells. Fig. 5 shows that the mutant VirA protein having either single amino acid change at position 210 ($VirA^{E}210^{V}$: indicating VirA with amino acid change from E to V at position 210) or that at position 211 ($VirA^{G}211^{D}$) did not respond to glucose. In contrast, enhancement by glucose was observed in *Agrobacterium* having either $VirA^{Q}229^{L}$ or $VirA^{E}235^{V}$. Note that *vir* gene expression in the $VirA^{E}235^{V}$ strain incubated in the presence or the absence of glucose was in the similar levels to that in a wild type strain. Therefore, it is likely that the region extending from TM1 to amino acid position 229 is responsible for enhancement by sugars, but the amino acid region between position 229 and TM2 is not crucial.

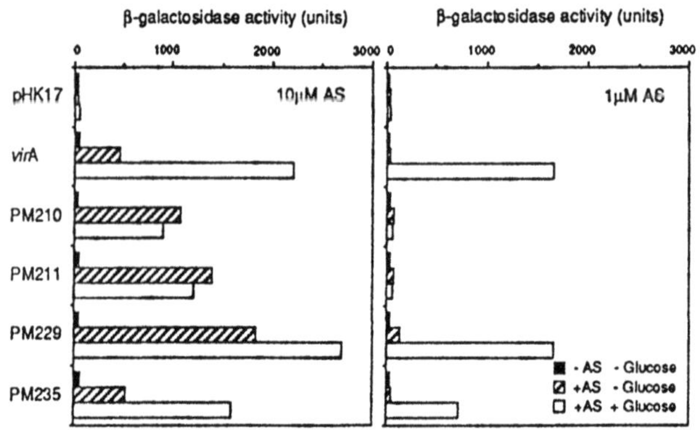

Fig. 5. Effects of glucose on *virB* gene expression in the base substitution mutants of *virA*. Activity of ß-galactosidase was assayed as described in Fig. 4.

3.3. MUTATIONS IN THE CHVE GENE SUPPRESS THE SUGAR-NON-RESPONSIVE PHENOTYPE OF THE VIRA MUTANT

In order to understand interaction between VirA and ChvE proteins, we attempted to create suppressor mutations of the *chvE* gene. To this end, we first of all disrupted the *chvE* gene in the chromosome of the *Agrobacterium* strain ΔA/Blac carrying $VirA^{E}210^{V}$ in

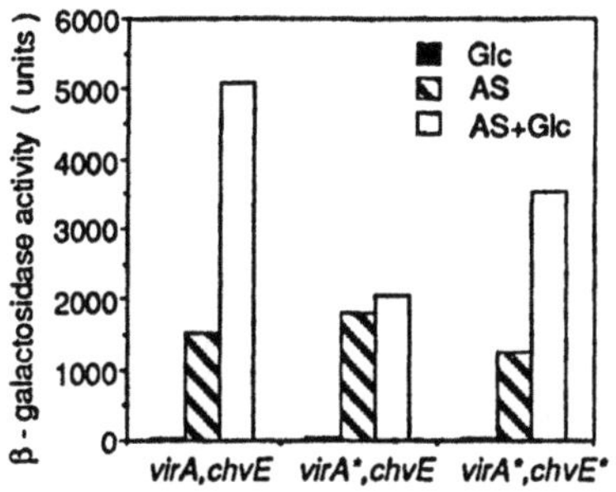

Fig. 6. Effects of the *chvE* mutation on *virB* expression in the *virA* mutant ($^{E}210^{V}$). 10 µM AS and 10 mM glucose were used. Activity of ß-galactosidase was assayed as described in Fig. 4. Asterisks indicate mutated genes.

the plasmid. The strain thus created had the base substitution mutation in the *virA* gene and the deletion in the *chvE* gene (designated as $A^{E}210^{D}$/Blac/ΔE). Plasmid DNA having the *chvE* gene was treated with hydroxylamine and transformed into cells of *Agrobacterium* $A^{E}210^{V}$ /Blac/ΔE. Bacterial cells were plated on agar medium containing 10 µM AS, 10 mM glucose and X-gal and incubated (colonies of *Agrobacterium* $A^{E}210^{V}$/Blac carrying a wild-type *chvE* gene in the plasmid showed faint blue under these incubation conditions). Out of 38,000 colonies, 11 showed higher intensity of blue color than that carrying the wild-type *chvE* . These colonies were classified into two groups (group A and group B) on the basis of the intensity of blue color.

Plasmid DNAs were purified from the bacteria exhibiting blue color. The DNA fragment containing the *chvE* region was isolated and reinserted into other vector DNA. The resulted construct was again introduced into cells of $A^{E}210^{V}$/Blac/ΔE. Induction of *vir* gene expression by AS in the cells carrying each construct was biochemically examined in the presence or the absence of glucose. As shown in Fig. 6, enhancement by glucose was restored in cells carrying either group A or B plasmid (data with group B plasmid was not shown). When the DNA fragment carrying a wild-type *chvE* gene was similarly introduced into $A^{E}210^{V}$/Blac/ΔE cells, no sugar enhancement was observed (Fig. 6). These results indicate that the mutation(s) in the *chvE* gene which was probably generated by treatment with hydroxylamine suppresses the sugar-non-responsive mutation of *virA*. Since the extent of suppression by the group A differed from that of suppression by the group B, mutations in these groups are supposed to occur in different sites.

3.4. AGROBACTERIUM-MEDIATED TRANSIENT GENE EXPRESSION IN PLANT CELLS IS USEFUL FOR STUDYING PROCESSES OF T-DNA TRANSFER

A Procedure for detecting transfer of T-DNA from Agrobacterium to plant nuclei. A procedure for assaying expression levels of genes that are transiently introduced into plant cells is currently used for estimating activity of a promoter of interest, since results can be rapidly obtained. However, this procedure can be useful for only protoplasts in case of plant cells. It has been shown that a certain type of tobacco suspension culture cell lines are able to be directly transformed by co-culturing with *Agrobacterium* harboring binary vector plasmid (Ann 1985). We here examined whether transient expression of the GUS reporter gene can be detected by using the co-culture procedure.

To detect expression of the reporter gene specifically in plant cells, we used binary plasmid pIG221HM (Ohta et al., 1990), which carries the 35S promoter-linked GUS gene with the intron sequence in the GUS coding region (it was referred to as P35S/INTGUS) so that normal GUS protein from this construct can be synthesized only in plant cells. BY-2 tobacco suspension culture cells were co-cultured with various strains of *Agrobacterium* harboring pIG221HM and GUS activity was assayed. As shown in Fig.

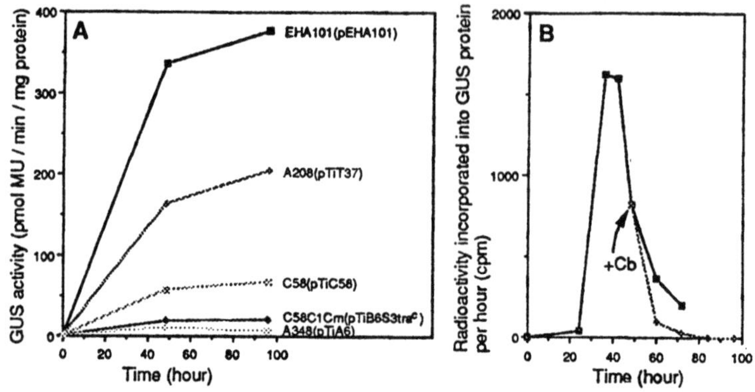

Fig. 7. (A) Activity of GUS detected after co-culturing BY-2 tobacco suspension cultured cells with the *Agrobacterium* strain carrying pIG221HM (35S/INTGUS). Co-culture and measurement of GUS activity were carried out as described in Materials and Methods. (B) Rate of GUS protein synthesis. An aliquot of the co-culture was labeled with [^{35}S] methionine for one hours and the reaction was quenched by chilling. Proteins were extracted and GUS protein was immunoprecipitated with antisera against ß-glucuronidase. Precipitated proteins were fractionated on SDS-polyacrylamido gel. Radioactivity of ^{35}S in the region corresponding to the GUS protein band was measured. Cb, 200 µg/ml carbenicillin.

7A, activity of GUS was detected in all the samples up to 48 hours after co-culturing, although the level of activity differed from one strain to another. They scarcely increased by further co-culturing. Subsequently, we examined whether the activities of GUS we observed were due to transient expression of the 35S/INTGUS before it would be integrated into tobacco nuclear DNA. We measured the rate of synthesis of GUS protein by pulse-labeling aliquots of the co-cultured sample of the EHA101 strain with [^{35}S] methionine. Fig. 7B shows that the rate of GUS synthesis was maximum between 36 and 42 hours after co-culturing, then decreased rapidly. When carbenicillin was added to the co-culture at 48 hours, the degree of decrease was exaggerated. Such transient synthesis cannot be explained by assuming expression of 35S/INTGUS stably resident in tobacco chromosomes. It is most likely that the GUS protein observed was synthesized from the 35S/INTGUS molecules that had been transferred from *Agrobacterium* cells to tobacco nuclei but not yet integrated into the chromosomal DNA. In addition, the present results suggest that a large number of 35S/INTGUS molecules can be first transferred to plant nuclei, then only a limited portion of them can be integrated into nuclear DNA. It has been reported that when *Agrobacterium* harboring the GUS gene in a binary vector was inoculated on leaf discs, the similar pattern of transient expression was observed (Janssen

Table 1. Comparison of transient activity of GUS to the frequency of transformation by various Agrobacterium strains

		Transformation frequency (number of Km^r calli / 0.1ml culture)		GUS activity (pmol MU / min / mg protein)	
Strain	Ti plasmid	+pIG221HM	-pIG221HM	+pIG221HM	-pIG221HM
EHA101	pEHA101	494 ±203	0	338 ±202	1.5 ± 0.70
A208	pTiT37	229 ± 19	0	164 ± 55	1.1 ± 1.3
C58	pTiC58	86 ± 75	0	58 ± 42	0.80± 0.10
C58C1Cm	pTiB6S3tra^c	25 ± 8	0	19 ± 9.3	0.73± 0.16
A348	pTiA6	15 ± 1	0	12 ± 2.6	1.6 ± 0.74

Table 2. Effects of *vir* mutations on transient activity of GUS

Strain	Mutation	GUS activity (pmol MU / min / mg protein)
A348 / pAO416VG , pIG221HM	Wild type	171
A348 / pIG221HM	Wild type	15.0
A348 / pAO416VG	Wild type	1.0
A348mx226 / pAO416VG , pIG221HM	*virA*	0.59
A348mx238 / pAO416VG , pIG221HM	*virB*	0.31
A348mx355 / pAO416VG , pIG221HM	*virD4*	0.79
A348mx341 / pAO416VG , pIG221HM	*virE2*	8.6

and Gardner 1989; Vancanneyt et al. 1990).

As shown in Fig. 7A, the level of GUS expression varied from one *Agrobacterium* strain to another: the highest GUS activity was found in the co-culture with EHA101, which is known to carry the disarmed supervirulent Ti plasmid. There seems to be a parallel correlation between the level of GUS activity and the frequency of transformation of BY-2 by *Agrobacterium* strains used (Table 1). These results imply that the level of GUS activity reflects amounts of 35S/INTGUS molecules in tobacco nuclei which must have been transferred from *Agrobacterium*. Therefore, the efficiency of T-DNA transfer can be roughly measured by assaying GUS activity after co-culturing BY-2 tobacco cells.

VirB, VirD4 and VirE2 Proteins Are Responsible for Transfer of T-DNA. We examined mutants of *vir* genes for abilities to transfer 35S/INTGUS to plant nuclei. A series of mutants, however, were created with the A348 strain harboring pTiA6 (Stachel and Nester 1986), which has been shown to be inefficient for T-DNA transfer in the present study. To increase the efficiency of T-DNA transfer, we introduced into the A348 strain pAO416VG containing the DNA region from *virG* to *virC* from supervirulent Ti plasmid pTiBo542 which is shown to be responsible for the supervirulent characteristics and the high level of AS-inducible expression of *vir* genes (Jin et al. 1987). As shown in Table 2, the efficiency of 35S/INTGUS transfer increased 10-fold, when the A348 strain harboring pAO416VG and pIG221HM was used for co-culturing.

We introduced pAO416VG and pIG221HM into A348 derivative strains with Tn*3*-HoHo1 insertions in *virA, virB, virD4* and *virE2,* respectively, and co-cultured BY-2 cells with each of these *vir* mutants. Table 2 summarizes activity of GUS detected at 72 hours after co-culturing. When co-cultured with mutants of *virA, virB* and *virD4,* GUS activity was in the background level . Significant activity was detected in the co-culture with the *virE2* mutant, although the level of activity decreased 20-fold.

4. Discussion

The present results of deletion analyses of the *virA* gene reveal that the large region of the periplasmic portion of VirA protein is responsible for enhancement by sugars. Taken together with the results of experiments with base substitution mutations, the responsible region is likely present between the site corresponding to the *Pst*I site near TM1 (the 1st *Pst*I site) and amino acid position 211. The periplasmic region close to TM2 may not be involved in sugar enhancement.(Fig. 5).

Roles of the periplasmic region for AS and sugar response, however, seem to be rather complex. When the region between the 1st *Pst*I site and the *Pma*CI site was deleted (DM106), only enhancement by glucose was abolished (Fig. 4). When deletion was extended to the 2nd *Pst*I site near TM2 (DM178), the induction level by AS markedly increased without sugars. Deletion of two amino acid residues encoded by the sequence between *Pma*CI and *Dra*II cleavage sites also resulted in increase in the induction level by AS without sugars (Fig. 4). Therefore, the region encoded by the DNA sequence between the *Pma*CI site and the 2nd *Pst*I site seems to have a function to repress *vir* induction by AS in addition to that responsible for sugar enhancement.

Genetic analysis with suppressor mutants of *chvE* provides evidence that the ChvE protein directly interacts with the VirA protein, probably with its periplasmic region. The results of deletion analysis show that the large region in the periplasmic domain of the VirA protein is required for sugar enhancement. It, therefore, can be speculated that a particular stereostructure of the periplasmic domain is necessary for interaction with the ChvE protein or the ChvE/sugar complex.

We also have demonstrated that transient gene expression is observed when tobacco BY-2 cells are co-culturing with *Agrobacterium* cells and that this transient expression system is useful for studying the process of T-DNA transfer. When BY-2 cells were co-cultured with *Agrobacterium* having a mutation in either *virA, virB, virD4* or *virE2*, transient expression was abolished or lowered (Table 2). T-DNA processing has been shown to normally take place in cells of *virB* and *virD4* mutants when they are incubated with AS or plant exudates (Yanofsky et al. 1986; Yamamoto et al. 1987; Veluthambi et al. 1987; Stachel et al. 1987), although they are not able to induce crown gall tumor. Therefore, the results described above suggest that products of *virB* and *virD4* are involved in the process of transfer of T-DNA from the bacterial cells to plant nuclei. Previous observations that some of VirB proteins and VirD4 protein are localized in inner membranes of *Agrobacterium* cells (Christie et al. 1989; Ward et al. 1990; Okamoto et al., 1991) support this suggestion.

Since only the low level of GUS activity was found in the co-culture with the *virE2* mutant, VirE2 protein is also required for efficient transfer of T-DNA, but it seems not to be essential. This result is consistent with the observation that *virE* mutants incite attenuated tumor on tobacco plants (Stachel and Nester 1986). VirE2 protein was shown to bind T-strand and to be localized in plant nuclei when it was synthesized in plant cells (see Introduction). Based on these results, it is proposed that VirE2 protein could facilitate nuclear transport of T-strand (Citovsky et al. 1992). Our observation is in line with this idea. Occurrence of the reduced level of transient expression (Table 2), however, suggests that T-DNA is transported into plant nuclei without VirE2 protein, which may be mediated by other protein(s) that may have the function similar to that of VirE2. Most likely candidate for that is VirD2 protein that has been also shown to bind to T-DNA molecules and to be localized to plant nuclei (Herrera-Estrella et al. 1990; Howard et al. 1992).

The present results suggest that a large number of T-DNA molecules are transiently transferred to plant nuclei during co-culture. Our previous estimation has shown that T-DNA can be transferred into at least 5% of BY-2 cells co-cultured with EHA101 cells (Onouchi et al., 1991). Therefore, the experimental system presented here can be applicable for analysis of structures of T-DNA molecules after transferred to plant cells.

Acknowledgements. We are grateful to Dr. E. W. Nester for the gift of *Agrobacterium* strain 348 and its derivatives. We also thank Drs. M.-D. Chilton and W. S. Chilton for their helpful discussions about correlation between sugar structures and activity and Dr. Imae for his helpful discussions about sugar binding proteins. This research was supported in part by a Grant-in-Aid for General Scientific Research, by a Grant-in-Aid for Scientific Research on Priority Area from The Ministry of Education, Science and Culture of Japan and a grant from the Ministry of Agriculture, Forestry, and Fisheries of Japan.

REFERENCES

Albright, L. M., Huala, E. and Ausubel, F. M. (1989) 'Prokaryotic signal transduction mediated by sensor and regulator protein pairs', Annu. Rev. Genet. 23, 311-336.

An, G. (1985) 'High Efficiency Transformation of Cultured Tobaccco Cells', Plant Physiol. 79, 568-570.

Cangelosi, G. A., Ankenbauer, R. G. and Nester, E. W. (1990) 'Sugars induce the *Agrobacterium* virulence genes through a periplasmic binding protein and a transmembrane signal protein', Proc. Natl. Acad. Sci. USA 87, 6708-6712.

Citovsky, V., Zupan, J., Warnick, D. and Zambryski, P. (1992) 'Nuclear Localization of *Agrobacterium* VirE2 Protein in Plant Cells', Science, 256, 1802-1805.

Cornish, A., Greenwood, J. A. and Jones, C. W. (1989) 'Binding-protein-dependent sugar transport by *Agrobacterium radiobacter* and *A. tumefaciens* grown in continuous culture', J. Gen. Microbiol., 135, 3001-3013.

Herrera-Estrella, A., Montagu, M. V. and Wang, K. (1990) 'A bacterial peptide acting as a plant nuclear targeting signal: The amino-terminal portion of *Agrobacterium* VirD2 protein directs a ß-galactosidase fusion protein into tobacco unclei', Proc. Natl. Acad. Sci. USA 87, 9534-9537.

Hohn, B., Koukolíková-Nicolá, Z., Bakkeren, G. and Grimnsley, N. (1989) '*Agrobacterium*-mediated gene transfer to monocots and dicots', Genome 31, 987-993.

Hood, E. E., Helmer, G. L., Fraley, R. T. and Chilton, M.-D. (1986) 'The Hypervirulence of *Agrobacterium tumefaciens* A281 Is Encoded in a Region of pTiBo542 Outside of T-DNA', J. Bacteriol. 168, 1291-1301.

Howard, E. A.; Zupan, J. R., Citovsky, V. and Zambryski, P. C. (1992) 'The VirD2 Protein of A. tumefaciens Contains a C-Terminal Bipartite Nuclear Localization Signal: Implications for Nuclear Uptake of DNA in Plant Cells', Cell 68, 109-118.

Huang, Y., Morel, P., Powell, B. and Kado, C. I. (1990) 'VirA, a coregulator of Ti-specifield virulence genes, is phosphorylated in vitro', J. Bacteriol. 172, 1142-1144.

Ito, W., Ishiguro, H. and Kurosawa, Y. (1991) 'A general method for introducing a series of mutations in cloned DNA using the polymerase chain reaction', Gene 102, 67-70.

Janssen, B.-J. and Gardner, R. C. (1989) 'Localized transient expression of GUS in leaf discs following cocultivation with *Agrobacterium*', Plant Mol. Biol. 14, 61-72.

Jefferson, R. A., Burgess, S. M. and Hirsh, D. (1986) 'ß-Glucuronidase from *Escherichia coli* as a gene-fusion marker', Proc. Natl. Acad. Sci. USA 83, 8447-8451.

Jin, S., Komari, T., Gordon, M. P. and Nester, E. W. (1987) 'Genes responsible for the supervirulence phenotype of Agrobacterium tumefaciens A281', J. Bacteriol. 169, 4417-4425.

Jin, S., Roitsch, T., Ankenbauer, R. G., Gordon, M. P. and Nester, E. W. (1990) 'The VirA Protein of *Agrobacterium tumefaciens* Is Autophosphorylated and Is Essential for *vir* Gene Regulation', J. Bacteriol. 172, 525-530.

Kado, C. I. (1991) 'Molecular mechanisms of crown gall tumorigenesis', Critical Reviews in Plant Sci. 10, 1-32.

Leroux, B., Yanofsky, M. F., Winans, S. C., Ward, J. E., Ziegler, S. F. and Nester, E. W. (1987) 'Characterization of the *virA* locus of *Agrobacterium tumefaciens*: A transcriptional regulator and host range determinant', EMBO J. 6, 849-856.

Machida, Y., Okamoto, S., Matsumoto, S., Usami, S., Yamamoto, A., Niwa, Y., Jeong, S. D., Nagamine, J., Shimoda, N., Machida, C. and Iwahashi, M. (1989) 'Mechanisms of crown gall formation: T-DNA transfer from *Agrobacterium* to plant cells', Bot. Mag. Tokyo 102, 331-350.

Melchers, L. S., Maroney, M. J., den Dulk-Ras, A., Thompson, D. V., van Vuuren, H. A. J., Schilperoort, R. A. and Hooykaas, P. J. J. (1990) 'Octopine and nopaline strains of *Agrobacterium tumefaciens* differ in virulence; molecular characterization of the *virF* locus', Plant Mol. Biol. 14, 249-259.

Melchers, L. S., Regensburg-Tuink, A. J. G., Bourret, R. B., Sedee, N. J. A., Schilperoort, R. A. and Hooykaas, P. J. J. (1989) 'Membrane topology and functional analysis of the sensory protein VirA of *Agrobacterium tumefaciens*', EMBO J. 8, 1919-1925.

Morel, P., Powell, B. S., Rogowsky, P. M. and Kado, C. I. (1989) 'Characterization of the *virA* virulence gene of the nopaline plasmid, pTiC58, of *Agrobacterium tumefavciens*', Mol. Microgiology 3, 1237-1246.

Melchers, L. S., Thompson, D. V., Idler, K. B., Neuteboom, S. T. C., De Maagd, R. A., Schilperoort, R. A. and Hooykaas, P. J. J. (1987) 'Molecular characterization of the virulence gene *virA* of the *Agrobacterium tumefaciens* octopine Ti plasmid' Plant Mol. Biol. 9, 635-645.

Nishiguchi, R., Takanami, M. and Oka, A. (1987) 'Characterization and sequence determination of the replicator region in the hairy-root-inducing plasmid pRiA4b', Mol. Gen. Genet. 206, 1-8.
Ohta, S., Mita, S., Hattori, T. and Nakamura, K. (1990) 'Construction and Expression in Tobacco of a ß-glucuronidase (GUS) Reporter Gene Containing an Intron Within the Coding Sequence', Plant Cell Physiol. 31, 805-813.
Okamoto, S., Toyoda-Yamamoto, A., Ito, K., Takebe, I. and Machida, Y. (1991) 'Localization and orientation of the VirD4 protein of *Agrobacterium tumefaciens* in the cell mumbrane', Mol. Gen. Genet. 228, 24-32.
Onouchi, H., Yokoi, K., Machida, C., Matsuzaki, H., Oshima, Y., Matsuoka, K., Nakamura, K. and Machida, Y. (1991) 'Operation of an efficient site-specific recombination system of *Zygosaccharomyces rouxii* in tobacco cells', Nucleic Acids Res. 19, 6373-6378.
Petit, A., Tempé, J., Kerr, A., Holster, M., Van Montagu, M. and Schell, J. (1978) 'Substrate induction of conjugative activity of *Agrobacterium tumefaciens* Ti plasmids', Nature 271, 570-571.
Rogowsky, P. M., Close, T. J., Chimera, J. J. and Kado, C. I. (1987) 'Regulation of the *vir* genes of *Agrobacterium tumefaciens* plasmid pTiC58' 169, 5101-5112
Sciaky, D., Montoya, A. L. and Chilton, M. D. (1978) 'Fingerprints of *Agrobacterium* Ti plasmids', Plasmid 1, 238-253.
Shimoda, N., Toyoda-Yamamoto, A., Nagamine, J., Usami, S., Katayama, M., Sakagami, Y. and Machida, Y. (1990) 'Control of expression of *Agrobacterium vir* genes by synergistic actions of phenolic signal molecules and monosaccharides', Proc. Natl. Acad. Sci. USA 87, 6684-6688.
Spencer, P. A. and Towers, G. H. N. (1989) 'Specificity of signal compounds detected by *Agrobacterium tumefaciens*', Phytochemistry 27, 2781-2785.
Stachel, S. E. and Nester, E. W. (1986) 'The genetic and transcriptional organization of the *vir* region of the A6 Ti plasmid of *Agrobacterium tumefaciens*', EMBO J. 5, 1445-1454.
Stachel, S. E. and Zambryski, P. C. (1986) '*Agrobacterium tumefaciens* and the susceptible plant cell: A novel adaptation of extracellular regognition and DNA conjugation', Cell 47, 155-157.
Stachel, S. E., Messens, E., Van Montagu, M. and Zambryski, P. (1985) 'Identification of the signal molecules produced by wounded plant cells that activate T-DNA transfer in *Agrobacterium tumefaciens*', Nature 318, 624-629.
Stachel, S. E., Nester, E. W. and Zambryski, P. C. (1986) 'A plant cell factor induces *Agrobacterium tumefaciens vir* gene expression', Proc. Natl. Acad. Sci. USA 83, 379-383.
Stachel, S. E., Timmerman, B. and Zambryski, P. (1987) 'Activation of *Agrobacterium tumefaciens vir* gene expression generates multiple single-stranded T-strand molecules from the TiA6 T-region: requirement for 5' *vir*D gene products', EMBO J. 6, 857-863.
Usami, S., Okamoto, S., Takebe, I. and Machida, Y. (1988) 'Factor inducir *Agrobacterium tumefaciens vir* gene expression is present monocotyledonous plants', Proc. Natl. Acad. Sci. USA 85, 3748-3752.
Vancanneyt, G., Schmidt, R., O'Connor-S., Willmitzer, L. and Rocha-S., M. (1990) 'Construction of an intron-containing marker gene: Splicing of the intron in transgenic plants and its use in monitoring early events in *Agrobacterium*-mediated plant transformation', Mol. Gen. Genet. 220, 245-250.
Veluthambi, K., Jayaswal R. K. and Gelvin, S. B. (1987) 'The virulence genes *A*, *G* and *D* mediate the double-stranded border cleavage of the T-DNA from the *Agrobacterium tumefaciens* Ti plasmid', Proc. Natl. Acad. Sci. USA 84, 1881-1885.
Ward, J. E., Dale, E. M., Nester, E. W. and Binns, A. N. (1990) 'Identification of a VirB10 protein aggregate in the inner membrane of *Agrobacterium tumefaciens*', J. Bacteriol. 172, 5200-5210.
Winans, S. C., Kerstetters, R. A., Ward, J. E. and Nester, E. W. (1989) 'A protein required for transcriptional regulation of *Agrobacterium* virulence genes spans the cytoplasmic membrane', J. Bacteriol. 171, 1616-1622.
Yamamoto, A., Iwahashi, M., Yanofsky, M. F., Nester, E. W., Takebe, I. and Machida, Y. (1987) 'The promoter proximal region in the *virD* locus of *Agrobacterium tumefaciens* is necessary for the plant-inducible circularization of T-DNA', Mol. Gen. Genet. 206, 174-177.
Yanofsky, M. F., Porter, S. G., Young, C., Albright, L. M., Gordon, M. P. and Nester, E. W. (1986) 'The *virD* locus of *Agrobacterium tumefaciens* encodes a site specific endonuclease', Cell 47, 471-477.
Zambryski, P. C. (1992) 'CHRONICLES FROM THE AGROBACTERIUM-PLANT CELL DNA TRANSFER STORY', Annu. Rev. Plant Mol. Biol. 43, 465-490.

DETECTION OF AVIRULENT MUTANTS OF *AGROBACTERIUM TUMEFACIENS* IN CROWN-GALL TUMORS PRODUCED *IN VITRO*.

C. BÉLANGER[1], M.L. CANFIELD[2], L.W. MOORE[2] and P. DION[1].
Département de phytologie, Faculté des sciences de l'agriculture et de l'alimentation, Université Laval, Québec, Québec, Canada, G1K-7P4, [1] and *Department of Botany and Plant Pathology, Oregon State University, Corvallis, Oregon 97330, USA.* [2]

ABSTRACT. Avirulent strains of *Agrobacterium* with the ability to catabolize opines have been described. In many strains, this property is encoded on a plasmid showing some homology with Ti plasmids. Recent studies (see accompanying abstract by Fortin *et al.*) showed that some *A. tumefaciens* strains had the capacity to mutate to avirulence under *vir* inducing conditions. The mutants thus produced were altered in their Ti plasmid. This suggested that, under some circumstances, avirulent mutants could perhaps be generated from virulent *A. tumefaciens* cells as a consequence of tumorigenesis. To test this possibility, seven *A. tumefaciens* isolates were inoculated on cherry, pear, blackberry and apple plantlets, which had been grown *in vitro* under axenic conditions. The same conditions of *in vitro* culture were maintained following bacterial inoculation. After one month, 100 bacterial clones recovered from each tumor were tested for virulence on tomato seedlings. Out of the seven strains so tested, only the biotype 2, nopaline-type strain D10B/87 produced avirulent mutants. These mutants were detected on apple tumors, but not on tumors of the other host plants tested. Depending on the experiment, between 30 and 99% avirulent D10B/87 clones were recovered from apple tumors. All of the 15 avirulent clones which were analyzed further had retained the capacity for nopaline catabolism. Of these, 14 were mutated in the Ti plasmid, and the remaining in the chromosome. No avirulent D10B/87 were recovered from apple tumors in greenhouse studies. We conclude that, under specific circumstances, mutation of *A. tumefaciens* to avirulence may accompany tumorigenesis.

Introduction

The soil microbe *Agrobacterium tumefaciens* is the causal agent of crown gall disease, which affects most dicotyledonous plants (DeCleen and DeLey, 1976). *A. tumefaciens* harbors a large plasmid called the Ti plasmid which is responsible for virulence (Van Larebeke et al., 1974). The Ti plasmid is divided in three important regions. One of these, called the T-DNA, is transferred and integrated into the plant genome. Once integrated, this region codes for proteins involved in the production of phytohormones which are responsible for tumor development. New compounds called opines are synthetised by enzymes coded by a second part of the T-DNA. Opines produced by transfomed cells are catabolised by the bacterium and this activity is catalized by products of genes present in a second important region of the Ti plasmid (Reviewed by; Kado, 1991; Ream,

E. W. Nester and D. P. S. Verma (eds.), Advances in Molecular Genetics of Plant-Microbe Interactions, 97–101.

1989). In nature, many *Agrobacterium* strains unable to produce disease, and this belonging to the species *A. radiobacter*, nevertheless exhibit the capacity to catabolize opines. In some cases, this ability is conferred by plasmids showing some homology with the Ti plasmid (Merlo and Nester, 1977). These observations suggest that mutations in virulent agrobacteria can lead to the appearance of avirulent opine-utilizing strains. These mutations would inactivate virulence without interfering with opine catabolism functions.

Yet another region of the Ti plasmid, called the virulence (*vir*) region, is induced by phenolic compounds produced at plant wounds (Stachel et al., 1985; Spencer and Towers, 1988). *vir* genes are responsible for the production of proteins which promote production of the single stranded T-DNA from the Ti plasmid, and transfer of this T-DNA from the bacterium to the plant cell. One of the *vir* genes (*virD*) codes for an endonuclease acting specifically on border sequences flanking the T-DNA (Yanofsky et al., 1986). We have hypothesized that under *vir* inducing conditions, avirulent mutants could be produced as a result of endonuclease or other type of genetic activity promoting rearrangements of the Ti plasmid. The reality of a such phenomenon has been demonstrated recently by studying the effect of *vir* phenolic inducers on the nopaline-type strain *A. tumefaciens* C58 (C. Fortin, unpublished).

In the present work, we have studied the occurrence of Ti plasmid rearrangements following tumorigenesis on various types of plantlets.

Materials and Methods

Bacterial strains. The *Agrobacterium tumefaciens* strains which were used in this study are listed in Table 1. For greenhouse studies, a spontaneous rifampicin and streptomycin resistant mutant of D10B/87 was used.

TABLE 1. *Agrobacterium tumefaciens* strains used in this study.

Strain	Host of origin	Biotype	Opine Utilization
C58	Cherry	I	Nopaline
B49C/83	Apple	II	Nopaline, Mannopine
D10B/87	Apple	II	Nopaline
I11/85	Cherry	II	Nopaline
I22/85	Cherry	II	Nopaline
B209B/85	Blackberry	ND(a)	ND
B230/85	Blackberry	ND	ND

(a) ND, not determined

Plant inoculations. Sterile plantlets of cherry, pear, blackberry and apple were cultivated on tissue culture media variously modified from Murashige and Skoog to meet the specific requirements of the different plant species. *In vitro*-grown plants were maintained at 25°C with a 16h photoperiod. Greenhouse-cultivated apple planlets came from *in vitro* culture and were placed under greenhouse conditions two weeks before inoculation. Plantlets were inoculated by piercing their stem with a needle previously dipped in a bacterial colony. After inoculation, *in vitro* and greenhouse grown plantlets were returned to their respective incubation conditions.

Recovery of bacteria from crown gall tumors. Following tumor development (one to three months), the tumor tissue was macerated in 2 ml of saline solution for 30 min. Dilutions were plated on rich medium. For greenhouse studies rifampicin (50 mg/ml), streptomycin (250 mg/ml) and cycloheximide (20 mg/ml) were added to the medium used for bacterial recovery. From each tumor, one hundred bacterial colonies were selected for further testing.

Determination of virulence and nopaline utilization. Bacterial virulence was tested in the greenhouse on tomato plants (Bonnie Best or Vendor) by stem inoculation three to four weeks after sowing. The capacity for nopaline utilization was examined in liquid AT medium containing nopaline (800 mg/L). After 72h of incubation at 28°C under shaking (175 rpm), development of turbidity indicated a positive result.

Ti plasmid transfer. The Ti plasmid was transfered from a donor *Agrobacterium* strain to the recipient strain C58C1rs. This recipient had been cured of the Ti plasmid and was resistant to streptomycin and rifampicin. Selection was made on AT medium containing nopaline (800 mg/L), rifampicin and streptomycin. The transconjugants were tested for virulence as described above.

Results

Seven strains of *Agrobacterium tumefaciens* listed in Table 1 were inoculated on *in vitro*-grown plants of cherry, pear, apple and blackberry and then returned to conditions of *in vitro* culture. After one month, the tumors were excised and the bacteria were recovered. One hundred clones of each bacterial strain - plant combination were tested for virulence.

Out of the seven strains tested in this manner, only one, D10B/87, produced avirulent mutants. Furthermore, these avirulent mutants were obtained only on apple rootstocks. Tumors of cherry, pear or blackberry contained only virulents clones of D10B/87. In two independant inoculations of strain D10B/87 on apple rootstocks, 90 and 35%, respectively, of avirulent mutants were recovered. All of the clones, virulent or avirulent, were able to utilize nopaline as the sole source of carbon and nitrogen (Table 2).

These initial tests had been done with Mark apple rootstocks. In order to see if the recovery of mutants was possible only from Mark apple plantlets, other types of apple rootstocks were also inoculated with D10B/87. Three months after inoculation, tumors had become visible on the various rootstocks and bacterial clones were recovered from each of these. Bacteria were collecded either from the internal tissue of the tumor (with the gall surface removed and following maceration as described above) or else from the tumor surface. In the last case, bacteria were recovered simply by washing the surface of the gall with saline solution and transferring portion of this solution to selective medium as described above. Tumors formed by the three different apple rootstocks contained between 68 and 99% of avirulent mutants and most of the clones recovered had retained the capacity for nopaline catabolism (Table 2). No obvious difference in the proportion of avirulent mutants was observed between the tumor inner tissue and the tumor suface. These results indicated that avirulent mutants of D10B/87 were not recovered only from Mark tumors but that they occurred also in tumors of other rootstocks. Furthermore, it was apparent that mutant production was not a direct consequence of the tissue maceration used for isolation of bacteria. Inoculation of strain D10B/87 in tissue culture medium in the absence of plantlets did not induce mutagenesis.

In vitro culture may alter various aspects of plant metabolism (Ibrahim, 1987; Hegedus and Phan, 1983), including metabolism of phenolic compounds and photosynthesis. In addition, chemical composition of the plant is likely to be affected. Thus, an experiment was performed to compare

TABLE 2. Properties of bacteria recovered from surface or inner tissue of tumors induced on various apple rootstocks.

Apple rootstock	Plant growth conditions	Proportion (%) of: nopaline-utilizing clones surface	inner tissue	avirulent clones surface	inner tissue
Mark	In vitro	ND (a)	100	ND	90
Ottawa3	In vitro	100	100	99	98
P106	In vitro	100	100	68	95
BUD116	In vitro	99	100	82	90
Ottawa3	Greenhouse	100	100	0	0
Malling7	Greenhouse	100	100	0	0

(a) ND, not determined

the rate of mutant recovery from apple tumors produced on plants grown *in vitro* or else in the greenhouse. Strain D10B/87 was inoculated on two different apple cultivars grown in the greenhouse. After three months, bacteria were harvested from tumors and tested for virulence. No mutants were obtained under these conditions (Table 2).

Fifteen avirulent mutants obtained from tumors on *in vitro*-grown Mark apple rootstocks were further analyzed and compared to the wild type D10B/87. Eckhardt gel analysis showed that strain D10B/87 harbored two plasmids. The larger of these was transferred to the recipient *Agrobacterium* C58C1rs to establish that this particular plasmid encoded nopaline catabolism and virulence. This pTiD10B was of about the size of pTiC58. No deletions in the pTi were detected in the fifteen mutants analyzed by Eckhardt gel (data not shown).

The plasmidic or chromosomal localization of the mutations to avirulence was also examined by conjugative transfer of the Ti plasmid from the D10B/87 mutants to the avirulent C58C1rs recipient. The resulting transconjugants were tested for virulence on tomato plants. Transfer of the Ti plasmid from the wild type D10B/87 to C58C1rs produced a virulent transconjugant. However, transfer of the Ti plasmid from fourteen out of the fifteen avirulent mutants analyzed failed to confer virulence on the C58C1rs recipient (Table 3). This indicated that, in the case of these fourteen mutants, the mutation to avirulence resided on the Ti plasmid.

TABLE 3. Localization of the mutation to avirulence of D10B/87 clones recovered from apple tumors produced *in vitro*.

Clone of D10B/87	Virulence of clone	Virulence of transconjugant	Mutation on Ti plasmid
14/15 avirulent mutants	-	-	Yes
1/15 avirulent mutant	-	+	No

Conclusions

In this work, avirulent mutants were obtained following tumorigenesis under conditions of *in vitro* culture. Out of seven strains tested, only one of these, strain D10B/87, produced mutants on various types of apple rootstocks. This strain did not yield mutants on three other plant species. Most of the avirulent mutants retained the capacity to catabolize nopaline. Mutagenesis was limited to conditions of *in vitro* culture. This may correspond to alterations in plant physiology or plant composition which would be induced by *in vitro* culture. Out of fifteen mutants analyzed, fourteen were mutated in pTiD10B. Results from this stydy suggest that, under particular conditions, tumor induction by some strains of *A. tumefaciens* can be accompanied by the massive production of avirulent mutants.

References

DECLEEN, M. and DELEY, J. (1976) 'The host range of crown gall', Bot. Rev. 42, 389-466.

HEGEDUS, P. and PHAN, C.T. (1983) 'Action des phénols sur les malformations observées chez les porte-greffes de pommiers M-26 et O-3 cultivés *in vitro*', Rev. Can. Biol. Exptl. 42, 33-38.

IBRAHIM, K.R. (1987) 'Regulation of synthesis of phenolics', in F.Constabel and I.K. Vasil (eds.), Cell culture and somatic cell genetics of plants. Vol.4: Cell culture in phytochemistry, Academic Press, London, pp. 77-95.

KADO, C.I. (1991) 'Molecular mechanisms of crown gall tumorigenesis', Crit. Rev. Plant Sci. 10, 1-32.

MERLO, D.J. and NESTER, E.W. (1977) 'Plasmids in avirulent strains of *Agrobacterium*', J. Bact. 129 ,76-80.

REAM, W. (1989) '*Agrobacterium tumefaciens* and interkingdom genetic exchange', Annu. Rev. Phytopathol. 27, 583-618.

SPENCER, P.A., and TOWERS, G.H.N. (1988) 'Specificity of signal compounds detected by *Agrobacterium tumefaciens*', Phytochem. 27, 2781-2785.

STACHEL, S.E., MESSENS, E., VAN MONTAGU, M. and ZAMBRYSKI, P. (1985) 'Identification of the signal molecules produced by wounded plant cells that activate T-DNA transfer in *Agrobacterium tumefaciens*', Nature 318, 624-629.

VAN LAREBEKE, N., ENGLER, G., HOLSTER, M., VAN DEN ELSACKER, S., ZAENEN, I., et al. (1974) 'Large plasmid in *Agrobacterium tumefaciens* essential for crown gall inducing activity', Nature 252, 169-170.

YANOFSKY, M.F., PORTER, S.G., YOUNG, C., ALBRIGHT, L.M., GORDON, M.P. and NESTER, E.W. (1986) 'The *virD* operon of *Agrobacterium tumefaciens* encodes a site-specific endonuclease', Cell 47, 471-477.

VIRB4 IS AN INNER MEMBRANE-ASSOCIATED PROTEIN ESSENTIAL FOR T-DNA TRANSFER FROM *AGROBACTERIUM TUMEFACIENS* TO PLANTS.

K. SHIRASU[1], Z. KOUKOLÍKOVÁ-NICOLA[2], B. HOHN[2] AND C. I. KADO[1].

1-Department of Plant Pathology
University of California, Davis
California 95616, U.S.A.
2-Friedrich Miescher-Institut
P.O. BOX 2543
CH-4002 Basel, Switzerland

ABSTRACT. *Agrobacterium tumefaciens* is unique in its ability to transfer bacterial DNA to plant cells. The virulence regulon of the Ti plasmid of *A. tumefaciens* contains 24 genes organized in six *vir* operons that are involved in the DNA transfer process. The *virB* operon consists of 11 genes, all of which have been characterized as to their primary structure. *virB4* and *virB11* encode proteins that possess a potential ATP-binding site and are localized in the inner membrane fraction of *A.tumefaciens*. In this study, we focus on the product of *virB4* which is the largest gene in the operon. Purified VirB4 protein was shown to possess ATPase activity. To test the essentiality of *virB4* in the T-DNA transfer process, non-polar mutants were constructed using a Tn*5pvirB* transposon in which the *virB* promoter faces outward. Several independent mutants were isolated and the precise insertion sites in *virB4* were confirmed by sequence analysis. These mutants were tested for T-DNA transfer ability by agroinfection and tumorigenicity by inoculation in *Brassica* and *Datura*. All mutants were agroinfection negative and tumorigenicity negative. These data strongly suggest that *VirB4* is essential for T-DNA transfer from *A. tumefaciens* to plants.

1. Introduction

Genes of the 28.6 kb *vir* regulon of the Ti plasmid of *Agrobacterium tumefaciens* are responsible for the processing of T-intermediates that are transferred to plants and integrated into the host chromosome (for recent reviews see (Kado (1991); Winans (1992)). This processing is initiated by an endonuclease encoded by *virD1* and *virD2* which causes strand- and site-specific cleavage at the 25 bp sequences flanking each side of the T-DNA. Following the cleavage reaction, both single- and double-stranded species of T-intermediates have the VirD2 protein covalently attached to their 5'-termini. Nuclear targeting sequences present near the carboxyl end of VirD2 permit this protein to accumulate in nuclei of transgenic tobacco bearing the *virD2* gene. Facilitation of T-intermediate processing is apparently mediated by the single-stranded binding VirE2 protein. In vivo, the VirD2 capped single-stranded T-intermediate species is thought to be coated along its length by VirE2 protein to protect against host and plant nucleases. By its presence in transgenic tobacco, a VirE2 protein also functions exocellularly

E. W. Nester and D. P. S. Verma (eds.),
Advances in Molecular Genetics of Plant-Microbe Interactions, 103–107.

(Citovsky, et al. (1992)). This interaction may be taking place at the interface between the *A. tumefaciens* cell (donor) and the plant cell (recipient). At the interface is the requirement of DNA export proteins. Of the *vir* genes, the prime candidates would be proteins that interact with the bacterial and plant membranes. Eight of eleven VirB proteins possess domains likely to associate with the membranes. In this paper, we have introduced non-polar mutations in *virB4* and show that VirB4 protein is required for virulence and T-DNA transfer.

2. Procedure

2.1. BACTERIAL STRAINS, PLASMIDS, AND MEDIA

A. tumefaciens LBA4301 Rec$^-$, pTi$^-$, Rif^r was maintained in medium 523 or induction medium (8 g casein hydrolysate, 4 g yeast extract, 0.2 g $Mg_2SO_4 \cdot 7H_2O$, 0.57 $K_2HPO_4 \cdot 3H_2O$, 10g glucose, 9.76 g MES, pH 5.5, per liter). *Escherichia coli* BL21 (DE3) (Studier, et al. (1990)) was grown in 2 x YT medium at 37°C. The Tn*5pvirB* delivering vector and pBM plasmids are described elsewhere (K. Shirasu, Z. Koukolíková-Nicola, B. Hohn, and C. I. Kado, submitted).

2.2. VIRULENCE AND AGROINFECTION ASSAY

Three-week-old *Datura stramonium* and turnip (*Brassica rapa* var. Just Right) were inoculated as described (Steck, et al. (1990)). pEAP42 which carries 1.4 genomes of Cauliflower mosaic virus (CaMV) in the T-DNA was introduced into the strain LBA4301 containing pBM plasmids by triparental mating using the helper strain HB101 (pRK2013) (Ditta, et al. (1980)). These strains were tested for agroinfection ability on three week old turnip plants (Grimsley, N. (1989)).

2.3 MEMBRANE FRACTIONATION AND IMMUNO BLOT ANALYSIS

Freshly grown (5 ml) *A. tumefaciens* was inoculated into 50 ml of medium 523 with appropriate antibiotics and incubated overnight at 30°C. Cells were inoculated into 500 ml of induction medium. Acetosyringone was added to final concentration 40 µg/ ml when $O.D._{600}$ reached 0.3. The cells were collected by centrifugation after incubation for 12 hours. Cells were fractionated into outer membrane, cytoplasmic membrane, periplasmic, and cytoplasmic components by using De Maagd and Lugtenberg's method (De Maagd and Lugtenberg (1986)) NADH oxidase and alkaline phosphatase activities were measured as cytoplasmic- and periplasmic-markers, respectively. Proteins from these fractions were separated by SDS-PAGE and blotted onto nitrocellulose membrane by standard methods. VirB4 specific antiserum was used to detect VirB4 protein with ECL detection kit (Amersham, IL).

2.4. VIRB4 PROTEIN PURIFICATION AND ATPASE ASSAY

VirB4 protein was overproduced under the control of the bacteriophage T7 promoter in *E.coli* as described (Shirasu, K., et al., (1990)). Protein purification scheme is described elsewhere (K. Shirasu et al., submitted). The assay measures the production of free phosphate from [γ-^{32}P]ATP (6000 Ci/mmol : Amersham, IL) in 40 mM HEPES/KOH, pH 7.6, 50 µg of bovine serum albumin /ml, 11 mM magnesium acetate.

3. Results

3.1. VIRB4 IS ESSENTIAL FOR T-DNA TRANSFER

Five Tn*5pvirB* insertional mutants (pBM1110, pBM1123, pBM1125, pBM1130, and pBM1159) in *virB4* were isolated. Precise insertion sites were determined by sequencing. These mutants were tested for T-DNA transfer ability by agroinfection assay. Only pBM1070 which contains Tn*5pvirB* in the vector alone showed CaMV symptom, while none of *virB4* mutants showed symptom (Table 1). The polarity of these mutants were tested for tumor formation using *virB4* polar mutant pJK190 and *virB5* polar mutant pJK104 (Rogowsky, et al. (1990)).

TABLE 1. Tumor formation and agroinfection by *virB* mutants

plasmid	pBM1110	pBM1123	pBM1125	pBM1130	pBM1159	pBM1070
Tumor formation						
+ pJK190	-	-	-	-	-	+
+ pJK104	-	+	+	+	-	+
Agroinfection						
+ pEAP42	-	-	-	-	-	+

pBM1110, pBM1123, pBM1125, pBM1130, and pBM1159 are Tn*5pvirB* insertional mutants in *virB4*. pBM1070 contains Tn*5pvirB* in the vector region of pUCD2619. pJK190 and pJK104 are *virB4* and *virB5* polar mutants, respectively. pEAP42 is a CaMV delivery vector.

3.2. VIRB4 IS AN INNER MEMBRANE ASSOCIATED PROTEIN

A. tumefaciens LBA4301 with and without pUCD2614 (plasmid containing the entire *vir* region) (Rogowsky, et al. (1990)) were fractionated into four distinct cellular components after acetosyringone induction. The outer and inner membranes were separated by density gradient centrifugation, whereas the periplasmic and cytoplasmic fraction were isolated by lysozyme-EDTA treatment. The VirB4 protein was detected only in the inner membrane fraction as determined by the use of VirB4 antiserum and SDS/PAGE.

3.3. THE VIRB4 PROTEIN HAS ATPASE ACTIVITY

From the sequence analysis, *virB4* as well as *virB11* contain a putative nucleotide binding site (Thompson, et al. (1988)). Purified VirB11 protein was reported to possess ATPase activity (Christie, et al. (1989)). We show here that VirB4 also has ATPase activity. VirB4 protein was overexpressed in *E. coli* and purified to homogeneity as determined by SDS/PAGE factionation. The ATPase activity of purified VirB4 protein was detected by measuring the hydrolysis of [γ-^{32}P]ATP into free phosphate and ADP on thin-layer chromatography. The ATPase activity was inhibited by proteinA-sepharose purified anti-VirB4 immunoglobulin by at least 65%, whereas the immunoglobulins from preimmune serum showed no inhibition. The addition of poly dA(dT) or poly dT did not stimulate the activity.

4. Discussion

The gene products of the *virB* operon have been proposed to be located on the bacterial membrane from sequence analysis (Shirasu, et al. (1990); Ward, J. E., et al. (1988); Thompson, et al. (1988); Kuldau et al. (1990)) and from early biochemical studies (Engstrom, et al. (1987)). Indeed, VirB11 was found to associate with the inner membrane, whereas VirB10 aggregates in the inner membrane. In this study we found that VirB4 associates with the inner membrane even though it does not contain any significant membrane spanning region nor obvious signal peptide at N-terminal end. This suggests that VirB4 might not cross the inner membrane but associate with it like a peripheral protein. This raises the question as to the function of the protein and its requirement for the T-DNA transfer. Therefore, we have constructed non-polar mutants in *virB4* gene and also purified the protein. Non-polar mutants in *virB4* abolished not only tumor formation ability but agroinfection ability. This strongly suggests that VirB4 is essential for the T-DNA transfer. We found that purified VirB4 protein bears ATPase activity, which may be used to generate energy for either T-DNA transfer or for the assembly of the DNA transfer apparatus in the bacterial membrane. Recently, the *virB* operon has been shown to share the homology with the *trw* operon from broad-host-range IncW plasmid R388 (de la Cruz, F. personal communication; Kado et al. (1992)). The *trw* operon located in PILw region is required for W pilus synthesis and assembly. According to the sequence comparison analysis, all genes except *virB1* are homologous to the equivalents in *trw* operon with 51 % similarity on the average. For example, VirB4 has 54% similarity and 32% identity with TrwK which is the third gene in the *trw* operon. This strongly suggests that the gene products of the *virB* operon are also involved in pilus synthesis and assembly in *Agrobacterium* cell. Analogous to bacterial conjugation, it is very likely that *Agrobacterium* use the pilus for interkingdom DNA transfer. We are now investigating the possibility that the gene products of *virB* form an interkingdom sex pilus.

5. Acknowledgments

This work supported by NIH grant (GM45550) and a NIH training Grant in Molecular and Cell Biology.

6. References

Christie, P. J., Ward, J. J., Gordon, M. P., and Nester, E. W. (1989). 'A gene required for transfer of T-DNA to plants encodes an ATPase with autophosphorylating activity', Proc Natl Acad Sci U S A, 86, 9677-9681.

Citovsky, V., Zupen, J., Warnick, D., and Zambryski, P. (1992). 'Nuclear localization of *Agrobacterium* VirE2 protein in plant cells', Science, 256, 1802-1805.

De Maagd, R. A. and Lugtenberg, B. (1986). 'Fractionation of *Rhizobium leguminosarum* cells into outer membrane, cytoplasmic membrane, periplasmic, and cytoplasmic components', J. Bacteriol, 167, 1083-1085.

Ditta, G., Stanfield, S., Corbin, D., and Helinski, D. R. (1980). 'Broad host range DNA cloning system for Gram-negative bacteria: Construction of a gene bank of *Rhizobium melioti*', Proc. Natl. Acad. Sci. USA, 77, 7347-7351.

Engstrom, P., Zambryski, P., Van, M. M., and Stachel, S. (1987). 'Characterization of *Agrobacterium tumefaciens* virulence proteins induced by the plant factor acetosyringone', J Mol Biol, 197, 635-645.

Grimsley, N., Hohn, B., Ramos, C., Kado, C. I., and Rogowsky, P. (1989). 'DNA transfer from *Agrobacterium* to *Zea* mays or *Brassica* by agroinfection is dependent on bacterial virulence functions', Mol Gen Genet, 217, 309-316.

Kado, C. I. (1991). 'Molecular mechanisms of crown gall tumorigenesis', Crit Rev Plant Sci, 10, 1-32.

Kado, C. I., Shirasu, K., Koukolíková-Nicola, Z., Hohn, B., de la Cruz, F., and Lin, T-s. (1992) The *virB* and *virD* operons encode functions essential for T-DNA transmission to plants.' EMBO workshop on Promiscuous Plasmids.

Kuldau, G. A., De, V. G., Owen, J., McCaffrey, G., and Zambryski, P. (1990). 'The *virB* operon of *Agrobacterium tumefaciens* pTiC58 encodes 11 open reading frames', Mol Gen Genet, 221, 256-266.

Rogowsky, P. M., Powell, B. S., Shirasu, K., Lin, T. S., Morel, P., Zyprian, E. M., Steck, T. R., and Kado, C. I. (1990). 'Molecular characterization of the *vir* regulon of *Agrobacterium tumefaciens*: complete nucleotide sequence and gene organization of the 28.63-kbp regulon cloned as a single unit', Plasmid, 23, 85-106.

Shirasu, K., Morel, P., and Kado, C. I. (1990). 'Characterization of the *virB* operon of an *Agrobacterium tumefaciens* Ti plasmid: nucleotide sequence and protein analysis', Mol Microbiol, 4, 1153-1163.

Steck, T. R., Lin, T. S., and Kado, C. I. (1990). '*VirD2* gene product from the nopaline plasmid pTiC58 has at least two activities required for virulence', Nucleic Acids Res, 18, 6953-6958.

Studier, F. W., Rosenberg, A. H., Dunn, J. J., and Dubendorff, J. W. (1990). 'Use of T7 RNA polymerase to direct expression of cloned genes', Methods Enzymol, 185, 60-89.

Thompson, D. V., Melchers, L. S., Idler, K. B., Schilperoort, R. A., and Hooykaas, P. J. (1988). 'Analysis of the complete nucleotide sequence of the *Agrobacterium tumefaciens virB* operon', Nucleic Acids Res, 16, 4621-4636.

Ward, J. E., Akiyoshi, D. E., Regier, D., Datta, A., Gordon, M. P., and Nester, E. W. (1988). 'Characterization of the *virB* operon from an *Agrobacterium tumefaciens* Ti plasmid', J Biol Chem, 263, 5804-5814.

Winans, S. C. (1992). 'Two-way chemical signaling in *Agrobacterium*-plant interactions', Microbiol Rev, 56, 12-31.

The *Rhizogenes* tale: modification of plant growth and physiology by an enzymatic system of hydrolysis of phytohormone conjugates.

Spena, A., Estruch, J.J., Hansen, G., Langenkemper, K., Berger, S. and Schell, J..

MPI für Züchtungsforschung, 5000 Köln 30, FRG.

Abstract

More than 60 years ago, *Agrobacterium rhizogenes* , at that time called *Phytomonas rhizogenes*, was identified as the aethiological agent of the plant disease called hairy-root or woolly knot (Riker, 1930). Since then, the study of the hairy-root disease, and of the related hairy-root syndrome, has contributed to our knowledge of plant biological processes. The possible application in stimulating rooting of cuttings, proposed by Riker (1930) and patented by Tepfer and Lambert (1985), still wait for large commercial exploitation.

In the early seventies, tobacco plants regenerated from hairy-roots were found to be phenotypically altered (Ackermann, 1977). These alterations were transmitted to the F1 and F2 progeny (Ackermann, 1977). On the basis of these results it was consequently inferred that "Man muss daher annehmen, dass unter dem Einfluss von *A. rhizogenes* eine genetische Veränderung stattgefunden hat, die sich manifestiert, wenn mann über die Bildung von

E. W. Nester and D. P. S. Verma (eds.),
Advances in Molecular Genetics of Plant-Microbe Interactions, 109–124.

"Tumorwurzeln" wieder ganze Pflanzen induziert. Es ist offen, ob es sich dabei um genetisches Material aus den Bakterien handelt, zumal alle bisher bekanntgewordenen neueren Untersuchungen keine fremden DNA-Sequenzen in den infizierten Geweben nachweisen konnten" (Ackermann, 1977). In other words, it was becoming evident that the hairy-root syndrome and consequently the hairy-root disease were due to a genetic modification of the plant genome. Once Bacon wrote: "Solomon saith: there is no new thing upon the earth. So that as Plato had an imagination that all knowledge was but remembrance; so Solomon giveth his sentence, that all novelty is but oblivion" (Francis Bacon, Essays, LVIII). This might be one of the reasons why, news (Chilton et al., 1982; White et al., 1982; Willmitzer et al., 1982; Costantino et al., 1984) that *Agrobacterium rhizogenes* , as previously shown to be the case for *Agrobacterium tumefaciens*, was introducing genetic information into plant cells did not fully take advantage of the data on hairy-root plants. However, considering the fact that somaclonal variation may take place in plant regenerants, it was felt necessary to provide further and more extensive data to support the observation and interpretation of Ackermann. Prompted by this necessity, the comprehensive and detailed study of Tepfer (1984) clearly reminded us that the hairy-root syndrome is due to genetic transformation. Thus, the time was ripe for identifying and studying the genes responsible of the hairy-root disease and of the hairy-root syndrome.

In 1985, White et al. (1985) defined the *rol* genes by insertional mutagenesis. Four loci were identified by their role in the hairy-root disease and called *rol* A, B, C and D (*rol* = root locus). When DNA sequence analysis assigned to the *rol* genes four open reading frames in the T_L-DNA or Ri plasmid A4 (Slightom et al., 1986), the experimental question became: what is the biochemical function of these plant (onco)genes? In the last few years, several hypotheses have been proposed and reiterated extensively in the scientific literature, but one has been largely forgotten. The forgotten one was formulated in the same article defining the *rol* genes. Indeed, White et al. (1985) wrote that " The Ri T_L-DNA could functionally complement a mutation in the *tmr* locus of pTiA6NC... The genes of the T_L-DNA may be involved in the synthesis or regulation of substances with cytokinin-like effects". The concept was repeated by Taylor et al., (1985), when they said that: "One or more of these loci appear to have a cytokinin-like function, based on the ability of the *A. rhizogenes* T_L-DNA to complement mutations in the *A. tumefaciens tmr* gene". Tenaciously, but at that time in vain, White and Sinkar (1987) reiterated in a review article: "One function of the *rol* genes involves alterations in cytokinin synthesis or cytokinin-like effect".

So, on the basis of microbial genetic analysis, at least one of the T_L-DNA genes was postulated to have a cytokinin-like effect. In the same line of evidence, tobacco plants transgenic for the *rol* C gene, either with its own promoter (Oono et al., 1987) or under the control of the 35S promoter of cauliflower mosaic virus (Schmülling et al., 1988) showed a reduction in apical dominance and a bushy phenotype. These types of alterations were also interpreted as being indicative of cytokinin-like effects due to "synthesis of an auxin-antagonising agent such as cell-autonomous cytokinin" (Schmülling et al., 1988). Moreover, somatic mosaics for the expression of the *rol* C gene showed pale green sectors (Spena et al., 1989), characterised by stomata opening and cytokinin content higher than in neighbouring non-expressing tissue (Estruch et al. 1991a). Stomata opening can be caused by cytokinins (Mathysse and Scott, 1984), and administration of cytokinins has cell-autonomous effects in the mesophyll (Mothes and Engelbrecht, 1961). Consequently, these findings could also be interpreted as consistent with the hypothesis of a *rol* C mediated alteration of cytokinin activity. When the rol C peptide turned out to be localised in the plant cytosol (Estruch et al., 1991b), the increased membrane sensitivity to auxin of *rol* C and 35S-*rol* C protoplasts (Maurel et al., 1991) appeared to be a likely consequence of the intracellular function of the rol C protein.

Since the early eighties, it has been proposed that phytohormone activity can be regulated by enzymatic systems of conjugation of phytohormones and hydrolysis of phytohormone conjugates (Cohen and Bandurski, 1982; Letham and Palni, 1983). Moreover, *Pseudomonas savastanoi*, another bacterial plant pathogen, contains an indoleacetic lysine synthetase, encoded by the *iaa* L gene, which is able to conjugate IAA with lysine and other aminoacids such as ornithine (Kosuge et al., 1984; Glass and Kosuge, 1986; Roberto et al., 1990), and consequently to form less active or inactive IAA-conjugates. When the gene coding for the indolacetyl-lysine synthetase was expressed in tobacco, potato and *Arabidopsis* plants, it caused phenotypical alterations. Conferring new capacity for conjugating IAA obviously provoked changes in plant morphological and physiological processes. This finding showed that phytohormone content and activity can be modulated by enzymatic systems of hormone conjugation (Romano et al., 1991; Spena et al., 1991).

Similarly and in agreement with the hypothesis for the role of phytohormone conjugates in the modulation of phytohormone activity (Cohen and Bandurski, 1982; Letham and Palni, 1983), the rol C protein of the plant pathogen *Agrobacterium rhizogenes* expressed both in bacteria and plants, was found to have a β-glucosidase activity able to hydrolyse cytokinin N- and O-glucosides (Estruch et al., 1991). Several glucosidases are able to hydrolyse O-glucosides, however the rol C protein is, so far, the only one able to hydrolyse cytokinin N-glucosides (Table 1) which are biologically inactive forms of cytokinins (Laloue, 1977).

The finding that the rol C protein is a glucosidase able to release free active cytokinins from inactive cytokinin-glucosides and to alter plant developmental and physiological processes represents, together with the data obtained with plants transgenic for the 35S-*iaa* L gene, genetic evidence that plant physiological processes can be altered by enzymatic sytems of hydrolysis and conjugation of phytohormone-conjugates and phytohormones, respectively.

Whilst the phenotypic alterations caused by the expression of the *rol* C gene were interpreted as resembling a cytokinin effect, expression of the *rol* B gene caused, in transgenic plants, plant tissues and plant cells, biological effects indicative of an increased auxin biological activity (Cardarelli et al., 1987; Spena et al., 1987; Schmülling et al., 1988; Maurel et al., 1991). Leaves of hairy-root tobacco plants transgenic for the T_L-DNA were found to contain more than double the normal amount of IAA (Spano' et al., 1988). However, this finding was interpreted as not indicative of an alteration in IAA content but of an increased auxin sensitivity of plant cells transformed by *A. rhizogenes*. Plant biologist have been influenced by Trewavas' view on the limiting factors in plant growth (Trewavas, 1982). However, this provocative view should be reconsidered in the light of experimental evidence showing that alterations in phytohormone content within one order of magnitude causes physiological and developmental alterations in transgenic plants (Medford et al., 1989; Schmülling et al., 1989; Romano et al., 1991; Spena et al., 1991; Estruch et al., 1991c; Hocart et al., 1992). Tobacco anthers transgenic for the *rol* B gene were found and interpreted to have an IAA content up to 4 fold higher than control tissue (Spena et al., 1992). This increase in IAA content can well explain the increased sensitivity to auxin, measured as membrane depolarisation, of *rol* B transgenic protoplasts (Maurel et al., 1991) as a consequence of an altered intracellular auxin metabolism. This interpretation is also supported by the finding that the rol B protein is a cytosolic protein (unpublished results). Lastly, the rol B protein, expressed either in plants or in bacteria, has β-glucosidase activity able to hydrolyse indoxyl-glucoside in the presence of UTP (Estruch et al., 1991d). Indole and IAA are competitive inhibitors of the indoxyl-β-glucosidase activity of the rol B enzyme, but not of that one of emulsin. However, none of the IAA-conjugates tested is a substrate for the rol B enzyme (Table 2). Consequently, barrying the presence of untested forms of auxin conjugates which are substrates for the rol B enzyme, the increase in IAA content can not be explained by a direct release of IAA from IAA-conjugates. Thus, it still remains to be elucidated which indolic-glucoside(s) is the substrate *in planta* and how the end product(s) of the reaction lead to an increase in auxin content and activity. Whatever the mechanism, the rol B enzyme, able to hydrolyse indoxyl-glucoside and to increase auxin content and activity in transgenic tissue, is part of the strategy used by *A. rhizogenes* to alter phytohormone metabolism and activity.

Altogether, the study of the biological function of the *iaa* L, *rol* C and *rol* B genes has shown that phytopathogens manipulate phytohormone content and activity by using enzymatic systems of conjugation/hydrolysis of hormone and hormone-conjugates. Is this a peculiar strategy used by plant pathogens to modify plant physiological processes, or is it a mimicking of physiological modes of regulation ?

Recalling Jacob, we should remember: "What is disease, save an exaggeration or deficiency of certain processes which occur in a healthy animal" (or plant)?..."If knowledge of the physiological state was obviously necessary for the interpretation of pathological conditions, the study of pathological conditions also provide a precious instrument to study biological functions" (Jacob, 1974). In this respect, it is interesting to note that Mok et al. (1992) have identified zeatin-O-glycosyltransferases which appear to be differentially expressed in tissues of *Phaseolus vulgaris*. The article of Mok et al. (1992) ended by saying: "The model of *in vivo* control of plant development resulting from selective expression of specific metabolic enzymes is worthy of further investigation".

What then, is the hairy-root disease ? It is the result of a genetic transformation event caused by the transfer of genetic information from the bacteria to the plant cell. Two of the genes transferred with the T_L-DNA from the bacteria to the plant cell code for β-glucosidases able to increase intracellular auxin content and activity (i.e. rol B) and to release free active cytokinins from their inactive glucosides (i.e. rol C). Auxins trigger root formation, consequently, we envisage that formation of roots will be mainly triggered by the auxin increase caused by the expression of the *rol* B gene. In this regard, it is important to note that the *gus* gene under the control of the *rol* B promoter is expressed preferentially in the initial cells of meristems (Altamura et al., 1991). Expression of the *rol* C gene, with a different specificity of expression (Schmülling et al., 1989), would then allow further root growth. This is because auxin doses optimal for triggering root formation inhibit root growth, and optimal root formation and growth is obtained by a combination of auxin and cytokinin (Wightman et al., 1980). Consequently, the cytokinin effect caused by *rol* C might antagonise the inhibitory effects of auxin on root growth, allowing further root growth and branching. Other genes are coded by the T_L-DNA of *A. rhizogenes* A4 and by the T-DNA of other strains. Among them, much research has been dedicated to the *rol* A gene of *A. rhizogenes*. It is still to premature to address the biochemical function(s) of the *rol* A and other T-DNA encoded oncogenes. However, it would not be completely surprising if some of them do turn out to code for enzymes able to alter phytohormone metabolism and activity.

CONCLUSIONS

The study of the hairy-root disease has not only brought us information of heuristic significance, but also genetic tools of possible applied interest. The *rol* genes, together with other genes of microbial origin (for recent review: Spena et al., 1992), can be used to manipulate plant growth. The type of morphological and developmental alterations will reflect both the type of gene used, the specificity of expression of the promoter driving the expression of the gene, the stability and activity of the encoded enzyme and the availability of substrates (Spena et al., 1992). Furthermore, such genes can be used for the establishment and growth of roots for the production of secondary metabolites (Hamill et al., 1987; Scragg, 1992). Root cultures transgenic for single *rol* genes might well have different amounts of secondary metabolites. Last and probably least, the *rol* C gene can also be useful as visual cell autonomous marker to monitor transposon excision in somatic tissue (Spena et al., 1989).

Acknowledgments. We thank Tony Michael for critical reading of this manuscript. Our work described in this comunication has been supported by an EEC Bridge grant to A.Spena. J.J.Estruch and G. Hansen are holders of long-term BRIDGE fellowships.

REFERENCES

ACKERMANN C: **Pllanzen aus *Agrobacterium rhizogenes*** Tumoren an ***Nicotiana tabacum.*** *Plant Sci Lett* 1977, 8:23-30.

ALTAMURA MM, ARCHILLETTI T, CAPONE I, COSTANTINO P: **Histological analysis of expression *Agrobacterium rhizogenes rol* B-GUS gene fusions in transgenic tobacco.** *New Phytol* 1991, 118: 69-78.

CARDARELLI M, MARIOTTI D, POMPONI M, SPANO L, CAPONE, I, COSTANTINO P: ***Agrobacterium rhizogenes* T-DNA genes capable of inducing hairy root phenotype.** *Mol Gen Genet* 1987, **209**:475-480.

CHILTON MD, TEPFER D, PETIT A, CASSE-DELBART F, TEMPE J : ***Agrobacterium rhizogenes* inserts T-DNA into genomes of the host plant root cells.** *Nature* 1982, **295**:432-434.

COHEN JD, BANDURSKI RS: **Chemistry and Physiology of the Bound Auxins.** *Annu Rev Plant Physiol* 1982, **33**:403-430.

COSTANTINO P, SPANO` L, POMPONI, M, BENVENUTO, E, ANCORA, G: **The T-DNA of *Agrobacterium rhizogenes* is transmitted through meiosis to the progeny of hairy root plants.** J Mol Appl Genet 1984, 2: 465-470.

ESTRUCH JJ, CHRIQUI D, GROSSMANN K, SCHELL J, SPENA A: **The Plant Oncogene *rolC* is Responsible for the Release of Cytokinins from Glucoside Conjugates.** *EMBO J* 1991a, **10**:2889-2895.

ESTRUCH JJ, PARETS-SOLER, A, SCHMÜLLING, T., SPENA, A. **Cytosolic localization in transgenic plants of the *rolC* peptide from *Agrobacterium rhizogenes* .** *Plant Mol. Biol.* 1991b, **17**: 547-550.

ESTRUCH JJ, PRINSEN E, VAN ONCKELEN H, SCHELL J, SPENA A: **Viviparous Leaves Produced by Somatic Activation of an Inactive Cytokinin-Synthesizing Gene.** *Science* 1991c, **254**:1364-1367.

ESTRUCH JJ, SCHELL J, SPENA A: **The Protein Encoded by the *rolB* Plant Oncogene Hydrolyses Indole Glucosides.** *EMBO J* 1991, **10**:3125-3128.(d).

GLASS NL, KOSUGE T: **Cloning of the Gene for Indoleacetic Acid-Lysine Synthetase from *Pseudomonas syringae* subsp. *savastanoi.*** *J Bacteriol* 1986, **166**:598-603.

HAMILL JD, PARR AJ, RHODES MJC, ROBINS RJ, WALTON NJ: **New routes to plant secondary metabolites.** *BioTechnology* 1987, 8:800-804.

HOCKART CH, LETHAM DS, WANG J, CORNISH E, PARKER CW: **Control of cytokinin levels by inhibitors of metabolism, symbiosis and genetic manipulation .** In Progress in Plant Growth Regulation edited by Karssen C.M., Van Loon L.C. and Vreugdenhil, D., Dordrecht/Boston/London: Kluwer Ac. publishers, 1992, p. 607-616.

JACOB F: **The logic of life. A history of heredity.** p.123 Pantheon ed. New York, 1974.

KOSUGE T, COMAI L, GLASS NL: **Virulance Determinants in Plant-Pathogen Interactions.** In Plant Molecular Biology edited by B. Goldberg, New York: Alan R. Liss, Inc., 1983, pp 167-177.

LALOUE M: **Cytokinins: 7-Glucosylation is Not a Prerequisite of the Expression of Their Biological Activity.** *Planta* 1977 **134:** 273-275.

LETHAM DS, PALNI LMS: **The biosynthesis and metabolism of cytokinins.** *Annu Rev Plant Physiol* 1983, **34**:163-197.

MATHYSSE AG, SCOTT TK: **Functions of hormones at the whole plant level of organisation.** In: *Encyclopedia of Plant Physiology* TK Scott, ed. pp. 219-243.

MAUREL C, BARBIER-BRYGOO H, SPENA A, TEMPE J, GUERN J: **Single *rol* genes from the *Agrobacterium rhizogenes* T_L-DNA alter some of the cellular responses to auxin in *Nicotiana tabacum.*** *Plant Physiol* 1991, **97**:212-216.

MEDFORD JI, HORGAN R, EL-SAWI Z, KLEE HJ: **Alterations of Endogenous Cytokinins in Transgenic Plants Using a Chimeric Isopentenyl Transferase Gene.** *The Plant Cell* 1989, **1**:403-413.

MOK DWS, MOK MC, MARTIN MC, BASSIL NV, LIGHTFOOT DA: **Zeatin metabolism in *Phaseolus* : Enzymes and genes.** In: *Current Plant Sciences and Biotechnology in Agriculture. Progress in Plant Growth Regulation.* CM Karssen, LC Van Loon, D Vreugdenhil eds. Vol 13, pp 597-606. Kluwer Academic Publishers, Dordrecht 1992.

MOTHES K, ENGELBRECHT L : **Kinetin-induced directed transport of substances in excised leaves in the dark.** *Phytochemistry* 1961, **1**:58-62.

OONO BY, HANDA T, KANAYA K, UCHIMIYA H: **The T_L-DNA gene of Ri plasmid responsible for dwarfness of tobacco plants.** *Jpn J Genet* 1987, **62**:501-505.

ROBERTO FF, KLEE H, WHITE F, NORDEEN R, KOSUGE T: **Expression and fine structure of the gene encoding N^{ε}-indole-3-acetyl-L-lysine synthase from *Pseudomonas savastanoi.*** *Proc Natl Acad Sci USA* 1990, **87**:734-738.

ROMANO CP, HEIN MB, KLEE HJ: **Inactivation of auxin in tobacco transformed with the indoleacetic acid-lysine synthetase gene of *Pseudomonas savastanoi.*** *Genes & Development* 1991, **5**:438-446.

RICKER AJ: **Studies on infectious hairy-root of nursery apple trees.** *J Agric Res* 1930, **41**:507-540.

SCRAGG AH: **Large scale plant cell culture: methods, applications** and products. *Current Opinion Bitechnol* 1992, 3:105-109.

SCHMÜLLING T, SCHELL J, SPENA A: **Single genes from *Agrobacterium rhizogenes* influence plant development.** *EMBO J* 1988, 7:2621-2629.

SCHMÜLLING T, BEINSBERGER S, DE GREEF J, SCHELL J, VAN ONCKELEN H, SPENA A: **Construction of a Heat-Inducible Chimaeric Gene to Increase the Cytokinin Content in Transgenic Plant Tissue.** *FEBS Lett.* 1989, **249**:401-406.

SCHMÜLLING T, SCHELL J, SPENA A: **Promotors of the *rol* A, B and C genes of *Agrobacterium rhizogenes* are differentially regulated in transgenic plants.** *The Plant Cell* 1989, 1:665-670.

SLIGHTOM JL, DURAND-TARDIF M, JOUANIN L, TEPFER D: **Nucleotide sequence analysis of T_L-DNA of *Agrobacterium rhizogenes* agropine type plasmid.** *J Biol Chem* 1986, **261**:108-121.

SPANO L, MARIOTTI D, CARDARELLI M, BRANCA C, COSTANTINO P: **Morphogenesis and auxin sensitivity of transgenic tobacco with different complements of Ri T-DNA.** *Plant Physiol* 1988, **87**:479-483.

SPENA A, SCHMÜLLING T, KONCZ C, SCHELL J: **Independent and synergistic activity of *rol A, B* and *C* loci in stimulating abnormal growth in plants.** *EMBO J* 1987, **7**: 3891-3899.

SPENA A, AALEN RD, SCHULZE, RD: **Cell-autonomous behavior of the *rolC* gene of *Agrobacterium rhizogenes* during leaf development: A visual assay for transposon excision in transgenic plants.** *Plant Cell* 1: 1157-1164.

SPENA A, PRINSEN E, FLADUNG M, SCHULZE SC, VAN ONCKELEN H : **The indoleacetic acid-lysine synthetase gene of *Pseudomonas syringae* subsp. *savastanoi* induces developmental alterations in transgenic tobacco and potato plants.** *Mol Gen Genet* 1991,**227**:205-212.

SPENA A, ESTRUCH, JJ, SCHELL JJ: **On microbes and plants: new insights in phytohormonal research.** *Curr. op. Biotechnology* 1992, 3, 159-163

SPENA A, ESTRUCH JJ, PRINSEN E, NACKEN W, VAN ONCKELEN H, SOMMER H: **Anther-specific expression of the *rol* B gene of *Agrobacterium rhizogenes* increases IAA content in anthers and alters anther development and whole flower growth.** *Theor Appl Genet* 1992,in press.

TAYLOR BH, AMASINO RM, WHITE FF, HUFFMAN GA, GORDON MP, NESTER EW: **Transcription of *Agrobacterium rhizogenes* A4 T-DNA.** *Mol Gen Genet* 1985, 201:554-557.

TEPFER D: **Transformation of several species of higher plants by *Agrobacterium rhizogenes* : sexual transmission of the transformed genotype and phenotype.** *Cell* 1984, 37:959-967.

TEPFER D, LAMBERT C: **Composition a base d'inoculum bacterien et application de cette composition au bouturage et au marcottage des plantes, en particulier les pommiers porte-greffes.** Patent no 85/16691. Priority 12 Nov 1985.

TREWAVAS AJ: **Growth substance sensitivity: the limiting factor in plant development.** *Physiol. Plant.* 1982, **55**: 60-72.

WIGHTMAN F, SCHNEIDER EA, THIMANN KV: **Hormonal Factors Controlling the Initiation and Development of Lateral Roots II. Effects of Exogenous Growth Factors on Lateral Root Formation in Pea Roots.** *Physiol Plant* 1980, **49**:304-314.

WHITE FF, GHIDOSSI G, GORDON MP, NESTER EW: **Tumor induction by *Agrobacterium rhizogenes* involves the transfer of plasmid DNA to the plant genome.** Proc Natl Acad Sci USA 1982, **79**: 3193-3197.

WHITE FF, SINKAR VP: **Molecular analysis of root induction by *Agrobacterium rhizogenes*.** In: *Plant DNA infectious agents, Plant gene research* (T Hohn, J Schell eds.). Vol 4, pp 149-175. Springer-Verlag Wien Press, Berlin 1987.

WHITE FF, TAYLOR BH, HUFFMAN GA, GORDON MP, NESTER EW: **Molecular and Genetic Analysis of the Transferred DNA Regions of the Root-Inducing Plasmid of *Agrobacterium rhizogenes*.** *J Bacteriol* 1985, **164**:33-44.

WILLMITZER L, SANCHEZ-SERRANO J, BUSCHFELD E, SCHELL J: **DNA from *Agrobacterium rhizogenes* is transferred to and expressed in axenic hairy root plant tissue.** *Mol Gen Genet* 1982, **186**:16-22.

TABLE 1:
RELATIVE SPECIFICITIES OF CYTOKININ-GLUCOSIDES TESTED AS SUBSTRATES OF THE ROLC ENZYME

N-glucosides	**Substrates***
Benzyladenine-9-glucoside	++
Kinetin-9-glucoside	++
Zeatin-9-glucoside	++
Dihydrozeatin-9-glucoside	++
Isopentenyl-adenosine-9-glucoside	+++
Benzyladenine-7-glucoside	+
Zeatin-7-glucoside	+
Dihydrozeatin-7-glucoside	+
Benzyladenine-3-glucoside	+
O-Glucosides	
Zeatin-O-glucoside	+
Dihydrozeatin-O-glucoside	+

* The enzymatic activity was measured as release of glucose. 100% activity corresponds to 0.3 nmol/h.mg protein. (+++: 100%; ++: 70% - 80%; +: 20% - 50%).

TABLE 2:

INDOLIC CONJUGATES TESTED AS SUBSTRATES OF THE ROLB ENZYME

COMPOUNDS	SUBSTRATES
ETHERS	
Indoxyl-β-glucoside	+
Indoxyl-β-glucuronide	-
ESTERS	
Indole-3-acetyl-1-O-β-glucopyranose	-
Indole-3-acetyl-myo-inositol	-
Indole-3-acetic acid-methylester	-
Indole-3-acetic acid-ethylester	-
Indole-3-acetic acid-n-hexylester	-
Indole-3-acetic acid-pentafluorobenzylester	-
AMIDES	
Indole-3-acetic acid-alanine	-
Indole-3-acetic acid-glycine	-
Indole-3-acetic acid-aspartate	-
Indole-3-acetic acid-phenylalanine	-

THE *AGROBACTERIUM TUMEFACIENS* TRANSCRIPTIONAL ACTIVATOR OccR CAUSES A BEND AT A TARGET PROMOTER THAT IS PARTIALLY RELAXED BY A PLANT TUMOR METABOLITE

L. WANG, K. CHO, J. D. HELMANN, AND S. C. WINANS
Section of Microbiology
Cornell University
Ithaca, New York 14853

ABSTRACT. Crown gall tumors incited by *A. tumefaciens* release compounds called opines, which serve as nutrient sources and as signal molecules for the colonizing bacteria. One such opine, octopine, induces expression of a cognate catabolism operon (*occQ*) via a LysR-type protein called OccR. Here we describe biochemical properties of complexes formed between OccR and operator DNA. This regulatory system was reconstituted *in vitro* using purified OccR protein and *Escherichia coli* RNA polymerase. OccR binds with high affinity to a single site overlapping the divergent *occQ* and *occR* promoters. Octopine increases the gel mobility of OccR-DNA complexes, relaxes an OccR-incited DNA bend, and shortens the DNase I footprint of OccR. This operator contains a 6 bp dyad symmetry in its left half and an unrelated 4 bp dyad symmetry in its right half. A 19 bp region including the right dyad is necessary and sufficient for high affinity OccR binding, while a larger region including both the left and right dyads is necessary for DNA bending and transcriptional repression.

1. Introduction

Agrobacterium tumefaciens strains direct transformed plant cells to synthesize a number of secondary metabolites called opines which are a source of nutrients for these bacteria (Tempe and Petit, 1982). Utilization of one such opine, octopine, requires the products of the *occQ* operon, which is activated by the OccR protein in response to octopine. OccR shows extensive amino acid similarity with members of the large LysR family of transcriptional regulators and has properties similar to many members of this family (Habeeb *et al.*, 1991, Henikoff *et al.*, 1988). These proteins function in ways that are somewhat atypical for prokaryotic regulators. Although most LysR proteins are activated by specific ligands, these molecules do not greatly affect the affinity of the proteins for their target promoters. They must therefore alter other properties of these proteins in order to effect transcriptional activation. In an effort to understand OccR-mediated transcriptional control, we have studied the physical properties of OccR-DNA complexes in the presence and absence of octopine.

2. Results

2.1. Reconstitution of regulated *occQ* and *occR* expression in vitro

OccR protein was purified to virtual homogeneity by conventional chromatography. To determine whether all proteins and DNA sites essential for regulated expression of the *occQ* and

E. W. Nester and D. P. S. Verma (eds.),
Advances in Molecular Genetics of Plant-Microbe Interactions, 125–129.

occR promoters had been identified, we attempted to reconstitute regulated expression of these genes *in vitro*. A cloned 226 bp PCR fragment predicted to direct synthesis of a 108 base *occQ* transcript and a 72 base *occR* transcript was used as a DNA template. In the presence of OccR and octopine, an RNA product of approximately 108 bases was synthesized. Reactions containing OccR protein but not containing octopine did not produce detectable levels of this transcript, whereas low levels of this transcript were detected in reactions lacking OccR whether octopine was present or absent. In the absence of OccR protein, a second RNA product of approximately 72 bases was also synthesized. Addition of OccR protein abolished synthesis of this transcript in the presence or absence of octopine while octopine alone had no effect. Further evidence for the identity of these transcripts was provided by truncation of the left or right ends of the DNA template. Truncation of the left end reduced the length of the 108 bp transcript, while truncation of the right end reduced the length of the 72 bp transcript. Therefore, OccR effectively regulates the transcription of these two promoters *in vitro*.

2.2. OccR binds to a specific site in the *occQ-occR* intergenic region

OccR protein specifically retarded the migration of a DNA fragment containing the *occQ* and *occR* promoters in polyacrylamide gels (Wang *et al.*, 1992). The apparent dissociation constant was 1.4 nM OccR monomers. In the presence of 40 μM octopine, the apparent dissociation constant increased to 2.7 nM, indicating that octopine slightly weakens the binding of OccR to DNA. In both cases, binding of OccR to its target is second order with respect to OccR concentration, with a Hill coefficient of 2.0 ± 0.1, indicating that OccR protein bound to its operator has a 2-fold higher oligomeric state than it does in solution. OccR protected a region of approximately 59 bp against DNase I digestion. The presence of octopine in the binding buffer shortened this footprint by 8 bp at the end closest to the *occQ* gene. Furthermore, certain unprotected bases within this footprint showed hypersensitivity to cleavage in the absence of octopine but not in its presence. These results provide further evidence that octopine significantly alters the interactions between OccR and the *occQR* operator.

2.3. OccR incites an inducer-relaxed bend in the *occQ-occR* intergenic region.

Octopine causes a slight but reproducible increase in the gel mobility of DNA-OccR complexes (Wang *et al.*, 1992), suggesting that it might cause a conformational change in either OccR or the target DNA, an hypothesis confirmed by the DNase I protection assays. To determine whether this altered conformation might involve a DNA bend, we used the plasmid pBend3 (Kim *et al.*, 1990), to perform circular permutation assays. These assays are based upon the fact that a bend at the middle of a DNA molecule slows the mobility of that fragment more severely than the same bend at the end of the molecule (Wu and Crothers, 1984). pBend3 contains a large number of restriction endonuclease cleavage sites arranged in two tandem sets, with unique sites at the center for introduction of the fragment of interest. A 112 bp PCR product containing the OccR binding site was introduced into this plasmid. The resulting plasmid was individually digested with each of 6 different endonucleases, creating fragments that had the same 215 bp sequence in a permuted order. These fragments were incubated with OccR in the presence or absence of octopine and size-fractionated on native gels. Complexes formed in the absence of octopine exhibited a strongly position-dependent mobility, indicative of a static DNA bend (Wu and Crothers, 1984). The center of this bend lies at the middle of the footprinted region. Complexes formed in the presence of octopine showed a weaker position-dependent mobility, indicating that octopine partially relaxes this bend. All complexes formed in the presence of octopine also migrated more quickly than the equivalent complexes formed in the absence of octopine, again consistent with a differential DNA bend. This operator does not show any sequence-directed bend in the absence of OccR. The angles of these

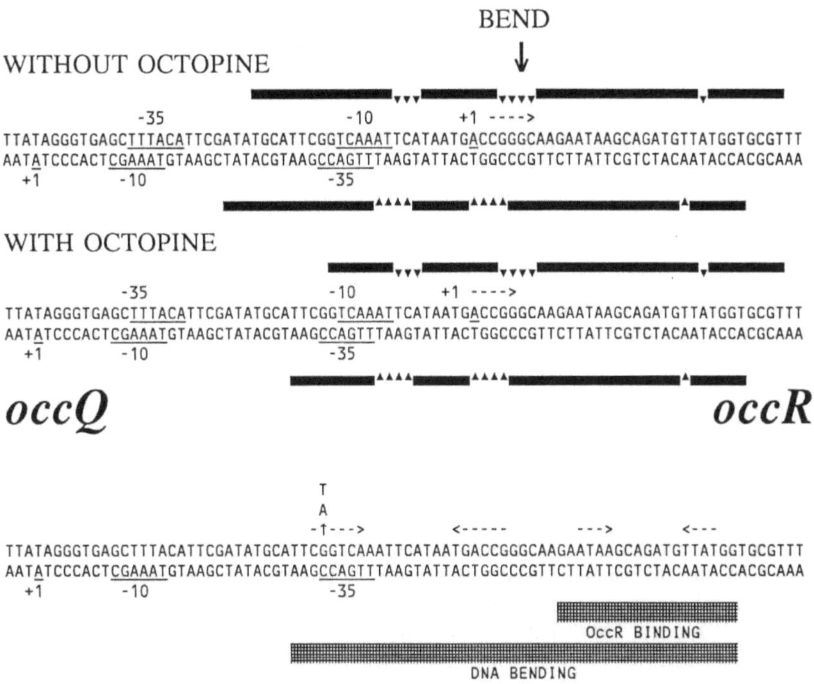

Figure 1. The *occQR* operator. Solid bars and arrowheads represent DNase I footprint. Shaded bars represent sequences needed for OccR binding or DNA bending. A point mutation that abolishes OccR-mediated repression of *occQ* is also shown.

DNA bends were conservatively estimated at 62° without octopine and 46° with octopine (Wang *et al.*, 1992).

2.4. Resections and point mutants of the *occQR* operator

We have made a series of resections of the *occQR* operator using cloned PCR amplified DNA fragments, and tested the resulting plasmids for OccR binding and OccR-mediated DNA bending. A 19 bp region at the right half of the footprinted region (containing the dyad symmetry ATAAN$_7$TTAT) is sufficient for OccR binding, while deletion of any part of this region abolishes high affinity, stable binding. In contrast, almost the entire footprinted region is required for high-angle DNA bending. The left half of the footprint contains a second dyad symmetry (CGGTCAN$_9$TGACCG), and deletions removing any part of this sequence abolished bending. We have also obtained an interesting point mutation in the *occQR* operator. This mutation stimulates expression of the *occQ* promoter in the absence of OccR and does not alter the activation of *occQ* by OccR plus octopine. Significantly, this mutant promoter is also activated (rather than repressed) by OccR in the absence of octopine. The mutation is a GC to TA transversion that simultaneously "improves" the -35 of the *occQ* promoter and disrupts the left half of the left dyad symmetry. The elevated OccR-independent expression is probably a consequence of the improved -35 region. The fact that OccR can activate but cannot repress this promoter could be due to an inability of OccR to interact with operator DNA in the left dyad.

3. Discussion

We have described the physical properties of OccR-DNA complexes in the presence and absence of octopine. Synthesis of the *occR* and *occQ* transcripts is regulated by OccR protein *in vitro* in a fashion similar to that observed *in vivo*. OccR protein binds specifically to a region just upstream of the *occQ* promoter and just downstream of the *occR* promoter. The right half of this protected region is necessary and sufficient for OccR binding, while almost the whole protected region is necessary for DNA bending. Octopine also increases the gel mobility of OccR-DNA complexes, relaxes a DNA bend incited by OccR, and shortens the region protected by OccR. It seems plausible that these three phenomena might all reflect the same conformational change in the protein-DNA complex.

There are several other examples of proteins that cause other environmentally responsive DNA distortions. AraC protein causes a repression loop in the *araBAD* promoter region that is relieved by arabinose (Lobell and Schleif, 1990). OmpR probably causes a similar repression loop in the *ompF* promoter (Slauch and Silhavy, 1991). The MerR protein of Tn*501* induces a -19° underwinding of its operator. Addition of the inducing ligand mercury increases this underwinding to -52°, thereby inciting a ligand-responsive DNA twist (Ansari *et al.*, 1992). LysR, AraC, and MerR all appear to be evolutionarily unrelated. In addition, several regulatory proteins have been reported to cause fixed angle bends at operator sites (Long *et al.*, 1991, Schultz *et al.*, 1991; Verrijzer *et al.*, 1991; Zwieb *et al.*, 1989, Kerppola and Curran, 1991).

Acknowledgments

We thank S. Adhya for providing plasmid pBend3, and C. Fuqua and N. Mantis for helpful discussions and critical evaluation of this manuscript. These studies were supported by N.I.H. grant #1 R29 GM2893-01.

References

Ansari, A. Z., Chael, M. L., and O'Halloran, T. V. (1992). Allosteric underwinding of DNA is a critical step in positive control of transcription of Hg-MerR. Nature 355:7-9.

Habeeb, L. F., Wang, L., and Winans, S. C. (1991). Transcription of the octopine catabolism operon of the Agrobacterium tumor-inducing plasmid pTiA6 is activated by a LysR-type regulatory protein. Mol. Plant-Microbe Interac. 4:379-385.

Henikoff, S., Haughn, G. W., Calvo, J. M., and Wallace, J. C. (1988). A large family of bacterial activator proteins. Proc. Natl. Acad. Sci., USA 85: 6602-6606.

Kerppola, T. K., and Curran, T. (1991). Fos-Jun heterodimers and Jun homodimers bend DNA in opposite orientations: implications for transcription factor cooperativity. Cell 66:317-326.

Kim, J., Zwieb, C., Wu, C., and Adhya, S. (1989). Bending of DNA by gene regulatory proteins: construction and use of a DNA bending vector. Gene 85:15-23.

Lobell, R. B., and Schleif, R. F. (1990). DNA looping and unlooping by AraC protein. Science 250, 528-532.

Long, S. R., R. F. Fisher, J. Ogawa, J. Swanson, D. W. Ehrhardt, E. M. Atkinson, and J. S. Schwedock. (1991). *Rhizobium meliloti* nodulation gene regulation and molecular signals. *in* Advances in Molecular Genetics of Plant-Microbe Interactions. H. Hennecke and D. P. S. Verma, (eds) Kluwer, Dordrecht.

Schultz, S. C., Shields, G. C., and Steitz, T. A. (1991). Crystal structure of a CAP-DNA complex: the DNA is bent by 90°. Science 253:1001-1007.

Slauch, J. M. and T. J. Silhavy. (1991). *cis*-acting *ompF* mutations that result in OmpR-dependent constitutive expression. J. Bacteriol. 173:4039-4048.

Tempe, J., and Petit, A. (1982). Opine utilization by Agrobacterium. p. 451-459., in G. Kahl and J. Schell (eds), Molecular Biology of Plant Tumors. Academic Press, New York.

Verrijzer, C. P., J. A. W. M. van Oosterhout, W. W. van Weperen, and P. C. van der Vliet. (1991). POU proteins bend DNA via the POU-specific domain. EMBO J. 10:3007-3014.

Wang, L., J. D. Helmann, and S.C. Winans. 1(992). The *A. tumefaciens* transcriptional activator OccR causes a bend at a target promoter, which is partially relaxed by a plant tumor metabolite. Cell 69:659-667.

Wu, H.-M., and D. M. Crothers. (1984). The locus of sequence-directed and protein induced DNA bending. Nature 308:509-513.

Zwieb, C., Kim, J., and Adhya, S. (1989). DNA bending by negative regulatory proteins: Gal and Lac repressors. Genes Dev. 3:606-611.

Section 3 / *Rhizobium*-Plant Interactions: *Rhizobium* Side

RHIZOBIUM NODULATION FACTORS: VARIATIONS ON A THEME

N. DEMONT, P. ROCHE, H. AURELLE, F. TALMONT, D. PROME, and J. C. PROME
LPTF, CNRS, 205 Route de Narbonne, 31077 Toulouse Cedex, France
N. P. J. PRICE, B. RELIC and W. J. BROUGHTON
LBMPS, Université de Genève, 1 Chemin de l'Impératrice, Chambesy/Genève, Switzerland
F. DEBELLE, M. Y. ARDOUREL, F. MAILLET, C. ROSENBERG
G. TRUCHET and J. DENARIE
LBMRPM, CNRS-INRA, B.P. 27, 31326 Castanet-Tolosan Cedex, France

ABSTRACT. Nodulation (*nod*) genes of *Rhizobium*, which control infection, nodulation and host specificity, determine the production of extracellular lipo-oligosaccharidic Nod factors. The Nod factors of various rhizobia, including the narrow host-range temperate *R. meliloti* and the broad host-range tropical *Rhizobium* sp. NGR234 belong to the same chemical family: they are mono-N-acylated and substituted chitin oligomers. The common *nodC* gene codes for a protein which has homology with chitin synthases and may be involved in the synthesis of the chitin backbone. The species-specific *nod* genes determine the decoration of these core molecules by encoding substitutions with specific groups at particular sites. For example in *R. meliloti* (i) the *nodH* and *nodPQ* genes are involved in the O-sulfation of the lipo-oligosaccharides on the carbon 6 of the reducing glucosamine; (ii) the *nodL* gene is required for the O-acetylation of the carbon 6 of the non-reducing terminal glucosamine; (iii) the *nodFE* genes specify the synthesis of the specific C16:2 N-acyl chain. *Rhizobium* sp. NGR234 secretes a large family of Nod factors carrying a variety of substituents. The non-reducing end is N-acylated with vaccenic or palmitic acids, is N-methylated, and carries varying numbers of carbamoyl groups. The reducing glucosamine residue is substituted on position 6 with 2-O-methyl-L-fucose which may be acetylated, or sulfated, or non-substituted. The O-acetylation and the O-sulfation are mutually exclusive: thus this broad host range strain secretes a variety of charged and uncharged Nod factors. We also present data which strongly suggest that the host range *nod* genes mediate host specificity by determining the type of substitutions decorating the lipo-oligosaccharidic Nod factors.

1. Introduction

The *Rhizobium*-legume symbiosis is an attractive system for dissecting the molecular basis of host specificity because of the specificity of these associations and the well studied genetics of a number of the bacterial partners (Young and Johnston, 1989; Dénarié and Roche, 1991; Fisher and Long, 1992). For example the narrow host range *Rhizobium meliloti* strains elicit the formation of nitrogen-fixing nodules on species of only three genera, *Medicago*, *Melilotus* and *Trigonella*, whereas the broad host range *Rhizobium* sp. NGR234 is currently known to nodulate more than sixty different legumes genera, as well as the non-legume *Parasponia andersonii*. The *R. meliloti nod* genes, which determine infection, nodulation, and host specificity have been studied in detail and are classified in two groups, the regulatory and structural *nod* genes. The three regulatory *nodD* genes encode DNA-binding proteins that activate the transcription of the other *nod* operons in the presence of plant phenolic signals present in seed and root exudates. These genes determine a first level of host specificity, since the NodD proteins are activated by specific phenolic compounds whose nature and abundance may vary according to the legume host. The structural *nod* genes are divided into common and species-specific. Common *nodABC* genes are structurally and

E. W. Nester and D. P. S. Verma (eds.),
Advances in Molecular Genetics of Plant-Microbe Interactions, 133–141.

functionally conserved among all rhizobia and are absolutely required for both root hair curling and infection, for eliciting mitosis in the root cortex, and for nodule formation. The species-specific *nod* genes are the major determinants of the specificity of infection and nodulation (Dénarié et al., 1992).

In *R. meliloti*,the common *nodABC* genes are involved in the production of lipo-oligosaccharidic extracellular Nod factors, which are tetra- or pentamers of chitin (N-acetyl-glucosamine linked in β, 1-4), sulfated at the reducing end and O-acetylated and N-acylated at the non-reducing end (Lerouge et al., 1990; Roche et al., 1991a; Roche et al., 1991b; Schültze et al., 1992). These purified molecules exhibit biological activities on alfalfa at low concentrations: hair deformations (10^{-8} M- 10^{-11} M), cortical cell divisions and nodule organogenesis (10^{-9} M- 10^{-6} M)(Roche et al., 1991b; Truchet et al., 1991). During the last two years, we have addressed the question of the mechanisms used by rhizobia to determine host specificity via Nod factors, using two approaches. Firstly the study of mechanisms by which the species-specific *nod* genes of *R. meliloti* make the NodRm factors plant-specific; secondly the study of the structure of Nod factors from various rhizobia having different host ranges, particularly from the broad host-range strain *Rhizobium* sp. NGR234.

2. Functions of *R. meliloti* host-range genes

2. 1. GENERATION OF NOD FACTOR OVERPRODUCING STRAINS

The low levels of extracellular Nod factors produced by *R. meliloti* 2011 make the structural analysis of these molecules exceedingly difficult. To improve the yield of these factors we have increased the gene dosage of the transcriptional activator genes *syrM* and *nodD3* by introducing the pMH682 plasmid in the strains of interest (Fig. 1). The introduction in the wild-type strain of the pMH682 plasmid resulted in an increase of NodRm factor production by a factor of at least 100-fold. After butanol extraction of the sterile supernatants of *nod*-induced cultures, and fractionation by reverse phase high performance liquid chromatography (RP-HPLC) the NodRm factors were studied by mass spectrometry and nuclear magnetic resonance. The major factors were found to be sulfated lipo-oligosaccharides with four or five glucosamine residues, which are O-acetylated (NodRm-IV(Ac,S) and NodRm-V(Ac,S)) or not (NodRm-IV(S) and NodRm-V(S)). In all these factors the major N-acyl chain is C16:2 with double bonds in positions 2 and 9 (Roche et al., 1991b).

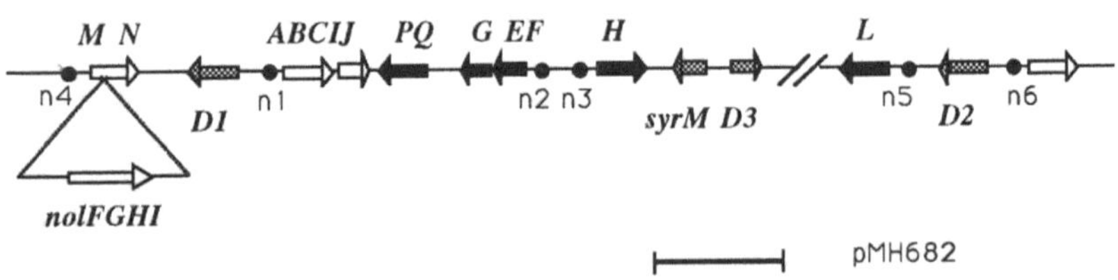

Fig. 1 Genetic map of the nodulation (*nod* and *nol*) genes of *R. meliloti*, after Baev et al. (1990) and Dénarié et al. (1992)

To study the role of various *nod* genes in the production of Nod factors our strategy was to construct overproducing strains carrying mutations in the *nod* genes of interest and to study the structure of the Nod metabolites produced by such mutants.

2. 2. *nodH* and *nodPQ* genes

nodH is a major host range gene. *R. meliloti nodH* mutants have lost the ability to elicit root hair curling (Hac), infection thread formation (Inf) and nodule formation (Nod) on the homologous host alfalfa but have acquired the ability to infect and nodulate a non-homologous host common vetch. The *nodH* gene is involved in modifying the plant specificity of extracellular Nod factors (Faucher et al., 1988; Banfalvi et al., 1989). To examine the role of *nodH* in the chemical modification of the lipo-oligosaccharidic factors, we purified and characterized the Nod factors present in the supernatants of cultures of overproducing *nodH*$^-$ strains. Butanol extracts of a *nodH* mutant culture were fractionated by RP-HPLC. The inactivation of the *nodH* gene resulted in a clear change in the HPLC profile: the peaks (fraction I) corresponding to Nod factors found in the *nodH*$^+$ control strain were not detectable, and new peaks (fraction II) corresponding to more hydrophobic compounds appeared. The chemical study of these new compounds revealed that they were similar to the Nod factors found in the corresponding *nodH*$^+$ strain, except that the O-sulfate group was no longer present on the C6 of the terminal reducing glucosamine. The following molecules were identified: NodRm-IV(Ac), NodRm-IV, NodRm-V(Ac) and NodRm-V. It can be concluded that the *nodH* gene is necessary for the 6-O-sulfation of the various lipo-oligosaccharide precursors (Roche et al., 1991b).

*nodQ*1$^-$ mutants exhibit an extended host-range, since they still infect and nodulate alfalfa but they are also able to infect common vetch. The *nodQ1* gene is involved in modifying the plant specificity of Nod factors (Faucher et al., 1989). The HPLC profiles of extracts from strains having a Tn5 insertion in *nodP1* or *nodQ1* showed two sets of peaks having the same retention times than the fractions I and II observed in the wild-type and the *nodH*$^-$ mutants respectively. Chemical studies showed that these fractions contained the sulfated and non-sulfated Nod factors already described. Thus, the *nodP1Q1* genes are also involved in lipo-oligosaccharide sulfation but, in contrast to *nodH*$^-$ mutants, *nodP1*$^-$ and *nodQ1*$^-$ mutants exhibit a leaky phenotype. What could be the cause of this leakiness? Schwedock and Long (1989) have found a reiteration of the *nodP1Q1* genes, called *nodP2Q2*, which is located on the second megaplasmid of *R. meliloti.* Overproducing strains carrying these mutations produced also both sulfated and non-sulfated factors. Double mutants *nodQ1Q2* produced only very reduced quantities of sulfated Nod factors. Thus both copies of *nodPQ* genes are functional and are able to specify the sulfation of Nod factors (Roche et al., 1991b).

What could be the biochemical functions of the NodH, NodP and NodQ proteins in the sulfation pathway of the *R. meliloti* Nod factors? The nodPQ genes share homology with ATP-sulfurylase and APS-kinase and specify the synthesis of PAPS, an activated sulfate donor (Schwedock and Long 1990; Long et al., this volume). NodH shares homology with mammalian sulfotransferases and its function might be to specifically transfer a sulfuryl group from PAPS to the carbon 6 of the reducing sugar of tetra- and pentamers of the lipo-oligosaccharide precursors (Roche et al., 1991b).

On alfalfa, a homologous host of *R. meliloti*, purified sulfated NodRm factors elicited root hair deformations (10^{-8}-10^{-11} M) and nodule formation (10^{-7} M), whereas the non-sulfated NodRm factors were not active. In contrast, on common vetch the non-sulfated factors elicited root hair initiation (10^{-9}-10^{-11} M) and the thick and short root phenotype (10^{-9} M), but at these concentrations the sulfated factors were not active. There is a striking correlation among *R. meliloti* strains between the production of sulfated or non-sulfated Nod factors, on the one hand, and the host specificity of infection and nodulation, on the other. This correlation indicates that the *nodH* and *nodPQ* genes specify the host range, the alfalfa specificity, by determining the sulfation of lipo-oligosaccharidic Nod factors (Roche et al., 1991b).

2. 3. *nodFE* and *nodL* genes

R. meliloti nodFE mutants are altered in the ability to elicit the formation of infection threads on alfalfa roots, but are able to induce root hair deformations on a non-homologous host, white clover (Debellé et al., 1986). Overproducing *R. meliloti* strains carrying *nodFE* mutations produce extracellular Nod factors which have N-acyl substituents different from those found in the "wild-type" strain. The C16:2 chain cannot be detected and the major substituents are vaccenic acid (C18:1) and various (ω-1)-hydroxylated fatty acids with chains ranging from 18 to 24 carbons. Thus the *R. meliloti nodFE* genes seem to be involved in the synthesis of the characteristic C16:2 chain with double bonds in positions 2 and 9. In the absence of this specific C16:2 fatty acid, rhizobial acyl-transferase(s) would N-substitute the non-reducing terminal glucosamine residue with fatty acids predominant in the cell compartment where the synthesis of the Nod factors occurs. In *R. leguminosarum* bv. *viciae nodFE* mutants the specific highly unsaturated C18:4 is not found and the major N-acyl substituent is vaccenic acid (Spaink et al. 1991). The fact that NodF is homologous to acyl carrier proteins (Shearman et al;, 1986) and NodE to β-ketoacyl synthases (Spaink et al., 1989) support the hypothesis of a role of NodF and NodE in the synthesis of the Nod acyl chain.

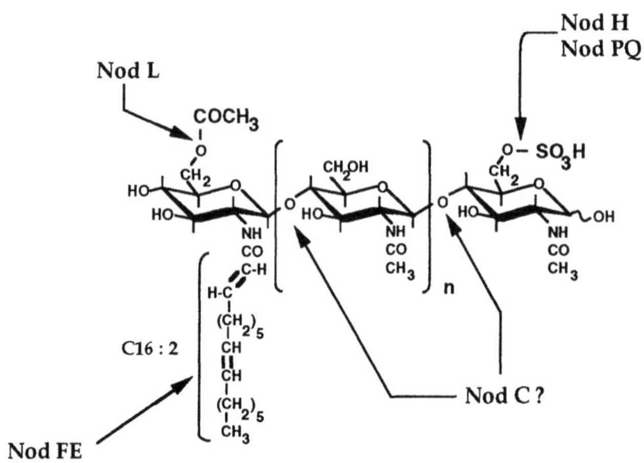

Fig. 2 Structures of the major Nod factors of *R. meliloti* and role of *nod* genes in their biosynthesis

R. meliloti nodL⁻ mutants exhibit a slight nodulation delay on alfalfa. *nodL⁻* mutants secrete Nod factors which lack the O-acetyl group at the non-reducing end. NodL shares homology with bacterial acetyl transferases (Downie, 1989). We can thus hypothesize that in *R. meliloti* the function of *nodL* is to determine the O-acetylation of the carbon 6 of the terminal non-reducing glucosamine residue of the Nod factors. A similar role of NodL has already been described in *R. leguminosarum* bv. *viciae* (Spaink et al. 1991).

Thus the role of the various host range *nod* genes of *R. meliloti* is to decorate the core molecule with specific substituents at specific sites (see Fig. 2). These decorations result in the production of Nod signal molecules which have a higher activity on the homologous host. The role of the common *nodABC* genes would be to determine the synthesis of the core molecule. NodC shares homology with yeast chitin synthases and might be involved in the creation of the β,1-4 linkage between the glucosamine residues (Atkinson and Long, 1992; Debellé et al., 1992).

3. **Towards a comparative "Nod-factorology"**

To address the question of the strategies adopted by various rhizobia to determine host range of infection and nodulation we have initiated the study of the structure of Nod factors from various *Rhizobium* species.

3. 1. *RHIZOBIUM* SP. NGR234: A BROAD HOST-RANGE STRAIN

The *Rhizobium* sp. NGR 234 strain is particularly interesting since it has a unique broad host range (Lewin et al., 1990). The sterile supernatant of cultures induced by the plant flavonoid apigenin elicited clear root hair deformations on *Macroptilium* and *Vigna*. The supernatant of a mutant deleted of the *nodABC* region did not induce root hair deformations, indicating that NGR234 secretes Nod factors. The introduction of extra-copies of the NGR234 regulatory gene *nodD1* increased Nod factor production 5 to 10 times as estimated from the amount of radioactivity present in thin layer chromatography (TLC) spots. Nod factors were prepared from the supernatant of large-scale cultures of an overproducing strain and were purified by solid-phase extraction followed by preparative HPLC. Two major fractions (A and B) were collected. Fraction A co-eluted with radioactive material from ^{35}S-sulfate labelled culture, whereas fraction B was not labelled, suggesting that NGR234 produces both sulfated and non-sulfated Nod factors.

Biochemical and chemical studies revealed that NGR234 produces a family of NodNGR factors which are mono-N-acylated chitin pentamers carrying a variety of substituents. The terminal non-reducing glucosamine is N-acylated with vaccenic or palmitic acids and is also N-methylated. Moreover this residue can be O-carbamoylated on positions 3 and 4. The reducing N-acetyl-glucosamine residue is substituted on position 6 with 2-O-methyl fucose which may be 3-O-sulfated or 4-O-acetylated, or non-substituted (Fig. 3). The O-acetylation and the O-sulfation seem to be mutually exclusive, which leads to the production of a mixture of sulfated and non-sulfated Nod factors. The combinations of these various substitutions result in a family of different charged and non-charged molecules. It is reasonable to hypothesize that this diversity of structures is involved in the control of broad host range (Price et al., submitted).

Although methods have yet to be devised to separate and purify all members of the NodNGR family, two groups of factors, sulfated and non-sulfated, could be separated by RP-HPLC and were tested for biological activity with root hair deformation assays. Both groups of NodNGR factors were active on homologous tropical hosts, at concentrations as low as 10^{-11} M on

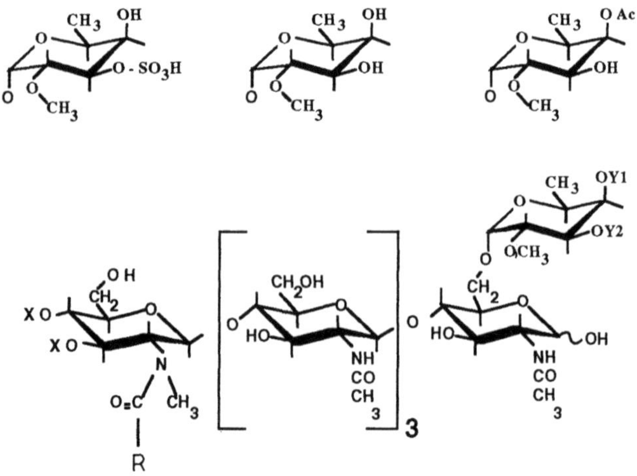

x = carbamoyl

R1 = vaccenic acid (C18:1)
R2 = palmitic acid (C16:0)

Fig. 3 Structures of the major Nod factors of *Rhizobium* sp. NGR234

Macroptilium and 10^{-12} M on *Vigna*. Interestingly the sulfated NodNGR molecules are also active on alfalfa root hairs at concentrations down to 10^{-11} M and approximately 10,000 times more active than the non-sulfated ones. In contrast the non-sulfated NodNGR factors were more active on common vetch than the sulfated ones (Price et al., submitted). Thus, taken as a whole, the NodNGR factors exhibit hair-deforming activities on a broad range of hosts. These data also confirm that the presence or absence of a sulfate group is a major element in the recognition of Nod factors by host plants. It is interesting to note that the sulfated NodNGR factors are active on alfalfa root hairs in spite of the numerous structural differences between NodNGR factors and the NodRm factors of *R. meliloti*. For example, in the NodRm factors, the sulfate group is located on carbon 6 of the reducing glucosamine residue, while in nodNGR factors the sulfate group is bound to the methyl fucose.

3. 2. OTHER SPECIES

Nod factors have been purified from other rhizobia, and have always been found to share a common core which is a chitin oligomer (tetra- or penta for major compounds), mono-N-acylated on the non-reducing end. Nod factors differ by the type and location of substituents. Spaink et al.

(1991) have shown that in the Nod factors of *R. leguminosarum* bv. *viciae* the reducing glucosamine is not sulfated and the non-reducing end is O-acetylated and N-acylated by a highly unsaturated fatty acid (C18:4).

Studies of the Nod factors produced by *R. tropici* and *R. leguminosarum* bv. *phaseoli* have been initiated because these two genetically distinct species induce nitrogen-fixing nodules on French beans but have also clear differences in their host range. *R. tropici* produces a mixture of sulfated and non-sulfated factors and the N-atom at the non-reducing end is doubly substituted by a methyl and an acyl group (R. Poupot, E. Martinez and J. C. Promé, personal communication). *Rhizobium fredii* is genetically quite distant from *Bradyrhizobium japonicum*, however these two species nodulate soybeans. *B. japonicum* 110 produces Nod factors carrying a methyl fucose substitution on the reducing glucosamine (G. Stacey, personal communication). Interestingly *R. fredii* secretes Nod factors that are substituted by a fucose residue which can be methylated or not (M. P. Bec and J. C. Promé, personal communication). The stem-nodulating *Azorhizobium caulinodans* secretes Nod factors which are also mono-acylated chitin oligomers but exhibit different decorations (P. Mergaert, M. Holsters and J. C. Promé, personal communication). These variations are likely to be responsible for the variety of rhizobia host-ranges.

4. **Conclusions and perspectives**

On the basis of genetical and cytological results we had previously proposed a model for a function of *Rhizobium nod* genes: the common *nodABC* genes would determine the production of Nod factor precursors and the host range *nod* genes would modify these precursors to make plant-specific extracellular signals (Faucher et al., 1988 and 1989). The genetic and biochemical results summarized in this paper, and results obtained by other laboratories (Spaink et al. 1991), fully support this model. The common *nodABC* genes determine the production of mono-N-acylated chitin oligomers and the species-specific *nod* genes decorate these core molecules by specifying their substitution by various groups at specific positions. These variations on a theme constitute a molecular basis for the host specificity on the bacterial side. Further studies will analyze in details the different strategies that various *Rhizobium*, *Bradyrhizobium* and *Azorhizobium* strains use to synthesize the specific Nod signals which will allow the recognition of their hosts. In fact, given the difficulties of the genetic analysis of "exotic" rhizobia on the one hand and the relative chemical homogeneity of the Nod factors on the other hand, it might be simpler to start by the determination of the structures of the major Nod factors and then try to identify the species-specific *nod* genes which specify the decoration characteristic of the strain, rather than identifying all the *nod* genes and studying subsequently their function. Analyzing the various mechanisms which determine rhizobial specificity of infection and nodulation will probably not only help understanding the *Rhizobium*-legume symbiosis but also contribute to our understanding of the very important developmental biology problem of cell-cell recognition.

If the molecular basis of host specificity on the bacterial side begins to be understood, the mechanisms by which the host plants recognize these signal molecules is completely unknown. Receptors are likely to be involved in the recognition and amplification of these signals and probably recognize the both ends of the molecules which carry the specific substitutions. However preliminary results obtained in various laboratories suggest that the Nod factor recognition is a complicated process which is not an all-or-none response at the level of membrane receptors of root hairs (see the paper of T. Bisseling and co-workers in this book). Thus to study the specific recognition of Nod factors we need biological assays to study the plant responses at the gene level and at the cellular level (Nap and Bisseling 1990; Pichon et al., sous presse). In a near future we will study the structure-function relationships of Nod factors by using a variety of Nod factors and studying their activity on the expression of different plant genes in different tissues. We hope thus to decipher the different levels at which differences in the signal structures are recognized by the plant.

Acknowledgements

We are very grateful to Sharon Long for providing bacterial strains prior to publication. We thank François Couderc and Jean Roussel (CRBGC-CNRS, Toulouse) and the NMR Unit of the Laboratoire de Chimie de Coordination CNRS Toulouse, for their help in physicochemical measurements. This work was partly supported by a grant from the Conseil Régional Midi-Pyrénées, by the Erna och Victor Hasselblads Stiftelse and by the Université de Genève.

References

Atkinson, E. M. and Long, S. R. (1992) Homology of *Rhizobium meliloti* NodC to polysaccharide polymerizing enzymes. Mol. Plant-Microbe Interact., in press.

Baev, N., Endre, G., Petrovics, G., Banfalvi, Z., and Kondorosi, A. (1991) Six nodulation genes of *nod* box locus 4 in *Rhizobium meliloti* are involved in nodulation signal production: *nodM* codes for D-glucosamine synthetase. Mol. Gen. Genet. 228, 113-124.

Banfalvi, Z. and Kondorosi, A. (1989) Production of root hair deformation factors by *Rhizobium meliloti* nodulation genes in *Escherichia coli*: *HsnD (NodH)* is involved in the plant host-specific modification of the NodABC factor. Plant Mol. Biol. 13, 1-12.

Debellé, F., Rosenberg, C., Vasse, J., Maillet, F., Martinez, E., Dénarié, J., and Truchet, G. (1986) Assignment of symbiotic developmental phenotypes to common and specific nodulation (*nod*) genetic loci of *Rhizobium meliloti*. J. Bacteriol. 168, 1075-1086.

Debellé, F., Rosenberg, C. and Dénarié, J. (1992) The *Rhizobium, Bradyrhizobium* and Azorhizobium NodC proteins are homologous to yeast chitin synthases. Mol. Plant-Microbe Interact., in press.

Dénarié, J., Debellé, F. and Rosenberg, C. (1992) Signaling and host range variation in nodulation. Ann. Rev. Microbiol., 46, 497-531.

Dénarié, J. and Roche, P. (1991) *Rhizobium* nodulation signals. In Molecular Signals in Plant-Microbe Communication. D.P.S. Verma, ed. (Boca Raton: CRC Press) pp. 295-324.

Downie, J. A. (1989) The *nodL* gene from *Rhizobium leguminosarum* is homologous to the acetyl transferase encoded by *lacA* and *cysE*. Molec. Microbiol. 3, 1649-1651.

Faucher, C., Camut, S., Dénarié, J., and Truchet, G. (1989) The *nodH* and *nodQ* host range genes of *Rhizobium meliloti* behave as avirulence genes in *R. leguminosarum* bv. *viciae* and determine changes in the production of plant-specific extracellular signals. Mol. Plant-Microbe Interact. 2, 291-300.

Faucher, C., Maillet, F., Vasse, J., Rosenberg, C., van Brussel, A.A.N., Truchet, G., and Dénarié, J. (1988) *Rhizobium meliloti* host range *nodH* gene determines production of an alfalfa-specific extracellular signal. J. Bacteriol. 170, 5489-5499.

Fisher, R. F. and Long, S. R. (1992) Rhizobium-plant signal exchange. Nature 357, 655-660.

Horvath, B., Bachem, C.W.B., Schell, J., and Kondorosi, A. (1987) Host-specific regulation of nodulation genes in *Rhizobium* is mediated by a plant signal, interacting with the *nodD* gene product. EMBO J. 6, 841-848.

Lerouge, P., Roche, P., Faucher, C., Maillet, F., Truchet, G., Prome, J.C., and Dénarié, J. (1990) Symbiotic host-specificity of *Rhizobium meliloti* is determined by a sulphated and acylated glucosamine oligosaccharide signal. Nature 344, 781-784.

Lewin, A., Cervantès, E., Wong, C. H. and Broughton, W. J. (1990) *nodSU*, two new *nod* genes of the broad host-range *Rhizobium* strain NGR234 encode host-specific nodulation of the tropical tree *Leucaena leucocephala*. Mol. Plant-Microbe Interact., 3, 317-326.

Nap, J.P. and Bisseling, T. (1990) Developmental biology of a plant-prokaryote symbiosis: The legume root nodule. Science 250, 948-954.

Pichon, M., Journet, P. E., Dedieu, A., de Billy, F., Truchet, G. and Barker, D.G.(1992) *Rhizobium meliloti* elicits transient expression of the early nodulin gene *ENOD*12 in the differentiating root epidermis of transgenic alfalfa. Plant Cell, in press.

Price, N. P. J., Relic, B., Talmont, F., Lewin, A., Promé, D., Pueppke, S. G., Maillet, F., Dénarié, J., Promé, J. C. and Broughton, W. J. Broad host-range of *Rhizobium* sp. NGR234 is based on a family of carbamoylated and fucosylated nodulation signals that are O-acetylated or sulphated. Submitted.

Roche, P., Lerouge, P., Ponthus, C., and Promé, J.C. (1991a) Structural determination of bacterial nodulation factors involved in the *Rhizobium meliloti*-alfalfa symbiosis. J. Biol. Chem. 266, 10933-10940.

Roche, P., Debellé, F., Maillet, F., Lerouge, P., Faucher, C., Truchet, G., Dénarié, J., and Promé, J. C. (1991b) Molecular basis of host specificity in *Rhizobium meliloti*: *nodH* and *nodPQ* genes encode the sulfation of lipo-oligosaccharide signals. Cell 67, 1131-1143.

Schültze, M., Quiclet-Sire, B., Kondorosi, E., Virelizier H., Glushka, J. N., Endre, G., Géro, S. D. and Kondorosi, A. (1992) *Rhizobium meliloti* produces a family of sulfated lipo-oligosaccharides exhibiting different degrees of plant host specificity. Proc. Natl. Acad. Sci. USA, 89, 192-196.s

Schwedock, J. and Long, S.R. (1990) ATP sulphurylase activity of the NodP and NodQ gene products of *Rhizobium meliloti*. Nature 348, 644-647.

Shearman, C. A., Rossen, L., Johnston, A. W. B., Downie, J. A. (1986) The *Rhizobium* gene *nodF* encodes a protein similar to acyl carrier protein and is regulated by *nodD* plus a factor in pea root exudate. EMBO J. 5, 647-652.

Spaink, H. P., Sheeley, D. M., van Brussel, A. A. N., Glushka, J., York, W. S., Tak, T., Geiger, O., Kennedy, E. P., Reinhold, V. N. and Lugtenberg, B. J. J. (1991) A novel highly unsaturated fatty acid moiety of lipo-oligosaccharide signals determine host specificity of Rhizobium. Nature 354, 125-130.

Spaink, H. P., Weinman, J., Djordjevic, M. A., Wijffelman, C. A., Okker, J. H., and Lugtenberg, B. J. J. (1989) Genetic analysis and cellular localization of the Rhizobium host specificity-determining NodE protein. EMBO J. 8, 2811-2818.

Truchet, G., Roche, P., Lerouge, P., Vasse, J., Camut, S., De Billy, F., Promé, J.C., and Dénarié, J. (1991) Sulphated lipo-oligosaccharide signals of *Rhizobium meliloti* elicit root nodule organogenesis in alfalfa. Nature 351, 670-673.

Young, J. P. W. and Johnston, A. W. B. (1989) The evolution of specificity in the legume-*Rhizobium* symbiosis. Trends Ecol. Evol. 4, 331-349.

CONTROL OF NODULE INDUCTION AND PLANT CELL GROWTH BY NOD FACTORS

[1]E. KONDOROSI, [1]M. SCHULTZE, [1]A. SAVOURE, [1]B. HOFFMANN, [2]D. DUDITS, [1]M. PIERRE, [1]L. ALLISON, [1]P. BAUER, [2]G.B. KISS and [1,2] A. KONDOROSI
[1]Institut des Sciences Végétales CNRS, 91198 Gif-sur-Yvette, France, [2]Biological Research Center, Hung. Acad. Sci. 6701, Szeged, Hungary

ABSTRACT. We have shown that *R. meliloti* cells induced for their *nod* genes excrete a family of structurally related lipo-oligosaccharides, the Nod factors. These molecules evoke root hair deformation, cortical cell division, nodule induction and expression of an early nodulin gene, *Nms-8b*, to different extents. Using various approaches we have demonstrated that the NodRm-IV(S) factor can act as plant growth regulator not only on *Medicago* root cells but also on cells grown in suspension. The factor was shown to stimulate the cell cycle; increased transition of cells from G1 to G2 phase was observed. In addition, perturbation of the hormonal balance in the plant cells was found to affect nodulation drastically, suggesting that the Nod factors may act in conjunction or *via* plant hormones to reprogram cells for nodule organogenesis.

1. Introduction

During the last five years it has become evident that nodule development is governed by signal exchanges between rhizobia and their host plants. The first signal in the symbiotic pathway comes from the host plant which induces the expression of *nod* genes in the bacterium in conjunction with the constitutively produced activator protein NodD (Kondorosi, 1991 for review). These molecules are products of the phenyl-propanoid pathway (primarily flavonoids and isoflavones). For a given *Rhizobium* species or strain several flavonoids of the host plant may serve as *nod* gene inducers, albeit they may activate the *nod* genes with varying efficiency. Other flavonoids exuded by the host plant roots have no effect on *nod* gene expression or are inhibitors. Thus, *nod* gene activation is controlled not only by the inducing flavonoids but also by the inhibitors, indicating that the ratio of inducing and inhibitory molecules is important for *nod* gene activation. Besides the plant signal and NodD, additional positive and negative regulatory factors controlling *nod* gene expression were identified and all these factors together provide a highly complex regulation of *nod* genes.

The activated *nod* genes produce return signals which evoke root hair curling and meristematic cell division (Faucher et al., 1988; Banfalvi and Kondorosi, 1989; Truchet et al., 1991). The first Nod signal molecule was purified from *R. meliloti* and identified as a lipo-oligosaccharide, a sulfated β-1,4 linked tetrasaccharide of *D*-glucosamine where three amino groups are acetylated. At the non-reducing end the sugar unit carries a C16:2 unsaturated fatty acid while the sugar moiety at the reducing end contains a sulfate group (Lerouge *et al*., 1990). The presence of the sulfate group was shown to be essential for host specific action on *Medicago* (Roche et al., 1991).

In this paper we will discuss our recent results which indicate that *R. meliloti* produces a family of Nod factors exhibiting different degrees of host specificity. We showed that the Nod factors differ in their ability to evoke root hair deformation, nodule induction or expression of the early nodulin gene, *Nms-8b*.

Early studies indicated that, by changing the hormonal balance in the root, nodule-like structures may appear on roots (Allen et al., 1953). It is possible that Nod factors induce nodule formation by changing the hormonal balance in the root. In order to get some insight into the mode of action of the Nod factors we studied the possible correlation between the changes of hormonal balance and nodulation of *Medicago*. These studies are summarized in the second part of the Results.

E. W. Nester and D. P. S. Verma (eds.),
Advances in Molecular Genetics of Plant-Microbe Interactions, 143–150.

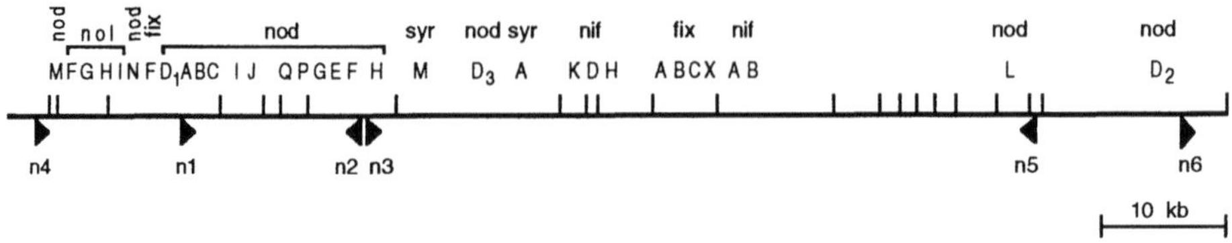

Figure 1. Organization of nodulation genes in *R. meliloti.* For ref. see Kondorosi et al. (1991); on *nodL*: Baev and Kondorosi (1992).

Finally, based on the ability of the Nod factors to induce cortical cell division, the possible role of Nod factors in controlling the cell cycle was addressed.

2. Results

2.1. NOD FACTORS OF *R. MELILOTI*

In *R. meliloti* strain 41 the *nod* genes are located on about a 100 kb segment of the symbiotic megaplasmid (Fig. 1, Banfalvi et al., 1981; Kondorosi et al., 1991). Until now we have identified more then 20 *nod* genes in this strain. Many of them are likely to be involved in the production or export of the Nod signal. The flavonoid-inducible *nod* transcriptional units are preceded by the highly conserved *nod* box sequences (Rostas et al., 1986) which provide common sites for the coordinated expression of *nod* transcriptional units forming the *nod* regulon (Kondorosi, 1989, 1991).

2.1.1. *Identification of a family Nod factors in* R. meliloti. In order to test the existence of other Nod factors in *R. meliloti,* strains were constructed which expressed the *nod* genes at high level and over-produced the Nod factors (Schultze et al., 1992). This was achieved by introducing the *syrM-nodD3* gene region or, in addition to *syrM* and *nodD3*, further *nod* genes on plasmids into the wild type or $NolR^-$ derivatives of *R. meliloti.* In this way, the Nod factor production could be increased at least 1000 fold. The Nod factors were purified by following the protocol of Lerouge et al. (1990). Due to the high level of factors in the bacterial supernatants, in some cases supernatants fractionated by analytical reverse phase C_{18} HPLC revealed several peaks corresponding to various Nod factors which has facilitated the purification and further analysis of factors produced by the overexpressing "wild type" or *nod* mutant strains.

Structural analysis of the purified compounds was carried out by ^{13}C-NMR, ^{1}H-NMR spectrometry as well as by FAB mass spectroscopy. One of the compounds was identical to the compound NodRm-1 reported by Lerouge et al. (1990) which has been renamed as NodRm-IV(C16:2,S) (Roche et al., 1991). In addition, a number of related compounds were identified as shown in Fig. 2 (Schultze et al., 1992). These compounds are penta-, tetra- or trisaccharides of *N*-acetyl-*D*-glucosamine. (The detailed chemical structure determination studies on the trisaccharides are in progress). The sulfate group was found always at the reducing end sugar which was shown to be essential for activity on *Medicago* (Truchet et al., 1991). Depending on the *nod* genes provided on plasmid, Nod factors without the sulfate groups were also produced. The acyl chains on the various compounds were C:16 fatty acids with 2 or 3 double bonds. In these preparations the Nod factors were not *O*-acetylated.

The biological activity of the pentasaccharides was tested and compared to that of NodRm-IV(C16:2,S) using the root hair deformation (Had) assay on the host plants *Medicago sativa* and *Melilotus albus* and on the non-host plant *Vicia sativa.* On *M. sativa* pentasaccharides containing fatty acids with 2 or 3 double bonds were about 100 fold less active than the tetrasaccharide NodRm-IV(C16:2,S). On *M. albus* these differences were not pronounced and on *V. sativa* the pentasaccharides were 10 fold more active. Furthermore, on *M. sativa* the root hair deformation was blocked at increased concentrations of either factors and the activity of the pentasaccharides exhibited a rather narrow concentration range (Schultze et al., 1992). These data suggest that, in addition to the presence of the sulfate group, the length of the oligosaccharide chain is important for host specific recognition and activity. Changes in the acyl chain had less influence, at least on the tested plants. However, saturation of the acyl chain of NodRm-IV(C16:2,S) by hydrogenation reduced the activity at least tenfold, indicating that the unsaturated fatty acids are important for biological activity (Truchet et al., 1991; Schultze et al., 1992).

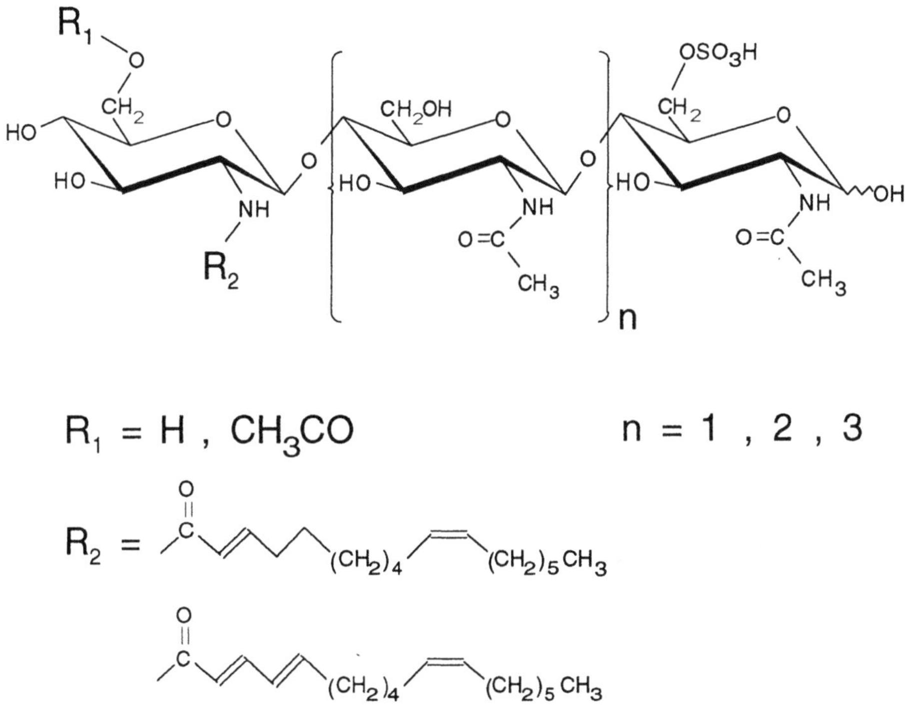

Figure 2. Nod factor family produced in *R. meliloti.* Data are from Schultze et al. (1992) and Truchet et al. (1991).

The Nod factors were able to induce nodule-like structures on *M. sativa.* Nodulation requires at least 100 fold higher concentrations of the Nod factors than the Had assay. Again, the pentasaccharides were two orders of magnitude less active than the tetrasaccharides.

Infection of the pea with *R. leguminosarum* bv. *viciae* was shown to induce the expression of the early nodulin gene *PsENOD12* in pea root hairs (Nap and Bisseling, 1990). From *Medicago sativa* we have isolated 2 genes, *Nms-8a* and *Nms-8b*, homologous to *PsENOD12* and showed that one class of these genes *Nms-8b* was induced at high level by NodRm-IV (C16:2,S), while the Nod factor from *R. leguminosarum* bv. *viciae* (Spaink et al., 1991), NodRlv-V(Ac,C18:4) (kindly provided by H. Spaink, Leiden), NodRm-V(C16:2,S) or *N*-acetyl-glucosamine tetrasaccharides activated this gene at a rather low level. These data indicate that the high level of expression of *Nms-8b* requires the *Medicago* specific Nod factor. At the same time, related glucosamine oligosaccharides, even without the fatty acid chain could have some inducing activity. Since ENOD12 is a cell wall protein belonging to the family of proteins where several of them are involved in cell wall synthesis in defense reactions, the action of the Nod factors resemble that of elicitors inducing these plant responses.

2.1.2. *The possible involvement of Nod factors in modular signalling.* The first two signalling steps between *Rhizobium* and its plant host show certain similarities. As discussed in the Introduction, the plant signals represent a family of molecules with different host specificity. In *R. meliloti* the likely receptors, the different NodD proteins, exhibit distinct flavonoid specificities.

The return signals of the initial steps of *Rhizobium*-plant interactions, the Nod factors, are again a family of related compounds with differing sensitivity to the plants, acting again in a host dependent manner. While the plant signals are recognized by the different NodD proteins (Györgypal et al., 1991), the Nod signals are likely to be recognized by specific plant receptors.

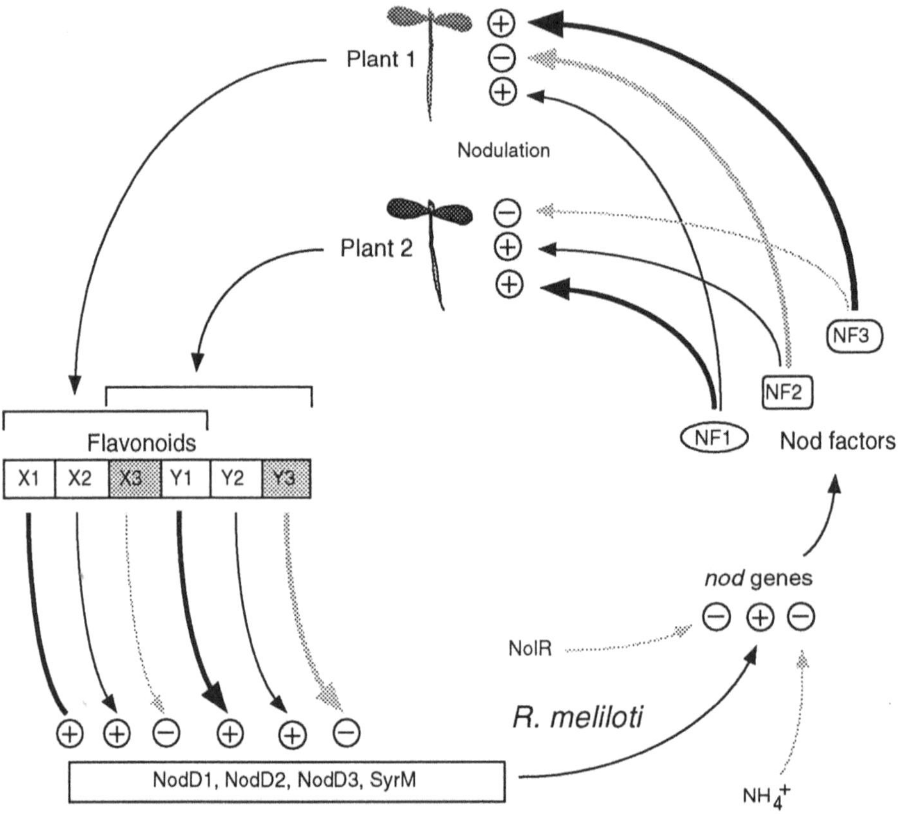

Figure 3. Control of nodule induction: a model for modular signalling.

These receptors, however, have not been identified yet. A modular and specific signal recognition may provide the molecular basis for the highly host specific symbiotic nodule induction (Fig. 3). Matching of plant signals with the bacterial receptor and the bacterial signals with the plant receptors, respectively, are prerequisites for the development of a symbiotic association.

2.2. INFLUENCE OF AUXIN-SENSITIVITY AND HORMONAL BALANCE ON NODULATION OF *MEDICAGO*

Early and more recent observations that auxin-transport inhibitors can induce the formation of empty nodule-like structures on *Medicago* roots (Allen et al., 1953; Hirsch et al., 1989) or that nodule-like structures may appear even spontaneously in the absence of *R. meliloti*, albeit at a low frequency (Dart, 1977), indicate that the plants harbour the essential components of the nodule developmental pathway which can be activated probably by changing the hormonal balance in the root. Apparently, the Nod signals seem to be required for high stimulation of this pathway. They may interact with auxins, cytokinins or other related hormones or might act directly as host specific plant hormones. To elucidate these possibilities we studied how the increased auxin-sensitivity of the host plant or other perturbations of the endogenous hormone balance affect nodule induction and development.

2.2.1. Auxin-sensitivity correlates with nodulation ability. We compared the nodulation kinetics of two genetically related lines of *Medicago varia* differing in their auxin-sensitivity (Bögre et al., 1990). It was found that the A2 plant, selected as an embryogenic line, was about tenfold more sensitive to auxins than R15 and formed significantly more nodules when infected with *R. meliloti* and nodulation started earlier.

In another study, four *Medicago* lines two of them with good embryogenic capabilities, were compared for spontaneous nodulation ability. They were found to have distinct abilities for spontaneous nodulation in the absence of *Rhizobium* and combined nitrogen. Again, a direct correlation was found between the frequency of spontaneous nodulation and auxin sensitivity. The highest frequency of spontaneous nodule formation was observed on the A2 line which exhibited the highest auxin-sensitivity and the best embryogenic properties.

These data provide further support to (but still do not prove) the idea that during nodulation the rhizobial partner alters the hormonal balance in the root cells.

2.2.2. Perturbation of hormone balance affects nodule development. The *rolABC* genes of *Agrobacterium rhizogenes* have been reported to alter the auxin-sensitivity and the endogenous active hormone ratios (Shen et al., 1989; Estruch et al., 1991a and b). If changes in the hormonal balance are involved in nodule induction by *Rhizobium,* introduction of the *rol* genes into *Medicago* should affect nodule formation on *Medicago* roots. To test this assumption, we transformed two embryogenic *Medicago* lines (A2 and RA3) with different combinations of the *rolABC* genes which were expressed either from their own or from the cauliflower mosaic virus 35S promoter. We found that depending on the genes introduced, the presence of *rol* genes affected the development of the plant and/or nodulation by *R. meliloti*. In general, the development of roots was positively influenced by the presence of the *rolB* genes as expected from the work of Schmülling et al. (1988) and in most cases a negative effect of the *rolC* genes was observed.

The nodulation ability of transgenic plants was affected more drastically. Plants containing the *rolABC* or particularly the *rolB* genes were nodulated faster and the number of nodules were much higher than on the non-transformed plants. Interestingly, in the rapidly developing nodules of RA3 plants expressing *rolC* from the 35S-promoter discontinuity of the cell walls of nodule cells in the symbiotic zone was observed. The presence of *rolB* caused delay in nodulation, affected the development of bacteroids and resulted in a faster senescence. High expression of *rolC* in RA3 halted symbiosome development in the interzone region.

Our data suggest that both spontaneous nodulation ability and nodule induction by *Rhizobium* correlates with the auxin-sensitivity and probably with the embryogenic properties of the *Medicago* lines. It is likely that the mode of action of the Nod signals is connected to the signal transduction pathway(s) of plant hormones (auxins ?).

2.3. CONTROL OF CELL CYCLE BY NOD SIGNALS

Purified Nod factors from *R. meliloti*, particularly NodRm-IV(S), can induce the formation of nodule-like structures indicating that Nod factors act as mitogens by inducing cortical cell divisions. In order to elucidate the (direct or indirect) mode of action of Nod factors on the cell cycle in *Medicago,* we studied the effects of purified Nod factors on root cells and on cells grown in suspension. Studying the effects of Nod factors on intact roots is technically rather difficult, since only a limited number of cells in the root are induced for cell division. Therefore, the effect of the Nod factors is often masked by the mass of the non-responding root tissue. In order to circumvent these technical difficulties, we have used cell suspensions. The experiments were carried out with three types of cell suspensions, i) single cell suspension culture from A2 (2n=8c), ii) line RA3 as a microcallus cell suspension (MCS) and iii) embryo cultures (EC) from RA3.

2.3.1.. Addition of Nod factors to cell suspensions and alfalfa seedlings stimulate expression of specific genes. Cell suspensions at different stages of growth were treated with NodRm-IV(S) at different concentrations, samples were taken at different time intervals and RNA was isolated and translated *in vitro*. By two dimensional gel analysis of the *in vitro* translation products we found that under certain conditions several new spots appeared on the gels and a number of spots became more or less intensive than the corresponding spots on the control gels.

The same samples were used for Northern blot analysis with the early nodulin *Nms-8* (recently isolated in the laboratory) and leghemoglobin (*Lb*) probes (Kiss et al., 1987), with the histone *H3c-1*, a S-phase specific gene probe (Wu et al., 1989) and with *Msc27*, a gene highly expressing in *Medicago* cell suspensions (Hirt et al., 1991). In none of the NodRm-IV(C16:2,S)-treated samples the expression of *Nms-8* or *Lb* was detected. However, 5 hrs after treatment a significant enhancement of the expression of the *H3c-1* gene was observed.

In parallel experiments, 2D gel analysis of the *in vitro* translation products as well as Northern blot and PCR analyses were performed with *M. sativa* seedlings treated with NodRm-IV(S) or infected with *R. meliloti* and with nodules of different age. Upon factor or *R. meliloti* treatment the appearence of new spots, enhancement or diminishment of certain spots on the gels were observed. In the Northern and PCR experiments the expression of *Nms-8b* was detected in the NodRm-IV(S)-treated roots as well as in the nodules. In young nodules, the expression of *Nms-8b* was particularly strong. *Lb* expression was not found in non-treated or in treated roots but was detectable in 7 day old nodules. Moreover, the *H3C-1* probe showed expression in all material but was strongly stimulated in young roots and nodules.

Expression of leghemoglobin and histone *H3c-1* was studied by *in situ* hybridization of 14 and 21 day old nodule sections. The *Lb* probe showed high expression in the zone of nodule cells invaded with bacteroids, as expected. The expression of the *H3c-1*, however, was restricted to the meristematic zone and was localised in few cells of the invasion zone.

2.3.2. NodRm-IV(S) stimulates transition of Medicago *cells from phase G1 to G2* . The possible effects of the NodRm-IV(S) on the cell cycle was studied by flow cytometry on A2 cell cultures consisting of single cells. For the determination of the DNA content of DAPI stained nuclei, A2 cells cultured for 10 old days in hormone-free medium were used. The cells were treated for various times and with various concentrations of NodRm-IV(S). (The flow cytometrical analyses were performed in collaboration with S. Brown and D. Marie). The experiments under these varying conditions were carried out in several repetitions. The effects of Nod factors on the number of nuclei in the different phases of cell cycle were studied under several conditions which allowed to find conditions where strong effects of the Nod factor were observed. Thus, after 15 hours of incubation the number of nuclei in the G2 phase increased by more than 50%.

3. Conclusions

We have shown that *R. meliloti* produces a family of Nod factors with different degrees of activity on the host and non-host plants. Further studies on these factors may contribute to the better understanding of the structural-functional relationship with respect of host specificity, complementing related work carried out in other laboratories (see also Chapters of this volume).

Using different approaches we provided evidence that the Nod factors affect gene expression not only in the intact root of *Medicago* but, to a limited extent, also in cultured *Medicago* cells. Our studies suggest that the Nod factor acts on the cell cycle, at least promotes transition of cells from G1 to G2. Studies are in progress to identify the mode of action of Nod factors on the control of cell cycle which involves the identification of cell cycle regulators in *Medicago* specifically responding to Nod factors and testing responses of various regulators of the cell cycle to the application of Nod factors in cell suspensions or in the roots.

4. References

Allen, E.K., Allen, O.N. and Newman, A.S. (1953) 'Pseudonodulation of leguminous plants induced by 2-bromo-3,5-dichlorobenzoic acid', Am. J. Bot. 40, 429-435.

Baev, N. and Kondorosi, A. (1992) 'Nucleotide sequence of the *Rhizobium meliloti nodL* gene located in locus n5 of the *nod* regulon', Plant Mol. Biol. 18, 843-846.

Banfalvi, Z., Sakanyan, V., Koncz, C., Kiss, A., Dusha, I and Kondorosi, A. (1981) 'Location of nodulation and nitrogen fixation genes on a high molecular weight plasmid of *R. meliloti*", Mol. Gen. Genet. 184, 318-325.

Banfalvi, Z. and Kondorosi, A. (1989) 'Production of root hair deformation factors by *Rhizobium meliloti* nodulation genes in *Escheria coli* : HsnD (NodH) is involved in the plant host specific modification of the NodABC factors', Plant Mol. Biol. 13, 1-12.

Bögre, L., Stefanov, I., Abraham, M., Somogyi, I. and Dudits, D. (1990) 'Differences in responses to 2,4-dichlorophenoxy acetic acid (2,4-D) treatment between embryogenic and non-embryogenic lines of alfalfa, in Nijkamp, H.J.J., Van der Plas, L.H.W., Van Aartrijk, J. (eds.), 'Progress in Plant Cellular and Molecular Biology', Kluwer Academic Publishers, 427-436.

Dart P. (1977) 'Infection and development of leguminous nodules', in Hardy, R.W.F., Silver, W.S. (eds.), A Treatise on Dinitrogen Fixation, J. Wiley and Sons, Inc. 367-472.

Estruch, J. J., Chriqui, D., Grossman, K., Schell, J. and Spena, A. (1991a) 'The plant oncogene *rolC* is responsible for the release of cytokinins from glucoside-conjugates', EMBO J. 10, 3125-3128.

Estruch, J.J., Schell, J. and Spena, A. (1991b) 'The protein encoded by the *rolB* plant oncogene hydrolyses indole glucosides', EMBO J. 10, 3125-3128.

Faucher, C., Maillet, F., Vasse, J., Rosenberg, C., Van Brussel, A.A.N., Truchet, G. and Dénarié, J. (1988) '*Rhizobium meliloti* host range *nodH* gene determines production of an alfalfa-specific extracellular signal', J. Bacteriol. 170, 5489-5499.

Györgypal, Z., Kiss, G.B. and Kondorosi, A. (1991) 'Transduction of plant signal molecules by the *Rhizobium* NodD protein', BioEssays 13, 575-581.

Hirt, H., Pay, A., Györgyey, J., Bako, L., Nemeth, K., Bögre, L., Schweyen, R.J., Heberle-Bors, E. and Dudits, D. (1991) 'Complementation of a yeast cell cycle mutant by an alfalfa cDNA encoding a protein kinase homologous to $p34^{cdc2}$, Proc. Natl. Acad. Sci. U.S.A. 88, 1636-1640.

Hirsch, A.M., Bhuvaneswari, T.V., Torrey, J.G. and Bisseling, T. (1989) 'Early nodulin genes are induced in alfalfa root outgrowths elicited by auxin transport inhibitors', Proc. Natl. Acad. Sci. U.S.A. 86, 1244-1248.

Kiss , G.B., Végh, Z. andVincze, E. (1987) 'Nucleotide sequence of a cDNA clone encoding leghemoglobin III (LbIII) from *Medicago sativa',* Nucl. Acid. Res. 15, 3620.

Kondorosi, A. (1989) '*Rhizobium*-Legume Interactions: Nodulation Genes', in Kosuge T., Nester E.W. (eds.), Plant-Microbe Interactions, McGraw-Hill P. Co. N.Y, Vol. III 383-420.

Kondorosi, A. (1991) 'Regulation of nodulation genes in rhizobia', in Verma, D.P.S. (ed.), Molecular Signals in Plant-Microbe Communications, CRC Press, Boca Raton, 325-340.

Kondorosi, A., Kondorosi, E., John, M., Schmidt, J. and Schell, J. (1991) 'The role of nodulation genes in bacterium-plant communication', in Setlow J.K. (ed.), Genetic Engineering, 13, Vol. 13, 115-136.

Lerouge, P., Roche, Faucher, C., Maillet, F., Truchet, G., Promé, J.C. and Dénarié, J. (1990) 'Symbiotic host-specificity of *Rhizobium meliloti* is determined by a sulphated and acylated glucosamine oligosaccharide signal', Nature 344, 781-784.

Nap, J.P. and Bisseling, T. (1990) 'Developmental biology of a plant-prokaryote symbiosis: the legume root nodule', Science 250, 948-954.

Roche, P., Debellé, F., Maillet, F., Lerouge, P., Faucher, C., Truchet, G., Dénarié, J. and Promé, J.C. (1991) 'Molecular Basis of Symbiotic host specificity in *Rhizobium meliloti*: *nodH* and *nodPQ* genes encode the sulfation of lipo-oligosaccharides signals', Cell 67, 1131-1143.

Rostas, K., Kondorosi, E., Horvath, B., Simoncsits, A. and Kondorosi, A. (1986) 'Conservation of extended promoter regions of nodulation genes in *Rhizobium*' , Proc. Natl. Acad. Sci. U.S.A .83, 1757-1761.

Schultze, M., Quiclet-Sire, B., Kondorosi, E., Virelizier, H., Glushka, J.N., Endre, G., Géro, S.D., Kondorosi, A. (1992) '*Rhizobium meliloti* produces a family of sulfated lipo-oligosaccharides exhibiting different degrees of plant host specificity', Proc. Natl. Acad. Sci. U.S.A. 89, 192-196.

Schmülling, T., Schell, J. and Spena, A. (1988) 'Single genes from *Agrobacterium rhizogenes* influence plant development', EMBO J. 7, 2621-2629.

Shen , W.H., Petit, A., Guern, A., and Tempé, J. (1988) 'Hairy roots are more sensitive to auxin than normal roots', Proc. Natl. Acad. Sci. U.S.A. 85, 3417-3421.

Spaink , S.P., Sheeley, D.M., van Brussel, A.A.N., Glushka, J., York, W.S., Tak, T., Kennedy, E.P., Reinhold, V.N. and Lugtenberg, B.J.J. (1991) 'A novel highly unsaturated fatty acid moiety of lipo-oligosaccharide signals determine host specificity of *Rhizobium*' Nature 354, 125-130.

Truchet, G., Roche, Lerouge, P., Vasse, J., Camut, S., de Billy, F., Promé, J.C. and Dénarié, J. (1991) 'Sulfated lipo-oligosaccharide signals of *Rhizobium meliloti* elicit root nodule organogenesis in alfalfa', Nature 351, 670-673.

Wu, S.C., Bögre, L. and Dudits, D. (1989) 'Polyadenylated H3 histone transcripts and H3 histone variants in alfalfa', Nucl. Acids Res. 17, 3057-3063.

RHIZOBIAL LIPO-OLIGOSACCHARIDE SIGNALS: THEIR BIOSYNTHESIS AND THEIR ROLE IN THE PLANT

H.P. SPAINK, A. AARTS, G.V. BLOEMBERG, J. FOLCH, O. GEIGER, H.R.M. SCHLAMAN, J.E. THOMAS-OATES*, A.A.N. VAN BRUSSEL, K. VAN DE SANDE, P. VAN SPRONSEN, A.H.M. WIJFJES, AND B.J.J. LUGTENBERG
*Leiden University, Institute of Molecular Plant Sciences, Nonnensteeg 3, 2311 VJ Leiden, The Netherlands and * University of Utrecht, Bijvoet Center for Biomolecular Research, Sorbonnelaan 16, 3584 CA Utrecht, The Netherlands.*

ABSTRACT. All tested strains of the genera *Rhizobium*, *Bradyrhizobium* and *Azorhizobium* are able to produce lipo-oligosaccharide signals upon activation of the flavonoid-inducible *nod* genes. *R.leguminosarum* was chosen as a model system to study the biochemical function of the Nod proteins in the synthesis of the lipo-oligosaccharides. In *R.leguminosarum* bv. *viciae*, the operons *nodABC* and *nodFEL* encode for the Nod proteins which are minimally required for synthesis of the wild type lipo-oligosaccharides and have therefore been studied in most detail. The NodE protein determines host-specificity by its involvement in the synthesis of highly unsaturated fatty acid moieties. The role of the lipo-oligosaccharides in the host plant will be emphasized. Results are presented which tempt us to believe that lipo-oligosaccharides, similar to those produced by rhizobia, are naturally occurring in the uninfected plant

1. Introduction

Bacteria belonging to the genera *Rhizobium*, *Bradyrhizobium* and *Azorhizobium*, collectively called rhizobia, are able to invade the roots of their host plants where they trigger the formation of a new organ called the root nodule. In these root nodules a differentiated form of the rhizobia, the bacteroid, is able to fix nitrogen into ammonia, which then can be utilized by the plant. Rhizobia produce certain lipo-oligosaccharides (modified chitin oligomers) after induction of nodulation (*nod*) gene transcription by flavonoids of the host plant (reviewed in [6,7,9,29]). The induction of *nod* gene transcription requires the rhizobial NodD protein, a transcriptional regulatory protein which belongs to the LysR family [13,22]. The common nodulation genes *nodABC* are essential for the biosynthesis of the lipo-oligosaccharides (also called Nod metabolites), consistent with their pivotal role in the infection and

E. W. Nester and D. P. S. Verma (eds.),
Advances in Molecular Genetics of Plant-Microbe Interactions, 151–162.

nodulation processes. Several *nod* genes which were shown to be involved in the determination of the host specificity of nodulation also appear to be involved in the production of lipo-oligosaccharide signals. Most noteworthy are the genes *nodQ* and *nodH*, which are major determinants of host specificity in *R.meliloti*, and the gene *nodE*, which was shown to be the major determinant of host specificity in the *R.leguminosarum* biovars *viciae* and *trifolii* [6,7,9,25,29].

The lipo-oligosaccharides can elicit various responses in the roots of the host plants in nanomolar concentrations. Examples are: the induction of nodulin gene expression [12,21], the production of new flavonoids [18], deformation [14,27] and membrane depolarization [8] of root hairs, formation of nodule meristems [28,30] and induction of pre-infection thread structures [31]. These discernible effects of the lipo-oligosaccharides on the plant root emphasize the importance of this new class of signal molecules as a tool to study plant morphogenesis.

2. Properties of rhizobial lipo-oligosaccharides

After the initial discovery of lipo-oligosaccharides produced by *R.meliloti* strain 2011 [14], the structures of lipo-oligosaccharides produced by various other rhizobial species have been published [6,29] or reported at symposia (Table 1). The structures of these molecules all conform to the basic structure which is shown in Figure 1. The sugar backbone of the lipo-oligosaccharides can be described as a short fragment of chitin which is invariably substituted with a fatty acyl chain at the amino group of the non-reducing terminal sugar. In accordance with the similarity to chitin, it has been reported that the rhizobial signals can be degraded by chitinase [19]. The nature of the fatty acyl chain is variable between the producing rhizobial strains. The lipo-

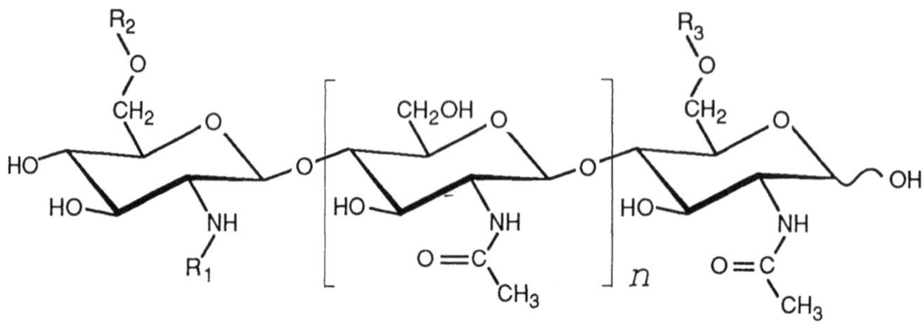

Figure 1. Basic structure of rhizobial lipo-oligosaccharides. The oligosaccharide backbone, which is identical to chitin, can vary in length between 3 and 5 sugar units (n=1,2,3). The R1 moiety is always a fatty acyl residue, which either is a special highly unsaturated fatty acid or a common bacterial fatty acyl residue. R2 and R3 can be hydrogen or a different substitution (see Table 1).

TABLE 1. Chemical properties of rhizobial lipo-oligosaccharides. Other substituents comprise *O*-acetyl groups (R2, Fig.1), novel sugar moieties or other novel substitutions. The data are according to Dénarié and Roche [6], E. Martinez (pers. comm.), N.Price (pers. comm.), M.Holsters (pers. comm.), R.Carlson and G.Stacey (pers.comm.) and data presented in this manuscript.

Bacterial Strain	sulfate (R3)	special lipid (R1)	other substituents
R.l. bv. *viciae*	no	C18:4	yes
R.l. bv. *trifolii*	no	yes	yes
R.l. bv. *phaseoli*	no	?	?
R.meliloti 2011	yes	C18:2	yes
R.meliloti AK41	yes	C18:2 C18:3	no
B.japonicum USDA135	no	C18:1	yes
B.japonicum USDA110	no	C18:1	yes
R.tropicii CIAT899	yes or no	yes	yes
NGR234	yes or no	no	yes
R.loti	no	no	?
Azorhizobium	no	no	yes
R.galega	no	?	?

oligosaccharides produced by several bacterial strains contain fatty acids which are common constituents of the bacterial phospholipids. Other rhizobial strains, like *R.meliloti* [14,24], *R.leguminosarum* biovars *viciae* [27], *R.leguminosarum* biovar *trifolii* [A.Aarts, unpublished], and *R.tropicii* [J.Folch, unpublished] produce Nod metabolites which contain a multi-unsaturated fatty acyl chain. The additional double bonds of these fatty acyl moieties (in comparison with common acyl chains of phospholipids), which are always in the *trans* configuration and conjugated to the carbonyl group, account for the observed UV absorption maxima [14,24,27]. In several strains a mixture of lipo-oligosaccharides is produced which contain variations in their fatty acyl moiety. In *R.meliloti* this moiety can either be a fatty acyl chain with two double bonds (designated as C16:2) or a C16:3 acyl chain [24] whereas in *R.leguminosarum* biovar *viciae* either a C18:4 acyl chain or a C18:1 acyl chain, the common *cis*-vaccenic acid, is present [27]. Also in *R.leguminosarum* biovar *trifolii* the fatty acyl moiety can either be a common or highly unsaturated species, the structure of the latter being variable but different from the C18:4 moiety found in biovar *viciae*.

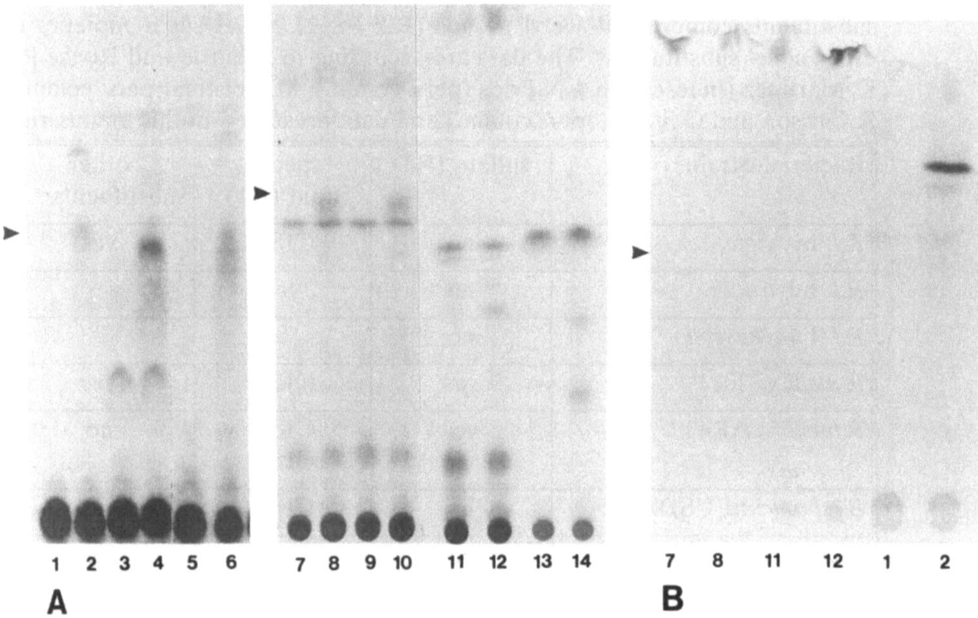

Figure 2. Analysis of Nod metabolites produced by several rhizobia by thin layer chromatography (TLC). The spent growth medium of the cells, grown in the presence of ^{14}C-labelled acetate (panel A) or ^{35}S-labelled sulfate (panel B), was extracted with n-butanol and applied on reverse phase (ODS: 100% octadecyl silanization) TLC plates [26,28]. The plates were developed in 50% acetonitrile/water and exposed to X-ray film. Several strains contained the plasmid pMP280 in which the transcriptional regulatory *nodD* gene of *R.leguminosarum* biovar *viciae* was cloned [28]. Used strains were *R.tropicii* strain CIAT899.pMP280 (lanes 1 and 2); *R.leguminosarum* biovar *phaseoli* strain CE3.pMP280 (lanes 3 and 4); *Rhizobium* strain NGR234 (lanes 5 and 6); *R.loti* strains E1R.pMP280 and B1R.pMP280 (unpublished isolates) (lanes 7,8,9 and 10) and NZP2037.pMP280 (lanes 11 and 12); *R.galega* strain 1261R.pMP280(lanes 13 and 14). The odd and even numbered samples were derived from bacteria grown in the absence or presence of the inducer naringenin, respectively. Arrowheads indicate the position of the compound NodRlv-V (C18:4, Ac) [27] which was used as a reference sample on each TLC plate.

The structure of the Nod metabolites produced by various other rhizobial strains is still under investigation. The analysis of these compounds by thin layer chromatography (TLC) has been a very valuable method in these characterizations [26,28]. Examples for the Nod metabolites of several strains which are presently under investigation in our laboratory are given in Figure 2. In several strains the production of these metabolites could be greatly enhanced by the introduction of a multicopy plasmid which contains the transcriptional regulatory *nodD* gene of *R.leguminosarum* biovar *viciae* (Figure 2).

3. The biosynthesis of the lipo-oligosaccharides

The biosynthesis of the lipo-oligosaccharides involves the translational products of the *nod* genes and they have therefore been referred to as Nod metabolites. Many of the initial indications for the function of the Nod proteins have been inferred from homologies with other proteins of known function (for reviews see [6,7,9,29]). These homologies suggested that the NodC protein might be an *N*-acetylglucosamine-transferase [3], involved in the synthesis of the oligosaccharide backbone of the Nod metabolites. For NodF and NodE a function was suggested in the synthesis of the highly unsaturated lipid moiety. The presence on NodF of a 4'-phosphopantetheine prosthetic group, which can function as a carrier for acyl chains during fatty acid biosynthesis, supports this suggestion [10]. Several genes were indicated to be involved in the presence of substituents on the basic structures, namely the NodPQ and NodH proteins for the sulfate moiety and the NodL gene for the *O*-acetyl moiety. We have been involved in the further biochemical analysis of the function of the Nod proteins using *R.leguminosarum* biovar *viciae* as a model system because its symbiotic properties are genetically and phenotypically well characterized. In the Sym plasmid-cured strain LPR5045 that we have used the presence of the operons *nodABC(IJ)* and *nodFEL* appeared to be sufficient for the production of all major wild type lipo-oligosaccharide signals [26,27] and therefore the genes in these operons have been studied in most detail. The *nodIJ* genes have been suggested to be involved in transport of the Nod metabolites on the basis of homology [7]. However, mutations in *nodI* and *nodJ* appeared to have no influence on the amount of radio-labelled Nod metabolites secreted in the growth medium [28]. These findings seem to be in contrast and therefore the influence of the *nodIJ* genes was further investigated. As shown in Figure 3, the presence of the *nodIJ* genes apparently does influence the amount of excreted Nod metabolites when only the *nodABC* and *nodFE* genes are present on a high copy number plasmid. Since the effect of the presence of *nodIJ* is only relatively small, the role of these genes in the transport of Nod metabolites remains disputable.

Efforts to set up *in vitro* synthesis systems for Nod metabolites have so far led to a better understanding of the role of the NodL protein. This protein, as suggested by its sequence homologies, indeed appears to be a transacetylase involved in the addition of the *O*-acetyl moiety. An important finding was that the NodL protein appears to use the complete unacetylated lipo-oligosaccharides as a substrate [G. Bloemberg, unpublished]. With the developed *in vitro* system, using radiolabelled acetyl-CoA as the acetyl donor, it was possible to obtain radiolabelled derivatives of

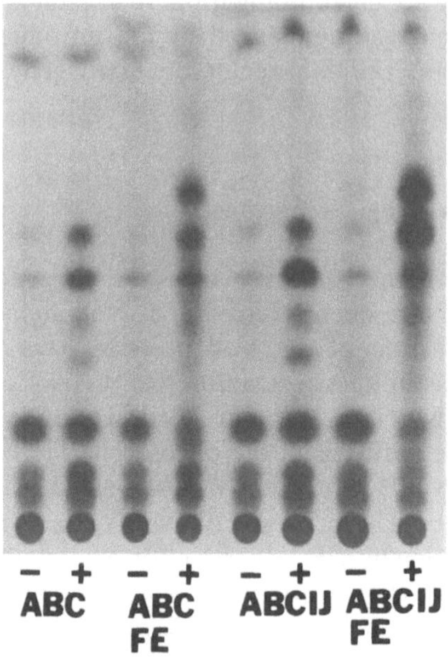

Figure 3. Effect of the *nodIJ* genes on the amount of radiolabelled Nod metabolites present in the spent growth medium. Nod metabolites were labelled and analyzed as described in Figure 2. Plasmids used were pMP292 and pMP247, IncP plasmids containing the cloned *nodABC* genes and *nodABCIJ* genes, respectively, and pMP1254, an IncQ plasmid containing the *nodFE* genes of *R.leguminosarum* biovar *viciae* [10,26,27]. Strains used were: LPR5045.pMP292 (lanes 1 and 2); LPR5045.pMP292.pMP1254 (lane 3 and 4); LPR5045.pMP247 (lanes 5 and 6); LPR5045.pMP247.pMP1254 (lanes 7 and 8). On odd and even lanes the samples derived from bacteria grown in the absence (-) or presence (+) of naringenin, respectively, were applied.

the lipo-oligosaccharides of various rhizobial species which can be very useful for further studies on the role of these signals in the plant host. For the NodF and NodE proteins it has not yet been possible to set up an *in vitro* synthesis system for the fatty acid groups. This is mainly due to the complexity of the fatty acid biosynthesis system, which involves many proteins encoded by chromosomal loci [29].

Recently, we found that the highly unsaturated fatty acid moiety of lipo-oligosaccharides of various rhizobial species is also present in the phospholipids of the cell membrane of naringenin-induced bacteria [O. Geiger, unpublished]. This finding could have important implications since it raises the possibility that either during the biosynthesis of lipo-oligosaccharides the phospholipids act as intermediates or that these phospholipids which contain a multi-unsaturated acyl moiety have a signal function themselves. In any case this finding gives us an important tool in the study of the biochemical function of the NodF and NodE proteins.

4. The role of the rhizobial lipo-oligosaccharides

4.1 EFFECTS ON THE HOST PLANT

The effect of the lipo-oligosaccharides on plant roots which is most obviously relevant to the symbiotic interaction is the induction of nodule meristems [27,30]. In *Vicia sativa* plants the nodule meristems which are induced by the lipo-oligosaccharides do not develop further to full-grown nodules and never go beyond the stage where small bumps on the roots are visible. This result raises the question whether other signals from the bacterium are required to obtain full-grown root nodules. Ethylene appears to be a powerful antagonist of the nodule meristem formation as illustrated by the reversability of the negative effect of visible light on this process by ethylene inhibitors [A. Van Brussel, unpublished]. Besides their role in the formation of the root nodule meristem, lipo-oligosaccharides also seem to be involved in the infection process, as suggested by the induction of pre-infection thread structures in the outer cortex of *Vicia* roots by the mitogenic lipo-oligosaccharides in the absence of bacteria [31]. These pre-infection thread structures are characterized by the occurrence of radially aligned cytoplasmic bridges in the outer cortex. This gives the impression of strands which cross the outer cortex. The formation of these structures, which are indistinguishable from those observed after infection with *R.leguminosarum* biovar *viciae* bacteria, always precedes the formation of infection threads and they are therefore named pre-infection thread structures [31]. The formation of cytoplasmic bridges in vacuolated cells is preceded by polarization of the cell in which the nucleus moves to the centre of the cell just like in cells which are going to divide [1]. The process of preinfection thread formation can therefore be interpreted as being the result of activation of the cell cycle as is the case with the formation of the nodule meristem in the inner cortex [17,31]. The final result apparently is determined by the position of the cells in the cortex.

Are the rhizobial lipo-oligosaccharides also involved in the later events of the symbiosis, like nodule maturation and nitrogen fixation? This possibility is suggested by the positive effect of flavonoid-independent transcription activator (FITA) *nodD* genes on symbiotic nitrogen fixation [22]. However, recent results indicate that the transcription of the *nod* genes is turned off in the bacteroid stage [22]. Several results indicated that the positive effect of FITA *nodD* genes on nitrogen fixation is more likely the result of an alternative role of the *nodD* gene in nitrogen fixation [H. Schlaman, unpublished]. At present we favour the idea that the lipo-oligosaccharide signals are only important during the early stage of the symbiosis.

4.2 DO THE RHIZOBIAL LIPO-OLIGOSACCHARIDES RESEMBLE ENDOGENOUS SIGNAL MOLECULES OF THE PLANT?

It has been shown that oligosaccharides can have various effects on plant morphogenesis [4]. The resemblance of the rhizobial signals to chitin together with recent indications of the involvement of chitinases in plant embryogenesis [5] gives new ground for discussion about the possibility that similar molecules occur in uninfected plants. Indeed the occurrence of various classes of chitinases in plants and their constitutive expression in the pistels of flowers [11,15,16] has been the cause for much speculation as to their function and their possible substrates in the plant [5]. The results of Benhamou [2] suggest that chitin derivatives occur in secondary plant cell walls of various plant species. The results of Schmidt *et al.* [23] who have shown that transgenic plants containing either the *nodA* or *nodB* genes are disturbed in their normal development could be explained by an interference with the synthesis of hypothetical chitin derivatives in plants. Also suggestive is the high level of similarity, both in primary and secondary structure, of NodC with the DG42 protein of the frog *Xenopus leavis* [3,20] which is transiently expressed during embryogenesis. Since NodC is probably involved in the synthesis of the sugar backbone of Nod metabolites this could mean that chitin-like molecules are even involved as signals in vertebrate animals.

4.3 LIPOPHILIC COMPOUNDS DEGRADABLE BY CHITINASE OCCUR IN *LATHYRUS* PLANTS

To test the hypothesis that lipo-oligosaccharides analogous to the rhizobial signals occur in plants, we have performed radioactive labelling studies with uninfected plants. The approach which was taken is very similar to the one used for the detection of the rhizobial Nod metabolites (Figure 2): *Lathyrus* plants were incubated in a solution of ^{14}C-labelled acetate and subsequently extracted with n-butanol. These extracts were analyzed in several TLC systems known to separate the Nod metabolites (Figure 4). Flowering plants were used as a test system for the following reasons: (i) The development of the flower (and subsequently the embryo) is a complex developmental system which probably involves various unknown signal molecules, (ii) the ENOD12 gene, an important early nodulin gene, is expressed in flowers [21] and, (iii) several chitinases are expressed constitutively in particular parts of flowers [11,15,16]. *Lathyrus odoratus* plants were used because they develop many flowers which are easy to fractionate and because the host range of this plant for rhizobia is very similar to that of *Vicia* which is an important test plant for bio-assays. As can be seen in Figure 4, the flowering *Lathyrus* plants contain many lipo-philic compounds which migrate similarly to the rhizobial lipo-oligosaccharides on TLC plates. To test whether compounds related to chitin were present in the n-butanol extracts, these were subjected to incubation with a chitinase which is able to degrade the rhizobial lipo-oligosaccharides (Figure 4, panel B). The results (Figure 4, panel C) show that

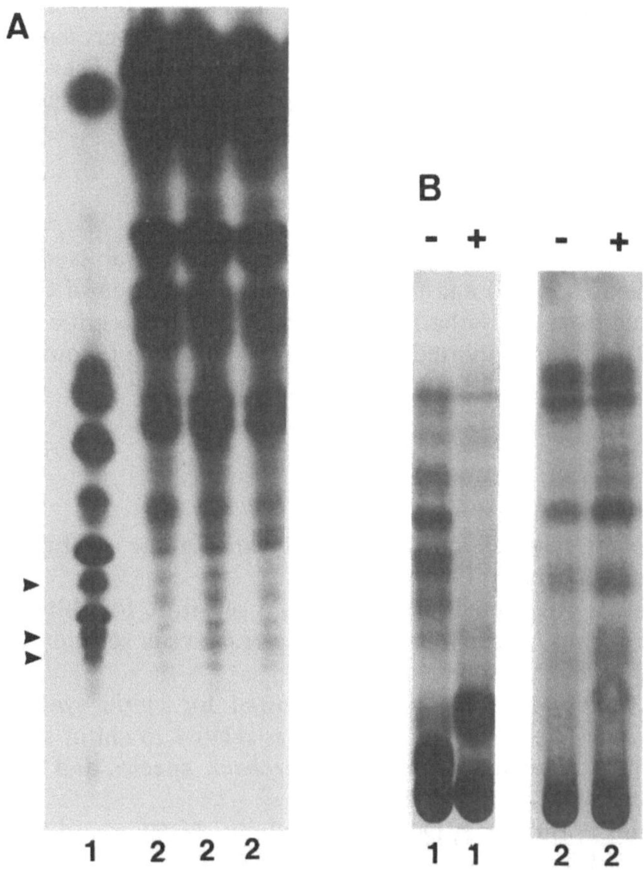

Figure 4. TLC analysis of n-butanol extracts of flowering *Lathyrus* plants. n-butanol extracts were prepared from gnotobiotically grown *Lathyrus odoratus* plants which, after amputation of the roots, were incubated with their stems in a solution of ^{14}C-labelled acetate for five days (sample number 2). The extracts were analyzed on silica TLC plates (panel A: running solvent n-butanol/acetic acid/water) or ODS-TLC (panels B and C) as described in [28] and Figure 2. TLC analysis was also performed after treatment of the n-butanol extracts with chitinase (from *Streptomyces griseus*, 18 hr, 0.2 units, from Sigma Chemical Company, St. Louis, MO, USA). As a control an n-butanol extract was used of the spent growth medium of naringenin-induced *R.leguminosarum* biovar *viciae* (sample number 1). A plus (+) is used to indicate the chitinase treated samples wherase a minus (-) is used to indicate the control to which only buffer without chitinase was added. The positions of the Nod metabolites of this strain are indicated by arrowheads.

although no disappearance of spots is apparent there are at least three new spots appearing after the chitinase treatments. Two of these spots migrate very similarly to the lipo-oligosaccharide signals of *R.leguminosarum* biovar *viciae*. These results therefore are very suggestive for the presence of lipophilic chitin derivatives in uninfected plants.

5. Acknowledgements

The research described in this work was supported by the Royal Netherlands Academy of Arts and Sciences (HPS) and the Netherlands Foundation of Chemical Research with financial aid from the Netherlands Organization for Scientific research (GVB, HRMS and JT-O). JF was supported by an EMBO short term fellowship. We thank Lena Suominen for providing us with the *R.galega* strain.

6. References

1. Bakhuizen R: The plant cytoskeleton in the *Rhizobium*-legume symbiosis. Ph.D. Thesis, Leiden University (1988).
2. Benhamou N, Asselin A: Attempted localization of a substrate for chitinases in plant cells reveals abundant *N*-acetyl-D-glucosamine residues in secondary walls. Biol Cell 67:341-350 (1989).
3. Bulawa CE: CSD2,CSD3 and CSD4, genes required for chitin synthesis in *Saccharomyces cerevisiae*: the CSD2 gene product is related to chitin synthases and to developmentally regulated proteins in *Rhizobium* species and *Xenopus laevis*. Mol Cell Biol 12:1764-1776 (1992).
4. Darvill AG, Albersheim P, Bucheli P, Doares S, Doubrava N, Eberhard S, Gollin DJ, Hahn MG, Marfa-Riera V, York WS, Mohnen D: Oligosaccharins: plant regulatory molecules. In: Lugtenberg BJJ (ed) Signal Molecules in Plants and Plant-Microbe Interactions, NATO ASI Series, pp. 41-48. Springer-Verlag, Berlin (1989).
5. De Jong AJ, Cordewener J, Lo Schiavo F, Terzi M, Vandekerckhove J, Van Kammen A, De Vries S: A carrot somatic embryo mutant is rescued by chitinase. Plant Cell 4:425-433 (1992).
6. Dénarié J, Roche P: *Rhizobium* nodulation signals. In: Verma DPS (ed) Molecular Signals In Plant-Microbe Communications,pp. 295-324. CRC Press, Boca Raton (1992).
7. Downie JA: A *nod* of recognition. Curr Opinion Biol 1:382-384 (1991).
8. Ehrhardt DW, Atkinson EM, Long SR: Depolarization of alfalfa root hair membrane potential by *Rhizobium meliloti* Nod factors. Science 256:998-1000 (1992).
9. Fisher RF, Long SR: *Rhizobium*-plant signal exchange. Nature 357:655-660 (1992).

10. Geiger O, Spaink HP, Kennedy EP: Isolation of *Rhizobium leguminosarum* NodF nodulation protein: NodF carries a 4'-phosphopantetheine prosthetic group. J Bacteriol 173:2872-2878 (1991).
11. Harikrishna K, Jampates-Beale R, Charles S: A basic chitinase is expressed at high levels in the transmitting tissue of tomatoes. In: Halick R.B. (ed) Molecular Biology of Plant Cell Growth and Development, Program and abstracts, Department of Biochemistry, Tucson, Arizona USA, abst. 511 (1991).
12. Horvath B, Franssen H, Heidstra R, Kardailsky I, Lados M, Meshi T, Moerman M, Mylona P, Novak K, Vijn I, Yang W-C, Bisseling T: Mechanisms involved in early nodulin gene expression. In Hallick RB (ed) Molecular Biology of Plant growth and Development , Program and Abstracts. Tucson, Department of Biochemistry, University of Arizona, abstract 127 (1991).
13. Kondorosi A: Regulation of nodulation genes in rhizobia. In: Verma DPS (ed) Molecular Signals In Plant-Microbe Communications,pp. 325-340. CRC Press, Boca Raton (1992).
14. Lerouge P, Roche P, Faucher C, Maillet F, Truchet G, Promé JC, Dénarié J: Symbiotic host-specificity of *Rhizobium meliloti* is determined by a sulphated and acylated glucosamine oligosaccharide signal. Nature (London) 344:781-784 (1990).
15. Lotan T, Ori N, Fluhr R: Pathogenesis-related proteins are developmentally regulated in tobacco flowers. Plant Cell 1:881-887 (1989).
16. Neale AD, Wahleithner JA, Lund M, Bonnet HT, Kelly A, Meeks-Wagner DR, Peacock WJ, Denis ES: Chitinase, ß-1,3-glucanase, osmotin, and extensin are expressed in tobacco explants during flower formation. Plant Cell 2:673-684 (1990).
17. Rae AL, Bonfante-Fasolo P, Brewin N: Structure and growth of infection threads in the legume symbiosis with *Rhizobium leguminosarum*. Plant Journal 2:385-395 (1992).
18. Recourt K, Van Tunen AJ, Mur LA, Van Brussel AAN, Lugtenberg BJJ, Kijne JW: Activation of flavonoid biosynthesis in roots of *Vicia sativa* subsp. *nigra* plants by inoculation with *Rhizobium leguminosarum* biovar *viciae*. Plant Molec. Biol. 19:411-420 (1992).
19. Roche P, Lerouge P, Ponthus C, Promé JC: Structural determination of bacterial nodulation factors involved in the *Rhizobium meliloti*-alfalfa symbiosis. J. Biol. Chem. 266:10933-10940 (1991).
20. Sandal NN, Marcker KA: Some nodulin and Nod proteins show similarity to specific animal proteins. In: Gresshoff PM, Roth LE, Stacey G, Newton WE (eds) Nitrogen Fixation: Achievements and Objectives, pp.687-692. Chapman and Hall, New York (1990).
21. Scheres B, Van de Wiel C, Zalensky A, Horvath B, Spaink HP, Van Eck H, Zwartkruis F, Wolters A-M, Gloudemans T, Van Kammen A, Bisseling T: The ENOD12 gene product is involved in the infection process during the pea-

Rhizobium interaction. Cell 60:281-294 (1990).

22. Schlaman HRM, Okker RJH, Lugtenberg BJJ: Regulation of nodulation gene expression by NodD in rhizobia. J Bacteriol 174 (in press).
23. Schmidt J, John M, Wieneke U, Stacey G, Röhrig H, Schell J: Studies on the function of *Rhizobium meliloti* nodulation genes. In: Hennecke H, Verma DPS (eds)Advances in Molecular Genetics of Plant-Microbe Interactions, pp. 150-155. Kluwer Academic, Dordrecht (1991).
24. Schultze M, Quiclet-Sire B, Kondorosi E, Virelizier H, Glushka JN, Endre G, Géro SD, Kondorosi A: *Rhizobium meliloti* produces a family of sulfated lipo-oligosaccharides exhibiting different degrees of plant host specificity. Proc Natl Acad Sci USA 89:192-196 (1992).
25. Schwedock J, Long SR: ATP sulphurylase activity of the *nodP* and *nodQ* gene products of *Rhizobium meliloti*. Nature (London) 348:644-647 (1990).
26. Spaink HP, Geiger O, Sheeley DM, van Brussel AAN, York WS, Reinhold VN, Lugtenberg BJJ, and Kennedy EP: The biochemical function of the *Rhizobium leguminosarum* proteins involved in the production of host specific signal molecules. In: H.Hennecke,and D.P.S.Verma (eds) Advances in Molecular Genetics of plant- microbe interactions., pp. 142-149. Kluwer Academic Publishers, Dordrecht/Boston (1991).
27. Spaink HP, Sheeley DM, Van Brussel AAN, Glushka J, York WS, Tak T, Geiger O, Kennedy EP, Reinhold VN, Lugtenberg BJJ: A novel highly unsaturated fatty acid moiety of lipo-oligosaccharide signals determines host specificity of *Rhizobium*. Nature (London) 354:125-130 (1991).
28. Spaink HP, Aarts A, Stacey G, Bloemberg GV, Lugtenberg BJJ, Kennedy EP: Detection and separation of *Rhizobium* and *Bradyrhizobium* Nod metabolites using thin layer chromatography. Mol Plant Microbe Interact 5:72-80 (1992).
29. Spaink HP: Rhizobial lipo-oligosaccharides: Answers and questions. Plant Mol. Biol. (in press).
30. Truchet G, Roche P, Lerouge P, Vasse J, Camut S, De Billy F, Promé J-C, Dénarié J: Sulphated lipo-oligosaccharide signals of *Rhizobium meliloti* elicit root nodule organogenesis in alfalfa. Nature (London) 351:670-673 (1991).
31. Van Brussel AAN, Bakhuizen R, Van Spronsen P, Spaink HP, Tak T, Lugtenberg BJJ, Kijne J: Induction of pre-infection thread structures in the host plant by lipo-oligosaccharides of *Rhizobium*. Science (in press).

NodO: A NODULATION PROTEIN THAT FORMS PORES IN MEMBRANES

M. J. SUTTON*, E. J. A. LEA+, S. CRANK*, R. RIVILLA*, A. ECONOMOU*, S. GHELANI*, A. W. B. JOHNSTON+ and J. A. DOWNIE*
*John Innes Institute and +University of East Anglia,
NORWICH NR4 7UH, UK

ABSTRACT. The *Rhizobium leguminosarum* biovar *viciae nodO* gene encodes a secreted protein which is involved in the nodulation of peas and vetch. In the absence of the *nodE* gene, mutants of *R.l.* bv. *viciae* nodulate inefficiently (about 50% of normal) and this residual nodulation is greatly decreased in *nodE* mutants that also contain a mutation in *nodO*. The NodO protein was purified from the supernatant of a strain of *R. leguminosarum* induced for high-level expression of *nodO*. The purified protein had no detectable cellulase, pectinase or protease activity. However, when NodO was added to lipid bilayers, it was found to form ion channels that allowed the passage of Na^+ or K^+. The ion channels appear to have a major peak of conductivity of about 1000 pS (in 1 M NaCl) and were stable for 1-5s. It is proposed that NodO plays a role in the nodulation process by directly interacting with plant plasma-membranes and forming ion channels.

1. Introduction

In *Rhizobium leguminosarum* biovar *viciae*, twelve nodulation (*nod*) genes have been identified that are under the control of the transcriptional activator encoded by *nodD*. These twelve genes are present in four operons: *nodABCIJ*, *nodFEL*, *nodMNT* and *nodO* (see Economou et al., 1990).

It is clear that the *R.l.* bv. *viciae nodABC* and *nodFEL* gene products are involved in the biosynthesis of nodulation signals that are oligomers of N-acetyl glucosamine carrying an N-linked acyl group (Spaink et al., 1991; Spaink et al., this volume). The *nodM* gene encodes a glucosamine synthase that provides glucosamine precursors used in the synthesis of these signals (Marie et al., 1992).

In the absence of the *nodE* gene, the host-specific nodulation signals (which carry a C18:4 acyl group) are not made and only signals with a C18:1 acyl group are formed. The signals with the C18:4 acyl group can, induce nodule meristem formation when added to *Vicia sativa* plants, whereas the signals with the C18:1 acyl chain cannot (Spaink et al., 1991).

A deletion analysis (Downie and Surin, 1990) revealed that in the absence of the *nodFE* genes the *nodO* gene region was essential for the reduced level of nodulation of peas or vetch observed with *nodE* mutants.

E. W. Nester and D. P. S. Verma (eds.),
Advances in Molecular Genetics of Plant-Microbe Interactions, 163–167.

The *nodO* gene encodes a Ca^{2+}-binding protein that is secreted into the growth medium supernatant of strains of *R.l.* bv. *viciae* induced for *nod* gene expression (de Maagd et al., 1989; Economou et al., 1990). The NodO protein shows limited homology to secreted haemolysins and to some proteases (Economou et al., 1990) and shares a similar mechanism of secretion to that of haemolysin and the protease (PrtB) secreted by *Erwinia chrysanthemi* (Scheu et al., 1992).

It appears that the NodO function must be complementary to the secreted lipooligosaccharide nodulation signals. In this work, the NodO protein has been purified in a biologically active form and been shown to form ion channels in artificial bilayers. It is proposed that NodO functions during nodulation by inducing ion fluxes across the plasma membrane of legume root cells.

Results

EFFECTS ON NODULATION OF *nodO nodE DOUBLE MUTANTS*

rreviously (Downie and Surin, 1990) it was shown that although Tn*5* insertions in any one of the *nodF, nodE nodL, nodM, nodN, nodT* or *nodO* genes did not block nodulation, deletion of these genes (plus a 8 kb flanking region) did block nodulation. This deletion mutant could be complemented for nodulation of vetch by a cosmid clone (pIJ1088) carrying *nodO* (plus the flanking genes *rhiABCR*) but not by a similar plasmid carrying a mutation in *nodO*. These experiments revealed a clear role for *nodO* in nodulation but did not exclude the possibility that genes other than *nodO* on pIJ1088 might contribute to its ability to complement the deletion mutant. Indeed, subsequent experiments (Cubo et al., 1992) have indicated a role for the *rhiABCR* genes in nodulation.

To determine if *nodO* alone accounted for the residual nodulation seen with a *nodE* mutant (about 50% of normal), a double mutant was constructed carrying both the *nodE68::*Tn*5* and *nodO93*::Tn*3HoHo1* alleles on the symbiotic plasmid pRL1JI. When this double mutant was tested for nodulation on peas, the *nodO* mutation significantly reduced the level of nodulation (as compared with a *nodE* mutant) but interestingly, some nodulation was observed (about 5% of normal levels).

2.2 MECHANISM OF NodO SECRETION

NodO is secreted via a mechanism that does not use an N-terminal transit sequence (de Maagd et al., 1989; Economou et al., 1990). The NodO secretion mechanism must be analogous to that involved in the secretion of haemolysin and the *Erwina chrysantheum* protease PrtB, since the haemolysin secretion genes (*hlyBD* and *tolC*) or the protease secretion genes (*prtDEF*) could confer an *Escherichia coli* the ability to secrete NodO (Scheu et al., 1992). Significantly, we have shown that a wide range of strains from the genera *Rhizobium*, *Bradyrhizobium*, *Azorhizobium* and *Agrobacterium* have the ability to secrete NodO when the cloned *nodO* gene is transferred to them.

If these secretion systems are analogous to the haemolysin and PrtB (protease) secretion systems then it could be predicted that *Rhizobium* strains might be able to secrete the protease PrtB. A clone carrying the *prtB* gene expressed from a vector promoter was constructed and transferred to *R. leguminosarum*. The transconjugant was indeed found to be able to secrete the protease PrtB into the growth medium supernatant. Strains of *R. leguminosarum* either carrying or lacking the symbiotic plasmid pRL1JI had the ability to secrete the protease indicating that genes that encode the secretion apparatus are present elsewhere than on pRL1JI.

It was considered possible that the genes involved in *nodO* secretion might include the *nodT* gene, the product of which shows sone homology to the *prtF* gene product. However, strains of *R. leguminosarum* lacking *nodT* retained the ability to secrete NodO. Significantly, we have identified a gene homologous to *nodT* in a strain of *R. leguminosarum* lacking a symbiotic plasmid. Therefore, it is possible that two sets of genes may encode two related secretion systems both of which can mediate NodO secretion.

2.3 PURIFICATION OF NodO

Previous strategies used to purify the NodO protein (de Maagd et al., 1989; Economou et al., 1990) would have been unlikely to result in a preparation that retained biological activity. To optimise NodO purification a strain of *R. leguminosarum* was constructed that lacked exopolysaccharide and expressed high levels of *nodO* to facilitate the large scale purification of NodO from the growth medium supernatant of a culture induced for *nodO* expression.

A purification protocol, involving ion exchange chromatography and size fractionation chromatography, was devised. This protocol yielded large amounts of NodO in a form that was likely to retain biological activity. The protein profile of the final column chromatography step is shown (Fig. 1). Whereas the predicted molecular weight of NodO is 30,002, the native NodO protein ran with an apparent Mr of about 67,000 on this (non-denaturing) column. Therefore, we conclude that NodO probably purifies as a dimeric protein.

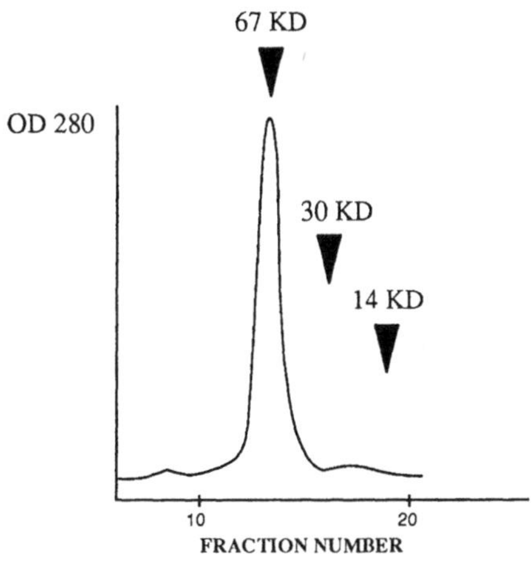

Figure 1. Purified NodO protein runs as a dimer. A fraction containing NodO protein (approx. 5mg) eluted from a Pharmacia FPLC MONO Q ion exchange column, was applied to a Pharmacia Superose 12 FPLC column and the elution of protein was monitored at OD 280. The NodO protein eluted at Mr = 67,000. The arrows indicate the elution positions of the standard proteins, Bovine Serum Albumin (67 kD), Carbonic Anhydrase (30 kD), and Lysozyme (14 kD), used to calibrate the column.

2.4. BIOLOGICAL PROPERTIES OF NodO

Since NodO functions synergistically with the nodulation signal molecules, it was thought possible that NodO might bind to the lipooligosaccharide signals. However, no such binding could be detected between C^{14}-labelled lipooligosaccharide nodulation signals and the pure NodO protein.

An alternative hypothesis was that NodO might facilitate the "access" of the nodulation factors to the legume root cells. Such an effect was thought possible if NodO were found to have pectinase, cellulase or protease activity (especially since NodO had limited homology to some proteases). However, extensive tests of potential NodO enzymic activity using a variety of carbohydrate, cell wall or protein substrates revealed no cellulase, hemicellulase, pectinase or protease activity.

NodO also has some limited homology to the Ca^{2+}-binding domain of the haemolysins secreted by haemolytic strains of *E. coli* (Economou et al., 1990). These haemolysins induce Ca^{2+}-dependent ion efflux from red blood cells (Boehm et al., 1990) and when added to lipid bilayers induce the formation of ion channels (Ludwig et al., 1991).

The NodO protein from the purified sample as shown in Fig. 1 was used in lipid bilayer experiments to determine if it induced ion fluxes. The lipid bilayer was formed using asolecithin (soybean lipids) in the presence of Ca^{2+} and 1M NaCl. Such bilayers had no ion conductance. When NodO protein was added on either side of the lipid bilayer (i.e. the compartments containing the anode or the cathode) specific ion channels were formed. These pores had a typical conductance of about 1000 pS. In addition, secondary pores of about 450 pS were formed. These ion channels opened and closed at irregular intervals but on average remained open for periods of 1-5 sec. It is thought that the different sizes of pores may reflect different numbers of NodO proteins in an aggregate pore (e.g. NodO dimers or tetramers).

Increasing the Ca^{2+} concentration decreased the current flow by reducing the number of channels which opened. This closure of preformed channels indicates that Ca^{2+} may play a key role in NodO function and in nodulation. In general terms, the characteristics of NodO-induced pores were found to be similar to those formed by haemolysin in lipid bilayers (Ludwig et al., 1991).

3. Discussion

It appears that the primary function of NodO is to mediate ion flux across membranes in a manner similar to that of haemolysin. Such a role in bacterial-plant communication for a bacterially made protein is novel and has broad implications with regard to other plant-microbe interactions. It is evident that if NodO does induce pores in plant membranes, the net effect is unlikely to induce cell death since *nod* genes (including *nodO*) are expressed in infection threads (Schlaman et al., 1991) but this does not kill those cells.

Since the pure lipooligosaccharide nodulation factors can induce nodule meristem formation in the absence of bacteria, (Spaink et al., 1991; Truchet et al., 1991) it is evident that NodO must have a complementary function to the nodulation factors. Such a role might be to contribute toward infection thread growth by causing a specific ion flux across the plasma membrane. In this regard, it is particularly significant that the lipooligosaccharide nodulation factors play a role in ion movement in the alfalfa root hairs (Ehrhardt et al., 1992). The complementary roles of these two signals in nodulation will be of fundamental importance in our understanding of infection and nodule formation.

4. References

Boehm, D.F., Welch, R.A. and Snyder, I.S. (1990) Calcium is required for binding of *Escherichia coli* hemolysin (HlyA) to erythrocyte membranes. Inf. and Imm., 58, 1951-1958.

Cubo, T., Economou, A., Murphy, G., Johnston, A.W.B. and Downie, J.A. (1992) Molecular characterization and regulation of the rhizosphere-expressed genes *rhiABCR* that can influence nodulation by *Rhizobium leguminosarum* biovar viciae. J. Bacteriol., 174,

de Maagd, R.A., Rao, A.S., Mulders, I.H.M., Goosen-de Roo, L., van Loosdrecht, M.C.M., Wijffelman, C.A. and Lugtenberg, B.J.J. (1989) Isolation and characterization of mutants of *Rhizobium leguminosarum* bv. *viciae* 248 with altered lipopolysaccharides: possible role of surface charge of hydrophobicity in bacterial release from the infection thread. J.Bacteriol., 171, 1143-1150.

Downie, J.A. and Surin, B.P. (1990). Either of two *nod* gene loci can complement the nodulation defect of a *nod* deletion mutant of *Rhizobium leguminosarum* bv. *viciae*. Mol.Gen.Genet., 222, 81-86.

Economou, A., Hamilton, W.D.O., Johnston, A.W.B. and Downie, J.A. (1990) The *Rhizobium* nodulation gene *nodO* encodes a Ca^{2+}-binding protein that is exported without N-terminal cleavage and is homologous to haemolysin and related proteins. EMBO J., 9, 349-354.

Ehrhardt, D.W., Atkinson, E.M. and Long, S.R. (1992) Depolarization of alfalfa root hair membrane potential by *Rhizobium meliloti* Nod factors. Science, 256, 998-1000.

Ludwig, A., Schmid, A., Benz, R. and Goebel, W. (1991). Mutations affecting pore formation by haemolysin from *Escherichia coli*. Mol.Gen.Genet., 226, 198-208.

Marie, C., Barny, M-.A and Downie, J.A (1992). *Rhizobium leguminosarum* has two glucosamine synthases, GlmS and NodM, required for nodulation and development of nitrogen-fixing nodules. Mol. Micro., 6, 843-851.

Scheu, A.K., Economou, A., Hong, G.F., Ghelani, S., Johnston, A.W.B. and Downie, J.A. (1992). Secretion of the *Rhizobium leguminosarum* nodulation protein NodO by haemolysin-type systems. Mol. Micro., 6, 231-238.

Schlaman, H.R.M., Horvath, B., Vijgenboom, E., Okker, R.J.H. and Lugtenberg, B.J.J. (1991). Suppression of nodulation gene expression in bacteroids of *Rhizobium leguminosarum* biovar *viciae*. J. Bacteriol., 173, 4277-4287.

Spaink, H.P., Sheeley, D.M., Van Brussel, A.A.N., Glushka, J., York, W.S., Tak, T., Geiger, O., Kennedy, E.P., Reinhold, V.N. and Lugtenberg, B.J.J. (1991). A novel, highly unsaturated, fatty acid moeity of lipo-oligosaccharide signals determines host specificity of *Rhizobium leguminosarum*. Nature, 354, 125-130.

Truchet, G., Roche, P., Lerouge, P., Vasse, J., Camut, S., de Billy, F., Prom', J-C. and D'nari', J. (1991). Sulphated lipo-oligosaccharide signals of *Rhizobium meliloti* elicit root nodule organogenisis in alfalfa. Nature, 351, 670-673.

NONFLAVONOID INDUCERS OF *nod* GENES IN *RHIZOBIUM MELILOTI*: APPARENT NODD2 ACTIVATORS RELEASED NATURALLY FROM ALFALFA SEEDS ADD NEW DIMENSIONS TO RHIZOSPHERE BIOLOGY

D.A. PHILLIPS, C.M. JOSEPH and C.A. MAXWELL
Department of Agronomy & Range Science,
University of California, Davis, CA 95616

ABSTRACT. New data show that two betaines, trigonelline and stachydrine, are major components of alfalfa (*Medicago sativa* L.) seed rinse and induce transcription of nodulation (*nod*) genes by activating the regulatory NodD2 protein in *Rhizobium meliloti*. The nonflavonoid nature of these compounds broadens our understanding of natural ecochemical factors that plants use to regulate rhizosphere microbes.

1. Introduction

Molecular data show that plants use chemical signals to regulate specific microbial processes in the rhizosphere [15, 23]. Nodulation (*nod*) and virulence (*vir*) genes are induced in *Rhizobium* and *Agrobacterium* by particular flavonoids and simple phenolics at concentrations far below those required for inducing catabolic systems that supply energy for microbial growth [13, 22]. Some *nod* gene inducers and closely related flavonoids also promote bacterial or fungal processes such as chemotaxis, spore germination, and growth [16]. Thus the classical concept that plants control rhizosphere microbes by supplying carbon and nitrogen substrates [18] can be expanded to suggest that powerful regulatory molecules from plants probably structure microbial communities on and near plant roots through effects on general rhizosphere competence, root colonization, and infection by both symbiotic and pathogenic microbes (Figure 1).

Little is known about the ecology of plant signals to microbes. Studies with alfalfa (*Medicago sativa* L.) show that structurally different flavonoids are released from seeds and roots (5-hydroxy vs. 5-deoxy) [8, 13], and thus one can speculate that different signals are present in hypothetical seed and root zones around a young seedling. Because the 5,7-dihydroxyl substitution pattern of seed flavonoids can enhance growth rate of *R. meliloti* [7], these structural variations may increase the number of rhizobia near seeds. As yet, however, no ecochemical zones have been demonstrated by the direct isolation of *nod* gene inducers from soil. Moreover, there is little understanding of the distances these compounds move in soil.

As a first step toward studying the ecological significance of plant signals to microbes, we have searched for natural products from alfalfa that regulate gene transcription in soil microbes. The availability of *nodC-lacZ* reporter fusions in *R. meliloti* has helped identify four flavonoid *nod* gene inducers, in addition to luteolin [15], which are released naturally from alfalfa: 4',7-dihydroxyflavone, 4',7-dihydroxyflavanone, 4,4'-dihydroxy-2'-methoxychalcone (MCh), and chrysoeriol [8, 13]. Identifying a wide array of transcriptional signals released from alfalfa seedlings under controlled conditions will help future investigations into the movement and fate of these compounds in rhizosphere soil.

E. W. Nester and D. P. S. Verma (eds.),
Advances in Molecular Genetics of Plant-Microbe Interactions, 169–173.

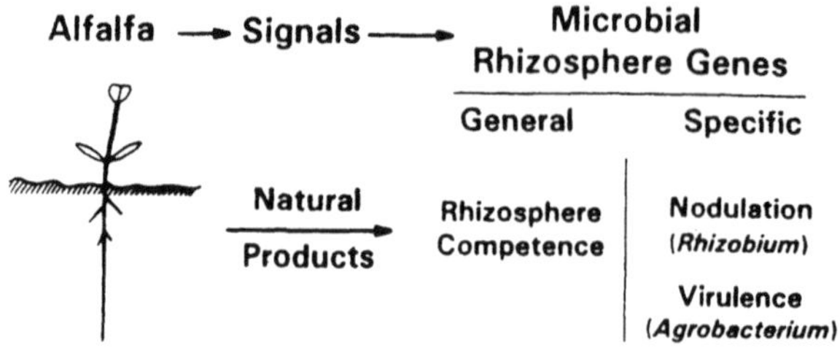

Figure 1. Plant regulation of microbial genes and processes in the rhizosphere.

R. meliloti contains three different *nodD* regulatory genes, *nodD1*, *nodD2*, and *nodD3*, that affect the rate at which this bacterium nodulates alfalfa [5, 11]. All *nod* gene inducers thus far reported from alfalfa have been identified by their capacity to induce transcription of a *nodC-lacZ* fusion in *R. meliloti* strains containing multiple copies of *nodD1* [16]. MCh, the most powerful of those NodD1 activators, is the only known compound that also induces *nod* genes strongly with extra copies of *nodD2* [9]. Luteolin does not activate NodD2 [6, 9, 10], but an unidentified substance in alfalfa seed rinse apparently does [6]. Because MCh is not present in seed rinse [13] and because the seed flavonoid chrysoeriol is only a weak activator of NodD2 [9], we have searched for a strong NodD2 activator in alfalfa seed rinse. Details of data summarized in this report are published elsewhere [17].

2. Materials and Methods

Seed rinse was prepared from surface-sterilized `Moapa 69' alfalfa seeds by soaking 4 h in water. *R. meliloti* strains containing a *nodC-lacZ* translational fusion on pSym with extra copies of either *nodD1* (JM57pRmJ30) or *nodD2* (JM57pRmM137) were generously supplied by Dr. S.R. Long, Stanford University. Transcription of *nod* genes was measured as β-galactosidase activity [13]. Seed rinse was passed through a C_{18} cartridge to remove flavonoids, applied to a cation-exchange HPLC column, and eluted in a linear gradient from water to 40:60 (v/v) methanol:water containing 3 mM HNO_3. Fractions containing NodD2-dependent *nod*-gene-inducing activity were determined by assaying with *R. meliloti* JM57pRmM137. One-dimensional ^{1}H-NMR, two-dimensional ^{1}H-NMR (COSY), ^{13}C-NMR, FAB-MS, high resolution MS, and UV/visible absorbance data were collected on the purified fraction showing NodD2-dependent *nod*-gene-inducing activity. Authentic trigonelline was purchased; stachydrine and MCh were synthesized by published procedures [2, 14].

3. Results

The major NodD2-activating fraction from alfalfa seed rinse was retained in water on a cation-exchange column, and a single fraction with an absorbance maximum at 264 nm was eluted by 40% methanol containing 3mM HNO_3. ^{1}H-NMR, ^{13}C-NMR, and MS data indicated that the NodD2-activating fraction contained two compounds: trigonelline and stachydrine [1, 3, 20, 21]. These zwitterionic betaines are, respectively, the *N*-methyl derivative of nicotinic acid and the *N,N*-dimethyl derivative of proline (Figure 2). Each seed released large quantities of these betaines relative to flavonoid *nod* inducers: trigonelline, 8.6 nmole, and stachydrine, 16.4 nmole, versus the previously reported value for luteolin, 0.3 nmole [8].

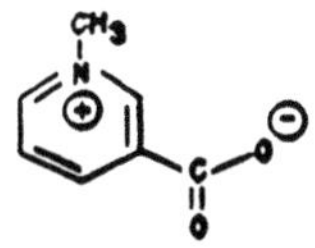

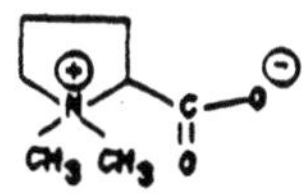

Trigonelline Stachydrine

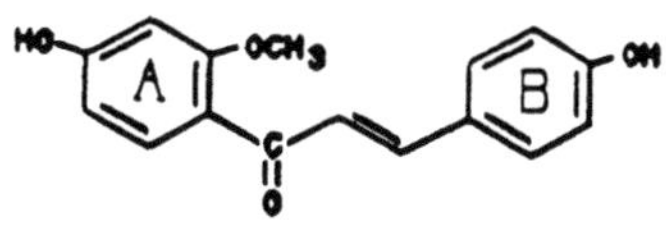

4,4'-Dihydroxy-2'-methoxychalcone

Figure 2. Structures of apparent NodD2 activators released by alfalfa.

Confirmation that trigonelline and stachydrine induce transcription of a *nodC-lacZ* fusion by apparently activating NodD2 was obtained in direct bioassays with authentic compounds (Table 1). The concentration required for half-maximum induction (I_{50}) by both trigonelline and stachydrine was four orders of magnitude greater than that of MCh (*i.e.* 10 μM vs. 1 nM). The small, but significant induction by trigonelline in strain JM57pRmJ30 may have been caused either by a slight activation of NodD1 or by the presence of a single *nodD2* gene on pSym in that strain.

TABLE 1. Induction of *nodC-lacZ* in *R. meliloti* strains containing extra copies of *nodD1* (JM57pRmJ30) or *nodD2* (JM57pRmM137). Values are mean ± SE; a background of 25 to 27 units was subtracted from each.

Compound:	Trigonelline		Stachydrine		MCh	
Concentration	NodD1	NodD2	NodD1	NodD2	NodD1	NodD2
			(units of ß-galactosidase activity)			
1 nM	-	-	-	-	56.8 ± 0.5	26.3 ± 1.8
10 nM	-	-	-	-	97.9 ± 3.0	32.2 ± 0.9
1 μM	-	0.4 ± 0.2	-	9.5 ± 1.0	110 ± 3.6	42.1 ± 0.4
10 μM	-	35.0 ± 1.6	-	25.8 ± 0.4	-	-
50 μM	1.1 ± 0.5	63.7 ± 1.7	0.0	30.6 ± 0.4	-	-
100 μM	5.5 ± 0.2	91.4 ± 1.4	0.0	37.4 ± 0.8	-	-
1 mM	9.5 ± 1.7	96.2 ± 0.1	0.0	36.6 ± 0.9	-	-

4. Discussion

The presence of trigonelline and stachydrine in alfalfa seed rinse and their capacity to induce *nod* gene transcription through NodD2 add new dimensions to alfalfa rhizosphere biology. First, these data offer direct evidence that nonflavonoids can be signals from a legume to its rhizobial symbiont. Thus future attempts to find regulatory factors in soil cannot be restricted to the phenolic fraction. Second, the excellent solubility of these betaines in water may allow them to diffuse more easily than flavonoids in soil. This potentially larger ecochemical zone could be restricted in some soils where the positive charges on these molecules may impair movement. Third, the possibility that both compounds serve as methyl donors suggests that methylation might be examined as a biochemical mechanism for regulating soil microbes. Fourth, the existence of compounds that act with NodD2 but not NodD1 opens the possibility that these proteins regulate some genes jointly (*e.g. nod*) and others separately. The fact that *R. meliloti* carries genes for degrading millimolar concentrations of trigonelline and stachydrine [4] does not prove that these compounds are significant sources of energy in the rhizosphere. It seems equally likely that rhizobia benefit by removing these plant signals from the environment. When existing betaines have been degraded, rhizobia can sense the presence of a new alfalfa seedling.

Both trigonelline and stachydrine have been reported previously in alfalfa [12, 24]. Although trigonelline occurs in seeds of many other legumes [25], data reported here are ecologically important because they establish its presence in alfalfa seed rinse. Trigonelline was previously described as a very weak inducer of *nod* gene transcription in *R. meliloti* [19], but there was no indication of how the signal operated. Although we interpret our results as showing that trigonelline and stachydrine activate NodD2 protein, all assay strains contained a single copy of each *nodD* gene on pSym. Eliminating all other interpretations will require studies with strains containing major deletions in those *nodD* genes. The naturally occurring betaines identified in this report should contribute to those future experiments because they clearly function differently from MCh and luteolin.

It is unclear whether trigonelline and/or stachydrine induce microbial genes involved in general rhizosphere competence separately from their effects on *nod* genes (Figure 1). The relatively large amount of these compounds on seeds suggests they are important, but alternatively, their rather high I_{50} for *nod* gene induction could have favored alfalfa genotypes that released large quantities. Addressing this and other questions raised by these data may help foster a new understanding of rhizosphere biology.

5. Acknowledgements

This work was supported by USDA, NRICGP award 91-37305-6513 and grant US-1884-90 from the U.S.-Israel Binational Agricultural Research and Development Fund (BARD). Address of C.A.M.: Plant Biology Division, The Samuel Roberts Noble Foundation, P.O. Box 2180, Ardmore, OK 73402.

6. References

1. Ahman, V.U., Basha, A., Rahman, A. (1975) 'Identification and C-13 N.M.R. spectrum of stachydrine from *Cadaba fruticosa*', Phytochemistry 14, 292-293.
2. Carlson, R.E., Dolphin, D.H. (1982) '*Pisum sativum* stress metabolites: two cinnamylphenols and a 2'-methoxychalcone', Phytochemistry 21, 1733-1736.
3. Ghosal, S., Dutta, S.K. (1971) 'Alkaloids of *Abrus precatorius*', Phytochemistry 10, 195-198.
4. Goldmann, A., Boivin, C., Fleury, V., Message, B., Lecoeur, L., Maille, M., Tepfer, D. (1991) 'Betaine use by rhizosphere bacteria: genes essential for trigonelline, stachydrine, and carnitine catabolism in *Rhizobium meliloti* are located on pSym in the symbiotic region', Mol. Plant-Microbe Inter. 4, 571-578.

5. Göttfert, M., Horvath, B., Kondorosi, E., Putnoky, P., Rodriguez-Quinones, F., Kondorosi, A. (1986) 'At least two different *nodD* genes are necessary for efficient nodulation on alfalfa by *Rhizobium meliloti*', J. Mol. Biol. 191, 411-420.
6. Györgypal, Z., Iyer, N., Kondorosi, A. (1988) 'Three regulatory *nodD* alleles of diverged flavonoid-specificity are involved in host-dependent nodulation by *Rhizobium meliloti*', Mol. Gen. Genet. 212, 85-92.
7. Hartwig, U.A., Joseph, C.M., Phillips, D.A. (1991) 'Flavonoids released naturally from alfalfa seeds enhance growth rate of *Rhizobium meliloti*', Plant Physiol. 95, 797-803.
8. Hartwig, U.A., Maxwell, C.A., Joseph, C.M., Phillips, D.A. (1990) 'Chrysoeriol and luteolin released from alfalfa seeds induce *nod* genes in *Rhizobium meliloti*', Plant Physiol. 92, 116-122.
9. Hartwig, U.A., Maxwell, C.A., Joseph, C.M., Phillips, D.A. (1990) 'Effects of alfalfa *nod* gene inducing flavonoids on *nodABC* transcription in *Rhizobium meliloti* strains containing different *nodD* genes', J. Bacteriol. 172, 2769-2773.
10. Honma, M.A., Asomaning, M., Ausubel, F.M. (1990) '*Rhizobium meliloti nodD* genes mediate host-specific activation of *nodABC*', J. Bacteriol. 172, 901-911.
11. Honma, M.A., Ausubel, F.M. (1987) '*Rhizobium meliloti* has three functional copies of the *nodD* symbiotic regulatory gene', Proc. Natl. Acad. Sci., USA 84, 8558-8562.
12. Jones, G.P., Naidu, B.P., Starr, R.K., Paleg, L.G. (1986) 'Estimates of solutes accumulating in plants by ^{1}H-nuclear magnetic resonance spectroscopy', Aust. J. Plant Physiol. 13, 649-658.
13. Maxwell, C.A., Hartwig, U.A., Joseph, C.M., Phillips, D.A. (1989) 'A chalcone and two related flavonoids released from alfalfa roots induce *nod* genes of *Rhizobium meliloti*', Plant Physiol. 91, 842-847.
14. Musich, J.A., Rapoport, H. (1977) 'Reaction of *O*-methyl-*N,N'*-diisopropylisourea with amino acids and amines', J. Org. Chem. 42, 139-141.
15. Peters, N.K., Frost, J.W., Long, S.R. (1986) 'A plant flavone, luteolin, induces expression of *Rhizobium meliloti* nodulation genes', Science 233, 977-980.
16. Phillips, D.A. (1992) 'Flavonoids: Plant signals to soil microbes', Rec. Adv. Phytochem. 26, 201-231.
17. Phillips, D.A., Joseph, C.M., Maxwell, C.A. (1992) 'Trigonelline and stachydrine released from alfalfa seeds activate NodD2 protein in *Rhizobium meliloti*', Plant Physiol. 99, In press.
18. Rovira, A.D. (1969) 'Plant root exudates', Bot. Rev. 35, 35-57.
19. Schmidt, J., John, M., Wieneke, U., Krüssmann, H.-D., Schell, J. (1986) 'Expression of the nodulation gene *nodA* in *Rhizobium meliloti* and localization of the gene product in the cytosol', Proc. Natl. Acad. Sci., USA. 83, 9581-9585.
20. Sciuto, S., Chillemi, R., Piattelli, M. (1988) 'Onium compounds from the red alga *Pterocladia capillacea*', J. Nat. Prod. 51, 322-325.
21. Smith, G.M., Pettigrew, G.W. (1980) 'Identification of *N,N*-dimethylproline as the N-terminal blocking group of *Crithidia oncopelti* cytochrome c_{557}', Eur. J. Biochem. 110, 123-130.
22. Spencer, P.A., Towers, G.H.N. (1988) 'Specificity of signal compounds detected by *Agrobacterium tumefaciens*', Phytochemistry 27, 2781-2785.
23. Stachel, S.E., Messens, E., Van Montagu, M., Zambryski, P. (1985) 'Identification of the signal molecules produced by wounded plant cells that activate T-DNA transfer in *Agrobacterium tumefaciens*', Nature 318, 624-629.
24. Steenbock, H. (1918) 'Isolation and identification of stachydrine from alfalfa hay', J. Biol. Chem. 35, 1-13.
25. Tramontano, W.A., McGinley, P.A., Ciancaglini, E.F., Evans, L.S. (1986) 'A survey of trigonelline concentrations in dry seeds of the Dicotyledoneae', Environ. Expt. Bot. 26, 197-205.

ROLE IN NODULATION AND GENETIC REGULATION OF EXOPOLYSACCHARIDE SYNTHESIS IN *RHIZOBIUM MELILOTI*

JOHN A. LEIGH, LAURIE BATTISTI, CHI CHANG LEE,
DAVID A. OZGA, HANGJUN ZHAN, and SABINA ASTETE
Department of Microbiology, SC-42
University of Washington
Seattle, WA 98195
USA

ABSTRACT. Genetic evidence has shown that *Rhizobium meliloti* exopolysaccharide is required for root nodule invasion on alfalfa. We have found that a specific form of the *R. meliloti* exopolysaccharide, succinoglycan, can rescue nodule invasion by exopolysaccharide deficient (*exo*) mutants. The active form of succinoglycan was present in a low molecular weight fraction, consisted of four repeat units, and was relatively anionic. A non-succinylated form of succinoglycan was not active, nor were low molecular weight exopolysaccharides from other species. In other work, we have determined that a spontaneous mutation in the *exoS* locus of *R. meliloti* decreases *exo* gene expression. We have also identified genes that regulate the expression of *muc* genes involved in the synthesis of the second *R. meliloti* exopolysaccharide, EPSb. *mucR*::Tn*5* mutants showed decreased *exo* gene expression as well as increased *muc* gene expression. A Tn*5* insertion in a new locus, *muc8*, decreased *muc* gene expression. Finally, a *phoR* or *phoB*-like gene in *R. meliloti* seems to be necessary for the induction of EPSb synthesis by low phosphate concentrations. A model for the genetic regulation of exopolysaccharide synthesis in *R. meliloti* is presented.

1. Introduction

Genetic and phenotypic analyses of *Rhizobium meliloti* mutants (Exo-) that are defective in exopolysaccharide (EPS) synthesis have shown that EPS is necessary for nodule invasion on alfalfa (11, 13). *exo* mutants form nodules that are round and white and contain no bacteroids. Infection threads are found but do not penetrate beyond the nodule cortex (21). Root hair curling is delayed (21). Nodules contain only two of the known nodulins (4, 14). Finally, structures suggestive of a plant defense response form (15). Either of two *R. meliloti* EPSs, succinoglycan or EPSb, is sufficient for nodule invasion (7, 25). Both EPSs are acidic heteropolysaccharides consisting of repeating units. The repeat unit of succinoglycan is a particular octasaccharide containing 1-carboxyethylidene (pyruvate), succinyl, and acetyl substituents (1), while the repeat unit of EPSb is a disaccharide containing 1-carboxyethylidene (pyruvate) and acetyl substituents (8, 12).

E. W. Nester and D. P. S. Verma (eds.),
Advances in Molecular Genetics of Plant-Microbe Interactions, 175–181.

Much of our work in recent years has addressed whether the *R. meliloti* EPSs play specific or non-specific roles. Specific roles would include serving as a ligand that is recognized by a plant receptor, thereby inducing a plant cellular response leading to infection tread penetration or suppressing a plant defense response. Non-specific roles would include mediating attachment to the plant root, forming a morphological component of the infection thread such as the fibrillar infection thread matrix, or coating the bacterial cell and preventing a plant defense response from blocking invasion. A role that might require an intermediate degree of specificity would be to chelate divalent cations by virtue of the multivalent anionic nature of the EPS.

To determine which models, specific or non-specific, could be correct, we have studied the effect of altering the chemical structure of succinoglycan on its ability to promote nodule invasion. Early evidence in favor of specificity in the *R. meliloti*-alfalfa system was that *exoH* mutants, which produced succinoglycan lacking the usual succinyl substituent, had the same nodulation phenotype as *exo* mutants that produced no succinoglycan (10). Additional evidence for specificity was found in the *Rhizobium* sp. strain NGR234-*Leucaena* and *R. leguminosarum* bv. *trifolii*-clover systems (5). In these systems, EPSs harvested from the wild type strains were found to rescue the nodule development defects of Exo^- mutants on the appropriate host plants. Only EPS harvested from the wild type strain of the same species as the Exo^- mutant had this ability; EPS from the other species did not.

In another line of work, we have studied the genetic regulation of EPS synthesis. Three classes of mutants with increased levels of succinoglycan synthesis have been reported, *exoR*::Tn*5* (17), *exoS*::Tn*5* (6), and *exoX*::Tn*5* (16, 24). EPSb synthesis occurred only in certain mutants or under certain conditions. *mucR*::Tn*5* (25) and *expR* (7) mutants produced EPSb. In addition, *mucR*::Tn*5* produced much less succinoglycan than wild type (25). Finally, low phosphate concentrations activated EPSb synthesis in the wild type strain Rm1021, and this synthesis depended on an apparent regulatory network that also activated alkaline phosphatase (PhoA) gene expression as do PhoR and PhoB in *E. coli* (23).

2. Exogenous Addition of Low Molecular Weight Succinoglycan Rescues Nodule Invasion by *R. meliloti exo* Mutants on Alfalfa.

We harvested succinoglycan from culture supernatants of wild type *R. meliloti* Rm1021 and tested its ability to rescue the nodule invasion defect of *exo* mutants. The succinoglycan was routinely added to the plant root one day before inoculation or at the time of inoculation with an *exo* mutant. Succinoglycan was first separated into high molecular weight (HMW) and low molecular weight (LMW) fractions by Biogel A5 chromatography (9). LMW succinoglycan, but not HMW succinoglycan, rescued nodule invasion. LMW succinoglycan rescued nodule invasion at concentrations as low as 1 μM. The rescue of nodule invasion on alfalfa by the addition of LMW succinoglycan has been observed independently by Urzainqui and Walker (19) and constitutes a biological assay for the nodule invasion-promoting activity of EPS on alfalfa (2).

3. Only a Specific Form of LMW Succinoglycan Promotes Nodule Invasion.

We fractionated LMW succinoglycan into six subfractions by QAE anion exchange chromatography. Each subfraction consisted of a different form of succinoglycan, and differed from the others in repeat unit multiplicity and/or the density of negative charges stemming presumably from the carboxylated substituents. Repeat unit multiplicities of one (monomer), three (trimer), and four (tetramer) were observed. When tested singly, only the peak eluting last in an increasing concentration gradient of KCl (QAE peak 6) promoted nodule invasion in the bioassay. QAE peak 6 consisted of tetrameric succinoglycan with a relatively high negative charge density.

Heterologous EPSs did not promote nodule invasion on alfalfa. LMW EPS from *R. leguminosarum* bv. *trifolii* and LMW EPS from *R.* sp. strain NGR234 failed to promote invasion. In addition, LMW non-succinylated succinoglycan obtained from culture supernatant of an *exoH* mutant failed to promote nodule invasion. This observation was consistent with the nodulation phenotype of *exoH* mutants.

These results suggest that the function of succinoglycan in nodule invasion is specific. However, the degree of specificity is still unresolved. One possibility is that the exact chemical structure of succinoglycan is crucial for its function. Another possibility is that only the number of negative charges on a molecule of a certain size is important.

4. Discovery of a Spontaneous Mutant Allele of *exoS* that Decreases *exo* Gene Expression.

In *R. meliloti* three genetic classes of mutants overproduce succinoglycan. *exoR*::Tn*5* mutants (17) and an *exoS*::Tn*5* mutant (6) have increased levels of *exo* gene expression, while *exoX*::Tn*5* mutants (16, 24) appear to alter succinoglycan synthesis posttranslationally. *exoR*::Tn*5* mutants overproduced succinoglycan more severely than *exoS*::Tn*5* or *exoX*::Tn*5* mutants. We observed that bacterial suspensions obtained from young colonies of *exoR*::Tn*5* mutants formed empty nodules on alfalfa. However, older colonies of *exoR*::Tn*5* mutants contained sectors of pseudorevertants that produced normal levels of succinoglycan, and these pseudorevertants did colonize nodules even in the presence of a vast excess of *exoR*::Tn*5* overproducers. Pseudorevertants, obtained either from colonies or from nodules, contained spontaneous mutations that decreased succinoglycan synthesis and nearly always mapped to the *exoS* locus. These mutations, called *exoS**, were recessive and belonged to the same genetic complementation group as the overproducing mutation, *exoS*::Tn*5*. We determined the effect of *exoS** on the expression of genes for succinoglycan synthesis (*exo* genes) using an *exoY-lacZ* fusion. ß-galactosidase activities were at least ten-fold higher in an *exoR*::Tn*5* mutant than in the wild type background, and only two to three times higher in an *exoR*::Tn*5* - *exoS** pseudorevertant than in the wild type background. Similar results were obtained with an *exoP-phoA* fusion measuring alkaline phosphatase activity. Therefore, the *exoS** mutation decreased *exo* gene expression even in an *exoR*::Tn*5* background.

5. *mucR* Regulates the Expression of Genes for Succinoglycan Synthesis Positively and the Expression of Genes for EPSb Synthesis Negatively.

mucR::Tn*5* mutants produce EPSb instead of succinoglycan (25). We determined if *mucR*::Tn*5* affected the expression of succinoglycan synthesis genes (*exo* genes) and the expression of EPSb synthesis genes (*muc* genes) accordingly. ß-galactosidase activity arising from an *exoY-lacZ* fusion was four to six fold lower in a *mucR*::Tn*5* background than in a wild type background, while ß-galactosidase activity arising from a *muc-lacZ* fusion was four to six fold higher in a *mucR*::Tn*5* background than in a wild type background. Therefore, *mucR* regulates succinoglycan synthesis and EPSb synthesis inversely at the gene expression level.

6. At Least One Gene in the *muc* Cluster is an Activator of *muc* Gene Expression.

EPSb synthesis occurs in Rm1021 containing EPSb synthesis genes on the recombinant plasmid pMuc (25). Glazebrook and Walker (7) identified two loci on a cosmid containing the same region that were required for this stimulation of EPSb synthesis. Similarly, we isolated two Tn*5* insertions in pMuc that eliminated EPSb synthesis. One of these insertions, *muc8*::Tn*5*, consistently decreased *muc* gene expression. Thus, *muc-lacZ* expression was three to four fold lower in the presence of pMuc *muc8*::Tn*5* than in the presence of pMuc. Therefore, *muc8* may encode an activator of *muc* gene expression.

7. Model of Genetic Regulation of EPS Synthesis in *R. meliloti*

Figure 1 shows a model for the genetic regulation of succinoglycan and EPSb synthesis. ExoC (28) is a phosphoglucomutase required for the synthesis of UDP-glucose, a precursor of lipopolysaccharide (LPS) and periplasmic cyclic ß-(1,2) glucan as well as succinoglycan and EPSb (9, 25). ExoB (3) is a galactose epimerase required for the synthesis of UDP-galactose, a precursor of LPS, succinoglycan, and EPSb (9, 25). MucR regulates succinoglycan synthesis and EPSb synthesis inversely, by regulating *exo* gene expression positively and *muc* gene expression negatively. ExoR and ExoS regulate *exo* gene expression, while ExoX regulates succinoglycan synthesis posttranslationally, probably by complexing with ExoY (16, 24). Muc8 activates *muc* gene expression. Finally, gene products equivalent to *E. coli* PhoR and PhoB may activate *muc* gene expression in response to low phosphate concentrations (23).

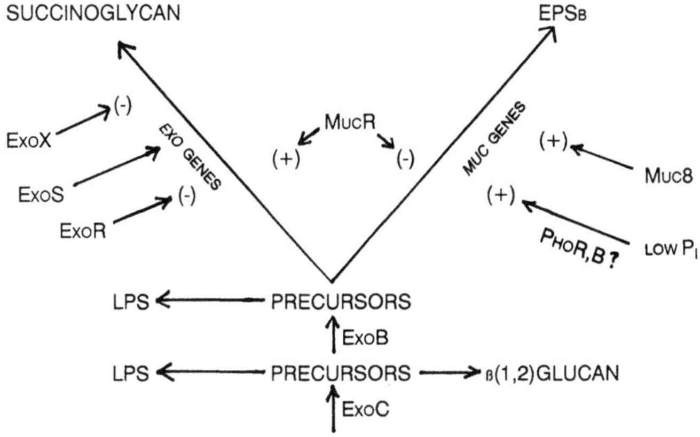

Figure 1. Model for the genetic regulation of EPS synthesis in *R. meliloti*.

Acknowledgements

This work was supported by Public Health Service grant GM39785 from the National Institutes of Health.

References

1. Aman, P., McNeil, M., Franzen, L., Darvill, A. G., Albersheim, P. 1981. Structural elucidation, using HPLC-MS and GLC-MS, of the acidic polysaccharide secreted by *R. meliloti* strain 1021. Carbohydr. Res. 95:263-282.
2. Battisti, L., Lara, J. C., and Leigh, J. A. . 1992. Specific oligosaccharide form of the *Rhizobium meliloti* exopolysaccharide promotes nodule invasion in alfalfa. Proc. Natl. Acad. Sci. USA 89:5625-5629.
3. Buendia, A. M., Enenkel, B., Köplin, R., Niehaus, K., Arnold, W., Pühler, A. 1991. The *Rhizobium meliloti exoZ/exoB* fragment of megaplasmid 2: ExoB functions as a UDP-glucose 4-epimerase and ExoZ shows homology to NodX of *Rhizobium leguminosarum* biovar *viciae* strain TOM. Molecular Microbiol. 5:1519-1530.
4. Dickstein, T., Bisseling, T., Reinhold, V. N., Ausubel, F. 1988. Expression of nodule-specific genes in alfalfa root nodules blocked at an early stage of development. Genes and Development 2:677-687.
5. Djordjevic, S. P., Chen, H., Batley, M., Redmond, J. W., Rolfe, B. G. 1987. Nitrogen fixation ability of exopolysaccharide synthesis mutants of *Rhizobium* sp. NGR234 and *Rhizobium trifolii* is restored by the addition of homologous exopolysaccharides. J. Bacteriol. 169:53-60.

6. Doherty, D., Leigh, J. A., Glazebrook, J., Walker, G. C. 1988. *Rhizobium meliloti* mutants that overproduce the *R. meliloti* acidic Calcofluor-binding exopolysaccharide. J. Bacteriol. 170:4249-4256.
7. Glazebrook, J, Walker, G. C. 1989. A novel exopolysaccharide can function in place of the Calcofluor-binding exopolysaccharide in nodulation of alfalfa by *Rhizobium meliloti*. Cell 56: 661-672.
8. Her, G. R., Glazebrook, J., Walker, G., and Reinhold, V. N. 1990. Structural studies of a novel exopolysaccharide produced by a mutant of *Rhizobium meliloti* strain Rn1021. Carbohydr. Res. 198:305-312.
9. Leigh, J. A., Lee, C. C. 1988. Characterization of polysaccharides of *Rhizobium meliloti exo* mutants that form ineffective nodules. J. Bacteriol. 170: 3327-3332.
10. Leigh, J. A., Reed, J. W., Hanks, J. F., Hirsch, A. M., Walker, G. C. 1987. *Rhizobium meliloti* mutants that fail to succinylate their Calcofluor-binding exopolysaccharide are defective in nodule invasion. Cell 51:579-587.
11. Leigh, J.A., Signer, E. R., Walker, G. C. 1985. Exopolysaccharide-deficient mutants of *Rhizobium meliloti* that form ineffective nodules. Proc. Natl. Acad. Sci. USA 82:6231-6235.
12. Levery, S. B., Zhan, H., Lee, C. C., Leigh, J. A., Hakomori, S. 1991. Structural analysis of a second acidic exopolysaccharide of *Rhizobium meliloti* that can function in alfalfa root nodule invasion. Carbohydr. Res. 210:339-347.
13. Long, S., Reed, J. W., Himawan, J., Walker, G. C. 1988. Genetic analysis of a cluster of genes required for synthesis of the Calcofluor-binding exopolysaccharide of *Rhizobium meliloti*. J. Bacteriol. 170:4239-4248.
14. Norris, J. H., Macol, L. A., Hirsch, A. M. 1988. Nodulin gene expression in effective alfalfa nodules and nodules arrested at three different stages of development. Plant Physiol. 88:321-328.
15. Pühler, A., Arnold, W., Buendia-Claveria, A., Kapp, D., Keller, M., Niehaus, K., Quandt, A., Roxlau, A., and Weng, W. M. 1991. The role of the *Rhizobium meliloti* exopolysaccharides EPS I and EPS II in the infection process of alfalfa nodules. In H. Hennecke and D. P. S. Verma (eds.), Advances in Molecular Genetics of Plant-Microbe Interactions, Vol. 1, Klewer Academic Publishers, Dordrecht, pp. 189-194.
16. Reed, J. W., Capage, M., Walker, G. C. 1991. *Rhizobium meliloti exoG* and *exoJ* mutations affect the ExoX-ExoY system for modulation of exopolysaccharide production. J. Bacteriol. 173:3776-3788.
17. Reed, J. W., Glazebrook, J., Walker, G. C. 1991. The *exoR* gene of *Rhizobium meliloti* affects RNA levels of other *exo* genes but lacks homology to known transcriptional regulators. J. Bacteriol. 173:3789-3794.
18. Reuber, T. L., Long, S., Walker, G. C. 1991. Regulation of *Rhizobium meliloti exo* genes in free-living cells and in planta examined by using Tn*phoA* fusions. J. Bacteriol. 173: 426-434.
19. Urzainqui, A., Walker, G. C. 1992. Exogenous suppression of the symbiotic deficiencies of *Rhizobium meliloti exo* mutants. J. Bacteriol. 174:3403-3406.
20. Uttaro, A. D., Cangelosi, G. A., Geremia, R. A., Nester, E. W., Ugalde, R. A. 1990. Biochemical characterization of avirulent *exoC* mutants of *Agrobacterium tumefaciens*. J. Bacteriol. 172:1640-1646.

21. Yang, C., Signer, E. R., Hirsch, A. M. 1991. Nodules initiated by *R. meliloti* exopolysaccharide (*exo*) mutants lack a discrete, persistent meristem. Plant Physiol. 98:143-151.

22. Zhan, H., Gray, J. X., Levery, S. B., Rolfe, B. G., Leigh, J. A. 1990. Functional and evolutionary relatedness of genes for exopolysaccharide synthesis in *Rhizobium meliloti* and *Rhizobium* sp. strain NGR234. J. Bacteriol. 172:5245-5253.

23. Zhan, H., Lee, C. C., Leigh, J. A. 1991. Induction of the second exopolysaccharide (EPSb) in *Rhizobium meliloti* SU47 by low phosphate concentrations. J. Bacteriol. 173:7391-7394.

24. Zhan, H., Leigh, J. A. 1990. Two genes that regulate exopolysaccharide production in *Rhizobium meliloti*. J. Bacteriol. 172:5254-5259.

25. Zhan, H., Levery, S. B., Lee, C. C., and Leigh, J. A. 1989. A second exopolysaccharide of *Rhizobium meliloti* strain SU47 that can function in root nodule invasion. Proc. Natl. Acad. Sci. U.S.A. 86:3055-305

OXYGEN REGULATION OF NITROGEN FIXATION GENE EXPRESSION IN RHIZOBIUM MELILOTI

J. BATUT, P. de PHILIP, J.M. REYRAT, F. WAELKENS and P. BOISTARD
LBMRPM, CNRS-INRA
BP 27
31326 Castanet-Tolosan Cedex
France

ABSTRACT. The major signal inducing nitrogen fixation gene expression in *R. meliloti* is microaerobiosis. The pathway from oxygen sensing to transcriptional activation of the intermediate regulatory genes *nifA* and *fixK* is being studied by a combination of *in vivo* and *in vitro* approaches. The protein responsive to oxygen, FixL, is a hemoprotein kinase that belongs to the sensor class of two-component regulatory proteins. The sensing part of FixL is its central non-conserved domain. This domain prevents activation of FixJ by FixL *in vivo* under aerobic conditions. Anaerobic conditions enhance the rate of autophosphorylation of purified FixL *in vitro* and strongly promote FixJ phosphorylation. An *in vitro* transcription assay was developed in order to study the mechanism by which FixJ, or its phosphorylated form, activates transcription of *nifA* and *fixK*. Potential binding sites for FixJ have been identified on the *fixK* promoter. FixJ-activated promoters display a consensus sequence at position -35.

1. Introduction

Rhizobium meliloti is an aerobic soil bacterium able to fix N2 in symbiosis with a limited range of legumes including alfalfa. Symbiotic nitrogen fixation by *Rhizobium* is a complex process that requires differentiation of both partners. Rhizobia induce the formation on plant roots of genuine new organs, nodules, whose structure and physiology is specialized for nitrogen fixation. Inside the nodule, bacteria, which do not fix nitrogen in the free-living state, differentiate into nitrogen-fixing bacteroids. Many plant and microbial genes need to be coordinately expressed (or repressed) for a successful nitrogen-fixing interaction to occur.

Signals are exchanged between the two partners during the differentiation process and also likely later on in the mature N2-fixing nodule. Two kinds of signals, flavonoids and lipo-oligosaccharides synthesized by the plant and the bacteria respectively, have been identified that control the onset of nodule organogenesis (reviewed by Fisher and Long (1992)). At a later stage in nodule development, bacterial genes involved in nitrogen fixation (*nif* and *fix* genes) start being expressed whereas they are silent in the free-living aerobic state.

We and others have found that microaerobiosis is probably the key signal responsible for eliciting expression of nitrogen fixation genes in the nodule. The

E. W. Nester and D. P. S. Verma (eds.),
Advances in Molecular Genetics of Plant-Microbe Interactions, 183–191.

aim of this paper is to summarize the evidence for this and our knowledge about the mechanism by which oxygen regulates nitrogen fixation gene expression.

2. Results

2.1. MICROAEROBIOSIS INDUCES TRANSCRIPTION OF NITROGEN FIXATION GENES IN R.MELILOTI

All *nif* and *fix* genes whose expression has been studied are under the control of either of two regulatory genes, *nifA* and *fixK*, both of which encode transcriptional activators (Figure 1). Expression of *nifA* and *fixK* is itself under the control of the *fixLJ* regulatory operon. Null mutations in *fixL* or *fixJ* prevent symbiotic activation of *nifA* and *fixK* and hence of all genes downstream in the regulatory cascade (David et al. (1988), Batut et al.(1989)).

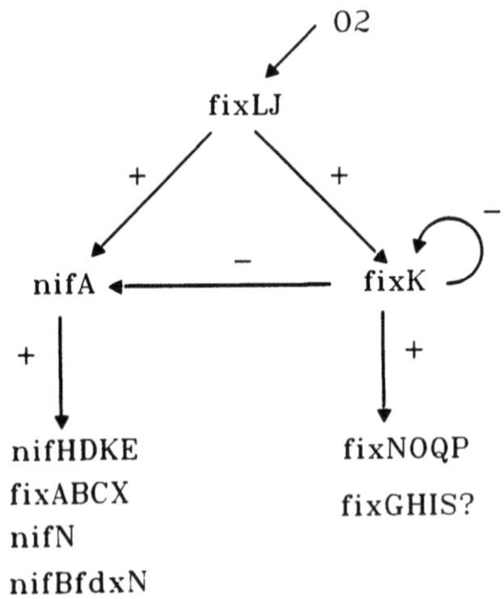

Figure 1: Cascade regulation of nitrogen fixation genes in *R.meliloti*

We and Ditta et al.(1987) have found that *nif* and *fix* gene expression could be induced at the transcriptional level *ex planta* under microaerobic (c.a. 2% oxygen) conditions (however nitrogen fixation could not be demonstrated under such conditions). Microaerobic expression of *nif* and *fix* genes also depends on *fixLJ* (David et al.(1988)).

Since the central part of the nodule is microaerobic (see Witty and Minchin (1990)), it is likely that a drop in free oxygen concentration is the signal responsible for symbiotic activation of the *fixLJ* regulatory cascade.

2.2. FIXL MEDIATES MICROAEROBIC INDUCTION OF NITROGEN FIXATION GENES (de Philip et al.(1990))

How is microaerobiosis sensed by *Rhizobium*? From sequence inspection, the FixLJ proteins were shown to belong to the two-component family of regulatory proteins (David et al.(1988)). FixL is homologous to sensor proteins whereas FixJ is homologous to regulator proteins. Consequently we hypothesized that FixL might be responsible for oxygen sensing.

In order to test this view, we reconstituted the genetic circuitry of *nifA* and *fixK* expression in the heterologous host *E. coli*. This offered two advantages: first, a limited number of genes was brought from *Rhizobium*; second, *fixL* and *fixJ* could be expressed independently under the control of appropriate promoters. We found that *fixJ*, provided it was overexpressed, drove *nifA* and *fixK* expression at the genuine promoters in *E. coli* in the absence of *fixL*. Expression in that case was independent of the oxygen status of the culture. Regulation by oxygen was restored when *fixL* was provided in addition to *fixJ* thereby demonstrating that *fixL* mediates the response to oxygen.

2.3. THE CENTRAL NON-CONSERVED DOMAIN OF FIXL IS RESPONSIBLE FOR OXYGEN-SENSING (de Philip et al.(1992))

FixL, as well as several other sensor proteins, is organized in three structural domains: an amino(N)-terminal hydrophobic domain that, supposedly, anchors the protein to the membrane; a central non-conserved domain and a carboxy(C)-terminal domain homologous to the kinase domain of sensor proteins (Figure 2).

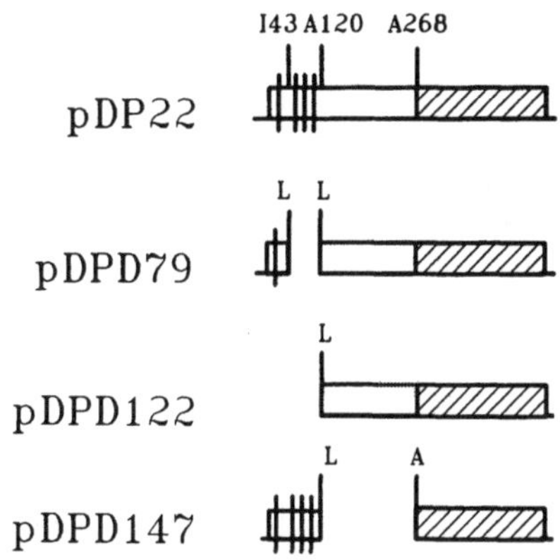

Figure 2: Structure of FixL and deleted derivatives

In order to probe the structure-function relationship of FixL, we engineered precise deletions of one or the other of the two non-conserved domains of FixL. The ability of the truncated FixL proteins to promote *nifA* and *fix*K expression was investigated in *E.coli*. Because we anticipated that the FixL mutant proteins may have a reduced activity as compared to wild-type, we cloned *fixL* under the control of the strong IPTG-inducible lac promoter whereas *fixJ* was expressed from a weak constitutive promoter. Under those conditions *fixJ* cannot activate target gene expression in the absence of *fixL* (data not shown). Regulation by oxygen was observed whatever the level of expression of *fixL* (Table 1).

Table 1: Expression of nifA (pCHK57) and fixK (pJJ5) lacZ fusions in the presence of FixJ (pDP8) and wild-type FixL or truncated derivatives. (A and B represent two different series of experiments)

STRAIN	[IPTG] mM	ß-GALACTOSIDASE UNITS Aerated culture	Microaerobic culture
A)			
TG1(pDP22, pDP8, pJJ5)	0	2	120
	2	5	206
TG1(pDPD79, pDP8, pJJ5)	0	12	29
	2	4	211
TG1(pDPD147, pDP8, pJJ5)	0	5	3.5
	2	215	230
TG1(pDP22, pDP8, pCHK57)	0	0.2	37
	2	0.4	63
TG1(pDPD79, pDP8, pCHK57)	0	1.8	2.4
	2	1.5	13
TG1(pDPD147, pDP8, pCHK57)	0	0.3	0.2
	2	86	71
B)			
TG1(pDP22, pDP8, pJJ5)	0	8	85
	2	1	92
TG1(pDPD122, pDP8, pJJ5)	0	4	12
	2	0.2	74

We found that FixL proteins lacking part or all of the N-terminal domain (encoded by pDPD79 and pDPD122 respectively, Fig.2) did not differ significantly from the wild-type FixL protein in their ability to promote microaerobic induction of *nifA* and *fixK* expression. This suggests that the intrinsic ability of FixL to sense oxygen is preserved in the absence of the N-terminal hydrophobic domain. Therefore transmembrane signalling as has been described for other sensor proteins does not apply to FixL. However the truncated FixL proteins needed to be overexpressed, as compared to wild-type FixL protein, in order to show activity. The reason for this is not clear but may suggest that the N-terminal domain is necessary for full activity in *R.meliloti* in which *fixL* is poorly expressed.

On the other hand we found the central domain of FixL to be essential for oxygen-regulation *in vivo*. A FixL protein, encoded by pDPD147 (Fig.2), precisely deleted of the central domain, drove *nifA* and *fixK* expression similarly in *E.coli* under aerobic and microaerobic conditions (Table 1). Again overexpression was essential to detect the activity of the mutant protein.

We concluded from these results that 1. the central domain of FixL, but not the N-terminal domain, is essential for oxygen response 2. under aerobic conditions the central domain phenotypically represses the ability of FixL to activate FixJ .

2.4. ANAEROBIOSIS STIMULATES FIXJ PHOSPHORYLATION IN VITRO

Since the activity of the soluble FixL protein encoded by pDPD122 appeared to be regulated by oxygen *in vivo*, we have purified this protein to about 90 % homogeneity. We confirm previous finding by Gilles-Gonzalez et al (1991) that FixL is a hemoprotein with kinase activity.

When purified FixL and FixJ (>95% pure) were incubated together in the presence of γ 32P-ATP , about 5 to 10 fold more phosphorylated FixJ was generated under anaerobic than under aerobic conditions (Figure 3). Therefore microaerobic conditions stimulate *nifA* and *fixK* expression *in vivo* and phosphorylation of FixJ *in vitro*. This strongly suggests that phosphorylation of FixJ is indeed the mechanism that connects oxygen sensing to gene expression.

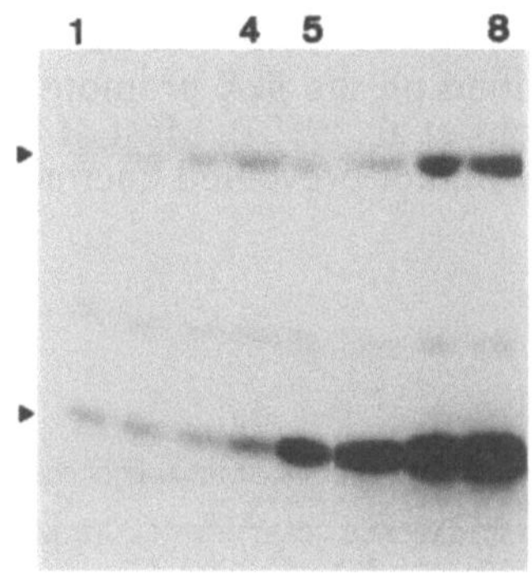

Figure 3: Kinetics of phosphorylation of FixL and FixJ from γ32P- ATP. Lanes 1 to 4: aerobic conditions, lanes 5 to 8: anaerobic conditions. Lanes 1,5: 10 min.; 2,6: 20 min.; 3,7: 40 min.; 4, 8: 80 min incubation at 28°C.

One reason for the concentration of FixJ-phosphate to increase under anaerobic conditions is a slight (2 to 3 fold) enhancement of the rate of FixL autophosphorylation under these conditions (data not shown). Similar results have been reported recently by Monson et al (1992). In addition, preliminary data point to an increased turn-over of FixJ-phosphate under aerobic conditions. This could account for the discrepancy between the stimulatory effect of anaerobiosis on FixL autophosphorylation (2 to 3 fold) and that observed on FixJ phosphorylation (5 to 10 fold). We are currently testing whether FixL is a phosphatase which dephosphorylates FixJ under aerobic conditions.

2.5. FIXJ IS A TRANSCRIPTIONAL ACTIVATOR OF NIFA AND FIXK EXPRESSION

FixJ is homologous, due to its N-terminal domain, to the regulator proteins of the two-component family, most of which are transcriptional activators (Stock et al. (1989)).

In order to conclusively establish FixJ as a transcriptional activator, we have set up an *in vitro* transcription assay using purified components. Transcripts of *nifA* and *fixK* with the correct size were obtained in the presence of *E.coli* RNA-polymerase-sigma70 holoenzyme and pure FixJ (data not shown). This demonstrates that FixJ is indeed a transcriptional activator and that the *nifA* and *fixK* promoters are direct targets for FixJ. According to *in vitro* experiments demonstrating that FixJ phosphorylation is enhanced under anaerobic conditions, it is very likely that phosphorylated FixJ is the physiologically active form of FixJ.

FixJ has been shown to consist of two functional domains (Kahn and Ditta (1991)). A N-terminal regulatory domain shared by other regulator proteins and a very short (74 aminoacids) C-terminal domain which is able by itself (i.e. in the absence of the N-terminal regulatory domain) to activate transcription of *nifA* and *fixK*. This activator domain also carries a putative helix-turn-helix DNA-binding motif. Mutations of the exposed residues of the recognition helix severely affected *nifA* activation thus suggesting that the DNA-binding motif is functional (Kahn and Ditta (1991)).

Three regions were identified on the *fixK* promoter that are essential for *in vivo* activation by *fixLJ* (Waelkens et al.(1992)). Most of the mutations in the -35, -45 and -60 regions of the *fixK* promoter prevented microaerobic expression in both *R.meliloti* and *E.coli*.

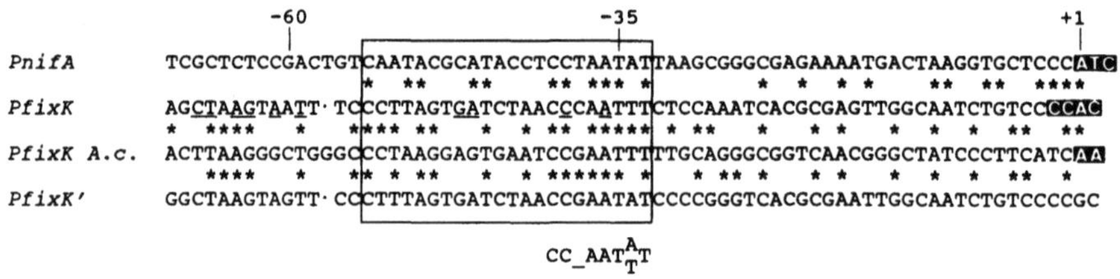

Figure 4: Alignment of FixJ-dependent promoters. PfixKAc is the *fixK* promoter from *A. caulinodans* (Kaminski and Elmerich (1991)). fixK' is the second functional copy of *fixK* found in *R.meliloti*. Nucleotides in PfixK in which mutations have a promoter-down effect are underlined.

Alignment of four FixJ-dependent promoters (Figure 4) and comparison of our data on PfixK with data obtained on PnifA (Virts et al.(1988)) suggest the following pattern of organization for FixJ-activated promoters. A consensus sequence CCNAATA/TT is conserved at position -35 and might be part of a larger motif extending up to position -54. This suggests the possibility that FixJ or its phosphorylated form may recognize the -35 region of these promoters.

An additional element, necessary for activation by FixJ, was found at position -60 on PfixK. Apparently, this element is lacking in the *nifA* promoter and may therefore account for the different behaviours of *nifA* and *fixK* promoters. Specifically, the -60 region allows activation of *fixK* by high levels of FixLJ in aerobic conditions in *E.coli*. A *fixK* promoter mutated at -60 behaves as a *nifA* promoter and always requires microaerobic conditions for activation by *fixLJ*. Interestingly the -60 region does not display homology to the -35,-54 region.

3. Discussion

3.1 FROM OXYGEN SENSING TO GENE EXPRESSION

Among the numerous two-component systems that are under study, the FixLJ system offers the interesting properties that the primary signal is clearly identified as oxygen and that the complete circuitry from oxygen sensing by FixL to transcriptional activation of *nifA* and *fixK* by FixJ can be reconstituted *in vitro* with purified components. These, in combination with *in vivo* studies and the analysis of mutants, should allow a detailed analysis of each step of the transduction pathway.

Based on available data, we propose the following model (Figure 5).

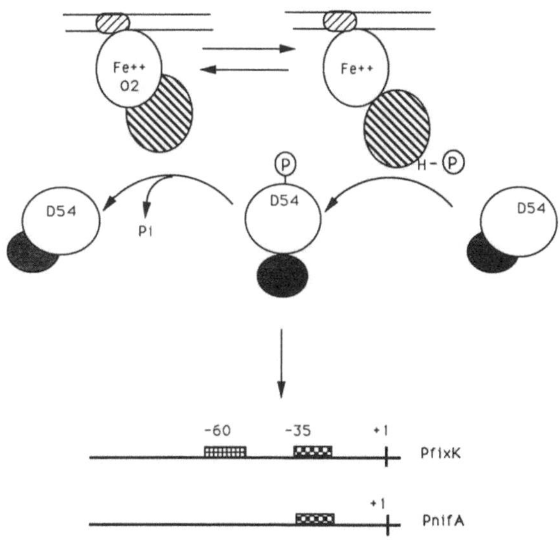

Figure 5: Model for oxygen-regulated activity of FixL and FixJ.

The limited rate of FixL autophosphorylation under aerobic as compared to anaerobic conditions is consistent with a repressive action of the oxygen-sensing central domain of FixL on the kinase activity carried by the C-terminal domain. Monson et al (1992) recently demonstrated by heme-specific staining that the heme moiety is indeed bound to the central domain of FixL and suggested that molecular oxygen binds to heme. Conceivably, the binding of oxygen to heme may induce a conformational change that leads the central domain to inhibit the kinase activity carried by the C-terminal domain. Possibly, the same conformational change may stimulate a phosphatase activity of FixL. If so, FixL would both promote FixJ phosphorylation under microaerobic conditions and FixJ dephosphorylation under aerobic conditions. Such a dual role of FixL would be consistent with *in vivo* data demonstrating that FixL both promotes microaerobic expression and prevents aerobic expression of *nifA* (de Philip et al. (1990)). According to the mechanism proposed for CheY and CheB (Lukat et al. (1992)), FixJ could phosphorylate itself at an aspartate residue (D54) using phosphorylated FixL as phospho-donor. Kahn and Ditta (1991) have shown that FixJ has a modular structure and that the N-terminal regulatory domain of FixJ inhibits the transcriptional activity carried by the C-terminal domain. They proposed that this inhibition would be relieved upon phosphorylation of the N-terminal domain of FixJ. The C-terminal of FixJ would then interact with *nifA* and *fixK* promoters and turn on transcription by a mechanism that remains to be elucidated.

3.2 OXYGEN IS A MAJOR REGULATOR OF SYMBIOTIC NITROGEN FIXATION

In *R.meliloti* as in other obligate aerobes, oxygen may have opposite effects on nitrogen fixation. On one hand it is necessary to generate, by oxidative phosphorylation, the energy and reducing power required for nitrogen fixation. On the other hand, at too high concentration, oxygen inactivates some key proteins such as NifA or the nitrogenase complex. We and others have demonstrated that oxygen is a key regulator of nitrogen fixation gene expression *ex planta* and, probably, *in planta* as well. Presumably the plant and the microsymbiont both participate in lowering free oxygen concentration in the nodule (Witty and Minchin (1990)).

In addition to *R.meliloti*, microaerobic induction of nitrogen fixation gene expression has been demonstrated in *R. leguminosarum*, *Bradyrhizobium japonicum* and *Azorhizobium caulinodans*. *fixLJ* genes have been described so far in *B. japonicum* (Anthamatten and Hennecke (1991)) and *A. caulinodans* (Kaminski and Elmerich (1991)). Although significant differences in the regulatory pathways have been described, a microaerobic control of nitrogen fixation gene expression, mediated at least in part by *fixLJ*, might be a conserved feature in Rhizobia.

While oxygen is likely to be a key regulator of nitrogen fixation gene expression in the *R.meliloti*-alfalfa symbiosis, other regulatory signals may exist. Conceivably, signals may repress nitrogen fixation gene expression under microaerobic but otherwise inappropriate conditions.

Acknowledgements

We thank M. David and T. Finan for critical reading of the manuscript. P.P. and J.M.R. were supported by grants from the French Ministère de la Recherche et de la Technologie. F.W. was the recipient of a fellowship from the D. Collen Research Foundation (Leuven, Belgium). This work was funded in part by a grant from the EEC in the framework of the BRIDGE Programme.

References

Anthamatten D and Hennecke H (1991) "The regulatory status of the *fixL*- and *fixJ*-like genes in *Bradyrhizobium japonicum* may be different from that in *Rhizobium meliloti*" Mol. Gen. Genet. 225, 38-48.

Batut J, Daveran-Mingot M-L, David M, Jacobs J, Garnerone A-M and Kahn D (1989) "*fixK*, a gene homologuous with *fnr* and *crp* from *Escherichia coli*, regulates nitrogen fixation genes both positively and negatively in *Rhizobium meliloti*" EMBO J. 8, 1279-1286.

David M, Daveran M-L, Batut J, Dedieu A, Domergue O, Ghai J, Hertig C, Boistard J and Kahn D (1988) "Cascade regulation of *nif* gene expression in *Rhizobium meliloti*" Cell 54, 671-683.

de Philip P, Batut J and Boistard P (1990) "*Rhizobium meliloti* FixL is an oxygen sensor and regulates *R. meliloti nifA* and *fixK* genes differently in *Escherichia coli*" J. Bacteriol. 172, 4255-4262.

de Philip P, Soupène E, Batut, J and Boistard, P (1992) "Modular structure of the FixL protein of Rhizobium meliloti" Mol. Gen. Genet.,in the press.

Ditta G, Virts E, Palomares A and Kim C H (1987) "The *nifA* gene of *Rhizobium meliloti* is oxygen regulated" J Bacteriol. 169, 3217-3223.

Fisher R and Long S R (1992) "Rhizobium-plant signal exchange" Nature 357, 655-660.

Gilles-Gonzalez M A, Ditta G S and Helinski D R (1991) "A haemoprotein with kinase activity encoded by the oxygen sensor of *Rhizobium meliloti*" Nature 350, 170-172.

Kahn D and Ditta G (1991) "Modular structure of FixJ: homology of the transcriptional activator domain with the -35 binding domain of sigma factors" Mol. Microbiol. 5, 987-997.

Kaminski P A and Elmerich C (1991) "Involvement of *fixLJ* in the regulation of nitrogen fixation in *Azorhizobium* caulinaudans" Mol. Microbiol. 5, 665-673.

Lukat G S, McCleary W, Stock A M and Stock J B (1992) "Phosphorylation of bacterial response regulator proteins by low molecular weight phosphodonors" Proc. Natl. Acad. Sci. USA 89, 718-722.

Monson E K, Weinstein M, Ditta G S and Helinski D R (1992) "The FixL protein of *Rhizobium meliloti* can be separated into a heme-binding oxygen-sensing domain and a functional C-terminal kinase domain" Proc. Natl. Acad. Sci. USA 89, 4280-4284.

Stock J B, Ninfa A J and Stock A M (1989) "Protein phosphorylation and regulation of adaptative responses in bacteria" Microbiol. Rev. 53, 450-490.

Virts E L, Stanfield S W, Helinski D R and Ditta G S (1988) "Common regulatory elements control symbiotic and microaerobic induction of *nifA* in *Rhizobium meliloti*" Proc. Natl. Acad. Sci. USA 85, 3062-3065.

Waelkens F, Foglia A, Morel J B, Fourment J, Batut J and Boistard P (1992) "Molecular genetic analysis of the *Rhizobium meliloti fixK* promoter: identification of sequences involved in positive and negative regulation" Mol. Microbiol. 6, 1447-1456.

Witty J F and Minchin F R (1990) "Oxygen diffusion in the legume root nodule", in P M Gresshoff, L E Roth, G Stacey and W E Newton (eds), Nitrogen fixation: achievements and objectives, Chapman and Hall, New-York, London, pp. 285-292.

OXYGEN SENSING AND PROTEIN PHOSPHORYLATION BY THE TWO-COMPONENT REGULATORY FIXLJ SYSTEM FROM *RHIZOBIUM MELILOTI*.

A. F. LOIS, M. WEINSTEIN, E. K. MONSON, G. DITTA AND D. R. HELINSKI
Dept. of Biology, University of California, San Diego
La Jolla, CA 92093
USA

ABSTRACT. We have been studying the signal transduction mechanism mediated by the *Rhizobium meliloti* FixL and FixJ proteins in response to oxygen. FixL is a hemoprotein kinase that can sense oxygen tension and can respond to this signal by modifying its kinase activity. *In vitro* studies with a soluble truncated version of FixL (FixL*) demonstrated that low oxygen tension specifically augments the autophosphorylating and not the transfer activity to FixJ. FixL* also contains a phosphatase activity that is repressed under anaerobic conditions. Regulation of the phosphatase activity by oxygen is dependent on the phosphorylation state of the FixL* protein. These results demonstrate that oxygen regulates two opposing activities in the FixL protein, kinase and phosphatase, in a reciprocal manner and suggest a specific mechanism by which the oxygen tension within the nodule regulates transcription of the nitrogen fixation genes.

INTRODUCTION

Rhizobium bacteria infect alfalfa plants and become nitrogen fixing symbionts within a specialized structure in the root called the nodule. A number of bacterial genes, termed *nod*, *nif* and *fix*, are required for a successful nodulation and subsequent nitrogen fixation (for review see Long (1989)). The *nif* and *fix* genes are expressed during the later stages of nodulation and are involved directly in the process of nitrogen fixation. The expression of the plant gene leghemoglobin during this stage of nodulation produces a microaerobic environment that is required to prevent inactivation of the oxygen sensitive nitrogenase enzyme and indirectly signals the onset of further maturation of the bacterium (Appleby (1984)). The regulation of the nitrogen fixation (*nif, fix*) genes in *Rhizobium meliloti* is ultimately controlled by a pair of proteins, FixL and FixJ in response to the oxygen tension within the nodule (Ditta *et al.* (1987), Virts *et al.* (1988), David *et al.* (1988)).

FixL is a hemoprotein with kinase activity capable of transfering a phosphate group to the FixJ protein (Gilles-Gonzalez *et al.* 1991, Lois, A.F., Weinstein, M., Ditta, G. and Helinski, D.R., manuscript submitted for publication). Both, FixL and FixJ, belong to a family of bacterial two-component signal transducing systems that sense a number of environmental signals through phosphorylation-dephosphorylation reactions and induce subsequent modification of gene expression (Stock *et al.* (1989)). Bacterial systems that belong to this protein family are usually composed of two proteins: a sensor that has kinase/phosphatase activity and a regulator protein that is the substrate for the kinase and is usually a transcriptional activator. Phosphorylation of the regulator protein by the sensor produces an active transcription factor (Ninfa (1991)). The FixL protein from *R. meliloti* has homology in its C-terminal end with the sensor class of these proteins and FixJ, in its N-terminal region to the regulator class (David *et al.* (1988).

Recent studies in our laboratory have demonstrated that a soluble version of FixL (FixL*) possess oxygen-regulated kinase activity. Upon incubation of FixL* with labeled ATP, Monson

E. W. Nester and D. P. S. Verma (eds.),
Advances in Molecular Genetics of Plant-Microbe Interactions, 193–197.

et al. (1992) reported a three fold enhancement of FixL autophosphorylation under anaerobic conditons in comparison to aerobic conditions. In the same work, it was also shown that the accumulation of FixJ phosphate also increases under anaerobic conditions when both FixL and FixJ are incubated with ATP. These data strongly suggested that FixL senses oxygen tension directly and that the phosphorylation state of FixJ may be important in regulating the expression of nitrogen fixation genes during anaerobiosis within the nodule structure.

More recent work in our laboratory has shown that FixL also contains phospho-FixJ dephosphorylating activity (Lois, A.F., Weinstein, M., Ditta, G. and Helinski, D.R., manuscript submitted for publication). The phosphatase activity appears to respond to the oxygen concentration of the reaction in an opposite manner as the kinase activity. Other two component sensor kinases (EnvZ and NtrB) have also been shown to contain phosphatase activity but little is known about the role of this activity during signal transduction (for review see Ninfa (1991)).

In this study we attempted to understand in greater detail the signal transduction mechanism mediated by the FixLJ system in response to oxygen. We were specifically interested in determining whether the phosphatase activity present in FixL played a role during oxygen sensing and signal transduction.

MATERIALS AND METHODS

Autophosphorylation. Autophosphorylation assays were carried out with 3 μg of FixL* in 15 ml volume of reaction buffer [50mM Tris pH 8.0, 50mM KCl, 0.8mM $MgCl_2$, 0.8mM ATP, 1.2mM $CaCl_2$, 1mM dithiothreitol, 5% glycerol and 10mCi [γ-^{32}P]-ATP (6,000 Ci/mmole)] for the indicated times.

Phosphatase Assay. An affinity column to bind FixL was made by coupling FixL antiserum to cyanogen bromide activated Sepharose CL-4B (Pharmacia) according to manufacturer's specifications. To isolate [^{32}P]-labeled FixJ-phosphate, FixL*362 (Lois, A.F., Weinstein, M., Ditta, G. and Helinski, D.R., manuscript submited for publication) (4μg) was incubated with FixJ (6μg) in the autophosphorylation reaction buffer at 25°C under anaerobic conditions for 1 hr. The reaction mixture was then passed 3 consecutive times through a 150ml FixL affinity column in FixL buffer. The phosphatase assay was carried out by incubating [^{32}P]-labeled FixJ-phosphate (0.6mM) with FixL* (0.3mM) in the autophosphorylation reaction buffer either without ATP, or with unlabeled FixL*-phosphate (0.3mM) in the reaction buffer with ATP.

Anaerobic conditions were established by placing tubes containing the reaction mixes in a dessicator jar under house vacuum for 4 minutes. The jar was filled with nitrogen gas and the vaccuum extraction was repeated twice. All subsequent manipulations of the deoxygenated samples were done under a stream of nitrogen.

Gel Electrophoresis. SDS polyacrylamide gel electrophoresis was run according to Laemmli (1970) using 4% stacking and 8% separating gels. Gels containing labeled proteins were dried and exposed to x-ray films. For quantitation, autoradiograms were scanned using a laser densitometer and the data were processed using software from the Ambis Corporation (San Diego, CA).

RESULTS AND DISCUSSION

FixL from *R. meliloti* is a bifunctional protein that catalyzes both the synthesis and degradation of FixJ-phosphate, an important regulatory protein that is involved in the regulation of the nitrogen fixation genes. The reaction catalyzed by the FixL protein appears to follow a ping-pong mechanism of phosphorylation whereby the kinase is first autophosphorylated and can then transfer the phosphate moiety to the second component, the regulator FixJ (Lois, A.F., Weinstein, M., Ditta, G. and Helinski, D.R., manuscript submitted for publication). These results are in agreement with findings obtained from studying the kinase reaction of other two component systems (Stock *et al.* (1989), Ninfa (1991)).

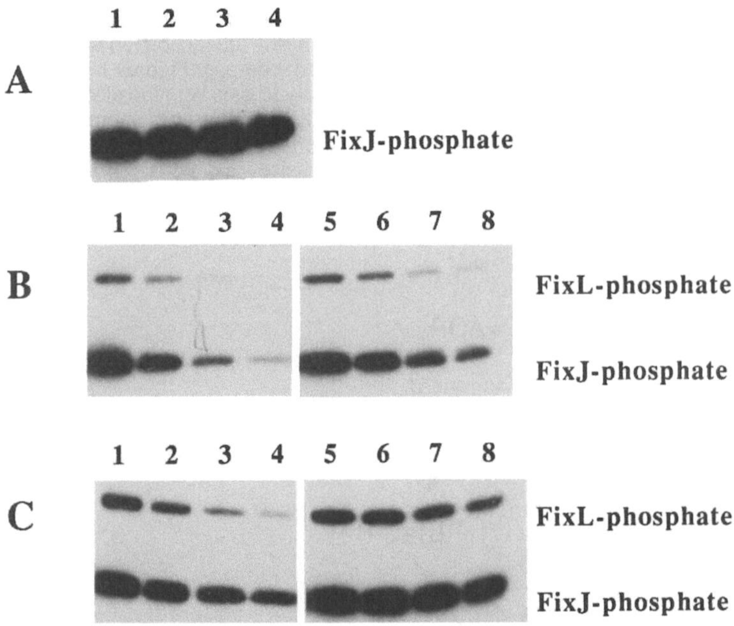

Figure 1. Phosphatase reaction catalyzed by FixL. [^{32}P]-FixJ-phosphate was incubated in reaction buffer (A) with FixL* (B) or with FixL*-phosphate (C) under aerobic (lanes 1-4) and anaerobic (lanes 5-8) conditions. Lanes 1, 5 (5 min), lanes 2, 6 (10 min), lanes 3, 7 (20 min), lanes 4, 8 (30 min). At the indicated times reactions were terminated with SDS. The labeled proteins were separated by gel electrophoresis and detected by autoradiography.

Our studies of the catalytic activities of a soluble version of FixL (FixL*) revealed that low oxygen tension specifically augments the autophosphorylating activity (Monson *et al.*, (1992), Lois, A.F., Weinstein, M., Ditta, G. and Helinski, D.R., manuscript submitted for publication). These studies provided direct evidence for a mechanism whereby the oxygen concentration within the nodule can modulate gene expression in *R. meliloti*. The phosphorylated form of FixJ is thought to be the active form of the protein capable of inducing transcription from the target promoters, *nif*A and *fix*K.

More recent studies in our laboratory indicate that, in addition to the kinase acitivity, FixL* also contains a FixJ-phosphate dephosphorylating activity (Lois, A.F., Weinstein, M., Ditta, G. and Helinski, D.R., manuscript submitted for publication). When labeled FixJ-phosphate is incubated with excess FixL*, the phosphate moiety is rapidly removed (Figure 1B). In this assay, the FixL protein becomes phosphorylated as shown by the presence of the upper band in the autoradiogram (see discussion below). Oxygen depletion has a small but reproducible effect in the rate of phosphate removal from FixJ-phosphate (compare in Figure 1B, lanes 1-4 with lanes 5-8). A low oxygen tension appears to have the effect of reducing the phosphatase activity.

FixL*-phosphate also has a phosphatase activity under aerobic conditons (Figure 1C). Under anaerobic conditions, the phosphatase activity of FixL*-phosphate is severely repressed. Thus, oxygen causes reciprocal changes in the enzymatic activities of FixL*. Under low oxygen tension, the kinase activity is activated and the phosphatase activity is substantially reduced. These data suggest that the net physiological effect of a reduction in the oxygen tension appears to be an increase in the level of FixJ-phosphate intracellularly.

As shown in Figure 1, FixL* becomes rapidly labeled when incubated with [^{32}P]-FixJ-phosphate. When FixL* is preincubated with ATP prior to and during the incubation with FixJ-phosphate, equivalent or even greater labeling of FixL is observed (compare Figures 1 B and 1C). Since the presense of ATP results in most of the kinase catalytic sites being loaded with phosphate (Lois, A.F., Weinstein, M., Ditta, G. and Helinski, D.R., manuscript submitted for publication), the inability of ATP to inhibit labeled FixL formation from [^{32}P]-FixJ-phosphate suggest the presence of two phosphoacceptor sites, one for the kinase , and a different site for the phosphatase activity. It is interesting to note that the rate of labeling of FixL from FixJ-phosphate substantially different than the rate of labeling from ATP.

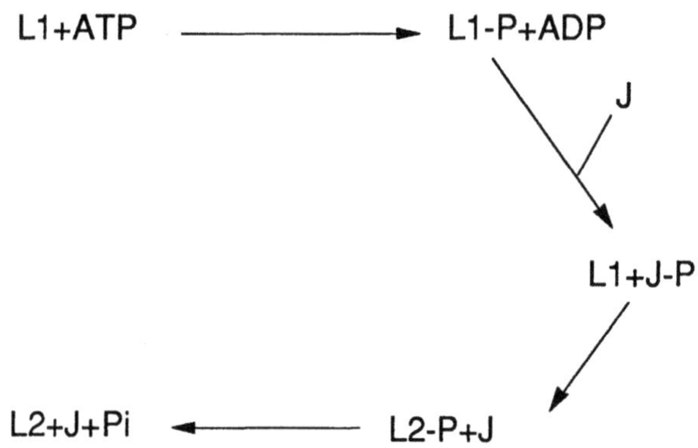

Figure 2. Proposed scheme for the reaction mechanism catalyzed by the FixL protein. L1 and L2 represent different catalytic sites in FixL. J denotes the FixJ protein.

The results of the investigation of the FixL kinase/phosphatase reaction are consistent with the basic catalytic mechanism shown in Figure 2. This scheme is put forward as a working hypothesis which requires much additional study. In the presence of ATP, FixL becomes phosphorylated at the kinase active site. The phosphoryl group is then rapidly transfered to FixJ. The phosphate from FixJ-phosphate can then be transfered to a second site on FixL where it is subsequently hydrolyzed to form inorganic phosphate. Under conditions where the phosphatase activity is inhibited (anaerobic), the phosphoryl group is not hydrolyzed and becomes stable. In the overall scheme oxygen would affect both the kinase and phosphatase activities driving the reaction in the direction of either FixJ-phosphate formation or the non-phosphorylated form of FixJ.

REFERENCES

Appleby, C. A. (1984) Leghemoglobin and *Rhizobium* respiration. Annu. Rev. Plant Physiol. 35:443-478.

David, M., M.-L. Daveran, J. Batut, A. Dedieu, O. Domergue, J. Ghai, C. Hertig, P. Boistard, and D. Kahn. (1988) Cascade regulation of nif gene expression in *Rhizobium meliloti.* Cell 54:671-683.

Ditta, G., E. Virts, A. Palomares, and C.-H. Kim. (1987) The *nif* A gene of *Rhizobium meliloti* is oxygen regulated. J. Bacteriol. 169:3217-3223.

Gilles-Gonzalez, M.A., G.S. Ditta, and D.R. Helinski. (1991) A haemoprotein with kinase activity encoded by the oxygen sensor of *Rhizobium meliloti*. Nature (London) 350:170-172.

Laemmli, U. K. (1970) Cleavage of Sturctural Proteins during the Assembly of the head of bacteriophage T4. Nature (London) 227:680-685.

Long, S.R. (1989) *Rhizobium*-legume nodulation: Life together in the underground. Cell 56:203-214.

Monson, E.K., Weinstein, M., Ditta, G.S. and D.R. Helinski. (1992) The FixL protein of *Rhizobium meliloti* can be separated into a heme-binding oxygen-sensing domain and a functional C-terminal kinase domain. Proc. Natl. Acad. Sci. USA 89:4280-4284.

Ninfa, A.J. (1991) Protein phosphorylation and the regulation of cellular processes by the homologous two-component regulatory systems of bacteria in J.K. Setlow (ed.), Genetic Engineering, Vol. 13, Plenum Press, New York, pp. 39-72.

Stock, J.B., A.J. Ninfa, and A. M. Stock. (1989) Protein phosphorylation and regulation of adaptive responses in bacteria. Microbiol. Rev. 53:450-490.

Virts, E.L., W.W. Stanfield, D.R. Helinski, and G.S. Ditta. (1988) Common regulatory elements control symbiotic and microaerobic induction of *nif*A in *Rhizobium meliloti.* Proc. Natl. Acad. Sci. USA 85:3062-3065.

ACKNOWLEDGEMENTS

This work was supported by an NIH grant to G. Ditta. A. Lois was the recipient of a supplement to the above NIH grant.

GENETIC AND PHYSIOLOGIC REQUIREMENTS FOR OPTIMAL BACTEROID FUNCTION IN THE *BRADYRHIZOBIUM JAPONICUM* SOYBEAN SYMBIOSIS

H. HENNECKE, D. ANTHAMATTEN, M. BABST, M. BOTT, H.M. FISCHER, T. KASPAR, I. KULLIK, H. LOFERER, O. PREISIG, D. RITZ AND M. WEIDENHAUPT
Mikrobiologisches Institut, Eidgenössische Technische Hochschule, ETH-Zentrum, Schmelzbergstr. 7, CH-8092 Zürich, Switzerland

ABSTRACT. It is crucial for endosymbiotic legume root-nodule bacteria to generate metabolic energy by respiration at a very low partial pressure of oxygen. This physiologic condition also permits derepression of genes for nitrogen fixation and other anaerobic processes. Here we report on our progress in the characterization of genes that are essential for respiration in *Bradyrhizobium japonicum*, the soybean root-nodule bacterium. Furthermore, we present the state of the art of the complex genetic regulatory network that controls gene expression in response to the presence or absence of oxygen.

1. Introduction

The conversion of free-living *B. japoncium* cells to endosymbiotic bacteroids is accompanied by a dramatic physiologic switch in the bacterium, i.e. from an aerobic to a microaerobic mode of life. It is not known precisely at which stage during symbiotic development this switch occurs. It may happen already as early as during bacterial propagation in the infection thread, or later, when the bacteria are released from the infection thread into the nodule cell cytoplasm. Alternatively, the bacteria may transiently live through a gradient of increasing oxygen deprivation as root-nodule development progresses. It is now well established that the oxygen concentration within the infected nodule tissue (3 to 30 nM) is by a factor of 10^4 to 10^5 lower as compared with standard aerobic conditions (~ 250 µM) (Layzell et al. 1990; Witty and Minchin 1990). The usual way of bacteria to cope with such extremes is by induction of different respiratory chains terminated by oxidases with different affinities for O_2. In a comparative spectroscopic analysis, Appleby (1969a, b) first showed that the hemoprotein composition in aerobically cultured *B. japonicum* differed from that in root-nodule bacteroids. In particular, cytochromes aa_3 and *o*, whose oxidase function in aerobic cells was demonstrated by photochemical action spectra, were absent in bacteroids. Keister and

E. W. Nester and D. P. S. Verma (eds.),
Advances in Molecular Genetics of Plant-Microbe Interactions, 199–207.

Marsh (1990) showed that several *B. japonicum* strains retained cytochromes *o* and aa_3 even during symbiosis, but it is unknown whether these oxidases are functional in the root-nodule environment. Thus, a hitherto unidentified high-affinity terminal oxidase is probably capable of reducing the O_2 that is provided by oxygenated leghemoglobin (Appleby 1984). In our laboratory we are searching for *B. japonicum* genes that code for components of the branched respiratory chain and for proteins involved in their biosynthesis. The analysis of mutants with lesions in those genes should help answer the question which of them are essential for a functional symbiosis.

Another consequence of the extremely low oxygen concentration in endosymbiosis is the induction of bacterial genes that are normally not expressed in aerobiosis. The nitrogen fixation genes are the most prominent among them. In the past years, we have described three regulatory elements (the NifA, FixK and FixLJ systems) that lead to activation of target genes only at low, but not at high, O_2 concentrations. In this article, we describe how these regulatory systems may be connected in a complex network, and report on some new genes that are controlled by them.

2. Genes for Respiration

More than 20 *B. japonicum* genes have been identified, which are involved in respiration. These are compiled in Table 1. Aerobically grown *B. japonicum* possesses a respiratory chain which resembles that occurring in mitochondria (Thöny-Meyer et al. 1989; Bott et al. 1990, 1991). In this chain, the electrons are transferred from ubiquinol via the cytochrome bc_1 complex, a membrane-bound cytochrome *c* (CycM) and an aa_3-type cytochrome *c* oxidase to O_2. Two lines of genetic evidence suggest that *B. japonicum* possesses two additional respiratory chains. (i) Genes for a second heme/copper-type cytochrome *c* oxidase were found in an operon, *coxMNOP*, in which *coxN* codes for subunit I and is thus very similar, but not identical, to *coxA* (Table 1). It is not yet known under which growth conditions this new oxidase functions. It was speculated that it might be equivalent to cytochrome *o* (Bott et al. 1992), but biochemical support for this assertion is still missing. A *B. japonicum coxN* mutant exhibited no obvious defects in free-living, aerobic growth or in root-nodule symbiosis with soybean. This shows that the CoxMNOP-oxidase is not essential for respiration in the N_2 fixing bacteroid. (ii) Mutants that are defective in the genes (*fbcFH*) for the cytochrome bc_1 complex have a Fix^- phenotype. Based on this finding it was suggested that the bc_1 complex is a branchpoint at which electrons can be transferred to a novel, symbiosis-specific high-affinity terminal oxidase operating in microaerobiosis (Thöny-Meyer et al. 1989; Bott et al. 1990). This implies that, like all other bc_1-dependent respiratory chains, this new respiratory chain is composed of a new cytochrome *c* plus a new cytochrome *c* oxidase. Further support for this assumption is given as follows.

TABLE 1. *Bradyrhizobium japonicum* genes involved in respiration

Genes	Known or proposed function of gene products	Symbiotic phenotype of mutants	Reference
fbcFH	Ubiquinol-cytochrome *c* oxidoreductase (cytochrome bc_1 complex)	Fix^-	Thöny-Meyer et al. (1989)
cycM	Membrane-bound 20 kDa cytochrome *c*	Fix^+	Bott et al. (1991)
coxA	Subunit I of aa_3-type cytochrome *c* oxidase	Fix^+	Bott et al. (1990); Gabel and Maier (1990)
coxMNOP	Subunits II, I, IIIa and IIIb of alternative heme/copper cytochrome *c* oxidase	Fix^+	Bott et al. (1992)
cycA	Cytochrome c_{550}	Fix^+	M. Bott, unpublished
cycB	Cytochrome c_{552}	Fix^+	Rossbach et al. (1991)
cycC	Cytochrome c_{555}	Fix^+	Tully et al. (1991)
cycV,W,X	Components of an ATP-dependent, membrane-bound transport system (possibly for heme), necessary for biogenesis of all *c*-type cytochromes	Fix^-	Ramseier et al. (1991)
cycH	Protein possibly involved in cytochrome *c* heme lyase activity, necessary for formation of *c*-type cytochromes except cytochrome c_1	Fix^-	D. Ritz, unpublished
tlpA	Membrane-bound thioredoxin-like protein, necessary for biogenesis of cytochrome aa_3	Fix^-	H. Loferer, unpublished
fixNOQP	Components of a putative bacteroid oxidase complex containing a diheme cytochrome *c* (FixP)	Fix^-	O. Preisig and D. Anthamatten, unpublished

The *cycV*, *cycW* and *cycX* genes were discovered which code for an ATP-dependent transport system (heme transport?). Mutations in these genes led to the complete absence of the entire set of membrane-bound and soluble *c*-type cytochromes and to a Fix⁻ phenotype (Ramseier et al. 1991). While the Fix⁻ phenotype corroborates the requirement of a *c*-type cytochrome in a symbiotic respiratory chain, it can already be explained simply by the absence of cytochrome c_1. More compelling in this respect is the Fix⁻ phenotype of the *cycH* mutant (Table 1) in which cytochrome c_1 is still synthesized whereas the formation of most, if not all, of the other cellular *c*-type cytochromes is severely affected. This suggest that, in addition to cytochrome c_1, yet another cytochrome *c* is essential for symbiosis. Finally, the discovery of the *tlpA* gene (Table 1) is of interest in this context. The *tlpA* gene codes for a membrane-bound thioredoxin-like protein that is required for the formation of cytochrome aa_3, a heme/copper-containing cytochrome *c* oxidase. While cytochrome aa_3 itself is not essential for symbiosis (see *coxA* mutant in Table 1), the Fix⁻ phenotype of a *tlpA* mutant may be interpreted to mean that the TlpA protein is also required for the formation of a symbiosis-specific, heme/copper-type cytochrome *c* oxidase.

In the search for a cytochrome *c* that forms part of a bc_1-dependent, symbiosis-specific respiratory chain, we first focussed on the analysis of three soluble *c*-type cytochromes that were purified from soybean root-nodule bateroids, namely cytochromes c_{550}, c_{552} and c_{555} (Appleby et al. 1991). The genes for cytochromes c_{550} and c_{552}, *cycA* and *cycB* (Table 1), were cloned and sequenced in our laboratory, whereas the cytochrome c_{555} gene (*cycC*) was found by others (Tully et al. 1991). Mutations in either *cycA*, *cycB* or *cycC* led to a Fix⁺ phenotype (Table 1) and even a *cycB/cycC* double mutant or a *cycA/cycB/cycC* triple mutant was not affected in symbiotic N_2 fixation (M. Bott and H. Loferer, unpublished results). This precludes a role of cytochromes c_{550}, c_{552} and c_{555} in a bacteroid-specific, microaerobic respiratory chain.

Recently we identified a symbiotically essential gene region consisting of four open reading frames that are located upstream of, and in divergent orientation to, the regulatory *fixLJ* genes. Insertion mutations in these genes led to a Fix⁻ phenotype, and the respective mutants characteristically lacked a 32 kDa membrane-bound cytochrome *c* which was induced in the *B. japonicum* wild type only when cells had been grown microaerobically or anaerobically under conditions of nitrate respiration. Based on interspecies hybridization, it is likely that the four ORFs correspond to the *Rhizobium meliloti fixNOQP* gene cluster (mentioned by Boistard et al. 1991) whose sequence is not yet published. Our sequence analysis of the fourth open reading frame (*fixP*-like) of *B. japonicum* predicts it to code for a diheme cytochrome *c* equivalent to the aforementioned 32 kDa protein. In conclusion, this 32 kDa cytochrome *c* is a candidate to function in the electron transfer from the bc_1 complex to the bacteroid oxidase. Whether or not the *fixN*, *fixO* and *fixQ* genes code for other constituent proteins of the bacteroid-specific respiratory chain is currently under investigation.

3. Genetic Regulatory Network

With the recent discovery of the *fixLJ*-like genes and one of two postulated *fixK*-like genes (Anthamatten and Hennecke 1991; Anthamatten et al. 1992), the model of the genetic regulatory pathways for *nif* and *fix* genes in *B. japonicum* (Fischer et al. 1991) could be extended substantially. An updated version is shown in Figure 1. The model now accommodates two, largely independent regulatory cascades that respond to the cellular oxygen or redox status: the FixLJ/FixK cascade and the NifA cascade (left and right, respectively, in Figure 1). A connection between these two cascades exists by the fact that $rpoN_1$, one of two σ^{54} genes, is regulated by oxygen in a FixLJ-dependent fashion (Kullik et al. 1991). Although this connection may be physiologically useful, it is certainly not essential for symbiotic N_2 fixation because a mutant with an $rpoN_1$ gene disruption still has a Fix^+ phenotype owing to the existence of $rpoN_2$.

While the NifA protein controls the nitrogenase genes and other accessory nitrogen fixation genes, the FixLJ/FixK cascade is more likely concerned with microaerobic or anaerobic respiratory processes of which some may also be necessary for symbiotic nitrogen fixation, though only indirectly. For example, there is preliminary evidence to suggest that the *fixNOQP* genes, just as in *R. meliloti* (Boistard et al. 1991), could be among the controlled target genes (O. Preisig und D. Anthamatten, unpublished data), which would explain the Fix^- phenotype of *fixL* and *fixJ* mutants. Other target genes are probably involved in denitrification, because *fixL* and *fixJ* mutants (in contrast to the wild type) are unable to grow anaerobically with nitrate as the terminal electron acceptor.

The model shown in Figure 1 assumes that *B. japonicum* has two *fixK*-like homologs, both of which are FixLJ-regulated. However, only one of the two genes has been characterised so far (Anthamatten et al. 1992). There were essentially two observations which led us to postulate the existence of a second *fixK* homolog. (i) Although the identified *fixK* gene was shown to be regulated by FixLJ, a *fixK* mutant did *not* have the phenotypic characteristics (Fix^-; no denitrification) of *fixL* and *fixJ* mutants. This result is intelligible, if a second *fixK* gene can functionally substitute for the mutated one. (ii) When a plasmid with a constitutively expressed (i.e. unregulated) *fixK* gene was introduced into a *fixJ* mutant, anaerobic growth with nitrate was restored. This result is explicable, if the *B. japonicum* wild type has two FixLJ-regulated *fixK* genes; none of the two is expressed in a *fixJ* mutant, whereas the deregulated expression of one of the two *fixK* genes suffices to compensate for the *fixJ* defect. We thus hope to be able to find a second *fixK*-like gene in the near future.

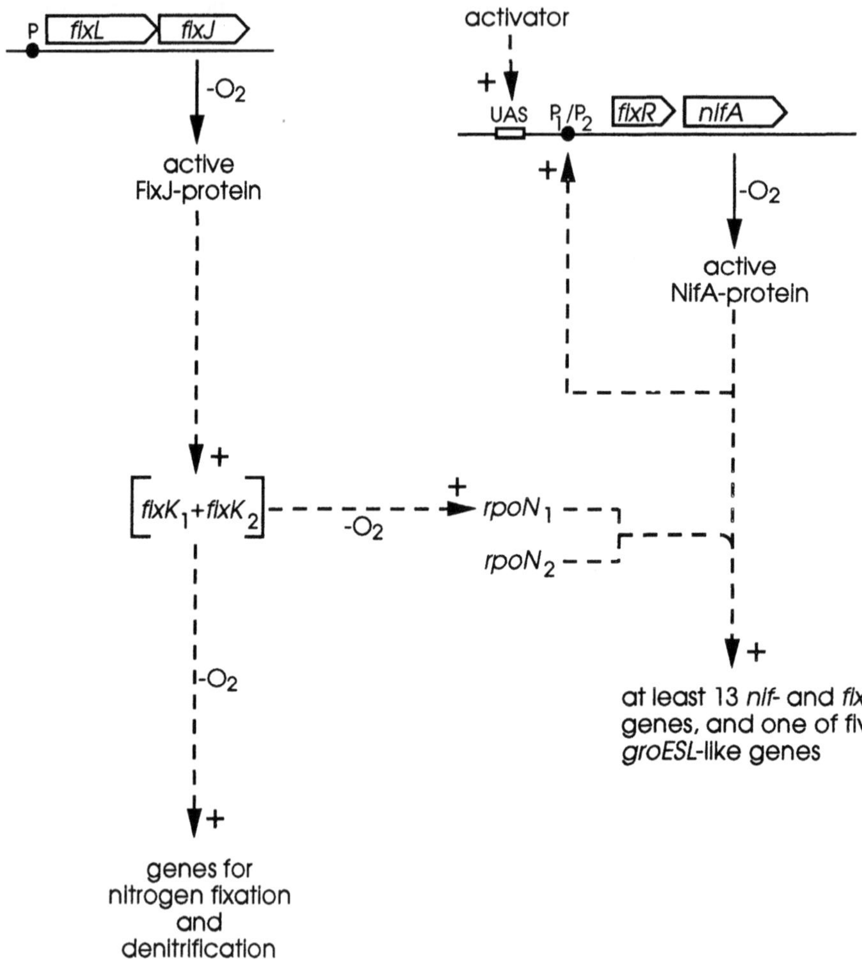

Figure 1. Model for the control of genes involved in symbiotic nitrogen fixation and other anaerobic processes in *Bradyrhizobium japonicum*. Dashed lines denote regulatory routes. The symbol (+) means positive control (gene activation).

At present we cannot rule out the possibility that there are FixLJ-regulated genes which are not activated via FixK (not considered in Figure 1). In the case of *rpoN*$_1$, however, there is a typical 'anaerobox' located upstream of the promoter region, which could serve as a binding site for FixK. In fact, deletion analyses have shown that this sequence is necessary for maximal *rpoN*$_1$ gene expression (I. Kullik, unpublished data), which implies that the FixLJ-dependent *rpoN*$_1$ gene regulation goes via FixK.

How the NifA cascade works has been described previously (Fischer et al. 1991; Morett et al. 1991; and further references therein). The novelty reported here concerns the identification of additional, NifA-activated target genes. Using a promoter-probe vector system, a new *B. japonicum* gene region was found which was transcribed in a NifA-dependent manner. The transcription start site was mapped and shown to be preceded by a -24/-12-type promoter (recognized by the σ^{54}-RNA polymerase) and a NifA binding site. Yet, the gene transcribed from this hitherto unidentified NifA-dependent promoter (*ndp*) is not essential for symbiotic nitrogen fixation because a deletion of the *ndp* region does not result in a Fix$^-$ phenotype (M. Weidenhaupt, unpublished data).

In connection with the NifA regulatory circuit, we recently explored another attractive new area of research. Two *B. japonicum* proteins with apparent molecular weights of 12000 and 58000, whose synthesis is controlled by NifA, were identified by N-terminal sequence analysis as homologs of the bacterial chaperonins GroES and GroEL, respectively. In the course of searching for their corresponding structural genes, the surprising discovery was made that *B. japonicum* has five *groESL*-like gene regions. These five regions were cloned and partially sequenced (H.M. Fischer, M. Babst and T. Kaspar, unpublished results) which revealed that all of them were very similar, but not identical. A comparative mutational and transcriptional analysis of all five gene regions is now in progress. It will be of interest to see whether or not the specific derepression of a particular *groESL*-like homolog is another prerequisite for optimal bacteroid function.

Acknowledgements

The research reported in this article was supported by grants from the Swiss National Foundation for Scientific Research, the Swiss Federal Institute of Technology and the Human Frontier Science Program. We acknowledge G. Acuña and J. Sanjuan for help in some preliminary aspects of this work and P. Brouwer for technical assistance. We are grateful to D. Keister for sending a mutated *cycC* clone and D. Kahn for providing the predicted amino acid sequence of the *R. meliloti* FixP protein. H. Paul is thanked for typing the manuscript.

References

Anthamatten, D and Hennecke, H. (1991) The regulatory status of the *fixL*- and *fixJ*-like genes in *Bradyrhizobium japonicum* may be different from that in *Rhizobium meliloti*. Mol. Gen. Genet. 225, 38-48.

Anthamatten, D., Scherb, B. and Hennecke, H. (1992) Characterization of a *fixLJ*-regulated *Bradyrhizobium japonicum* genes sharing similarity with the *Escherichia coli fnr* and *Rhizobium meliloti fixK* genes. J. Bacteriol. 174, 2111-2120.

Appleby, C.A. (1969a) Electron transport systems of *Rhizobium japonicum.* 1. Hemoprotein P-450, other CO-reactive pigments, cytochromes and oxidases in bacteroids from N_2 fixing root nodules. Biochim. Biophys. Acta 172, 71-87.

Appleby, C.A. (1969b) Electron transport systems of *Rhizobium japonicum.* 2. *Rhizobium* hemoglobin, cytochromes and oxidases in free-living (cultured) cells. Biochim. Biophys. Acta 172, 88-105.

Appleby, C.A. (1984) Leghemoglobin and *Rhizobium* respiration. Annu. Rev. Plant Physiol. 35, 443-478.

Appleby, C.A., James, P. and Hennecke, H. (1991) Characterization of three soluble *c*-type cytochromes isolated from soybean root nodule bacteroids of *Bradyrhizobium japonicum* strain CC705. FEMS Microbiol. Lett. 83, 137-144.

Boistard, P., Batut, J., David, M., Fourment, J., Garnerone, A.M., Kahn, D., de Philip, P., Reyrat, J.M. and Waelkens, F. (1991) Regulation of nitrogen fixation genes in *Rhizobium meliloti.* In H. Hennecke and D.P.S. Verma (eds.) Advances in Molecular Genetics of Plant-Microbe Interactions, Kluwer Academic Publishers, Dordrecht, pp. 195-202.

Bott, M., Bolliger, M. and Hennecke, H. (1990) Genetic analysis of the cytochrome *c*-aa_3 branch of the *Bradyrhizobium japonicum* respiratory chain. Mol. Microbiol. 4, 2147-2157.

Bott, M., Ritz, D. and Hennecke, H. (1991) The *Bradyrhizobium japonicum cycM* gene encodes a membrane-anchored homolog of mitochondrial cytochrome *c*. J. Bacteriol. 173, 6766-6772.

Bott, M., Preisig, O. and Hennecke, H. (1992) Genes for a second terminal oxidase in *Bradyrhizobium japonicum.* Arch. Microbiol. in press.

Fischer, H.M., Anthamatten, D., Kullik, I., Morett, E., Acuña, G. and Hennecke, H. (1991) Complex regulatory network for *nif* and *fix* gene expression in *Bradyrhizobium japonicum.* In H. Hennecke and D.P.S. Verma (eds.) Advances in Molecular Genetics of Plant-Microbe Interactions, Kluwer Academic Publishers, Dordrecht, pp. 203-210.

Gabel, C. and Maier, R.J. (1990) Nucleotide sequence of the *coxA* gene encoding subunit I of cytochrome aa_3 of *Bradyrhizobium japonicum.* Nucleic Acids Res. 18, 6143.

Keister, D.L. and Marsh, S.S. (1990) Hemoproteins of *Bradyrhizobium japonicum* cultured cells and bacteroids. Appl. Environ. Microbiol. 56, 2736-2741.

Kullik, I., Fritsche, S., Knobel, H., Sanjuan, J., Hennecke, H. and Fischer, H.-M. (1991) *Bradyrhizobium japonicum* has two differentially regulated, functional homologs of the σ^{54} gene (*rpoN*). J. Bacteriol. 173, 1125-1138.

Layzell, D.B., Hunt, S., Moloney, A.H.M., Fernando, S.M. and del Castillo, L.D. (1990) Physiological, metabolic and developmental implications of O_2 regulation in legume nodules. In P.M. Gresshoff, L.E. Roth, G. Stacey and W.E. Newton (eds.) Nitrogen Fixation: Achievements and Objectives, Chapman and Hall, New York, pp. 21-32.

Morett, E., Fischer, H.M. and Hennecke, H. (1991) Influence of oxygen on DNA binding, positive control, and stability of the *Bradyrhizobium japonicum* NifA regulatory protein. J. Bacteriol. 173, 3478-3487.

Ramseier, T.M., Winteler, H.V. and Hennecke, H. (1991) Discovery and sequence analysis of bacterial genes involved in the biogenesis of *c*-type cytochromes. J. Biol. Chem. 266, 7793-7803.

Rossbach, S., Loferer, H., Acuña, G., Appleby, C.A. and Hennecke, H. (1991) Cloning, sequencing and mutational analysis of the cytochrome c_{552} gene (*cycB*) from *Bradyrhizobium japonicum* strain 110. FEMS Microbiol. Lett. 83, 145-152.

Thöny-Meyer, L., Stax, D. and Hennecke, H. (1989) An unusual gene cluster for the cytochrome bc_1 complex in *Bradyrhizobium japonicum* and its requirement for effective root nodule symbiosis. Cell 57, 683-697.

Tully, R.E., Sadowsky, M.J. and Keister, D.L. (1991) Characterization of cytochrome c_{550} and c_{555} from *Bradyrhizobium japonicum*: cloning, mutagenesis and sequencing of the c_{555} gene (*cycC*). J. Bacteriol. 173, 7887-7895.

Witty, J.F. and Minchin, F.R. (1990) Oxygen diffusion in the legume root nodule. In P.M. Gresshoff, L.E. Roth, G. Stacey and W.E. Newton (eds.) Nitrogen Fixation: Achievements and Objectives, Chapman and Hall, New York, pp. 285-292.

Section 4 / Bacterial-Plant Interactions: Bacterial Side

DETERMINANTS OF SPECIFICITY IN THE INTERACTION OF PLANTS WITH BACTERIAL PATHOGENS

N.T. KEEN, J.J. SIMS, S. MIDLAND, M. YODER, F. JURNAK, H. SHEN, C. BOYD, I. YUCEL, J. LORANG AND J. MURILLO
Departments of Plant Pathology and Biochemistry
University of California
Riverside, California, 92521
USA

ABSTRACT. General and specific virulence mechanisms have been defined in plant pathogenic bacteria over the last few years and several global regulator genes identified. Recent information also suggests that the *hrp* cluster of genes required for pathogenicity may provide a secretion function for virulence factors similar to those occurring in bacterial pathogens of vertebrates. A *hrp* gene protein from *Erwinia amylovora* has also been shown to function as an elicitor of the hypersensitive response in several plants. Additional avirulence genes have recently been cloned and characterized from bacterial pathogens that are cosmopolitan—that is, they function in several plant species when expressed in normal pathogens of those plants. One avirulence gene, *avr*D, causes *P. syringae* cells to secrete a specific elicitor that interacts with the cognate plant disease resistance gene.

1. Introduction

Considerable progress has occurred in our understanding of factors determining plant-pathogen specificity—that is, mechanisms predicating whether the pathogen multiplies rapidly and the plant is susceptible to disease or the pathogen multiplies slowly and the plant is resistant. We will discuss mechanisms accounting for basic pathogenicity and virulence in bacterial pathogens as well as *hrp* gene and avirulence gene-specified elicitors that trigger active plant defense responses. Because of space constraints, we will concentrate only on information generated in the last two years but may nevertheless inadequately address important recent work.

2. Pathogenicity and General Virulence Mechanisms

Bacterial plant pathogens contain large clusters of *hrp* genes that are required for pathogenicity on host plants and for the capacity to elicit the hypersensitive reaction in non-host plants (see Lindgren et al., 1986; Willis et al., 1991). A major unsolved issue has been the biochemical functions of *hrp* loci. Recent information from several laboratories indicates that certain *hrp* genes facilitate the extracellular export of one or more proteins from the pathogen. H-C Huang, S. Hutcheson and A. Collmer showed at the Interlaken meeting that *hrpH* from *Pseudomonas syringae* pv. *syringae* exhibited

E. W. Nester and D. P. S. Verma (eds.),
Advances in Molecular Genetics of Plant-Microbe Interactions, 211–220.

considerable homology to the *pulD* secretory protein from *Klebsiella*. Huang et al. (1992) also noted that two *hrp* proteins of *P. syringae* pv. *syringae* were either exported or membrane associated. Boucher and associates showed by sequence data that genes in three different transcriptional units of the *P. solanacearum hrp* cluster had significant homology to known protein export genes from *Yersinia* and *Klebsiella* (C. Boucher, personal communication). Furthermore, all three of the *P. solanacearum hrp* genes hybridized to DNA from all the major groups of bacterial plant pathogens, indicating a conserved role in protein export. Wei et al. (1992) recently isolated a protein elicitor of the plant HR from *E. coli* cells expressing the *hrp* gene cluster of *Erwinia amylovora*. This protein, encoded by the *hrpN* gene, was found to be associated with the membrane fraction of bacterial cells.

Cloned *hrp* clusters have been shown to elicit the HR in heterologous host plants (e.g. Bauer and Beer, 1991; Huang et al., 1988), indicating that they may have been adapted to prevent initiating the HR on normal host species. Isolation of the protein elicitor by Wei et al. (1992) supports this idea as well as the finding of HR-inducing DNA sequences outside both the *hrpS* and *hrpL* ends of the *hrp* clusters of, respectively, *P.s. tomato* (Lorang et al., 1990) and *P.s. phaseolicola* (J. Mansfield, personal communication). All of these results suggest that *hrp* gene clusters may be a determinant of host range in plant pathogenic bacteria.

Studies in the last few years have established the roles of various toxins as general virulence factors of certain bacterial plant pathogens and efforts are underway to understand their biosynthesis (e.g. Kinscherf et al., 1991; Mo and Gross, 1991; Young et al., 1992). Extracellular polysaccharides are important virulence factors (e.g. Coplin et al., 1992; Denny and Baek, 1991) as is the production of plant growth hormones by certain pathogens. Recently, Manulis et al. (1991) studied IAA production in a gall-forming strain of *Erwinia herbicola* and showed that while the bacterium possesses two pathways for the synthesis of IAA, the functional pathway during pathogenesis is the indole acetamide pathway dissected genetically by Kosuge and co-workers in *Pseudomonas savastanoi* (e.g. Comai and Kosuge, 1982). Another group of virulence factors in bacterial pathogens are the pectic enzymes (Denny et al., 1992; Ried and Collmer, 1988; Rogriguez-Palenzuela et al., 1991).

The X-ray diffraction analysis of a pectic enzyme, PelC from *Erwinia chrysanthemi* (Tamaki et al., 1988), has recently been completed (M. Yoder and F. Jurnak, manuscript in preparation). The 38 kDa protein is folded into one domain with a unique 'corkscrew' motif. The backbone is a large cylindrical helix comprised of parallel B strands with occasional polypeptide loops protruding from and folding over the core cylinder. The active site region has been tentatively identified on one external side of the cylinder. Protruding loops, containing amino acids invariant in all pectate lyases, are spatially clustered together near the putative Ca^{2+} binding site. The unique structure of PelC suggests that folding occurs after the protein is secreted through the inner membrane and that it may represent a common topology with other secreted proteins which share sequence similarity. The structural results should facilitate further investigation of basic questions such as the mechanism of catalysis, interactions with the metal cofactor and the saccharide substrate, the ability of the protein to penetrate the higher plant cell wall and identification of amino acids involved in targeting the protein for extracellular secretion by the *out* genes (He et al., 1991).

The regulation of virulence factors has received a great deal of attention in pathogens of vertebrates and has also been studied in plant pathogenic bacteria. For instance, the group of Tim Denny has shown that several virulence factors in *Pseudomonas solanacearum* including EPS and cellulase are coordinately regulated by *phcA*, a global regulator gene (Brumbley and Denny, 1990). Similarly, Murata et al. (1991) have characterized a global regulatory gene, *aepA*, of genes encoding pectolytic, cellulolytic and proteolytic enzymes in *Erwinia carotovora*. A gene, *lemA*, was also shown to regulate toxin production, extracellular enzymes and virulence in two pathovars of *Pseudomonas syringae* (Barta et al., 1992; Hrabek et al., 1992). The single *lemA* protein appeared to contain both members of a two component regulatory system. Finally, the *hrpS* and *hrpL* regulatory genes have been shown to be environmental sensors in *P. syringae* pv. *phaseolicola* that control *hrp* and several other genes (Rahme et al., 1992), including certain avirulence genes, as will be discussed later.

3. Host Range Factors

The range of plant species and cultivars infected by bacteria is determined by factors which act positively (specific virulence factors) or negatively (avirulence genes). Unlike more general virulence factors such as pectic enzymes or toxins such as coronatine which affect a wide range of plant species, the specific virulence factors condition high level virulence by the producing bacterium on a limited range of plant species and/or cultivars. There are currently few recognized examples of specific virulence factors, but mutational studies have indicated the occurrence of such genes in certain bacterial pathogens (Arlat and Boucher, 1991; De Feyter and Gabriel, 1991; Waney et al., 1991). Swarup et al. (1992) made the important finding that a gene from *Xanthomonas citri* which appears responsible for water-soaking and canker formation on citrus leaves also functions as an avirulence gene in other plants. Furthermore, this gene is a member of the *avrBs3* family of avirulence genes to be described later. As yet, however, we have no indications as to the virulence mechanism conferred by this or other genes involved with specific virulence.

Several avirulence genes have been cloned and characterized from bacterial pathogens that restrict the pathogen host range on plants. Some of these determine specificity at the plant cultivar level (e.g. Jenner et al., 1991), but others have been shown to function in heterologous bacterial taxa to elicit defense reactions on the normal hosts of these bacteria (Carney and Denny, 1990; Fillingham et al., 1992; Kobayashi et al., 1990; Ronald et al., 1992; Whalen et al., 1991). This indicates that these genes may restrict the range of plant species parasitized and implies that different plant species contain resistance genes with conserved recognitional specificities.

Bonas et al. (1989) cloned *avrBs3* from *Xanthomonas campestris* pv. *vesicatoria* and showed that it was an unusual avirulence gene encoding a large protein with several internal repeats. This gene has recently gained greater stature with the finding by several groups that avirulence genes yielding distinct specificities are homologous to *avrBs3* (Canteros et al., 1991; DeFeyter and Gabriel, 1991; J. Leach, personal communication). As noted above, Swarup et al. (1992) have also shown that *pthA*, a gene from *X. citri* believed to function as a specific virulence factor in citrus, also belongs to the *avrBs3* gene family and behaves as an avirulence gene in non-host plants. While the basis of the differential specificities possessed by the various *avrBs3* family members is not clear, Herbers et al. (1992) have shown that altering the number of repeats in *avrBs3* changes the avirulence specificity. These results may indicate that the *avrBs3* protein product or processed peptides emanating from it may function directly as elicitors of the plant HR, unlike the situation discussed later for *avrD* in *P. syringae*.

Several avirulence genes have been cloned from bacterial pathogens of *Arabidopsis thaliana* and tomato (Canteros et al., 1991; Debener et al., 1991; Dong et al., 1991; Ronald et al., 1992; Whalen et al., 1991), plants in which cloning of the complementary disease resistance genes can be approached by map based chromosome walking or gene tagging strategies. Several groups are pursuing such projects and it is probable that the structures of plant disease resistance genes will be revealed in the near future. While it has long been proposed that disease resistance genes encode components of receptors which specifically recognize avirulence gene-specified pathogen elicitors, no direct evidence is currently available to test such hypotheses.

Recently, specific elicitors have been identified from pathogens expressing particular avirulence genes (Culver and Dawson, 1992; Keen et al., 1990; Van den Ackerveken, 1992) and there are clear indications that specific elicitors are produced by other pathogens expressing defined avirulence genes [e.g. Chen and Heath (1990) with *Uromyces vignae*; Knogge et al. (1991) with *Rynchosporium secalis*; Schoelz et al. (1986) with cauliflower mosaic virus; D. Baulcombe and collaborators with potato virus X, personal communication].

The *avrD* gene, originally cloned from *Pseudomonas syringae* pv. *tomato* (Kobayashi et al., 1990), also directs production of a specific elicitor (Keen et al., 1990). This gene occurs on an indigenous plasmid of ca. 83 kb (J. Murillo, manuscript in preparation) and is the first of at least five open reading frames which appear to comprise an operon (Kobayashi et al., 1990). Data base searches have not yielded known genes with high homology to ORF 1 (*avrD*). Open reading frames 2 and 3 of the *avrD* operon also do not exhibit significant homology with known genes, but Roche et al. (1991) showed that ORF 4 shares significant homology with *nodH*, the sulfotransferase gene of *Rhizobium meliloti* which transfers a sulfate group to the nodulation signal molecule. We have recently observed that ORF 5 of the *avrD* operon exhibits considerable homology with amino transferase genes of *Bacillus* sp. and *E. coli* (*bioA*) that are involved in biotin synthesis (Gloeckler et al., 1990). As yet, the relationship of these putative functions for ORF4 and ORF5 to *avrD* are not clear.

AvrD is the only bacterial avirulence gene for which we understand some aspects of its biochemical interaction with the cognate plant disease resistance gene. Expression of *avrD* in a range of Gram negative bacteria, including *E. coli*, causes them to secrete a discrete elicitor molecule which is specifically recognized by plants carrying the matching *Rpg4* resistance gene (Keen and Buzzell, 1991). The *avrD* elicitor has recently been isolated and its structure determined as two homologous gamma lactones (Fig. 1, J. Sims, S. Midland, M. Stayton, M. Smith, E. Mazzola, and N. Keen, manuscript in preparation). These molecules have been given the trivial names syringolides 1 and 2. They have not been previously encountered but have similarity to the butyrolactones isolated from *Streptomyces* sp. which serve as autoinducers of sporulation and antibiotic production (Sakuda et al., 1992). Inspection of the structures of the *avrD* elicitors suggests that, like the butyrolactones, they may biosynthetically result from the initial condensation of a β-keto fatty acid (β-keto octanoic and β-keto decanoic acids for syringolides 1 and 2, respectively) with the sugar xylulose. It is assumed that the *avrD* protein catalyses the first, committed step leading to the syringolides, but this activity has thus far not been demonstrated *in vitro*.

AvrD is of particular interest because it is expressed at much higher level when bacteria are inoculated into plants than when they are grown in rich laboratory culture media. H. Shen and N. Keen (manuscript in preparation) observed that *avrD* utilizes a sigma 54 requiring promoter. It contains the GG-10 bp-GC motif characteristic of such

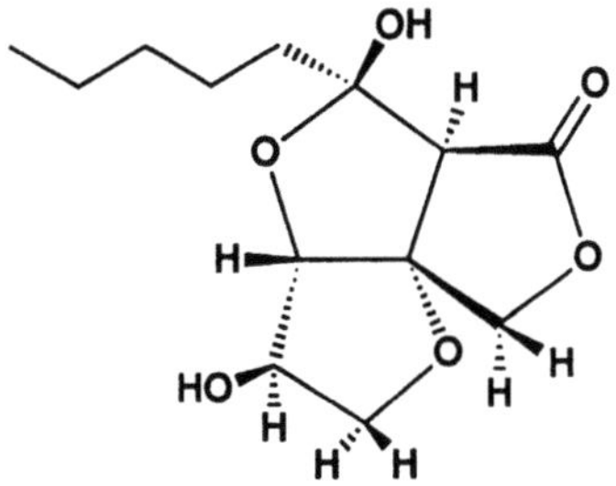

Syringolide 1

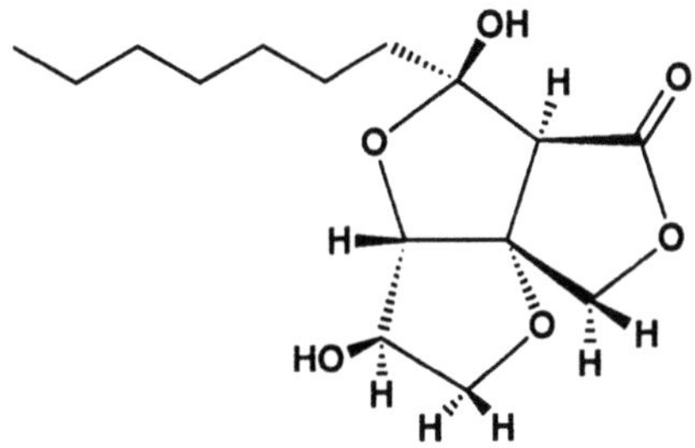

Syringolide 2

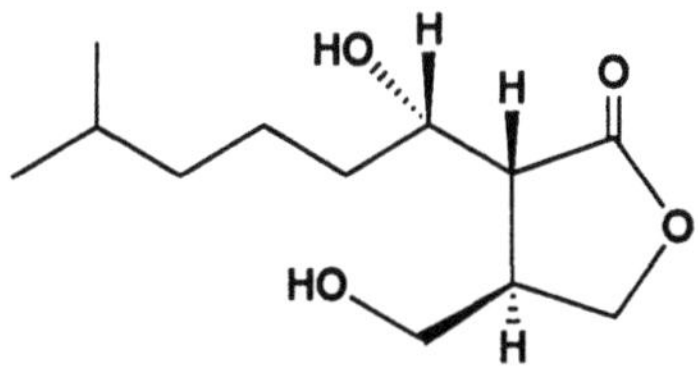

Virginiae butanolide A

Fig. 1. Structures of syringolides 1 and 2 (upper), produced by Gram negative bacteria expressing *avrD* and virginiae butanolide A (lower), a signal molecule produced by *Streptomyces* species (Sakuda et al., 1992).

promoters and also requires an upstream element assumed to function as an activator. Significantly, an upstream element (TGGAACC-15 or 16 bp-CCAC) occurs in the promoter regions of at least 9 different avirulence genes cloned from various *Pseudomonas syringae* pathovars. The *avrD* promoter also requires the *hrp* regulatory genes *hrpS* and *hrpL* for full activity, similar to previous results for *avrB* obtained by Huynh et al. (1989). Also similar to *avrB*, the expression of *avrD* is low when bacteria are grown in rich culture media and is strongly repressed in minimal media by certain repressing carbon sources, pH values above 6.5 or high concentrations of nitrogen sources.

4. Future Considerations

Further work should provide elucidation of biochemical functions for several of the specific virulence factors that have been identified in plant pathogenic bacteria as well as for the elusive *hrp* cluster of pathogenicity genes. It will be particularly interesting to study how these factors, for instance the elicitor protein described by Wei et al. (1992), have been adapted to the particular plant species attacked by a bacterium. With the successful crystallization and X-ray diffraction analyses of pectate lyase proteins, we are in the lucrative position of being able to address plant-pathogen interactions at the atomic level. Doubtless, additional virulence-related proteins will be crystallized in the future to extend these possibilities. Thus far, *avrD* is the only bacterial avirulence gene shown to direct production of a discrete elicitor molecule. It is not clear whether failures to observe similar signal molecules with other avirulence genes results from the relative instability of elicitors or whether these interactions are more complex. With the *avrBs3* gene family, it is appealing to speculate that the protein products or processed proteins function as elicitors. The observation that certain *avrBs3* family members also appear to possess virulence-related functions (De Feyter et al., 1991; Swarup et al., 1992) fosters the speculation that these proteins or processing products may mimic natural plant growth regulators. It is noteworthy that a large plant gene with repetitive sequences reminiscent of *avrBs3* has recently been shown to direct production of a low molecular weight signal molecule for systemic induction of protease inhibitor genes in tomato (Pearce et al., 1991).

With few exceptions (Kearney and Staskawicz, 1990; Swarup et al., 1992), mutation of avirulence genes has not been shown to impair pathogen virulence or other detectable phenotypes. In view of the shared regulatory elements present in several *P. syringae* avirulence genes and the multi-gene nature of the *avrBs3* gene family in *X. campestris* strains, it is possible that even structurally unrelated avirulence genes such as those in the *P. syringae* pathovars may contribute to a common biological function, such that mutation of all of them would be required for loss of function. Work with avirulence genes indicates that plant disease resistance genes, when they are cloned, should reveal of wealth of information regarding plant-pathogen specificity. With the recent cloning of the protein *hrp* elicitor (Wei et al., 1992), it is appealing to consider cloning the plant gene encoding the recognitional mechanism.

5. References

Arlat, M., and Boucher, C. (1991) 'Identification of a *dsp* DNA region controlling aggressiveness of *Pseudomonas solanacearum*', Molec. Plant-Microbe Inter. 4, 211-213.

Bauer, D.W. and Beer, S.V. (1991) 'Further characterization of an *hrp* cluster of *Erwinia amylovora*', Molec. Plant-Microbe Inter. 4, 493-499.

Barta, T.M., Kinscherf, J.G., and Willis, D.K. (1992) 'Regulation of tabtoxin production by the *lemA* gene in *Pseudomonas syringae*', J. Bacteriol. 174, 3021-3029.

Bonas, U., Stall, R.E., and Staskawicz, B. (1989) 'Genetic and structural characterization of the avirulence gene *avrBs3* from *Xanthomonas campestris* pv. *vesicatoria*', Molec. Gen. Genet. 318, 127-136.

Brumbley, S.M., and Denny, T.P. (1990) 'Cloning of *phcA* from wildtype *Pseudomonas solanacearum* a gene that when mutated alters expression of multiple traits that contribute to virulence', J. Bacteriol. 172, 5677-5685.

Canteros, B., Minsavage, G., Bonas, U., Pring, D., and Stall, R. (1991) 'A gene from *Xanthomonas campestris* pv. *vesicatoria* that determines avirulence in tomato is related to *avrBs3*', Molec. Plant-Microbe Inter. 4, 628-632.

Carney, B.F., and Denny, T.P. (1990) 'A cloned avirulence gene from *Pseudomonas solanacearum* determines incompatibility on *Nicotiana tabacum* at the host species level', J. Bacteriol. 172,4836-4843.

Chen, C.Y., and Heath, M.C. (1990) 'Cultivar-specific induction of necrosis by exudates from basidiospore germlings of the cowpea rust fungus', Physiol. Molec. Plant Pathol. 37, 169-178.

Comai, L., and Kosuge, T. (1982) 'Cloning and characterization of *iaaM*, a virulence determinant of *Pseudomonas syringae* pv. *savastanoi*', J. Bacteriol. 149, 40-46.

Coplin, D.L., Frederick, R.D. and Majerczak, D.R. (1992) 'New pathogenicity loci in *Erwinia stewartii* identified by random Tn*5* mutagenesis and molecular cloning', Molec. Plant-Microbe inter. 5, 266-268.

Culver, J.N., and Dawson, W.O. (1992) 'Tobacco mosaic virus elicitor coat protein genes produce a hypersensitive phenotype in transgenic *Nicotiana sylvestris* plants', Molec. Plant-Microbe Inter. 4, 458-463.

De Feyter, R., and Gabriel, D.W. (1991) 'At least six avirulence genes are clustered on a 90-kilobase plasmid in *Xanthomonas campestris* pv. *malvacearum*', Molec. Plant-Microbe Inter. 4, 423-432.

Debener, T., Lehnackers, H., Arnold, M., and Dangl, J.L. (1991) 'Identification and molecular mapping of a single *Arabidopsis thaliana* locus determining resistance to a phytopathogenic *Pseudomonas syringae* isolate', Plant J. 1, 289-302.

Denny, T.P., and Baek, S.-R. (1991) 'Genetic evidence that extracellular polysaccharide is a virulence factor of *Pseudomonas solanacearum*', Molec. Plant-Microbe Inter. 4, 198-206.

Denny, T.P., Carney, B.F., and Schell, M.A. (1992) 'Inactivation of multiple virulence genes reduces the ability of *Pseudomonas solanacearum* to cause wilt symptoms', Molec. Plant-Microbe Inter. 3:293-300.

Dong, X., Mindrinos, M., Davis, K.R., and Ausubel, F.M. (1991) 'Induction of *Aradidopsis* defense genes by virulent and avirulent *Pseudomonas syringae* strains and by a cloned avirulence gene', Plant Cell 3, 61-72.

Fillingham, A.J., Wood, J., Bevan, J.R., Crute, I.R., Mansfield, J.W., Taylor, J.D. and Vivian, A. (1992) 'Avirulence genes from *Pseudomonas syringae* pathovars *phaseolicola* and *pisi* confer specificity towards both host and non-host species', Physiol. Molec. Plant Pathol. 40, 1-15.

Gloeckler, R., Ohsawa, I., Speck, D., Ledoux, C., Bernard, S., Zinsius, M., Villeval, D., Kosou, T., Kamogawa, K., and Lemoine, Y. (1990) 'Cloning and characterization of the *Bacillus sphaericus* genes controlling the bioconversion of pimelate into dethiobiotin', Gene 87, 63-70.

He, S.Y., Lindeberg, M., Chatterjee, A.K., and Collmer, A. (1991) 'Cloned *Erwinia chrysanthemi out* genes enable *Escherichia coli* to selectively secrete a diverse family of heterologous proteins to its milieu', Proc. Natl. Acad. Sci., USA 88, 1079-1083.

Herbers, K., Conrads-Strauch, J. and Bonas, U. (1992) 'Race specific plant resistance to bacterial spot disease determined by repetitive motifs in a bacterial avirulence protein', Nature 356, 172-174.

Hrabek, E.M. and Willis, D.K. (1992) 'The *lemA* gene required for pathogenicity of *Pseudomonas syringae* pv. *syringae* on bean is a member of a family of two-component regulators', J. Bacteriol. 174, 3011-3020.

Huang, H.-C., Schuurink, R., Denny, T.P., Atkinson, M.M., Baker, C.J., Yucel, I., Hutcheson, S.W., and Collmer, A. (1988) 'Molecular cloning of a *Pseudomonas syringae* pv. *syringae* gene cluster that enables *Pseudomonas fluorescens* to elicit the hypersensitive response in tobacco plants', J. Bacteriol. 170:4748-4756.

Huang, H.-C., Hutcheson, S.W., and Collmer, A. (1992) 'Characterizaion of the *hrp* cluster from *Pseudomonas syringae* pv. *syringae* 61 and Tn*phoA* tagging of genes encoding exported or membrane-spanning Hrp proteins', Molec. Plant-Microbe Inter. 4, 469-476.

Huynh, T.V., Dahlbeck, D., Staskawicz, B.J. (1989) 'Bacterial blight of soybean: regulation of a pathogen gene determining host cultivar specificity', Science 245, 1374-1377.

Jenner, C., Hitchin, E., Mansfield, J., Walters, K., Betteridge, P., Teverson, D., and Taylor, J. (1991) Gene for-gene interactions between *Pseudomonas syringae* pv. *phaseolicola* and *Phaseolus*', Molec. Plant-Microbe Inter. 4, 553-562.

Kearney, B., and Staskawicz, B.J. (1990) 'Widespread distribution and fitness contribution of *Xanthomonas campestris* avirulence gene *avrBs2*', Nature 346, 385-386.

Keen, N.T. and Buzzell, R.I. (1991) 'New disease resistance genes in soybean against *Pseudomonas syringae* pv. *glycinea*; evidence that one of them interacts with a bacterial elicitor', Theor. Appl. Genet. 81, 133-138.

Keen, N.T., Tamaki, S., Kobayashi, D., Gerhold, D., Stayton, M., Shen, H., Gold, S., Lorang, J., Thordal-Christensen, H., Dahlbeck, D., and Staskawicz, B. (1990) 'Bacteria expressing avirulence gene D produce a specific elicitor of the soybean hypersensitive reaction', Molec. Plant-Microbe Inter. 3,112-121.

Kinscherf, T.G., Coleman, R.H., Barta, T.M., and Willis, D.K. (1991) 'Cloning and expression of the tabtoxin biosynthetic region from *Pseudomonas syringae*', J. Bacteriol. 173, 4124-4132.

Knogge, W., Hahn, M., Lehnackers, H., Rupping, E., Wevelsiep, L. (1991) 'Fungal signals involved in the specificity of the interaction between barley and *Rhynchosporium secalis*', in H. Hennecke and D.P.S. Verma (eds.), Advances in Molecular Genetics of Plant-Microbe Interactions, Vol 1, Kluwer Acad. Publ., Dordrecht, pp. 250-253.

Kobayashi, D.Y., Tamaki, S.J., and Keen, N.T. (1990) 'Molecular characterization of avirulence gene D from *Pseudomonas syringae* pv. *tomato*', Molec. Plant-Microbe Inter. 3, 94-102.

Lindgren, P., Peet, R.C., and Panopoulos, N.J. (1986) 'Gene cluster of *Pseudomonas syringae* pv. "*phaseolicola*" controls pathogenicity on bean plants and hypersensitivity on nonhost plants', J. Bacteriol. 168, 512-522.

Lorang, J.M., Boucher, C.A., Dahlbeck, D., Staskawicz, B and Keen, N.T. (1990) 'An avirulene function from *Pseudomonas syringae* pv. *tomato* is located within a *hrp* cluster', Abstract 28, Amer. Phytopathol. Soc. meeting, Grand Rapids, Michigan.

Manulis, S., Valinski, L., Gafni, Y., and Hershenhorn, J. (1991) 'Indole-3-acetic acid biosynthetic pathways in *Erwinia herbicola* in relation to pathogenicity on *Gypsophila paniculata*', Physiol. Molec. Plant Pathol. 39, 161-172.

Mo, Y.-Y., and Gross, D.C. (1991) 'Expression *in vitro* and during plant pathogenesis of the *syrB* gene required for syringomycin production by *Pseudomonas syringae* pv. *syringae*', Molec. Plant-Microbe Inter. 4, 28-36.

Murata, H., McEvoy, J.L., Chatterjee, A., Collmer, A., and Chatterjee, A.K. (1991) 'Molecular cloning of an *aepA* gene that activates production of extracellular pectolytic, cellulolytic and protolytic enzymes in *Erwinia carotovora* subsp. *carotovora*', Molec. Plant-Microbe. Inter. 4, 239-246.

Pearce, G., Strydom, D., Johnson, S. and Ryan, C.A. (1991) 'A polypeptide from tomato leaves induces wound-inducible proteinase inhibitor proteins', Science 253, 895-898.

Rahme, L.G., Mindrinos, M.N., and Panopoulos, N.J. (1992) 'Plant and environmental sensory signals control the expression of *hrp* genes in *Pseudomonas syringae* pv. phaseolicola', J. Bacteriol. 174, 3499-3507.

Ried, J.L., and Collmer, A. (1988) 'Construction and characterization of an *Erwinia chrysanthemi* mutant with directed deletions in all of the pectate lyase structural genes', Molec. Plant-Microbe Inter. 1:32-38.

Roche, P., Debelle, F., Maillet, F., Lerouge, P., Faucher, C., Truchet, G., Denarie, J., and Prome, J-C. (1991) 'Molecular basis of symbiotic host-specificity in *Rhizobium meliloti: nodh* and *nodPQ* genes encode the sulfation of lipo-oligosaccharide signals', Cell 67, 1131-1144.

Rodriguez-Palenzuela, P., Burr, T.J. and Collmer, A., (1991) 'Polygalacturonase is a virulence factor in *Agrobacterium tumefaciens* biovar 3', J. Bacteriol. 173, 6547-6552.

Ronald, P.C., Salmeron, J.M., Carland, F.M., and Staskawicz, B.J. (1992) The cloned avirulence gene *avrPto* induces disease resistance in tomato cultivars containing the *Pto* resistance gene', J. Bacteriol. 174, 1604-1611.

Sakuda, S., Higashi, A., Tanaka, S., Nihira, T., Yamada, Y. (1992) 'Biosynthesis of virginiae butanolide A, a butyrolactone autoregulator from *Streptomyces*', J. Amer. Chem. Soc. 114, 663-668.

Schoelz, J., Shepherd, R.J., and Daubert, S. (1986) 'Region VI of cauliflower mosaic virus encodes a host range determinant', Molec. Cell. Biol. 6, 2632-2637.

Swarup, S., Yang, Y., Kingsley, M.T., and Gabriel, D.W. (1992) 'An *Xanthomonas citri* pathogenicity gene, *pthA*, pleiotropically encodes gratuitous avirulence on nonhosts', Molec. Plant-Microbe Inter. 5, 204-213.

Van den Ackerveken, G.F.J.M., van Kan, J.A.L., and DeWit, P.J.G.M. (1992) 'Molecular analysis of the avirulence gene *avr9* of the tomato pathogen *Cladosporium fulvum* fully supports the gene-for-gene hypothesis', Plant J. 2, 359-366.

Waney, V.R., Kingsley, M.T., and Gabriel, D.W. (1991) '*Xanthomonas campestris* pv. *translucens* genes determining host-specific virulence and general virulence on cereals identified by Tn*5-gusA* insertion mutagenesis', Molec. Plant-Microbe Inter. 4, 623-627.

Wei, Z-M., Laby, R.J., Zummof, C.H., Bauer, D.W., He, S.Y., Collmer, A., and Beer, S.V. (1992) 'Harpin, elicitor of the hypersensitive response produced by the plant pathogen *Erwinia amylovora*', Science 257, 85-88.

Whalen, M.C., Innes, R.W., Bent, A.F., and Staskawicz, J.B. (1991) 'Identification of *Pseudomonas syringae* pathogens of *Arabidopsis* and a bacterial locus determining avirulence on both *Arabidopsis* and soybean', Plant Cell 3, 49-59.

Willis, D.K., Rich, J.J., and Hrabek, E.M. (1991) '*hrp* genes of phytopathogenic bacteria', Molec. Plant-Microbe Inter. 4, 132-138.

Young, S.A., Park, S.K., Rodgers, C., Mitchell, R.E., and Bender, C.L. (1992) 'Physical and functional characterization of the gene cluster encoding the polyketide phytotoxin coronatine in *Pseudomonas syringae* pv. glycinea', J. Bacteriol. 174, 1837-1843.

A FAMILY OF AVIRULENCE GENES FROM *XANTHOMONAS ORYZAE* PV. *ORYZAE* IS INVOLVED IN RESISTANT INTERACTIONS IN RICE

JAN E. LEACH, CHRISTOPHER HOPKINS, AILAN GUO, SEONG-HO CHOI, MARK MAZZOLA, MARIETTA RYBA-WHITE, AND FRANK F. WHITE
Department of Plant Pathology
Kansas State University
Throckmorton Hall
Manhattan, Kansas 66506-5502
U.S.A.

ABSTRACT. *Xanthomonas oryzae* pv. *oryzae*, the bacterial blight pathogen of rice, contains at least three avirulence genes (*avrxa5, avrXa7*, and *avrXa10*) that are related to *avrBs3*, an avirulence gene from the pepper pathogen *X. campestris* pv. *vesicatoria*. The gene *avrXa10*, which specifies the resistant phenotype on rice cultivars with the bacterial blight resistance gene *Xa-10*, like *avrBs3*, contains a 102-bp repeated sequence, although *avrXa10* contains fewer copies than *avrBs3* (15.5 vs. 17.5). Each 102-bp repeat contains a six-nucleotide variable region; the order of the repeats within *avrXa10* is distinct from *avrBs3*. The number of repeat copies in *avrXa7* was estimated at 25. The physiological events associated with resistant interactions were investigated using the cloned avirulence genes and near-isogenic rice cultivars with single bacterial blight resistance genes. Resistance in rice to *X. oryzae* pv. *oryzae* is characterized by a light-dependent increase in the extracellular activity of three peroxidases, lignin deposition, and a decrease in the rate of bacterial multiplication. The timing of these events is dependent on the specific avirulence gene-resistance gene interaction, that is, the responses associated with resistance occur later in interactions involving *avrxa5-xa-5* and *avrXa7-Xa7* than those between *avrXa10* and *Xa-10*.

Introduction

Bacterial blight, caused by *Xanthomonas oryzae* pv. *oryzae* (Ishiyama) Dye (Swings et al. 1990), is the most serious bacterial disease of rice (Ou 1985). Races of *X. oryzae* pv. *oryzae* have been defined using rice cultivars containing different resistance genes (Mew 1987). Resistant interactions between *X. oryzae* pv. *oryzae* and rice involve the light-dependent increase in the extracellular activity of three peroxidases, lignin deposition, host cell death, and a decrease in the rate of bacterial multiplication (Reimers and Leach 1991; Reimers et al. 1992; Guo et al. unpublished).

Race specific interactions such as those observed between *X. oryzae* pv. *oryzae* and rice are thought to follow the gene-for-gene model (Ellingboe 1976; Flor 1955), which predicts that incompatible interactions are the consequence of positive functions encoded by avirulence genes in the pathogen and corresponding resistance genes in the host. Avirulence genes have been cloned from several different species and pathovars of *Xanthomonas* (Bonas et al. 1989;

E. W. Nester and D. P. S. Verma (eds.),
Advances in Molecular Genetics of Plant-Microbe Interactions, 221–230.

Canteros et al. 1991; De Feyter and Gabriel 1991; Gabriel et al. 1986; Minsavage et al. 1990; Ronald and Staskawicz 1988; Swanson et al. 1988; Swarup et al. 1992; Whalen et al. 1988). The sequences of two avirulence genes *avrBs3* and *avrBsP*, both from *X. campestris* pv. *vesicatoria*, are very similar, and both contain a 102-bp sequence in the coding region that is repeated 17.5 and six times, respectively (Bonas et al. 1989; Canteros et al. 1991). DNA sequences related to *avrBs3* from *X. campestris* pv. *vesicatoria* were detected in seven additional pathovars of *X. campestris* that cause disease on diverse dicotyledonous plants (Bonas et al. 1989). Recently, homologs of *avrBs3* from *X. campestris* pv. *malvacearum* and *X. campestris* pv. *citri* have been shown to have avirulence activity (Swarup et al. 1992).

We are interested in the physiological and molecular mechanisms by which *X. oryzae* pv. *oryzae* specifically elicits resistance in rice plants. We have found that some avirulence genes *X. oryzae* pv. *oryzae* are members of a multiple gene family with sequence similarity to *avrBs3*. Further, the timing of the physiological responses correlated with resistant interactions involving these avirulence genes is dependent on the specific *avr* gene-resistance gene combination.

Results and Discussion

HYBRIDIZATION OF *X. ORYZAE* PV. *ORYZAE* GENOMIC DNA WITH *AVRBS3*

The internal *Bam*HI fragment from *avrBs3* (Bonas et al. 1989) hybridized with multiple *Bam*HI fragments in DNA from strains of *X. oryzae* pv. *oryzae* under high stringency conditions (data not shown). For example, approximately 12 DNA fragments from the race 2 strain PXO86 of *X. oryzae* pv. *oryzae* hybridized with *avrBs3*, and the fragments ranged in size from 2.8 to greater than 12 kb. DNA from over 100 strains of *X. oryzae* pv. *oryzae* from various geographic areas contained multiple fragments that hybridized with *avrBs3*. The only exception was DNA from the U.S.A. strains (Jones et al. 1989), which did not hybridize to *avrBs3* under these conditions.

IDENTIFICATION OF AVIRULENCE GENES FROM *X. ORYZAE* PV. *ORYZAE*

Identification of Avirulence Genes. Forty-three clones from a genomic library of strain PXO86 hybridized with the 3.3 kb *avrBs3 Bam*HI fragment. Six of the clones, when mobilized into the virulent recipient strain PXO99^{A}, conferred an avirulent phenotype when inoculated to cultivars containing either *xa-5*, *Xa-7*, or *Xa-10*. Only clone pXO6-33 specified avirulence to more than one genotype; PXO99^{A}(pXO6-33) was avirulent to cultivars of rice with both *Xa-10* and *xa-5* resistance genes.

The clones containing *avrXa7* (pXO29-29) and *avrXa10* (pXO5-15) were subjected to further analysis by transposon insertional mutagenesis, and the positions of the insertions were determined by restriction digestion mapping. Insertions that resulted in inactivation of *avrXa7* and *avrXa10* were located either within the 4.1-kb *Bam*HI fragment of pXO29-29 and the 3.1-kb *Bam*HI fragment of pXO5-15, respectively or immediately adjacent to these regions (Fig. 1). Mutations in the 3.7-kb *Bam*HI fragment of pXO29-29 (A7L) and the 4.5-kb *Bam*HI fragment of pXO5-15 (A10L), which hybridize with *avrBs3*, and are adjacent to the respective *avr* gene, had no effect on avirulence activity in the tested cultivars (Fig. 1).

The two avirulence genes from *X. campestris* pv. *vesicatoria*, *avrBs3* and *avrBsP*, are not adjacent to one another, nor are tandem, nonfunctional sequences present (Bonas et al. 1989; Canteros et al. 1991). Clone pXO6-33, which encodes both *avrxa5* and *avrXa10* activities,

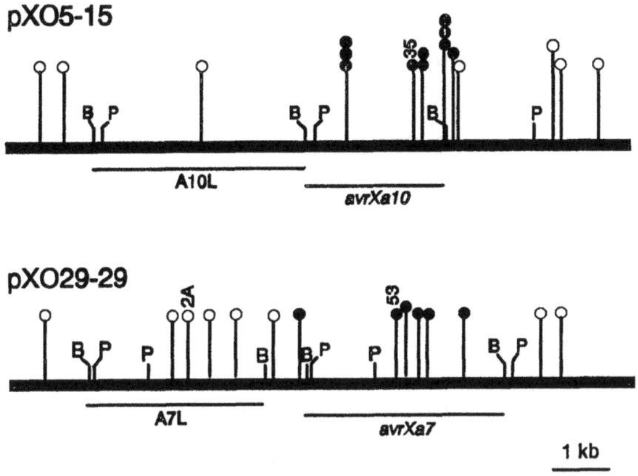

Figure 1. Partial restriction map of clones with *avrXa7* and *avrXa10* activity. Positions of transposon Tn*5*-B20 insertions that do (●) and do not (○) inactivate the avirulence phenotypes are shown. Transposon insertions used for marker exchange and physiological studies are indicated. B, *Bam*HI; P, *Pst*I. *Bam*HI fragments that hybridized with the 3.3-kb *Bam*HI fragment of *avrBs3* are indicated below the restriction maps with horizontal lines.

contains two *Bam*HI fragments similar in size to those present in pXO5-15, which encodes only *avrXa10* activity. We are investigating if pXO5-15 and pXO6-33 are overlapping clones and, if so, if *avrxa5* activity is encoded by the homolog A10L and was affected during cloning. Alternatively, the *avrXa10* gene may be represented in more than one copy on the genome or the same gene may encode both *avrxa5* and *avrXa10* activities in pXO6-33.

Bacterial Growth in planta. In the compatible interactions, bacterial numbers increased steadily until 48 h after inoculation to approximately 1×10^8 cfu/leaf, after which the rate of multiplication decreased and bacterial numbers did not substantially increase (Fig. 2). In incompatible interactions, the multiplication rate decreased between 24-48 (*Xa-10*) or 48-72 h (*Xa-7*) (approx. 5×10^7 cfu/leaf). Multiplication rates of the transconjugants carrying the *avr* genes on pXO5-15 and pXO29-29 were similar to those observed for strain PXO86 in rice cultivars with the corresponding resistance genes. If the cosmid-borne avirulence gene was inactivated by insertion of Tn*5*, the multiplication rate of transconjugants was similar to that of the virulent parental strain PXO99^A. Growth curves were similar for all strains in IR24, which carries no resistance genes (data not shown).

CHARACTERIZATION OF THE AVIRULENCE GENES

Restriction Analysis of Repeat Domains. The 3.3-kb *Bam*HI fragment of *avrBs3* is internal to the gene and contains the entire repeat domain. The *Bam*HI fragments internal to *avrXa7* and *avrXa10* as well as the adjacent *avrBs3*-hybridizing, *Bam*HI fragments with unknown activity (A7L and A10L, Fig. 1) were analyzed for the presence of a domain with a directly

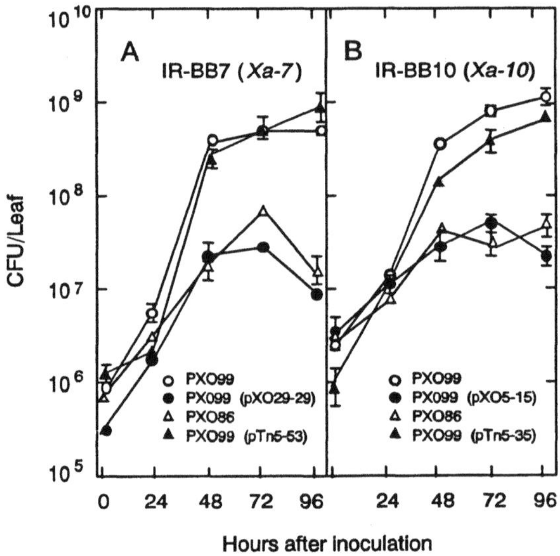

Figure 2. Time course of bacterial growth in leaves of rice cultivars (A) IR-BB7 and (B) IR-BB10. Leaves were infiltrated with bacterial suspensions and were sampled for 96 h following infiltration. (A) IR-BB7 infiltrated with *X. oryzae* pv. *oryzae* $PXO99^A$ (○); PXO86 (△); $PXO99^A$(pXO29-29) (•); $PXO99^A$(pTn*5*-53) (▲). (B) IR-BB10 infiltrated with *X. oryzae* pv. *oryzae* $PXO99^A$ (○); PXO86 (△); $PXO99^A$(pXO5-15) (•); $PXO99^A$(pTn*5*-35) (▲). Values are means from three repetitions, and vertical bars represent $\pm$1 SEM. Similar results were obtained in three independent experiments.

repeated 102-bp sequence analogous to that found in *avrBs3*. The presence and number of direct repeated sequences in each plasmid were demonstrated by partial digestion with *Bal*I and Southern blot analysis using pBSavrBs3 as a probe. Each direct repeat from *avrBs3* contains a unique *Bal*I site; thus, a partial digestion with *Bal*I created a ladder of 102-bp units. We estimated that *avrXa7* and *avrXa10* contained at least 25 and 15 bands, respectively, corresponding to the number of repeats. The adjacent *avrBs3*-homologous sequences A7L and A10L had 17 and 19 bands, respectively. Based on comigration with *avrBs3*, the tested *avrBs3*-hybridizing fragments from *X. oryzae* pv. *oryzae* contain repeats of approximately 102 bp in size.

Nucleotide Sequence of avrXa10. The nucleotide sequence and deduced amino acid sequence of the *X. oryzae* pv. *oryzae avrXa10* gene are very similar to those of *avrBs3*. The *avrXa10* sequence contains a 3306-bp open reading frame which would code for a 1102-amino acid, 116-kd protein; *avrBs3* is predicted to encode a 1163-aa protein.

From sequence analysis, the repeat of *avrXa10* was found to be 102-bp in length, which is the same length as that of *avrBs3*. The repeat is present in 15.5 directly repeated copies as compared to 17.5 copies in *avrBs3*. The sequence of each copy of the 102-bp repeat in *avrXa10* is almost identical to the other copies with the exception of a six-base pair region

```
                   VARIABLE
                    REGION
avrXa10            \  /
   1  LTPDQVVAIASNIGGNQALETVQRLLPVLCQAHG
   2  ...........HG..K..................
   3  ........S..NI..K...A...........D..
   4  ...........HG..K...............D..
   5  ...........NI..K...............D..
   6  ...........NI..K...............D..
   7  ..........NNN..K...............T..
   8  ...........HD..K...............D..
   9  ...........NI..K...A..............
  10  ...........HD..K............V..D..
  11  ..........NNN..K...............D..
  12  ...A.......HG..K...............D..
  13  ...V.......NS..K...............D..
  14  ...V.......NG..K...A...........D..
  15  ...V.......HD..K...............D..
  16  ...........NG..K...

avrBs3
   1  ...E.......HD..K..................
```

Figure 3. Amino acid sequence of the *avrXa10* repeat domain. The repeat number is indicated on the left and the two amino acid variable region is indicated above. The first repeat of *avrBs3* (Bonas et al. 1989) is below for comparison.

(starting at nucleotide number 34), which we hereafter refer to as the variable region. The variable region alternates between six possible nucleotide sequences (AATATT, CATGGC, AATAAC, CACGAT, AATAGC, AATGGC), and there is no discernable order in the position of any variable region within the sequence. The 102-bp repeat represents a 34-amino acid repeat domain in the predicted protein sequence (Fig. 3). Most of the nucleotide variations do not result in amino acid substitutions, with the exception of the variable region which represents codons 12 and 13 in the protein repeat domain. Codon 12 encodes either asparagine or histidine, while codon 13 encodes isoleucine, glycine, asparagine, aspartate, or serine. No arrangement for the repeat domain is apparent from the amino acid sequence, except that histidine at codon 12 is only present with glycine or aspartate at codon 13.

The fourth codon of the *avrBs3* repeat encodes either glutamate or glutamine compared to aspartate, alanine, or valine for the *avrXa10* repeat, and the thirty-second codon in *avrBs3* is for alanine compared to aspartate and alanine in *avrXa10*. In addition to the amino acid combinations present in the variable region of *avrBs3*, *avrXa10* contains the combinations histidine-glycine (repeat 2) and arginine-arginine (repeats 7 and 11).

PHYSIOLOGICAL RESPONSES DURING RESISTANT INTERACTIONS

Effect of Light on Resistant Interactions. The resistant response in interactions between rice and *X. oryzae* pv. *oryzae* is dependent on light (Guo, Reimers and Leach, unpublished). In the absence of light, or if tissues are exposed to less than 8 h light after inoculation, interactions that would normally result in resistance are altered, that is, the bacterial populations reach high levels (Fig. 4, PXO86, 24 h dark), and tissues become watersoaked, similar to susceptible interactions (Fig. 4, PXO99, 24 h light or dark). Furthermore, the accumulation of peroxidase and lignin that are correlated with resistance are not observed in the absence of light (data not shown). The presence of the plasmid containing *avrXa10* (pXO5-15) in a normally virulent strain ($PXO99^A$) is sufficient to alter the growth rate in the *Xa-10* cultivar, as described above, if adequate light is provided [Fig. 4, $PXO99^A$(pXO5-15), 24 h light]. However, in the absence of light, the effects of the cloned avirulence gene are not observed, and the tissues are susceptible [Fig. 4, $PXO99^A$(pXO5-15), 24 h light].

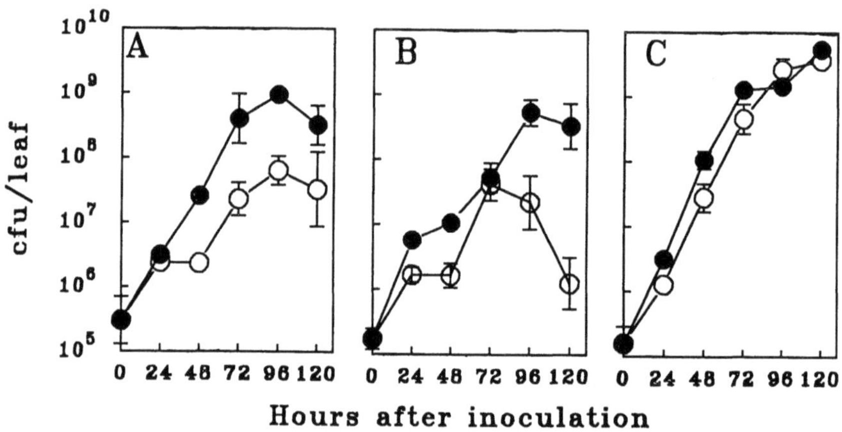

Figure 4. Time course of bacterial multiplication in leaves of a rice cultivar containing *Xa-10*. Leaves were infiltrated with PXO86 (A), PXO99(pXO5-15) (B), and PXO99 (C). Immediately after infiltration, plants were incubated in complete darkness for 24 h (•), or in light for 24 h (○). After the first 24 h treatment, the plants were returned to the normal 12 h day/12 h night cycle in a growth chamber.

Timing of Resistant Responses Varies Between Specific Avirulence and Resistance Gene Combinations. The phenotypes of the reactions varied with the resistance gene involved in the interaction. Interactions involving *Xa-10* resulted in a dark brown color throughout the infiltrated site within 24-48 h, whereas with *xa-5* and *Xa-7*, a dark ring formed around the perimeter of the watersoaked site at 48 h, and the tissue within the site became tan to brown after 72 h. Compatible interactions in both combinations remained watersoaked in the infiltration site through 5 days, after which time the watersoaked lesion had spread, and the leaf had wilted.

Incompatible interactions between *X. oryzae* pv. *oryzae* and rice containing the *Xa-10* resistance gene are characterized by an increase in the activity of an extracellular cationic peroxidase and a decrease in the rate of bacterial multiplication (Barton-Willis et al. 1989; Reimers et al. 1992; Reimers and Leach 1991). The timing of both these events corresponds with the phenotypic reaction observed on the plant leaf. However, the timing of incompatible interactions between different avirulence gene-resistance gene combinations varies.

Incompatible interactions involving near-isogenic cultivars carrying *Xa-10* were characterized by an increase in the extracellular cationic peroxidase within 16-24 h, which coincided with a decrease in the rate of bacterial multiplication (beginning at 24 h) (Fig. 5, Reimers et al. 1992). In incompatible interactions with *avrXa7*, the increase in cationic peroxidase activity and the decrease in the rate of bacterial multiplication were not observed until 48 h. The timing and intensity of the browning response characteristic of resistance also was delayed in incompatible interactions with *avrXa7* as compared with *avrXa10*. The browning response and accumulation of cationic peroxidase activity are delayed in interactions involving *avrxa5* and *xa-5* (data not shown). Previously, differences in interactions between PXO86 (a wild-type race 2 strain carrying *avrxa5*, *avrXa7*, and *avrXa10*) and plants with *xa-5*, *Xa-7*, or *Xa-10* resistance genes could have been explained as the effect of other genes in PXO86. However, we have demonstrated that the timing of the different interactions seen when *avrxa5*, *avrXa7*, and *avrXa10* are in the same genetic background (PXO99^A) is the same as that observed in the wild-type strain (PXO86) (Figs. 2 and 5). We therefore conclude that the timing of the physiological and phenotypical reactions is dependent on the specific avirulence-resistance gene combination.

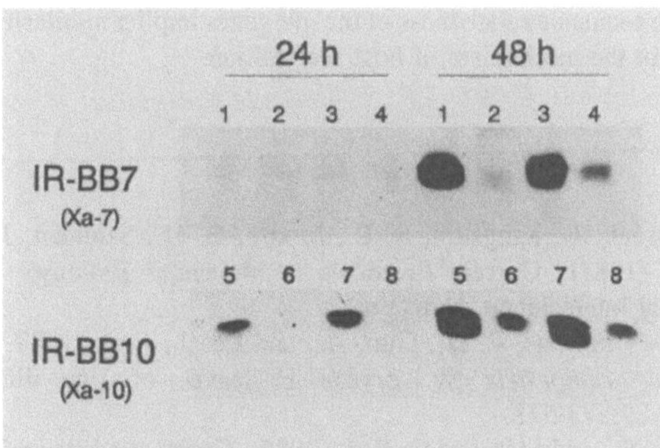

Figure 5. Peroxidase activity in extracellular fluids from rice leaves extracted at 24 and 48 h after infiltration with bacterial suspensions. Rice cultivars IR-BB7 and IR-BB10 were infiltrated with suspensions of *X. oryzae* pv. *oryzae* PXO86 (1 and 5); PXO99^A (2 and 6); PXO99^A(pXO29-29) (3); PXO99^A(pTn5-53) (4); PXO99^A(pXO5-15) (7); PXO99^A(pTn5-35) (8). Extracellular proteins (0.2 μg/lane) were separated in a nondenaturing, cathodic gel (Thomas and Hodes 1981). Peroxidase activity was detected with a mixture of the substrates guaiacol and 3-amino-9-ethylcarbazole.

Summary

We have identified clones from *X. oryzae* pv. *oryzae* that contain three avirulence genes (*avrxa5, avrXa7, avrXa10*), which control bacterial elicitation of resistance in rice cultivars carrying the *xa-5, Xa-7,* and *Xa-10* resistance genes, respectively. All of these genes were found on cosmid clones containing sequences homologous to avrBs3, an avirulence gene from the pepper pathogen *X. campestris* pv. *vesicatoria* (Bonas et al. 1989). Two of the genes, *avrXa7* and *avrXa10,* were localized by transposon mutagenesis to regions of the clone which hybridized to *avrBs3.* The identification of a 102-bp repeated sequence within the active regions of *avrXa7* and *avrXa10* and other closely linked homologous sequences along with the DNA sequence of *avrXa10* indicates that these genes are most likely homologs of *avrBs3.* The third gene (*avrxa5*) is contained within a clone that hybridized with *avrBs3* and is likely to be an *avrBs3*-homolog. The *avrBs3*-homologs in *X. oryzae* pv. *oryzae,* therefore, represent a family of genes of which some are involved in the induction of gene-specific resistance in rice. In contrast, *avrBs3* and the *avrBs3*-like gene *avrBsP* are the only related avirulence genes found in *X. campestris* pv. *vesicatoria* (Canteros et al. 1991).

The cloning of avirulence genes and the development of cultivars near-isogenic for resistance to *X. oryzae* pv. *oryzae* has added a new dimension to our investigations into the physiological responses associated with resistance in rice. Responses that are induced in such "near-isogenic" interactions are clearly the result of the interplay between avirulence genes and corresponding resistance genes.

Our work adds to the growing list of related genes that function to confer avirulence, and extends the types of plants with which they interact from dicots (pepper, Bonas et al. 1989; tomato, Canteros et al. 1991; and cotton, Swarup et al. 1992) to a monocot (rice). Although nothing is known about how the avirulence genes function to confer resistance, it is tempting to speculate that the sequence relatedness of the *avr* genes implies similarity in both avirulence gene function and in the mechanism of host recognition.

References

Ausubel, F. M., Brent, R., Kingston, R. E., Moore, D. D., Seidman, J. G., Smith, J. A., and Struhl, K. (1987) Current Protocols in Molecular Biology. Greene Publishing Associates-Wiley Interscience, New York.

Barton-Willis, P. A., Roberts, P. D., Guo, A., and Leach, J. E. 1989. Growth dynamics of *Xanthomonas campestris* pv. *oryzae* in leaves of rice differential cultivars. Phytopathology 79:573-578.

Bonas, U., Stall, R. E., and Staskawicz, B. J. 1989. Genetic and structural characterization of the avirulence gene *avrBs3* from *Xanthomonas campestris* pv. *vesicatoria.* Mol. Gen. Genet. 218:127-136.

Canteros, B., Minsavage, G., Bonas, U., Pring, D., and Stall, R. 1991. The cloning and characterization of a gene from *Xanthomonas campestris* pv. *vesicatoria* that determines avirulence in tomato. Mol. Plant-Microbe Interact. 4:628-632.

De Feyter, R., and Gabriel, D. W. 1991. At least six avirulence genes are clustered on a 90-kilobase plasmid in *Xanthomonas campestris* pv. *malvacearum.* Mol. Plant-Microbe Interact. 4:423-432.

Ellingboe, A. H. 1976. Genetics of host-parasite interactions. Pages 761-778 in: Encyclopedia of Plant Physiology, vol. 4. R. Heitefuss and P. H. Williams, eds. Springer-Verlag, Berlin.

Flor, H. H. 1955. Host-parasite interaction in flax rust-its genetics and other implications. Phytopathology 45:680-685.

Gabriel, D., Burges, A., and Lazo, G. 1986. Gene-for-gene interactions of five cloned avirulence genes from *Xanthomonas campestris* pv. *malvacearum* with specific resistance genes in cotton. Proc. Natl. Acad. Sci. USA 83:6415-6419.

Herbers, K., Conrads-Strauch, J., and Bonas, U. 1992. Race-specificity of plant resistance to bacterial spot disease determined by repetitive motifs in a bacterial avirulence protein. Nature 356:172-174.

Jones, R., Barnes, L., Gonzalez, C., Leach, J., Alvarez, A., and Benedict, A. 1989. Identification of low-virulence strains of *Xanthomonas campestris* pv. *oryzae* from rice in the United States. Phytopathology 79:984-990.

Knoop, V., Staskawicz, B., and Bonas, U. 1991. Expression of the avirulence gene *avrBs3* from *Xanthomonas campestris* pv. *vesicatoria* is not under the control of *hrp* genes and is independent of plant factors. J. Bacteriol. 173:7142-7150.

Mew, T. W. 1987. Current status and future prospects of research on bacterial blight of rice. Annu. Rev. Phytopathol. 25:359-382.

Minsavage, G. V., Dahlbeck, D., Whalen, M. C., Kearney, B., Bonas, U., Staskawicz, B. J., and Stall, R. E. 1990. Gene-for-gene relationships specifying disease resistance in *Xanthomonas campestris* pv. *vesicatoria*-pepper interactions. Mol. Plant-Microbe Interact. 3:41-47.

Ou, S. H. 1985. Rice Diseases, 2nd Ed. Commonw. Mycol. Inst., Kew, Surrey, U.K.

Reimers, P. J., Guo, A., and Leach, J. E. 1992. Increased activity of a cationic peroxidase associated with incompatible interactions between *Xanthomonas oryzae* pv. *oryzae* and rice (*Oryza sativa*). Plant Physiol., in press.

Reimers, P. J., and Leach, J. E. 1991. Race-specific resistance to *Xanthomonas oryzae* pv. *oryzae* conferred by bacterial blight resistance gene *Xa-10* in rice (*Oryza sativa*) involves accumulation of a lignin-like substance in host tissues. Physiol. Mol. Plant Pathol. 38:39-55.

Ronald, P. C., and Staskawicz, B. J. 1988. The avirulence gene *avrBs1* from *Xanthomonas campestris* pv. *vesicatoria* encodes a 50-kD protein. Mol. Plant-Microbe Interact. 1:191-198.

Simon, R., Quandt, J., and Klipp, W. 1989. New derivatives of transposon Tn*5* suitable for mobilization of replicons, generation of operon fusions and induction of genes in Gram-negative bacteria. Gene 80:161-169.

Swanson, J., Kearney, B., Dahlbeck, D., and Staskawicz, B. 1988. Cloned avirulence gene of *Xanthomonas campestris* pv. *vesicatoria* complements spontaneous race-change mutants. Mol. Plant-Microbe Interact. 1:5-9.

Swarup, S., Yang, Y., Kingsley, M. T., and Gabriel, D. W. 1992. An *Xanthomonas citri* pathogenicity gene, *pthA*, pleiotropically encodes gratuitous avirulence on non-hosts. Mol. Plant-Microbe Interact. 5:204-213.

Swings, J., Van den Mooter, M., Vauterin, L., Hoste, B., Gillis, M., Mew, T. W., and Kersters, K. 1990. Reclassification of the causal agents of bacterial blight (*Xanthomonas campestris* pv. *oryzae*) and bacterial leaf streak (*Xanthomonas campestris* pv. *oryzicola*) of rice as pathovars of *Xanthomonas oryzae* (ex Ishiyama 1922) sp. nov., nom. rev. Int. J. Syst. Bacteriol. 40:309-311.

Thomas, J. M., and Hodes, M. E. 1981. A new discontinuous buffer system for the electrophoresis of cationic proteins at near-neutral pH. Anal. Biochem. 118: 194-196.

Whalen, M. C., Stall, R. E., and Staskawicz, B. J. 1988. Characterization of a gene from a tomato pathogen determining hypersensitive resistance in non-host species and genetic

analysis of this resistance in bean. Proc. Natl. Acad. Sci. USA 85:6743-6747.

FURTHER CHARACTERIZATION OF GENES ENCODING EXTRACELLULAR POLYSACCHARIDE OF *Pseudomonas solanacearum* AND THEIR REGULATION

MARK A. SCHELL, TIMOTHY P. DENNY, STEVEN J. CLOUGH, AND JIANZHONG HUANG
Departments of Microbiology and Plant Pathology
University of Georgia, Athens, GA 30602

ABSTRACT. *Pseudomonas solanacearum* produces an acidic, nitrogen-rich, extracellular polysaccharide (EPS) that is required for wilting and killing of infected plants. Biosynthesis of EPS is partially encoded by the 18-kb *eps* locus; additional loci near *eps* are also required for EPS production, but only under specific growth conditions. All these loci are transcriptionally controlled by an interacting regulatory network involving the products of at least five distinct regulatory loci: *phcA*, *phcB*, *vsrA*, *vsrB*, and *xpsR*. This network, which also regulates other virulence genes, may control transcription in response to various environmental signals, such as nutritional status and cell density.

1. Introduction

Pseudomonas solanacearum is a major plant pathogen that causes a lethal wilting disease of several hundred diverse species of plants, including the important crops peanut, potato, tomato, and banana [1]. It is especially a problem in tropical and subtropical environments where the severity and incidence of the disease are on the rise. *P. solanacearum* is a soil-borne pathogen that infects susceptible plants through the roots and then spreads through the vascular system multiplying to nearly 10^{11} cells per plant, often wilting and killing it in less than two weeks [2, T. P. D. and M. A. S., unpublished].

Many, if not all, virulent *P. solanacearum* strains produce an abundance of a noncapsular, extracellular polysaccharide slime (EPS) in amounts nearly equivalent to the cell mass. This EPS is heterogenous, but analysis of EPS from one strain [3] shows it is composed of two major fractions, each representing 40% of its total mass. One fraction (EPS I) is a high-molecular-weight, acidic polysaccharide composed of N-acetyl galactosamine and derivatives thereof, with a trimeric repeat unit of N-acetyl galactosamine, 2-N-acetyl-2-deoxy-L-galacturonic acid, and 2-N-acetyl-4-N-(3-hydroxybutanoyl)-2,4,6-trideoxy-D-glucose (Fig. 1). The latter two components and high nitrogen content of the EPS trimer are relatively rare in prokaryotic EPSs [4]. However an analogous trimer is found in the O-antigen in the lipopolysaccharide of *P. aeruginosa* strain Hab [3]. The other major fraction (EPS III) is an uncharged, uncharacterized, mostly non-glycosidic fraction that is rich in mannose. A slightly acidic, rhamnose-rich polymer (EPS II) and glucan are found as minor (< 10%) components of EPS.

E. W. Nester and D. P. S. Verma (eds.),
Advances in Molecular Genetics of Plant-Microbe Interactions, 231–239.

The virulence of various EPS-deficient *P. solanacearum* mutants in tomato and eggplant has been investigated [5-8]. All these mutants have nonmucoid colony morphology and produce reduced levels of extracellular polymeric galactosamine; however, which EPS fractions they are missing has not been reported. In general, such mutants grow in planta like wild type, but only slowly or partially wilt, and often do not kill the host. To date EPS appears to be the most important proven virulence factor in wilt disease caused by *P. solanacearum* [6,8]. Although EPS probably causes wilting by plugging xylem vessels thereby reducing water flow, non-nitrogenous EPSs such as dextrans could theoretically do the same [9]. The unusually high nitrogen content of *P. solanacearum* EPS I and the exceedingly complex regulation of its synthesis described later implies that EPS may perform very important, as yet unknown functions. EPS could serve as a storage polymer for later use as a carbon and nitrogen source. Purified EPS appears to be utilized as sole carbon, nitrogen, and energy source by several bacteria, including *P. solanacearum*, albeit poorly [M. A. S., unpublished].

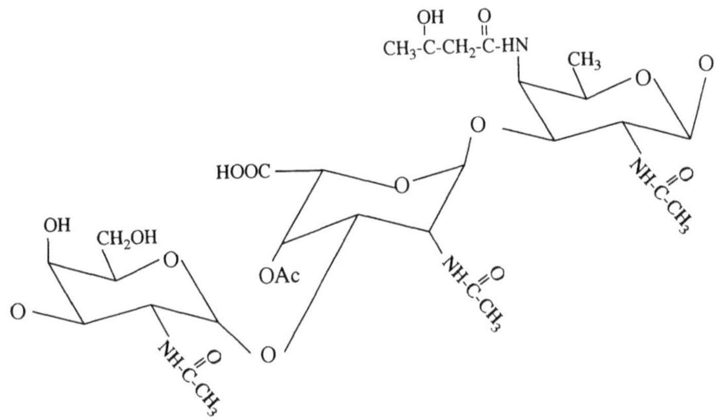

Figure 1. Trimeric repeat unit of EPS I, the acidic galactosamine-rich extracellular polysaccharide of *P. solanacearum* encoded by *eps*. Structure taken from [3]. OAc, O-acetyl.

P. solanacearum also produces many extracellular proteins (EXPs) [10], and it is likely that many are important in disease. However, the involvement in virulence of only three EXPs which degrade components of plant cell-walls: the 43-kDa endoglucanase, Egl [11], the 52-kDa endopolygalacturonase PglA, and the 80-kDa exopolygalacturonase, PglB [12, M.A.S. and J.H., unpublished; C. Allen personal communication], has been studied. The majority of evidence [8] suggests that once the pathogen enters the plant vascular system, the absence of one or even two of these enzymes slows the rate of wilting and death by anywhere from 25% to 75%, and that these enzymes are not absolutely required for wilt disease or normal growth in planta. However, most virulence analyses have used stem-inoculation, which bypasses the natural root infection process where such enzymes may be more important. In one study [8], Egl appeared to be much more important in root infection (and/or soil survival) than PglA. *P. solanacearum* produces a 44-kDa pectinmethylesterase [13] and many other uncharacterized EXPs, including an EPS-related 28-kDa EXP [10]. The role of these in wilt disease is presently under active study in our labs.

Other than *eps* and *hrp* genes [14], the only other genes shown to be required for wilt production by *P. solanacearum* are those encoding virulence gene regulators which when inactivated, simultaneously and differentially affect EPS production, EXP production, and/or other physiological systems [15, M.A.S., J.H., and T.P.D., in preparation]. The characterization of these virulence gene regulators and others [16; C. Allen, personal communication] has just begun. Here we will present new data about the organization of *eps* and also reveal its complex regulation by a sensory network controlling production of many other virulence factors of *P. solanacearum*.

2. Strains and methods.

Strains or methods for their construction are described elsewhere [5,8]. Transposon mutagenesis of plasmids in *E. coli* or the genome of *P. solanacearum* with Tn*phoA* or Tn*lacZ* has been described [17-19]. Marker exchange was by electroporation of *P. solanacearum* with pUC-based plasmids or genomic DNA of *P. solanacearum* [M.A.S. unpublished]. pRK415 derivatives were transfered from *E. coli* to *P. solanacearum* by conjugative mobilization [11, 12]. For EPS analysis liquid cultures were grown shaking at 30°C, centrifuged twice at 15,000 x g for 20 min, and the culture supernatant dialyzed extensively against water. EPS was quantified by measuring hexosamine with the Elson-Morgan reaction [15] or hexose with anthrone reagent.

3. Results

3.1. FURTHER CHARACTERIZATION OF THE *eps* LOCUS.

Region I of the *P. solanacearum* genome was originally defined and subcloned by Denny and Baek [5] and shown to be a 9 kb segment required for EPS synthesis and virulence. Using transposon mutagenesis of flanking segments and marker exchange, we have found that the genomic region required for EPS synthesis is actually a larger 18-kb locus we now call *eps*. Transposon insertion throughout *eps* (Fig. 2; 19A, 130, 129, Z7) inactivates EPS production (i.e. causes loss of mucoid colony morphology). Insertions in and around *rgnII* (Region II [5]) conditionally reduce EPS synthesis (see 3.2 below). Most insertions in 5-kb regions either upstream of *eps* or downstream of *rgnII* do not affect the EPS+ phenotype, limiting contiguous EPS production genes to the 25-kb region shown in Fig. 2. Another region of the *P. solanacearum* genome is required for EPS production; however it is also necessary for synthesis of the lipopolysaccharide O-antigen and may possibly be involved in synthesis of polysaccharide precursors [6].

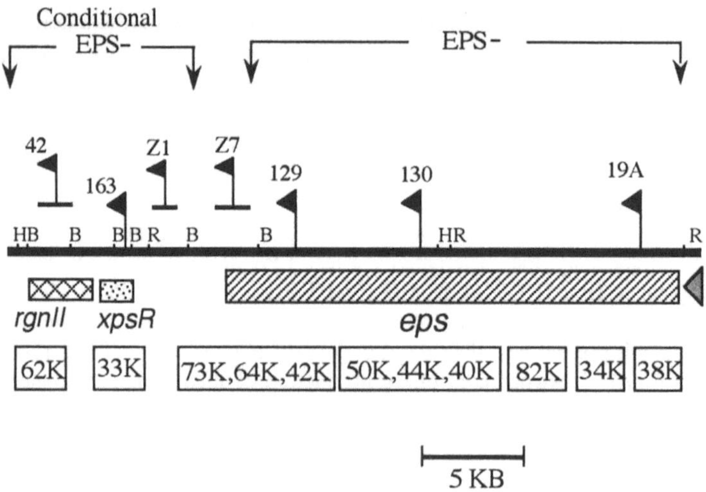

Figure 2. Map of cloned genomic region of *P. solanacearum* involved in EPS biosynthesis. Flags show positions and transcription direction of *eps::*Tn*3-HoHo1* (*lacZ*) insertions (129: AW1-129; 130: AW1-130; 42: AW1-42; ref.5) and Tn*5lacZB20* insertions (19A, Z1, Z7, 163, unpublished). Exact insertion sites and transcription directions of 42, Z1, and Z7 are unkown, but underline indicates limits. Hatched boxes show extent of loci; open boxes below show encoded polypeptides and sizes from maxicell analysis in approximate order. *rgnII* = Region II [5]. R-*Eco*R1, H-*Hind*III; B-*Bam*HI.

Orgambide et al [3] showed that anion-exchange chromatography in 7M urea divides the EPS of *P. solanacearum* GMI1000 largely into a neutral and a galactosamine-rich acidic subfractions, each representing approximately 40% of the total weight of EPS. Using the same fractionation methods on total EPS from our wild type strain AW and two *eps* insertion mutants (Fig. 2, 19A and 129), we found that inactivation of *eps* blocks production of the acidic-galactosamine rich EPS subfraction (EPS I), but not the neutral subfraction (EPS III) (Table 1). Thus, we beleive that the *eps* locus encodes proteins specifically involved in production of the nitrogen rich EPS I (Fig. 1). The 18-kb size of *eps* is consistent with the amount of genetic information needed to synthesize and polymerize the unusual components of EPS I. In comparison, synthesis of *Xanthomonas campestris* EPS is encoded by >16 kb region containing 12 genes/polypeptides tentatively identified as glycosyl transferases, acylases, polymerases, and a ketalase [4].

TABLE 1. Fractions of extracellular polysaccharide (EPS) produced by *P. solanacearum* strain AW and *eps* mutants

		Amounts Produced*			
Fraction	Charge†	Wild type		*eps*::Tn*lacZ*	
		Hex	Hex-NH_2	Hex	Hex-NH_2
EPS I	Strongly acidic	26	650	38	12
EPS II	Weakly acidic	<4	<4	7	<4
EPS III	Neutral	90	<4	85	<4
Total	-	140	730	150	16

†Strongly acidic: elutes from DEAE-Sephacel with 0.5 M NaCl; weakly acidic: elutes with 0.05 M NaCl; neutral: unretained by DEAE-Sephacel.
*$\mu g/10^{10}$ cells: Hex = polymeric (>8 kDa) hexose; Hex-NH_2 = polymeric hexosamine. *eps*::Tn*lacZ*: AW1-129 or AW-19A (see Fig. 2) gave largely the same results.

Maxicell, Tn*phoA*, and DNA sequence analysis of fragments from *eps* have identified at least ten encoded polypeptides (some membrane-associated) that are probably involved in EPS I biosynthesis (Fig. 2); overexpression of some of these polypeptides by *E. coli* in glucose-free media markedly inhibits growth. We are now further characterizing these and the remaining *eps* gene products for function and localization. Based on the structure of EPS I we are attempting to correlate *eps* gene products with putative enzymatic activities that could polymerize EPS I, modify it, or make its precurors. Radiotracer studies are also in progress with EPS-deficient mutants to help elucidate the EPS biosynthetic pathway and correlate it with individual *eps* gene products.

3.2. EFFECTS OF COMPLEX NITROGEN ON EPS PRODUCTION/REGULATION.

Transposon insertions in and on one side of *rgnII* [5] (Fig. 2; Z1, 163, 42) cause only a conditional loss of EPS production which is dependent on growth medium; whereas EPS production by wild type is growth-medium insensitive (Table 2). Insertions in *rgnII* decrease EPS production by 15-fold in BG medium, but have no effect on EPS levels in minimal medium. Addition of peptone to minimal medium causes a 10-fold decrease in EPS I production by *rgnII* mutants, indicating that it contains a component(s) which lowers EPS levels. Since *rgnII* does not regulate *eps* (see 3.3), our working hypothesis is that *rgnII* encodes an EPS biosynthetic enzyme(s) which can be replaced by another, isofunctional enzyme, whose synthesis or activity is reduced by peptone. Investigation of the nature of the peptone component(s) responsible is under way; several amino acids have been implicated.

TABLE 2. Influence of growth medium on EPS I production by various *P. solanacearum* mutants

	EPS Production† in:			
Genotype*	MIN	BG	MIN/CAA	MIN/PEP
Wild type	450	510	660	625
rgnII::Tn*5*	525	40	690	70
xpsR::Tn*phoA*	nt	20	175	nt
epsZ1::Tn*lacZ*	150	125	700	nt

†μg polymeric (>8 kDa) hexosamine per 10^{10} cells; MIN, basal salts medium + 1% sucrose [11]; BG, 1% glucose + 1% peptone + 0.1% CAA + 0.1% yeast extract; CAA, 0.2% casamino acids; PEP, 0.5% peptone; nt, not tested.
*See Fig. 2.

Insertions in the *xpsR* locus near *rgnII* (Fig. 2; 163) cause only a 3.5-fold decrease in EPS levels in minimal medium with casamino acids, but a 25-fold decrease in peptone-containing (BG) medium (Table 2). Insertions upstream in a 2-kb *Bam*HI-*Eco*RI fragment (Z1; Fig. 2) reduce EPS levels in minimal or BG medium to 25% of wild type; addition of casamino acids to the medium restores wild type EPS levels to Z1 mutants. These results suggest that nitrogen status may be an important physiological factor regulating EPS synthesis; this would not be surprising given the relatively high nitrogen content of *P. solanacearum* EPS I. Transcriptional regulation of *eps* in response to nitrogen status may be mediated by *xpsR* and/or some of other regulatory genes described below.

3.3. A COMPLEX NETWORK REGULATES PRODUCTION OF EPS

Transposon mutagenesis and subcloning identifed five distinct, non-allelic loci in *P. solanacearum* (*phcA*, *phcB*, *xpsR*, *vsrA*, and *vsrB*) which when inactivated cause 90% reduction in EPS synthesis, loss of ability to kill or wilt host plants, and dramatic effects on production of other virulence-related factors and extracellular proteins [J.H, T.P.D., and M.A.S. in preparation]. Except for *vsrA* and *xpsR* mutants which are very similar, mutations at each locus confer a unique phenotype. Whereas *phcA* and *vsrB* mutants grow like wild type in planta and produce minor symptoms (e.g. yellowing and stunting), *xpsR* or *vsrA* mutants appear to grow much more poorly in planta, and often cause no overt disease symptoms.

Inactivation of *vsrA*, *xpsR*, or *vsrB* caused >90% reduction in transcription of *eps* as measured with an *eps::lacZ* reporter (130; Fig. 2); inactivation of *phcA* [5,15] caused a 99% reduction of *eps* transcription measured with the same reporter. *phcA* and *vsrA* (but not *vsrB*) are also required for full expression of the *rgnII::lacZ* reporter (Fig. 2; 42). This suggests that transcription of *eps* is controlled by *phcA*, *vsrA*, *xpsR*, and *vsrB*, and that *phcA* and *vsrA* additionally control *rgnII* transcription. Inactivation of *rgnII* did not dramatically affect expression of the *eps-130::lacZ* reporter. However a 6 kb fragment containing *rgnII* and *xpsR* in trans on pQF44 [5] restored wild type EPS production and expression of the *eps-130::lacZ* reporter in *vsrA* mutants, and partially suppressed (i.e. increased 20-fold) lowered *eps::lacZ* expression caused by the *phcA* mutation. While *phcA* appears to be a LysR-type transcriptional activator [S. Brumbley, B. Carney, and T. P. Denny; submitted], Tn*phoA* mutagenesis experiments suggest that *vsrA* and *vsrB* may be exported proteins, possibly membrane sensors involved in signal transduction for virulence gene regulation. Investigations of the molecular nature of these regulatory products, the signals they recognize, and their interactions is in progress.

Insertional inactivation of the complex *phcB* locus located 12 kb downstream of *phcA* produces a phenotype nearly identical in all aspects (e.g. loss of EPS production and *eps::lacZ* expression, altered EXP production, etc.) to that of *phcA* mutants [15], except that the *phcB* phenotype can be completely (and reversibly) restored to wild type by exposure to cell-free culture supernatants or vapors of log-phase cultures of *phcB*+ strains [S. Clough, M.A.S., and T.P.D., in preparation]. A limited number of very specific, pure chemicals (e.g. fatty acid derivatives), but not ethylene, can also suppress the *phcB* mutant phenotype. We hypothesize that this may signify the presence of a *phcA*/*phcB*-mediated extracellular cell-cell signalling system that regulates virulence gene expression. This system may be analogous to the cell-density dependent bioluminescence (*lux*) gene regulation system of *Vibrio fischeri* [20]. In support of this we have found that *eps::lacZ* expression (specific activity per cell) is 50-fold less in low density log-phase cultures compared to fully grown log-phase cultures. At present we are attempting to correlate the cell-density dependent regulation of *eps* with *phcB* and the volatile signal molecule to confirm this hypothesis.

4. Conclusions

Host-pathogen interactions are dynamic and complex. Therefore it is not suprising that virulence gene expression is modulated by multicomponent, interacting, and environmentally-responsive systems [21]. The 18-kb *eps* locus of *P. solanacearum* encoding at least ten proteins required for synthesis of an unusual nitrogen-rich polysaccharide virulence factor is also highly regulated by an unusually complex system. While the regulatory network has five known components, recent data strongly suggest additional components remain undiscovered. Suppression of *phcA* and *vsrA* mutations by a locus on pQF44 implies that some network components interact at the transcriptional or posttranslational level. Mutations in *vsrB*, *phcA*, *phcB*, *vsrA*, or *xpsR* cause different, simultaneous positive and negative effects on levels specific extracellular proteins (some known or suspected virulence factors), indicating that the network regulates targets other than *eps* and may crosstalk with other virulence regulators such as *pehR* [16] or *epsR* [L. Sequieria, personal communication].

Our working model for the regulatory network is that *phcA* (which regulates multiple virulence determinants [15]), is a master switch which determines when virulence gene expression is appropriate. Its ability to transcriptionally activate virulence genes appears to depend on a signal molecule that accumulates with increasing cell density and requires *phcB* for its synthesis. It is likely that *phcA* controls virulence genes indirectly via an underlayer of intermediate transcriptional regulators such as *vsrA*, *xpsR* or *vsrB* working below the global level of *phcA*. The interdependence of *xpsR* and *vsrA* and nitrogen-conditional effects on EPS levels caused by their inactivation, suggests that they or other network components modulate *eps* expression in response to nitrogen status. Elucidation of the signals and interactions in the network controlling *P. solanacearum* virulence genes will undoubtedly further our understanding of strategies employed by both plant and animal pathogens.

Acknowledgements. Some work described was supported by grants from the University of Georgia Biotechnology Program and the National Science Foundation (DMB-89-04472 and DCB-91-17544). We thank Rosemary Wood for help in preparation of the manuscript.

5. References

1. Hayward, A. C. 1991. Biology and epidemiology of bacterial wilt caused by *Pseudomonas solanacearum*. Annu. Rev. Phytopathol. 29:65-87.
2. Wallis, F. M. and Truter, S. J. 1978. Histopathology of tomato plants infected with *Pseudomonas solanacearum*, with emphasis on ultrastructure. Physiol. Plant Pathol. 13:307-316.
3. Orgambide, G., Montrozier, H., Sevin, P., Roussel, J., Trigalet-Demery, D., and Trigalet, A. 1991. High heterogeneity of the exopolysaccharides of *Pseudomonas solanacearum* strain GMI1000 and the complete structure of the major polysaccharide. J. Biol. Chem. 266:8312-8321.
4. Sutherland, I. W. 1991. Microbial Polysaccharides. Marcel Dekker, NY, NY.
5. Denny, T. P. and Baek, S.-R. 1991. Genetic evidence that extracellular polysaccharide is a virulence factor of *Pseudomonas solanacearum*. Molec. Plant-Microbe Interact. 4:198-206.
6. Kao, C. C. and Sequeira, L. 1991. A gene cluster required for the coordinated biosynthesis of lipopolysaccharide also affects virulence of *Pseudomonas solanacearum*. J. Bacteriol. 173:7841-7847.
7. Denny, T. P., Makini, F. W., and Brumbley, S. M. 1988. Characterization of *Pseudomonas solanacearum* Tn*5* mutants impaired in production of extracellular polysaccharide. Molec. Plant-Microbe Interact. 1:215-223.

8. Denny, T. P., Carney, B. F., and Schell, M. A. 1990. Inactivation of multiple virulence genes reduces the ability of *Pseudomonas solanacearum* to cause wilt symptoms. Molec. Plant-Microbe Interact. 3:293-300.
9. Van Alfen, N. K. 1989. Reassessment of plant wilt toxins. Annu. Rev. Phytopathol. 27:533-550.
10. Schell, M. A. 1987. Purification and characterization of an endoglucanase from *Pseudomonas solanacearum*. Appl. Environ. Microbiol. 53:2237-2241.
11. Roberts, D P., Denny, T. P., and Schell, M. A. 1988. Cloning of the *egl* gene of *Pseudomonas solanacearum* and analysis of its role in phytopathogenicity. J. Bacteriol. 170:1445-1451.
12. Schell, M. A., Roberts, D. P., and Denny, T. P. 1988. Analysis of the *Pseudomonas solanacearum* polygalacturonase encoded by *pglA* and its involvement in phytopathogenicity. J. Bacteriol. 170:4501-4508.
13. Spök, A., Stubenrauch, G., Schorgendorfer, K., and Schwab, H. 1991. Molecular cloning and sequencing of a pectinesterase gene from *Pseudomonas solanacearum*. J. Gen. Microbiol. 137:131-140.
14. Arlat, M., Gough, C. L., Barber, C. E., Boucher, C., and Daniels, M. J. 1991. *Xanthomonas campestris* contains a cluster of *hrp* genes related to the larger *hrp* cluster of *Pseudomonas solacearum*. Mol. Plant-Microbe Interact. 4:593-602.
15. Brumbley, S. M. and Denny, T. P. 1990. Cloning of *phcA* from wild-type *Pseudomonas solanacearum*, a gene that when mutated alters expression of multiple traits that contribute to virulence. J. Bacteriol. 172:5677-5685.
16. Allen, C., Huang, Y., and Sequeira, L. 1991. Cloning of genes affecting polygalacturonase production in *Pseudomonas solancearum*. Mol. Plant-Microb. Interact. 4:147-154.
17. Simon, R., Quandt, J., and Klipp, W. 1989. New derivatives of transposon Tn5 suitable for mobilization of replicons, generation of operon fusons and induction of genes in Gram-negative bacteria. Gene 80:161-169.
18. Stachel, S. E., An, G., Flores, C., and Nester, E. W. 1985. A Tn*3lacZ* transposon for the random generation of ß-galactosidase gene fusions: application to analysis of gene expression in *Agrobactorium*. EMBO J. 4:891-898.
19. Manoil, C., Mekalanos, J., and Beckwith, J. 1990. Alkaline phosphatase fusions: sensors of subcellular localization. J. Bacteriol. 172:515-518.
20. Baldwin, T. O., Devine, J. H., Heckel, R. C., Lin, J.-W., and Sadel, G. S. 1989. The complete nucleotide sequence of the *lux* regulon of *Vibrio fischeri* and the *luxABN* region of *photobacterium leiognathi* and the mechanism of control of bacterial bioluminescence. J. Biolumin. Chemilumin. 4:326-341.
21. Miller, J. F., Mekalanos, J., and Falkow, S. 1989. Coordinate regulation and sensory transduction in the control of bacterial virulence. Science 243:916-919.

Physiological and genetic regulation of a pectate lyase structural gene, pel-1 of Erwinia carotovora subsp. carotovora strain 71

Asita Chatterjee, Yang Liu, Hitoshi Murata, Thouraya Souissi[1/], and Arun K. Chatterjee*
Department of Plant Pathology, 108 Waters Hall, University of Missouri-Columbia, Columbia, MO 65211

[1/]Present address: Department of Agronomy, 210 Waters Hall, University of Missouri-Columbia, Columbia, MO 65211

Abstract. Of the various exoproteins secreted by Erwinia carotovora subsp. carotovora strain 71 (Ecc71), Pel-1 is the major pectate lyase species with tissue macerating activity. To analyze pel-1 expression, a pel1-lacZ operon fusion, created by MudI1734 insertion, was placed by homologous recombination within the chromosome of the LacZ⁻ derivative of Ecc71. pel1-lacZ expression was inhibited by a negative regulator gene, repN and activated by plant extracts during late exponential growth and by the products of the regulator genes aepA, aepB and aepH*. Since these regulatory genes control, in addition to pel-1 expression, the production of polygalacturonase, cellulase, protease and the other Pels, we propose that they constitute a global network and control extracellular protein production in Ecc71 by modulating the levels of transcripts.

[Introduction]

Erwinia carotovora subsp. carotovora (Ecc) elicits post-harvest decay (= soft rot) in a wide variety of plant products, thereby causing severe economic losses throughout the world. The production of pectolytic enzymes, especially the pectate lyases (Pels), by the pathogen is the most critical factor in the development of the disease. All of the Ecc strains and most other soft-rot Erwinia spp. produce a number of Pel species that are readily distinguished by their isoelectric points (pIs) [2, 25, 28]. Ecc strain 71, the model organism in our laboratory, produces at least five Pel species: the three most basic Pel species [Pel-1, pI 10; Pel-2, pI 9.8; and Pel-3, pI 9.4] are secreted outside the bacterial cell, whereas Pel-4 (pI 8.0) and Pel-5 (pI 6.6) are predominantly localized in the periplasm [31, 34]. The specific contributions of the periplasmic and secreted Pel species to pathogenicity and pectate catabolism and the molecular basis for the differential Pel localization await clarification.

The observation that Pels can elicit tissue maceration has generated

E. W. Nester and D. P. S. Verma (eds.),
Advances in Molecular Genetics of Plant-Microbe Interactions, 241–251.

considerable interest in genes specifying these enzymes. Structural genes for a number of pectinases and out genes specifying secretion functions have been cloned and sequenced [4, 8, 9, 10, 12, 13, 16, 19, 22, 26, 27, 29, 30, 33 and references cited therein]. In contrast, molecular and genetic aspects of the regulation of Pels and other extracellular enzymes await an in depth analysis. In E. chrysanthemi, two negative regulator genes [kdgR and ptlR] have been identified [23, 33]. The kdgR product has been shown to repress the expression of pel as well as the expression of other genes involved in pectate catabolism [18, 23]. Despite the occurrence of KdgR binding sites in several pel and peh genes of Ecc [10, 26], there is no direct evidence for a regulatory role of KdgR in this bacterium.

In Ecc, there is now substantial evidence for a coordinate regulation of exoenzyme production. For example, mutants have been isolated that are pleiotropically deficient in the production of Pel, Peh, Cel and Prt [1, 17, 20]. In addition, regulatory genes (aepA, activator of extracellular protein production [17]; exp, exoprotein production [20]) have been cloned that restore enzyme production in most such mutants. We have now extended these findings by isolating two additional activator genes [aepB, $aepH^{*}$] and a negative regulator gene [repN = negative regulator of extracellular protein production] whose products control the production of Pel, Peh, Cel and Prt in Ecc71. Using a pel1-lacZ operon fusion, we have examined the effects of these regulatory genes and several physiological parameters on pel-1 expression. Our data demonstrate that (a) pel-1 expression is dependent upon the bacterial growth phase; (b) substances known to stimulate exoenzyme production also stimulate pel-1 expression; and (c) the products of aep and repN genes affect pel-1 expression. Our findings support the hypothesis that these genes regulate production of secreted proteins by modulating transcription of the target genes.

[Material and Methods]

BACTERIAL STRAINS AND MEDIA.

Bacterial strains and plasmids are described in the text. The strains were maintained on appropriate agar media as described [16]. The compositions of various media and procedures for the agar plate assays of enzymatic activities have been published [6, 16, 17]. For quantitation of enzymatic activities, bacterial cultures were grown in SYG (= salts-yeast extract-glycrol) medium or in the SYG medium supplemented with pectate or crude extracts of plant tissues [17].

ISOLATION OF $AepA^-$, $AepB^-$ and $AepH^{*}$ STRAINS BY ETHYL MENTHANSULFONATE (EMS) MUTAGENESIS.

A procedure described in Miller [15] was used in EMS mutagenesis of the Pel1-LacZ strain, AC5030. Bacterial cells were incubated for 3 hours with EMS, resulting in >95% killing. Single colonies were screened for enzymatic activities as described below.

CONSTRUCTION OF pelI-lacZ FUSIONS.

The mini-Mu-lacZ element, MudI1734 [3], was used as described [14, 17] for the construction of pelI-lacZ transcriptional fusions in pAKC228, a pBR329 derivative containing the Ecc71 $pel\text{-}1^{+}$ DNA. The location and orientation of the mini-Mu-lacZ insertion in pAKC619, which carries a pelI-lacZ fusion, were determined by restriction analysis. To construct an Ecc71 derivative carrying a chromosomal copy of pelI-lacZ DNA, pAKC619 was transferred to the $LacZ^{-}$ strain, AC5006. Isolates resulting from marker exchange were obtained using standard procedures [35]. The exchange of pel-1 with pelI-lacZ in the strain, AC5030 was confirmed by examining the Pel profile and by Southern hybridization.

DNA TECHNIQUES.

Standard procedures described in our previous publications were used [4, 14, 16]

Results

EFFECTS OF PLANT EXTRACTS AND GROWTH PHASE ON pelI-lacZ EXPRESSION.

To determine if the expression of pelI-lacZ responds to substances known to stimulate Pel production [5, 7, 32], we grew AC5030 in the salts-yeast extract-glycerol (SYG) medium containing various supplements and assayed the cultures for ß-galactosidase activity. ß-galactosidase levels in AC5030 grown in celery extract, potato extract and carrot extract were 2058, 1910 and 1200 Miller units, respectively. In contrast, in SYG + pectate, AC5030 produced 575 units of ß-galactosidase which was about two-fold higher than the level in SYG.

To assess the effect of growth phase upon the expression of pelI-lacZ, AC5030 was grown in the celery extract medium and the levels of ß-galactosidase were determined at prescribed intervals. The highest rate of ß-galactosidase synthesis commenced during late exponential growth after the culture had reached a Klett value of about 200 (Fig. 1).

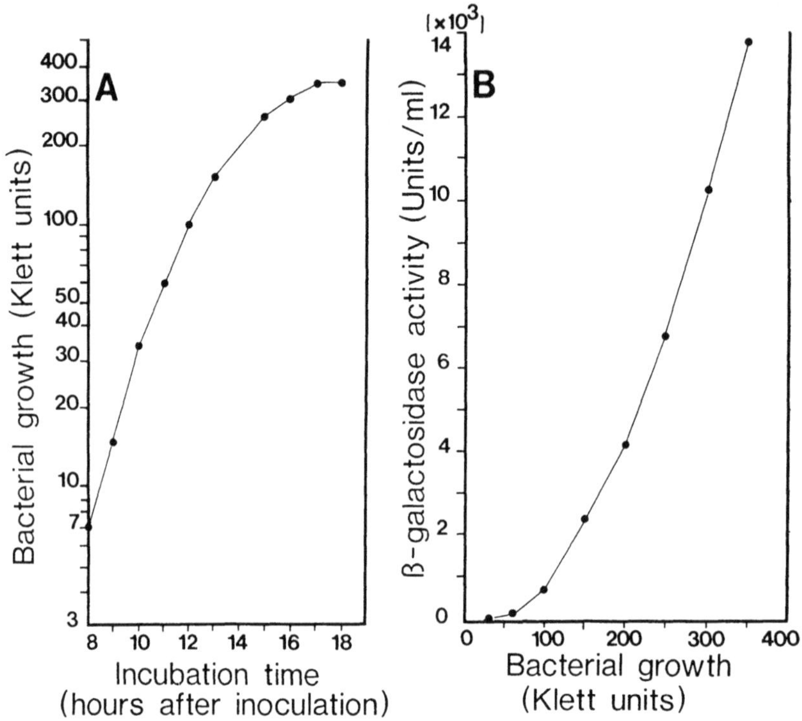

Figure 1. Production of ß-galactosidase by AC5030, a LacZ$^-$, Pel-1$^-$ and pel1-lacZ derivative of Ecc71, in a celery extract medium. The bacterium was grown at 28°C in a shaker and, at desired intervals, samples of cultures were removed and assayed for ß-galactosidase activity. The data for incubation periods beyond the initial eight hours are shown.

ISOLATION OF MUTANTS ALTERED IN pel1-lacZ EXPRESSION AND EFFECTS OF THE ACTIVATOR GENES.

Mutants altered in pel1-lacZ expression were isolated by EMS mutagenesis of AC5030. The mutagenized cells were screened on MacConkey-Lactose agar for ß-galactosidase activity or on nutrient-gelatin agar for protease activity. We identified two classes of mutants: one class produced very low levels of ß-galactosidase, pectate lyase, protease and cellulase, while the other class overproduced these enzymes. The data presented in Table 1 show that ß-galactosidase was essentially non-inducible in AC5040 and AC5043, the two representative class I mutants. In contrast, AC5034, a representative of class II mutants, produced a high basal level of ß-galactosidase; however, pel1-lacZ expression was barely inducible in this strain with celery extract (Table 2).

Table 1. Expression of pell-lacZ in AC5030 and its mutants pleiotropically defective in extracellular enzyme production and effects of cloned regulatory genes on ß-galactosidase production[a/].

Bacterial[b/] strains	Plasmid[c/]	Relevant genotypes[d/]	ß-galactosidase (Miller units) SYG + CE
AC5030	pSF6	$aepAB^+/aepAB^-$	1119
	pAKC264	$aepA^+/aepA^+$	1912
	pAKC637	$aepB^+/aepB^+$	1035
AC5040	pSF6	$aepA^-/aepA^-$	109
	pAKC264	$aepA^-/aepA^+$	637
	pAKC637	$aepA^-/aepB^+$	131
AC5043	pSF6	$aepB^-/aepB^-$	274
	pAKC264	$aepB^-/aepA^+$	229
	pAKC637	$aepB^-/aepB^+$	1005

[a/]Bacteria were grown in minimal salts-yeast extract glycerol medium containing celery extract (= SYG + CE) [17] Cultures were grown to a Klett value of about 350 in a shaker at 28 °C. ß-galactosidase activity was determined using samples of whole cultures, i.e., cells and culture medium.
[b/]AC5030 [$aepA^+$, Pell-LacZ], AC5040 [aepA, Pell-LacZ], AC5043 [aepB, Pell-LacZ].
[c/]pSF6 = the cloning vector, pAKC264 = pSF6 + $aepA^+$ DNA, pAKC637 = pSF6 + $aepB^+$ DNA.
[d/]Chromosomal genotype/plasmid genotype.

As previously reported [17], the aepA plasmid, pAKC264, was isolated from the library of Ecc71 genome cloned into the cosmid, pSF6. This cosmid complemented most of our pleiotropic mutants with the exception of AC5043 (Table 1). This mutant was designated $AepB^-$. To isolate the $aepB^+$ DNA, a genomic library of AC5034 was constructed in pSF6 using standard procedures. The library was mobilized into AC5043 and transconjugants screened for the restoration of extracellular enzyme production. By screening 1200 clones, we identified a cosmid, pAKC637, which complemented AC5043 (Table 1). This cosmid did not complement any of our $AepA^-$ mutants tested; the data for one $AepA^-$ mutant are shown in Table 1. The lack of cross complementation of $AepA^-$ and $AepB^-$ mutants by the cognate genes indicates that an interaction between AepA and AepB leads to the stimulation of enzyme production in Ecc71 and its derivatives. Our findings also reveal that aepA and aepB are not clustered on the Ecc71 chromosome. Southern blot hybridization and nucleotide sequence data have demonstrated that aepA and aepB do not share significant homology.

For the isolation of the gene responsible for enzyme overproduction (see above and Table 2 for the phenotype of AC5034), the library of AC5034 was mobilized into AC5006, and transconjugants were screened for the production of high levels of pectolytic and proteolytic enzymes on appropriate agar media. We identified several cosmids that in agar plate assays appeared to cause an overproduction of Pel, Peh, Cel and Prt. Further characterization of the cosmids revealed two classes: one class of cosmids resembled the AepA⁺ cosmid pAKC264 in that they stimulated the production of ß-galactosidase and extracellular enzymatic activities in AC5030 when grown in SYG+celery extract but not in SYG medium.

The cosmid, pAKC636, representing the other class also caused stimulation of extracellular enzyme production (Table 2). However, pAKC636 caused a significant increase in the basal level of ß-galactosidase in AC5030 (Table 2). A further three-fold increase in the level of ß-galactosidase was noted when the bacteria were grown in SYG + celery extract medium (Table 2). These phenotypes resembled those of the overproducing mutants described above. The cosmid pAKC636 restored enzyme production in AepA⁻ as well as in AepB⁻ mutants (data not shown).

Table 2. Expression of pell-lacZ in AepH* strain AC5034 and the parent strain, AC5030, carrying the aepH* plasmid pAKC636[a/].

Strains	Plasmids	ß-gal activity in (Miller units)	
		SYG	SYG+CE
AC5034	pSF6	3192	4868
AC5030	pSF6	417	1117
AC5030	pAKC636	1101	3192

[a/] Refer to Footnotes of Table 1 for the details of bacterial growth conditions and ß-galactosidase (ß-gal) assays.

[b/] pSF6 = the cloning vector, pAKC636 = pSF6 + aepH* DNA.

The aepH* gene, responsible for the overproduction of extracellular enzymes, has been localized to a 1,400 bp DNA segment by deletion analysis and by saturation mutagenesis with MudI. The nucleotide sequence of aepH* has very high homology (i.e., ca. > 99% identity) with aepA. Whether of not aepH* is a mutated allele of aepA remains an open issue.

repN INHIBITS pell-lacZ EXPRESSION.

As stated above, in E. chrysanthemi KdgR represses the production of Pel and several other pectate catabolic enzymes [23]. To ascertain if there is an analogous negative regulatory gene controlling extracellular enzyme production in Ecc71, we mobilized the library of AC5034 into AC5006, and scored for the effects on Pel, Peh, Cel and Prt production. We identified a cosmid, pAKC642 which severely suppressed enzyme production in Ecc71; the levels of Pel, Peh, Cel and Prt were reduced by 10-fold, 80-fold, 4-fold and 40-fold, respectively. Likewise, pAKC642 prevented the induction of pell-lacZ expression in AC5030 with celery extract. This cosmid affected exoenzyme levels but not the levels of periplasmic or cytoplasmic proteins. By subcloning, the "repressor" function has been localized to a 1.2 kb DNA segment. The nucleotide sequence of this DNA does not share homology with kdgR. This observation and the differences in phenotypes elicited by pAKC642 in Ecc and by kdgR in E. chrysanthemi indicate that these two genes are structurally and functionally different.

Discussion

The expression of pel-1 is activated during late exponential growth. While the physiological basis for this response in Erwinia is not yet apparent, our findings and those with E. chrysanthemi [11, 24], show that transcription of the pel genes is stimulated during the later stages of growth. It is, however, unclear what limits transcription of pel and other genes specifying extracellular degradative enzymes during early exponential growth or how the bacterium makes the choice between the expression of the housekeeping genes and those genes that encode degradative enzymes. A better understanding of this phenomenon should result from the analysis of mutants in which transcription of pel and other genes is no longer growth phase dependent, as is the case with the AepH* strain, AC5034 (data not shown). The aepH* allele cloned from AC5034 allows constitutive production of Pel, Peh and Cel in Ecc strain 193. Moreover, Pel and Peh production in E. coli in LB broth is also activated by an aepH* plasmid. Further analysis of this gene and its product may help us understand the molecular basis for constitutive expression of pel and the other genes specifying exoenzymes.

Recent studies of Collmer and associates revealed that, like the expression of pel genes [11, 24], the expression of the out genes of E. chrysanthemi are activated during the late exponential growth. These observations raise the possibility that genes specifying the export machinary and the genes whose products are secreted may be coordinately expressed. The possibility should also be considered that the export [SEC and OUT] systems [21], by some sort of feed back mechanism, may control transcription of the structural genes for pel and the other secreted proteins.

We have discovered several genes whose products affect pel-1 transcription. Three positive regulator genes are aepA, aepB and aepH*. The gene repN specifies a negative regulator which suppresses pel-1 expression, again by affecting transcription. How these aep and repN products control pel transcription is not yet known. Sequences of aepA and aepH* bear no homology to other prokaryotic regulator gene sequences present in the GenBank data base, suggesting that extracellular protein production is regulated by the activities of several proteins via a novel mechanism. In this context, we should note that the repN DNA has no sequence homology with kdgR, the negative regulator of pel and other pectate catabolic genes of E. chrysanthemi [18, 23]. This observation, of course, does not rule out a function of KdgR in Ecc71. In fact, the presence of a putative KdgR binding sites upstream of the translational start site of pel-1 (Fig. 1) would suggest an interaction with a KdgR-like protein. We should point out that putative KdgR binding sites have also been detected upstream of peh and other pel genes of several E. carotovora strains [10, 26].

In summary, we have discovered genes whose products specifically control exoprotein production in Ecc. The data presented here document that these gene products affect transcription of pel-1. It remains to be determined if the other target genes are similarly affected. Our ongoing studies should help resolve this issue as well as produce an understanding of the mechanisms by which these regulatory genes activate or suppress exoenzyme production.

Acknowledgements

This research was supported by grant 87-CRCR-1-2504 from the U.S. Department of Agriculture and a grant from the Food for the 21st Century program of University of Missouri-Columbia. This article is journal series 11718 of the Missouri Agricultural Experiment Station.

References

1. Beraha, L., and Garber, E.D. (1971) Avirulence and extracellular enzymes of Erwinia carotovora. Phytopathol. Z. 70:335-344.
2. Boccara, M., Vedel, R., Lalo, D., Lebrun, M.-H. and Lafay, J. F. (1991) Genetic diversity and host range in strains of Erwinia chrysanthemi. Mol. Plant-Microbe Interact. 4:293-299.
3. Castilho, B.A., Olfson, P., and Casadaban, M.J. (1984) Plasmid insertion mutagenesis and lac gene fusion with mini-Mu bacteriophage transposons. J. Bacteriol. 158:488-495.
4. Chatterjee, A., McEvoy, J.L., Chambost, J.P., Blasco, F. and Chatterjee. A.K. (1991) Nucleotide sequence and molecular characterization of pnlA, the structural gene for damage-inducible pectin lyase of Erwinia carotovora subsp. carotovora 71. J. Bacteriol. 173:1765-1769.

5. Chatterjee, A.K., McEvoy, J.L., Murata, H. and Collmer, A. (1991) Regulation of the production of pectinases and other extracellular enzymes in the soft-rotting Erwinia spp. In: Molecular strategies of pathogens and host plants (eds: S.S. Patil, S. Ouchi, D. Mills, C. Vance, Springer-Verlag, p.45-55.
6. Chatterjee, A.K., Ross, L.M., McEvoy, J.L. and Thurn, K.K. (1985) pULB113, an RP4::mini-Mu plasmid, mediates chromosomal mobilization and R-prime formation in Erwinia amylovora, Erwinia chrysanthemi, and subspecies of Erwinia carotovora. Appl. Environ. Microbiol. 50:1-9.
7. Chatterjee, A.K., and Vidaver, A.K. (1986). Genetics of pathogenicity factors: application to phytopathogenic bacteria. In: Advances in plant pathology. Vol. 4. (eds: D. Ingram, and P.H. Williams) Academic Press, London. p. 1-244/
8. He, S.Y., Lindeberg, M., Chatterjee, A.K., and Collmer, A. (1991) Cloned Erwinia chrysanthemi out genes enable Escherichia coli to selectively secrete a diverse family of heterologous proteins to its milieu. Proc. Natl. Acad. Sci. USA 88:1079-1083.
9. Hinton, J.C.D., Gill, D.R., Lalo, D., Plastow, G.S. and Salmond, G.P.C. (1990) Sequence of the peh gene of Erwinia carotovora: homology between Erwinia and Plant enzymes. Mol. Microbiol. 4:1029-1036.
10. Hinton, J.C.D., Sidebotham, J.M., Gill, D.R. and Salmond, G.P.C. (1989) Extracellular and periplasmic isoenzymes of pectate lyase from Erwinia carotovora subspecies carotovora belong to different gene families. Mol. Microbiol. 3:1785-1795.
11. Hugouvieux-Cotte-Pattat, N., Reverchon, S., Condemine, G. and Robert-Baudouy, J. (1986) Regulatory mutations affecting the synthesis of pectate lyase in Erwinia chrysanthemi. J. Gen. Microbiol. 132:2099-2106.
12. Ji, J., Hugouvieux-Cotte-Pattat, N. and Robert-Baudouy, J. (1989) Molecular cloning of the outJ gene involved in pectate lyase secretion by Erwinia chrysanthemi. Mol. Microbiol. 3:285-293.
13. Lei, S-P., Lin, H-C, Wang, S-S. and Wilcox, G. (1988) Characterization of the Erwinia carotovora pelA gene and its product pectate lyase A. Gene 62:159-164.
14. McEvoy, J.L., Murata, H. and Chatterjee, A.K. (1990) Molecular cloning and characterization of an Erwinia carotovora subsp. carotovora pectin lyase gene that responds to DNA-damaging agents. J. Bacteriol. 172:3284-3289.
15. Miller, J.H. (1972) Experiments in Molecular Genetics. Cold Spring Harbor Laboratory, Cold Spring Harbor, NY.
16. Murata, H., Fons, M., Chatterjee, A., Collmer, A. and Chatterjee, A.K. (1990) Characterization of transposon insertion Out mutants of Erwinia carotovora subsp. carotovora defective in enzyme export and of a DNA segment that complements out mutations in E. carotovora subsp. carotovora, E. carotovora subsp. atroseptica, and Erwinia chrysanthemi. J. Bacteriol. 172:2970-2978.

17. Murata, H., McEvoy, J.L., Chatterjee, A., Collmer, A. and Chatterjee, A.K. (1991) Molecular cloning of an aepA gene that activates production of extracellular pectolytic, cellulolytic, and proteolytic enzymes in Erwinia carotovora subsp. carotovora. Mol. Plant-Microbe Interact. 4:239-246.
18. Nasser, W., Reverchon, S. and Robert-Baudouy, J. (1992) Purification of the KdgR protein, a major repressor of pectinolysis genes of Erwinia chrysanthemi. Mol. Microbiol. 6:257-265.
19. Ohnishi, H., Nishida, T., Yoshida, A., Kamio, Y. and Izaki, K. (1991) Nucleotide sequence of pnl gene from Erwinia carotovora Er. Biochem. Biophys. Res. Commun. 176:321-327.
20. Pirhonen, M., Saarilahti, H., Karlsson, M.-B., and Palva, E.T. (1991) Indentification of pathogenicity determinants of Erwinia carotovora subsp. carotovora by transposon mutagenesis. Mol. Plant-Microbe Interact. 4:276-283.
21. Pugsley, A.P. (1988) Protein secretion across the outer membrane of Gram-negative bacteria, in R.C. Das and P.W. Robbin (eds.), Protein transfer and organelle biogenesis. Academic Press, Inc., San Diego, CA. p. 607-652.
22. Reverchon, S., Huang, Y., Bourson, C. and Robert-Baudouy, J. (1989) Nucleotide sequences of the Erwinia chrysanthemi ogl and pelE genes negatively regulated by the kdgR gene product. Gene 85:125-134.
23. Reverchon, S., Nasser, W. and Robert-Baudouy, J. (1991) Characterization of kdgR, a gene of Erwinia chrysanthemi that regulates pectin degradation. Mol. Microbiol. 5:2203-2216.
24. Reverchon, S. and Robert-Baudouy, J. (1987) Regulation of expression of pectate lyase genes pelA, pelD, and pelE in Erwinia chrysanthemi. J. Bacteriol. 169:2417-2423.
25. Ried, J.L. and Collmer, A. (1986) Comparison of pectic enzymes produced by Erwinia chrysanthemi, Erwinia carotovora subsp. carotovora, and Erwinia carotovora subsp. atroseptica. Appl. Environ. Microbiol. 52:305-310.
26. Saarilhati, H.T., Heino, P., Pakkanen, R., Kalkkinen, N. Palva, I. and Palva, E.T. (1990) Structural analysis of the pehA gene and characterization of its protein product, endopolygalacturonase, of Erwinia carotovora subspecies carotovora. Mol.Microbiol. 4:1037-1044.
27. Tamaki, S.J., Gold, S., Robeson, M., Manulis, M. and Keen, N.T. (1988) Structure and organization of the pel genes from Erwinia chrysanthemi EC16. J. Bacteriol. 170:3468-3478.
28. Thurn, K.K., Barras, F., Kegoya-Yoshino, Y. and Chatterjee, A.K. (1987) Pectate lyases of Erwinia chrysanthemi:PelE-like polypeptides and pelE homologous sequences in strains isolated from different plants. Physiol. Mol. Plant Pathol. 31:429-439.
29. Trollinger, D., Berry, S., Belser, W. and Keen, N.T. (1989) Cloning and characterization of a pectate lyase gene from Erwinia carotovora EC153. Mol. Plant-Microbe Interact. 2:17-25.

30. Van Gijsegem, F. (1989) Relationship between the pel genes of the pelADE cluster in Erwinia chrysanthemi strain B374. Mol. Microbiol. 3:1415-1424.
31. Willis, J.W., Engwall, J.K. and Chatterjee, A.K. (1987) Cloning of genes for Erwinia carotovora subsp. carotovora pectolytic enzymes and further characterization of the polygalacturonases. Phytopathology 77:1199-1205.
32. Yang, Z., Cramer, C.L. and Lacy, G.H. (1989) System for simultaneous study of bacterial and plant gene expression in soft rot of potato. Mol. Plant-Microbe Interact. 2: 195-201.
33. Yankovsky, N.K., Bukanov, N.O., Gritzenko, V.V., Evtushenkov, A.N., Fonstein, M.Y. and Debabov, V.G. (1989) Cloning and analysis of structural and regulatory pectate lyase genes of Erwinia chrysanthemi ENA49. Gene 81:211-218.
34. Zink, R.T. and Chatterjee, A.K. (1985) Cloning and expression in Escherichia coli of pectinase genes of Erwinia carotovora subsp. carotovora. Appl. Environ. Microbiol. 49:714-717.
35. Zink, R.T., Engwall, J.K., McEvoy, J.L. and Chatterjee, A.K. (1985) recA is required in the induction of pectin lyase and carotovoricin in Erwinia carotovora subsp. carotovora. J. Bacteriol. 164:390-396.

OXIDATIVE STRESS RESPONSE IN *Xanthomonas oryzae.*

Skorn Mongkolsuk[1,2], Sangpen Chamnongpol[2], Niwat Supsamran[2] and Siritida Rabibhadana[2]

Laboratory of Biotechnology, Chulabhorn Research Institute[1] and Department of Biotechnology, Faculty of Science, Mahidol University[2], Rama 6 Rd, Bangkok 10400, Thailand.

ABSTRACT. *Xanthomonas oryzae (Xoo)* oxidative stress responses were investigated by monitoring the protective enzymes, catalases (KATs), superoxide dismutases (SODs) and an important DNA repairing protein (RECA). In *Xoo*, paraquat, a superoxide generator was a potent inducer of KATs whereas SOD and RECA remained largely uninduced. In contrast, H_2O_2 did not induce either enzymes but induced RECA. Using paraquat induction of catalases as the criteria, it was possible to dividing *Xanthomonas spp* into two groups namely; *Xoo, X.c.malvacearum, X.c.phaseolin* were inducible while *X.c.campestris, X.c.citri* and *X.maltophilia* were uninducible. The *Xoo kat* gene was cloned using PCR and gene isolation from the genomic library techniques. The cloned *kat* was subsequently expressed in an *E.coli kat* mutant.

INTRODUCTION

Xanthomonas is an important family of bacterial plant pathogens known to effect virtually every economically important crops. *Xanthomonas oryzae oryzae (Xoo)* and *oryzicola* are major rice bacterial pathogens causing leaf blight and leaf streak diseases respectively. Currently, the molecular mechanisms of pathogenicity and plant microbes interactions with respect to *Xanthomonas* and rice remain largely unknown. The studies on *avr, hrp* and *path* genes are the major focused of molecular plant microbe interaction research and has resulted in the molecular cloning and analysis of *avr* genes from several races of *Xoo*.(1) These findings will significantly contribute to our understanding of the disease at the molecular level.

Our laboratory has taken a different approach to the problem and focused on the oxidative stress responses in *Xoo*. In plant pathogen interaction, an incompatible interaction would lead to localization and eventual destruction of the microbes at the infection site. An increased in superoxide anions production, within the first few hours of infection has been associated with the incompatible interaction(2). This reaction was not observed or delayed in wounding alone or compatible interaction. Recent evidence suggests that H_2O_2 and peroxides productions may be important reactions in the initial stages of plant defense response(3). The peroxides are released to polymerize cell wall components and also generated as a product of lignification process. These observations suggest that superoxides and peroxides anions may have active roles in the plant defense processes. The invading pathogens would encounter the host generated oxidative stresses. Thus, the

E. W. Nester and D. P. S. Verma (eds.),
Advances in Molecular Genetics of Plant-Microbe Interactions, 253–257.

pathogens must posses mechanisms to protect and overcome these stresses in order to survive. Our laboratory is interested in the molecular mechanism of oxidative stress protection in *Xanthomonas* species.

Paraquat and Peroxide Induction of *Xanthomonas* KAT and SOD.

The most well studied oxidative stress responses were carried out in enterobacteriaceae. Paraquat, a potent superoxide generator and H_2O_2 are known to induce SODs and KATs respectively. These enzymes confer some protection against oxidative stresses and may be involved in the bacterial survival under the stressful conditions. The enzymes catalyze the following reactions:

$$1)\ 2O_2^- + 2H^+ \xrightarrow{\textbf{SOD}} H_2O_2 + O_2$$

$$2)\ 2H_2O_2 \xrightarrow{\textbf{KAT}} O_2 + 2H_2O$$

We have investigated the effects of oxidative stress on locally isolates strains of *Xoo* and *Xanthomonas spp* using these enzymes and RECA as indicators.

Xanthomonas strains were grown in rich complex media (SB). Inducers were added to early log phase cultures and growth continue for three hours. The cultures were collected and lysed by sonication. Total KAT and SOD activities were assayed according to publish protocols. The results showed that paraquat was a potent inducer of KAT with the induction levels varied from 3 to 10 fold in all of the *Xoo* and *Xo pv oryzicola* strains tested (8 different isolates). However, total SOD activities showed only a marginal increased at very high concentrations (500 to 1000 μM) of paraquat. The induction of KAT by superoxides anions have previously been reported in *E.coli* and *B.subtilis*. The likely mechanism of the induction is that according to the equations 1 and 2, the superoxides anions can be converted to H_2O_2 via the action of SOD and the H_2O_2 in turn induced KAT. Thus, the *Xanthomonas* induction experiment was repeated using H_2O_2 as the inducer. The results in Table 1 indicate that H_2O_2 is not an inducer of KAT in all of the *Xanthomonas* strains tested. Thus, the paraquat induction of KAT may be operated through other mechanisms, possibly as a result of cellular damages.

Paraquat and H_2O_2 inductions of SOD and KAT were tested in serveral *Xanthomonas* pathovars. The results(Table 1) indicate that with respect to paraquat induction of KAT, *Xanthomonas spp* can be divided into two classes namely inducible and non inducible. H_2O_2 did not induce KAT or SOD in all strains tested which implied that there may be other oxidative stress protective mechanisms existed in various *Xanthomonas*. The results also illustrate that there are basic differences in the oxidative stress responses in *Xanthomonas* and other bacteria thus far investigated.

Table 1, Paraquat and H_2O_2 induction of *Xanthomonas* KAT

	Strains	Paraquat[a] Induction	H_2O_2[a] Induction
	X.o.oryzae	YES	NO
1	*X.o.oryzicola*	YES	NO
	X.c.malvacearum	YES	NO
	X.c.phaseolin	YES	NO
	X.c.campestris	NO	NO
2	*X.c.citri*	NO	NO
	X.maltophilia	NO	NO

[a] All strains were grown in a SB medium and induced with 100μM of either paraquat or H_2O_2. The positive induction(YES) indicated an increased of at least three fold whereas the negative induction(NO) indicated less than two fold increased in KAT specific activities over the non induced level.

Expression of *Xoo* RECA under Oxidative Stress.

RECA is an important multi-functions protein involved in the regulation of DNA repair and recombination processes. Reagents such as paraquat and peroxides which induced oxidative stress are also known to cause damages to DNA. We have cloned the *Xoo recA* gene by complementation of an *E.coli recA* mutant and showed that *Xoo* RECA protein cross reacted with an anti *E.coli* RECA antibody(5), (S.Mongkolsuk,manuscript submitted). The expression of *Xoo recA* under stressful conditions was monitored with anti *E.coli* RECA antibody to evaluate the effects of oxidative stress on DNA repair process. Paraquat did not induce RECA under conditions tested whereas a treatment with 100 μm H_2O_2 caused more than 5 fold increased in the amount of RECA protein.

Molecular Cloning of *Xoo kat* Gene.

Serveral attempts to clone the *Xoo kat* gene by a complementation of H_2O_2 sensitive phenotype of an *E.coli katG, katE* double mutants were unsuccessful. Furthermore, there was no crossed hybridization between *Xoo* genomic DNA and the cloned bacterial *kat* heterologous probe. Thus, an alternative approach using PCR was attempted. This approach makes used of the regions of highly conserved amino acids sequence among the cloned *kat* genes to synthesize degenerate oligonucleotide primers.(4)

	117 126	372 381
HPII	**FD hERIPER iV**	**KmVLNRNPDN**
RLC	**FDRERIPERVV**	**KLVLNRNP aN**
BLC	**FDRERIPERVV**	**KLVLNRNP vN**
HKC	**FDRERIPERVV**	**KLVLNRNP vN**
SCC	**FDRERvPERVV**	**t I t L t e N vDN**
CTC	**FDRERIPERVV**	**K f tLN eNP kN**
ZMC	**FDRERIPERVV**	**RLVLNRN iDN**
CON.	**FDRERIPERVV**	**KLVLNRNPDN**

Fig.1 Highly Conserved KAT Amino Acids Sequences Used to Make *kat* Degenerate PCR Primers. Numbering are from *E.coli katE* sequence(HPII), RLC, rat liver, BLC,bovine liver, HKC, human kidney, SCC, *S.cerevisiae*, CTC, *C.tropicalis*, ZMC, maize(4). CON is the amino acids sequences used for the synthesis of degenerate 5' and 3' *kat* oligonucleotide primers.

In the PCR reaction with a *Xoo* genomic DNA as the template and *kat* degenerate primers, a band corresponded to the expected molecular weight of *kat* PCR product was detected which was subsequently excised from the gel and used to probe *Xoo* genomic library. After screening 10,000 plaques with the probe, 10 positive clones were isolated, purified and the restriction enzymes analysis performed. Eight of the clones contained a common 7.2 kb BamHI fragment(Fig 2) which hybridized to the PCR probe. This fragment was subsequently sub-cloned into pBluescripts in both orientations and tested for KAT activities in the strain, UM 2 (*katG,katE4* mutants).

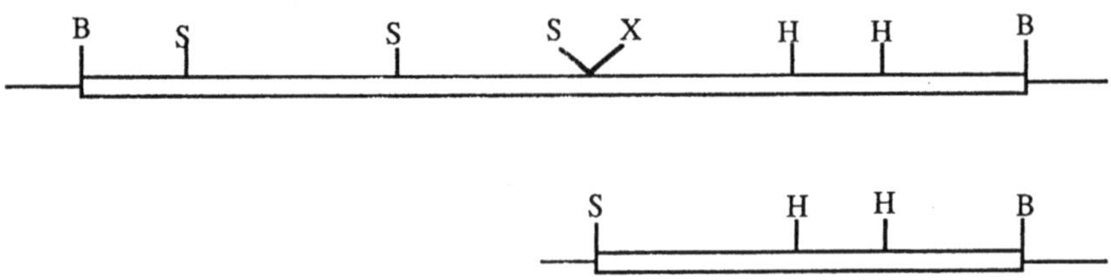

Fig.2 A restriction map of *Xoo* 7.2 kb fragment containing the *kat* gene and the subsequent *SalI* deletion.

KAT activity was not detectable in any of the clones tested in a UM2 mutant. Only after the *SalI* deletion was performed on one of the clone that a low level of KAT activity was detectable in a UM2 mutant (S.Mongkolsuk, manuscript submitted). We are currently characterizing the *kat* gene at nucleotide level and using it as a probe to determine its expression in response to stresses.

Acknowledgment. This work was partially supported by grants from UNDP THA/88/019/A/01/99 and a Rockefeller Rice Biotechnology.

References.

1 Kelemu, S. and Leach,J. (1990) 'Cloning and Characterization of an avirulance gene from *Xanthomonas oryzae*.', Mol.Plant-Microbe interact 3, 59-65.

2 Adam, A., Farkas, T., Somlyai, G., Hevesi, M. and Kiraly Z. (1989) 'Consequence of O_2^- generation during a bacterially induced hypersensitive reaction in Tobacco: deterioration of membrane lipids.', Physiol and Mol Plant Pathol 34, 13-26

3 Dixon R.A. and Lamb, C. (1990) 'Molecular communication in interactions between plants and microbial pathogens.', Annu.Rev.Plant Physio. 41, 339-67

4 von Ossowski I, Mulvey, M.R., Leco, P.A., Bory A.and Lowen, P.C. (1991) ' Nucleotide Sequence of *E.coli katE* which encodes catalase HPII.', J.Bacteriol 173, 514-520

5 Marrero,R. and Yasbin, R,E., (1988) 'Cloning of *B.subtilis recE* gene and functional expression of *recE* in *B.subtilis*.', J.Bacteriol. 170, 335-344

INVOLVEMENT OF *PSEUDOMONAS SOLANACEARUM hrp* GENES ON THE SECRETION OF A BACTERIAL COMPOUND WHICH INDUCES A HYPERSENSITIVE-LIKE RESPONSE ON TOBACCO.

S. GENIN, C.L. GOUGH, M. ARLAT, C. ZISCHEK,
F. VAN GIJSEGEM, P. BARBERIS and C.A. BOUCHER
Laboratoire de Biologie Molèculaire des Interactions Plantes Microorganismes
CNRS-INRA
B.P. 27
31326 - Castanet Tolosan Cedex
France

ABSTRACT. Based on DNA sequencing of 20kb of the left hand of the *Pseudomonas solanacearum hrp* gene cluster, 19 open reading frames (ORFs) with a high coding probability have been identified. One of these ORFs codes for a positive regulator which controls the expression of at least 5 *hrp* transcription units in addition to the expression of yet unidentified genes adjacent to the *hrp* gene cluster. Five other ORFs code for putative proteins which share homology with pathogenicity genes from the animal and human pathogens *Yersinia enterocolitica, Y. pestis* and *Shigella flexneri*. These homologies led us to look in the supernatant of bacteria cell cultures for a *hrp* gene dependant bacterial factor which is able to induce a hypersensitive-like response on tobacco. Preliminary experimental data suggest that such a factor does exist and that it is a heat resistant protein and that it is released in the culture medium via a *hrp* gene encoded secretion machinery. Cross-hybridizations of a *P.solanacearum hrp* gene with the *hrp* gene clusters of *P. syringae* pv. *phaseolicola* and *Erwinia amylovora* are also presented.

1. Introduction

Plant pathogenic bacteria have the ability to infect and to multiply within living plants leading to the occurence of typical disease symptoms. The aptitude to colonize such a particular ecological niche relies on specific biochemical functions encoded by pathogenicity genes of the pathogen, which have to be coordinately expressed during the infection process. Genetic studies of the determinism of such plant-bacteria interactions have been undertaken in *Pseudomonas solanacearum* (Boucher *et al.*, 1992), *P. syringae* (Rahme *et al.*,1991) , *Xanthomonas campestris* (Arlat *et al.*, 1991, Bonas *et al.*, 1991, Daniels *et al.*, 1984) and *Erwinia amylovora* (Barny *et al.*, 1990, Wei *et al.*, 1992), which are representatives of the vast majority of plant pathogenic bacteria. These studies have revealed that numerous bacterial genes are involved in the control of pathogenicity and have enlightened the crucial importance of a particular group of genes called *hrp* genes (Willis *et al.*,1991). In addition to controlling the aptitude of the bacteria to elicit disease on host plants, these genes are also required for the induction of the hypersensitive response (HR) on resistant or non host plants. In most of the organisms studied, the majority of the *hrp* genes are clustered in a region of about 25kb, although in *E.amylovora* this cluster covers approximately 40 kb (Wei *et al.*, 1992). Based on DNA cross hybridization and/or on heterologous complementation, there appear to be two groups of phytopathogenic bacteria within each of which *hrp* gene conservation has been shown. One group includes *P.solanacearum* and *X.campestris* pathovars (Arlat *et al.*, 1991a) ; the other includes

E. W. Nester and D. P. S. Verma (eds.),
Advances in Molecular Genetics of Plant-Microbe Interactions, 259–266.

pathovars of *P.syringae* (Lindgren *et al.,* 1988) and *E.amylovora* (Laby and Beer 1990). Until recently homology had not been found between members of these two groups.
With the exception of the *hrpS* locus of *P.syringae* which codes for a transcriptional activator of other *hrp* genes (Grimm and Panopoulos 1989), no biochemical function has been yet assigned to *hrp* genes in any bacterial species.
In this paper we firstly present, DNA sequence data from the *P.solanacearum hrp* gene cluster, and explain how the interpretation of this data led us to speculate on the function of these genes. Secondly we show that *hrp* genes are involved in the production of (an) extracellular protein(s) which induces a HR-like reaction following infiltration into tobacco leaves.

2. Results

2.1 GENETIC ORGANIZATION OF THE *P.solanacearum hrp* GENE CLUSTER

We have recently proposed that the *P.solanacearum hrp* gene cluster is organized in a minimum of six transcriptional units (Arlat *et al.,* 1991b, Figure 1).

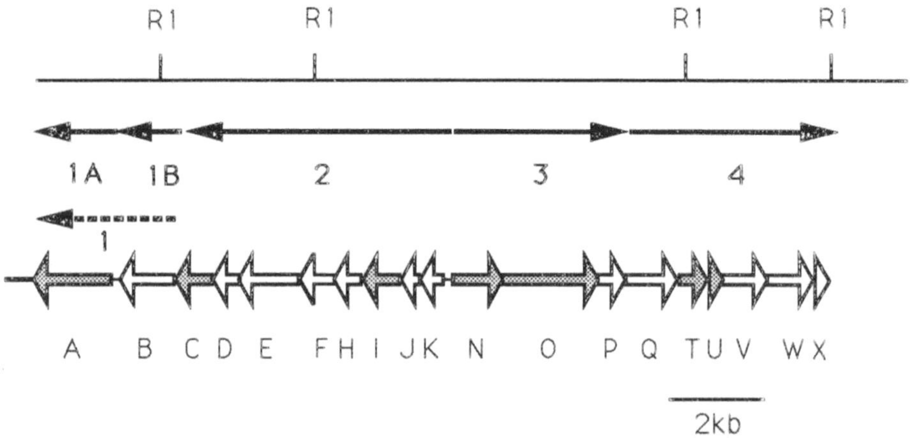

Figure 1: Genetic organization of the *hrp* gene cluster. The top line is an *Eco*R1 (R1) restriction map of the left end of the cluster. Solid arrows show the individual transcription units, with the original transcription unit 1 (dashed arrow) being split into two (1A and 1B). Open arrows at the bottom of the figure represent the individual open reading frames for the putative Hrp proteins. When a protein is predicted to have transmembrane domain(s) the corresponding arrow is shaded.

DNA sequencing of a stretch of 20 kb covering the four transcriptional units located at the left hand end of the cluster has been achieved leading to the identification of 19 open reading frames (ORFs), which have a high coding probability as deduced from codon usage. The relative position of these ORFs within the cluster together with the proposed nomenclature for the corresponding genes is presented in Figure 1. This organization is in agreement with the previously proposed transcriptional organization of this region with the exception that, as shown in Figure 1, preliminary genetic data

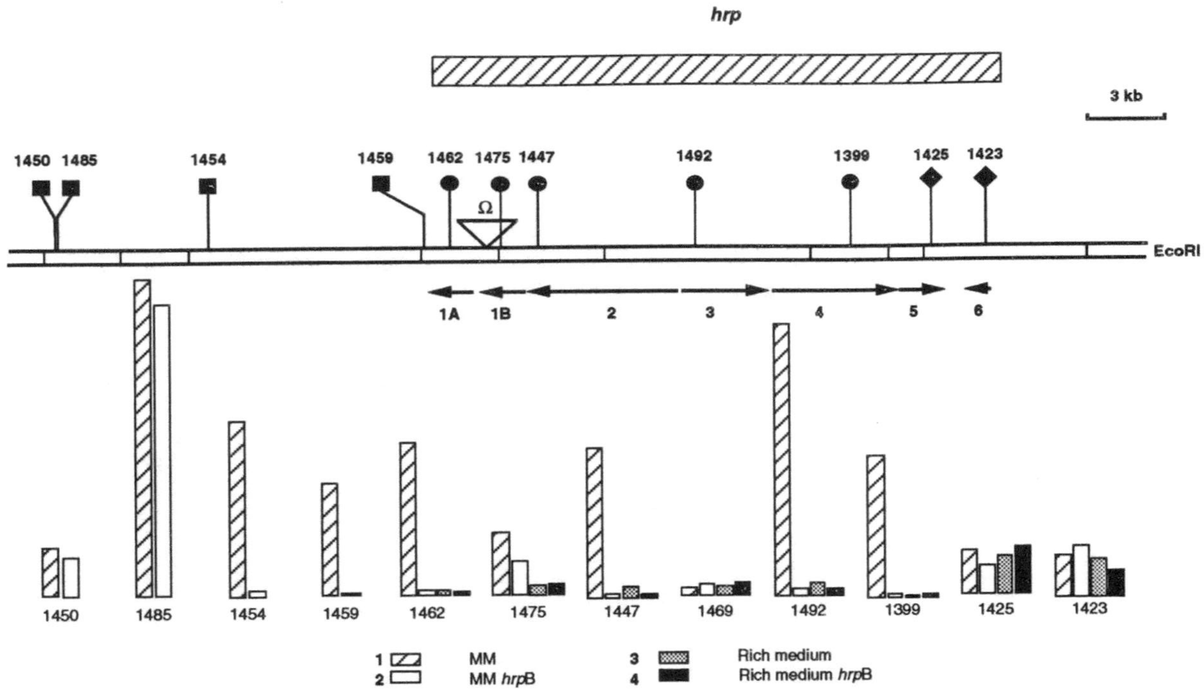

Figure 2: Regulation of hrp gene expression by hrpB. The top of the figure represents the Eco R1 restriction map of the hrp gene cluster and of the flanking region together with the localization of the reporter lacZ gene fusions which have been generated using the transposon Tn5-B20 (Arlat et al.1992). The triangle above the map shows the location of the insertion of the omega interposon leading to the hrpB mutation. Histograms represent the level of expression of each gene fusion under inducing condition in the wild-type (1) versus a hrpB mutant background (2) and in rich medium in the wild-type (3) versus in a hrpB mutant background (4) repectively.

indicate that transcription unit 1 probably corresponds to two monocistronic operons (1A and 1B in Figure 1). This assumption is also supported by the presence of a non coding at least 85 bp stretch of DNA between the 3' end of *hrpB* and the 5' end of *hrpA*. Analysis of the amino acid (AA) sequence of the corresponding putative proteins revealed that 7 of them (Figure 1) are predicted to have transmembrane alpha helix domain(s) and could therefore be integral membrane proteins.

2.2 *hrpB* CODES FOR A POSITIVE REGULATOR

The *hrpB* gene has a coding capacity for a 477 amino acid (AA) peptide. This putative peptide shows significant similarity to several procaryotic transcriptional activators (Genin *et al.*, 1992) including the AraC protein of *Escherichia coli* (Wallace *et al.*, 1980), the XylS protein of *P. putida* (Mermod *et al.*, 1987) and the VirF protein of *Yersinia enterocolitica* (Cornelis *et al.*, 1989) . The *hrpB* gene product therefore belongs to a different family of bacterial regulators from the previously described HrpS protein of *P.syringae* (Grimm and Panopoulos 1989). Genetic evidence demonstrates that the *hrpB* gene product actually acts as a positive regulator of the expression, in minimum medium, of transcription units 2, 3 and 4 of the *hrp* gene cluster (Figure 2) (Genin *et al.*, 1992).
If, as previously discussed, the original transcription unit 1 actually has to be split into two transcription units, 1A and 1B, then *hrpB* would also positively regulate the expression of *hrpA* (Figure 2). The *hrpB* gene also controls the expression of additional genes located to the left of the *hrp* gene cluster, for which no involvement in bacteria/plant interactions has yet been identifed (Genin *et al.*, 1992). Finally, we have also established that expression of *hrp*B is partly self-activated and dependent on other regulatory factor(s).
Introduction of *hrpB* on the multicopy plasmid pLAFR3 into *P.solanacearum* also resulted in an increased level of transcription of transcription unit 5 when bacteria were grown under inducing conditions.

2.3 SEVERAL *hrp* GENES SHARE HOMOLOGY WITH GENES GOVERNING PATHOGENICITY IN MAMMALIAN PATHOGENS

In addition to *hrp*B discussed above at least five other genes have been shown to share homology with pathogenicity genes of animal and human pathogens (Gough *et al.*, 1992)

- *hrpA* codes for a putative protein of 568 AA which is 34% identical over its entire length to the YscC protein of *Y.enterocolitica* (Michiels *et al.*, 1991b),
- *hrpE* codes for a putative 439 AA protein which is 44% identical to the putative Spa47 protein of *Shigella flexneri* (Venkatesan *et al.*, 1992)
- *hrpI* codes for a putative lipoprotein of 269 AA which is 35% identical over its entire length to the YscJ lipoprotein of *Y. enterocolitica* (Michiels *et al.*, 1991b),
- *hrpO* codes for a putative protein of 690 AA which is 43% identical to the LcrD protein from *Y.pestis* (Plano *et al.*, 1991) and 32% identical over 189 AA to a partial ORF located 5' of the *spa15* locus of *S.flexneri* (Andrews and Maurelli, unpublished, GenBank accession M91664).,
- *hrpT* codes for a putative protein of 218 AA which is 40% identical to the putative protein encoded by the *spa24* gene of *S.flexneri* (Venkatesan *et al.*, 1992).

In the corresponding organisms, the structural genes for these proteins have been shown to map on virulence plasmids (Cornelis *et al.*, 1989, Halle 1991) and to be (or suspected to be) involved in the secretion of extracellular proteins (called Yops for *Yersinia* and Ipas for *S.flexneri*). In both *Yersinia*

and *Shigella*, these proteins have been shown to be crucial for pathogenicity and host cell invasion (Goguen *et al.*, 1984, Michiels *et al.*, 1991b, Halle 1991).
Based on the functional role of the *hrp* gene homolgues in these different animal pathogens, we hypothesized that *hrp* genes are involved in the secretion of macromolecule(s) which are directly involved in the elicitation of a plant response. We now present experimental data, which tend to support this hypothesis.

2.4. PRESENCE OF A HR-LIKE INDUCING FACTOR IN BACTERIAL CULTURE SUPERNATANTS

A necrogenic activity was found in the supernatant of a culture of the wild-type strain GMI1000 grown in minimum medium, which is a condition known to induce *hrp* gene expression (Arlat *et al.*, 1991b). Infiltration of a 20 fold concentrate of the non-dialysable (MW >10,000) fraction of the supernatant into tobacco leaf parenchyma resulted in the development of a necrotic lesion, which remained strictly localized to the infiltrated tissues. This reaction was highly reminiscent of the HR induced by live bacteria, both in terms of the kinetics of development and of the aspect of the lesions. This activity was not found in the supernatant of seven *hrp* mutants tested which were deleted of most of the *hrp* gene cluster or which carried a mutation in the proximal part of transcription units 1A-4. It was retained for two mutants mapping in transcription units 5 and 6.

This activity was heat resistant (100°C, 8mn) and sensitive to proteinase K and was therefore assumed to be due to the presence of one or more heat resistant protein(s).

2.5. *hrp* GENES ARE INVOLVED IN TRANSMEMBRANE TRANSIT OF THE HR-LIKE INDUCING FACTOR(S)

Although no HR-like inducing activity could be detected in the supernatant of any of the *hrp* mutants tested, a similar activity was found in the sonicated cell lysates from most *hrp* mutants tested as well as from the wild-type strain . Only the 2 *hrpB* mutants tested had no intracellular activity. Since these mutants were deficient in *hrp* functions it is clear that this intracellular activity does not result from an unspecific toxic effect of the lysate, but rather that it is *hrp* gene specific. As for the extracellular activity, the intracellular one was found to be heat resistant and sensitive to proteinase K. This is an indication that both intra and extra cellular activities are due to the same protein(s). If this is true, our data establish that in a similar way to their *Yersinia* or *Shigella* counterparts, several *hrp* genes are involved in the transit through the bacterial cell envelopes of factor(s) which are HR-like inducing factor(s) in the case of *P.solanacearum*.
The requirement of a functional *hrpB* gene for the production of the intracellular activity suggests that the gene(s) coding for the active compound(s) is (are) part of the *hrp* gene regulon.

2.6 *hrpO* HOMOLOGUES ARE PRESENT IN *E.amylovora* AND *P.syringae*

Since certain *hrp* genes have their counterparts in animal pathogens, we thought that these genes could also be conserved in plant pathogenic bacteria. In order to test this hypothesis DNA probes corresponding to internal fragments of genes *hrpA, hrpI* and *hrpO* were hybridized to cosmid clones carrying the *hrp* gene clusters of *X.campestris* pv. *campestris* (Arlat *et al.*, 1991a), *P.syringae* pv.

phaseolicola (Lindgren *et al.,* 1986) and *E.amylovora* (Barny *et al.,* 1990). As expected from previous data (Arlat *et al.,* 1991a, Boucher *et al.,* 1987), these experiments confirmed the presence of sequences homologous to all three probes within the *X.campestris hrp* gene cluster. The *hrpO* gene probe also hybridized to the *hrp* gene clusters of *E.amylovora* and *P.syringae.*

3. Discussion and Conclusions :

We have shown that in *P.solanacearum hrp* genes are involved in the production and transit through the bacterial cell envelope of a compound which mimicks the ability of the live bacteria to induce an HR following infiltration into tobacco leaves.

Although the role of *hrp* genes in coding for a secretion system in *P.solanacearum* is not yet definitively established, experimental data favour this hypothesis since most *hrp* mutants tested which were devoid of the extracellular HR-inducing activity, had retained a similar heat resistant, proteinase K sensitive activity in the cell lysates. This is a good indication that they could be affected in a secretion process. In addition, the striking similarities observed between *hrp* genes from *P.solanacearum* and certain *Yersinia* and *Shigella* genes which have been shown (or have been postulated) to be involved in the secretion of essential pathogenicity determinants, further support this assumption. Moreover, additional homologies have been found for putative Hrp proteins which again suggest the involvement of *hrp* genes in governing secretion :

HrpA shares homology with PulD of *Klebsiella pneumoniae* (d'Enfert *et al.,* 1989), and protein pIV of filamentous bacteriophage I2-2 (EMBL accession X14336),

HrpO shares homology with FlhA from *Bacillus subtilis* (EMBL accession No X63698) and with FlbF of *Caulobacter crescentus* (Sanders *et al.,* 1992),

HrpQ shares homology in its C terminal half with the FliN protein of *S.typhimurium* (Kihara *et al.,* 1989), and the MotD protein from *E. coli*.(Malakooti *et al.,*1989).

A common feature of all these proteins which share homology with Hrp proteins, is their (possible) involvement in transmembrane transit of various macromolecules. *hrp* mutants retaine the ability of the wild-type parent to secrete cellulolytic and pectinolytic enzymes. Therefore it is assumed that, as in *Yersinia* and *Shigella*, the *hrp* gene secretion pathway of *P.solanacearum* might be specific for pathogenicity proteins.

With the exception of HrpE, all the Hrp proteins for which we have identified homologous proteins in other organisms, are proteins which are predicted to be located in the membrane and thus are good candidates for participating in a secretion process. HrpE shares homology with membrane associated ATPases and has a typical ATP binding site motive and a typical catalytic domain, suggesting that this protein might be required in energizing the secretion machinery. This has already been postulated for the Spa47 protein of *S.flexneri* (Venkatesan *et al.*, 1992).

Until now no homology has been found for any of the putative hydrophylic Hrp proteins (except for HrpE dicussed above). This raises the question of what could be the functions of these proteins. One possibility is that certain of them could be secreted through the *hrp* machinery and be involved in plant-bacteria interactions.

In *Yersinia,* the Yop specific secretion system does not rely on the presence of a classical signal peptide on the preprotein but rather on a yet unidentified signal (Michiels and Cornelis, 1991a). Therefore, based on the amino acid sequence of the Hrp proteins it is not presently possible to predict whether some of them could be secreted.

Our results established that transcription of the region adjacent to left of the *hrp* gene cluster is dependant of *hrpB* and therefore part of the *hrp* gene regulon. This suggests that the corresponding genes might be involved in controlling plant bacteria interactions. Until now they have been shown to be neither necessary for HR induction on tobacco nor for disease expression in tomato but the possibility that they might be required in interactions with other plants remains open.

The reported DNA homology of the *P. solanacearum hrpO* gene with the *hrp* gene cluster from *P.syringae* and *E.amylovora* is the first indication that *hrp* genes might be conserved between these organisms and suggests that these plant pathogenic bacteria might share a common machinery for secretion of pathogenicity determinants.

Preliminary characterization of an active compound present in cell culture supernatants has been acheived. This compound was active on tobacco, where it mimicked the activity of the wild-type strain to induce an HR. Since it was sensitive to proteinase K it is assumed to be of a proteinaceaous nature. The question remains as to the activity observed on tobacco was due to a single factor or whether different factors with a similar activity are involved. Whatever the answer, we are on the way to identifying a key bacterial component mediating plant-bacteria interactions.

Whether this factor is specific for the interaction with tobacco or whether it could be involved in the interaction with other plants and particularly with host plants remains to be established.

References

Arlat, M., Gough, C.L., Barber, C.E., Boucher C. and Daniels, MJ. 1991 *Xanthomonas campestris* contains a cluster of *hrp* related to the larger *hrp* cluster of *Pseudomonas solanacearum*. Mol. Plant Microbe Interact. 4 : 593-601.

Arlat, M., Gough, C.L., Zischek, C., Barberis, P.A., Trigalet, A., and Boucher, C.A. (1992). Transcriptional organisation and expression of the large *hrp* gene cluster of *Pseudomonas solanacearum*. Mol. Plant-Microbe Interact. 5 : 187-193.

Barny, M.A., Guinebretiére, M.H., Marcais, B., Coissac, A., Paulin, J.P and Laurent, J. (1990). Cloning of a large gene cluster involved in *Erwinia amylovora* CFBP1430 virulence.Mol. Microbiol. 4 : 777-787.

Bonas, U., Shulte, R., Fenseleau, S., Misavage, G.V. and Staskawicz, B.J. (1991). Isolation of a gene cluster from *Xanthomonas campestris* pv. *vesicatoria* that determine pathogenicity and the hypersensitive response on pepper and tomato. Mol. Plant-Microbe Interact. 4 : 81-88.

Boucher, C.A., Gough, C.L. and Arlat, M. (1992). Molecular genetic of pathogenicity determinants of *Pseudomonas solanacearum* with special emphasis on hrp genes. Annu. Rev. Phytopathol. 30 : 443-461.

Boucher, C.A., Van Gijsegem, F., Barberis, P.A., Arlat, M. and Zischek, C. (1987). *Pseudomonas solanacearum* genes controlling both pathogenicity on tomato and hypersensitivity on tobacco are clustered. J. Bacteriol. 169 : 5626-5632.

Cornelis, G.R., Biot, T., Lambert de Rouvroit, C., Michiels, T., Mulder, B., Sluiters, C., Sory, M.P., Van Bouchaute, M. and Vanooteghem, J.C. (1989). The *Yersinia* Yop regulon. Mol. Microbiol. 3 : 1455-1459.

Cornelis, G., Sluiters, C., Lambert de Rouvroit, C. and Michiels, T. (1989). Homology between VirF, the transcriptional activator of the *Yersinia* virulence regulon, and AraC, the *Escherichia coli* arabinose operon regulator. J. Bacteriol. 171 : 254-262.

Daniels, M.J., Barber, C.E., Turner, P.C., Sawczyc, M.K., Byrde, R.J.W. and Fielding, A. H. (1984). Cloning of genes involved in pathogenicity of *Xanthomonas campestris* pv. *campestris* using the broad host range cosmid pLAFR1. EMBO J. 3 : 3323-3328.

d'Enfert, C., Reyss, I., Wandersman, C., and Pugsley, A.P. (1989). Protein secretion by Gram-negative bacteria. J. Biol. Chem. 264 : 17462-17468.

Genin, S. ,Gough, C.L., Zischek, C. and Boucher C.A. (1992). The *hrpB* encodes a positive regulator of *hrp* genes from *Pseudomonas solanacearum*. Molecular Microbiology (In press).

Goguen, J.D., Yother, J. and Straley S.C. (1984). Genetic analysis of the low calcium response in *Yersinia pestis* Mud1(Ap*lac*) insertion mutants. J. Bacteriol. 160 : 842-848.

Gough, C.L., Genin, S., Zischek, C. and Boucher, C.A. (1992). *hrp* genes of *Pseudomonas solanacearum* are homologous to pathogenicity determinants of animal pathogenic bacteria and are conserved among plant pathogenic bacteria. Mol. Plant Microbe Interact. (In press).

Grimm and Panopoulos (1989). The predicted product of a pathogenicity locus of *Pseudomonas syringae* pv. *phaseolicola* is homologous to a highly conserved domain of several procaryotic regulatory proteins. J. Bacteriol. 171 : 5031-5038

Halle, T.L. (1991). Genetic basis of virulence in *Shigella* species. Microbiological Reviews. 55 : 206-224.

Kihara, M., Homma, M., Kutsukake, K. and Macnab, R. (1989). Flagellar switch of *Salmonella typhimurium* : gene sequences and deduces protein sequences. J. Bacteriol. 171 : 3247-3257.

Laby, R.J. and Beer, S.V. (1990). The *hrp* gene cluster of *Erwinia amylovora* shares DNA homology with other bacteria. Phytopathology 80 : 1038-1039.

Lindgren, P.B., Panopoulos, N.J., Staskawicz, B.J. and Dahlbeck,D. (1988). Gene required for pathogenicity and hypersensitivity are conserved and interchangeable among pathovars of *Pseudomonas syringae.* Mol. Gen. Genet. 211 : 4999-5006.

Malakooti, J., Komeda, Y. and Matsumara, P. (1989). DNA sequence analysis, gene product identification, and localization of flagellar motor components of *Escherichia coli.* J. Bacteriol. 171 : 2727-2734.

Mermod, N., Ramos, J.L., Bairoch, A. and Timmis, K.N. (1987). The *xylS* gene positive regulator of TOL plasmid pWWO : identification, sequence analysis and over production leading to constitutive expression of *meta* cleavage operon. Mol. Gen. Genet. 207 : 349-354.

Michiels, T. and Cornelis, G. (1991). Secretion of hybrid proteins by the *Yersinia* export system. J. Bacteriol. 173 : 1677-1685.

Michiels, T., Vanooteghem, J.-C., Lambert de Rouvroit, C., China, B., Gustin, A., Boudry, P., and Cornelis, G.R. (1991). Analysis of *virC*, an operon involved in the secretion of Yop proteins by *Yersinia enterocolitica*. J. Bacteriol. 173 : 4994-5009.

Plano, G.V., Barve, S.S., and Straley, S.C. (1991). LcrD, a membrane-bound regulator of the *Yersinia pestis* low-calcium response. J. Bacteriol. 173 : 7293-7303.

Rahme, L.G., Mindrinos, M.N., and Panopoulos, N.J. (1991). Genetic and transcriptional organization of the *hrp* cluster of *Pseudomonas-syringae* pv. *phaseolicola*. J. Bacteriol. 173 : 575-586.

Sanders, L.A., Van Way, S. and Mullin, D.A. (1992). Characterization of the *Caulobacter crescentus flbF* promoter and identification of the inferred FlbF product as a homolog of the LcrD protein from the *Yersinia enterocolitica* virulence plasmid. J. Bacteriol. 174 : 857-866.

Venkatesan, M.M., Buysse, J.M. and Oaks, E.V. (1992). Surface presentation of *Shigella flexneri* invasion plasmid antigens required the products of the *spa* locus. J. Bacteriol. 174 : 1990-2001

Wallace, R.G., Lee, N. and Fowler, A.V. (1980). The *araC* gene of *Escherichia coli* : transcriptional and translational start-points and complete nucleotide sequence. Gen. 12 : 179-190.

Wey, Z.M., Sneath, B.J. and Beer, S.V. (1992). Expression of *Erwinia amylovora hrp* genes in response to environmental stimuli. J. Bacteriol. 174 : 1875-1882.

Willis, D.K., Rich, J.J., and Hrabak, E.M. (1991). *hrp* Genes of phytopathogenic bacteria. Mol. Plant Microbe Interact. 4 : 132-138.

PSEUDOMONAS SYRINGAE PV. PHASEOLICOLA-PLANT INTERACTIONS: HOST-PATHOGEN SIGNALLING THROUGH CASCADE CONTROL OF HRP GENE EXPRESSION

W. Miller, M. N. Mindrinos[1], L. G. Rahme[1], R. D. Frederick[2], C. Grimm[3], R. Gressman[4], X. Kyriakides[4], M. Kokkinidis[4] and N. J. Panopoulos.
Department of Plant Pathology, University of California, Berkeley, CA, USA. [1]Present address: Department of Molecular Biology, Massachusetts General Hospital, Boston, MA; [2]Department of Biological Sciences, Purdue University, Lafayette, IN; [3]Institute of Botany, University of Vienna, Austria; [4]Institute of Molecular Biology and Biotechnology, Foundation for Research and Technology-Hellas, and Department of Biology, University of Crete, Heraklio, Greece.

ABSTRACT. The current catalogue of hrp genes in P. syringae pv. phaseolicola includes seven contiguous operons (hrpL, hrpAB, hrpC, hrpD, hrpE, hrpF and hrpRS) found in a 22 kb cluster, and at least three genes unlinked to this cluster, namely, the bicistronic locus hrpM and the newly discovered genes hrpT, and hrpQ. The first six operons in the cluster are positively controlled by hrpS and depend on a functional ntrA gene for full induction. The hrpS and hrpL genes undergo dramatic induction at the infection site, evidently in response to plant signals, both in compatible and incompatible interactions. The hrpT and hrpQ genes control an early step(s) in a signalling process in which transcriptional activation of hrpS occurs in response to plant signal(s). In addition to plant signal control, the hrpAB, hrpC, hrpD, and hrpE operons are regulated by pH and medium osmolarity in vitro. Expression of the hrpT and hrpM genes is not controlled by any known genes or signals. The hrpR gene encodes a protein that is structurally similar to HrpS and is also involved in hrp regulon function but its role is not clear. Based on computer predictions, structural modelling, homology and Tn-phoA analysis, at least four Hrp proteins appear to be membrane associated and one of them is predicted to be a lipoprotein. It appears that environmental signalling and expression of the operons in the hrp cluster is mediated by a cascade involving hrpT and hrpQ as "upstream" elements, hrpS as a transcriptional activator acting in conjunction with σ^{54}, and hrpL and hrpR playing additional, although functionally not understood roles in the signalling process.

1. Introduction

1.1. THE HRP GENES OF PSEUDOMONAS SYRINGAE PV. PHASEOLICOLA

The hrp ("harp") genes (Lindgren et al. 1986) control the interactions of many phytopathogenic bacteria with both susceptible and resistant plants (rev. in Willis et al, 1991). In the bean halo blight pathogen, Pseudomonas syringae pv. phaseolicola, the majority of known hrp genes form a large contiguous cluster that spans approximately 22 kb, is located in the bacterial chromosome, and is genetically

E. W. Nester and D. P. S. Verma (eds.),
Advances in Molecular Genetics of Plant-Microbe Interactions, 267–274.

organized into seven complementation groups: hrpL, hrpAB, hrpC, hrpD, hrpE, hrpF and hrpRS (Rahme et al. 1992). There are at least three additional hrp loci in this bacterium that are physically distinct from this cluster: hrpM (Frederick 1989), and two newly identified genes, hrpT and hrpQ, described in this report.

1.2. GENETIC CONTROL OF HRP GENE EXPRESSION: THE "HARP" REGULON

The seven operons in the hrp cluster form a complex regulon in which the hrpRS locus transcriptionally controls the expression of the six other operons in the hrp cluster, but not hrpM. The hrpRS locus spans ≈2 kb and can potentially encode two similarly sized (each ≈34 kD), structurally related proteins, HrpS and HrpR (Grimm and Panopoulos, 1989; Grimm and Panopoulos, unpubl. data). These proteins resemble in their primary sequence the highly conserved "central" domain of several well known prokaryotic transcriptional regulators ("NtrC subfamily"). HrpS is essential for the expression of hrpL, hrpAB, hrpC, hrpD, hrpE and hrpF, based on the fact that an hrpS::Tn5 mutant is unable to induce any of these operons in vitro or in planta (Grimm and Panopoulos, 1989; Mindrinos, Rahme and Panopoulos, unpubl. data). The analysis of the role of hrpR and its product (HrpR) in the expression of the other hrp operons has been complicated by the difficulty of separating polar effects of transposon insertions in the hrpR region on the expression of the hrpS gene located immediately downstream and our present lack of knowledge of the transcriptional organization of the hrpRS operon and its promoter elements. Insertion mutations in hrpR block the induction of the hrpS controlled operons (Mindrinos, Rahme and Panopoulos, unpubl. data). Furthermore, analysis of site directed mutants in the putative ATP binding site of HrpR indicated that it is involved in the control of expression of the hrp operons in conjunction with HrpS (Fellay et al. 1991; Fellay et al. in preparation). Our results suggest that the two structurally similar products of hrpRS may interact physically, possibly through heterodimer formation, in regulating the expression of hrpD and other target operons. Most proteins possessing the conserved central domain of NtrC promote transcription of genes that possess -24/-12 type promoters by the RNA polymerase-σ^{54} holoenzyme (Kustu et al. 1989). A requirement for a functional ntrA gene (encoding σ^{54}) for the expression of hrpS controlled operons in P. s phaseolicola has been established (Fellay et al. 1991, 1992).

1.3. REGULATION OF HRP GENES BY PLANT AND ENVIRONMENTAL SENSORY SIGNALS

Expression of the hrp genes in P. s. phaseolicola in vitro and in planta is regulated by at least three types of signals, but to a different degree (Rahme et al. 1992). Under defined in vitro conditions: a) the expression of hrpAB, hrpC and hrpD is downregulated to a similar degree by osmolarity, pH and type of carbon source; b) the expression of hrpE is strongly affected by pH and carbon substrate and to a small extent by osmolarity; c) hrpF is not substantially affected by any of these factors. The expression levels of hrpS and hrpL in vitro are very low, compared to the other hrp genes. However, in plant leaves the expression of both genes are dramatically increased (1,000- to 2,000-fold); a similar response, but of smaller amplitude, is also observed with the other genes in the hrp cluster. Thus, the seven operons in the hrp cluster are controlled by a complex signalling mechanism and a plant signal(s) evidently is necessary for maximum induction. The behavior of hrpS and hrpL also suggested that an additional gene(s) exists that regulates hrpS transcription.

2. Results and Discussion

2.1. IDENTIFICATION OF NEW HRP GENES THAT CONTROL HRPS TRANSCRIPTION

To substantiate the above hypothesis and to further analyze the signalling mechanisms controlling hrp gene expression we sought to identify mutants in which hrpS expression in planta is either decreased or no longer occurs. A strain carrying an hrpS::inaZ fusion (inaZ is a promoterless ice nucleation gene terminally positioned in transposon Tn3-Spice [Lindgren et al. 1989]) was mutagenized by insertion of another transposon, Tn-plac (Miller and Simon 1992). The resulting mutants were screened for ice nucleation activity after co-cultivation with a tobacco cell suspension culture. This screening led to the identification of two mutants, Q1 and T24, that showed a 15- to 20-fold lower expression of hrpS (assayed as inaZ activity), compared to the parental strain (Table 1). The mutations were subsequently shown by Southern blot analysis

Table 1. Effect of mutations in the hrpQ and hrpT genes on the expression of hrpS in tobacco tissue culture (TS) and in bean and tobacco leaves.

	Tobacco cell culture		Tobacco leaves		Bean leaves	
	log INA units	fold decrease	log INA units	fold decrease	log INA units	fold decrease
G94	-3.1	--	-2.1	--	-0.9	--
Q1	-5.4	14	-3.1	3	-2.9	10
T24	-5.5	17	-3.7	6	<-4.1	>39
G94 (-TC)	-7.2	≈100	N.A.	N.A.	N.A.	N.A.

G94 is the hrpS::inaZ marker exchange mutant used in the original screening; Q1 and T24 carry the G94 insertion and each has an additional mutation elsewhere in the genome. TC: tobacco cells. Expression levels were measured by droplet freezing assays at -9 degrees Celsius. INA units (defined as the number of ice nuclei/cell) are given in logarithms; thus, smaller negative values represent greater activity; Because the signal generated by the InaZ protein at -9 °C is proportional to the square power of InaZ monomer concentration (see Lindgren et al. 1989), the fold-decrease in hrpS expression in the Q1 and T24 mutants is obtained as the square root of the difference in INA levels (e. g. a difference of 2 and 3 log INA units represents a 10- and 33-fold change in expression level, respectively).

to be located outside of the hrp cluster at two distinct chromosomal locations and provisionally identified two new genes, designated hrpQ and hrpT. In Red Kidney bean leaves, hrpT mutants showed a 40-fold reduction in hrpS expression, whereas hrpQ mutants showed a 10-fold reduction; neither mutant caused disease symptoms. In tobacco leaves, hrpT and hrpQ mutants showed a 3- to 6-fold reduction in hrpS expression and a weak (hrpQ) or no (hrpT) hypersensitive response. Cosmids from a wild-type P. s. phaseolicola library have been isolated which complemented the hrpT mutation. Tn3-Spice and Tn-phoA mutagenesis of a hrpT-complementing cosmid revealed that the hrpT locus is approximately 1.7 kb in length and is transcribed at moderate levels in both complex media and in planta.

Table 2. Phenotypic complementation of G94 and the Q1 and T24 mutants with plasmids carrying the hrpSR operon in bean and tobacco leaves.

Strain	Hypersensitive response (Tobacco leaves)	Disease symptoms (Bean leaves)
G94*	-	-
G94/pBBGL1	+	+
G94/pBBGL6**	+	+
Q1/pBBGL1**	-	+/-
Q1/pBBGL6	+	+/-
T24/pBBGL1	-	-
T24/pBBGL6	+	-

* G94 is a *hrpS::inaZ* marker exchange mutant; Q1 and T24 carry the G94 insertion and each has an additional mutation elsewhere in the genome. ** The plasmids are depicted in Figure 1b.

The pBBGL1 and pBBGL6 plasmids referred to in the above table and in the text. Prs stands for native promoter of the hrpRS operon, Plac for the lac promoter in the vector (solid bars). Arrows show the direction of hrpRS transcription.

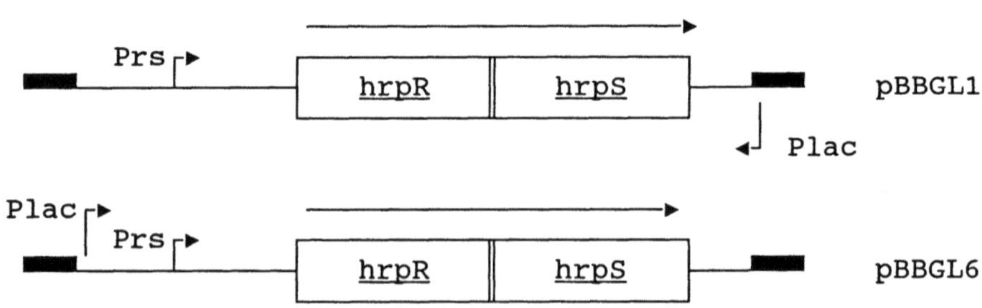

2.2. HRP PHENOTYPE OF HRPT AND HRPQ MUTANTS

The original hrpT and hrpQ mutants each contained two insertion mutations, one in hrpS (the "reporter" mutation), which alone would lead to an Hrp phenotype, and the Tn-plac insertions at the respective chromosomal locations. To determine whether the hrpT and hrpQ mutations alone lead to loss of pathogenicity on bean and ability to elicit HR on tobacco, we restored hrpS function in the double mutants by reverse marker exchange, using a cosmid that had an intact hrpS gene to replace the chromosomal hrpS::inaZ insertion allele. The resulting strains were non-pathogenic and unable to elicit HR. Additionally, we introduced into the T24 and Q1 strains two plasmids, one of which expressed the hrpRS operon under its native promoter (pBBGL1) and the other under the E. coli lac promoter (pBBGL6). Both plasmids fully restored the Hrp^+ phenotype in the original hrpS::inaZ strain (Table 2) - thereby assuring that the lac promoter provided appropriate expression of the hrpRS operon for pathogenesis and HR elicitation. We reasoned that plasmid pBBGL6 would bypass the transcriptional block caused by the hrpT and hrpQ mutations on the plasmid borne hrpRS operon. Therefore, by disengaging hrpS expression from hrpT/hrpQ control this plasmid should restore the Hrp^+ phenotype in the double mutants. Plasmid pBBGL6, but not pBBGL1, restored the ability to ability to elicit HR in the double mutants (Table2). Neither plasmid restored pathogenicity in the hrpT mutant; however, both plasmids partially restored pathogenicity in the hrpQ mutant. These findings suggested that hrpT and hrpQ may control additional genes that are specifically involved in pathogenesis but not for HR. However, the mutations could lead to loss of pathogenesis in an indirect way (see below).

2.3. AUXOTROPHIC DEFICIENCIES OF HRPT AND HRPQ MUTANTS

Besides having reduced transcription of hrpS, the hrpT and hrpQ mutants are auxotrophs for tryptophan and methionine, respectively. The metabolic blocks of these mutants have been identified on the basis of auxanographic tests. Specifically, the hrpT and hrpQ mutants behave as if they were blocked in the first step of the tryptophan biosynthesis and in the last step in methionine biosynthesis, respectively. The amino acid requirements of these mutants are not directly responsible for their inability to elicit HR, based on the results of restoration experiments with the hrpRS plasmids discussed above. However, these requirements may be responsible for the inability of these plasmids to fully restore pathogenesis in the hrpT and hrpQ mutants.

2.4. MEMBRANE ASSOCIATED HRP-ENCODED PROTEINS AND THEIR POSSIBLE FUNCTIONS

Approximately half of the 22 kb region from P. syringae pv. phaseolicola and the hrpM locus have been sequenced (Grimm and Panopoulos, 1989, Grimm et al, unpublished data,; Frederick 1991; Mindrinos et al, unpublished data). Based on computer predictions, homology to other proteins, and Tn-phoA analysis, at least four Hrp proteins appear to be membrane associated. One of these is encoded by the hrpM locus. Based on a recent report concerning the synonymous locus of Pseudomonas syringae pv. syringae (Loubens et al. 1992), hrpM is structurally similar to the Escherichia coli mdoH locus which is involved in periplasmic glucan biosynthesis. Two other putative membrane peptides are encoded by the hrpD locus (Frederick 1989). A fourth protein (Grimm and Panopoulos, unpublished data), encoded by the hrpF locus, resembles the YscJ lipoprotein of Yersinia enterocolitica, which is thought to be involved in Yop protein export (Michiels et al, 1991). This protein contains a typical lipoprotein signal sequence (Wu and Tokunaga 1986) and two hydrophobic domains positioned in the N- and C-terminal segments of the predicted polypeptide.

2.5. STRUCTURE/FUNCTION ANALYSIS OF THE HRPS PROTEIN

To define the structural/functional features which the HrpS protein shares with its structural relatives we have attempted to purify it from E. coli carrying a plasmid in which the hrpS gene is expressed under the control of the phage T7 promoter. Overexpressed HrpS overwhelmingly forms inclusion bodies. After solubilization with 8 M urea at pH 9.5, dialysis and Sephadex G75 filtration the protein was highly purified (>90%). The solubilized protein aggregated when its concentration in solution exceeds 1-2 mg/ml. During gel filtration the freshly purified protein tended to form dimers and to some extent higher oligomers. The dimers disappeared after about one week and were replaced by higher oligomers or by fragments smaller than the monomer. This is the first evidence, although indirect, that an oligomeric structure may be the form of this protein that is involved in hrp promoter control, as is generally the case with other bacterial transcriptional activators, including NtrC subfamily members (Weiss et al. 1991).

As mentioned earlier, hrpS and hrpR encode ca. 34 kD proteins that are structurally related to each other (≈60% amino acid sequence identity, Grimm et al. 1992) as well as to the highly conserved "central" domain of the NtrC subfamily of transcriptional regulators in bacteria (Gross et al. 1991, Gussin et al. 1986). Similar to these proteins, both HrpS and HrpR posses a C-terminal helix-turn-helix motif that may be involved in DNA binding. However, the two Hrp proteins differ in an important respect from other NtrC subfamily members, namely, the virtual lack of long N-terminal domains, relative to the central domain core. If the ATP binding motif (GxxGXGKT), which is highly conserved in all members of the NtrC family, is used as a common reference point, the predicted N-terminal domain of HrpS and HrpR are only ≈36 and 38 amino acids, respectively, compared to 167 or 168 in NtrC and >200 in NifA proteins. Since the N-terminal domain in NtrC and other similar proteins modulates activator function, its absence in the two Hrp proteins suggests that transcriptional control of hrpS assumes greater importance in hrp regulon control. This suggestion is in line with the dramatic induction which hrpS undergoes in planta (Rahme et al. 1992) and the regulation of hrpS transcription by hrpT and hrpQ.

3. Conclusions and Speculations

The hrpT and hrpQ genes reported in this study constitute novel elements in what appears to be a signalling cascade that mediates plant and other environmental signal transduction to the pathogen and leads to hrp gene activation in the course of infection. The isolation and properties of the hrpT and hrpQ mutants provide strong evidence that a plant signal molecule(s) is recognized by the bacterium.

The metabolic blocks identified in the hrpT and hrpQ mutants are catalyzed by enzymes that are composed of two different subunits (TrpE plus TrpG and MetE plus MetH, respectively). Experiments are in progress to determine which subunit in each pair is the target of, or is otherwise interfered with, in the hrpT and hrpQ mutants. In the case of hrpT we speculate that: a) if the gene encodes a housekeeping anthranilate synthase subunit, it is likely trpE (based on the comparable size of the hrpT and trpE coding regions); b) hrpT could be a substrate ambiguous TrpE-like protein that serves more than one metabolic function; c) hrpT is a protein that activates trpE. We consider it unlikely that the anthranilate synthase function of TrpE, the tryptophan biosynthetic intermediates, or tryptophan itself are the relevant elements in hrpS induction, since the gene is only minimally expressed in wild type ($hrpT^+$) cells under in vitro conditions.

In regard to hrpQ, the position of the metabolic block in the methionine biosynthetic pathway suggests that the gene is: a) either metE or metH, b) a metE- or metH-like protein that somehow interacts with the methionine biosynthetic activity of MetE and/or MetH, c) a regulatory element similar to metR, which positively regulates metE and metH expression in E. coli (Weissbach and Brot, 1991).

REFERENCES

Fellay, R., Rahme, L. G., Mindrinos, M. N., Frederick, R. D., Pisi, A. and Panopoulos, N. J. (1991), "Genes and signals controlling the Pseudomonas syringae pv. phaseolicola-plant interaction", in Proc. 5th Internat. Sympos. on the Molecular Genetics of Plant-Microbe Interactions. Klewer Academic Publishers, Dordrecht, The Netherlands, pp. 45-52.

Fellay, R., Rahme, and N. J. Panopoulos (1992). In preparation.

Frederick, R. D. (1989), Ph. D. thesis, University of California.

Grimm, C. and Panopoulos, N. J. (1989) "The predicted protein product of a pathogenicity locus from Pseudomonas syringae pv. phaseolicola is homologous to a highly conserved domain of several procaryotic regulatory proteins", J. Bacteriol. 171, 5031-5038.

Gross, R., Arico, B. and R. Rapuoli (1989) "Families of bacterial signal-transducing proteins", Mol. Microbiol. 3, 1661-1667.

Gussin, G. N., Ronson, C. W. and F. M. Ausubel (1986) "Regulation of nitrogen fixation genes", Ann. Rev. Genet. 20, 567-591.

Kustu, S., Santero, E., Keener, J., Pophan, D. and Weiss, D. (1989) "Expression of σ^{54} (ntrA)-dependent genes is probably united by a common mechanism", Microbiol. Rev. 53, 367-376.

Lindgren, P. B., Peet, R. C. and Panopoulos, N. J. (1986), "Gene cluster of Pseudomonas syringae pv. phaseolicola controls pathogenicity of bean plants and hypersensitivity on non-host plants", J. Bacteriol. 168, 512-522.

Lindgren, P. B., Frederick, R., Govindarajan, N. J., Panopoulos, N. J., Staskawicz, B. J. and Lindow, S. E. (1989), "An ice nucleation reporter gene system: identification of inducible pathogenicity gene in Pseudomonas syringae pv. phaseolicola", EMBO J. 8, 1291-1301.

Loubens, I., Richter, G., Mills, D. and Bohin, J.-P. (1992), "The hrpM locus of Pseudomonas syringae pv. syringae complements a defect in periplasmic glucan biosynthesis in Escherichia coli K-12, Sixth Internat. Sympos. on Plant-Microbe Interact., July 11-16, Seattle, Washington, Abstr. No. 192.

Michels, T., Vanooteghem, J.-C., Rouvroit, C.-L., de, China, B., Gustin, A., Boudry, P. and Cornellis, G. R. (1991), "Analysis of the virC operon involved in the secretion of Yop proteins by Yersinia enterocolitica", J. Bacteriol. 173, 4994-5009.

Miller, W., and Simons, R. (in preparation).

Rahme, L. G., Mindrinos, M. N. and Panopoulos, N. J. (1991) "Genetic and transcriptional organization of the hrp cluster of Pseudomonas syringae pv. phaseolicola", J. Bacteriol. 173, 575-586.

Rahme, L. G., Mindrinos, M. N. and Panopoulos, N. J. (1992), "Plant and environmental sensory signals controlling the expression of hrp genes in Pseudomonas syringae pv. phaseolicola", J. Bacteriol. 174, 3499-3507.

Weiss, D.S., Klose, K. E., Hoover, T. R., North, A. K., Porter, S. C., Wedel, A. B. and Kustu, S. (1991) "Prokaryotic transcriptional enhancers. In Transcriptional Regulation", in McKnight, S. L. and Yamamoto, K. R. (eds.), Cold Spring Harbor Laboratory, Cold Spring Harbor, N. Y., in press.

Weissbach, H. and Brot, N. (1991) "Regulation of methionine synthesis in Escherichia coli", Mol. Microbiol. 5, 1593-1597.

Willis, D. K., Rich, J. J. and Hrabak, E. M. (1991) "hrp genes of phytopathogenic bacteria" Mol. Plant-Micr. Interact. 4, 132-138.

Wu, H. C. and Tokunaga, M. (1986) "Biogenesis of lipoproteins in bacteria", in Protein secretion and export in bacteria, Wu, H. C. and Tai, P. C. (eds.), Springer-Verlag, pp.127-157.

MOLECULAR GENETIC ANALYSIS OF *hrp* AND AVIRULENCE GENES OF *XANTHOMONAS CAMPESTRIS* PV. *VESICATORIA*

ULLA BONAS, JUTTA CONRADS-STRAUCH, STEFAN FENSELAU, TORSTEN HORNS, KAI WENGELNIK AND RALF SCHULTE

Institut für Genbiologische Forschung Berlin GmbH, Ihnestr. 63, 1000 Berlin 33, Federal Republic of Germany

ABSTRACT. *Xanthomonas campestris* pv. *vesicatoria* (*Xcv*) causes bacterial spot disease in pepper and tomato. The *hrp* genes (hypersensitive reaction and pathogenicity) are organized in six loci clustered in a 25-kb region. *hrp* expression is suppressed in complex and minimal media and is induced in the plant or in plant culture filtrates. The *hrpF* locus can also be induced in a defined minimal medium without any plant-derived molecule. Sequence analysis revealed striking similarities (up to 70%) between putative Hrp proteins and proteins from other bacteria, including the mammalian pathogens *Shigella* and *Yersinia*. We hypothesize that *hrp* gene products are involved in transport of molecules important for the interaction of *Xcv* with the plant. Induction of the HR in pepper and tomato requires *hrp* genes and a particular avirulence gene in *Xcv*. The avirulence genes *avrBs3* and *avrBs3-2* are highly homologous to each other but differ in specificity. Both genes express repetitive proteins containing 17.5 copies of a nearly identical repeat motif. Sequence and genetic analyses of *avrBs3* alleles have demonstrated that the presence and position of certain repeats determine specificity of the Avr-protein.

Introduction

Xanthomonas campestris pathovar (pv.) *vesicatoria* (*Xcv*) is the causal agent of bacterial spot disease of pepper and tomato. After infection of a plant with *Xcv* two different types of reactions can be observed. In the susceptible plant, the infection gives rise to watersoaked lesions (compatible interaction). Is the plant resistant, a hypersensitive response (HR) is induced (incompatible interaction). The HR is a local defense reaction accompanied by a rapid necrosis of the infected tissue, leading to restriction of bacterial growth. It has been shown that the incompatible interactions between *Xcv* and pepper cultivars are examples of gene-for-gene systems. This means that a resistance gene in the plant is matched by a corresponding avirulence gene in the particular race of the pathogen (Minsavage et al. 1990).

We are interested in the molecular mechanisms determining the interaction between susceptible host plants and for the induction of the HR on resistant host and nonhost plants. The term *hrp* has been introduced by Lindgren and coworkers (1986) and,

E. W. Nester and D. P. S. Verma (eds.),
Advances in Molecular Genetics of Plant-Microbe Interactions, 275–279.

since then, operationally defines sets of genes in a number of different phytopathogenic bacteria based on their mutant phenotype (see review by Willis et al. 1991). To study the incompatible interaction we focus on the analysis of the *avrBs3* avirulence gene family (Bonas et al. 1989).

Results and Discussion

hrp GENES

The hrp region of Xcv. Recently, a chromosomal region containing the *hrp* gene cluster from *X. c. vesicatoria* has been isolated by complementation of nonpathogenic NTG mutants (Bonas et al. 1991). Defined insertion mutants were generated by transferring a large number of transposon Tn*3-gus* insertions, generated in the cloned wild type *hrp* region, into the genome of *Xcv* strain 85-10 by marker gene exchange. Most insertions into a 25-kb region eliminated both pathogenicity and the ability to induce the HR on resistant host plants and on tobacco. In addition, the nonpathogenic mutants have lost the ability to grow in the plant tissue whereas bacterial growth *in vitro* is as wild-type. Complementation studies showed that the *hrp* region of *Xcv* contains at least six different complementation groups (loci), designated *hrpA* to *hrpF*. A region of ca. 4 kb between *hrpE* and *hrpF* does not appear to be essential for pathogenicity (Bonas et al. 1991).

Expression of the hrp loci is regulated by environmental conditions. To determine the conditions for expression of the *hrp* loci we used the β-glucuronidase (*gus*) gene as a reporter gene. The transposon Tn*3-gus* which was used for insertion mutagenesis of the cloned *hrp* region contains a promoterless *gus* gene. The β-glucuronidase (GUS)-activity of *Xcv* strains harboring derivatives of the plasmids pXV2 and pXV9, carrying a Tn*3-gus* insertion in one of the *hrp* loci, was determined after bacterial growth in minimal or in complex medium (NYG) (Schulte and Bonas 1992a). In no case was significant GUS-activity detected. However, after growth of the bacteria in the plant, for each *hrp* locus at least one insertion with plant-inducible GUS-activity was found. We also examined various plant extracts for their ability to induce expression of the *hrp* genes, however, only culture filtrates from tomato, pepper or tobacco suspension cultures were found to induce expression of all *hrp* loci. The culture filtrate from tomato suspension cultures, tomato conditioned medium (TCM), gave the highest activities, comparable to those obtained in the pepper plant. MS medium, the basal medium for the plant suspension cultures, did not induce *hrp* expression. The putative plant factor(s) present in TCM is organic, heat-stable, hydrophilic, and smaller than 1000 Dalton (Schulte and Bonas, 1992a).

In order to mimic the conditions within the intercellular space of the plant tissue and to test whether a plant factor is really needed for induction we searched for a defined medium that would allow both growth of the bacteria and *hrp* gene induction. Since the *Xcv* merodiploid strain carrying pF312, with a Tn*3-gus* insertion in the *hrpF* locus, gives the highest inducible GUS-activity this strain was used as a test strain. We designed the medium XVM1, which supports bacterial growth and induces *hrpF* to high activities (Schulte and Bonas, 1992b). This medium is a basal salt medium, low in phosphate and sodium chloride concentration, and contains 10 mM sucrose and 2 µg/ml methionine as organic compounds. Interestingly, both sucrose (or fructose) and methionine are required for *hrpF* gene induction. High concentrations of phosphate, sodium chloride, and organic nitrogen were found to suppress gene induction. The other *hrp* loci are less well or not induced under these conditions. Whether additional (plant?) factors are required or the XVM1 medium contains some components in suppressing concentration is not clear (Schulte and Bonas, 1992b).

The gene(s) regulating *hrp* gene expression has not been identified yet. Since the sequences of the putative *hrp* promoters share some homology with sigma-54 regulated promoters, *hrp* gene expression might be sigma-54 dependent. To test this, we are currently isolating sigma-54 mutants of *Xcv*. An effect of sigma-54 mutants on gene expression has been reported previously for *hrp* genes of *P. s. phaseolicola* (Fellay et al. 1991).

Function of hrp genes. As a first step towards an understanding of what the function of *hrp* genes might be, the nucleotide sequence of the entire *hrp* cluster will be determined. Recently, we discovered striking sequence similarities to known proteins for some of the putative Hrp proteins encoded within the 10-kb left hand part of the cluster. Three of the 12 putative proteins, designated HrpA1, HrpB3, and HrpC2, are similar to the YscC, YscJ, and LcrD proteins, respectively, of subspecies of the mammalian pathogen *Yersinia* (Fenselau et al. 1992). For example, HrpC2 is 70% similar (46% identical) to LcrD throughout its entire length. The YscC, YscJ, and LcrD proteins of *Yersinia* are encoded by genes localized on a 70-kb virulence plasmid and are essential for secretion of the so-called Yop (Yersinia outer proteins) proteins. The Yop proteins are encoded by genes on the same 70-kb plasmid and are important virulence factors (Cornelis et al. 1989). Furthermore, the HrpB6 protein shares 65% sequence similarity with the FlaA, FliI, and Spa47 proteins from *B. subtilis*, *Salmonella typhimurium*, and *Shigella flexneri*, respectively. These related proteins are putative ATPases; they share similarity throughout their entire lengths and not only in the highly conserved regions for nucleotide or magnesium binding. Interestingly, the HrpB6 related proteins are involved in transport of proteins, e. g. the Spa47 protein is necessary for the export of virulence proteins (Fenselau et al. 1992).

Based on the sequence similarities we hypothesize that the Hrp proteins mentioned above and possibly others play a role in the transport of molecules essential for the interaction with the plant, and that they might be part of a specialized transport apparatus. These findings indicate for the first time that the fundamental determinants of pathogenicity may be conserved among bacterial pathogens of plants and animals.

ANALYSIS OF *avrBs3* ALLELES

Structure of avrBs3 alleles. The avirulence gene *avrBs3* was isolated from *Xcv* race 1 and is responsible for the induction of the HR on pepper cultivar ECW-30R (Bonas et al. 1989). Southern blots of total genomic DNA of different strains of *Xcv*, probed with the internal *Bam*HI fragment of the *avrBs3* gene, revealed the presence of a 15 kb *Eco*RI fragment homologous to *avrBs3* in a number of strains. As described for *avrBs3*, the homologous sequence is also plasmid-borne. We have isolated the 15 kb *Eco*RI fragment and found that it contains an avirulence gene. In contrast to *avrBs3*, the HR induction occurs on tomato but not on pepper. Since this gene is highly homologous to *avrBs3*, it is considered to be allelic to *avrBs3* and was, therefore, designated *avrBs3-2* (Bonas et al., submitted).

The nucleotide sequence of *avrBs3-2* was determined. The gene encodes a protein of 1160 amino acids which is 97% identical to AvrBs3. Interestingly, the internal regions of both *avrBs3-2* and *avrBs3* contain 17.5 copies of a nearly identical 102-bp repeat motif. Basically the same repeat units are present in both genes, with a variable region at amino acid positions 12 and 13. The organization of the repeats, however, is different, and most likely is the reason for the different specificities. This is corroborated by the fact that new avirulence specificities could be generated by deleting repeat units in the original *avrBs3* gene (Herbers et al. 1992).

The *avrBs3-2* sequence is 100% identical to the corresponding region of the 1.7 kb sequence of the *avrBsP* gene which was isolated from a different strain of *Xcv* and also displays avirulence activity on tomato but not on pepper (Canteros et al. 1991). We assume that *avrBs3-2* and *avrBsP* are identical genes. The *pthA* gene from *X. citri* appears also to belong to the family of *avrBs3*-related genes; the N-terminal region is identical to *avrBs3* and *avrBs3-2* and it contains the same kind of basic repeat motif of 102 bp (Swarup et al. 1992). The *pthA* gene has avirulence activity on bean and cotton when it is introduced into heterologous pathovars of *Xanthomonas*.

Expression and activity of the avrBs3-2 gene. Expression of the *avrBs3-2* gene is constitutive, as was described for the *avrBs3* gene (Knoop et al. 1991). To our surprise, derivatives of *avrBs3-2* with deletions into the C-terminal region remain active in inducing the HR on tomato. The shortest active derivative encodes a protein which consists of the N-terminal 288 amino acids plus 3.5 repeat units. In contrast, deletions into the C-terminus of the *avrBs3* gene abolish avirulence activity. The role of *avrBs3-2* and the other members of the gene family in eliciting the HR remains elusive.

Acknowledgements

We thank I. Balbo and M. Gutschow for excellent technical assistance. This research was supported by grants from the Bundesministerium für Forschung und Technologie, the Deutsche Forschungsgemeinschaft and the European community to U. B.

References

Bonas, U., Stall, R.E., and Staskawicz, B. 1989. Genetic and structural characterization of the avirulence gene *avrBs3* from *Xanthomonas campestris* pv. *vesicatoria*. Mol. Gen. Genet. 218:127-136.

Bonas, U., Schulte, R., Fenselau, S., Minsavage, G.V., Staskawicz, B.J., and Stall, R.E. 1991. Isolation of a gene cluster from *Xanthomonas campestris* pv. *vesicatoria* that determines pathogenicity and the hypersensitive response on pepper and tomato. Mol. Plant-Microbe Interact. 4:81-88.

Canteros, B., Minsavage, G., Bonas, U., Pring, D., and Stall, R. 1991. A gene from

Xanthomonas campestris pv. *vesicatoria* that determines avirulence in tomato is related to *avrBs3*. Mol. Plant-Microbe Interact. 4:628-632.

Cornelis, G. R., Biot, T., Lambert de Rouvroit, C., Michiels, T., Mulder, B., Sluiters, C., Sory, M.-P., Van Bouchaute, M., and Vanooteghem, J.-C. 1989. The Yersinia *yop* regulon. Mol. Microbiol. 3:1455-1469.

Fellay, R., Rahme, L. G., Mindrinos, M. N., Frederick, R. D., Pisi, A., and Panopoulos, N. J. 1991. Genes and signals controlling the *Pseudomonas syringae* pv. *phaseolicola*-plant interaction. In Advances in Molecular Genetics of Plant-Microbe-Interactions. Vol. 1. H. Hennecke and D. P. S. Verma, eds. (Dordrecht, The Netherlands: Kluwer Academic Publishers), pp. 45-52.

Fenselau, S., Balbo, I., and Bonas, U. 1992. Determinants of pathogenicity in *Xanthomonas campestris* pv. *vesicatoria* are related to proteins involved in secretion in bacterial pathogens of animals. Mol. Plant-Microbe Interact. 5 (in press).

Herbers, K., Conrads-Strauch, J., and Bonas, U. 1992. Race-specificity of plant resistance to bacterial spot disease determined by repetitive motifs in a bacterial avirulence protein. Nature 356:172-174.

Knoop, V., Staskawicz, B. J., and Bonas, U. 1991. The expression of the avirulence gene *avrBs3* from *Xanthomonas campestris* pv. *vesicatoria* is not under the control of *hrp* genes and is independent of plant factors. J. Bacteriol. 173:7142-7150.

Lindgren, P. B., Peet, R., and Panopoulos, N. J. 1986. Gene cluster of *Pseudomonas syringae* pv. "*phaseolicola*" controls pathogenicity on bean plants and hypersensivity on nonhost plants. J. Bacteriol. 168:512-522.

Minsavage, G. V., Dahlbeck, D., Whalen, M. C., Kearney, B., Bonas, U., Staskawicz, B. J., and Stall, R. E. 1990. Gene-for-gene relationships specifying disease resistance in *Xanthomonas campestris* pv. *vesicatoria*-pepper interactions. Mol. Plant Microbe Interact. 3:41-47.

Schulte, R., and Bonas, U. 1992a. The expression of the *hrp* gene cluster from *Xanthomonas campestris* pv. *vesicatoria* that determines pathogenicity and hypersensitivity on pepper and tomato is plant-inducible. J. Bacteriol.174:815-823.

Schulte, R., and Bonas, U. 1992b. A *Xanthomonas* pathogenicity locus is induced by sucrose and sulfur-containing amino acids. Plant Cell 4:79-86.

Swarup, S., Yang, Y., Kingsley, M. T., and Gabriel, D. W. 1992. A *Xanthomonas citri* pathogenicity gene, *pthA*, pleiotropically encodes gratuitous avirulence on nonhosts. Mol. Plant-Microbe Interact. 5: 204-213.

Willis, D. K., Rich, J. J., and Hrabak, E. M. 1991. *hrp* Genes of phytopathogenic bacteria. Mol. Plant-Microbe Interact. 4:132-138.

ARE HARPINS UNIVERSAL ELICITORS OF THE HYPERSENSITIVE RESPONSE OF PHYTOPATHOGENIC BACTERIA?

STEVEN V. BEER, ZHONG-MIN WEI, RON J. LABY, SHENG YANG HE, DAVID W. BAUER, ALAN COLLMER AND CATHY ZUMOFF
Department of Plant Pathology
Cornell University
Ithaca, NY 14853 USA

ABSTRACT. The entire *hrp* gene cluster of *Erwinia amylovora* strain Ea321 had been cloned previously on a single cosmid, designated pCPP430. This cosmid is well expressed in *Escherichia coli* and Gram-negative bacteria harboring it strongly elicit the hypersensitive response (HR) on non-host plants. Cell-free sonicates of *E. coli* DH5α(pCPP430) and Ea 321 caused collapse of tobacco leaf tissue within 12 hrs after infiltration of the intercellular spaces. From the HR-eliciting sonicates, we purified a 44 kD, heat-stable, cell-surface-associated protein (harpin) by ion-exchange chromatography and reverse-phase chromatography. Purified harpin caused collapse of leaf tissue that was indistinguishable from the collapse caused by DH5α(pCPP430) or wild-type *E. amylovora*. Harpin also elicited the K^+/H^+ exchange reaction in tobacco cell suspensions. A polyclonal antibody raised in rabbits in response to injections of harpin may prove useful in the diagnosis of the fire blight pathogen.

The structural gene encoding harpin (*hrpN*) was identified on pCPP430 with the aid of a synthesized oligonucleotide probe corresponding to the ninth to fifteenth amino acids of the NH_2- terminus of harpin. This gene hybridized with genomic DNA of *E. amylovora*, *E. chrysanthemi*, *E. stewartii* and *Pseudomonas solanacearum*. No consistent hybridization was observed with genomic DNA of *P. syringae* and *Xanthomonas campestris*. Evidence is accumulating that harpin-like proteins are the active products of *hrp* genes of *Erwinia* species. Based on compelling genetic evidence with respect to *E. amylovora*, harpin also plays a crucial role in disease.

1. Introduction

Bacteria that cause necrotic symptoms in their host plants generally elicit a plant defense reaction known as the hypersensitive response (HR) when suspensions are infiltrated into the intercellular spaces of leaves of non-host plants (Klement, 1982). Recently, the factor responsible for elicitation of the HR by the fire blight pathogen, *Erwinia amylovora*, was isolated, characterized and named harpin (Wei *et al.*, 1992). Harpin is a product of the *hrp* gene cluster of *E. amylovora*. The entire *hrp* gene cluster had been cloned and is well-expressed in *Escherichia coli* (Beer *et al.*, 1991). *hrp* genes have been defined as those that function in both pathogenicity to host plants and elicitation of the HR in non-host plants (Lingdren *et al.*, 1986). They have been described also from many other bacteria that cause necrotic symptoms (Willis *et al.*, 1991). Whether other necrogenic plant pathogenic bacteria produce harpin-like elicitors of the HR is the subject of the work described here.

E. W. Nester and D. P. S. Verma (eds.),
Advances in Molecular Genetics of Plant-Microbe Interactions, 281–286.

1.1. CHARACTERISTICS OF HARPIN

Harpin was purified from *Escherichia coli* DH5α(pCPP430). Cells were grown in LB broth, pelleted and resuspended in buffer with phenylmethylsulfonyl fluoride, a potent protease inhibitor. Following sonication, harpin was purified by anion-exchange and reverse-phase chromatography. The final product, which moved as a single band in SDS polyacrylamide gels, was used to raise antibodies in rabbits that reacted specifically with harpin.

Harpin is a protein of approximately 44 kD in size, based on mobility in SDS-polyacrylamide gels. It is acidic (pI = 4.3), and its deduced amino acid sequence is markedly hydrophilic and contains 22% glycine. The protein is heat-stable, but highly sensitive to protease. Harpin is cell-surface associated, based on fractionation studies in which only whole cells and the membrane fraction reacted with anti-harpin antibodies.

The gene encoding harpin (*hrpN*) was localized to a 1.3 kb *Hind*III fragment of the *hrp* gene cluster of *E. amylovora* contained in the cosmid pCPP430 (Fig. 1). Localization was accomplished by hybridization of a synthesized oligonucleotide corresponding to the ninth to fifteenth amino acids from the NH_2- terminus of the protein. Sequence analysis revealed that the gene is 1155 bp in size and that it is not similar in nucleic acid or amino acid sequence to DNA of other organisms reported to GenBank, except for some glycine-rich plant cell wall proteins.

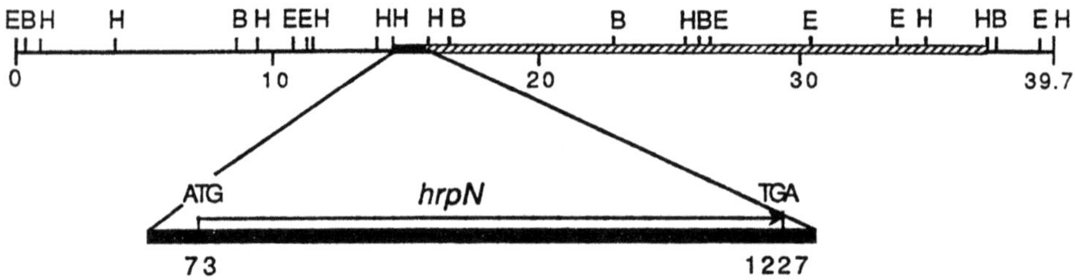

Figure 1. The *hrp* gene cluster of *Erwinia amylovora* indicating location and extent of *hrpN* within a 1.3 *Hind*III fragment. The cross-hatched portion plus *hrpN* is required for the Hrp phenotype.

Purified harpin elicited an HR in leaves of tobacco at a concentration $\geq$ 0.5 μM. HR also developed in leaves of other plants infiltrated with harpin including tomato, petunia, pepper and *Arabidopsis thaliana*. The time-course of HR development was similar to that elicited by metabolizing cells of DH5α(pCPP430) or Ea321. Harpin also elicited the K^+/H^+ exchange reaction (XR) (Atkinson *et al.*, 1985) in tobacco cell suspension culture (TCSC). However, the TCSC responded to harpin more rapidly and without the lag that is observed with bacteria. Harpin also is required by *E. amylovora* for pathogenicity on pear as well as elicitation of an HR in nonhosts. Mutagenesis of *hrpN* with Tn*5*tac1 resulted in a strain that failed to elicit the HR and was not pathogenic to immature pear fruits.

1.2. HYPOTHESIS AND APPROACH

We hypothesized that other necrogenic phytopathogenic bacteria produce harpin-like proteins that are secreted by *hrp* genes, elicit the HR in nonhosts, and are required for pathogenicity to hosts. To test the hypothesis, we used several tools developed during studies of harpin. Isolation and assay techniques similar to those used for harpin from *E. amylovora* were used with other bacteria. The *hrpN* gene was used as a hybridization probe and for sequence comparison with candidate *hrpN*-like genes. DNA of other bacteria also was used in attempts to functionally restore the *hrp* phenotype to a *hrpN*-minus mutant of *E. amylovora*, produced by marker-exchange mutagenesis of Ea321 with *hrpN*::Tn*5*tac1. Attempts were made to isolate harpin-like proteins and *hrpN*-like genes from *E. chrysanthemi*, *E. stewartii*, *P. syringae*, *P. solanacearum* and *X. campestris*. Collectively, these bacteria are representative of those responsible for the majority of agricultural losses due to phytopathogenic bacteria.

2. Studies on *Pseudomonas* Species

Pseudomonas syringae pv. *syringae* strain 61 was addressed initially because its entire *hrp* gene cluster had been cloned in cosmid pHIR11 (Huang *et al.*, 1988). It is expressed in *P. fluorescens* and *E. coli* and bestows on strains of these saprophytes the ability to elicit the HR in plants. Thus, pHIR11 has functional characteristics similar to those of pCPP430 of *E. amylovora.*

Cell-free preparations from *P. syringae* pv. *syringae*, prepared according to methods used with *E. amylovora*, failed to elicit the HR when infiltrated into tobacco leaves. When the preparations were resolved on SDS polyacrylamide gels and then reacted with anti-harpin antibodies in western blots, no reaction occurred. Nor did the *hrpN* coding sequence hybridize with digests of pHIR11 in southern blots. Furthermore, subclones of pHIR11 failed to complement the *hrpN*-minus mutant of *E. amylovora*. Finally, *hrpN* failed to complement the several mutated derivatives of pHIR11 for elicitation of the HR in tobacco. These results indicate that the harpin isolation methods that were successful with *E. amylovora* do not yield a protein from *P. syringae* similar to harpin in biological, immunological or genetic characteristics.

Although *hrp* genes have been identified from *P. solanacearum* and *X. campestris*, entire functional *hrp* gene clusters have not been cloned from these bacteria. Strain K60 of *P. solanacearum*, which has been studied extensively, was used after confirming its pathogenicity to tomato, pepper and tobacco. A genomic library of this strain was constructed in a plasmid vector. Hybridization of several members of the library occurred with *hrpN* in colony hybridizations. Further tests for hybridization of *hrpN* with southern-blotted DNA are in progress.

3. Studies on *Erwinia* Species

Functional *hrp* genes recently were identified in *E. chrysanthemi* strain EC16. The possible presence of harpin-like elicitors in this strain was explored. *hrpN* hybridized with cosmid clones containing DNA of EC16. The hybridizing portion of the cosmid was cloned, and initial sequencing showed significant similarity between *hrpN* and the hybridizing clone. However, no complementation of the *hrpN* mutant of *E. amylovora* was observed with the *hrpN* homolog from *E. chrysanthemi.* The *hrpN*-like gene was mutagenized with Tn*5*gus and then marker-exchanged into a strain of *E. chrysanthemi,* resulting in an HR-minus phenotype. Complementation of the *E. chrysanthemi* HR-minus mutant with *hrpN* of *E. amylovora* did not succeed.

Hybridization of *hrpN* with genomic DNA of *E. stewartii* strain DC283, digested with *Eco*R1, in Southern blots revealed a single hybridizing band of DNA. Similarly, *hrpN* hybridized with DNA of cosmid pES411 (Coplin *et al.*, 1992), which contains genes of the *wts* (watersoaking) cluster of *E. stewartii.* These hybridization data suggest that the Stewart's wilt pathogen has a *hrpN* homolog located in DNA already identified as being involved in pathogenicity.

4. Studies on *Xanthomonas*

A strain of *X. campestris* pv. *glycines* was selected from among 39 strains tested, based mainly on its pathogenicity characteristics. The strain elicited a strong HR on tobacco and was pathogenic on soybean, an important crop plant.

Cell-free preparations from *X. campestris* pv. *glycines*, prepared according to methods used with *E. amylovora,* failed to elicit the HR when infiltrated into tobacco leaves. When the preparations were resolved on SDS polyacrylamide gels and then reacted with anti-harpin antibodies in western blots, no reaction occurred. Hybridization of the 1.3 kb *Hind*III fragment from pCPP430, which contains *hrpN,* in colony hybridizations to a library containing xanthomonad DNA did not result in consistent hybridization. Furthermore, no member of cosmid library containing *X. campestris pv. glycines* DNA restored the Hrp phenotype to the *hrpN* mutant of *E. amylovora.* These results indicate that the methods that were successful with *E. amylovora* do not yield a protein from *X. campestris* pv. *glycines* similar to harpin in biological, immunological or genetic characteristics.

5. Conclusions and Speculations

Our search for harpin-like elicitors among phytopathogenic bacteria, other than *E. amylovora,* thus far, has given mixed results. Within the genus *Erwinia,* evidence for a *hrpN*-like gene has been found both in *E. chrysanthemi* and *E. stewartii.* The case is strongest for *E. chrysanthemi* in which sequence similarity has been found. Previous work by Laby and Beer (1992) has indicated lack of restriction fragment polymorphism in the *hrp* gene clusters among 22 strains of *E. amylovora* of diverse origin. Thus, harpin-like elicitors are expected to be found in all strains of the fire blight pathogen.

Our failure to detect harpin-like proteins or *hrpN*-like genes in single strains of *P. syringae* and *X. campestris* is indeed puzzling. Bacteria of both species elicit an HR which is quite similar to that elicited in nonhosts by *E. amylovora.*

Analysis of the *hrp* gene cluster of *P. syringae* strain 61 and comparison of it with the functionally homologous *hrp* gene cluster of *E. amylovora* has indicated a highly conserved and colinear region of DNA (Laby and Beer, 1992). Examination of available sequences confirm the close relationship. Based on homologies with DNA of other bacteria with known functions, it seems clear that *hrp* gene clusters include several genes that have a function in the translocation of proteins that are synthesized intracellularly without leader sequences (e.g., the "Yop" virulence proteins of *Yersinia* spp.). As an example, *hrpH* and *hrpI* of *P. syringae* (Huang *et al.*, in preparation) and *hrpI* of *E. amylovora* (Wei *et al.*, in preparation) are strikingly similar to each other and to genes of *Yersinia, Escherichia, Shigella* and others (Galán et al., 1992) that encode proteins thought to be involved in protein translocation (Fig. 2).

We conclude that proteinaceous elicitors of the HR probably are produced by pseudomonads and xanthomonads, but these elicitors differ substantially in primary structure from the *E. amylovora* harpin. Thus, we consider harpin to be the archetype of potentially diverse disease proteins that share a dependence on the products of the conserved *hrp* genes for secretion.

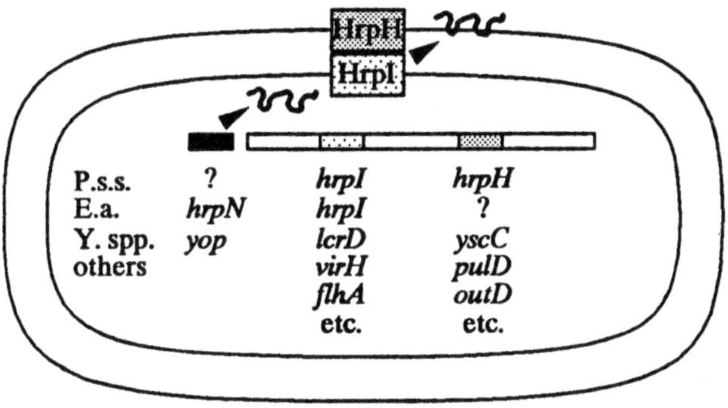

Figure 2. Cartoon illustrating relationships between genes encoding exported proteins (solid bar) of several bacteria and genes encoding secretion apparatus proteins (stippled bars). HrpI and HrpH are located in the inner and outer membranes, respectively (Huang et al., in preparation); their encoding genes are in conserved portions of the *hrp* clusters. Homologs of HrpI and HrpH, with apparent functions in protein translocation, are found in several bacteria, including *Yersinia* spp. (LcrD and YscC), *S. flexneri* (VirH), *E. coli* (FlhA), *Klebsiella oxytoca* (PulD) and *E. chrysanthemi* (OutD). Abbreviations: P.s.s. = *Pseudomonas syringae* pv. *syringae*, E.a. = *Erwinia amylovora*, Y. = *Yersinia*.

6. References

Atkinson, M. M., J. S. Huang, and J. A. Knopp. 1985. The hypersensitive reaction of tobacco to *Pseudomonas syringae* pv. *pisi*: activation of plasmalemma K^+/H^+ exchange mechanism. Plant Physiol. 79:843-847.

Beer, S. V., D. W. Bauer, X. H. Jiang, R. J. Laby, B. J. Sneath, Z.-M. Wei, D. A. Wilcox, and C. H. Zumoff. 1991. The *hrp* gene cluster of *Erwinia amylovora*, p. 53-60. *In* H. Hennecke and D. P. S. Verma (eds.), Proceedings 5th Int. Symp. Mol. Genetics Plant-Microbe Interaction. Interlaken, Switzerland.

Coplin, D. L., R. D. Frederick, and D. R. Majerczak. 1992. New pathogenicity loci in *Erwinia stewartii* identified by random Tn*5* mutagenesis and molecular cloning. Mol. Plant-Microbe Interact. 5:266-268.

Galan, J. E., C. Ginocchio, and P. Costeas. 1992. Molecular and functional characterization of the *Salmonella* invasion gene *invA*: Homology of InvA to members of a new protein family. Journal of Bacteriology 174:4338-4349.

Huang, H.-C., R. Schurrink, T. P. Denny, M. M. Atkinson, C. J. Baker, I. Yucel, S. W. Hutcheson, and A. Collmer. 1988. Molecular cloning of a *Pseudomonas syringae* pv. *syringae* gene cluster that enables *Pseudomonas fluorescens* to elicit the hypersensitive response in tobacco plants. J. Bacteriol. 170:4748-4756.

Klement, Z. 1982. Hypersensitivity, p. 149-177. *In* M. S. Mount and G. H. Lacy (ed.), Phytopathogenic Procaryotes, vol. 2. Academic Press, New York.

Laby, R. J., and S. V. Beer. 1992. Hybridization and functional complementation of the *hrp* gene cluster from *Erwinia amylovora* strain Ea321 and DNA of other bacteria. Mol. Plant-Microbe Interact. (In press).

Lindgren, P. B., R. C. Peet, and N. J. Panopoulos. 1986. Gene cluster of *Pseudomonas syringae* pv. "phaseolicola" controls pathogenicity on bean plants and hypersensitivity on nonhost plants. J. Bacteriol. 168:512-522.

Wei, Z-M., R. J. Laby, C. H. Zumoff, D. W. Bauer, S. Y. He, A. Collmer, and S. V. Beer. 1992. Harpin, elicitor of the hypersensitive response produced by the plant pathogen *Erwinia amylovora*. Science 257:85-88.

Section 5 / Fungal-Plant Interactions: Fungal Side

MOLECULAR CLONING AND FUNCTIONS OF AVIRULENCE AND PATHOGENICITY GENES OF THE TOMATO PATHOGEN *CLADOSPORIUM FULVUM*

PIERRE J.G.M. DE WIT, GUIDO F.J.M. VAN DEN ACKERVEKEN, PAUL M.J. VOSSEN, MATTHIEU H.A.J. JOOSTEN, TON J. COZIJNSEN, GUY HONEE, JOS P. WUBBEN, NADIA DANHASH, JAN A.L. VAN KAN, ROLAND MARMEISSE (*) AND HENK W.J. VAN DEN BROEK (*)

Agricultural University Wageningen
Department of Phytopathology
Binnenhaven 9
6709 PD Wageningen
The Netherlands

(*)
Agricultural University Wageningen
Department of Genetics
Dreijenlaan 2
6703 HA Wageningen
The Netherlands

ABSTRACT. Of the fungal tomato pathogen *Cladosporium fulvum* a number of *in planta* induced genes and their products have been isolated and characterized. They include: (i) putative pathogenicity genes of which two genes, (extracellular proteins) *ecp*1 and *ecp*2, have been cloned and sequenced; the function of those genes during pathogenesis is not known yet; effects of *ecp* gene disruption on pathogenicity are being studied, and: (ii) avirulence genes which encode race-specific elicitors interacting with the products of complementary resistance genes in the host plant, resulting in a hypersensitive response and other defense responses. Avirulence gene *avr*9 of *C. fulvum* is the first fungal avirulence gene that has been cloned. The regulation of this gene has been studied *in vitro* and *in planta*. *In vitro*, the gene is induced under low nitrogen conditions, whereas *in planta* the gene is highly expressed around the vascular tissue. Avirulent races carrying the *avr*9 gene become virulent on *Cf9* genotypes of tomato after disruption of *avr*9.

Strategies to obtain broad spectrum disease-resistant plants by transformation with a gene cassette containing both *avr*9 and the complementary resistance gene *Cf9* will be presented.

E. W. Nester and D. P. S. Verma (eds.),
Advances in Molecular Genetics of Plant-Microbe Interactions, 289–298.

1. Introduction

Cladosporium fulvum is a biotrophic fungal pathogen which enters tomato leaves through stomata, colonizes the intercellular spaces between mesophyll cells and is confined to the apoplast during the main part of the infection cycle. The interaction between *C. fulvum* and tomato is supposed to fit into a gene-for-gene relationship. Of the many defined resistance genes, *Cf2*, *Cf4*, *Cf5*, and *Cf9* are available in near-isogenic lines of the cultivar Moneymaker. These lines give a clear differential response to the presently known fungal races, resulting either in a compatible or an incompatible interaction. The *C. fulvum*-tomato interaction is an ideal model system to study communication between plant and pathogen, as the interface exchange of molecules is confined to the apoplast, from which washing fluids can easily be obtained (De Wit et al., 1991). In apoplastic fluids of *C. fulvum*-infected leaves, plant and fungal proteins occur. They are constitutively produced or specifically induced during the infection. Among the plant proteins the pathogenesis-related (PR) proteins prevail. The fungal proteins can be divided into two categories. The first consists of proteins which are involved in establishing or maintaining basic compatibility and may thus be considered as pathogenicity factors. Two of these proteins are the extracellular proteins ECP1 and ECP2 which are produced by all races of the fungus during colonization of the host plant. Their structure and regulation will be discussed in detail. The second category consists of proteins which are the products of avirulence genes (the so-called race-specific elicitors). They induce a hypersensitive response (HR) in genotypes of tomato carrying the corresponding genes for resistance. The structure and regulation of one of the race-specific elicitors and its encoding avirulence gene *avr9* (which interacts with the product of the *Cf9* resistance gene) will be discussed in detail. The potential of exploiting avirulence genes in molecular breeding for disease-resistant plants will be discussed briefly.

2. Putative Pathogenicity Genes *Ecp1* and *Ecp2*

C. fulvum colonizes the leaves of susceptible tomato plants without inducing defense responses. The way *C. fulvum* parasitizes tomato without the formation of specialized structures and the production of damaging hydrolytic enzymes raises questions about the mechanism by which *C. fulvum* is establishing or maintaining basic compatibility. To gain insight into the strategy *C. fulvum* uses to successfully infect its one and only host species, putative pathogenicity factors and their encoding genes need to be isolated and characterized. In compatible *C. fulvum*-tomato interactions several low molecular weight proteins accumulate in the extracellular space during the infection process. These proteins are not detected in healthy plants and incompatible *C. fulvum* - tomato interactions. Two such extracellular proteins, ECP1 (synonymous to P1, Joosten and De Wit, 1988) and ECP2 (Wubben et al., unpublished) have been purified and polyclonal antisera have been raised against these proteins. Western blot analyses indicated that ECP1 and ECP2 are not present in healthy plants, in *C. fulvum* grown *in vitro*, nor in incompatible tomato-*C. fulvum* interactions. The genes encoding ECP1 and ECP2 are highly expressed during colonization of tomato leaves. The cloning and sequencing of both *ecp* genes were carried out by Van den Ackerveken et al. (1992).

ECP1 was purified as described (Joosten and De Wit, 1988) and the N-terminal sequence of 36 amino acids was determined. A degenerated oligonucleotide probe, complementary to the putative mRNA sequence was designed. This probe gave a clear signal on northern blots, containing RNA from compatible *C. fulvum*-tomato interactions. To obtain a perfectly matching probe, RNA from a compatible *C. fulvum*-tomato interaction was sequenced using the oligonucleotide probe as primer. The fully homologous probe that was synthesized according to the obtained sequence was used to screen a cDNA library of a compatible *C. fulvum*-tomato interaction. One positive cDNA clone was obtained of which the insert encoded a protein containing the determined N-terminal amino acid sequence of ECP1. The cDNA clone was subsequently used to screen a genomic library of *C. fulvum*. The *ecp*1 gene occurs in a single copy in the genome of all *C. fulvum* races and contains two short introns.

Antiserum raised against ECP2 was used to screen a λgt11 expression cDNA library from *C. fulvum*-infected tomato leaves. One positive phage contained an insert which was shown to be of fungal origin and hybridized to a highly abundant mRNA only present in compatible *C. fulvum*-tomato interactions. The cDNA insert was subsequently used to screen a genomic library of *C. fulvum*. Several positive clones were obtained and a fragment, containing the *ecp2* gene, was subcloned and sequenced. The *ecp*2 gene is a single copy gene and contains one short intron. Comparison of both DNA and protein sequences with the GenBank, EMBL, PIR and SwissProt databases did not reveal significant homology of both *ecp*1 and *ecp*2 genes to any known sequence.

2.1. EXPRESSION OF *ECP*1 AND *ECP*2 *IN VITRO* AND *IN PLANTA*

Comparison of the promoters of the *ecp*1 and *ecp*2 genes might reveal some common sequences which could be involved in the regulation of those genes *in planta*. Transforming *C. fulvum* with *ecp* promoter-reporter gene fusions and inoculating the transformants onto tomato will tell us more about the expression of the *ecp* genes in space and time during pathogenesis. Until now, we have not been able to find conditions which induce one or both *ecp* genes *in vitro*. Low nitrogen concentrations which induce the avirulence gene *avr*9 *in vitro* are not inducive for the *ecp* genes. The role of the *ecp* genes in establishing and maintaining basic compatibility remains unclear. The high expression of the *ecp* genes *in planta* as compared to *in vitro* and the abundance of the *ECP* proteins in the extracellular space of the infected tomato leaf suggests a role in pathogenicity. The extracellular localization as observed by Wubben et al. (unpublished) suggests a function in the matrix which is present between the fungal hyphae and the host cell wall. Possibly, the *ECP* proteins are actively interfering with the metabolism or transport of nutrients within the tomato leaf. Gene disruptions by transformation and homologous recombination with a mutated *ecp* gene are essential to determine their role in pathogenicity. These experiments are currently being carried out and will be discussed in the next paragraph.

2.2. DISRUPTION OF THE *ECP*1- and *ECP*2-GENES

Disruption of the *ecp*2 gene was achieved in *C. fulvum* using a plasmid containing the *ecp*2 genomic sequence interrupted by a hygromycin resistance gene (HygR) (Marmeisse et al., unpublished). Preliminary pathogenicity tests of the transformants with an interrupted *ecp*2 gene showed that they were still pathogenic on tomato. However, the transformants still produced a protein which was recognized by ECP2 antiserum, but was larger than ECP2 in the wild type. To avoid the chance of getting transformants which still produce truncated or mutant proteins after gene disruption, future *ecp*2 and also *ecp*1 gene disruptions will be carried out with plasmid constructs in which the whole coding sequence has been removed and replaced by selectable markers such as the *Pyr*4 and the *Hyg*R gene, respectively.

3. Avirulence Genes

The gene-for-gene hypothesis, which was proposed more than fifty years ago, states that for each gene conditioning race-specific resistance in the host plant, there is a corresponding gene conditioning avirulence in a race of the pathogen (Flor, 1942). Gene-for-gene complementarity occurs most frequently in plant-pathogen interactions involving obligate and biotrophic pathogens which are highly specialized and have a narrow host range. Resistance in host plants against these pathogens is usually inherited as a monogenic trait and is based on early recognition of the pathogen leading to a quickly induced hypersensitive response (HR) which restricts the pathogen to the infection site by local death of a few host cells (De Wit, 1992; Keen, 1990). The prevailing model in which recognition in gene-for-gene systems has been described is the elicitor-receptor model which states that a specific receptor in the resistant host plant interacts with a molecule of the pathogen, the so-called race-specific elicitor, leading to the induction of HR accompanied by a cascade of other defence responses (Dixon and Lamb, 1990; Gabriel and Rolfe, 1990; Heath, 1991; Keen and Dawson, 1992; Scheel and Parker, 1990).

The interaction between *C. fulvum* and tomato has a gene-for-gene basis. Avirulent races of the fungus with their characteristic avirulence genes produce race-specific elicitors which are recognized by tomato genotypes carrying the complementary genes for resistance, leading to the induction of a hypersensitive response (HR) (De Wit and Spikman, 1982; Scholtens-Toma and De Wit, 1988).

In order to overcome this race-specific resistance, the pathogen must evade recognition of its race-specific elicitor by the plant. In theory evasion of recognition can be achieved in three ways: (i) the avirulence gene may be mutated in such a way that its product is no longer recognized by the host, (ii) the avirulence gene may be disrupted by a transposon, and (iii) the avirulence gene may be lost from the pathogen's genome. The mechanism of elicitor recognition, however, is not known yet. Receptors in the host plasma membrane might interact with race-specific elicitors, followed by signal transduction and defence gene activation (Dixon and Lamb, 1990; Lamb et al., 1989; Scheel and Parker, 1990).

Races virulent on *Cf9* genotypes of tomato which have evaded recognition by the host, do not produce the peptide elicitor (Scholtens-Toma et al., 1989) as they lack the entire avirulence gene *avr*9 encoding the elicitor (Van Kan et al., 1991). We have isolated a genomic clone containing the avirulence gene *avr*9 , and used it to transform a race of *C. fulvum*, virulent on tomato genotype *Cf9* (Van den Ackerveken et al., 1992).

3.1. TRANSFORMATION OF *C. FULVUM* WITH AVIRULENCE GENE *AVR*9

Race 2.4.5.9.11, which is virulent on tomato genotype *Cf9*, was used in transformation studies. The *avr*9 gene was transferred to this race by co-transformation with pAN7-1, containing the HygR marker. Six selected hygromycin resistant transformants were screened for integration of the *avr*9 gene by Southern blot analysis. Five out of six transformants (C1 and C3 to C6) were found to have multiple copies of the *avr*9 gene integrated into their genomes, whereas one transformant (C2) had only obtained the HygR marker. Two transformants were selected and analysed in detail, C2 (HygR, *avr*9$^-$) and C3 (HygR, *avr*9$^+$).

Southern blot analysis of transformants C2 and C3, and two wild type races of *C. fulvum* indicated that transformant C3 obtained several copies of the cloned *avr*9 gene. The wild type recipient race 2.4.5.9.11. and transformant C2 both lack the *avr*9 gene. Pathogenicity of the transformants C2 and C3 was tested on tomato seedlings of cultivar Moneymaker (MM, carrying no Cf-resistance genes) and genotype *Cf*9 (a near-isogenic line of MM carrying resistance gene *Cf*9). Symptoms were visible 14-20 days after inoculation. Transformants C2 and C3 were both still virulent on cultivar MM, indicating that pathogenicity was not affected by the transformation procedure. The pathogenicity of transformant C3 on genotype *Cf*9, however, was changed from virulent to avirulent, whereas transformant C2 remained virulent.

The elicitor encoded by *avr*9, which induces HR on *Cf*9 genotypes but not on MM, was produced by transformant C3, but not by transformant C2. These results clearly indicate that the cloned avirulence gene *avr*9 is responsible for cultivar-specificity on tomato genotype *Cf*9, and that no other genes or factors are required to restore avirulence in a race previously virulent on tomato genotype *Cf*9. Our studies on the race-specific *avr*9 elicitor peptide provide the first example of a primary avirulence gene product, which directly induces HR in cultivars carrying the matching resistance gene.

*Avr*9 gene disruption experiments proved that disruption of the *avr*9 gene in race 4 and race 5 (both avirulent on tomato genotype *Cf*9) leads to transformants which have become virulent on tomato genotype *Cf*9 (Marmeisse et al., 1992). This is further evidence for having cloned a true fungal avirulence gene which fits into the gene-for-gene concept. Presently, other race-specific elicitors inducing HR on tomato genotypes carrying other Cf-genes are being characterized in order to clone their encoding avirulence genes. The purification of the AVR4 elicitor (a protein of ca. 12 kD) has now been completed and the cloning of its encoding gene is in progress.

3.2. REGULATION OF THE EXPRESSION OF AVIRULENCE GENE *AVR*9

The avirulence gene *avr*9 is not expressed *in vitro* when the fungus is grown under optimal growth conditions i.e.: sufficiently high nitrogen and carbon sources. Initially we could only detect high expression of *avr*9 *in planta*. By testing a great array of different growth conditions it was found that low nitrogen concentrations were inducive for *avr*9 expression, irrespective of whether the source was nitrate, nitrite, ammonia or glutamine.

In order to follow the expression of the *avr*9 gene *in planta*, *avr*9 promoter-reporter gene fusion constructs were made to transform *C. fulvum* and transformants were used to inoculate tomato plants. To this end the *avr*9 promoter was ligated to the ß-glucuronidase (GUS) coding region. *In planta* the *avr*9 promoter-GUS transformants showed high GUS activity after the fungus had penetrated the plant through the stomata, while no GUS activity was detected in runner hyphae present on the leaf surface. High expression was clearly visible in fungal hyphae growing around the vascular tissue of tomato leaves. The significance of the high expression of *avr*9 around vascular tissue is still unclear, but it could be indicative of a metabolic function for the AVR9 elicitor during pathogenesis.

3.3. BIOLOGICAL FUNCTION OF THE *AVR*9 GENE

A clear biological function of the *AVR*9 peptide for *C. fulvum,* while growing on a susceptible plant, is still unknown. The *avr*9 gene is highly expressed when the fungus grows *in planta*, especially around the vascular tissue (Van Kan et al., 1991; Van den Ackerveken et al., unpublished) suggesting that the peptide is of physiological importance for the pathogen. Races of *C. fulvum* that have overcome resistance gene *Cf*9 by loss of the *avr*9 gene are still pathogenic, indicating that the *avr*9 gene might be dispensible. Nevertheless, the *Cf*9 resistance gene still provides good protection to *C. fulvum* in greenhouse-grown tomato cultivars, suggesting that the competitive ability of these new races is slightly reduced. To quantify the role of the *avr*9 elicitor in fitness, the presence or absence of a functional *avr*9 gene in the population should be studied in competition experiments.

3.4. PROCESSING OF THE AVR9 ELICITOR *IN VITRO* AND *IN PLANTA*

The avirulence gene *avr*9 encodes a precursor protein of 63 amino acids. In *C. fulvum*-infected plants, susceptible to races which carry the *avr*9 gene, the prevailing elicitor molecule occurring in the apoplast consists of 28 amino acids which originates from the carboxyl part of the precursor protein. *C. fulvum* transformed with an *avr*9 construct under control of the constitutive *gpd*-promoter of *Aspergillus nidulans* excretes AVR9 into the culture medium during growth *in vitro*. The predominantly occurring peptides are 32, 33 and 34 amino acids in length, but traces of smaller and larger peptides are also detected. However, the 28 amino acid peptide, which is the predominant form in infected leaf tissue, is undetectable or even absent *in vitro*. These results indicate that processing of the AVR9 elicitor *in vivo* and *in vitro* is different. Probably the first processing occurs between amino acids 23 and 24 resulting in cleavage of the signal peptide and excretion of a 40 amino acid peptide, which *in vitro* can be processed to peptides of 32 to 34 amino acids, while *in planta* the peptide is further processed, possibly by plant proteases, to a stable peptide of only 28 amino acids. All these peptides, however, induce HR on *Cf*9 genotypes of tomato. It is not known whether the larger peptides are intrinsically active or whether the 28 amino acid peptide is the only active form which needs to be generated from the larger ones by proteases occurring *in planta*.

4. Approaches To The Molecular Cloning Of The *Cf*9 Disease Resistance Gene

Further support for the elicitor-receptor model of the *avr*9-*Cf*9 system can be obtained by cloning the *Cf*9 resistance gene. It is tempting to speculate about the existence of a receptor (the putative product of the resistance gene *Cf*9) for the *AVR*9 elicitor in tomato. A plasma membrane receptor for the *AVR*9 peptide elicitor might be the primary product of the *Cf*9 resistance gene. Approaches usually employed to clone a disease resistance gene include (i) map based cloning, (ii) cloning by transposon or T-DNA tagging, (iii) cloning of genes encoding receptors for race-specific elicitors and (iv) cloning by functional supplementation. All approaches have their specific potentials and problems. The latter two approaches will be briefly discussed. An indirect approach to clone resistance genes, is cloning of the plant gene encoding the receptor for the race-specific *AVR*9 peptide elicitor (iii). However, nothing is known yet about the localization of the receptor for this race-specific elicitor. Also the fate of the peptide elicitor in tomato leaves during natural infection by *C. fulvum* is not known. *In situ* localization of the peptide elicitor with monoclonal antibodies might clarify this question. In future, binding studies will be carried out with radioactively labeled peptide elicitor. Characterization of the receptor creates the possibility to clone the encoding *Cf9* gene. Another approach is (iv) the functional cloning of the *Cf9* resistance gene of tomato by a slightly modified version of the method described by Keen (personal communication), to clone a bacterial resistance gene in soybean. Different constructs of the avirulence gene *avr9* are being expressed in *Cf9*$^-$ tomato and transformants are being assayed for extracellular or intracellular production of AVR9 race-specific elicitor. Transgenic tomato plants which highly express the AVR9 race-specific elicitor will be used to clone the corresponding resistance gene *Cf9* by function. This can most probably be achieved by transforming (with a particle gun) leaves of *Cf9*$^-$, *avr9*$^+$ plants with pools of cDNAs from *Cf9*$^+$ tomato and screening for appearance of necrotic spots which represent the expression of HR. The cDNA pools giving HR can be purified and single-cloned by repeatedly transforming *Cf9*$^-$, *avr9*$^+$ plants with a diluted positive pool. In this way clones which encode an important molecule representing the *Cf9* resistance gene product itself or, the receptor which is responsible for the induction of the HR 'machinery' can be obtained. Alternatively, *Cf9*$^-$, *avr9*$^-$ plants can be transformed in a similar way as described above with a 35S-cDNA expression library of *Cf9*$^+$ tomato after which the transformants can be indirectly screened for development of HR after treatment with the *avr9* elicitor.

It is expected that several resistance genes, fitting into the gene-for-gene model, will be cloned in the coming year(s). These cloned resistance genes will, in combination with the corresponding genes for avirulence, become very useful to produce transgenic disease-resistant plants as will be discussed in the following paragraphs.

5. Engineering Disease-Resistant Crop Plants Via Transformation With Avirulence Gene *Avr9*

Most strategies to obtain disease-resistant plants through genetic engineering are presently focused on introducing genes which encode potential antimicrobial PR proteins with chitinase or 1,3-ß-glucanase activities or other antimicrobial proteins. In nature most PR proteins are induced in plants upon attack by pathogens (Bowles, 1990; Bol et al., 1990; Joosten et al., 1989; Van Kan et al., 1992; Wubben et al., 1992). However, induction of PR proteins is just part of a large array of active defense responses. Thus, engineered plants, constitutively expressing one or a few genes encoding PR proteins, might possibly be protected against some pathogens but only slightly, or not at all, against others. Furthermore, the constitutive overexpression of just one type of defense gene holds the risk of selecting variants of pathogens that might regain pathogenicity on such engineered plants.

As indicated earlier, gene-for-gene or race-specific resistance is usually based on specific induction of HR by avirulent races, or more precisely, by race-specific elicitors produced by the avirulence genes of these races. The pathogen can avoid induction of HR by producing variants of race-specific elicitors that are no longer recognized by the resistant cultivar, or by losing the genes encoding race-specific elicitors as occurs with the *avr9* gene in *C. fulvum*. An important lesson learned from the pathogen's strategies to overcome host resistance is that resistance genes are not broken by pathogens but, rather, the induction of HR is avoided.

The great potential of HR, which is the most widely occurring and effective way of active resistance to pathogens in plants, can be fully exploited. The recent cloning of the fungal avirulence gene *avr*9 and the prospects of cloning the corresponding host *Cf9* resistance gene in the near future, create new ways to obtain genetically engineered disease-resistant crop plants. The two components needed for the induction of HR, the race-specific elicitor and the resistance gene-derived specific receptor, should be brought together into one plant. By this strategy the production of race-specific elicitors by the plant itself could overcome the evasion of HR induction by virulent pathogens.

5.1. THE TWO-COMPONENT SENSOR SYSTEM

By combining both the resistance gene and the complementary avirulence gene in its genome, the plant has the genetic ingredients for an HR when both genes are expressed. However, if the two components were produced continuously in transgenic plants, a devastating HR would be induced that would destroy not only the attacking pathogen, but also the plant which is obviously not desirable. Hence, either one component or both in coordination must be regulated. Ideally, one or both genes should be specifically and immediately activated locally upon attack by pathogens but not otherwise.

This requirement implies the need to isolate both avirulence and resistance genes, and promoters that regulate these genes specifically. The ultimate aim would be to have the avirulence gene, or both avirulence and resistance genes, under the control of pathogen-inducible promoters, which respond both quickly and very locally upon attack by pathogens. Candidate promoters have already been described in the literature. Promoters of plant genes involved in the strict local accumulation of PR proteins, phytoalexins or other antimicrobial compounds, are good candidates to be used in the two-component sensor system, in order to obtain broad-spectrum resistance against pathogens.

The two components under control of the selected promoters could be introduced as a cassette into any plant, provided that both genes can be expressed correctly in a biologically active form. To apply the two-component sensor system in unrelated crop plants the isolation of different avirulence and corresponding resistance genes is needed, as it is uncertain whether both types of genes will be active in completely unrelated plant species.

6. Conclusions and Future Prospects

Molecular characterization of gene-for-gene systems is still in its infancy. Of the many bacterial genes and one fungal avirulence gene cloned to date, none show homology to other characterized genes. One can only speculate about their primary functions for the pathogens. It is likely that the primary function of avirulence genes varies considerably. In some cases an avirulence gene may be so important for the pathogen that loss of it is detrimental. A property all avirulence genes have in common is the antigenicity of their direct or indirect products. Through these products the pathogens are recognized by the host, which responds by inducing an HR. The primary function of resistance genes is still a matter of speculation as none have yet been cloned. Without knowing the primary function of either avirulence and resistance genes, their secondary function (the induction of HR by their interacting products) could perhaps be exploited to engineer transgenic disease-resistant plants. Future experiments should tell us whether plants transformed with highly controlled avirulence-resistance gene cassettes, indeed become resistant to various pathogens. Regardless of the success or otherwise of this concept, such transgenic plants can teach us more about stimulus perception and signal transduction in gene-for-gene systems.

7. Acknowledgement

Part of these investigations were supported by the Foundation for Biological Research (BION) and the Foundation for Applied Research (STW), which are both subsidized by the Netherlands Organization for Scientific Research (NWO), and the EC-BRIDGE programme (project number: BIOT-CT90-0163).

8. References

Bol, J.F., Linthorst, H.J.M., Cornelissen, B.J.C. 1990. Plant pathogenesis-related proteins induced by virus infection. Annu. Rev. Phytopathol. 28, 113-138

Bowles, D.J. (1990) Defense-related proteins in higher plants. Annu. Rev. Biochem. 59, 873-907.

De Wit, P.J.G.M. and Spikman, G. (1982) Evidence for the occurence of race- and cultivar-specific elicitors of necrosis in intercellular fluids of compatible interactions between *Cladosporium fulvum* and tomato. Physiol. Plant Pathol. 21, 1-11.

De Wit, P.J.G.M., Van Kan, J.A.L., Van den Ackerveken, G.F.J.M., Joosten, M.H.A.J. (1991) Specificity of plant-fungus interactions: molecular aspects of avirulence genes. In "Advances in Molecular Genetics of Plant-Microbe Interactions", eds. H.Hennecke, D.P.S. Verma, Vol. 1, 233-41, Dordrecht; Kluwer Academic Publisher.

Wit, P.J.G.M. (1992) Molecular characterization of gene-for-gene systems in plant-fungus interactions and the application of avirulence genes in control of plant pathogens. Annu. Rev. Phytopathol. 30, 391-418.

Dixon R.A. and Lamb, C.J. (1990) Molecular communication in interactions between plants and microbial pathogens. Annu. Rev. Plant. Physiol. Plant Mol. Biol. 41, 339-367.

Flor, H.H. (1942) Inheritance of pathogenicity in *Melampsora lini* Phytopathol. 32, 653-669.

Gabriel, D.W. and Rolfe, B.G. (1990) Working models of specific recognition in plant-microbe interactions. Annu. Rev. Phytopathol. 28, 365-391.

Heath, M.C. (1991) The role of gene-for-gene interactions in the determination of host species specificity. Phytopathol. 81, 127-30

Joosten, M.H.A.J. and De Wit, P.J.G.M. (1988) Isolation, purification and preliminary characterization of a protein specific for compatible *Cladosporium fulvum* (syn. *Fulvia fulva*)-tomato interactions. Physiol. Mol. Plant Pathol. 33, 241-253.

Joosten, M.H.A.J. and De Wit, P.J.G.M. (1989) Identification of several pathogenesis-related proteins in tomato leaves inoculated with *Cladosporium fulvum* (syn. *Fulvia fulva*) as 1,3-ß-glucanases and chitinases. Plant Physiol. 89, 945-951.

Keen, N.T. (1990) Gene-for-gene complementarity in plant-pathogen interactions. Annu. Rev.

Genet. 24, 447-463.

Keen, N.T. and Dawson, W.O. (1992) Pathogen avirulence genes and elicitors of plant defense. In Plant Gene Research, Vol.8, Genes involved in plant defense, (Boller T. and Meins G., eds) Vienna/New York: Springer-Verlag. in press.

Lamb, C.J., Lawton, M.A., Dron, M. and Dixon, R.A. (1989) Signals and transduction mechanisms for activation of plant defences against microbial attack. Cell 56, 215-224.

Marmeisse, R., Van den Ackerveken, G.F.J.M., Van den Broek, H.W.J. and De Wit P.J.G.M. (1992) Disruption of the avirulence gene *avr*9 in isolates of the fungal tomato pathogen *C. fulvum* renders these isolates to become virulent on tomato cultivars carrying the corresponding *Cf*9 gene for resistance. Manuscript in preparation.

Scheel, D. and Parker, J.E. (1990) Elicitor recognition and signal transduction in plant defense gene activation. Z. Naturforsch. 45c, 569-575.

Scholtens-Toma, I.M.J. and De Wit, P.J.G.M. (1988) Purification and primary structure of a necrosis-inducing peptide from the apoplastic fluids of tomato infected with *Cladosporium fulvum* (syn. *Fulvia Fulva*). Physiol. Mol. Plant Pathol.33, 59-67.

Scholtens-Toma, I.M.J., De Wit, G.J.M. and De Wit P.J.G.M. (1989) Characterization of elicitor activities of apoplastic fluids isolated from tomato lines infected with new races of *Cladosporium fulvum*. Neth. J. Plant Pathol. 95 (Suppl.1), 161-168.

Van Kan, J.A.L., Van den Ackerveken, G.F.J.M. and De Wit, P.J. G.M. (1991) Cloning and characterization of cDNA of avirulence gene *avr*9 of the fungal pathogen *Cladosporium fulvum*, causal agent of tomato leaf mould. Mol. Plant-Microbe Interact. 4, 52-59.

Van Kan, J.A.L., Joosten, M.H.A.J., Sibbel-Wagemakers, C.A.M., Van den Berg-Velthuis, G.C.M. and De Wit, P.J.G.M. (1992) Differential accumulation of mRNAs encoding extracellular and intracellular PR proteins in tomato induced by virulent and avirulent races of *Cladosporium fulvum*. Plant Mol. Biol. in press.

Van den Ackerveken, G.F.J.M., Van Kan, J.A.L. and De Wit P.J.G.M. (1992) Molecular analysis of the avirulence gene *avr*9 of the fungal tomato pathogen *Cladosporium fulvum* fully supports the gene-for-gene hypothesis. The Plant Journal 2, 359-366.

Van den Ackerveken, G.F.J.M., Van Kan, J.A.L., Joosten, M.H.A. J. and De Wit, P.J.G.M. (1992) Characterization of two putative pathogenicity genes of the fungal tomato pathogen *Cladosporium fulvum*. Submitted

Wubben, J.P., Joosten, M.H.A.J. Van Kan, J.A.L. and de Wit, P.J.G.M. (1990) Subcellular localization of plant chitinases and 1,3-ß-glucanases in *Cladosporium fulvum* (syn. *Fulvia fulva*)-infected tomato leaves. Physiol. Mol. Plant Pathol. In press.

GENOME DYNAMICS AND PATHOTYPE EVOLUTION IN THE RICE BLAST FUNGUS

JOHN E. HAMER, NICHOLAS J. TALBOT AND MORRIS LEVY
Dept. Biological Sciences
Lilly Hall of Life Sciences
Purdue University, West Lafayette, IN 47907

ABSTRACT. Intraspecific variation is a common feature of many plant pathogenic fungi. Many damaging species are composed of highly selective host or cultivar-specific forms, whose ability to evolve new virulence types continues to hamper the best efforts of plant breeders and pathologists. In our laboratory we are using the rice blast fungus, *Magnaporthe grisea*, to examine the genetic basis for intraspecific variation. This fungus is notorious for its ability to evolve new virulence types and infect new hosts. In this chapter we summarize current research efforts to understand the genetic organization of the *M. grisea* genome, the molecular basis for variation within and between various host-specific forms and the structure and dynamics of rice infecting populations.

1. Introduction

Several recent reviews have appeared on the rice blast fungus *Magnaporthe grisea* and its potential as a "model system" for studies of host-parasite interactions (Valent, 1990; Valent and Chumley, 1991). In addition, another chapter in this volume deals with some of the recent molecular genetic approaches to identifying pathogenicity genes. The cytology of the infection process has also recently been reviewed (Howard *et al.*, 1991). Although a considerable effort has been devoted to the understanding of genes that affect the pathogenicity of *M. grisea*, for agricultural and economic purposes the operational unit of primary consideration is the fungal population. Consequently, we have focused this chapter on the evolutionary genetics and population biology of this organism.

M. grisea is a pathogen of monocots (Ou, 1985). The fungus can be easily isolated from lesions found on a large number of grass hosts, including millets, wheat, barley, rice, maize, rye grass, St. Augustine grass, and a variety of forage grasses and weeds such as crabgrass, goosegrass, and weeping lovegrass (Ou, 1985; Mackill and Bonman, 1986). To our knowledge the fungus has never been isolated from soils. The most important aspect of the biology of *M. grisea* is that isolates that infect rice cause rice blast disease, one of the most widespread and devastating of all crop diseases (Ou, 1980a). Not surprisingly, the vast majority of research on *M. grisea* has focused on the genetics of rice infecting strains. However, recent outbreaks of wheat blast in South America (Valent, 1990) and continuing problems in developing durably resistant rice varieties (Gupta and O'Toole, 1986), makes studies of the evolutionary and population biology of the entire species important. Several key questions need to be addressed. What is the genetic relatedness of various host specific

E. W. Nester and D. P. S. Verma (eds.),
Advances in Molecular Genetics of Plant-Microbe Interactions, 299–311.

forms ? How variable are populations ? How is intraspecific variation generated ? How can new information on the population biology of *M. grisea* be used to help lessen the impact of rice blast disease ? Although complete answers to these questions are still elusive, recent research has begun to clarifying many of these issues.

1.1 BACKGROUND

M. grisea isolates from various hosts clearly represent a single species. They are morphologically identical, are occasionally capable of mating and have similar nutritional requirements in axenic media (Ou, 1985). Mycelial colony morphology is highly variable as are mating ability and fertility (ascospore formation and viability) (Yaegashi and Udagawa, 1978; Leung and Williams, 1986; Valent *et al.*, 1986; Yaegashi and Yamada, 1986). A considerable body of evidence now shows that the rice infecting populations are, by and large, clonal and are monophyletic in origin (Leung and Williams, 1986; Hamer *et al.*, 1989; Levy *et al.*, 1991b; Notteghem and Silue, 1991). Although mating ability is more common among non-rice pathogens, extensive population studies on non-rice pathogens have not been reported. Revised taxonomic considerations have established the name *Pyricularia grisea* for the anamorph (asexual state) and the name *Magnaporthe grisea* for the teleomorph (sexual stage) (Rossman *et al.*, 1990).

Infection assays have historically been the principal means of deciphering intraspecific variation in *M. grisea.* In these assays some cross infectivity has been noted with isolates from various hosts (Ou, 1985; Mackill and Bonman, 1986). However, most isolates generally cause the severest symptoms on their host of origin and host specificity is rather narrow for any particular isolate. *M. grisea* isolates that infect rice have also been sorted into infection types or pathotypes (historically called races) based on infection assays involving a collection of differentially susceptible rice cultivars (Latterell *et al.*, 1965; Ling and Ou, 1969). Hundreds of races have been reported to exist worldwide, and attempts to employ a universal set of cultivars have met with only limited success (see Bonman *et al.*, 1986 for discussion). Thus, intraspecific variation within rice infecting populations has been difficult to quantify. Because of the phenotypic nature of the infection assay the genetic relationship between races cannot be determined.

Other techniques have been used to assess intraspecific variation and genetic relatedness in *M. grisea.* Some have met with success. Isozyme polymorphisms (Leung and Williams, 1985; Leung and Williams, 1986) and single copy restriction fragment length polymorphisms (RFLPs) were found to be rare in rice infecting populations (Lebrun *et al.*, 1990). However, neither of these techniques have been extensively applied to non-rice infecting populations. In both rice infecting and non-rice infecting populations, repetitive DNA sequences have proven to be useful probes for assessing genetic relatedness (Hamer, 1991; Dobinson *et al.*, 1992). These probes are obtained in the following way. A genomic library is constructed from a particular host-specific form (e. g., a rice pathogen). The library is screened with randomly labelled total genomic DNA from the rice pathogen in order to identify clones carrying repeated sequences. The library is counter screened with a second randomly labelled genomic DNA probe constructed from a different host-specific form (e. g., a weeping lovegrass pathogen). Clones that hybridize to both probes are common to both isolates and generally include the ribosomal DNA cistron (Dobinson, *et al.*, 1992). Clones that hybridize only to one probe are candidates for form-specific repeated sequences. In *M. grisea* , several different form-specific sequences appear to be transposable elements, though many others remain uncharacterized. These results are somewhat surprising given that transposable elements were not identified in the well studied filamentous fungus *Aspergillus nidulans*, and genetic mechanisms exist to remove or inactivate such elements in *Neurospora crassa* (Selker, 1990).

In the following sections we describe the organization of the *M. grisea* genome and the surprising variability discovered in mitochondrial and nuclear DNA organization. We next discuss the genetic organization, distribution and characterization of various form-specific repeated sequences. Finally we discuss how DNA probes derived from these sequences have been used to analyze the population genetics of rice blast disease causing isolates.

2. Genome organization

2.1 GENOME SIZE AND CHROMOSOMAL ORGANIZATION

M. grisea appears to have a genome size of approximately 38Mb. This estimate is based on the sum of the molecular weights of chromosomal-sized DNA molecules resolved by pulsed field gel electrophoresis (PFGE) and genomic reconstruction experiments based on the hybridization of single copy DNA sequences (Hamer *et al.*, 1989). This size is typical of most ascomycete fungi (Kinghorn, 1987).

The *M. grisea* genome is organized into six chromosomes which vary in size from 3 to 12Mb. PFGE analysis using contour clamped homogeneous electric fields (CHEF) shows that electrophoretic karyotypes vary widely among isolates of *M. grisea* due to chromosome length differences (Hamer *et al.*, 1989; Skinner *et al.*, 1991; Talbot *et al.*, 1992). In general, 4-6 bands are resolvable by this technique. This observation corresponds well to cytological studies of chromosomes during metaphase of ascospore mitosis which suggests the presence of six chromosomes (Leung and Williams, 1987).

Karyotypic variation in rice pathogens is not correlated with an isolate's pathotype and appears to be as variable between isolates from the same geographic location as those from different regions. Often isolates of the same pathotype and with similar genetic backgrounds (based on DNA fingerprinting) will have different karyotypes. This karyotype variability appears to result from deletions and translocations, based on the hybridization of RFLP markers to blots of CHEF separations and the allocation of dispersed repeated sequences to chromosomal bands (Talbot, *et al.*, 1992).

Karyotypic variation has been observed between isolates of numerous plant pathogenic fungi (Mills and McCluskey, 1990; Skinner *et al.*, 1991) and is in marked contrast to the uniform karyotypes observed in populations of saprotrophic fungi such as *N. crassa* (Perkins and Turner, 1988). This has prompted considerable speculation that karyotype variation may be a manifestation of a mechanism for the generation of new virulence types. Two examples have recently illustrated that chromosomal rearrangements can occasionally be associated with a change in virulence in plant pathogenic fungi (Miao *et al.*, 1992; Tzeng *et al.*, 1992). The observation in *M. grisea,* however, that variable karyotypes are not correlated with pathotype suggests that genomic rearrangements are generally neutral in effect, and may merely reflect the absence of sexuality in *M. grisea* populations.

Some rice pathogens also contain a second class of resolvable chromosomal DNA molecules. These are significantly smaller than the 4-6 large chromosomal bands and vary in size from 0.47Mb to 2.2Mb. They have been termed 'minichromosomes', or 'B-like chromosomes', by comparison to those found in higher eukaryotic species (Mills and McCluskey, 1990) and have been found in many plant pathogenic fungi, most notably the pea wilt pathogen, *Nectria haematococca.* (Miao *et al.*, 1991; Miao, *et al.*, 1992). *M. grisea* strains can contain 0-4 of these minichromosomes The minichromosome karyotype appears to be as variable between isolates as the larger chromosome karyotype. Minichromosomes appear to be supernumerary because they do not segregate consistently through crosses (Valent and Chumley, 1991). The presence of minichromosomes is not ubiquitous and, although common in isolates from the United States, is infrequent in isolates from China (Talbot, *et al.*, 1992). Minichromosomes contain the same classes of repeated DNA as the main chromosomes suggesting that they are derived from them, although no other evidence for this derivation or its mechanism is available. It remains possible that minichromosomes are artifacts of the chromosomal DNA extraction procedure, since they could represent the products of breakage at particularly fragile sites within chromosomes. The restriction, however, of minichromosomes to rice infecting field isolates and the reproducibility of their resolution makes this unlikely.

2.2 GENE ORGANIZATION

Gene structure in *M. grisea* appears to be typical of most ascomycete fungi. This assumption is constrained, however, by the paucity of genes that have been cloned and sequenced in the organism. There appears to be a significant G,C bias in codon usage in *M. grisea* based on the sequencing of the *CUT1* gene. Sweigard et al.(1992) found that of the 228 amino acid codons, 75 and 120 ended with G and C, respectively, while only 11 and 22 ended in A and T, respectively. A similar observation has been made in the sequencing of pNJT15, a putative pathogenicity gene (Talbot and Hamer, unpublished). The promotor region of the *CUT1* gene contained the characteristic TATAA sequence but no CCAAT sequence was observed. In the 3' non-coding region of *CUT1* no AATAAA polyadenylation signal was seen, although this has been observed in pNJT15 sequencing (Talbot and Hamer, unpublished). Splicing recognition sequences in introns of *CUT1* also match consensus sequences observed in other fungi.

2.3 RIBOSOMAL RNA GENE ORGANIZATION

Genes encoding the 17S, 5.8S and 26S ribosomal RNAs are located in tandemly repeated units at one locus in *M. grisea* occupying an 8-9kb region. An *Eco*RI RFLP, resulting from differences in the length of the repeat unit between rice and weeping love-grass infecting laboratory strains, has been used to map the rDNA repeat unit to linkage group A (Romao and Hamer, 1992). RFLP mapping showed no recombination within the rDNA repeat in 192 progeny analyzed confirming that meiotic recombination is suppressed at this locus, paralleling observations made in yeast.

The rDNA repeat has been cloned from the rice infecting isolate Guy-11 and partially sequenced (Lebrun,*et al.*, 1990). Two domains of the 26S RNA gene were chosen for sequencing as reliable phylogenetic indicators of intraspecies and intrageneric relatedness. These domains were sequenced from Guy-11 and from two *Eleusine* infecting isolates. In all cases the sequences were identical, providing evidence for the classification of these host-specific forms as a single species. Restriction mapping of the rDNA repeat showed no internal restriction site differences between rice-infecting isolates and detected only a few between other host-limited forms. Length polymorphisms identified with *Eco*RI were used to classify rice-infecting isolates into three types based on rDNA repeat unit length. These classes were not correlated with race or geographic location . Similar *Eco*RI RFLPs have been used to classify 54 isolates of *M. grisea* from 16 different grass hosts into four rDNA repeat length types (Borromeo *et al.*, 1992). This degree of intraspecies variability in rDNA repeat length is common in Ascomycetes (Garber *et al.*, 1988).

2.4 TELOMERES

The ends of chromosomes in *M. grisea* have been identified by hybridization with a 24-base oligonucleotide containing the telomere consensus sequence $(TTAGGG)_4$ (Orbach *et al.*, 1991; Romao and Hamer, unpublished results). Southern hybridization analysis detects an average of 12 restriction fragments, which conforms well to the estimated chromosomal number. Actual numbers of hybridizing bands vary between 9 and 15 in different isolates due, presumably, to co-migration of restriction fragments or the presence of internal telomeric repeats. Telomere-hybridizing restriction fragments have been shown to segregate in crosses allowing them to be mapped to the ends of linkage groups. Two-dimensional CHEF separation analysis has confirmed the presence of two hybridizing restriction fragments per separated chromosome (Farman and Leong, personal communication). Telomere-hybridizing restriction fragment lengths appear to be extremely variable between *M. grisea* isolates (Romao and Hamer, unpublished).

2.5 MITOCHONDRIAL GENOME ORGANIZATION

M. grisea has an average mitochondrial genome size of 35kb (Borromeo *et al.*, 1992). Restriction analysis of mitochondrial DNA (mtDNA) from Philippine isolates representing 17 host-limited forms has recently been performed. This revealed six mtDNA types, based on restriction analysis with three enzymes, among the 54 isolates examined. The most prevalent mtDNA type was shared by rice infecting isolates and isolates infecting 10 other grass species. Among the remaining five mtDNA types described, three types extracted from isolates infecting *Cyperus* spp. were sufficiently divergent to suggest the classification of these isolates as different species of *Pyricularia* (Borromeo,*et al.*, 1992). A more comprehensive study of mitochondrial DNA of isolates collected from diverse geographic locations suggests that there may be greater variation in mitochondrial types among host-limited forms of the fungus (B. Valent, personal communication).

3. Dispersed Repetitive Sequences

3.1. RICE PATHOGEN SPECIFIC SEQUENCES

The first dispersed repeated sequences identified in *M. grisea* were found to be highly conserved in isolates that cause rice blast disease (Hamer *et al.*, 1989). These sequences were, in fact, a collection of repeated sequences and were termed MGR for *Magnaporthe grisea* repeat. When subclones of these sequences from the rice pathogen genome are used as probes they hybridize to the genomic DNA of all strains of *M. grisea.* However, in Southern blots the intensity of hybridization (number of bands and intensity of the autoradiographic signal) is strikingly higher in rice pathogens. For example, a typical rice pathogen will contain 50-60 restriction fragments that hybridize to an MGR probe whereas a weeping lovegrass pathogen will contain only 1-2 fragments. Thus, either the MGR sequences in non-rice pathogens are present at a dramatically lower copy number, or the nucleotide sequence of MGRs in rice and non-rice pathogens has undergone substantial divergence. The former of these two possibilities now seems the most likely. Screens for repeated DNA in non-rice pathogens have not identified MGR-like sequences that hybridize faintly to the rice pathogen genome. Rather, these screens have identified repeated elements specific to populations of non-rice pathogens (see 3.2 below). Therefore the MGR sequences are ancestral to all host-specific forms of *M. grisea* and appear to have undergone amplification during the emergence of rice-specific pathogens. Table 1 summarizes relevant information on each of the repeated sequences currently known in rice pathogens of *M. grisea.*

TABLE 1. Dispersed repeated sequences in the genome of *M. grisea* rice pathogens.

Repeated Sequence	Estimated Copy Number	Characteristics
MGR586	50-60	< 1.5 kb in length, conserved in rice pathogens
MGR583, MGR604	60-100	> 5.0 kb in length, conserved in rice pathogens, homologous to mammalian LINE element [a]
PGR 1	72	< 1.0 kb in length, conserved in rice pathogens, similar to Alu sequences [b]
Fosbury	20-30	ca. 8.0 kb in length, terminally redundant, specific to rice pathogens, active transposable element.

[a] Valent and Chumley, 1991; [b] Leong *et al.*, 1991.

3.1.1. *MGR586*. This repeat, defined by the subclone pCB586 (Hamer *et al.*, 1989), was initially believed to be part of a larger repeated sequence. However, multiple clones containing only this isolated sequence have now been obtained (Givan, Dobinson and Hamer, unpublished results). MGR586 sequences appear to be well dispersed around the genome and at the level of RNA blot analysis do not appear to be transcribed (Hamer, unpublished results). High levels of restriction site polymorphism can be detected in rice pathogens using MGR586 as a Southern blot hybridization probe. Consequently, MGR586 has been used to construct DNA fingerprints of field isolates of *M. grisea* (see Section 4.0).

MGR586 sequences have also been used for genetic mapping in *M. grisea* (Hamer and Givan, 1990; Valent *et al.*, 1991; Romao and Hamer, 1992). By performing genetic crosses between rice and non-rice pathogen laboratory strains of similar genetic backgrounds, the segregation of many MGR586 RFLPs has been followed. In a recent study 57 MGR586 RFLPs in the genome of a laboratory strain pathogenic to rice were mapped (Romao and Hamer, 1992). Recombination was observed between each MGR586 sequence and a genetic map was constructed from the analysis of multi-point linkages. The MGR586 sequences are not clustered in the genome but, rather, appear to be randomly dispersed along all linkage groups. Consequently, a number of phenotypic traits could be mapped by simply detecting linkage to MGR586 sequences.

3.1.2. *MGR583, MGR604*. DNA sequence analysis of these repeated sequences has demonstrated that they encode an open reading frame with significant amino acid homology to the LINE family of poly(A) class retrotransposons (Valent and Chumley, 1991). These sequences hybridize to mRNAs of 7.5, 3.0 and 2.4 kbs in length (Hamer *et al.*, 1989). The copy number of this element is easily twice that of MGR586. Comparative studies in rice and non-rice pathogens suggests that amplification of this retroelement appears to have occurred within rice pathogens (Dobinson *et al.*, 1992). Similar episodes of LINE element amplification have been documented in certain mammalian taxa (Pascale *et al.*, 1990).

3.1.3. *PGR1*. This rice pathogen specific repeated DNA sequence appears related to the *Alu* family of repeats in that it contains sequences similar to canonical Pol III transcription signals (Leong *et al.*, 1991).

3.1.4. *Fosbury*. Recently, tetrad analysis of MGR586 segregation demonstrated the appearance of a new MGR586 RFLP that was present in only one of a pair of sister spores. Therefore, this polymorphism likely arose during the post-meiotic mitosis that occurs in all eight-spored tetrads. Molecular cloning demonstrated that this MGR586 polymorphism occurred as a result of the transposition of a new repeated sequence into a site adjacent to an MGR586 repeat. This new repeat is terminally redundant and has been termed *fosbury* (Shull and Hamer, unpublished results). *Fosbury* sequences appear to be dispersed and are specific to rice pathogens. No homology to *fosbury* sequences can be detected in any other host-specific form. Fine structure characterization of this transposon is underway.

3.2 REPEATED SEQUENCES IN OTHER HOST-SPECIFIC FORMS OF *M. GRISEA*.

The presence of specific repeated sequences in rice pathogens of *M. grisea* may have resulted from strong host selection due to the continual deployment of blast resistant rice genotypes. Consequently, non-rice pathogens might not be expected to contain these kinds of sequences and could be genetically more homogeneous. Recently, Dobinson et al. (1992) demonstrated that a subpopulation of *M. grisea* pathogenic to the the plant species *Eleusine coracana* (goosegrass) and *Eleusine indica* (finger millet), harbors a *Gypsy* class retrotransposon, which was named *grasshopper* (*grh*). *Grh* is present in approximately 40 copies in *Eleusine* pathogens from Japan, Nepal, India and the west African countries of Mali and Burkina Faso. Isolates from other areas in Africa, the Phillippines or South America did not contain *grh* sequences. All of the *Eleusine* pathogens examined in this study had identical mtDNA genotypes. Thus, *grh* is population specific and has been acquired by some horizontal transfer event sometime after the emergence of *Eleusine* pathogens.

These results, together with other analyses of mtDNAs in the species (see section 2.5), suggests that host-specific forms and even populations of *M. grisea* can be genetically quite distinct. The rather uniform isozyme patterns and morphology among different forms suggests that these difference in genome organization have arisen rather recently. One possible reason for this rapid pattern of genome evolution may have to do with the generally clonal organization of *M. grisea* populations. Neutral or even adaptive changes in genome structure (e. g., the amplification or intrusion of novel DNA sequences, chromosomal rearrangements, and/or minichromosomes) could readily persist within descendant clones. Infrequent or nonexistent use of the sexual cycle would permit a discontinuous array of genetic variation to exist within the species. Recently, it has been suggested on the basis of population level studies that many eukaryotic microorganisms may exist as clonal populations (Tibayrenc *et al.*, 1991). Although a considerable body of evidence suggests that *M. grisea* rice pathogen population are clonal, more analysis is required of other host-specific forms. If populations within host-specific forms are genetically distinct (as suggested by studies on the *Eleusine* pathogens), then reliably identifying clones becomes an important priority for controlling this pathogen.

Applications of DNA fingerprinting for Investigations of Population Dynamics and Genome Evolution

4.1 RATIONALE FOR GENEALOGICAL ANALYSIS OF RICE PATHOGENS

The distribution of MGRs and other repetitive genomic DNA sequences establishes that rice blast pathogens represent an evolutionarily independent and ecologically divergent branch of *M. grisea* (Hamer, 1991). The clearest dichotomy between rice and nonrice pathogens has been observed in the Philippines, where isolates from 16 weed species growing along rice fields infected by rice pathogens contain only one or a few *Eco*RI fragments that hybridize to the MGR586 probe (Borromeo,*et al.*, 1992). Such findings are typical and indicate a general lack of genetic exchange between rice and non-rice pathogens. Isolates of both genomic types which are capable of dually infecting rice and weeds have been observed infrequently (Bonman, *et al.*, 1986); Levy, Shajahan and Hamer, unpublished data). However, the field results indicate that neither form is an important inoculum source for alternate hosts.

RFLPs associated with MGR586 sequences provide a "DNA fingerprint" for assessing the genetic relatedness among rice pathogens. The 50-60 dispersed copies of MGR586 provide a large number of resolvable genetic polymorphisms for distinguishing isolate diversity. However, the RFLPs are not so hypervariable that they obscure the relatedness within clonal lineages (Levy *et al.*, 1991b). MGR-DNA fingerprints, thus, provide definition of rice pathogen population genetic structure as well as for the phylogenetic organization of pathotype diversity among clonal lineages. We have found this probe to be superior in efficiency and resolving power to other middle-repetitive or unique sequence probes.

4.2 LINEAGE -PATHOTYPE ASSOCIATIONS AMONG ARCHIVED USA ISOLATES

The utility of MGR-DNA fingerpints was first demonstrated in a study of 42 USA field isolates representing the eight major pathotypes observed throughout the southern rice belt of the USA from 1959-1988 (Levy *et al.*, 1991b). Although each isolate had a unique fingerprint, the overall variation was discretely partitioned into eight clusters, with each considered as a distinct genetic lineage (see Fig. 1). The average fingerprint similarity (proportion of shared RFLPs) within each lineage exceeded 90%. Variation between lineages was discontinuous, the average similarity ranging from 30-82%. Each of six lineages expressed a specific pathotype (International race; (Ling and Ou, 1969)). Two lineages expressed a pair of pathotypes, with each pair having similar virulence features. The same pathotype was convergently expressed by a pair of lineages in two cases. We have subsequently added 88 more archived isolates to the data base with no significant changes to the original findings (Fig. 1).

The results of multidecade sampling indicate that the USA rice blast population comprises a small number of geographically widespread clonal lineages. Each has been stably associated with very limited pathotype diversity, indicating persistently strong stabilizing selection by USA rice cultivars. Rare pathotypes are clonally derived from already identified lineages, differ generally by single virulence differences from the typical pathotypes within the lineage, and appear to occur recurrently during epidemics (Levy, Marchetti and Hamer, unpublished). The USA population structure defined by MGR-DNA fingerprinting does not support the view that rice blast pathogens are subject to promiscuous and continuous pathotype change (Ou, 1980b). Rather, it indicates that pathotype diversity and stability must be viewed in the context of the properties of the component clonal lineages in any population.

4.3 COLOMBIAN BLAST NURSERY REVEALS A STRATEGY FOR ANTI-LINEAGE RESISTANCE BREEDING

For most blast prone areas in Asia, Africa and South America disease control relies on the frequent introduction of new resistant cultivars that , usually, are selected for their resistance to the predominant pathotypes in a region. Resistance breakdown typically occurs in 2-3 years. Using MGR-DNA fingerprinting to define the clonal population structure of a

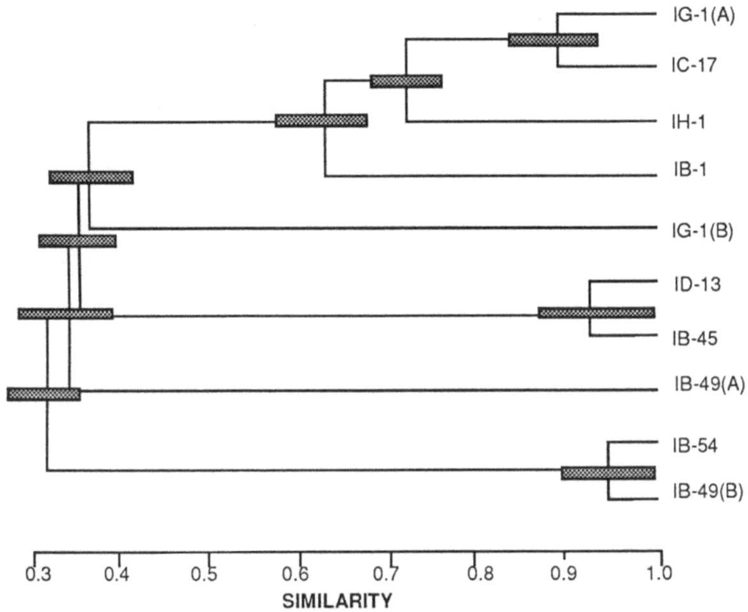

Figure 1. Phenogram of similarities among archived USA rice blast fungus pathotype groups. Based on consensus MGR-DNA fingerprints for 130 isolates, the error bars describe 95% confidence intervals on branch point values, using bootstrap data analysis and an unweighted pair-group with mathematical averaging clustering method. Pathotype names reflect International race designations (Ling and Ou, 1969). Each branch of the phenogram that is significantly different ($p < 0.05$) from the others is considered as a distinct lineage; isolates with pathotypes ID-13 and IB-45 are members of the same lineage as are those with pathotypes IB-54 and IB-49B.

pathotype-rich blast nursery in Colombia has suggested a novel breeding strategy that may lead to more durable blast resistance.

The 30-hectare Santa Rosa blast nursery in eastern Colombia exhibits chronic epidemic levels of disease and supports a blast population with more than 50 pathotypes. The spectrum of pathotypes is so broad that no known resistance genes or indigenous cultivar combinations would appear to be durably resistant at the site. MGR-DNA fingerprinting of 151 field isolates from 15 cultivars revealed that all isolate diversity was subsumed in only six distinct clonal lineages (Levy *et al.*, 1991a). The average fingerprint similarity within lineages was 92% or greater while the range between lineages was 37-85%. Each lineage was associated with a subset of cultivars and associated pathotypes. This hierarchical distribution of virulence features indicated that several cultivars, although severely infected in the field, were susceptible to only one lineage in the population and resistant to all others (Table 2). These relationships have been confirmed in virulence spectrum assays (Correa-Victoria and Zeigler, 1992). Consequently, the combined resistance features of certain cultivar combinations might collectively exclude all of the resident lineages.

A potential example of the efficacy of anti-lineage (rather than anti-pathotype) resistance breeding has been provided by the Colombian cultivar Oryzica Llanos 5, which has been durably blast resistant in the Santa Rosa region since its introduction in 1988. This cultivar combines the resistance ancestries found in the commercial cultivars Cica 9 and Oryzica 1, each of which is highly susceptible to different single lineages at Santa Rosa (Table 2). Currently this "lineage exclusion " resistance breeding strategy is being tested in a number of blast prone environments in Asia and South America.

TABLE 2. Field disease reactions to Santa Rosa rice blast fungus lineages. (R = resistant and S = susceptible.)

	Lineage					
Cultivar	SRL-1	SRL-2	SRL-3	SRL-4	SRL-5	SRL-6
IR8	S	S	R	S	S	S
Metica 1	R	R	S	S	S	S
Cica 8	R	R	R	S	S	S
Cica 9	S	R	R	R	R	R
Oryzica 1	R	R	R	R	R	S
Oryzica Llanos 5	R	R	R	R	R	R

4.4 INTERNATIONAL COMPARISONS

We are in the process of extending MGR-DNA fingerprint analysis to rice pathogen populations in China, Bangladesh, and the Philippines. Preliminary data indicates that each country contains a small of number (6-18) of distinct MGR-based lineages, most of which appear to be indigenous. However, we also have found several isolates in China that express 95% fingerprint similarity with the Colombian lineage, SRL-6, the most widespread on commercial cultivars in Colombia. This indicates a recent international movement of the pathogen and argues for greater hygiene controls on seed exchanges for breeding purposes. More importantly, we expect to identify several Chinese cultivars that are resistant to this lineage and may be very useful for future resistance breeding purposes in Colombia. Ultimately, we hope to construct an international atlas of MGR-based lineage distributions with which to target resistance sources specific for the rice pathogen populations in various regions.

5.0 Conclusions

At the level of genome organization, *Magnaporthe grisea* is a highly variable plant pathogenic fungus. Its potential to evolve new pathogenic forms (e. g. wheat blast pathogens) would also appear to be high. However, studies on a specialized population of *M. grisea*, the rice blast pathogens, have demonstrated a limited potential for pathotypic variation within clonal lineages. In other words, any one rice blast lineage would appear to have a defined cultivar range, even at sites promoting high disease pressure. We believe that an understanding of this genetic organization can be exploited for disease control purposes. Thus, in pathogens like *M. grisea*, it is imperative to accurately and reliably define population structures and their dynamics in space and time.

6.0 Literature Cited

Bonman, J. M., T. I. V. de Dios and M. M. Khin. (1986) 'Physiologic specialization of *Pyricularia oryzae* in the Philippines', Plant Dis. 70, 767-769.

Borromeo, E. S., R. J. Nelson, J. M. Bonman and H. Leung. (1992) 'Genetic differentiation among isolates of *Pyricularia grisea* infecting rice and weed hosts', Phytopathology,.in press.

Correa-Victoria, F. J. and R. S. Zeigler. (1992) 'Pathogenic variability in *Pyricularia grisea* at a rice blast "hot spot" breeding site in eastern Colombia', Plant Dis., in press.

Dobinson, K. F., R. E. Harris and J. E. Hamer. (1992) 'Distribution of a retroviral-like transposable element in *Magnaporthe grisea.*', Molec. Plant Path. Interact., in press.

Garber, R. C., B. G. Turgeon, E. U. Selker and O. C. Yoder. (1988) 'Organization of ribosomal RNA genes in the fungus *Cochliobolus heterostrophus*', Curr. Genet. 14, 573-582.

Gupta, P. C. and J. C. O'Toole.,(1986) Upland Rice: A Global Perspective, International Rice Research Institute, Los Banos, Philippines.

Hamer, J. E. (1991) 'Molecular probes for rice blast disease', Science 252, 632-633.

Hamer, J. E., L. Farrall, M. J. Orbach, B. Valent and F. G. Chumley. (1989) 'Host species-specific conservation of a family of repeated DNA sequences in the genome of a fungal plant pathogen', Proc. Natl. Acad. Sci. USA 86, 9981-9985.

Hamer, J. E. and S. Givan. (1990) 'Genetic mapping with dispersed repeated sequences in the rice blast fungus: Mapping the SMO locus', Mol. Gen. Genet. 223, 487-495.

Howard, R. J., T. M. Bourett and M. A. Ferrari. (1991) 'Infection by *Magnaporthe grisea*: an in vitro analysis, in K. Mendgen and D. E. Lesemann.(eds.), Electron Microscopy of Plant Pathogens, Springer-Verlag, Berlin, pp. 251-264 .

Kinghorn, J. (1987) Gene structure in eukaryotic microbes, IRL Press, Washington D.C.

Latterell, F. M., M. A. Marchetti and B. R. Grove. (1965) 'Co-ordination of effort to establish an international system for race identification in *Pyicularia oryzae*', in The Rice Blast Disease, Proc. Intl. Rice Res. Inst, Johns Hopkins Press, Baltimore, pp. 257-274.

Lebrun, M. H., M. P. Capy, N. Garcia, M. Dutertre, Y. Brygoo, J. L. Notteghem and M. Vales. (1990) 'Biology and genetics of *Pyricularia oryzae* and *Pyricularia grisea* populations: Current situation and development of RFLP markers', Rice Genetics Conference, Los Banos, Philippines.

Leong, S., M. Farman, A. Budde, P. Taylor, P. Tooley and H. Leung. (1991) 'Molecular analysis of pathogenesis in *Magnaporthe grisea.*', Fifth Annual Meeting of The International Program on Rice Biotechnology, Tucson, Arizona.

Leung, H. and P. H. Williams. (1985) 'Genetic analyses of electrophoretic enzyme variants, mating type, and hermaphroditism in *Pyricularia oryzae* Cavara',. Can. J. Genet. Cytol. 27, 697-704.

Leung, H. and P. H. Williams. (1986) 'Enzyme polymorphism and genetic differentiation among geographic isolates of the rice blast fungus', Phytopathology.76, 778-783.

Leung, H. and P. H. Williams. (1987) 'Nuclear division and chromosome behavior during meiosis and ascosporogenesis in *Pyricularia oryzae*', Can. J. Bot. **65,** 112-123.

Levy, M., F. J. Correa-Victoria, R. S. Zeigler, S. Xu and J. E. Hamer. (1991a) 'Organization of genetic and pathotype variation in the rice blast fungus at a Colombian "hot spot". Phytopathology 81, 1236-1237 (Abstract).

Levy, M., J. Romao, M. A. Marchetti and J. E. Hamer. (1991b) 'DNA fingerprinting with a dispersed repeated sequence resolves pathotype diversity in the rice blast fungus', The Plant Cell 3, 95-102.

Ling, K. C. and S. H. Ou. (1969) 'Standardization of the international race numbers of *Pyricularia oryzae*', Phytopathology 59, 339-342.

Mackill, A. O. and J. M. Bonman. (1986) 'New hosts of *Pyricularia oryzae*', Plant Dis. 70, 125-127.
Miao, V. P. W., S. F. Covert and H. D. VanEtten. (1992) 'A fungal gene for antibiotic resistance on a dispensable ("B") chromosome', Science 254, 1773-1776.
Miao, V. P. W., D. E. Matthews and H. D. VanEtten. (1991) 'Identification and chromosome locations of a family of cytochrome P-450 genes for pisatin detoxification in the fungus *Nectria haematococca*.', Mol. Gen. Genet. 226, 214-223.
Mills, D. and K. McCluskey. (1990) 'Electrophoretic karyotypes of fungi: the new cytology', Mol. Plant-Microbe Interact. 3, 351-357.
Notteghem, J. L. and D. Siluc. (1991) 'Distribution of the mating type alleles in *Magnaporthe grisea* populations pathogenic on rice', Phytopathology 82, 421-424.
Orbach, M., J. Sweigard, A. Walter, L. Farrall, F. Chumley and B. Valent. (1991) 'Strategies for the isolation of avirulence genes from the rice blast fungus *Magnaporthe grisea*.', 16th Fungal Genetics Conference, Pacific Grove, CA. (Abstract)
Ou, S. H. (1980a) 'A look at worldwide rice blast disease control', Plant Dis. 64, 439-445.
Ou, S. H. (1980b) 'Pathogen variability and host resistance in rice blast disease', Ann. Rev. Phytopathol. 18, 167-187.
Ou, S. H. (1985) Rice Diseases, Commonwealth Mycological Institute, Surrey, UK.
Pascale, E., E. Valle and A. V. Furano. (1990) 'Amplification of an ancestral mammalian L1 family of long interspersed repeated DNA occurred just before the murine radiation', Proc. Natl. Acad. Sci. USA 87, 9481-9485.
Perkins, D. D. and B. C. Turner. (1988) '*Neurospora* from natural populations: toward the population biology of a haploid eukaryote', Exp. Mycol. 12, 91-131.
Romao, J. and J. E. Hamer. (1992) 'Genetic organization of a repeated DNA sequence family in the rice blast fungus', Proc. Natl. Acad. Sci. USA 89, 5316-5320.
Rossman, A. Y., R. J. Howard and B. Valent. (1990) '*Pyricularia grisea*, the correct name for the rice blast disease fungus', Mycologia 82, 509-512.
Selker, E. U. (1990) 'Premeiotic instability of repeated sequences in *Neurospora crassa*', Ann. Rev. Gen. 24, 579-614.
Skinner, D. Z., A. D. Budde and S. A. Leong. (1991) 'Molecular karyotype analysis of fungi', in J. W. Bennet and L. L. Lasure (eds.), More gene manipulations in fungi, Academic Press Inc., San Diego., pp. 86-103.
Sweigard, J. A., F. G. Chumley and B. Valent. (1992) 'Cloning and analysis of CUT1, a cutinase gene from *Magnaporthe grisea*', Mol. Gen. Genet. 232, 174-182.
Talbot, N. J.., Y. Salch, M. Ma and J. E. Hamer. (1992) 'Karyotype variation within clonal lineages of the rice blast fungus', Appl. Environ. Microbiol., in review.
Tibayrenc, M., F. Kjellberg, J. Arnaud, B. Oury, S. F. Breniere, M-L. Darde and F. J. Ayala. (1991) 'Are eukaryotiv microorganisms clonal or sexual? A population genetics vantage', Proc. Natl. Acad. Sci. 88, 5129-5133.
Tzeng, T. H., L. K. Lynholm, C. F. Ford and C. R. Bronson. (1992) 'A restriction fragment length polymorphism map and electrophoretic karyotype of the fungal maize pathogen *Cochliobolus heterostrophus*', Genetics 130, 81-96.
Valent, B. (1990) 'Rice blast as a model system for plant pathology', Phytopathology 80, 33-36.
Valent, B. and F. G. Chumley. (1991) 'Molecular genetic analysis of the rice blast fungus, *Magnaporthe grisea*', Ann. Rev. Phytopathol.29, 443-467.
Valent, B., M. S. Crawford, C. G. Weaver and F. G. Chumley. (1986) 'Genetic studies of fertility and pathogenicity in *Magnaporthe grisea* (*Pyricularia oryzae*.', Iowa State J. Res. 60, 569-594.

Valent, B., L. Farrall and F. G. Chumley. (1991) '*Magnaporthe grisea* genes for pathogenicity and virulence identified through a series of backcrosses', Genetics 127, 87-101.

Yaegashi, H. and S. Udagawa. (1978) 'The taxonomical identity of the perfect state of *Pyricularia grisea* and its allies', Can. J. Bot. 56, 180-183.

Yaegashi, H. and M. Yamada. (1986) 'Pathogenic race and mating type of *Pyricularia oryzae* from Soviet Union, China, Nepal, Thailand, Indonesia and Colombia', Ann. Phytopath. Soc. Japan 52, 225-234.

MOLECULAR BASIS OF SPECIFICITY IN MAIZE LEAF SPOT DISEASE

J.D. WALTON
DOE-Plant Research Laboratory
Michigan State University
East Lansing, MI 48824
USA

ABSTRACT. The interaction between *Cochliobolus carbonum* and maize is mediated by a cyclic tetrapeptide known as HC-toxin. Isolates of the fungus that produce HC-toxin (called, by definition, race 1) show exceptional virulence on varieties of maize that are homozygous recessive at the nuclear *Hm* locus. In *C. carbonum*, production of HC-toxin, and hence virulence on maize of genotype *hm/hm*, is controlled by a single Mendelian genetic locus called *TOX2*. We have purified and characterized two enzymes that are involved in the synthesis of HC-toxin. Both enzyme activities are found only in isolates of the fungus that make HC-toxin and genetically co-segregate with *TOX2*. The gene encoding one of the enzymes was cloned. This gene is part of a 22-kb contiguous region of DNA that is found only in isolates of the fungus that make HC-toxin. Analysis of this region by cDNA mapping and sequencing has shown that it contains three genes, one of which constitutes a 15.7 kb open reading frame encoding a 570 kDa tetrapartite cyclic peptide synthetase.

1. Introduction

Until very recently the only factors known to mediate specificity in any plant/microbe interaction were the host-selective toxins. These small, secondary metabolites are known to be produced only by filamentous fungal pathogens. Host-selective toxins are virulence factors: disease develops only when a toxin-producing fungus invades a toxin-sensitive plant. In several cases diseases involving host-selective toxins are or have been serious problems in the U.S. and in Japan. Since the first host-selective toxin was described in the 1930's, much progress has been made on their structures, the genetics of production and response, and their modes of action. In the past few years there have been studies on the molecular biology and enzymology of their biosynthesis and on the molecular genetics and biochemistry of the host response.

Approximately fifteen diseases involving host-selective toxins have been definitively described. Chemically, the host-selective toxins are diverse, including cyclic peptides, polyketides, terpenoids, and compounds of uncertain origin. Some of the toxins are related, for example, AK-toxin, AF-toxin, and ACT-toxin produced by three pathotypes of *Alternaria alternata*, and T-toxin and PM-toxin produced by *Cochliobolus heterostrophus* race T and *Phyllosticta maydis*.

E. W. Nester and D. P. S. Verma (eds.),
Advances in Molecular Genetics of Plant-Microbe Interactions, 313–323.

Many of the toxins exist as families of closely related compounds that are active at different concentrations. In the case of helminthosporoside, the producing fungus *Cochliobolus sacchari* also makes related compounds that protect against the phytotoxic members of the family (Kohmoto and Otani, 1991).

The Ascomycete genus *Cochliobolus* (imperfect stage *Helminthosporium* or *Bipolaris*) contains at least four species that clearly use host-selective toxins as agents of pathogenicity (Table 1). *C. heterostrophus* race T is the causal agent of Southern Corn Leaf Blight and caused a severe epidemic of maize in the U.S. in 1970. *C. victoriae* caused severe epidemics on oats on the 1940's in the U.S. Apparently not all species and races of *Cochliobolus* make host-selective toxins, for example, there is no strong evidence that races 2, 3, and 4 of *C. carbonum*, race O of *C. heterostrophus*, and other important pathogens such as *C. sativus* make them. Host-selective toxins have also been reported from *C. miyabeanus*, an important disease of rice, and from other fungi with a *Helminthosporium* (sensu latu) imperfect stage such as *Pyrenophora tritici-repentis*, cause of tan spot of wheat. In common with most if not all other plant pathogenic fungi, *Cochliobolus* and its relatives make a number of non-specific phytotoxins. Even if not important in specificity some of these could be important in disease.

Table 1. Four host-selective toxins from *Cochliobolus* (*Helminthosporium*)

Species	Toxin (Gene)	Host (Gene)	Structure
C. heterostrophus	T-toxin (*TOX1*)	maize (T-cms)	C_{39}-C_{41} linear polyketide(s)
C. carbonum	HC-toxin (*TOX2*)	maize (*hm*/*hm*)	cyclic tetrapeptide
C. victoriae	victorin (*TOX3*)	oats (*Vb*/-)	cyclic pentapeptide(s)
C. sacchari	helminthosporoside or HS-toxin	sugarcane	galactosylated sesquiterpenoids(s)

The original establishment of the importance of host-selective toxins in disease relied above all on genetic manipulation, when possible, of the host and pathogen. *Cochliobolus*, unlike many other plant pathogenic fungi, can be crossed, and host plants such as maize and oats (but not sugarcane) are genetically tractable. It was thereby shown genetically for three *Cochliobolus* disease interactions that production of a species' characteristic host-selective toxin was correlated absolutely with virulence, and that for the plant sensitivity to a particular host-selective toxin was correlated to susceptibility to the producing fungus.

Even among just the *Cochliobolus* host-selective toxins there is great chemical diversity. T-toxin is a rather complex family of linear polyketides, all of which are host-selective. HC-toxin is a relatively simple cyclic tetrapeptide and although the fungus does make two related forms they occur only in minor amounts. Victorin is a family of five rather complex partially cyclic pentapeptides. Victorin is glyoxylated and chlorinated, and the fifth amino acid is cyclized to the second amino acid through an ether linkage (Kohmoto and Otani, 1991). Helminthosporoside (HS-toxin) from *C. sacchari* is a large family of galactosylated terpenoids.

In terms of activity and mode of action the *Cochliobolus* host-selective toxins also show great diversity. Victorin is active at less than 100 pg/ml and resistant oats as well as other plants do not respond even at concentrations one million-fold higher. Sensitivity to victorin and to *C. victoriae* is controlled by the dominant allele of the nuclear gene *Vb*. It was proposed some years ago that *Vb* might encode a victorin receptor, and that oats of genotype *vb/vb* are resistant to victorin by failing to recognize it. Evidence for this hypothesis was obtained by Wolpert et al. (1989) but Akimitsu et al. (1992) found that victorin bound equally well to resistant and susceptible oats. T-toxin is active at approx. 10 ng/ml and resistant plants respond at concentrations ten thousand-fold higher. Only maize with Texas male-sterile cytoplasm (Tcms) are susceptible to *C. heterostrophus* race T and sensitive to T-toxin. The utility of Tcms in hybrid seed production accounted for its wide distribution in 1970. Male-sterility and toxin sensitivity appear to be invariably linked for reasons that are not clear. Sensitivity to T-toxin (and male-sterility) result from a rearrangement of the mitochondrial DNA of Tcms maize resulting in production of a novel inner-membrane protein of 13 kDa. The 13 kDa protein confers sensitivity to T-toxin even when expressed in *E. coli* (Dewey et al., 1988).

Of the three well-studied *Cochliobolus* host-selective toxins, HC-toxin is the least active and least host-selective. It inhibits root growth of sensitive maize at approx. 1 μg/ml but resistant maize is similarly affected at concentrations approx. one hundred-fold higher. The mode of action of neither HC-toxin nor victorin is known. In constrast to victorin and T-toxin, the physiological response of cells and tissues to HC-toxin is subtle: protein and RNA synthesis are not inhibited, and protoplasts are not killed (Wolf and Earle, 1991).

One of most intriguing aspects of the host-selective toxins made by *Cochliobolus* is the classical genetics of their production. First, it appears that single, but different, genetic loci control host-specific toxin production in *C. heterostrophus*, *C. carbonum*, and *C. victoriae* (Table 1). These genes are called *TOX1*, *TOX2*, and *TOX3*, respectively (Bronson 1991). How could single loci control production of complex secondary metabolites? Second, this unusual genetic situation probably is important to the conclusion drawn from population genetic studies with *C. heterostrophus* that strongly suggest that race T arose once from race O and in the very recent past (Leonard, 1973).

2. *C. carbonum* race 1

C. carbonum was first described in the 1940's attacking certain maize inbreds. Reaction to the fungus was shown to be controlled by a single major gene, *Hm*, which maps on the long arm of chromosome 1. Resistance is completely dominant to susceptibility. Another gene, *Hm2*, also gives minor, and developmentally-modulated, resistance. Shortly thereafter, *C. carbonum* isolates showing weak pathogenicity on all maize lines (and no special virulence on maize of genotype *hm/hm*) was described and called race 2. More recently, races 3 and 4 of *C. carbonum* have been described. Inheritance of esistance to races 3 and 4, like to race 2, is polygenic.

Shortly after the discovery of the perfect stage, *Cochliobolus*, of what had previously been called *Helminthosporium*, it was shown that in crosses between race 1 and race 2 isolates the progeny were either race 1 or race 2, in a 1:1 ratio, that is, race 1 and race 2 differed by a single genetic locus. Hence, our level of understanding of the interaction between *C. carbonum* and maize in 1960 was at the same level as is our understanding of many other diseases today, which is restricted to a classical genetic understanding that the pathogenic phenotype of a single pathogen gene can be determined by the particular allele of a single host resistance gene, and vice versa.

A major advance in our understanding of this disease, and of host-selective toxins in general, came in 1965 with the report by Scheffer and Ullstrup (1965) that *C. carbonum* race 1 produced a low molecular weight compound, called HC-toxin, that selectively inhibits root growth of maize of genotype *hm/hm*. Shortly thereafter it was demonstrated that the single genetic locus differentiating race 1 from race 2 controlled production of HC-toxin. This gene is now called *TOX2*.

Some years later the structure of HC-toxin was elucidated by contributions from three groups (Walton, 1990). HC-toxin is a relatively simple cyclic tetrapeptide of structure cyclo(D-Pro-L-Ala-D-Ala-L-Aeo), where Aeo stands for 2-amino-9,10-epoxi-8-oxodecanoic acid. Determination of the structure allowed further studies on its biosynthesis, and hence the molecular genetics of *TOX2*, as well as on the function of the *Hm* gene. As a result of these studies the basis of specificity between *C. carbonum* race 1 and maize is now the best understood of any plant/microbe interaction.

This article will discuss only our studies on the biochemistry and molecular genetics of HC-toxin production. Results from this laboratory on the biochemical function of the maize *Hm* gene, which has involved a partly complementary and partly collaborative effort with S.P. Briggs and G.S. Johal at Pioneer Hi-Bred International, are presented in a chapter in this volume by R.B. Meeley and J.D. Walton.

3. Enzymology of HC-toxin Biosynthesis

C. carbonum race 1 must perforce contain genes that encode the enzyme or enzymes that make HC-toxin. *TOX2* could regulate some aspect of HC-toxin biosynthesis or could itself encode the enzymes that make HC-toxin. Other cyclic peptides are known to be synthesized by large, multifunctional enzymes. At least one, the cyclic hexidepsipeptide enniatin B, is synthesized by a single, large enzyme (Kleinkauf and von Döhren, 1987). In *Streptomyces*, antibiotic biosynthetic genes are typically clustered (Hopwood and Sherman, 1990). Hence, we reasoned, *TOX2* might encode a single, large biosynthetic enzyme, or *TOX2* might be a cluster of tightly linked genes. If *TOX2* were regulatory, it would likely regulate HC-toxin-synthesizing enzymes. Therefore, we began our approach to the cloning of *TOX2* through identification of the enzymes that synthesize HC-toxin.

All cyclic peptides so far examined are biosynthesized by a similar group of enzymes. These cyclic peptide synthetases all catalyze amino acid-dependent ATP/PP_i exchange, bind the amino acids as thioesters, and contain pantothenic acid as a co-factor. The amino acids are activated and bound independently of each other prior to peptide bond formation (Kleinkauf and von Döhren, 1987). Cyclic peptide synthetases are multifunctional and can catalyze, among other reactions, epimerization and N-methylation.

Extracts of *C. carbonum* race 1 contain ATP/PP_i exchange activity driven by D-ala, L-ala, and L-Pro. These three activities are present only in isolates that make HC-toxin and genetically co-segregate with *TOX2*. Enzyme activity is strong in mycelial mats that are 3 or 4 days old but falls to almost undetectable levels after 6 days, although HC-toxin production is known to continue for at least 18 days. The biological significance of our inability to extract the enzyme activity from old mats is unclear (one possibility is that the enzymes become associated with a particulate fraction of the cell), but could have practical significance for attempts to identify and extract toxin-biosynthetic enzymes from other fungi. Anion exchange or hydroxyapatite chromatography separates the L-Pro-activating activity from the Ala-activating activities. Hence, initially the biochemical data indicated that HC-toxin was not made by a single enzyme but that either *TOX2* was a cluster of biosynthetic enzyme genes or it regulated a series of genes (Walton, 1987). The two enzymes were purified and characterized (Walton and Holden, 1988). HTS-1, subunit M_r 220,000, activates L-Pro, binds it as a thioester, and epimerizes it to D-Pro. HTS-2, of subunit M_r 160,000, activates both L-Ala and D-ala, binds them as thioesters, and epimerizes L-Ala to D-Ala (but not vice versa). HTS-1 and HTS-2 have normal affinities for the substrates PP_i and ATP but low affinities for the amino acid substrates. For example, the K_m of HTS-2 for L-Ala is 100 mM (Walton and Holden, 1988). Because Aeo is not readily available, and information on its biosynthetic origin is limited (Wessel et al., 1988), we do not know if HTS-1 or HTS-2 have anything to do with its synthesis or activation.

4. Cloning and Analysis of Genes for HTS-1 and HTS-2

HTS-1 and HTS-2 were purified by conventional HPLC techniques and murine polyclonal antibodies prepared. The enzymes were digested with trypsin and peptides sequenced by automated Edman degradation. A cDNA library in λgt11 was prepared from polyA$^+$-enriched RNA from the tox$^+$ isolate SB111 and screened with anti-HTS-1 antibody. One cDNA thereby obtained hybridized to a synthetic oligonucleotide corresponding to a tryptic peptide from HTS-1. The DNA sequence of the cDNA confirmed that it encoded the entire tryptic peptide.

This first cDNA, pCC25, hybridizes only to DNA from tox$^+$ isolates and its presence genetically co-segregates with HC-toxin production and hence, by definition, *TOX2*. Starting from pCC25, we used overlapping genomic lambda clones (in λEMBL3) to map the DNA adjacent to pCC25 that was unique to tox$^+$ isolates. We discovered a total of 22 kb of contiguous genomic DNA that is found in all tox$^+$ and no tox$^-$ isolates examined. The 22-kb region is flanked on both ends by moderately repetitive DNA sequences that are common to tox$^+$ and tox$^-$ isolates but which, in some cases, show tox$^+$-associated polymorphisms. There are multiple copies of at least some of the tox$^+$-unique DNA sequences at the 3' end of the 22-kb region, but originally it appeared that there was just one copy of the DNA at the 5' end.

4.1 GENE DISRUPTION OF *TOX2*.

Integration of transforming DNA in *C. carbonum* is homologous, allowing one-step gene disruption. When the 22-kb tox$^+$-unique DNA was disrupted with a fragment of DNA near the 5' end, the transformants still made HC-toxin and were still pathogenic. Southern analysis of the transformants indicated that integration was homologous but that at least one copy of non-disrupted DNA remained. This led to the discovery that the 22-kb region exists in tox$^+$ isolates in two copies. These copies can be distinguished at the 3' end with any restriction enzyme that spans the junction between the tox$^+$-unique DNA and the flanking common DNA, but can be distinguished at the 5' end only with the enzyme ApaI, which cuts one copy 15 kb and the other copy 25 kb outside the tox$^+$-unique region. Using ApaI to analyze transformants, a strain of *C. carbonum* disrupted in both copies of the 22-kb region was engineered. This isolate could no longer make HC-toxin, lacked both HTS-1 and HTS-2 enzyme activities, and was no longer pathogenic (Panaccione et al., 1992). We also did a double disruption at the 3' end of the tox$^+$-unique 22-kb region. The resulting isolate did not make HC-toxin and was not pathogenic, but retained significant HTS-1 and HTS-2 activities, indicating that the 3' end of the 22-kb region was required for HC-toxin production but not HTS-1 or HTS-2 activity.

Both copies of the 22-kb region are functional, since isolates in which just one or the other copy was disrupted still make sufficient HC-toxin to be fully pathogenic. However, isolates with only one functional copy have less HTS-1 and HTS-2 enzyme activity (approximately 50% of wild-type levels) and make less HC-toxin.

The 5' gene disruptions were done in a region of the 22-kb region known to encode the enzyme HTS-1, and, as expected, the disruptions caused HTS-1 enzyme levels to fall. Unexpectedly, disruptions in the coding region for HTS-1 also caused levels of HTS-2 to fall, and by the same relative amount. This indicated that HTS-1 and HTS-2 levels were coordinately regulated in some way. A simple explanation for this finding, namely, that HTS-1 and HTS-2 are part of the same gene and of the same protein *in vivo*, came from DNA sequencing of the 22-kb region.

5. cDNA and Sequence Analysis of the Tox$^+$-unique 22-kb Region.

The 22-kb region is large enough to contain many genes. In addition to biosynthetic enzymes, these genes could encode proteins necessary for regulation, self-protection, and/or secretion of the toxin. Using sub-cloned genomic fragments as DNA probes to screen a cDNA library, six independent transcribed regions were found. Two of these lie at the extreme 5' and 3' ends of the 22-kb region and are transcribed in opposite directions. Although we have sequenced these cDNA's and the corresponding genomic DNA's, to date nothing is known about their possible function. From the middle of the 22-kb region four independent cDNA's, including pCC25, which was originally identified by immunoscreening, were obtained and sequenced (Fig. 1). All four are 1.0 to 1.5 kb in length, all are transcribed in the same direction, and all are polyadenylated, but, oddly, only the 3'-most cDNA contains a stop codon. Based on our sequencing of the 22-kb region (see below), we currently conclude that all but the 3'-most cDNA are artifacts due to premature transcriptional termination.

Figure 1. Schematic illustration of the structure of *HTS1*, a 15.7 kb genomic open reading frame encoding a tetrapartite cyclic peptide synthetase (HC-toxin synthetase). This ORF contains sequences encoding peptides derived from both the enzymes HTS-1 and HTS-2 (Walton and Holden, 1988).

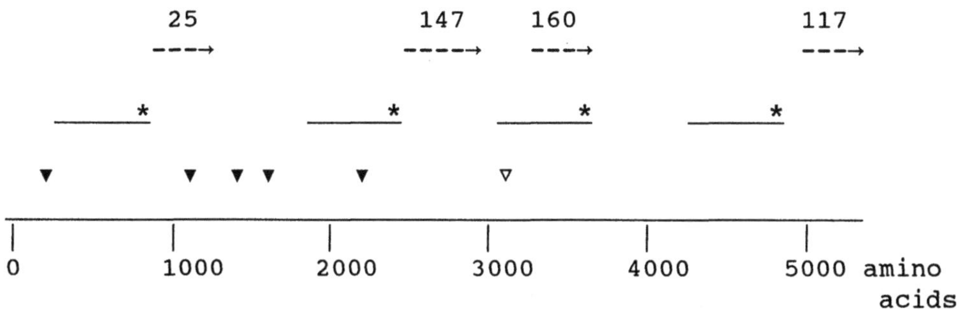

---→ Polyadenylated cDNA's (only 117 contains a stop codon)
____* Adenylate domain; * = putative amino acid-binding site
▼ Peptides from HTS-1 (L-Pro-activating) enzyme
▽ Peptide from HTS-2 (D- and L-Ala-activating) enzyme

Except for the two genes at the extreme ends of the 22-kb region, which appear to be real genes albeit of unknown function, all of the other cDNA's except the 3'-most one are embedded in a single open reading frame (ORF). From stop codon to stop codon the ORF is 15,693 nt in length. We call this ORF *HTS1*. Two ATG codons, occurring 48 and 54 nt from the start of the ORF, are in an excellent consensus context for fungal translational start sites. The predicted protein product contains 5215 amino acids (Fig. 1) and has a M_r of 574 kDa. This is potentially one of the largest polypeptides ever described - of proteins whose genes have been sequenced and are in the GENBANK/EMBL data base, it is second only to twitchin and its relatives, which are muscle structural proteins with no known enzymatic activity (Benian et al., 1989).

Two multifunctional cyclic peptide synthetase genes have been sequenced in the past two years - that encoding aminoadipyl-cysteinyl-valine synthetase (ACVS) from several prokaryotes and fungi, and gramicidin synthetase 2 (GS2) from *Bacillus subtilis*. The fungal ACVS genes contain no introns, similar to *HTS1*. ACVS activates three amino acids and GS2 activates four. Within the predicted ACVS amino acid sequence (the enzyme is about 410 kDa) contains three domains each of about 600 amino acids that are 35-45% percent identical. The predicted amino acid sequence for GS2 (an enzyme of 510 kDa) contains four such domains (Fig. 1). The sequence of *HTS1* also contains four domains. Therefore, we hypothesize that the protein product of *HTS1* is an enzyme capable of activating four amino acids.

Within the predicted amino acid sequence of *HTS1* are four peptides derived from HTS-1 and, surprisingly, a peptide of 19 amino acids derived from HTS-2, which had originally been purified as an enzyme distinct from HTS-1 (Walton, 1987; Walton and Holden, 1988)(Fig. 1). Apparently the single large ORF called *HTS1* encodes both HTS-1 and HTS-2. Is this protein processed in the cell or is its separation into two (and perhaps more) separate proteins an artifact of purification? During the several years we devoted to purifying HTS-1 and HTS-2, these two enzymatic activities did occasionally co-elute, at least in part, from some types of chromatography columns. However, co-elution was never consistent and could be explained on the basis of hydrophobic or electrostatic interaction, which would be expected for an enzymatic complex. To attempt to resolve this question, enzyme samples which had been prepared at different times, and using slightly modified protocols, were analyzed together by SDS-PAGE (3.2% acrylamide). This gel was blotted and the blot probed with antibodies raised against HTS-2 (the Ala-activating enzyme of reported M_r 160,000). In most preparations HTS-2 behaved as a single protein of 160,000, but in two preparations anti-HTS-2 antibodies reacted with proteins whose sizes were estimated to be at least 480 kDa. Thus, it appears that *HTS1* is translated into a single, large protein and that "HTS-2" comes from that protein. We think that this occurs as an artifact of purification, because a very similar situation has been found with other cyclic peptide synthetases. ATP/PP_i activity is known to be very resistant to even intentional proteolytic treatment, and in light of gene sequencing the M_r's of ACVS, GS2, and cyclosporin synthetase have also been re-evaluated upwards (MacCabe et al., 1991; Turgay et al., 1992).

The product of *HTS1* is apparently the central enzyme in the biosynthesis of HC-toxin. The evidence indicates that it activates three of the four amino acids in HC-toxin and epimerizes L-Ala and L-Pro to the D-forms. It presumably catalyzes peptide bond formation and cyclization. The presence of a fourth amino acid-activating domain strongly suggests that *HTS1* also activates Aeo (or a derivative) and incorporates it into the cyclic peptide. *HTS1* apparently does not, however, catalyze the synthesis of Aeo. This conclusion is based on the fact that when *HTS1* is inactivated by transformation-mediated gene disruption, a novel, epoxide-containing amino acid accumulates in the culture medium. This compound is not made by race 2 (tox^-) isolates, and its level is inversely proportional to the level of HC-toxin.

6. What is *TOX2*?

Despite identification and sequencing of a large region of DNA that is genetically linked to *TOX2* and is required for HC-toxin biosynthesis, and therefore composes at least in part this Mendelian gene, we still cannot claim to fully understand *TOX2*. Clearly, *TOX2* is not a simple gene in the normal sense. We know that *TOX2* is composed of two functional and very similar, if not identical, 22-kb regions. The core of this region is a single 15.7 kb ORF encoding a tetrapartite cyclic peptide synthetase. Establishment of the physical relationship between the two copies along the *C. carbonum* chromosome using pulsed-field gel electrophoresis is in progress. We estimate from limited chromosome walking that they are at least 20 kb apart. Therefore the minimum size of *TOX2* is at least 64 kb and it could be much larger. This large piece of DNA segregates as a single gene because there is no homologous DNA in the tox$^-$ isolates to permit crossing-over during meiosis.

There is probably additional DNA besides the 22-kb that is involved in HC-toxin biosynthesis and hence composes part of *TOX2*. First, probes such as pCC62 (at the 3'-most end of the 22-kb region) identify in addition to the two bands predicted for the two copies of the 22-kb region other tox$^+$-unique DNA (Panaccione et al., 1992). We are mapping these other DNAs and testing them for involvement in HC-toxin production. Second, when HC-toxin production is eliminated by disrupting *HTS1* (the 15.7 kb ORF), a novel, polar epoxide-containing metabolite (called PEC) accumulates in the culture medium. This compound is produced at low levels by wild-type race 1 (tox$^+$) isolates but not by race 2 (tox$^-$) isolates. This compound is likely a precursor or shunt-product of Aeo. Its accumulation suggests that Aeo is not synthesized by HTS and that therefore there must be other enzymes and genes necessary for HC-toxin synthesis. Again, if *TOX2* is really the only locus necessary for toxin synthesis, such genes must be a part of it.

7. Structure of *TOX2*: Implications for Evolution of Toxicogenesis in *Cochliobolus*

The natural history of *Cochliobolus* and its host-selective toxins poses some intriguing evolutionary questions. In two cases, the genes conferring sensitivity to host-selective toxins (and hence susceptibility to the producing pathogens) had been introduced into crop plants less than twenty years prior to the appearance on a devastating scale of species of *Cochliobolus* making toxins specific for those genes. The *Vb* gene conferring susceptibility to *C. victoriae* was introduced into this country from Uruguay in the 1930's (Walton, 1990). *C. victoriae* was first reported in the 1940's. Tcms was introduced into maize starting in the 1950's and *C. heterostrophus* race T was first reported in 1963. How, in a few years, did *Cochliobolus* evolve to produce complex secondary metabolites that permitted it to be a major pathogen on millions of plants throughout the U.S.?

A plausible explanation for the apparent rapid and recent evolution of toxicogenesis in *Cochliobolus* is that the *TOX* genes moved by horizontal gene transfer. The molecular structure of *TOX2* is consistent with a sudden, one-step acquisition of ability to produce HC-toxin. That T-toxin production and victorin are also controlled by single Mendelian "genes" suggests a similar situation might exist in these fungi. This raises two questions: where did the DNA for toxin production come from, and how did it move? In regards to the latter question, the occurrence of temporary anastomoses (hyphal fusions) is well-established in many filamentous fungi and can allow transfer of genetic material (e.g., Collins and Saville, 1990). In regards to the former question, a likely source of toxin-producing ability would be other organisms, especially fungi, that make related compounds. In fact, four unrelated filamentous fungi in addition to *C. carbonum* are known to make Aeo-containing cyclic tetrapeptides (Walton, 1990; Itazaki et al., 1991). Also, both *C. heterostrophus* and *Phyllosticta maydis*, unrelated pathogens of maize, make closely related host-selective toxins (Kohmoto and Otani, 1991). No known natural analogs of victorin are known, but this could be simply because they, unlike HC-toxin and its relatives, are not active in pharmacological assays.

Acknowledgments

I would like to thank the members of my laboratory who contributed to this work: Frank Holden, John Scott-Craig, Jean-Alain Pocard, Daniel Panaccione, and Carol Weiss. This work was supported by the U.S. Department of Energy (Division of Energy Biosciences), the U.S. National Science Foundation, and the M.S.U. Biotechnology Research Center.

References

Akimitsu, K., Hart, L.P., Walton, J.D., and Hollingsworth, R. (1992) 'Covalent binding sites of victorin in oat leaf tissues detected by anti-victorin polyclonal antibodies', Plant Physiol. 98, 121-126.

Benian, G.M., Kiff, J.E., Neckelmann, N., Moerman, D.G., and Waterston, R.H. (1989) 'Sequence of an unusually large protein implicated in regulation of myosin activity in C. elegans', Nature 342, 45-50.

Bronson, C.R. (1991) 'The genetics of phytotoxin production by plant pathogenic fungi', Experientia 47, 771-776.

Collins, R.A. and Saville, B.J. (1990) 'Independent transfer of mitochondrial chromosomes and plasmids during unstable vegetative fusion in Neurospora', Nature 345, 177-179.

Dewey, R.E., Siedow, J.N., Timothy, D.H., and Levings III, C.S. (1988) 'A 13-kilodalton maize mitochondrial protein in E. coli confers sensitivity to Bipolaris maydis toxin', Science 239, 293-295.

Hopwood, D.A. and Sherman, D.H. (1990) 'Molecular genetics of polyketides and its comparison to fatty acid biosynthesis', Ann. Rev. Genet. 24, 37-66.

Itazaki, H. et al. (1991) 'Isolation and structural elucidation of new cyclotetrapeptides, trapoxins A and B, having detransformation activities as antitumor agents', J. Antibiot. 43, 1524-1532.

Kleinkauf, H. and von Döhren, H. (1987) 'Biosynthesis of peptide antibiotics', Ann. Rev. Microbiol. 41, 259-289.

Kohmoto, K. and Otani, H. (1991) 'Host recognition by toxigenic plant pathogens', Experientia 47, 755-764.

Leonard, K.J. (1973) 'Association of mating type and virulence in Helminthosporium maydis, and observations on the origin of the race T population in the United States', Phytopathology 63, 112-115.

MacCabe, A.P. et al. (1991) 'δ-(L-α-aminoadipyl)-L-cysteinyl-D-valine synthetase from Aspergillus nidulans', J. Biol. Chem. 266, 12646-12654.

Panaccione, D.G., Scott-Craig, J.S., Pocard, J.-A., and Walton, J.D. (1992) 'A cyclic peptide synthetase gene required for fungal pathogenicity on maize', Proc. Natl. Acad. Sci. U.S.A., in press.

Scheffer, R.P. and Ullstrup, A.J. (1965) 'A host-selective toxic metabolite from Helminthosporium carbonum', Phytopathology 55, 1037-1038.

Turgay, K., Krause, M., and Marahiel, M.A. (1992) 'Four homologous domains in the primary structure of GrsB are related to domains in the superfamily of adenylate-forming enzymes', Mol. Microbiol. 6, 529-546.

Walton, J.D. (1987) 'Two enzymes involved in biosynthesis of the host-selective phytotoxin HC-toxin', Proc. Natl. Acad. Sci. U.S.A. 84, 8444-8447.

Walton, J.D. and Holden, F.R. (1988) 'Properties of two enzymes involved in the biosynthesis of the fungal pathogenicity factor HC-toxin', Mol. Plant-Micro. Inter. 1, 128-134.

Walton, J.D. (1990) 'Peptide phytotoxins from plant pathogenic fungi', in H. Kleinkauf and H. von Döhren (eds.), Biochemistry of Peptide Antibiotics, Walter de Gruyter, Berlin, pp. 179-203.

Wessel, W.L., Clare, K.A., and Gibbons, W.A. (1988) 'Biosynthesis of L-Aeo, the toxic determinant of the phytotoxin produced by Helminthosporium carbonum', Biochem. Soc. Trans. 16, 402-403.

Wolf, S.J. and Earle, E.D. (1991) 'Effects of Helminthosporium carbonum race 1 toxin on host and non-host cereal protoplasts', Plant Sci. 70, 127-137.

Wolpert, T.J. and Macko, V. (1989) 'Specific binding of victorin to a 100 kD protein from oats', Proc. Natl. Acad. Sci. U.S.A. 86, 4092-4096.

THE GENETICS OF DIMORPHISM IN THE SMUT FUNGI

J. KRONSTAD, A. YEE, G. BAKKEREN, S. GOLD,
K. BARRETT AND L. GIASSON
Biotechnology Laboratory, Departments of Microbiology and Plant Science
University of British Columbia, Vancouver, B.C.
Canada V6T 1Z3

ABSTRACT. Dimorphic growth in *Ustilago maydis* is regulated by mating-type loci called *a* and *b*. The *b* region encodes two polypeptides (*bE* and *bW*) which control the formation of the infectious dikaryon upon fusion of haploid, yeast-like cells of opposite *a* mating type. Given that there are at least 25 naturally occurring specificities at the *b* locus, and that the locus must be heterozygous to trigger filamentous growth, the molecular basis of self versus non-self recognition is of particular interest. The construction of recombinants between the *b1E* and *b2E* alleles identified a 30 to 48 amino acid region which determines specificity. In addition, hybridization and sequence analyses revealed homologs of the *bE* and *bW* genes in *U. hordei*, a smut thought to have only the *a* mating function. Genes have also been identified that may be regulated by the *b* locus and whose products influence cell morphology. One of these genes, called *rem1*, functions in the switch from yeast-like to mycelial growth.

1. Introduction

The smut fungi are basidiomycete phytopathogens that attack a variety of important crop species including corn, sorghum and small grain cereals (Fischer, 1953; Fischer and Holton, 1957; Christensen, 1963). Smut infections are generally distinguished by the presence of large masses of sooty black teliospores within the host plant. Some species, such as the corn smut fungus *Ustilago maydis*, incite galls on the host and sporulate within gall tissue. Other species, which infect small grain cereals (e.g., *Ustilago hordei*), do not incite galls but produce teliospores preferentially in floral tissue, thereby reducing the yield and quality of the grain (Agrios, 1988).

As a group, the smut fungi have basically similar life cycles in which cells exist in one of three types: diploid, haploid and dikaryotic (Fischer and Holton, 1957). Diploid teliospores are produced in infected tissue and germinate to give the four haploid products of meiosis, which are called basidiospores. Mating interactions between compatible haploid basidiospores yield dikaryotic cells which are capable of infecting the host plant, proliferating within host tissue and eventually sporulating to produce teliospores. Release of teliospores from infected tissue represents a major source of inoculum for subsequent infections.

One of the most interesting biological features of the smut fungi is the presence of at least two different mating systems within the group (Holton *et al.*, 1968). Many of the species, such as *U. hordei*, possess a bipolar mating system in which a single mating-type locus with two alternate forms, *MAT-1* and *MAT-2* (also called *a* and *A*), controls dikaryon formation (Thomas, 1991). Other smuts, such as *U. maydis*, display tetrapolar mating in which two loci control dikaryon formation (reviewed by Froeliger and Kronstad, 1990; Banuett, 1992). The tetrapolar smut fungi generally have one locus with two alternate sequences that is thought to control cell fusion (equivalent to *MAT*), and a second locus with multiple alternative forms that controls the establishment of the infectious dikaryon (Holton *et al.*, 1968). If the bunt fungi of the genus

E. W. Nester and D. P. S. Verma (eds.),
Advances in Molecular Genetics of Plant-Microbe Interactions, 325–333.

Tilletia are included in the smut group, then a third mating system (multiple alleles at a single locus) has been reported for *T. controversa* (Hoffman and Kendrick, 1965)

The tetrapolar mating system of *U. maydis* has been characterized in some detail due to recent efforts to isolate and sequence the mating-type genes. In this fungus, the locus which controls fusion is called *a* (*MAT*), and the locus that establishes the infectious dikaryon is called *b*. Haploid cells can fuse and form the infectious dikaryon only if they carry different sequences at the *a* and *b* loci. The *a* mating type locus has two alternative forms, *a1* and *a2*. These sequences have recently been shown to be idiomorphs (Froeliger and Leong, 1991), and to encode pheromones and pheromone receptors (Bolker *et al.*, 1992). The *b* region has at least 25 different alternative forms and appears to control events after cell fusion necessary for establishment of the filamentous, pathogenic dikaryon. The haploid cell type is not pathogenic. Two genes are present at *b*, *bE* and *bW*. The combination of the *bE* product (473 a.a.) from one version of the *b* locus with the *bW* product (626 a.a.) from another is believed to form a novel regulatory protein that triggers dikaryon formation (Gillissen *et al.*, 1992). The alignment of the predicted amino acid sequences of several alleles of *bE* revealed the presence of a variable N-terminal domain, a central homeodomain-like motif and a conserved C-terminal region (Kronstad and Leong, 1990; Schulz *et al.*, 1990). A similar organization was found for *bW*; however, it should be noted that *bE* and *bW* share sequence similarity only in the homeodomain region (Gillissen *et al.*, 1992).

It is interesting that the phytopathogenic smut fungi display a dimorphic growth habit similar to that found in some animal pathogens, e.g., *Candida albicans*. *Ustilago* species have the ability to switch between a yeast-like nonpathogenic phase (i.e., the haploid phase) and a filamentous pathogenic phase (i.e., the infectious dikaryon). This dimorphic switch is controlled primarily by the *b* genes. The finding that *bE* and *bW* contain homeodomain-like regions (Schulz *et al.*, 1990; Gillissen *et al.*, 1992), and the observation that disruption of *b* function blocks dikaryon formation (Kronstad and Leong, 1990; Gillissen *et al.*, 1992), suggests that the *b* products are regulatory proteins. The influence of the *b* locus can readily be detected because the mating of compatible haploids on rich medium containing activated charcoal (Puhalla, 1968) results in the formation of white aerial hyphae on the mixed colony. A similar phenotype is seen when diploid or haploid strains carrying two different versions of *b* are grown on the same medium. This "fuzzy" phenotype presumably indicates formation of the infectious, filamentous dikaryon and it provides a convenient assay to detect the activity of the *a* and *b* genes. For example, Day *et al.* (1971) employed this phenotype to isolate mutations at *b*, and Banuett (1991) used it to identify mutations that block the ability of haploid strains to mate; these mutations could conceivably be in genes necessary for fusion, dikaryon formation or both.

The primary goal of our work is to understand the genetics of formation of the infectious dikaryon in the smut fungi. We anticipate that this understanding will provide insight into fungal pathogenesis on plants, into the role of mating-type genes in pathogenesis, and into the regulation and mechanism of dimorphic growth in fungi. Here we describe progress in our analysis of the specificity region of the *bE* gene of *U. maydis* and in the characterization of a *b* locus in the bipolar smut *U. hordei*. In addition, we describe strategies that have proven successful for identifying other genes, besides the *a* and *b* mating-type genes, that play a role in formation of the infectious dikaryon.

2. Materials and Methods

The following strains of *U. maydis* were employed; 518 (*a2 b2*), 521 (*a1 b1*), 87 (*a2 b2 ad1-1 leu1-1*), 87-18 (*a2 b2 ad1-1 leu1-1 rem1-1*), P6D (*a2 b2 [a1 b1 phl^r]*). The strain P6D is phleomycin resistant (*phl^r*) *U. hordei* strains Uh112 (*MAT-1 ad1*) and Uh100 (*MAT-2 ad1*) were employed for the isolation of the *b* homologs. Other strains of *U. hordei*, *U. kolleri*, *U. avenae*, *U. bullata* and *U. nigra* were obtained from Dr. P.L. Thomas, Agriculture Canada, Winnipeg, Manitoba. All methods were employed as previously described (Kronstad and Leong, 1989, 1990; Sambrook *et al.*, 1989; Wang *et al.*, 1988).

3. Results and Discussion

3.1. THE *b* GENES OF *U. MAYDIS* AND *U. HORDEI*

3.1.1. *Recombinant alleles define a specificity region in the bE gene of U. maydis.* A series of hybrids between the *b1E* and *b2E* genes was constructed by an *in vivo* recombination strategy diagrammed in Figure 1A. DNA fragments carrying part of the *b1E* gene (truncated at various positions in the variable region) and a selectable marker for resistance to hygromycin B were introduced into a strain carrying a *b2E* allele (518, *a2 b2*). Targeted integration, which occurs at a high frequency in *U. maydis* (Kronstad *et al.*, 1989; Fotheringham and Holloman, 1989), generated recombinants between *b1E* and *b2E*. The transformants resulting from this experiment were screened for their mating specificity in a standard assay on rich medium containing activated charcoal (Puhalla, 1968; Holliday, 1974). The original strain used for transformation mates with *a2 b2* specificity such that it forms white aerial mycelium when mixed with strains carrying the *a1* mating specificity and any other *b* specificity besides *b2*. Three classes of transformants of this strain were found in the mating tests; I) those that mated with the original specificity (*a2 b2*); II) those with a novel *b* specificity (*a2 bx*); and III) those that had switched specificity from *b2* to *b1* (*a2 b1*). These results indicated that *bE* alleles with altered specificity could be generated by *in vivo* recombination. It should be noted that in these experiments, the hygromycin B marker was inserted into the *bW* orf; therefore, in the transformants carrying recombinant *bE* alleles, the *bW* orf is inactive. This allowed the analysis of the *bE* specificity of the transformants without complications arising from the activity of the *bW* gene.

The variable regions of the *bE* genes, from representatives of the three classes of transformants, were isolated by polymerase chain reaction and nucleotide sequence analysis was performed. Figure 1B shows a diagram of the recombination points found for transformants in each class. Class I transformants arose from recombination early in the *bE* orf, for example, at codons 28 or 39. Class II transformants had recombination points in the region between codons 40 and 79 and class III transformants had recombination points downstream of codon 87. Overall, these results demonstrate the presence of a region between codons 40 and 79 that determines allelic specificity for the *bE* gene. Recombination upstream of this region did not alter specificity while recombination downstream switched specificity from that of the resident *b2E* allele to that of the incoming *b1E* allele.

Current models for the molecular recognition events mediated by the *b* locus products postulate that the interaction between the *bE* gene product from one haploid parent (e.g. *b1E*) with the *bW* gene product from the other parent (e.g. *b2W*) combine to generate a regulatory activity that directly or indirectly influences the expression of genes required for formation of the infectious dikaryon (Gillissen *et al.*, 1992). It appears that the *bE* and *bW* products from one haploid parent (e.g., *b1E* and *b1W*) are unable to establish the regulatory activity. Given our discovery of a specificity domain in the variable region of the *bE* gene, we postulate that this region of the encoded polypeptide interacts with a corresponding domain in the polypeptide encoded by the *bW* gene (presumably also in the variable portion). The presence of a corresponding region in the *bW* gene is predicted by sequence alignments which reveal two clusters of hypervariable sequences within the variable region (Gillissen *et al.*, 1992; J. W. K., unpublished results). The prediction of our work is that recombinants between *b1W* and *b2W*, within the hypervariable region, should create alleles with specificity different from that of the parental alleles. This would be analogous to the results obtained from the construction of recombinant alleles for *bE*. In addition, it should be possible to find recombinant alleles of *bW* which match recombinant alleles of *bE* in terms of ability to interact and trigger formation of the filamentous cell type. If this turns out to be the case, our model predicts that this approach would lead directly to a molecular description of the amino acid residues that mediate recognition between *bE* and *bW*.

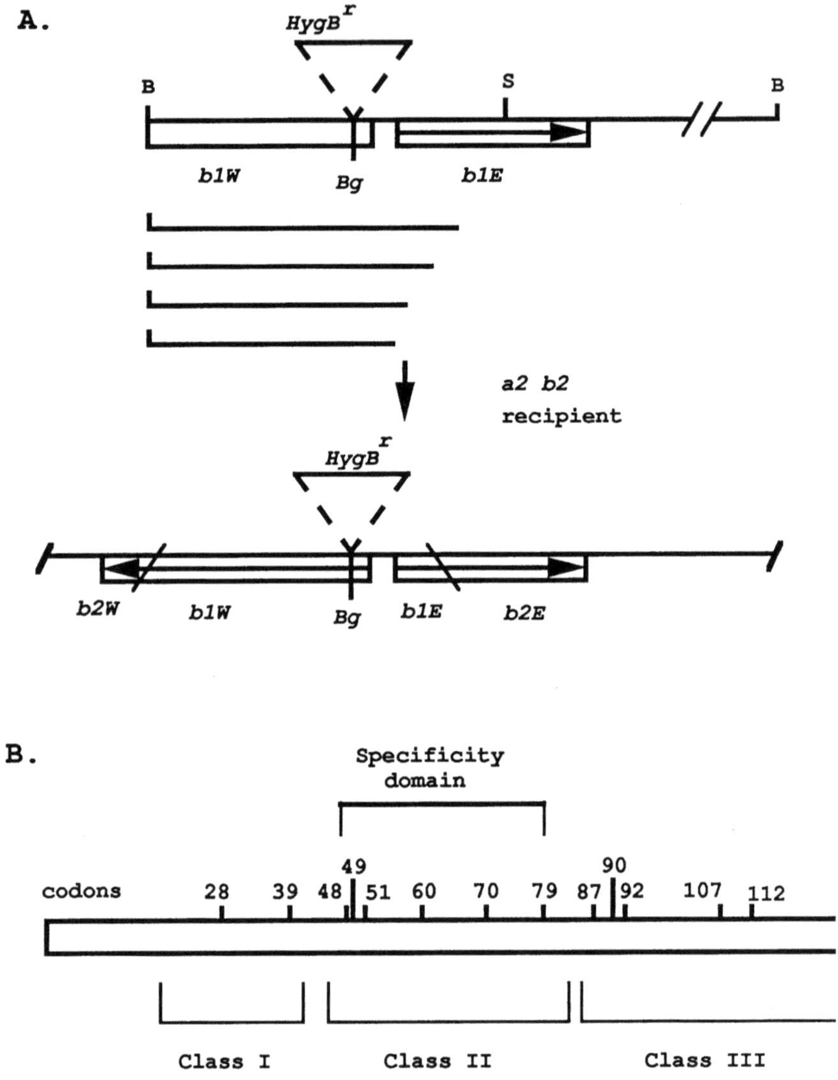

Figure 1. Construction and analysis of recombinant *bE* alleles.
(A) Strategy for the construction of recombinant alleles by targeted integration of truncated versions of *b1E* at the *b2* locus. Note that the hygromycin B resistance gene is inserted into the open reading frame of the *bW* gene. Restriction sites shown are: *Bam*HI, B; *Bgl*II, Bg; and *Sal*I, S.
(B) Recombination map of the *bE* open reading frame. Class I recombinants mate with *b2E* specificity, class II recombinants have specificity different from both *b1E* and *b2E* and class III recombinants have *b1E* specificity.

3.1.2. *The b genes are present in both tetrapolar and bipolar smut fungi.* Hybridization studies with the cloned *a* and *b* DNAs of *U. maydis* revealed that these sequences are present in other smut fungi with the tetrapolar mating system (having both the *a* and the *b* functions), as well as in some smut fungi with the bipolar mating system (Bakkeren *et al.*, in press). Previous genetic analyses suggested that the latter fungi have a single mating-type locus with two alternate forms. Therefore, it was particularly surprising to find that these fungi had sequences that hybridized with the *b* locus of *U. maydis*. We have initiated an analysis of the *b* genes in bipolar smut fungi by isolating cosmid clones that carry the *b* sequences from each mating-type (*MAT-1* and *MAT-2*) of *U. hordei*, a representative of the smuts with the bipolar mating system. The cosmid carrying DNA from the *MAT-1* strain (Uh112) was designated pa112 and the cosmid from the *MAT-2* strain (Uh100) was designated pA100.

Nucleotide sequence analysis of the *b*-like regions on the cosmids pa112 and pA100 revealed open reading frames analogous to the *bE* and *bW* genes of *U. maydis*. When the predicted amino acid sequences were aligned, the two *bE* alleles (one from each mating type) showed the same organization as was found for *U. maydis bE* (Kronstad and Leong 1990; Schulz *et al.*, 1990), i.e., variable amino termini, constant carboxy termini and a homeodomain-like motif. At the amino acid level, the two *bE* alleles from *U. hordei* were 81% identical to each other, and each gene was approximately 50% identical to the *b1E* allele of *U. maydis*. The *bW* alleles of *U. hordei* also possessed variable amino terminal regions, constant carboxy terminal regions and homeodomain-like motifs. These alleles were 74% identical to each other at the amino acid level and approximately 40% identical to the *b1W* allele of *U. maydis*. Overall, the sequencing results indicated that the *b* genes are structurally conserved between smut fungi with bipolar and tetrapolar mating systems.

Given the structural conservation, it was of interest to determine whether the sequences were functionally equivalent. Two tests of *b* allele function, based on colony morphology and pathogenicity (Kronstad and Leong, 1989), were performed. First, *U. hordei b* alleles were introduced into *U. maydis* haploid strains, e.g., *bE* and *bW* from *U. hordei* into *U. maydis*, and the resulting transformants were tested for their morphology on medium containing activated charcoal. The presence of two different *b* specificities in a single strain, either on an episomal vector or integrated, results in the formation of mycelial colonies; in contrast, haploid strains normally form yeast-like colonies. Placing either of the cosmids, pa112 or pA100, from *U. hordei* into *U. maydis* yielded transformants showing mycelial growth. Similarly, placing the cloned *b1E* and *b1W* alleles from *U. maydis* into strains of either mating-type of *U. hordei* also gave mycelial transformants. The transformants of *U. maydis*, which contained *U. hordei b* genes, were also injected into corn seedlings to determine whether the *U. hordei* and *U. maydis b* gene products would productively interact to yield a pathogenic cell type. These *U. maydis* transformants did give weak disease symptoms on corn seedlings indicating that the products of the *b* genes from *U. hordei* and *U. maydis* are capable of interaction.

Sequence analysis to date on the *bE* and *bW* variable regions from other bipolar smut fungi indicates that there may only be two alleles for each gene among this group of fungi. That is, the predicted amino acid sequences of the variable regions of *bE* and *bW* from 4 other bipolar smut fungi, *U. bullata*, *U. nigra*, *U. avenae* and *U. kolleri*, fall into two related groups. In one group, the sequences are almost identical to the genes from the *MAT-1* strain of *U. hordei*; in the other, they are almost identical to the sequences from the *MAT-2* strain. In contrast, a similar analysis of several *bE* genes from *U. maydis* revealed a large number of amino acid differences in the variable region (Kronstad and Leong, 1990; Schulz *et al.*, 1990). A more extensive survey of *bE* and *bW* variable regions from several isolates of *U. hordei* is currently underway to explore the extent of variability and to further establish the idea that only two alleles exist for *bE* and *bW* in the bipolar smut fungi.

3.2. OTHER GENES INVOLVED IN FORMATION OF THE INFECTIOUS DIKARYON.

3.2.1. *Isolation of constitutively mycelial mutants in U. maydis.* The ability of strains of *U. maydis* to form mycelial colonies on medium containing activated charcoal has proven to be a useful assay for the state of the *a* and *b* mating-type sequences (homozygous or heterozygous) and for the pathogenicity of a particular strain. In fact, it was this mycelial or "fuzzy" phenotype that originally provided the assay to clone the *b* locus (Kronstad and Leong, 1989). With this in mind, we set out to search for mutations that would cause normally yeast-like haploid strains of *U. maydis* to constitutively form mycelial colonies independent of the state of the *a* or *b* loci. To date, over 100 mutants have been isolated, following mutagenesis with ultraviolet light, which display a completely mycelial growth habit. This phenotype is in contrast to haploid or diploid strains carrying two different forms of the *b* locus. The latter strains generally grow with a yeast-like morphology in liquid and form yeast-like colonies which become covered with white aerial mycelium on solid medium. The mutation to a mycelial phenotype has been designated *rem1* for repressor of mycelial phenotype.

One of the mycelial haploid mutants, strain 87-18 (carrying *rem1-1*) has been characterized with respect to the distribution of nuclei in the hyphae and pathogenesis. Previous descriptions of the filaments that are thought to represent the infectious, filamentous dikaryon of *U. maydis* revealed that, in general, the cells are devoid of cytoplasm and nuclei except for those at the hyphal tips (Day and Anagnostakis, 1971). The mutant strain 87-18 differs in that cytoplasm and nuclei are present throughout the hyphae and each cell appears to have only one nucleus. Pathogenicity tests, performed by injecting mutant strains (carrying *rem1-1*), alone or with compatible haploid strains, into corn seedling revealed that the mutation did not allow haploid strains to be pathogenic by themselves. This is in contrast to weakly pathogenic haploid strains constructed by introducing a different set of *bE* and *bW* genes (e.g., *b1E, b1W*) into a haploid strain (e.g., *b2E, b2W*) (Kronstad and Leong, 1989). It is clear that mutation in the *rem1* gene (conferring the mycelial phenotype) is not sufficient to confer pathogenicity on a haploid strain. The haploid strains carrying the *rem1-1* mutation were capable of giving high levels of infection when paired with wild-type strains of compatible mating-type. However, mixtures of compatible strains, each carrying the *rem1-1* mutation, gave reduced levels of infection compared with infections caused by wild-type strains. Thus, the *rem1-1* mutation appears to reduce the ability of *U. maydis* to cause disease symptoms.

Stakman *et al.* (1943) reported the occurrence of "white" mutants of *U. maydis*, which have a similar phenotype to the *rem1* mutants described here. Interestingly, Stakman *et al.* (1943) reported that all of the white mutants were capable of causing disease symptoms but they were unable to produce teliospores in infected tissue.

The *rem1-1* mutation in strain 87-18 has been complemented by the introduction of a cosmid library and screening for transformants with a yeast-like morphology. Three transformants, all of which appear to contain the same cosmid, were obtained from a screen of approximately 2000 transformants. These cosmids, when introduced into strain 87-18, give transformants with a yeast-like morphology. Loss of the cosmid, due to growth in medium lacking hygromycin B, resulted in cells with the mycelial phenotype of the original mutant strain 87-18.

The region of the cosmid complementing the *rem1-1* mutation has been identified by Tn5 insertion mutagenesis. Tn5 insertions in an approximately 9 kb region block complementation upon transformation into strain 87-18. Insertion of the hygromycin B marker into this region and replacement of the genomic sequence with the disrupted sequence in haploid strains results in cells with a mycelial phenotype. This result suggests that the cosmid DNA does indeed carry the *rem1* gene, rather than a suppressor of the *rem1-1* mutation.

3.2.2. *Isolation of mutants blocked in formation of the infectious dikaryon.* One of the main problems in attempting to genetically dissect pathogenesis in the smut fungi is the requirement that two haploid partners must mate to form the infectious, dikaryotic cell type. Specifically, it is difficult to identify recessive mutations that block the ability of the cells to switch from budding to filamentous growth. To overcome this problem, we have constructed a haploid strain of *U. maydis* (P6D) that contains each of the *a* locus sequences and two different *b* locus sequences. This strain mimics diploid strains heterozygous at both *a* and *b* in that it forms mycelial colonies on charcoal medium and it is pathogenic (albeit weakly) upon injection into corn seedlings. With this strain, we have undertaken a screen for mutations that block the ability of the strain to form mycelial colonies on charcoal medium. Mutations have been generated by random insertion of an integrative plasmid vector and by treatment with ultraviolet light. A large number of mutants have been isolated (designated nf for nonfuzzy), and examples of the phenotypes of several of the mutants are shown in Figure 2. The phenotypes vary from almost as mycelial as the parental strain to completely yeast-like. In addition, one class of mutants has a phenotype of slow growth and abnormal cell shape. These mutants may have defects in cell wall biosynthetic functions that are involved in filamentous growth. Attempts are currently being made to complement two of the mutations that block the mycelial phenotype of P6D.

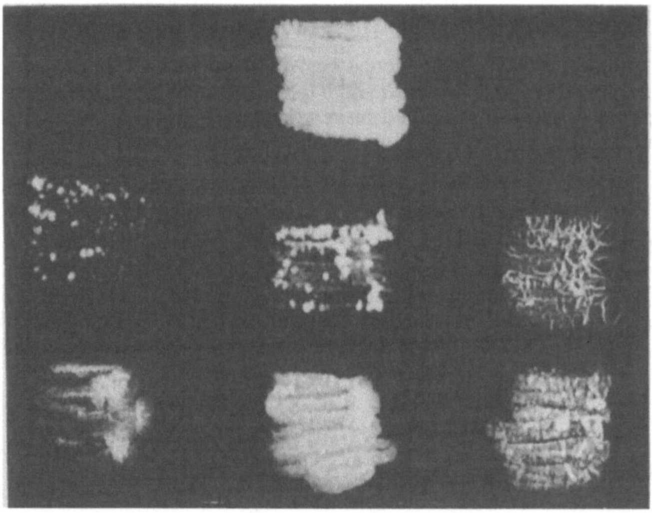

Figure 2. Colony morphology of the mycelial haploid strain P6D and nonfuzzy mutant derivatives. The colonies were grown on rich medium containing activated charcoal (Holliday, 1974) for 48 hours at 30°C.

4. Summary.

The analysis of the mating-type genes of *U. maydis* and *U. hordei* has provided an entry point to begin to understand the formation of the infectious, filamentous dikaryon from nonpathogenic yeast-like haploid strains. We have focused on the key step that triggers dikaryon formation; i.e., the recognition of the presence of products of different alleles of the *b* genes (*bE* and *bW*) in the same cell. Specifically, we have identified an approximately 40 amino acid region in the *bE* gene that we believe determines the specificity of the interaction of *bE* product with *bW*. What is needed now is a complementary analysis of *bW* to identify the corresponding region of specificity in this gene. This analysis will allow a detailed investigation of the specific amino acid residues that interact to mediate recognition.

Our work on *U. hordei* has revealed the surprising finding that smut fungi with bipolar mating systems, that is, without a genetically defined *b* locus, do indeed have *b* genes. Sequence analysis of these genes revealed that they have the same basic organization as the genes in the tetrapolar smut *U. maydis*. In addition, transformation experiments indicate that the sequences are functionally conserved when moved between species. The problem now is to explain why a *b* locus has not been identified genetically in smut species with bipolar mating systems. Our working model is that the *b* sequences are physically linked to the genetically defined *MAT* locus such that only a single mating-type function with two alternate forms, *MAT-1* and *MAT-2*, appears to be present.

A complete understanding of the regulatory control exerted by the *b* locus will require the identification and characterization of other genes whose expression may be regulated by *b* and whose products play additional regulatory or structural roles in dikaryon formation. We have identified a gene, *rem1*, which appears to play a central role in the switch between budding and filamentous growth. We envision that the product of the *rem1* gene establishes a site near one apex of the yeast-like haploid cell and that bud formation occurs at this site. Loss of the *rem1* function results in loss of the ability to bud; the fact that the resulting phenotype is completely filamentous suggests that this is the default growth morphology for *U. maydis.*

In contrast to the *rem1-1* mutation, which causes haploid cells to display a filamentous phenotype, other mutations have been identified which prevent expression of the mycelial phenotype induced by heterozygosity at the *a* and *b* loci. The characterization of these mutations and the corresponding genes will provide a more complete picture of the regulatory pathway controlled by mating-type and of the biochemical changes needed to change fungal cell morphology.

5. Acknowledgements

We thank Percy Thomas for providing strains of the smut fungi and Kande Williston for help with DNA sequencing. This work was supported by operating and strategic grants (to J.W.K.) and by fellowships (to A.Y. and L. G.) from the Natural Sciences and Engineering Research Council of Canada.

6. References

Agrios, G. N. 1988. Plant Pathology. Academic Press, Inc., San Diego.

Bakkeren, G., B. Gibbard, A. Yee, E. Froeliger, S. Leong and Kronstad, J. 1992. The *a* and *b* loci of *U. maydis* hybridize to DNAs from other smut fungi. Mol. Plant-Microbe Interaction. (in press).

Banuett, F. 1991. Identification of genes governing filamentous growth and tumor production by the plant pathogen *Ustilago maydis.* Proc. Natl. Acad. Sci. USA. 88: 3922-3926.

Banuett, F. 1992. *Ustilago maydis*, the delightful blight. Trends in Genetics. 8: 174-180.

Bolker, M., M. Urban and Kahmann, R. 1992. The *a* mating type locus of *U. maydis* specifies cell signalling components. Cell. 68: 441-450.

Christensen, J.J. 1963. Corn smut caused by *Ustilago maydis*. American Phytopathology Society Monograph. No. 2. American Phytopathology Society. St. Paul. pp. 1-41.

Day, P.R. and Anagnostakis, S.L. 1971. Corn smut dikaryon in culture. Nature New Biology. 231:19-20.

Day, P.R., Anagnostakis, S.L. and Puhalla, J. 1971. Pathogenicity resulting from mutation at the *b* locus of *Ustilago maydis*. Proc. Natl. Acad. Sci. USA. 68:533-535.

Fischer, G. W. 1953. Manual of North American Smut Fungi. Ronald Press, New York.

Fischer, G. W. and C. S. Holton. 1957. Biology and control of the smut fungi. Ronald Press, New York.

Fotheringham, S. and Holloman, W. K. 1989. Cloning and disruption of *Ustilago maydis* genes. Mol. Cell. Biol. 9: 4052-4055.

Froeliger, E. H. and Kronstad, J. W. 1990. Mating and pathogenesis in *Ustilago maydis*. Sem. Dev. Biol. 1: 185-193.

Froeliger, E.H. and Leong, S.A. 1991. The *a* alleles of *Ustilago maydis* are idiomorphs. Gene. 100: 113-122.

Gillissen, B., J. Bergemann, C. Sandmann, B. Schroeer, M. Bolker and Kahmann, R. 1992. A two-component regulatory system for self/non-self recognition in *Ustilago maydis*. Cell 68: 647-657.

Hoffman, J. A. and Kendrick, E. L. 1965. Compatibility relationships in *Tilletia controversa*. Phytopathology 55: 1061-1062.

Holliday, R. 1974. *Ustilago maydis*. in R. C. King (ed.), Handbook of Genetics, Vol 1. Plenum, New York, pp. 575-595.

Holton, C. S., J. A. Hoffman and Duran, R. 1968. Variation in the smut fungi. Ann. Rev. Phytopath. 6: 213-242.

Kronstad, J. and Leong, S.A. 1989. Isolation of two alleles of the *b* locus of *Ustilago maydis*. Proc. Natl. Acad. Sci. USA. 86: 978-982.

Kronstad, J. W., Wang, J., Covert, S. F., Holden, D. W., McKnight, G. L. and Leong, S. A. 1989. Isolation of metabolic genes and demonstration of gene disruption in the phytopathogenic fungus *Ustilago maydis*. Gene 79: 97-106.

Kronstad, J. and Leong, S.A. 1990. The *b* mating type locus of *Ustilago maydis* contains variable and constant regions. Genes and Development. 4: 1384-1395.

Puhalla, J. E. 1968. Compatibility reactions on solid medium and interstrain inhibition in *Ustilago maydis*. Genetics 60: 461-474.

Sambrook, J., Fritsch, E.F. and Maniatis, T. 1989. Molecular Cloning. 2nd ed. Cold Spring Harbor Press, USA.

Stakman, E.C., Kernkamp, M.F., Martin, W.J. and King, T.H. 1943. The inheritance of a white mutant character in *Ustilago zeae*. Phytopath. 33: 943-949.

Schulz, B., F. Banuett, M. Dahl, R. Schlessinger, W. Schafer, T. Martin, I. Herskowitz and Kahmann, R. 1990. The *b* alleles of *U. maydis*, whose combinations program pathogenic development, code for polypeptides containing a homeodomain-related motif. Cell 60: 295-306.

Thomas, P. L. 1991. Genetics of small-grain smuts. Ann. Rev. Phytopathol. 29: 137-148.

Wang, J., D. W. Holden and S. A. Leong. 1988. Gene transfer system for the phytopathogenic fungus *Ustilago maydis*. Proc. Natl. Acad. Sci. U.S.A. 85: 865-869.

GENETICAL AND FUNCTIONAL ORGANIZATION OF THE *a* MATING TYPE LOCUS OF *USTILAGO MAYDIS*

M. Bölker,[1] M. Urban,[1] S. Lauenstein,[1] R. Lurz[2] and R. Kahmann[1]
[1]Institut für Genbiologische Forschung Berlin GmbH, Ihnestr. 63, D-1000 Berlin 33, Germany and [2]Max-Planck-Institut für Molekulare Genetik, Ihnestr. 73, D-1000 Berlin 33, Germany

Introduction

The phytopathogenic fungus *Ustilago maydis* belongs to the Basidiomycetes and is the causal agent of corn smut (for review see Banuett, 1992). *U. maydis* can infect leaves, stems, tassels and ears of corn plants. Infection is followed by the formation of galls or tumors, which are filled with black masses of teliospores. The life cycle of *U. maydis* is characterized by a change in morphology: haploid sporidia are yeast-like, they grow vegetatively by budding and are nonpathogenic. The dikaryon, on the other hand, is filamentous and is able to infect corn plants. The dikaryotic stage is unstable on artificial media and requires the plant for sustained growth; karyogamy and spore formation also occur *in planta.* only The fusion of haploid cells and the switch between yeast-like and hyphal growth is controlled by two unlinked genetic loci, *a* and *b*. A successful mating reaction occurs only between strains carrying different alleles at both mating type loci. The mating reaction can be assayed on charcoal containing plates: only the filamentous growing dikaryon has a white fuzzy (Fuz^+ phenotype) appearance. The *b* locus exists in many different alleles and each allele codes for a pair of homeodomain proteins which are assumed to act as transcriptional regulators when appropriately combined (Gillissen et al., 1992). The *a* locus, which exists in two alleles (*a1* and *a2*), controls the fusion of haploid cells and, together with the *b* locus, the maintenance of filamentous growth (Banuett, 1989).

E. W. Nester and D. P. S. Verma (eds.),
Advances in Molecular Genetics of Plant-Microbe Interactions, 335–339.

Genetical organization of the *a* mating type locus

The *a* mating type locus is defined by a large region of DNA which is unique for each mating type. The region of sequence dissimilarity comprises 4.5 kb for *a1* and 8 kb for the *a2* allele (Froeliger and Leong, 1991; Bölker et al., 1992). The dissimilar sequences are flanked by identical sequences (Figure 1A). This overall structure in homologous and non-homologous sequences can also be visualized by electron microscopy of heteroduplex DNA generated from the cloned *a1* and *a2* alleles (Figure 1B): The regions specific for each allele remain single stranded indicating that there is no homology at the DNA level. By subcloning and mutational analysis two genes could be identified for each allele which are necessary for mating identity (Bölker et al., 1992): one codes for a putative pheromone (*mfa1* and *mfa2*) and the other for a presumptive pheromone receptor (*pra1* and *pra2*) (Figure 1A).

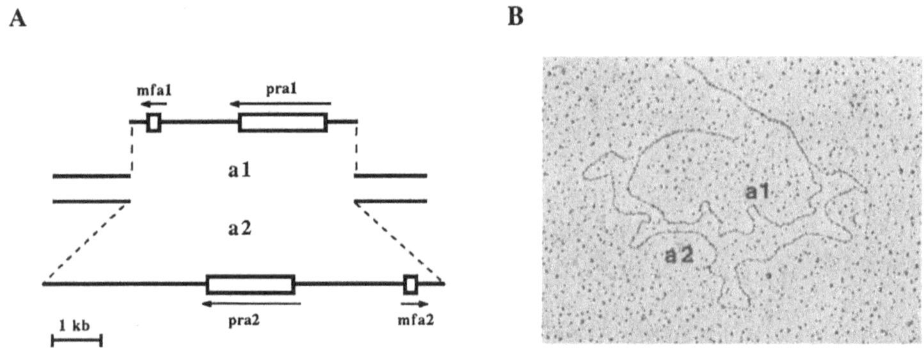

Figure 1: Genetical organization of the *a* mating type locus.
(A) Schematic drawing of the *a* locus. The regions specific for each allele are indicated. The genes identified within the *a* locus are indicated by open bars, arrows denote direction of transcription.
(B) Electron micrograph of heteroduplexes generated by melting and reannealing DNA fragments of the cloned *a1* and the *a2* alleles.

The assertion that the *a* locus encodes specific mating pheromones and their receptors rests on three lines of evidence: (i) The *mfa1* and *mfa2* genes encode short polypeptides (40 aa for *mfa1* and 38 for *mfa2*) that both contain a specific sequence motif (CAAX-box) which can serve as a signal for protein prenylation (Glomset et al., 1990). The same motif has been identified in the precursor peptides of several other fungal mating factors. (ii) Mutants in *mfa1* are mating deficient. This defect, however, can be complemented if the mating partner carries both the *a1* and *a2* allele, suggesting that the *mfa* product acts extracellularly and can be provided by the mating partner. (iii) The *pra1* and *pra2* genes display similarity with the yeast *STE3* gene, which has been identified as the receptor for the farnesylated yeast **a**-factor pheromone (Nakayama et al., 1985; Hagen et al., 1986). Sequence similarities are most prominent in the seven hydrophobic regions that comprise the potential membrane spanning domains.

The *a* locus thus contains the genes encoding the components involved in cell-cell recognition. Haploid *a1* cells secrete *a1* specific pheromone and respond to *a2* pheromone; *a2* cells secrete *a2* pheromone and respond to *a1* pheromone. The structural organization of the *a* mating type locus of *U. maydis* is reminiscent to that found in ascomycetous fungi. The functional role of the*U. maydis* *a* locus is, however, clearly distinct. In *Neurospora crassa* and *Podospora anserina* the genes contained within the mating type specific regions are thought to act as transcriptional regulators (Glass et al., 1990; Staben and Yanofsky, 1990; Debuchy and Coppin, 1992).

Role of the *a* locus for filamentous growth

The *a* locus, together with the *b* locus, controls the maintenance of filamentous growth. Diploid strains which carry two different *b* alleles but identical *a* alleles exhibit a Fuz$^-$ phenotype (Banuett, 1989). Transformation of these strains with either the pheromone or the receptor gene specific for the opposite mating type induces filamentous growth (Bölker et al., 1992). This implies that after cell fusion the pheromones serve to activate the receptors present on the same cell (autocrine response). This conclusion is further supported by the following experiment: A strain which carries a transposon insertion in the *pra1* receptor gene is completely mating deficient because it cannot be activated by the cognate pheromone. The strain still carries a functional *mfa1* gene, however, and can serve as donor for the *a1* pheromone. Co-cultivation of that strain with a diploid strain homozygous for *a2* and heterozygous for *b* (*a2/a2 b1/b2*) indeed induces filamentous growth of the diploid strain. This indicates that the ability to form filaments can be induced just by binding of the pheromone (provided that two different *b* alleles are present in the cell). It is likely therefore that also the filamentous growth of the dikaryon is maintained by an autocrine response of the pheromone receptors.

Regulation of the pheromone genes

The autocrine response proposed to be involved in control of filamentous growth implies that expression of the *a* specific genes should not be repressed in the dikaryon. This was demonstrated for the *mfa* genes by Northern analysis. We could not detect any differences in the expression of these genes in strains that were homozygous or heterozygous at either the *a* or *b* locus (not shown). Characteristic for both *mfa* promoters are short repeated sequence elements located about 300 bp upstream of their transcription start sites. It is the same 9 bp DNA consensus sequence ACAAAGGGA that is found in both promoters (Figure 2). To elucidate the role of these short repeats for the regulation of the *mfa* gene expression we have constructed a series of deletion constructs which were tested for their ability to confer mating type activity upon transformation (Figure 2). The results indicate that the *mfa1* gene retains its activity when at least 9 out of 11 of these repeats are present. The deletion of more or all repeats resulted for both *mfa* genes in a complete loss of activity. This demonstrates that these short repeats are essential components of the *mfa* gene promoters.

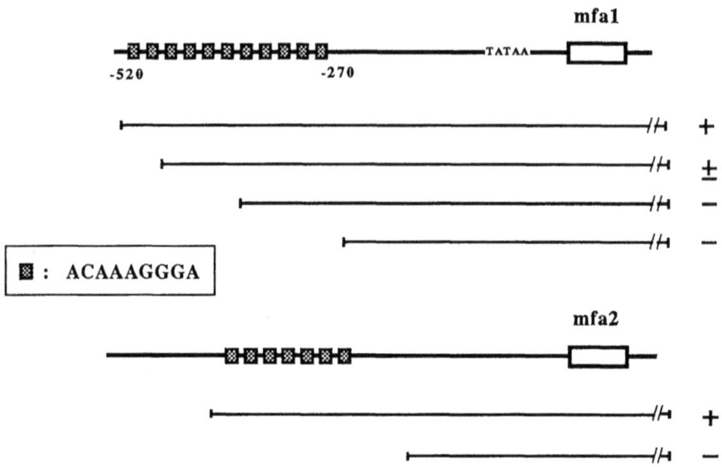

Figure 2: Deletion analysis of the *mfa* promoters.
The open bars indicate the *mfa* open reading frames, the shaded bars indicate the repeated DNA sequence elements. Successive deletion derivatives of the *mfa* promoters were created by using appropriate restriction sites for subcloning. The cloned fragments were transformed in haploid *U. maydis* strains of opposite mating type and transformants were screened for double mating activity. (+), (±) and (-) indicate the strength of double mating activity of transformants.

Signal transduction in *U. maydis*

The identification of pheromones and receptors as determinants of mating type specificity implies that pheromone stimulation of the receptors is the first step in the mating reaction. One of our future goals will be to elucidate the signal transduction pathway leading to the biological responses. Receptors containing seven transmembrane domains from other systems have been found to interact with GTP-binding proteins (reviewed in Kaziro et al., 1991). We have demonstrated for

U. maydis that the activation of either one receptor is sufficient to induce filamentous growth in strains heterozygous for *b*. This suggests that both receptors are likely to share a common signalling chain and probably interact with the same G protein. It will be interesting to determine the phenotype of mutations in the gene that encodes the α subunit of the G protein which is presumed to interact with the mating pheromone receptors. Such a gene has recently been cloned and experiments are currently performed to determine whether the disruption of this gene will interfere with mating or whether it will result in a constitutive response as in yeast.

References:

Banuett, F. (1989) 'Different *a* alleles of *Ustilago maydis* are necessary for maintenance of filamentous growth but not for meiosis', Proc. Natl. Acad. Sci. USA 86, 5878-5882.

Banuett, F (1992) '*Ustilago maydis*, the delightful blight', Trends Genet. 8, 174-180.

Bölker, M., Urban, M. and Kahmann, R. (1992) 'The *a* mating type locus of U. maydis specifies cell signaling components', Cell 68, 441-450.

Debuchy, R. and Coppin, E. (1992) 'The mating types of *Podospora anserina:* functional analysis and sequence of the fertilization domains', Mol. Gen. Genet. 233, 113-121.

Froeliger, E. H. and Leong, S. A. (1991) 'The *a* mating-tape alleles of *Ustilago maydis* are idiomorphs', Gene 100, 113-122.

Gillissen, B., Bergemann, J., Sandmann, C., Schroeer, B., Bölker, M. and Kahmann, R. (1992) 'A two-component regulatory system for self/non-self recognition in Ustilago maydis', Cell 68, 647-657.

Glass, N. L., Grotelueschen, J. and Metzenberg, R. L. (1990) '*Neurospora crassa A* mating-type region', Proc. Natl. Acad. Sci. USA 87, 4912-4916.

Glomset, J. A., Gelb, M. H. and Farnsworth, C. C. (1990) 'Prenyl proteins in eukaryotic cells: a new type of membrane anchor', Trends Biochem. Sci. 15, 139-142.

Hagen, D. C., McCaffrey, G. and Sprague, G. F. (1986) 'Evidence that the yeast *STE3* gene encodes a receptor for the peptide pheromone **a**-factor: gene sequence and implications for the structure of the presumed receptor', Proc. Natl. Acad. Sci. USA 83, 1418-1422.

Kaziro, Y., Itoh, H., Kozasa, T., Nakafuku, M. and Satoh, T. (1991) 'Structure and function of signal-transducing GTP-binding proteins', Annu. Rev. Biochem. 60, 349-400.

Nakayama, N. Miyajima, A. and Arai, K. (1985) 'Nucleotide sequence of *STE2* and *STE3*, cell type-specific sterile genes from *Saccharomyces cerevisiae*', EMBO J. 4, 2643-2648.

Staben, C. and Yanofsky, C. (1990) '*Neurospora crassa a* mating-type region', Proc. Natl. Acad. Sci. USA 87, 4917-4921.

Section 6 / *Rhizobium*-Plant Interactions: Plant Response

ROOT NODULE ORGANOGENESIS AND FORMATION OF THE PERI-BACTEROID MEMBRANE COMPARTMENT

DESH PAL S. VERMA, GUO-HUA MIAO, NA-GYONG LEE,
CHOONG-ILL CHEON AND ZONGLIE HONG
Department of Molecular Genetics and Biotechnology Center
The Ohio State University,
1060 Carmack Road,
Columbus, Ohio 43210-1002
USA

ABSTRACT. Early events in *Rhizobium*-plant interaction leading to nodule organogenesis require initiation of meristematic activity in root cortex. This step involves control of cell cycle genes in the host. We have demonstrated that two *cdc2* homologs of soybean are differentially regulated in root and shoot tissues and one of them, *cdc2-S5*, responds to bacterial infection following spot inoculation of root. In an attempt to understand the process of endocytosis and synthesis of peribacteroid membrane (PBM), we have isolated soybean and *Vigna* homologs of *YPT1* and *ras* genes known to be involved in vesicular transport of proteins in yeast. We believe that functional complementation of yeast vacuolar membrane protein transport mutants with nodule cDNA expression library may allow isolation of homologs of these genes which apparently play important roles in the synthesis of peribacteroid membrane in root nodules.

1. INTRODUCTION

Interaction of *Rhizobium* with its host triggers a cascade of events leading to induction of various nodulin genes that are essential for nodule organogenesis, development of infection, release of rhizobia from the infection thread, and establishment of a symbiotic state (Verma and Delauney, 1988). Although the root nodule structure can develop without bacterial infection (Truchet *et al.*, 1991), infection and organogenesis normally proceed together. Nodule organogenesis occurs under the control of specific nodulation (Nod) signal(s) produced by rhizobia (see Verma, 1992; Fisher and Long, 1992). Triggering of nodule ontogeny requires new cell division in the root cortical cells, and Nod factors appear to control initiation and/or regulation of plant cell division. Understanding the signal transduction pathways that control expression of host genes required for the initiation of cell division may not only reveal how the nodule organogenesis is regulated, but also how other plant organs are initiated in specific locations.

Formation of the PBM compartment is essential for endosymbiosis; failure to synthesize the PBM may evoke pathogenic responses in the host (Werner *et al.*, 1985; Djordjevic *et al.*, 1987). How bacteria enter the plant cell and how the plant forms a new subcellular compartment to house the bacteria are unknown. Recent studies on early interaction between host and *Rhizobium* have allowed the dissection of some of the events and the dissociation of the nodule organogenesis program from endocytosis of bacteria. The study of the molecular mechanism of *Rhizobium* endocytosis is thus vital to understanding of communication and interaction between the two organisms. Isolation of host genes involved in membrane biosynthesis and vesicle trafficking may provide invaluble insight into the endocytotic route(s) of rhizobia and the control of inter-membrane flow in the infected cell leading to the formation of PBM. Root nodules provide an ideal system for membrane biogenesis studies

E. W. Nester and D. P. S. Verma (eds.),
Advances in Molecular Genetics of Plant-Microbe Interactions, 343–352.

in plants because of a rapid membrane proliferation that occurs during endocytosis of rhizobia.

2. INITIATION OF CORTICAL MERISTEM AND NODULE ORGANOGENESIS

A precise exchange of molecular signals between rhizobia and host plant leads to the initiation of meristematic activity in the root cortex (see Verma, 1992a). This activity begins before the infection thread is formed and can occur in the absence of root hair curling (Libbenga and Harkes, 1973; Dudley *et al.*, 1987; Guinel and LaRue, 1991). These observations suggest that diffusible signal compounds produced by rhizobia are responsible for initiating the cascade of events in the host that lead to nodule organogenesis. These signal compounds are oligosaccharide molecules which in the case of *Rhizobium meliloti* have been identified (NodRm-1) as a sulfated and acylated oligo-*N*-acetylglucosamine (Lerouge *et al.*, 1990). Application of NodRm-1 to alfalfa roots at a very low concentration (10^{-11}M) initiates cortical cell division leading to the formation of nodule-like structures (Truchet *et al.*, 1991). These results suggest that Nod factors act on the host directly or by altering phytohormone levels which control cell division (Verma, 1992a). How the cell division is triggered by the nod signal molecules is not known, however. These signals may possibly interact with the cell cycle control mechanism so that resting cortical cells enter into mitosis. Some cells of the legume root cortex have a 4C DNA content and are arrested in the G_2 phase of the cell cycle (Gresshoff and Mohapatra 1981); these cells may act as foci for nodule meristem initiation. The cell cycle arrest in these cells has been suggested to be due to the translocation of 7-methyl nicotinate, a compound present in cotyledons of legumes (Evans and Van't Hof, 1973). Release of G2 arrest by nod signal(s) would provide a mechanism by which the host could respond rapidly to incoming signals from rhizobia. This would allow initiation of cell division without the need for DNA synthesis (see Verma, 1992a).

The basic mechanism for cell cycle control appears to be universal (Nurse, 1990). The signals triggering the cell division machinery, however, not only differ in different eukaryotes but also withen the different parts of the same orgnism (see Verma, 1992a). The product of the yeast *cdc25* gene encoding a tyrosine phosphatase that dephosphorylates $p34^{cdc2}$ protein kinase controls the activity of the maturation promoting factor (MPF), a complex of $p34^{cdc2}$ and cyclin B (Lewin, 1990), and this step is necessary for cells to enter into mitosis. Genetic studies have demonstrated that the *cdc25* gene product is a key regulator for cell division (Edgar and O'Farrell, 1989). Homologs of yeast *cdc25* have been isolated from animals (Sadhu *et al*, 1990), but no plant homolog of this gene has yet been identified. Because both external and internal signals can control cell division in plants, some of these signals are likely to be mediated by cdc25/RAS signal transduction pathways (see Malone, 1990). Apparently, these pathways converge on the same downstream gene(s) to initiate cell division.

Homologs of yeast *cdc2* gene have been isolated from legumes including alfalfa (Hirt *et al.*, 1991), pea (Feiler and Jacob, 1990), and soybean (Miao *et al.*, 1992a). We have identified two functional homologs of *cdc2* in soybean which appear to be differentially expressed in root and shoot meristems. Two degenerate oligonucleotide primers corresponding to the conserved sequence motifs were synthesized and used for PCR with the cDNAs reverse transcribed from soybean nodule total RNA. Using the amplified fragment as a probe, two positive clones (pCDC2-S5 and pCDC2-S6) were isolated by screening a soybean nodule cDNA library constructed in λZap II (Delauney and Verma, 1990). The two sequences show 90% identity with each other in the coding region. The both 5' and 3' non-coding regions of these clones are unique, however, suggesting that they represent two distinct genes encoding $p34^{cdc2}$ protein kinases. This was confirmed by Southern blot analysis of soybean genomic DNA probed with the *cdc2-S5*-and *cdc2-S6*-specific sequences.

Moreover, two copies of each of these genes in found in soybean genome, indicating that *cdc2-S5* and *cdc2-S6* are unlinked and non-allelic genes. In *Arabidopsis thaliana,* only a single gene encodes p34[cdc2] kinase (Ferreira *et al.*, 1991). The functionality of the soybean *cdc2-S5* and *cdc2-S6* was determined by the ability of these genes to rescue cdc28 mutation in *S. cerevisiae.* Transformants containing either soybean *cdc2-S5* or *cdc2-S6* on an expression vector grew at non-permissive temperature confirming that both sequences are *bona fide* soybean functional homologs of the yeast *cdc2/CDC28* genes. Growth characteristics of yeast cdc28 mutant harbouring either*cdc2-S5* or *cdc2-S6* differed (Figure 1), however, suggesting different efficiencies of these genes in rescuing the cdc28 mutation (Miao *et al.*, 1992a). Yeast cdc28 mutant cells containing soybean *cdc2-S5* gene grew faster than that harbouring *cdc2-S6* at restrictive temperature. The activity of cdc2/CDC28 protein kinase depends on the interaction with specific cyclins and other components of MPF (Lewin, 1990); the differences in abilities of the two soybean p34[cdc2] sequences to repair the cdc28 mutation imply different roles or regulation of these two genes in cell cycle control.

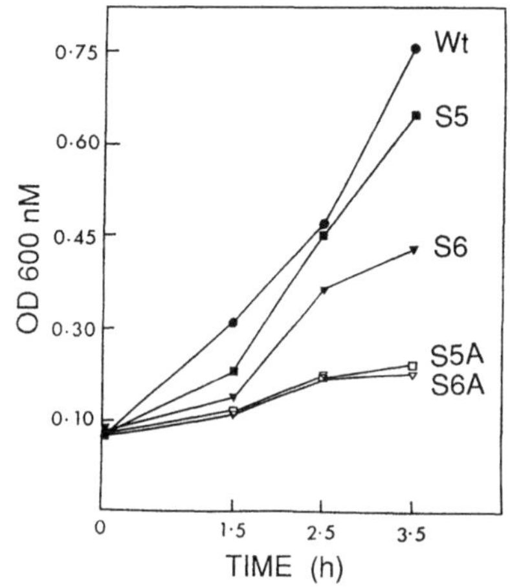

Figure 1. Functional complementation of *S. cerevisiae* cdc28 mutation by two soybean *cdc2/ CDC28* homologs and efficiency of complementation. Growth kinetics of cdc28 mutant cells harbouring various plasmid constructs grown at restrictive temperature. Wt, wildtype yeast cells containing pEMBLyex4 vector; S5, vector containing soybean *cdc2-S5*; S5A, same as S5 but in antisense orientation; S6, vector containing *cdc2-S6*; S6A, same as S6 but in antisense orientation (see Miao *et al.*, 1992a).

We determined the relative expression level of the two soybean *cdc2* transcripts in different tissues using RNA-PCR. Specific fragments corresponding to *cdc2-S5* and *cdc2-S6* were co-amplified (Miao *et al.*, 1992a). The transcripts of the two *cdc2* genes were higher in root and shoot meristem tissues and lower in differentiated leaf tissue (Figure 2). Thus, expression of the two *cdc2/CDC28* homologs of soybean appears to correlate with cell proliferation. The relative expression level of *cdc2-S6* was higher in shoot-derived organs, while the transcript of *cdc2-S5* was consistently higher in the root derived organs (Figure2). The possible role(s) of *cdc2-S5* and *cdc2-S6* in initiation of meristematic activity in cortex were studied by monitoring expression of each transcript following spot-inoculation with *Rhizobium* or auxin and cytokinin on soybean roots. Transcripts of *cdc2-S5* increased 3-4 fold after inoculation with *Bradyrhizobium japonicum* compared to mock inoculation (Figure 2). In contrast, the expression of *cdc2-S6* was not significantly affected. The enhanced *cdc2-S5* expression is *Rhizobium*-specific, as mock inoculation with *E. coli* did not increase the level of *cdc2-S5* transcript. The induction of *cdc2-S5* gene correlates with the timing of initiation of cortical cell division following inoculation with *Rhizobium* (Turgeon and Bauer, 1982). These findings suggest that cell division in host root cortex is induced by *Rhizobium via* activating genes involved in cell cycle control. Spot-inoculation with physiological concentration of α–naphthaleneacetic acid (NAA) on the root-elongation zone also enhanced expression of *cdc2-S5* and *cdc2-S6* genes. This demonstrated that up-regulation of plant *cdc2* genes in cells entering mitosis is possibly mediated by phytohormones. The plant growth regulator 2,4-dichlorophenoxyacetic acid also induces expression of alfalfa *cdc2* gene in cell suspension culture (Hirt *et al.*, 1991). A cyclin B cDNA has also been cloned from soybean (Hata *et al.*, 1991; our unpublished data), and this gene has been shown to be expressed at a very high level in early nodule development. Availability of plant cell cycle genes will help to elucidate the cell cycle control mechanism in legumes and reveal signal transduction pathways leading to cell division in the root cortex (Verma, 1992a).

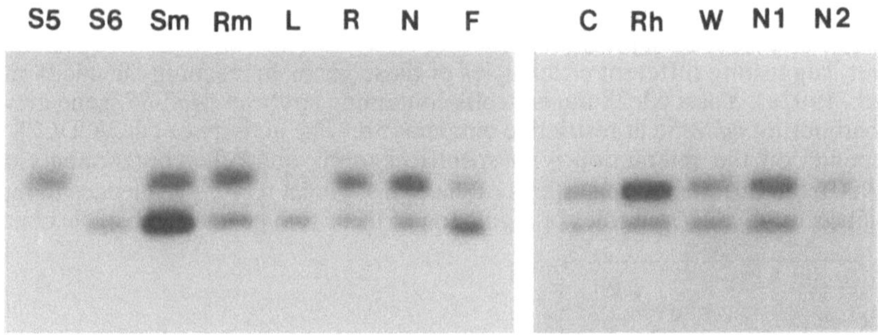

Figure 2. Differential expression of two soybean *cdc2* genes in different tissues and induction of these genes with *Rhizobium* and phytohormones. Soybean *cdc2-S5* and *cdc2-S6* sequences were amplified from total RNA from root using specific primers for S5 and S6 separately. RNA from shoot meristem (Sm), root meristem (Rm), mature leaf (L), total root (R), nodule (N) and flower (F) was used for RNA-PCR with mixed primers for both S5 and S6 sequences. RNA from three-day-old roots mock spot-inoculated with *E. coli* (C); *Bradyrhizobium* (Rh); water only (W); 0.5µg/ml of NAA (N1) and 2µg/ml NAA (N2). In lanes S5 and S6, primers for S5 and S6 were used, respectively, while samples in other lanes had mixture of both S5 and S6 primers (see Miao *et al.*, 1992a for details).

3. ENDOCYTOSIS OF *RHIZOBIUM* AND BIOGENESIS OF THE PERIBACTEROID MEMBRANE

The formation of a subcellular compartment to house the bacteria inside the infected cell is essential for symbiosis. Failure to form this membrane compartment or its disintegration renders the association pathogenic (Werner *et al.*, 1985). Many rhizobial mutants are known, including exo- and lipo-polysaccharide mutants (Gray and Rolfe, 1990; Noel, 1992), where bacterial entry does not take place but nodule morphogenesis proceeds to different stages; however, the specific interaction of these mutants with the host is not known. An interaction at the stage of endocytosis of bacteria in the host cell was observed in nodules formed by a *Tn*5 mutant of *B. japonicum* (Morrison and Verma, 1987). Infection by this strain arrests release of bacteria from the infection thread but allows differentiation to proceed, giving rise to nodules that appear normal but are largely devoid of bacteroids. Interaction between the plant membrane and bacteria is also suggested by the attachment of the PBM to components of the bacterial surface (Robertson *et al.*, 1985; Verma and Fortin, 1989; see also Brewin, 1991). In addition, an antibody prepared against the PBM proteins also recognizes a protein in purified bacteroids (Fortin *et al.*, 1985).

3.1 PBM Biogenesis

Early interaction of rhizobia with the host plant triggers proliferation of the membrane system that generates components of the PBM. In soybean root nodules, almost 30 times more membrane is synthesized in the form of PBM than plasma membrane (Verma *et al.*, 1978). Understanding biogenesis and the nature of this membrane may give insight into the basis of compatibility between these two organisms. PBM synthesis requires redirecting of some of the plasma membrane proteins and all of the PBM nodulins to this newly formed membrane compartment (Figure 3). All the signals necessary for the induction of the PBM nodulins are apparently transduced prior to the release of bacteria from the infection thread (Verma *et al.*, 1988), and these signals appear to vary for each PBM nodulin (Morrison and Verma, 1987). The PBM initially arises from the plasma membrane during endocytosis (Verma *et al.*, 1978), but it undergoes significant alterations and appears to be a mosaic membrane with properties common to both plasma and vacuolar membranes (see Verma, 1992b; Miao *et al.*, 1992b). Placing rhizobia in a subcellular compartment with lysosomal properties may allow the host cell to exert control over this invasion in the event nitrogen fixation does not occur and the bacteria need to be destroyed. In fact, in nodules formed by a *Tn5*-induced mutant of *Bradyrhizobium.* disintegration of bacteroids seem to occur leaving vesicles filled with fibrous material (Roth and Stacey, 1989). As in other endosymbiotic systems, in an effective root nodule the lysosomal function of the vacuole must be repressed (Oates and Touster, 1980, Roth and Stacey, 1989).

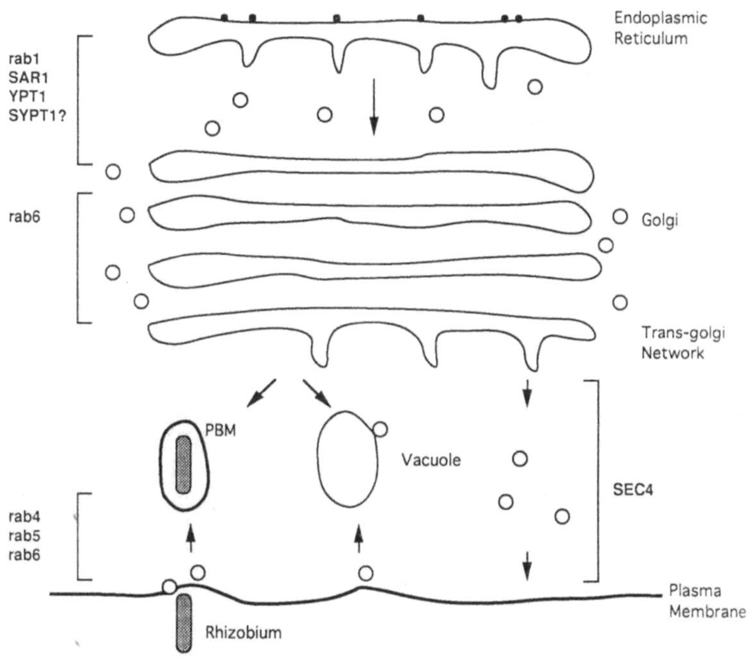

Figure 3. Postulated route of vesicular transport of endomembrane proteins in the infected cell of root nodule and the possible involvement of various gene products based on animal and yeast studies. SYPT1, soybean YPT1 homolog (C.-I Cheon and Verma, unpublished).

In an attempt to study biogenesis of PBM in more detail and to understand the interaction of specific membrane receptors with *Rhizobium*, we have initiated experiments on functional complementation of certain yeast mutants (Herman and Emr, 1990) that are blocked in membrane biosynthesis and intracellular vesicular transport. The yeast SEC4 protein has been suggested to be responsible for unidirectional transport of vesicles from Golgi to plasma membrane (Salminen and Novick, 1987). This protein binds GTP and has GTPase activity. Similarly, YPT1 controls an early step in vesicular transport between ER and Golgi (Segev *et al.*, 1988). Both Ca^{2+} and GTP are essential for vesicular transport (Hall, 1990). Mammalian homologs (rab1-7) of YPT1 and SEC4 have been isolated and shown to be located in the early and late endosomes, the intermediate compartment between ER and Golgi, suggesting that these proteins participate in the endocytosis and exocytosis pathways; rab5 was shown to be specifically involved in endosome fusion (Gorvel *et al.*, 1991). Thus, these small GTP binding proteins appear to be conserved and play some common roles in intracellular membrane trafficking in eukaryotes. Recently, a plant cDNA encoding a small GTP-binding protein (Rha1) which shows high homology with the members of *ras* family and 60% homology with mammalian rab5 protein has been isolated (Anuntalabhochai *et al.*, 1991). We have isolated, from soybean and *Vigna*, cDNAs encodingYPT1 and ras homologs, proteins involved in vesicular transport (see Figure 3). The soybean and *Vigna* homologs of *YPT1* and *ras* genes contained the conserved domains characteristic of GTP-binding proteins (Figure 4). Further characterization of the legume homologs of yeast genes involved in membrane biosynthesis and trafficking may allow delineation of the endocytotic route of rhizobia and the role of various pathways that contribute to the formation of PBM compartment able to support a foreign organism endosymbiotically.

```
                       G1                    G2                   G3
YPT1     MNSEYDYLFKLLLIGNSGVGKSCLLLRFSDDTYTNDYISTIGVDFKIKTVELDGKTVKLQIWDTAGQERFRT   72
SOYPT1   -SN-------------D-S-------------A--S-VDS-----------R----E----------------   72
V-RAS    -SLRRRT-I-VIVL-DT----TS-MNQYVHKKFSQQ-KA---A--VT-ELQI-DRL-T------------QS   72
BRL-RAS        M-I-VIIL-D-----TS-MNQYVNKKFS-Q-KA---A--LT-E-MV-DRL-TM-----------QS   66

                                                          G4
YPT1     ITSSYYRGSHGIIIVYDVTDQESFNGVKMW....LQEIDRYATSTVLKLLVGNKCDLK..DKRVVEYDVAKE 138
SOYPT1   --------A----------EM----N--Q-....-N------ND--C--------S--V.DN-.--DSLT--A 138
V-RAS    LGVAF---ADCCVLA---NVMK--DTLDNWHEEF-KQANPPDPRSFPFI-L---I-IDGGNS---SEKK--D 144
BRL-RAS  LGVAF---ADCCVL-F---APNT-KTLDSWRDEF-IQASPRDPENFPFVVL---I--E..N.-Q-ATKR-QA 135

                 G5
YPT1     F.ADANKMPFLETSALDSTNVEDAFLTMARQIKQSMSQQNLNETTQKKEDKGNVNLKGQSLTNTGGGCC*   206
SOYPT1   -.--ELGI-------K--I---Q------AE--KK-GS----T--G-SAE--S-QM---PIPQK-SN--*   206
V-RAS    WC-SKGNI-YF----KEDF--DA---CI-KAALANEHE-DIYFQGIPEAAVPENEQRSEHASRSDMEIP>   232
BRL-RAS  WCYSK-NI-YF----KEAI---Q--QTI--NALKQETEVE-YFNEFPEPI-LDK-ERAKASAESCSC*     204
```

Figure 4. Sequence comparison of soybean YPT1 and *Vigna ras* homologs (C.-I. Cheon, N.-G. Lee and D. P. S. Verma unpublished data) with YPT1 and BRL-*ras*.The amino acids sequences are deduced from the nucleotide sequences of cDNAs. The hyphen (-) indicates an identical amino acid and gaps (.) are generated to align the sequences. In boxes are the regions conserved among small GPT-binding proteins (Bourne *et al.*, 1991). SOYPT1: soybean homolog of YPT1, V-RAS: *Vigna* homolog of *ras*, BRL-RAS: rat homolog of *ras* (see Bucci *et al.*,1988).

3.2 Targeting of Nodulins to PBM

The movement of proteins from ER to Golgi and then to specific subcellular compartments requires specific topogenic information (see Chrispeels, 1991). Individual PBM nodulins appear to be targeted to the PBM by different mechanisms. The soybean PBM contains at least six major nodulins. Nodulin-24 has an amino terminus signal sequence that is cleaved co-translationally (Katinakis and Verma, 1985). The cleaved peptide is further processed post-translationally, which increases its molecular weight by 12kD; the nature of this modification is not known. *In vitro* co-translational processing and membrane protection experiments suggested that no part of nodulin-24 is exposed outside the vesicle and that it has no transmembrane regions; following signal removal, this peptide is free in the lumen of the ER (C.-I. Cheon and D. P. S. Verma, unpublished data). Attachment of this nodulin to the membrane occurs during its passage through the Golgi. Nodulin-24 forms an amphipathic α-helix that appears to face the bacteroid side of the PBM. In contrast, nodulin-

26 contains non-cleavable internal signal sequences. Thus these two PBM nodulins differ in their mode of targeting to the membrane. Nodulin-26 is glycosylated co-translationally as evidenced by ConA binding, and the glycosylation site faces the bacteroid. The orientation of the carboxyl and amino termini was determined by chemical cleavage mapping at cysteine residues of untreated and trypsinized *in vitro-* synthesized peptide (see Miao *et al.*, 1992b).

4. PHYSIOLOGICAL "INTERNALIZATION" OF BACTEROIDS

The space between the bacteria and PBM, known as the peribacteroid space, must be equilibrated with certain metabolites, including dicarboxylic acids that are used as a carbon source by bacteroids. Elimination of the concentration gradient between host and rhizobia physiologically internalizes bacteria and brings them into the closest possible association with the host cell. This step is apparently accomplished by opening specific channels in the PBM. Nodulin-26 (Fortin *et al.*, 1987) appears to be such a channel induced during nodule development. This nodulin was found to have homology with a group of proteins conserved from bacteria to mammals (Pao *et al.*, 1991; Verma, 1991). Three members of this group have experimentally been shown to form channels. The nature of the compounds transported through the nodulin-26 channel is although not defined, it may act as a water channel based on its homology with CHIP-28 (Preston *et al.,* 1992). Water activity may regulate osmoticum and control transport of metabolites such as dicarboxylic acids through the PBM in a manner that occurs in vacuole (Luttge *et al.*, 1982). Nodulin-26 is phosphorylated by a protein kinase located in the PBM (Miao *et al.*, 1992b) which suggests that transport through this channel may be an active process.

5. PERSPECTIVE

Endocytosis of *Rhizobium* in plant cell brings these two organism in a closest possible intracellular association. This requires development of a new subcellular comartment with many unique properties. Several nodulins are incorporated in the membrane enclosing the bacteroid, PBM. In an infected cell, the synthesis of PBM may amplify the basic membrane biogenesis machinery compared to that in a normal root cell (Verma *et al.*, 1978; Verma and Fortin, 1989). Because some of the steps in membrane biogenesis may be conserved between yeast and plants and yeast mutants blocked in vesicular transport are now available (Herman and Emr, 1990), functional complementation of these mutants may allow isolation of plant gene homologs and help to dissect PBM biosynthesis pathway. Finally, understanding subcellular compartmentalization, metabolite flow, and adaptation of various biosynthetic pathways may reveal how plant endosymbiotically houses a foreign organism and creats unique environment to sustain reduction and assimilation of dinitrogen.

6. ACKNOWLEDGMENTS

Work in the author's laboratory was supported by grants (DCB-8904101 and DCB-8819399) from the National Science Foundation.

7. REFERENCES

Anuntalabhochai, S., Terryn N, Van Montagu, M., Inze, D. (1991) Molecular characterization of an *Arabidopsis thaliana* cDNA encoding a small GTP-binding protein, Rha1. Plant J. **1**, 167-174.

Bourne, H. R., Sanders, D. A., McCormick, F. (1991) The GTPase superfamily: conserved structure and molecular mechanism. Nature **349**, 117-127.

Brewin, N. J. (1991) Development of the legume root nodule. Ann. Rev. Cell Biol. **7,** 191-226.

Bucci, C., Frunzio, R., Chiariotti, L., Brown, A. L., Rechler, M. M., Bruni, C. B. (1988) A new member of the *ras* gene superfamily identified in a rat liver cell line. Nucleic Acids Res. **16**, 9979-9993.

Chrispeels, M. J. (1991) Sorting of proteins in the secretory system. Ann. Rev. Plant Physiol. Plant Mol. Biol. **42**, 21-53.

Delauney, A.J. and Verma, D.P.S. (1990) Isolation of soybean Δ^1-pyrroline-5-carboxylate reductase cDNAs by functional complementation in *Escherichia coli*. Mol. Gen. Genet. **221**, 299-305.

Djordjevic, M. A., Gabriel, D. W. and Rolfe, B. G. (1987) *Rhizobium* - the refined parasite of legumes. Ann. Rev. Phytopathol. **25**, 145-168.

Dudley, M. E., Jacobs, T. W. and Long, S. R. (1987) Microscopic studies of cell divisions induced in alfalfa roots by *Rhizobium meliloti*. Planta **171**, 289-301.

Edgar, B. A. and O'Farrell, P. H. (1989) Genetic control of cell division patterns in the Drosophila embryo. Cell **57**, 177-187.

Evans, L. S., and Van't Hof, J. (1973) Cell arrest in G2 in root meristem: a control factor from the cotyledons. Exp. Cell Res. **82**, 471-473.

Feiler, H. S. and Jacobs, T. W. (1990) Cell division in higher plants: A *cdc2* gene, its 34 kD product, and histone H1 kinase activity. Proc. Nat. Acad. Sci. USA. **87**, 5397-5401.

Ferreira, P.C.G., Hemerly, A.S., Villarroel, R., Van Montagu, M.and Inzé, D. (1991) The *Arabidopsis* functional homolog of the $p34^{cdc2}$ protein kinase. Plant Cell **3**, 531-540.

Fisher, R. F. and Long, S. R. (1992) *Rhizobium*-plant signal exchange. Nature **357**, 655-660.

Fortin M. G., Morrison, N. A. and Verma, D. P. S. (1987) Nodulin-26, a peribacteroid membrane nodulin, is expressed independently of the development of peribacteroid compartment. Nucl. Acids. Res. **15**, 813-824.

Fortin, M. G., Zelechowska, M., and Verma, D. P. S. (1985) Specific targeting of membrane nodulins to the bacteroid-enclosing compartment in soybean nodules. EMBO J. **4**, 3041-3046.

Gorvel, J.-P., Chavrier, P., Zerial, M., Gruenberg, J. (1991) rab5 controls early endosome fusion *in vitro*. Cell **64**, 915-925.

Gray, J. X. anf Rolfe, B. G. (1990) Exopolysaccharide production in *Rhizobium* and its role in invasion. Mol. Microbiol. 4, 1425-1431.

Gresshoff, P. M. and Mohapatra, S. S. (1981) Legume cell and tissue culture. In Tissue culture of economically important crop plants, A. N. Rao, ed. (Singapore: Singapore Univ. Press), pp. 11-24.

Guinel, F. C. and LaRue, T. A. (1991) Light microscopy study of nodule initiation in *Pisum sativum* L. cv Sparkle and its low nodulating mutant E2 (*sym*5). Plant Physiol. **97**, 1206-1211.

Hall, A. (1990) The cellular functions of small GTP-binding proteins. Science **249**, 635-640.

Hata, S., Kouchi, H., Suzuka, I. and Ishii, T. (1991) Isolation and characterization of cDNA clones for plant cyclins. EMBO J. **10**, 2681-2688.

Herman, P.K.and Emr, S.D. (1990) Characterization of VPS34, a gene required for vacuolar protein sorting and vacuole segregation in *Saccharomyces cerevisiae*. Mol. Cell. Biol.**10**, 6742-6754

Hirt, H., Pay, A., Gyorgyey, J., Baco, L., Nemeth, K., Bogre, L., Schweyen, R. J., Heberle-Bors, E. and Dudits, D. (1991) Complementation of a yeast cell cycle mutant by an alfalfa cDNA encoding a protein kinase homologous to $p34^{cdc2}$. Proc. Nat. Acad. Sci. USA **88**, 1636-1640.

Katinakis, P. and Verma, D.P.S. (1985) Nodulin-24 gene of soybean codes for a peptide of the peribacteroid membrane and was generated by tandem duplication of an insertion element. Proc. Natl. Acad. Sci. USA **82**, 4157-4161 .

Lerouge, P., Roche, P., Faucher, C., Maillet, F., Truchet, G., Prome, J. C. and Denarie, J. (1990) Symbiotic host-specificity of *Rhizobium meliloti* is determined by a sulphated and acylated glucosamine oligosaccharide. Nature **344**, 781-784.
Lewin, B. (1990) Driving the cell cycle: M phase kinase, its partners, and substrates. Cell **61**: 743-752.
Libbenga, K.R.and Harkes P. A. A. (1973) Initial proliferation of cortical cells in the formation of root nodules in *Pisum sativum* L. Planta **114,** 17-28.
Lutge, U., Smith, J. A. C., Marigo, G. (1982) Membrane transport, osmoregulation and control of CAM. In Crassulacean Acid Metabolism (eds) I. P. Ting and M. Gibbs. American Society of Plant Physiol., Rockville, pp 69-91.
Malone, R. E. (1990) Dual regulation of meiosis in yeast. Cell **61**, 375-378.
Miao, G.-H, Hong, Z and Verma, D. P. S. (1992a) Two fuctional soybean genes encoding $p34^{cdc2}$ protein kinases are regulated by different plant developmental pathways. (Submitted)
Miao, G.-H, Hong, Z and Verma, D. P. S. (1992b) Topology and phosphorylation of soybean nodulin-26, an intrinsic protein of the peribacteroird membrane. J. Cell Biol. **118**, 481-490.
Morrison, N. and Verma, D.P.S. (1987) A block in the endocytosis of *Rhizobium* allows cellular differentiation in nodules but affects the expression of some peribacteroid membrane nodulins. Plant Mol. Biol. **9**, 185-196.
Noel, D. K. (1992) Rhizobial polysaccharides required in symbiosis with legumes. In Plant-Microbe communications (ed.) D. P. S. Verma, CRC Press, Boca Raton, Fl. pp. 340-357.
Nurse, P. (1990) Universal control mechanism regulating onset of M-phase. Nature **344**, 503-508.
Oates, P. J., Touster, O. (1980) In vitro fusion of *Acanthamoeba* phagolysosomes III. Evidence that cyclic nucleotides and vacuole subpopulations, respectively control the rate and extent of vacuole fusion in *Acanthamoeba* homogenates. J. Cell Biol. **85**, 804-819.
Pao, G.M., Wu, L.-F., Johnson, K. D., Hofte, H., Chrispeels, M. J. Sweet, G., Sandal, N. N., and Saier, M. H. (1991) Evolution of the MIP family of integral membrane transport proteins. Mol. Microbiol. **5**, 33-37
Preston GM, Carrol TP, Guggino WB, Agre P (1992) Appearance of water channels in *Xenopus* oocytes expressing red cell CHIP28 protein.. Science 256:385-387.
Robertson, J. G., Wells, B., Brewin, N. H., Wood, E., Knight, C. D. and Downie, J. A. (1985) The legume-*Rhizobium* symbiosis: A cell surface interaction. J. Cell Sci. Suppl. **2**, 317-331.
Roth, L. E. and Stacey, G. (1989) Cytoplasmic membrane systems involved in bacterium release in to soybean nodule cells as studied with two *Bradyrhizobium* mutant strains. Euro. J. Cell Biol. **49**, 24-32.
Sadhu, K., S. I. Reed, H. Richardson and P. Russell (1990) Human homolog of fission yeast cdc25 mitotic inducer is predominantly expressed in G2. Proc. Nat. Acad. Sci. USA **87**: 5139-5143.
Salminen, A. and Novick, P. J. (1987) A ras-like protein is required for post Golgi event in yeast secretion. Cell **49**, 527-538.
Segev, N., Mulholland, J.and Botstein, D. (1988) The yeast GPT-binding YPT1 protein and a mammalian counterpart are associated with the secretion machinery. Cell **52**, 915-924.
Truchet, G., Roche, P. Lerouge, P., Vasse, J., Camut, S., de Billy, F., Prome, J.-C. and Dénarié, J. (1991) Sulfated lipo-oligosaccharide signals of *Rhizobium meliloti* elicit root nodule organogenesis in alfalfa. Nature **351**, 670-673.
Turgeon, B.G.and Bauer, W.D. (1982) Early events in the infection of soybean by *Rhizobium japonicum.* Time course and cytology of the initial infection process. Can. J. Bot. **60**, 152-161.

Verma, D.P.S. (1991) Nodulin-26: A channel protein conserved from bacteria to mammals. In Transport and Receptor Proteins of Plant Membranes, D.T. Clarkson and D.T. Cooke, eds New York: Plenum Press p 113-117.
Verma, D.P.S. (1992a) Signals in root nodule organogenesis and endocytosis of *Rhizobium*. Plant Cell **4**: 373-382.
Verma, D.P.S. (1992b) Nodulin-26: A channel protein conserved from bacteria to mammals. In Transport and Receptor Proteins of Plant Membranes, D.T. Clarkson and D.T. Cooke, eds (New York: Plenum Press) (in press).
Verma, D.P.S. and Fortin, M.G. (1989) Nodule development and formation of the endosymbiotic compartment. In Cell Culture and Somatic Cell Genetics of Plants Vol VII: The Molecular Biology of Nuclear Genes. (Eds.) J. Schell and I.K. Vasil, Academic Press, New York, pp. 329-353.
Verma, D. P. S., Delauney, A. J., Guida, M., Hirel, B., Schafer, R.and Koh, S. (1988) Control of expression of nodulin genes. In Molecular Genetics of Plant-Microbe Interactions 1988. R. Palacios and D. P. S. Verma, eds (St Paul. MN: APS Press) pp. 315-320.
Verma, D.P.S., and Delauney, A. J. (1988) Root nodule symbiosis: Nodulins and Nodulin Genes. In Plant Gene Research : Temporal and Spatial Regulation of Plant Genes, D.P.S. Verma and R. Goldberg,eds. (New York: Springer Verlag) pp. 169-199.
Verma, D.P.S., Kazazian, V., Zogbi, V.and Bal, A.K. (1978) Isolation and characterization of the membrane envelope enclosing the bacteroids in soybean root nodules. J. Cell Biol. **78**, 919-936.
Werner, D., Mellor, R. B., Hahn, M. G. and Grisebach, H. (1985) Soybean root response to symbiotic infection glyceollin I accumulation in an ineffective type of soybean nodules with an early loss of the peribacteroid membrane. Z. Naturforsch **40c**, 179-181.

EARLY EVENTS IN ALFALFA NODULE DEVELOPMENT

A.M. HIRSCH, H.I. MCKHANN, M. LÖBLER, Y. FANG, B. NINER, S. ASAD, K. WYCOFF, P.T. ASMANN*, AND M. JACOBS*
Department of Biology
405 Hilgard Avenue
University of California
Los Angeles, CA 90024-1606 USA

*Biology Department
Swarthmore College
Swarthmore, PA 19081

ABSTRACT. The development of a nitrogen-fixing root nodule is a complex, multi-staged process. A large number of genes, from both the plant and bacteria, are specifically expressed during this interaction. We study the earliest stages of the alfalfa-*Rhizobium meliloti* interaction and have used a number of plant cDNA clones as molecular markers to examine nodule formation and nodule tissue differentiation. Some plant gene families, e.g., chalcone synthase, have members that are specifically induced after inoculation with *R. meliloti*, while other genes, e.g., the MsENOD12 gene, are expressed exclusively in nodules. We have determined that mRNAs of MsENOD2, another alfalfa early nodulin, are found in a number of different bacteria-free nodules, while MsENOD12 transcripts have been detected so far only in nodules having infection threads which elongate into the nodule interior. When phytohormones or chemicals that disturb the balance of endogenous plant hormones are applied to alfalfa roots, the ENOD2 gene is expressed. We have also determined that *R. meliloti* produces a molecule that competes with the auxin transport inhibitor *N*-1-(naphthyl)phthalamic acid (NPA) for its binding site. This molecule is synthesized by Nod$^-$ as well as by wildtype *R. meliloti*, and appears to require luteolin for maximum induction.

1. **Introduction.**

The earliest stages of alfalfa (*Medicago sativa* L.) nodule development -- recognition of the rhizobial cell, root hair deformation, penetration of the infection thread, and stimulation of cell divisions in differentiated root cortical cells -- require the expression of genes from both *Rhizobium meliloti* and the plant. Flavonoids present in alfalfa seed wash and root exudate induce the expression of *R. meliloti nod* genes (Maxwell et al., 1989; Hartwig et al., 1990), the product of which is NodRm-IV (S), a sulfated glycolipid (Lerouge et al., 1990). When NodRm-IV (S) is added to alfalfa roots at concentrations ranging from 10^{-7} to 10^{-11} M, responses varying from root hair deformation to cell divisions in the root cortex take place (Truchet et al., 1991).

E. W. Nester and D. P. S. Verma (eds.),
Advances in Molecular Genetics of Plant-Microbe Interactions, 353–364.

1.1 AUXIN TRANSPORT INHIBITORS PARTIALLY MIMIC RHIZOBIUM *NOD* GENES.

Some of the early stages of nodule development can be initiated by compounds that function as auxin transport inhibitors (ATIs). *N*-1-(naphthyl)phthalamic acid (NPA) and 2,3,5-triiodobenzoic acid (TIBA) induce pseudonodule formation on roots of clover, alfalfa and Afghanistan pea (Torrey, 1986; Hirsch et al., 1989; Scheres et al., submitted). Although the pseudonodules lack a discrete, persistent nodule meristem and peripherally located vascular bundles, they share a number of histological features with *Rhizobium*-induced root nodules including the presence of tissues comparable to nodule cortex, central zone, and nodule endodermis. They also contain transcripts for some early nodulin genes, which have been localized by *in situ* hybridization to specific nodule tissues (van de Wiel et al., 1990; Scheres et al., submitted). We concluded that NPA, by eliciting cell divisions and early nodulin gene expression, partially mimicked the effect of *Rhizobium nod* genes (Hirsch et al., 1989).

1.2. SOME EARLY NODULINS MAY BE DIAGNOSTIC FOR AN ALTERED HORMONE BALANCE IN THE ROOT.

The ATIs induce a change in the endogenous hormone balance of the root by blocking auxin transport (Rubery, 1990). The NPA-induced pseudonodules formed on alfalfa roots contain transcripts for at least two nodulin genes, MsENOD2 and Nms-30 (Hirsch et al., 1989), while pseudonodules formed on Afghanistan pea roots were found to contain PsENOD2 and PsENOD12 transcripts (Scheres et al., submitted). ENOD2, and ENOD12 in pea, may serve as molecular markers for cells with altered hormone levels. In support of this conclusion are experiments in which *Sesbania* roots have been treated with cytokinin for 48 h and shown to contain mRNAs of the early nodulin gene SrENOD2 (Dehio and de Bruijn, 1992).

1.3. NOD FACTOR APPLICATION IS HYPOTHESIZED TO ALTER ENDOGENOUS PHYTOHORMONE LEVELS.

Rhizobium meliloti Nod factor may cause a change in internal hormone balance after binding to a presumed receptor. Other responses to *Rhizobium* or to Nod factor may include alterations in the cytoskeletal components or Ca^{2+} levels. Treatment of alfalfa root hairs with Nod factor leads to a depolarization of the root hair membrane (Ehrhardt et al., 1992). Additional changes occurring after inoculation with rhizobia include the so-called Ini response (van Brussel et al., 1990), an increase in *nod* gene-inducing flavonoids. These and other subcellular changes brought about by interaction with *Rhizobium* lead to localized cell divisions, the formation of a nodule primordium, and the expression of early nodulin genes. Figure 1 illustrates a model that shows the interaction between host-derived flavonoids and the production of Nod factor, and the postulated responses that occur within the body of the root in response to Nod factor.

An alternative but not mutually exclusive explanation for some of the plant responses is that *R. meliloti* produces an ATI-like molecule which, rather than, or in addition to Nod factor, induces a change in endogenous phytohormone balance. This ATI-like molecule may be formed from: 1) the catabolism of plant-secreted flavonoids, compounds known to function as endogenous ATIs (Jacobs and Rubery, 1988); or 2) as a product of as yet uncharacterized *R. meliloti* genes that may be induced by flavonoids.

We have concentrated on testing the two following aspects of the model presented in Fig. 1:

A) Does inoculation with *R. meliloti* lead to a change in the phytohormone level and also to an alteration of the quantity and/or composition of flavonoids in the alfalfa root? The first possibility was examined by treating alfalfa roots directly with auxin or cytokinin and then looking for the expression of early nodulin genes. We have examined the second possibility by determining whether there was an increase in the levels of gene expression for enzymes important for the synthesis of flavonoids, in particular, chalcone synthase, the first committed step in flavonoid biosynthesis. An increase in flavonoids upon inoculation could not only serve to induce *Rhizobium nod* genes, but could also conceivably lead to an imbalance in the internal phytohormone levels because flavonoids are ligands for the NPA receptor (Jacobs and Rubery, 1988).

B) Does *Rhizobium* produce an ATI-like molecule? We have examined culture filtrates containing 1/10 seed wash with or without *R. meliloti* to determine whether *R. meliloti* cells produce a factor that competes with NPA for its binding site. *Rhizobium*-grown culture filtrates without seed wash were also tested to determine whether the ATI-like molecule was produced in the absence of flavonoids in the seed wash. We also examined the effect of *R. meliloti nod* gene mutants on the secretion of this ATI-like factor to determine whether *nod* genes are essential for producing a molecule that competes with radioactive NPA for its receptor. Lastly, we tested bacteria other than *R. meliloti* to determine whether the ability to produce such a molecule is specific to *R. meliloti*.

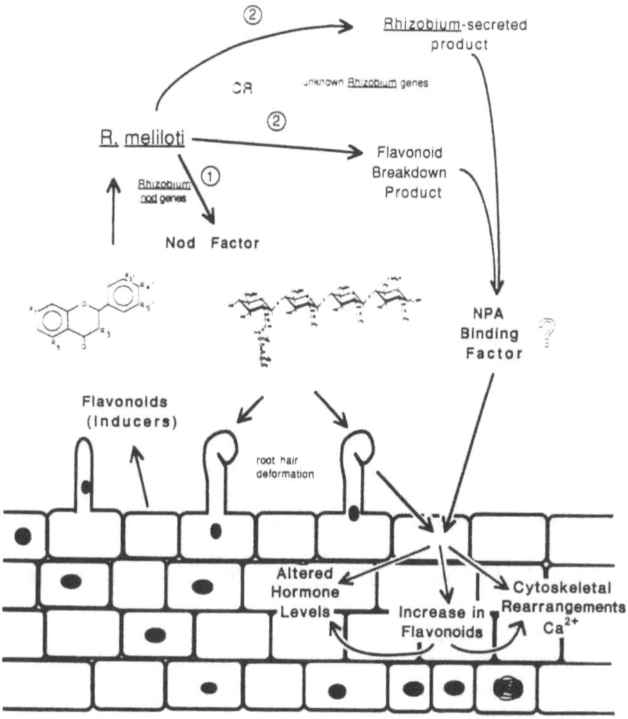

Figure 1. Legume roots secrete flavonoids which induce *Rhizobium nod* genes. Nod factors are produced (1) and these elicit root hair deformation. The level or type of flavonoids is proposed to increase within root cortical cells thereby causing a change in

plant hormone concentrations. Cytoskeletal components or Ca^{2+} levels could also change in response to Nod Factor. In addition, *R. meliloti* is proposed to produce a factor that functions as an auxin transport inhibitor (2). This factor may result from *Rhizobium*-mediated breakdown of flavonoids or as a newly secreted product from as yet uncharacterized rhizobial genes.

2. **Materials and Methods**

2.1. PLANTS AND GROWTH CONDITIONS

Alfalfa (*Medicago sativa* L. cv. Iroquois) were grown as described in Norris et al. (1988). Alternatively, after sterilization in 95% alcohol for 30-60 min. followed by full strength household bleach for 60 min. and copious rinses with sterile distilled water, seeds were placed on cheesecloth stretched over a metal screen suspended 2 cm over nitrogen-free Jensen's medium (Vincent, 1970). The cheesecloth, screen, and medium had been autoclaved previously within a polypropylene dish pan. After planting, the dish pan was covered with Saran wrap, and the plants were grown at 20°C under a 16 h light/8 h dark regime. When plants were inoculated, ca. 10^8 *R. meliloti* cells/ml were added 72 h after germination.

2.2 CONSTRUCTION OF A cDNA LIBRARY AND ISOLATION OF CLONES.

A cDNA library was constructed from poly(A)$^+$ RNA isolated from 21-d-old nodules in λgt11 using a cDNA synthesis kit and the λgt11 cloning kit from Amersham. The clones used to screen the library were pPsENOD12 (Scheres et al., 1990), a gift from Ton Bisseling, and pCHS1 from *Phaseolus vulgaris* (Ryder et al., 1984), a gift from Chris Lamb. The alfalfa ENOD2 cDNA clone A2ENOD2 (Dickstein et al., 1988) used in some experiments was a gift from Rebecca Dickstein.

2.3. TREATMENT OF ALFALFA ROOTS WITH PLANT HORMONES.

Seeds were sterilized, spread on moistened cheesecloth suspended over nitrogen-free Jensen's medium and germinated in the dark for 72 h. The seedlings were then transferred to fresh Jensen's medium containing either no phytohormone, 10^{-6} M NAA (α-naphthalene acetic acid), 10^{-5} M 2,4-D (2,4-dichlorophenoxyacetic acid) or concentrations of BAP (benzylaminopurine) ranging from 10^{-5} to 10^{-9} M. Some roots were inoculated with wildtype *Rhizobium meliloti* (Rm1021). After 2 or 4 d of treatment, roots and hypocotyls were harvested and separately frozen in liquid nitrogen. The roots were stored at -80°C until RNA was extracted.

2.4 RNA ANALYSIS AND *IN SITU* HYBRIDIZATION METHODS.

RNA was isolated from root or nodule tissue following the procedure of Goldberg et al. (1981), subjected to electrophoresis on a formaldehyde gel (Maniatis et al., 1982), and blotted onto Nytran membranes following the protocol provided by Schleicher and Schuell. Tissue was fixed and prepared for *in situ* hybridization as described by van de Wiel et al. (1990) and Bochenek and Hirsch (1990).

2.5. PREPARATION OF SEED WASH.

Seed wash prepared from alfalfa seeds as described by Mulligan and Long (1985) was filter-sterilized and stored at 4°C until use. At that time, the seed wash was added to RDM (*Rhizobium* Defined Medium; Vincent, 1970) at 1/10 volume. The diluted seed wash solutions were left uninoculated or inoculated with either *R. meliloti* wildtype or mutant strains (Table 1).

2.6 PREPARATION OF CULTURE FILTRATE.

To test whether a component from the seed wash or whether *Rhizobium* itself exhibited the ability to compete with NPA, bacteria were also grown in RDM without seed wash. In some cases, 10 μM luteolin was added to the culture medium. The bacterial strains tested are given in Table 1.

TABLE 1. Bacterial strains.

Strain	Characteristics	Source or reference
R. meliloti strains		
Rm1021	Nod$^+$Fix$^+$ wildtype SU47, Sm^r	Meade et al., 1982
Rm5610	SU47 *nodA*::Tn*5*	E.R. Signer
GMI2212	*nodH*::Tn*5*	J. Dénarié
GMI214	Δ 290 kb pSym1	J. Dénarié
R. leguminosarum strains		
128C53	wildtype, Nod$^+$Fix$^+$ on pea and vetch	F.M. Ausubel
TOM	wildtype, Nod$^+$Fix$^+$ on Afghanistan pea	T. Bisseling
Agrobacterium tumefaciens		
GMI9023	Δ pTi and cryptic plasmid	Rosenberg and Huguet, 1984

The bacteria were harvested by centrifugation after they had reached an O.D.$_{600}$ of 1.0 +/- 0.15. The supernatant was filter-sterilized and stored at 4°C until assayed for the ability to compete with NPA. In some experiments, after reaching an O.D.$_{600}$ of 1.0, the bacteria were transferred 1:10 to deposit-free Jensen's medium with or without 10 μM luteolin. After 48 h growth at 30°C, the bacteria were removed by centrifugation and the culture filtrate was filter-sterilized. These samples were then lyophilized, dissolved in sterile water as 10x concentrates, filter-sterilized, and stored at 4°C until assayed.

2.7 NPA BINDING ASSAY.

For measurements of specific NPA binding, a zucchini hypocotyl membrane fraction was prepared and treated as described by Jacobs and Rubery (1988). Replicate samples were analyzed for each treatment. NPA binding in the presence of culture medium alone (RDM or deposit-free Jensen's medium) served as the control for the assays; the controls were given a value of 100%. NPA binding activity in the presence of the test solutions was expressed as a percentage of the comparably treated control. Decreases in NPA binding activity in the presence of the test solution are referred to as evidence for an ATI-like molecule. Decrease in ligand-receptor interaction can be caused by compounds that change the conformation of the receptor, lowering its affinity, as well as by compounds competing directly with the ligand for binding. Our evidence is that the culture filtrate effect is not an effect on receptor affinity and thus represents competitive inhibition.

3. Results and Discussion

3.1 NODULE-SPECIFIC GENES

3.1.1. *Expression of ENOD2 after Treatment of Alfalfa Roots with Plant Hormones.*

Three-d-old alfalfa seedlings were grown with their roots suspended in Jensen's medium containing no phytohormone, medium containing either auxin or cytokinin, or inoculated with *R. meliloti*. The roots were examined under the light microscope immediately before harvest. Roots incubated with Rm1021 exhibited extensive root hair deformation while the untreated roots possessed only elongated, straight root hairs. The 2,4-D-treated roots were radially enlarged just proximal to the apical meristem and the root hairs were slightly deformed. The cytokinin-treated roots also exhibited slight root hair deformation, but the root itself was not distorted.

When RNA from the control and phytohormone-treated roots was probed with A2ENOD2, we observed a strong ENOD2 signal in the lanes that contained RNA isolated from the cytokinin-treated roots (data not shown). A weaker signal was observed in the RNA from auxin-treated roots and from *R. meliloti*-inoculated roots (Y. Fang and A.M. Hirsch, unpubl. results). We concluded that, like in *Sesbania* (Dehio and deBruijn, 1992) ENOD2 expression is most likely induced because of localized changes in hormone balance in treated roots.

3.1.2. *MsENOD12 is Not Expressed in Bacteria-free Nodules.* We had previously isolated a cDNA clone with homology to pPsENOD12 from the alfalfa cDNA library (Löbler and Hirsch, submitted). The DNA sequence of MsENOD12-1 is 509 base pairs (bp) long and contains an open reading frame of 336 bp. The encoded protein is proline-rich and is built of 14 repeats of the consensus motif pentamer PPIYK (proline-proline-isoleucine-tyrosine-lysine). There is limited sequence similarity between PsENOD12 and MsENOD12, except in the putative signal peptide. However, similar to the results of Scheres et al. (1990) for PsENOD12, we detected MsENOD12 transcripts in the invasion zone of nitrogen-fixing *Rhizobium*-induced nodules by *in situ* hybridization (Löbler and Hirsch, submitted).

We wanted to test whether MsENOD12 mRNAs were found in pseudonodules induced on alfalfa roots by NPA. The PsENOD12 gene, which is expressed in roots hairs 48 h after inoculation with *R. leguminosarum* bv. *viciae*, or after treatment with preparations containing Nod factors (Scheres et al., 1990) is also expressed within pea root hairs 48 h after NPA treatment (Scheres et al., submitted). In contrast, however, we could not detect MsENOD12 transcripts in NPA-induced pseudonodules. MsENOD12 transcripts are also not found in bacteria-free nodules that form spontaneously on alfalfa roots in the absence of *Rhizobium* (Hirsch et al., in press). Moreover, MsENOD12 mRNAs were not detected in bacteria-free alfalfa nodules elicited by *R. meliloti* exopolysaccharide (*exo*) mutants (Löbler and Hirsch, submitted). There was one exception, however. When alfalfa roots were inoculated with *R. meliloti exoH* mutants, which partially invade the developing nodule via infection threads that penetrate the central tissue of the nodule (Leigh et al., 1987), MsENOD12 transcripts were detected in the infected cells.

In conclusion, MsENOD12 appears to be expressed as part of the infection process, and does not appear to be a marker for altered hormone levels in alfalfa. This is in contrast to PsENOD12 which is expressed in NPA-induced pseudonodules formed on Afghanistan pea and also in pea root hairs treated with NPA for 48 h (Scheres et al., submitted).

3.2. SYMBIOSIS-ENHANCED GENES.

To test whether endogenous flavonoids increase in inoculated alfalfa roots, we screened the alfalfa cDNA library with a *P. vulgaris* cDNA clone for chalcone synthase (CHS). We reasoned that the levels of the mRNAs for enzymes involved in flavonoid biosynthesis should increase if an Ini response were taking place in inoculated alfalfa roots. Estabrook and Sengupta-Gopalan (1991) have isolated symbiosis-enhanced cDNAs for phenylalanine ammonia-lyase (PAL) and CHS from soybean.

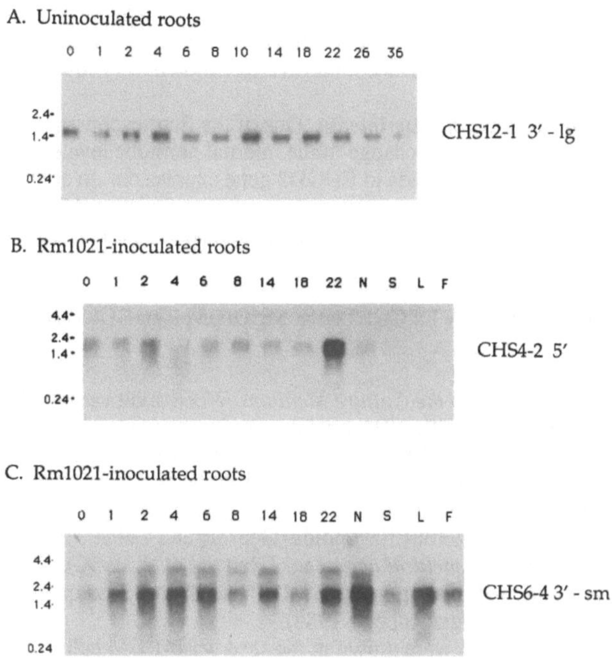

Figure 2. RNA transfer blot containing RNAs isolated from uninoculated (panel A) or Rm1021-inoculated (panels B, C) roots at varying times in days from the time of inoculation. N, nodules; S, stems; L, leaves; F, flowers.

We isolated four cDNA clones from the nodule library that hybridized to the CHS probe from bean. DNA sequence analysis demonstrated that the four cDNA clones were ca. 87-94% homologous to each other and differed primarily in their 3' untranslated ends.

When RNA from uninoculated roots was probed with any of the four alfalfa CHS cDNAs, a pattern similar to that of Fig. 2 (panel A) was obtained. The level of CHS gene expression appeared to be the same at any point in the time course. However, when RNA from Rm1021-inoculated roots was probed with any one of the CHS cDNAs corresponding to the 5' 250-350 bp of each clone, a pattern similar to Fig. 2 (panel B) emerged. Higher levels of gene expression were seen 22 d post-inoculation.

We made 3' specific probes that distinguished each of the four cDNA clones. When we examined RNA from roots harvested from 1-22 d after inoculation with Rm1021, we observed a slight increase in RNA accumulation 1-2 d and 6 d after inoculation with two of the CHS 3' probes (Fig. 2; panel C). When RNA was probed with the other two clones, there was an increase in CHS mRNA 1-2 d post-inoculation, but at much lower levels than with the putative symbiosis-enhanced clones. Moreover, the peak at 6 d was not observed when these other two clones were used as probes.

This increase in mRNA levels of the putative symbiosis-enhanced cDNAs at 1-2 d and at 6 d post-inoculation was borne out by additional analyses using RNAse protection (data not shown).

In conclusion, at least one CHS gene is specifically expressed 1-2 d and 6 d after inoculation with *R. meliloti* suggesting that it is symbiosis-enhanced. A second CHS gene appeared also to be symbiosis-enhanced, but its expression is very low.

3.2.1. *Conclusion for the first set of experiments.* One of the first responses to *Rhizobium* inoculation is likely to be a change in the internal hormone levels of the root. Treating alfalfa roots with cytokinin leads to ENOD2 gene expression. In addition, we detected an increase in CHS mRNA accumulation 1-2 and then 6 d after infection with *Rhizobium.* A change in phytohormone levels could be mediated by an increase in flavonoids in infected root cells.

3.3. DOES *R. MELILOTI* GROWN IN CULTURE MEDIUM PRODUCE AN ATI-LIKE COMPOUND?

3.3.1. *Adding Alfalfa Seed Wash to the Culture Medium.* When alfalfa seed wash was added at 1/10 volume to RDM, specific NPA binding was inhibited by 25-40% compared to RDM without seed wash (Fig. 3). The variation in percentage of NPA binding from one trial to another may be due to variations in the concentrations of seed wash flavonoids from experiment to experiment.

Alfalfa seed wash contains numerous flavonoids (Gehring and Geiger, 1980), including several which induce *R. meliloti nod* genes (Hartwig et al., 1990). Quercetin and quercetin conjugates are major components of alfalfa seed wash, and quercetin has been shown to be an ATI (Jacobs and Rubery, 1988).

When wildtype *R. meliloti* cells were grown in the seed wash-RDM culture medium, specific NPA binding was reduced by an additional 10-20% beyond that of seed wash alone (Fig. 3). This suggests that *R. meliloti* either alters components of the seed wash to produce a more effective ATI or makes an additional compound that competes with radioactive NPA for its receptor.

We examined the effects of *R. meliloti nod* mutants on the ability of the seed wash-culture medium supernatant to compete with radioactive NPA for binding to the NPA receptor (Table 2). Table 2 shows that the supernatant from an *R. meliloti nodA* mutant in the "SE"+ medium has the same effect on NPA binding as the "SE"+ supernatant derived from wildtype *R. meliloti.* Although the "SE"+ supernatant derived from the *nod* deletion mutant GMI214 appears to be not as effective, the values measured are not statistically different from supernatants derived from either wildtype or *nodA* mutant *R. meliloti.*

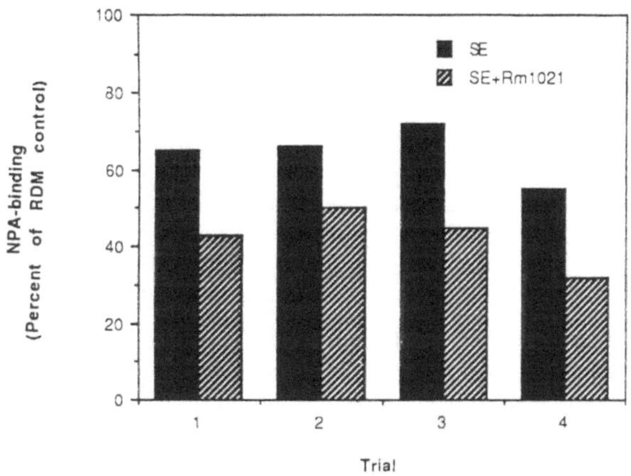

Figure 3. NPA-binding activity of alfalfa seed wash with or without wildtype *R. meliloti.*

TABLE 2. Effect of *R. meliloti* strains on an NPA competing factor.

Strain	% NPA Binding +/- S.D. SE-	% NPA Binding +/- S.D. SE+	Number of Assays
Rm1021	71.6 +/- 8.3	49.2 +/- 7.0	6
Rm5610	60.3 +/- 5.1	48.9 +/- 6.0	4
GMI214	72.2 +/- 7.7	59.1 +/- 4.0	6
Rl TOM	68.2 +/- 17.2	60.4 +/- 4.4	5
GMI9023	66.8 +/- 18.4	68.1 +/- 8.4	5

3.3.2. *Checking* R. meliloti *Culture Supernatant for an ATI-like Factor.* To test whether the flavonoids in seed wash were the source of the NPA competing factor, we grew *R. meliloti* in RDM without seed wash. In some experiments, the bacteria were transferred to deposit-free Jensen's medium with or without 10 µM luteolin and allowed to grow for 48 h. The culture filtrates were lyophilized, dissolved in sterile water to give 1x and 10x concentrations and analyzed for NPA binding activity. Figure 4 shows the results of the latter type of experiment. Jensen's medium in which Rm1021 had grown for 48 h has a low level of NPA-competing activity. However, the addition of 10 µM luteolin to the Jensen's medium in which the bacteria were grown significantly increased the amount of NPA-binding activity in the culture filtrate. Interestingly, 10x concentrated culture filtrate containing luteolin does not exhibit much more NPA binding than the 1x culture filtrate, suggesting that the NPA binding sites are saturated.

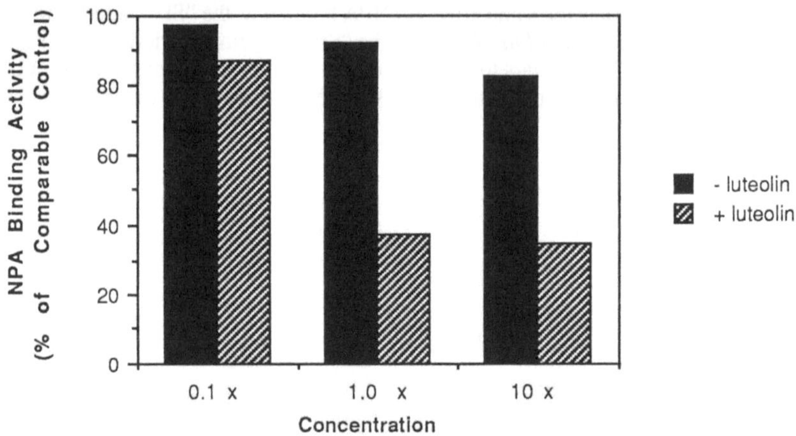

Figure 4. NPA binding activity of culture filtrate from wildtype *R. meliloti.* Methanol alone (-) or 10 μM luteolin dissolved in methanol (+) were added to deposit-free Jensen's medium immediately before Rm1021 cells were transferred from RDM.

We also examined the effects of wildtype, *nod* mutant *R. meliloti,* as well as non-*R. meliloti* strains on secretion of an ATI-like factor. Table 3 lists the NPA competitive ability of RDM culture filtrates prepared without luteolin. An additional 10-15% increase in NPA competitiveness was observed if 10 μM luteolin was included in the culture medium.

TABLE 3. NPA Competitive Ability of Culture Filtrates from Different Bacteria.

Treatment	Competitive Ability (% of control)	Number of Assays
Rm1021	79.2	6
Rm5610	79.7	6
GMI2212	78.1	6
GMI214	80.3	6
Rl TOM	77.1	2
Rl 128C53	80.7	2
GMI 9023	78.4	6

We conclude from these experiments that *R. meliloti* and other plant-interacting bacteria produce a factor that competes with radioactive NPA for its receptor. The production of this factor appears to be independent of the *R. meliloti nod* genes because culture filtrates derived from Nod$^-$ mutants can compete with NPA for binding to its receptor.

4. Acknowledgements

We thank Amy Kaizuka and Nu Usaha for their help in preparing the culture filtrates. We also thank Margaret Kowalczyk and Stefan Kirchanski for Fig. 1. This research was supported by NSF grants DCB 90-21587 and DCB 90-23888 and USDA Competitive Grant 91-37307-6603 to AMH, as well as NSF grant DCB 90-05774 and support from the Howard Hughes Medical Institute to MJ. Heather McKhann was supported by a USDA Training Fellowship and a California Biotechnology Fellowship. Shaheen Asad was sponsored by the USAID Pakistan Participant Training Program.

5. References

Bochenek, B. and A.M. Hirsch. (1990) *In-situ* hybridization of nodulin mRNAs in root nodules using non-radioactive probes. Plant Molec. Biol. Report. 8, 237-248.

Dehio, C. and F.J. deBruijn. (1992) The early nodulin gene SrENOD2 from *Sesbania rostrata* is inducible by cytokinin. The Plant J. 2, 117-128.

Dickstein, R., T. Bisseling, V.N. Reinhold, and F.M. Ausubel. (1988) Expression of nodule-specific genes in alfalfa root nodules blocked at an early stage of development. Genes Dev. 2, 677-687.

Ehrhardt, D.W., E.M. Atkinson, and S.R. Long. (1992) Depolarization of alfalfa root hair membrane potential by *Rhizobium meliloti* Nod factors. Science. 256, 998-1000.

Estabrook, E. and C. Sengupta-Gopalan. (1991) Differential expression of phenylalanine ammonia-lyase and chalcone synthase during soybean nodule development. The Plant Cell. 3, 299-308.

Goldberg, R.B., G. Hoschek, S.H. Tam, G.S. Ditta, and R.W. Breidenbach. (1981) Abundance, diversity, and regulation of mRNA sequences in soybean embryogenesis. Devel. Biol. 83, 201-217.

Gehring, E. and H. Geiger. (1980) Die Flavonoide der Samen von *Medicago* x *varia* Martyn c. v. Cardinal (Fabaceae). Z. Naturforsch. 35c, 380-383.

Hartwig, U.A., C.A. Maxwell, C.M. Joseph, and D.A. Phillips. (1990) Chrysoeriol and luteolin released from alfalfa seeds induce *nod* genes in *Rhizobium meliloti*. Plant Physiol. 92, 115-122.

Hirsch, A.M., T.V. Bhuvaneswari, J.G. Torrey, and T. Bisseling. (1989) Early nodulin genes are induced in alfalfa root outgrowths elicited by auxin transport inhibitors. Proc. Nat. Acad. Sci. (USA). 86, 1244-1248.

Hirsch, A.M., H.I. McKhann, and M. Löbler. (1992) Bacterial-induced changes in plant form and function. Inter. J. Plant Sci. In press.

Jacobs, M. and P.H. Rubery. (1988) Naturally-occurring auxin transport regulators. Science. 241, 346-349.

Leigh, J.A., J.W. Reed, J.F. Hanks, A.M. Hirsch, and G.C. Walker. (1987) *Rhizobium meliloti* mutants that fail to succinylate their calcofluor-binding exopolysaccharide are defective in nodule invasion. Cell. 51, 579-587.

Lerouge, R., P. Roche, C. Faucher, F. Maillet, G. Truchet, J.-C. Promé, and J. Dénarié. (1990) Symbiotic host-specificity of *Rhizobium meliloti* is determined by a sulphated and acylated glucosamine oligosaccharide signal. Nature. 344, 781-784.

Löbler, M. and A.M. Hirsch. MsENOD12, a proline-rich nodulin from alfalfa (*Medicago sativa* L.), is expressed in the invasion zone of *Rhizobium meliloti*-induced root nodules. Submitted for publication.

Maniatis, T., E.F. Fritsch, and J. Sambrook. (1982) Molecular Cloning, A Laboratory Manual, Cold Spring Harbor, NY.

Maxwell, C.A., U.A. Hartwig, C.M. Joseph, and D.A. Phillips. (1989) A chalcone and two related flavonoids released from alfalfa roots induce *nod* genes of *Rhizobium meliloti*. Plant Physiol. 91, 842-847.

Meade, H.M., S.R. Long, G.B. Ruvkun, S.E. Brown, and F.M. Ausubel. (1982) Physical and genetic characterization of symbiotic and auxotrophic mutants of *Rhizobium meliloti* induced by transposon Tn*5* mutagenesis. J. Bacteriol. 149, 114-122.

Mulligan, J.T. and S.R. Long. (1985) Induction of *Rhizobium meliloti nodC* expression by plant exudate requires *nodD*. Proc. Natl. Acad. Sci. (USA). 82, 6609-6613.

Norris, J.H., L.A. Macol, and A.M. Hirsch. (1988) Nodulin gene expression in effective nodules and in nodules arrested at three different stages of development. Plant Physiol. 88, 321-328.

Rosenberg, C. and T. Huguet. (1984) The pAtC58 plasmid is not essential for tumour induction. Molec. Gen. Genet. 196, 533-536.

Rubery, P.R. (1990) Phytotropins: receptors and endogenous ligands. Symp. Soc. Exp. Biol. 44, 119-146.

Ryder, T.B., C.L. Cramer, J.N. Bell, M.P. Robbins, R.A. Dixon, and C.J. Lamb. (1984) Elicitor rapidly induces chalcone synthase mRNA in *Phaseolus vulgaris* cells at the onset of the phytoalexin defense response. Proc. Natl. Acad. Sci. (USA). 81, 5724-5728.

Scheres, B., C. van de Wiel, A. Zalensky, B. Horvath, H. Spaink, H. van Eck, F. Zwartkruis, A.-M. Wolters, T. Gloudemans, A. van Kammen, and T. Bisseling. (1990) The ENOD12 gene is involved in the infection process during the pea-*Rhizobium* interaction. Cell. 60, 281-294.

Scheres, B., H.I. McKhann, A. Zalensky, M. Löbler, T. Bisseling, and A.M. Hirsch. The PsENOD12 gene is expressed at two different sites in Afghanistan pea pseudonodules induced by auxin transport inhibitors. Submitted for publication.

Torrey, J.G. (1986) Endogenous and exogenous influences on the regulation of lateral root formation, in M.B. Jackson (ed.). New root formation in plants and cuttings, Martinus Nijhoff Publishers, Dordrecht, pp. 31-66.

Truchet, G., P. Roche, P. Lerouge, J. Vasse, S. Camut, F. deBilly, J.-C. Promé, and J. Dénarié. (1991) Sulphated lipo-oligosaccharide signals of *Rhizobium meliloti* elicit root nodule organogenesis in alfalfa. Nature. 351, 670-673.

van Brussel, A.A.N., K. Recourt, E. Pees, H.P. Spaink, T. Tax, C.A. Wijffelman, J.W. Kijne, and B.J.J. Lugtenberg. (1990) A biovar-specific signal of *Rhizobium leguminosarum* bv. *viciae* induces increased nodulation gene-inducing activity in root exudate of *Vicia sativa* subsp. *nigra*. J. Bacteriol. 172, 5394-5401.

van de Wiel, C., J.H. Norris, B. Bochenek, R. Dickstein, T. Bisseling, and A.M. Hirsch. (1990) Nodulin gene expression and ENOD2 localization in effective, nitrogen-fixing and ineffective, bacteria-free nodules of alfalfa. The Plant Cell. 2, 1009-1017.

Vincent, J.D.. (1970) A Manual for the Practical Study of Root-Nodule Bacteria, ABP Handbook No. 15, Blackwell Scientific Publishers, Oxford.

RHIZOBIUM NOD METABOLITES AND EARLY NODULIN GENE EXPRESSION

Ton Bisseling, Henk Franssen, Renze Heidstra, Beatrix Horvath, Panagiotis Katinakis, Marja Moerman, Herman Spaink*, Ton van Bussel* and Irma Vijn

Department of Molecular Biology
Agricultural University
Dreijenlaan 3, 6703 HA Wageningen
The Netherlands

* University of Leiden
Nonnensteeg 3
2311 VJ LEIDEN
The Netherlands

• Agricultural University
Athens
Greece

INDEX. Nod factors
early nodulins
infection

1. INTRODUCTION

The *Rhizobium*-legume interaction results in the formation of a new plant organ, the root nodule, in which the bacteria are able to reduce atmospheric N_2. These root nodules have a very distinctive tissue organization, which allows the N_2 fixation process to occur [1]. In short the *Rhizobium*-legume interaction involves the following steps: It starts with the deformation and curling of root hairs. The rhizobia then invade the plant by means of a newly formed tube, called the infection thread. Meanwhile, cells in the root cortex start to divide and form the nodule primordium. Infection threads enter individual primordium cells, and bacteria are relased form the infection thread into the cytoplasm of the plant cells. The primordium cells then differentiate into the tissues that make up the mature nodule. *Rhizobium* thus induces three distinct developmental processes in its host (i) root hair deformation, (ii) infection thread formation, and (iii) induction of cell division followed by differentiation into the different nodule tissues.

Upon activation of the *nod* genes *Rhizobium* bacteria secrete Nod metabolites which are lipo-oligosaccharides and these Nod factors play a crucial role in the induction of the three induced plant developmental processes. The major factors secreted by *R.meliloti* are sulphated β-1,4-tetrasaccharides of D-glucosamine in which three amino

E. W. Nester and D. P. S. Verma (eds.),
Advances in Molecular Genetics of Plant-Microbe Interactions, 365–368.

groups are acetylated and one is N-acylated with an unsaturated fatty acid [2,3]. These factors, named NodRmIV-(S,C16:2) and NodRmIV-(Ac, S, C16:2), induce root hair deformation on alfalfa plants and furthermore, are the only *Rhizobium* factors required for the induction of cortical cell division and nodule formation [4].

R.leguminosarum bv. *viciae* makes several Nod factors and these molecules have a structure similar to NodRm factors since they are tetra- or pentasaccharides of D-glucosamine of which the terminal non-reducing sugar is N-acylated and the other sugar residues are N-acetylated. However, they lack the sulphate group and also the acyl group is different. *R.leguminosarum* bv. *viciae* Nod factors contain a C18 fatty acid chain with four (C18:4) or one (C18:1) unsaturated bond [5].

The synthesis of C18:4 requires a functional *nodE* gene whereas C18:1 containing Nod factors are still made by a *nodE*$^{-}$ mutant. *R.leguminosarum* bv. *viciae* Nod factors with either lipid deform vetch root hairs, but the induction of cell division is only induced by Nod factors containing the highly unsaturated C18:4 [5].

2. EARLY NODULIN GENES REGULATED BY NOD FACTORS

To study the relationship between the structure of Nod factors and its biological activity we searched for nodule specific (= nodulin) genes that are involved in infection or root cortical cell division. Two early nodulin sequences have been identified that are involved in the infection process [6,7]. *In situ* hybridization and PCR studies with infected pea roots have shown that the ENOD5 gene is expressed exclusively in root hairs and root cortical cells containing the infection thread tip [7]. The ENOD12 gene is also expressed in root cells containing the infection thread. However, this early nodulin gene is also expressed several cell layers in front of the infection thread tip [6,7]. Therefore, the mechanism by which *Rhizobium* elicits expression of these two early nodulin genes is likely to be different; the ENOD5 gene is induced by a factor with short range activity, whereas ENOD12 gene activation probably involves a diffusible compound.

Recently an early nodulin cDNA clone (ENOD40) has been identified representing a gene that is induced in dividing root cortical cells and is not expressed in root meristems (Yang *et al.*, submitted). An ENOD40 early nodulin clone was first isolated from a soybean nodule cDNA library. By *in situ* hybridization it was shown that the ENOD40 gene is induced in dividing root cortical cells. The expression of the ENOD40 gene is still detectable in mature nodules, but at this stage of development the expression mainly occurs in the pericycle of the nodule vascular bundle. Using the soybean ENOD40 clone as a probe, we isolated a pea ENOD40 cDNA clone. The pea ENOD40 gene is also expressed in dividing root cortical cells. Moreover, it is transcribed in the pericycle of the nodule vascular bundle. So the pattern of ENOD40 gene expression in determinate soybean and indeterminate pea nodule development is very similar.

To determine whether Nod factors induced the expression of early nodulin genes we used two different systems: 1.) Early nodulin expression was studied in root hairs for pea plants treated with Nod metabolites and 2.) Early nodulin transcripts were *in situ* localized in *Vicia sativa* nodule primordia induced by Nod factors.

3. EARLY NODULIN GENE INDUCTION IN PEA ROOT HAIRS

We studied the expression of ENOD5 and ENOD12 gene expression in root hairs of pea seedlings treated with purified Nod metabolites. ENOD40 gene expression was not studied since *Rhizobium* does not induce the expression of this gene in root hairs. Both ENOD5 and ENOD12 mRNA are detected by a semi-quantitative PCR assay. Such a PCR assay was used because only small quantities of root hair RNA were available and moreover both early nodulin genes are expressed at a low level.

We compared the inducing activity of Nod factors that only differ in their acyl moiety; NodRlvV (Ac, C18:4) and NodRlvV (Ac, C18:1). Both factors were applied to pea seedlings in a concentration of 10^{-8} M and ENOD5 and ENOD12 gene expression was studied. After 12, 24 and 48 hours NodRlvV (Ac, C18:4) induces ENOD12 gene expression within 12 hours and the ENOD12 mRNA level markedly decreases at 24 hours. The ENOD5 gene is also induced by this Nod factor, but expression first reaches a maximal level at 24 hours, after which the concentration of ENOD5 mRNA decreases. The C18:1 containing Nod factor also has the ability to induce both early nodulin genes in pea root hairs, but the induction occurs at a later time point than in NodRlvV (Ac, C18:4) treated plants.

We assume that the mechanism by which Nod factors induce early nodulin gene expression involves perception by a receptor and subsequently a transduction of the signal. Both Nod factors induce ENOD5 gene expression to a similar level but the kinetics of the induction is different. Therefore we think it is unlikely that the same receptor recognizes both Nod factors.

In situ hybridization studies on *Rhizobium* infected roots strongly suggested that the induction of the ENOD5 and ENOD12 genes involves different mechanisms, ENOD5 gene induction requiring intracellular signalling and ENOD12 gene activation involves intra- as well as intercellular signalling. Hence we were surprised that one single Nod factor can induce the expression of both early nodulins. We postulate that the same receptor mediates the induction of both early nodulins but different secondary signal molecules are finally involved in the induction of ENOD5 and ENOD12 gene expression.

4. EARLY NODULIN GENE INDUCTION IN VICIA NODULE PRIMORDIA

Nod factors induce cortical cell divisions in the roots of *Vicia sativa* [5]. These centers of mitotic activity result in the formation of small bumps, but root nodules are not formed. To determine at which step of development these bumps are blocked, we hybridized sections of these bumps with ^{35}S labeled antisense ENOD2 RNA. In *Rhizobium* induced nodule development, the ENOD2 gene is first expressed when the primordium differentiates into a central and peripheral tissues. The ENOD2 mRNA is specifically located in the nodule parenchyma [8]. In none of the sections of Nod factor induced primordia ENOD2 mRNA was detected. Therefore we conclude that these Nod factor induced primordia are blocked in development before a differentiation into nodule tissues takes place. This observation on early nodulin gene expression is consistent with e.g. the absence of vascular bundles in these Nod factor induced structures.

Sections of Nod factor induced structures were also hybridized with ENOD5, ENOD12 and ENOD40 probes. These preliminary studies showed that all 3 genes are expressed in the nodule primordium. The ENOD12 as well as the ENOD40 gene are expressed in *Rhizobium* induced primordia before it has been reached by the infection thread. Therefore the expression pattern of the ENOD12 and ENOD40 genes in Nod factor induced primordia is similar to that in *Rhizobium* formed primordia. In contrast, the ENOD5 gene is first expressed when a pea nodule primordium is penetrated by an infection thread. Therefore the observed expression of the ENOD5 gene in Nod factor induced *Vicia* primordia was rather unexpected. At the moment we do not understand this discrepancy in ENOD5 gene expression in the two types of nodule primordia. A possible explanation relates to the relatively high concentration (10^{-7} M) of Nod metabolites used to induce nodule primordia. On the other hand it can not be excluded that the *Vicia* ENOD5 gene expression pattern differs from that of the pea gene.

5. REFERENCES

1. Nap, J.P. and Bisseling, T. (1990) 'Developmental biology of a plant-prokaryote symbiosis: The legume root nodule', Science 250, 948-954.
2. Lerouge, P., Roche, P., Faucher, C., Maillet, F., Truchet, G., Promé, J.C. and Dénarié, J. (1990) 'Symbiotic host-specificity of *Rhizobium meliloti* is determined by a sulphated and acylated glucosamine oligosaccharide signal', Nature 344, 781-784.
3. Roche, P., Debellé, F., Maillet, F., Lerouge, P., Faucher, C., Truchet, G., Dénarié, J. and Promé, J.C. (1991) 'Molecular basis of symbiotic host specificity in *Rhizobium meliloti*: *nod*H and *nod*PQ genes encode the sulfation of lipo-oligosaccharide signals', Cell 67, 1131-1145.
4. Truchet, G., Roche, P., Lerouge, P., Vasse, J., Camut, S., De Billy, F., Promé, J.C. and Dénarié, J. (1991) 'Sulphated lipo-oligosaccharide signals of *Rhizobium meliloti* elicit root nodule organogenesis on Alfalfa', Nature 351, 670-678.
5. Spaink, H.P., Sheeley, D.M., van Brussel, A.A.N., Glushka, J., York, W.S., Tak, T., Geiger, O., Kennedy, E.P., Reinhold, V.N. and Lugtenberg, B.J.J. (1991) 'A novel highly unsaturated fatty acid moiety of lipo-oligosaccharide signals determines host specificity of *Rhizobium*', Nature 354, 125-120.
6. Scheres, B., van de Wiel, C., Zalensky, A., Horvath, B., Spaink, H., van Eck, H., Zwartkruis, F., Wolters, A.M., Gloudemans, T., van Kammen, A. and Bisseling, T. (1990) 'The ENOD12 gene product is involved in the infection process during pea-*Rhizobium* interaction', Cell 60, 281-294.
7. Scheres, B., van Engelen, F., van der Knaap, E., van de Wiel, C., van Kammen, A. and Bisseling, T. (1990) 'Sequential induction of nodulin gene expression in the developing pea nodule', The Plant Cell 8, 687-700.
8. Van de Wiel, C., Scheres, B., Franssen, H., van Lierop, M.J., van Lammeren, A., van Kammen, A. and Bisseling, T. (1990) 'The early nodulin transcript ENOD2 is located in the nodule parenchyma (inner cortex) of pea and soybean root nodules', EMBO J. 9, 1-7.

MECHANISMS OF CELL AND TISSUE INVASION BY *RHIZOBIUM LEGUMINOSARUM*: THE ROLE OF CELL SURFACE INTERACTIONS

N.J. BREWIN, S. PEROTTO, E.L. KANNENBERG, A.L. RAE, E.A.RATHBUN, M.M. LUCAS, I. KARDAILSKY, A. GUNDER, L. BOLAÑOS, N. DONOVAN, B.K. DRØBAK.
John Innes Institute, Colney Lane Norwich, NR47UH, UK

ABSTRACT. *Rhizobium* invades host cells and tissues as a result of a reorganisation of plant cell wall growth. This leads to the development of an intracellular tunnel, the infection thread, that traverses the host plant cell from one side to the other and provides a channel for the entry of rhizobial cells embedded in an extracellular matrix material secreted by the plant. Some of the components involved in cell surface interactions can be individually identified using monoclonal antibodies as molecular probes. The role of lipopolysaccharide (LPS) in cell and tissue invasion was investigated by examining pea nodules induced by mutants of *R. leguminosarum* with defects in LPS structure and biosynthesis. We conclude that the correct LPS structure is essential for "invasiveness" of plant cells and tissues; for avoidance of host defence responses; and for physiological adaptation to the endophytic microenvironment. Differentiation of released bacteria into nitrogen-fixing bacteroids is associated with further changes in cell surface composition, as revealed by *in situ* analysis of LPS epitope variation using monoclonal antibodies: these changes in epitope expression reflect variations in physiological and developmental status for bacteroids in different regions of the same pea nodule. Bacteroid differentiation is preceded by differentiation of the plant-derived peribacteroid membrane (pbm) which encloses the symbiosome compartment. Using a monoclonal antibody that identifies a group of plant membrane-associated inositol-containing glycolipids, we have identified a very early marker for the differentiation of pbm from plasma membrane.

1. Introduction

The *Rhizobium*-legume symbiosis is of major agronomic importance, especially in systems of sustainable agriculture: it is also a convenient model for the study of plant development and plant-microbe interactions. Our experimental system involves *R. leguminosarum bv viciae*, which has been very well characterised genetically. Among its hosts are the small-seeded hairy vetch *Vicia hirsuta* (convenient for cytological analysis) and the pea *Pisum sativum* (convenient for biochemical and genetic analysis). Thus, we have access to a wide range of plant and bacterial mutants affecting nodule development at various stages, as well as to some purified bacterial components known to have direct effects on plant cell morphogenesis. In addition, we have a range of monoclonal antibodies which identify individual cell surface components and allow us to monitor the differentiation of plant and bacterial surfaces associated with cell and tissue invasion

E. W. Nester and D. P. S. Verma (eds.),
Advances in Molecular Genetics of Plant-Microbe Interactions, 369–380.

during nodule development.

Rhizobium invades host cells and tissues as a result of a reorganisation of plant cell wall growth, which leads to the development of an intracellular tunnel, the infection thread, that traverses the host plant cell from one side to the other, thus providing a channel for the entry of rhizobial cells embedded in an extracellular matrix material secreted by the plant (Brewin, 1991). Recent genetic and biochemical analysis has shown that at least four kinds of *Rhizobium*-derived signal molecule are known to impinge on the plant cell surface during the initiation of root nodules on pea. These are:- the diffusible lipo-oligosaccharide "Nod-factor", which alone can stimulate root hair curling and cortical cell division (Spaink et al 1991); NodO, a haemolysin-type membrane channel protein (Sutton et al, this symposium volume); acidic extracellular polysaccharide, EPS (Gray & Rolfe, 1990); and lipopolysaccharide, LPS (Noel, 1991). The initiation of an infection thread in pea root hair cells involves the participation of live bacteria carrying appropriate extracellular polysaccharide and lipopolysaccharide components in their cell wall.

This paper will consider the processes of plant-microbe cell surface interaction that lead to cell and tissue invasion by *Rhizobium*. Our major experimental approach has involved the use of monoclonal antibodies as molecular probes to analyse the composition and tissue distribution of bacterial and plant cell surface components.

2. Results

We have examined the role of bacterial lipopolysaccharide, and also the involvement of plant cell surface components at successive stages of pea nodule development, both in the early stages of infection thread development and subsequently during the differentiation of nitrogen-fixing bacteroids within the "symbiosome" compartment.

2.1. STRUCTURE AND DEVELOPMENT OF INFECTION THREADS

Following inoculation with *Rhizobium*, root cortical cells frequently exhibit cytological features that are characteristic of mitotic reactivation. The first visible effect of a localised *Rhizobium* infection in the root cortex of pea or vetch is the development of centralised nuclei and transvacuolar cytoplasmic strands with "anticlinal" orientation (Kijne, 1991; Rae et al, 1992). As with the wounding response (Goodbody and Lloyd, 1990), the induction of transcellular strands is normally associated with nuclear reactivation and the onset of mitosis. This is also the case in the inner cortex of the *Vicia* root, where nuclear reactivation is followed by mitosis and the differentiation of the nodule primordium: moreover, the anticlinal orientation of these pre-mitotic transcellular strands predetermines the eventual orientation of cell plate formation. In the outer cortex, however, nuclear reactivation is not followed by mitosis, but instead it leads to the development of transcellular tunnels, namely the infection threads by which rhizobia gain entry into plant cells and tissues (Figure 1): in these cells, the orientation of the pre-mitotic transcellular strand predetermines the orientation of transcellular infection thread growth through the outer cortex. Thus, cell division in the inner cortex and infection thread development in the outer cortex appear to be alternative responses, although both may be consequences of a similar reorganisation of the cytoskeleton following exogenous application of Nod-factor.

Using the monoclonal antibody MAC 265 as a probe, we have identified a 95K plant

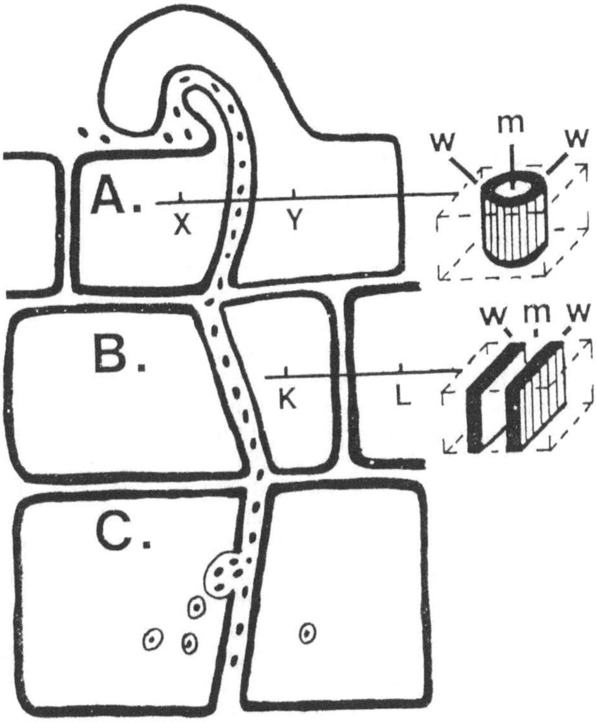

Figure 1. Diagram showing three stages of tissue and cell invasion by *Rhizobium*
A., root hair curling and infection thread initiation; B., cell-to-cell spread of rhizobia through the infection thread; C., release of rhizobia by endocytosis from an unwalled infection droplet into the intracellular compartment. Insets compare the topology of an infection thread, which serves as a trans-cellular tunnel, with that of a primary cell wall. The major differences are that the infection thread wall (w) is laid down as a cylinder rather than as a plate, and that instead of having a middle lamella (m) the infection thread contains in its lumen an intercellular matrix (m) similar in composition to that which is normally found in intercellular spaces. The bacteria embedded in plant-derived matrix glycoprotein grow and divide in the lumen of the infection thread.

glycoprotein secreted into the lumen of infection threads as an early response to *Rhizobium* infection (VandenBosch et al 1989). We have conducted a cytochemical analysis of the early stages of infection in *Vicia* roots in order to examine the basis for enhanced secretion of matrix glycoprotein and the initiation and organisation of infection thread development. The 95 K glycoprotein is secreted by root hair and cortical cells surrounding the initial infection site, and secretion continues in the meristematic zone at the apex of the developing nodule. Thus, the glycoprotein appears to be secreted by actively dividing cells, and may then accumulate in the apoplastic space, including the lumen of infection threads and intercellular spaces. Molecular probes for cell wall components showed that the infection thread wall contains polysaccharides similar to those of the normal primary cell wall, including pectins, xyloglucans and cellulose: therefore the infection thread can be thought of as an intracellular tunnel which conducts the apoplastic space (including extracellular matrix glycoprotein and embedded rhizobia) from one side of the cell to the other (Rae et al 1992). However, it has been noted that infection thread walls are more resistant to digestion than other cell walls (Higashi et al 1987), suggesting that additional modification of this wall may subsequently occur, perhaps by the introduction of phenolic compounds, or perhaps by incorporation of tissue-specific cell wall glycoproteins as has been hypothesised for ENOD12 (Scheres et al 1990).

Some indications are beginning to emerge about the physical properties of the plant matrix glycoprotein and how it might interact with invading rhizobia in the lumen of infection threads. We have observed that the attachment of this macromolecule to nitrocellulose sheets is affected by ionic strength and pH, perhaps indicating that the glycoprotein can adopt several alternative molecular conformations. Moreover, we have recently demonstrated that the matrix glycoprotein is capable of becoming physically attached to the surface of rhizobial cells, derived either from free-living culture or isolated from nodules. Although the molecular basis for this interaction is still unknown, the possible involvement of boron is being investigated, because of the long-standing report that boron is an essential mineral for nodule development (Brenchley & Thornton 1925), and because of the chemical properties of the borate ion which suggest a role in intra-molecular or inter-molecular cross-linking between glycoconjugates.

2.2. NODULE DEVELOPMENT INDUCED BY LPS-DEFECTIVE MUTANTS

Lipopolysaccharide (LPS) is a major component of the surface of *Rhizobium* (Carlson, 1984). Its important role in nodule development is suggested by the fact that mutants with a variety of modifications in the structure and biosynthesis of their LPS macromolecules are unable to establish a normal nitrogen-fixing symbiosis: these mutants induce the development of abnormal root nodules on peas and other legumes (Priefer, 1989; Stacey *et al.*, 1991; Rae *et al.*, 1991; Kannenberg et al., 1992). We analysed the development of these nodules and the fate of the LPS-defective mutant bacteria within them by using monoclonal antibodies and cytochemical techniques. The mutants fell into three general classes: severe mutants inducing an "empty nodule" phenotype; moderately severe mutants that delayed nodule development and reduced nitrogen fixation to less than 5% of the rate for wild-type nodules; and mildly disabled mutants which only slightly impaired the normal processes of nodule development and nitrogen fixation.

The most severe LPS-defective mutant failed to invade nodule tissue. Inoculation of pea seedling roots with strain B659 induced the development of empty nodule-like structures with peripheral vasculature, presumably because the mutant still secreted the Nod-factor that stimulates

Table 1.

CO-INOCULATION EXPERIMENTS INVOLVING STRAIN B659			
INOCULANT		NODULE PHENOTYPE	
Strain(s)	Characters	Morphology	Endophyte
B659	3841 *lps::Tn5*	Empty Nodule	-
16015	Δ *(nod,fix)*	Nod^-	-
B659+16015	Co-inoculation	Nod^+ Fix^-	16015 only
3841	wildtype *str*	Nod^+ Fix^+	3841

cortical cell division (Rae et al 1991). These "empty" nodules developed a rudimentary endodermis and the central (uninfected) tissue secreted quantities of extracellular matrix glycoprotein which accumulated in the intercellular spaces. However, in the absence of invading bacteria, no infection thread structures were seen. Interestingly, when this LPS-defective strain was used in a co-inoculation experiment with a non-nodulating mutant (cured of its symbiotic plasmid), the result was that the " Nod^- " strain was induced to invade the nodule tissue and differentiate into bacteroid-like cells (Table 1), although these bacteria obviously lacked the capacity for nitrogen fixation because they lacked the necessary *fix* genes. However, these same nodules were found to be virtually devoid of the other co-inoculant strain which had the severe LPS defect: this fact was determined by analysing nodule sections with a strain-specific monoclonal antibody, and by recovery of ex-nodule bacteria from surface sterilised nodules and checking for antibiotic resistance markers. Thus, the severely impaired LPS-defective mutant could not be rescued by co-inoculation and was apparently unable to survive in the endophytic environment. A further inference from this coinoculation experiment is that, because the bacteria within the nodule (strain 16015 or strain 8401 in a parallel experiment) lacked *nod* genes and were unable to produce Nod-factor, nodule meristematic activity must have been sustained by the exogenous non-invasive strain (B659), which still had a functional set of *nod* genes.

The second group of LPS-defective mutants induced nodules in which only a small proportion of the central nodule tissue was colonised by bacteria. Consequently, much of the central tissue was occupied by uninfected parenchyma, particularly underneath the nodule endodermis. In contrast to nodules formed by the wild-type strain, the nodules formed by these LPS mutants lacked a clear sequence of developmental zones arising from the apical meristem. Also, as already observed in other ineffective (Fix^-) nodules, starch was abundant and accumulated both in infected and in uninfected cells of the central nodule tissue and in the nodule parenchyma.

MAC 265, a monoclonal antibody which recognises an extracellular plant glycoprotein expressed in the lumen of infection threads (VandenBosch et al 1989), was used to follow by light microscopy the process of tissue invasion by bacteria. This antibody clearly revealed the infection threads and infection droplets which are formed in the invasion zone proximal to the nodule apex, where newly divided plant cells are actively colonised by wild-type rhizobia. When a longitudinal section from a nodule induced by an LPS-defective mutant strain was immunostained with this same antibody (MAC 265), large infection droplets and abnormal infection threads were commonly seen. Thus, the extracellular matrix glycoprotein appears to be overproduced by cells in contact with LPS-defective mutant bacteria: this phenomenon may be either a consequence or a cause of abnormal infection thread development. Enhanced secretion of matrix glycoprotein suggests the possible involvement of a plant defence response (with the matrix glycoprotein perhaps creating an extracellular barrier to reduce the progress of bacterial infection).

When early stages of nodule development were examined, a particularly notable feature of nodules containing LPS-defective mutants was a large amorphous invasion structure containing bacteria embedded in a matrix material that reacted with MAC 265 and seemed to be identical to the matrix glycoprotein that normally occupies the lumen of infection threads. Repeated observations indicate that this large invasion structure is derived from the primary infection thread that conveyed bacterial invasion into the root cortex from the infected root hair: this was established by taking serial sections which made it possible to trace the path of the large invasion structure outwards to the surface of the nodule where it was found to originate in the remnant of a curled root hair. These observations reinforce the point that the structure and growth of infection threads is radically altered when they harbour LPS-defective mutants.

Several lines of evidence indicated that the LPS-defective mutants induced some form of host defence response during nodule development, which is consistent with a recent report that transcription of chalcone synthase may be induced in host cells invaded by such mutants (Yang et al., 1992). In particular, the cells that surrounded the large amorphous structure typical of these Fix$^-$ nodules resulted in bright autofluorescence under UV light, indicating the presence of aromatic compounds. Furthermore, when nodule cryosections were stained with berberine / aniline and examined under UV-light, these cells were shown to have fluorescent cell walls, presumably indicating the presence of lignin or suberin. Moreover, the matrix material of these infection structures showed a blue fluorescence with aniline, perhaps indicating the deposition of callose.

The abnormal development and anatomy of nodules induced by LPS-defective mutants was often difficult to interpret without the use of cytological markers and molecular probes. However, by using different monoclonal antibodies, it was possible to locate all the bacteria within a nodule section, and to distinguish between intracellular and cytoplasmic locations for the bacteria in adjacent tissue sections from the same nodule. Immunostaining with MAC 57 (an anti-LPS antibody) was used to locate all the rhizobia; MAC 265 (which identifies the plant extracellular matrix glycoprotein) was used to identifiy the rhizobia present in infection threads and extracellular spaces; and MAC 206 (which identifies the plant peribacteroid membrane) was used to distinguish the rhizobia that have been released into the intracellular (symbiosome) compartment (Figure 2). Despite the abnormal development of infection threads and the relative low number of infected host cells in nodules containing LPS-defective mutants, these infected host cells nevertheless induced leghaemoglobin production and the endosymbiotic bacteria enclosed by peribacteroid membranes also induced the synthesis of nitrogenase. The presence of both of these proteins was detected by immunostaining with specific antisera. The population of nodule cells invaded by LPS-defective mutants was found to be highly heterogeneous. Next to viable infected cells with nitrogenase-containing bacteroids, other cells were found with clear signs of cytoplasmic disorganisation and collapse. This suggests that the signal eliciting the host defence

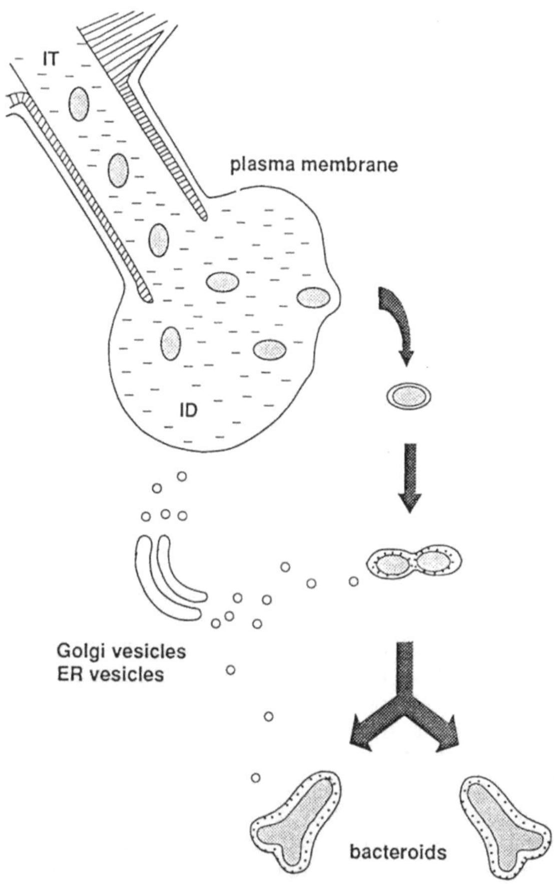

Figure 2. Stages of endocytosis and bacteroid differentiation.
Rhizobia within infection threads (IT) are bounded by plant cell wall and plant cell membrane. Uptake of bacteria into the cytoplasmic space takes place from infection droplets (ID), where host cell plasma membrane is not protected by plant cell wall. Endocytosis involves the elimination of plant matrix glycoprotein (recognised by antibody MAC 265) from the environment of rhizobia and its replacement by the surface glycocalyx of the plasma membrane (recognised by antibody MAC 206). Subsequent division and differentiation of rhizobia leads to the development of nitrogen-fixing bacteroids. The ensheathing plant membrane, which acquires new material by fusion of vesicles from the Golgi and endoplasmic reticulum (ER), becomes functionally specialised as the peribacteroid membrane. In pea nodules, the peribacteroid membrane divides in synchrony with dividing bacteroids so that only a single bacteroid is enclosed within each symbiosome unit.

response was very localised, and the host cell response was somewhat reminiscent of the hypersensitive response towards an invading pathogen.

The ultrastructural analysis of infected nodule cells also revealed clear differences between wild-type and LPS-defective bacteria. The mutant bacteria were always released into the plant cytoplasm surrounded by a peribacteroid membrane but, in contrast to wild-type bacteroids which are normally Y-shaped, the bacteroids formed by mutant strains were usually highly branched and much larger in size. Moreover, in nodules formed by LPS-defective mutants, several bacteroids were commonly seen inside the same peribacteroid membrane envelope even during the stage of most active division of wild-type bacteroids in the host cytoplasm, whereas in wild type nodule bacteroids were individually enclosed. Mutant bacteroids also showed premature senescence and induced the formation of apparently lytic vesicles in the host cell cytoplasm. From these observations of abnormal nodule development with LPS-defective strains, we conclude that the correct LPS structure is essential for invasiveness of plant cells and tissues; and for the avoidance of host defence responses.

From these observations of abnormal nodule development, it seems that LPS-defective mutants are a little closer to the borderline between symbiotic and pathogenic interactions with the host plant. We conclude that the correct LPS structure is essential for the avoidance of host defence responses; for the "fitness" of rhizobia as endophytes; and for "invasiveness" of plant tissues and cells.

2.3. LIPOPOLYSACCHARIDE EPITOPE VARIATION IN BACTEROIDS

In order to investigate the composition and structural variability of LPS for bacteria within the nodule, we isolated bacteroids from pea nodules by cell fractionation techniques and used this material as immunogen for rats. A range of monoclonal antibodies was isolated that reacted with different structural features (epitopes) associated with the lipopolysaccharide of *R. leguminosarum* strain 3841. Some of these antibodies were found to identify LPS epitopes whose expression *in vitro* was regulated by the physiological conditions of the growth medium, e.g., pH and oxygen concentration (Kannenberg & Brewin, 1989; Tao et al., 1992). We have now used these same monoclonal antibodies as probes to follow the expression of the corresponding LPS epitopes within the nitrogen-fixing pea nodule. This study has allowed us to investigate the physiological conditions in the peribacteroid fluid during the development of the symbiosome compartment. This was achieved by *in situ* immunolocalisation using the technique of immunogold staining with silver enhancement.

By taking serial median longitudinal sections and reacting each of them with an antibody of different specificity, it was possible to build up an interesting picture of LPS epitope variation for bacteroids in different regions of the same nodule. LPS epitope variation seemed to be governed by two different patterns (Figure 3). The first pattern (typified by MAC 281 epitope) showed a radial symmetry and seemed to reflect the local physiological conditions experienced by endosymbiotic bacteria within the nodule: this suggested that differences in oxygen availability might influence LPS epitope expression, presumably by affecting bacteroid respiration rate and hence perhaps by modifying the prevailing pH of the peribacteroid fluid. A second type of pattern for LPS epitope variation followed a linear axis of symmetry along the pea nodule from apex to base: these developmental changes appeared to be associated with endocytosis of bacteria and the progressive differentiation of the symbiosome compartment and the development of

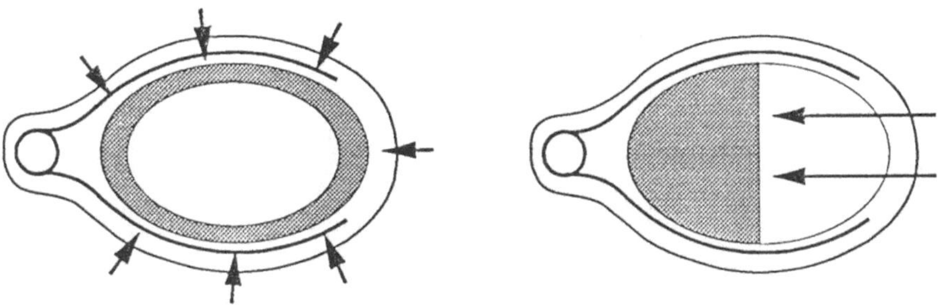

Figure 3. Patterns of LPS epitope variation in pea nodules.

nitrogen-fixation capacity in bacteroids. For one of these developmentally regulated LPS epitopes (identified by MAC 301), expression was closely linked to the expression of nitrogenase itself. However, expression of MAC 301 epitope was not dependent on the expression of nitrogenase because it was still found to be induced by bacteroids of Fix^- mutant strains that did not synthesise nitrogenase.

From these observations of LPS epitope variation in differentiating and nitrogen-fixing bacteroids, we suggest that the LPS of endosymbiotic rhizobia may play an important role in physiological adaptation to the microenvironmental conditions of the symbiosome compartment and may in fact govern the overall energetic efficiency of rhizobial strains as nitrogen-fixing endosymbionts.

2.4. DIFFERENTIATION OF THE PERIBACTEROID MEMBRANE

Within the maturing legume root nodule, Endosymbiotic bacteroids are individually enclosed by a plant-derived peribacteroid membrane, a differentiated form of plasma membrane (Perotto et al 1991), which apparently no longer is involved in synthesising cellulose or other components of the plant cell wall (Rae et al 1992). Instead, the peribacteroid membrane (pbm) becomes differentiated by acquiring nodule-specific proteins (nodulins) and transport functions associated with the specialised metabolism of this nitrogen-fixing "organelle". These developmental changes proceed in phase with the progressive differentiation of the intracellular rhizobia into nitrogen-fixing (bacteroid) forms.

We have recently identified a monoclonal antibody that recognises a new class of plant glycolipid membrane antigen. *In situ* immunostaining of pea nodule sections with this antibody reveals that the corresponding antigen is always present on the plasma membrane but it disappears from the peribacteroid membrane at a precise point in nodule differentiation, either

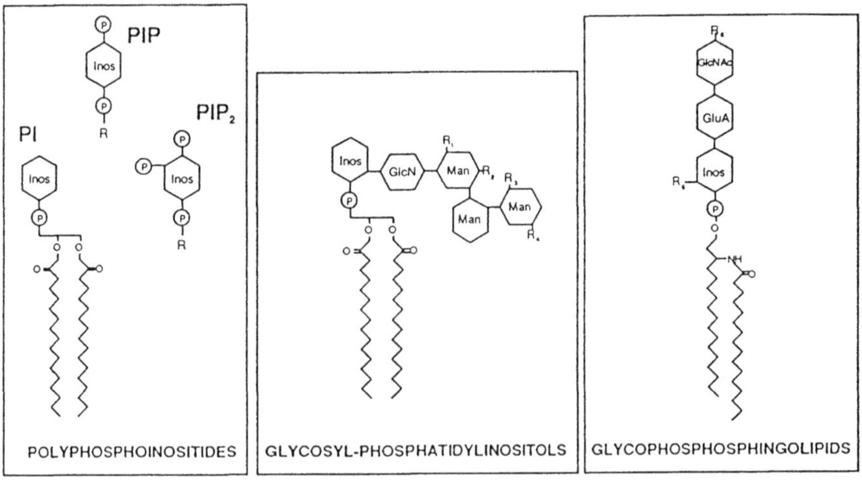

Figure 4. Inositol-containing lipids in plants.

synchronous with or slightly preceding the induction of leghaemoglobin. Thus, loss of the glycolipid antigen from the peribacteroid membrane coincides with differentiation of this membrane and the enclosed bacteria into nitrogen-fixing organelles. We therefore believe that we have identified a very early developmental switch, which may be relevant to symbiosome differentiation and perhaps to cell signal transduction pathways.

We have used various organic solvents to extract the glycolipid antigen from pea nodule membranes and we have examined the nature of the antigen by thin layer chromatography. Because we can identify a similar glycolipid component in carrot cell membranes, we have used carrot cell suspension cultures to perform incorporation studies with radiolabelled sugar precursors. We have thus demonstrated that the glycolipid contains inositol and glucosamine, probably as N-acetyl glucosamine since this sugar was found to inhibit antibody-antigen binding in ELISA assays. Moreover, the antibody reacts in dot immunoassays with phosphatidylinositol monophosphate (PIP), but not with phosphatidylinositol (PI) nor with phosphatidylinositol-bis-phosphate (PIP_2). Although further work is needed to characterise the chemical structure of this membrane glycolipid from pea nodules, the current experimental evidence suggests that it belongs to one of the groups of inositol-containing glycolipids shown in Figure 4, most probably the glycosyl-phosphatidylinositols. The properties of this group of "GPI" glycolipids has not been well characterised in plants. In animal systems, these molecules can act either as membrane anchors for Golgi-derived extracellular proteins or alternatively they can be cleaved as part of a signal transduction pathway, releasing diacyl glyceride and inositol phosphoglycan as intracellular messengers (Merida, 1992). Very little is known about the intracellular signals controlling differentiation of the peribacteroid membrane and of the consequent onset of nitrogen fixation. However, using a range of bacterial mutants in conjunction with molecular probes for *in situ* cytological analysis, it is now possible to embark on this investigation by analysing the synthesis and metabolism of this family of inositol-containing membrane glycolipids.

3. Conclusions

The analysis of abnormal nodule development induced by LPS-defective mutant strains and the analysis of epitope variation in bacteroids of wild type strains both suggest that LPS may have several important roles in colonising the intercellular spaces of plant tissues; in host cell invasion and in the process of nitrogen fixation itself. These functions include (i) initiation of infection threads; (ii) propagation and branching of infection threads; (iii) suppression of a host defence response; (iv) endocytosis; (v) bacteroid growth and cell division; (vi) physical interaction with the peribacteroid membrane; and (vii) physiological adaptation to the microenvironment of the peribacteroid fluid.

Perhaps these attributes can be grouped into two general categories, namely that of "invasiveness" and that of "fitness" as an endophyte. By "fitness" we imply that LPS structure is in some way adaptive to the extreme physiological conditions of the endophytic environment, either in the intercellular space or in the intracellular symbiosome compartment. By "invasiveness" we imply both the suppression of the host defence response and a physical interaction between bacterial LPS and a plant component such as the extracellular matrix glycoprotein or a component of the plant membrane glycocalyx present on the plasma membrane or the infection thread membrane (Perotto et al 1991).

In order to analyse these possible functions, it is now necessary to investigate aspects of the plant cell surface that might interact with bacterial lipopolysaccharide. These might include on the one hand the plant extracellular matrix glycoprotein which associates with the surface of rhizobia in infection threads and on the other hand components of the plant membrane glycocalyx which surrounds intracellular bacteroids as a peribacteroid membrane following endocytosis.

Further work is need to establish the significance of developmental regulation for glycosyl inositides on the peribacteroid membrane.

4. References

Brenchley, W. E., Thornton, H. G. (1925) The relation between the development, structure and functioning of the nodules on *Vicia faba*, as influenced by the presence or absence of boron in the nutrient medium, Proceedings Royal Society B XCVIII, 373-98.

Brewin, N. J. (1991) Development of the legume root nodule, Annual Review Cell Biology, 7, 191-226.

Carlson, R. W. (1984) The heterogeneity of *Rhizobium* lipopolysaccharide, Journal of Bacteriology, 158, 1012-1017.

Gray, J. X., and Rolfe, B. G. (1990) Exopolysaccharide production in *Rhizobium* and its role in invasion. Molecular Microbiology, 4, 1425-31.

Goodbody, K. C., and Lloyd, C. W. (1990) Actin filaments line up across *Tradescantia* epidermal cells, anticipating wound-induced division planes, Protoplasma, 157, 92-101.

Higashi, S., Kushiyama, K. and Abe, M. (1987) Electron microscopic observations of infection threads in driselase treated nodules of *Astragalus sinicus*, Canadian Journal of Microbiology 32, 947-952.

Kannenberg, E. L., and Brewin N. J. (1989) Expression of a cell surface antigen from *Rhizobium leguminosarum* 3841 is regulated by oxygen and pH, Journal Bacteriology, 171, 4543-4548.

Kannenberg, E. L., Rathbun, E. A., and Brewin, N. J. (1992) Molecular dissection of structure and function in the lipopolysaccharide of *Rhizobium leguminosarum* strain 3841 using monoclonal antibodies and genetic analysis, Molecular Microbiology (In Press).

Kijne, J.W. (1992) The *Rhizobium* infection process, in G. Stacey, R.H. Burris, and H.J. Evans (eds.) Biological Nitrogen Fixation, Chapman and Hall, pp 349-398.

Merida, I. (1992) Glycosyl-phosphatidylinositol: a novel mechanism of signal transduction, New Biologist 4, 207-211.

Noel, K.D. (1991) Rhizobial polysaccharides required in symbioses with legumes, in (D. P. S. Verma (ed.) Molecular Signals in Plant-Microbe Interactions, CRC Press Inc., Boca Raton, Fla., pp. 341-357.

Priefer, U.B. (1989) Genes involved in lipopolysaccharide production and symbiosis are clustered on the chromosome of *Rhizobium leguminosarum* biovar *viciae* VF39. Journal of Bacteriology, 171, 6161-6168.

Perotto, S., VandenBosch, K. A., Butcher, G. W., and Brewin, N.J. (1991) Molecular composition and development of the plant glycocalyx associated with the peribacteroid membrane of pea root nodules, Development, 112, 763-773.

Rae, A. L., Perotto, S., Knox, J. P., Kannenberg, E. L., and Brewin, N. J. (1991) Expression of extracellular glycoproteins in the uninfected cells of developing pea nodule tissue, Molecular and Plant-Microbe Interactions, 4, 563-570.

Rae, A. L., Bonfante-Fasolo, P., and Brewin, N. J. (1992) Structure and growth of infection threads in the legume symbiosis with *Rhizobium leguminosarum*, Plant Journal 2, 385-395.

Scheres, B., Van de Weil, C., Zalensky, A., Horvath, B., Spaink, H., Van Eck, H., Zwartkruis, F., Wolters, A. M., Gloudemans, T., van Kammen, A, and Bisseling, T. (1990) The ENOD12 gene product is involved in the infection process during the pea-*Rhizobium* interaction, Cell, 60, 281-94.

Spaink, H. P., Sheeley, D. M., van Brussel, A. A. N., Glushka, J., York, W. S., Tak, T., Geiger, O., Kennedy, E. P., Reinhold, V. N. and Lugtenberg, B. J. J. (1991) A novel highly unsaturated fatty acid moiety of lipo-oligosaccharide signals determines host-specificity of *Rhizobium*, Nature, 354, 125-130.

Stacey, G., So, J.-S., Roth, L. E., Bhagya Lakshmi, S. K., and Carlson, R. W. (1991) A lipopolysaccharide mutant of *Bradyrhizobium japonicum* that uncouples plant from bacterial differentiation, Molecular and Plant-Microbe Interactions 4, 332-340.

Sutton, M. J., Lea, E. J. A., Crank, S., Rivilla, R., Economou, E., Ghelani, S., Johnston, A. W. B., Downie, J.A. (1992) NodO: a nodulation protein that forms pores in membranes, this symposium volume.

Tao, H., Brewin, N. J., and Noel K. D. (1992) *Rhizobium leguminosarum* CFN42 lipopolysaccharide antigenic changes induced by environmental conditions, Journal of Bacteriology 174, 2222-2229.

VandenBosch, K. A., Bradley, D. J., Knox, J. P., Perotto, S., Butcher, G. W. and Brewin, N. J. (1989) Common components of the infection thread matrix and the intercellular space identified by immunocytochemical analysis of pea nodules and uninfected roots, EMBO Journal 8, 335-342.

Yang, W-C, Canter-Cramers, H. C. J., Hogendijk, P., Katanakis, P., Wijffelman, C. A., Franssen, H., Van Kammen, A., Bisseling, T. (1992) *In situ* localisation of chalcone synthase mRNA in pea root nodule development, Plant Journal 2, 143-151.

CONTROL OF INFECTION IN THE ALFALFA-*RHIZOBIUM MELILOTI* SYMBIOSIS

J. VASSE, F. de BILLY and G. TRUCHET
Biologie Moléculaire des Relations Plantes-Microorganismes, CNRS-INRA
31326 Castanet-Tolosan Cédex
France

1. Introduction

Among microorganisms which interact with plants, some are pathogenic and elicit either disease on a susceptible (compatible) host or disease resistance on a resistant (incompatible) plant. In the latter case, plant defence mechanisms are induced and a hypersensitive reaction, H.R., often occurs. The H.R. is a localized plant reaction resulting in the necrosis of the two partners and the arrest of the progression of the invader. In contrast to pathogenic associations, symbiotic interactions result in the ability of the two partners to establish a beneficial relationship. In the *Rhizobium*-legume symbiosis a series of events, recognition between symbionts, plant invasion by the bacterium and induction of plant cortical cell division, lead to the organogenesis of a new organ, the root nodule, in which molecular nitrogen is reduced to ammonia by the bacteria.

Nodulation is subjected to autoregulation, a finely balanced phenomenon genetically controlled by the plant which limits the number of nodules produced (Caetano-Anollés and Gresshoff, 1991a). Although evidence exists that various symbiotic steps, including the infection process, are feedback regulated, the autoregulatory mechanisms controlling root nodule formation are poorly understood. In our laboratory we are investigating autoregulation of nodulation by studying abortive infection in alfalfa inoculated with its natural symbiont *R. meliloti* strain RCR2011. Our results are briefly summarized below.

2. Results

This work was based initially on the finding that, 2 weeks after inoculation with the wild type strain, some root cortical cells looked different from adjacent cells by displaying a pigmentation which ranged from light yellow to black. Such cells, referred to as pigmented cells below, were seen either individually or in groups of 2 or 3 in the region of the root where nodules had developed. They were not seen in non-inoculated plants or in plants inoculated with strains having a mutation either in the common *nodA* gene or the host-specific *nodH* gene, both of which do not infect or nodulate alfalfa.

To determine if the appearance of pigmented cells was correlated with the infection process, we utilized a *R. meliloti* strain bearing the constitutively expressed chimaeric gene *hemA*::*lacZ* to visualize the location of the bacteria. Our analysis showed that

E. W. Nester and D. P. S. Verma (eds.),
Advances in Molecular Genetics of Plant-Microbe Interactions, 381–384.

infection threads were often located in the vicinity of pigmented cells. Looking at slightly cleared roots 2 weeks after inoculation, we found that, approximately 90% of the pigmented cells contained the extremity of an infection thread. The threads were often seen as coiled tubular structures. Quantitative experiments performed with plants collected at different times after inoculation showed that the ratio of the number of pigmented cells to the number of infection threads, increased progressively and significantly from day 7 to day 14, by which time around 25% of the infection threads were found to terminate in a pigmented cell.

Examination of whole plants by light microscopy, also allowed us to make two interesting findings. Firstly, we found that two different types of infection threads existed in the root area where nodules had developed. Whereas successful infection threads observed above a growing nodule were straight and regular in shape, infection threads which ended in a pigmented cell, generally had an anomalous aspect and were decorated by wall projections. In these threads, the chimaeric gene was not expressed in bacteria seen in the region of the thread lying in the pigmented cell. Secondly, we found that, in many cases, cell division had occurred in the inner cortical cells lying between the pigmented cells and the endodermis. These division centers displayed the cytological features of a nodule primordium but not those of a genuine nodule meristem.

Histological and ultrastructural studies confirmed that there was a close relationship between the occurrence of pigmented cells and the infection process. Light microscopy of transverse sections of pigmented cells confirmed what the examination of whole plants strongly suggested and provided more detailed observations: Pigmentation was restricted to individual or a few cortical cell(s); Infection threads were observed in the pigmented cells which displayed a star-like shape and a cytoplasmic content different from adjacent cells; Infection threads were decorated with wall appositions and; Cell divisions often occurred in the inner cortical cells located radially to the pigmented cells.

Ultrastructural investigations showed that both symbionts undergo necrosis in pigmented cells. The plant cell cytoplasm appeared highly disorganised and the bacteria no longer displayed the ultrastructural features characteristic of those seen in the distal region of the thread. Moreover, the thread cell walls appeared thickened by wall material appositions, while an amorphous material was observed inside the pigmented (necrotic) cells.

These results strongly suggested that the host plant might restrict the further development of some infection threads by eliciting a hypersensitive-like reaction. To examine this hypothesis further we carried out the following experiments.

Autofluorescent substances such as wall bound phenolics and phytoalexins accumulate in plant cells exhibiting a H.R. as a result of incompatible plant-pathogen interaction. We investigated whether an accumulation of such substances, also happened during abortive infection in alfalfa. Firstly, a fixation with potassium permanganate, an indicator of phenolic compounds, resulted in a strong staining of cortical cells previously detected as pigmented on whole plants. Secondly, fluorescence microscopy indicated the presence of autofluorescent compounds bound to the cell walls of both necrotic (pigmented) and adjacent cells and in the lumen of the necrotic cells. Looking at successful infection we found a strong autofluorescence at the infection site in the root hair cell. However, in contrast to infection threads terminating in necrotic cells which autofluoresced, no fluorescence was detected on successful threads examined during early infection or seen in the section of cortical cells located distal to a developing nodule.

Our last objective was to specify whether proteins involved in plant defence mechanisms could be immunodetected in necrotic cells in which infections aborted or, as a control, in cortical cells traversed by "successful" infection threads and observed distal to developing nodules. To address this question we used polyclonal antibodies directed against melon and alfalfa hydroxyproline-rich glycoproteins (HRGP), potato phenylalanine ammonia lyase (PAL) and chalcone synthase (CHS) and different tobacco pathogenesis related (PR) proteins such as chitinases, glucanases, thaumatin-like proteins and PR proteins of class 1C. Our results can be summarized as follows: HRGP were detected on walls of all cortical cells (non-inoculated, necrotic or located on developing nodules). They were also seen on the walls of infection threads and in the cytoplasm of necrotic cells. PAL and CHS were immunodetected in the cytoplasm of all invaded and adjacent cortical cells, as well as in the cytoplasm of the apical cells of growing nodules. Under the experimental conditions used, acidic chitinases PR-P and PR-Q were the only PR proteins detected in the lumen of necrotic cells. All the controls, involving adsorbed immune serum (HRGPs), preimmune serum (PR proteins) or omission of the first antibodies (PAL and CHS), were negative.

3. Discussion

To our knowledge, this work provides the first experimental evidence that a legume (alfalfa) can react to infection by a wild-type *Rhizobium* (*R. meliloti*) by eliciting a localized response displaying the cytological and biochemical features of a hypersensitive reaction (H.R.) similar to that only described so far in plant-pathogen interactions. We hypothesize that the H.R. response is one of the mechanisms by which the plant controls infection and therefore autoregulates nodulation.

It appears that autoregulatory mechanisms can impair various ontogenic steps in the alfalfa-*R. meliloti* symbiosis. Besides the infection process, the two ontogenic steps involving *Rhizobium*-induced cortical cell divisions, namely the induction of the nodule primordium (Caetano-Anollès and Gresshoff, 1991b) and the formation of the nodule meristem (this work), are also submitted to feedback control in alfalfa. It is worth mentioning that these two steps are induced on alfalfa by purified lipo-oligosaccharidic Nod factors produced by *R. meliloti* (Lerouge *et al.*, 1990; Truchet *et al.*, 1991). Moreover, we note that the core of rhizobial Nod factors is made of residues of N-acetyl-D-glucosamine β 1-4 linked, as in chitin (Lerouge *et al.*, 1990). Therefore, rhizobial Nod factors could both act as elicitors and as substrates for acidic chitinases which are specifically detected during abortive infection.

Our work raises many questions which deserve further investigation: How does the plant behave as a susceptible (compatible) or resistant (non-compatible) host towards the same invading bacterium? What is the primary stimulus which elicits the H.R. in the *Rhizobium*/legume interaction? Does *Rhizobium* possess pathogenicity-like genes? Answering these questions would lead to a better understanding of how autoregulation of nodulation takes place in symbiotic associations.

4. References

Caetano-Anollès, G. and Gresshoff, P.M. (1991a) "Plant genetic control of nodulation", Annu. Rev. Microbiol. 45, 345-382.

Caetano-Anollès, G. and Gresshoff, P.M. (1991b) "Alfalfa controls nodulation during the onset of *Rhizobium*-induced cortical cell division", Plant Physiol. 95, 366-373.

Lerouge, P., Roche, P., Faucher, C., Maillet, F., Truchet, G., Promé, J.C. and Dénarié, J. (1990) "Symbiotic host-specificity of *Rhizobium meliloti* is determined by a sulphated and acylated glucosamine oligosaccharide signal", Nature 344, 781-784.

Truchet, G., Roche, P., Lerouge, P., Vasse, J., Camut, S., de Billy, F., Promé, J.C. and Dénarié, J. (1991) "Sulphated lipo-oligosaccharide signals of *Rhizobium meliloti* elicit root nodule organogenesis in alfalfa", Nature 351, 670-673.

Acknowledgments. We are very grateful to M.T. Esquerré-Tugayé, K. Hahlbrock and B. Fritig for kindly providing us with the antibodies directed, respectively, against HRGPs, PAL and CHS and PR Ps. We also acknowledge J. Cullimore and T. Finan for reviewing the manuscript and A. Moisan for the statistical analysis of plant assays.

INDUCTION OF GENES ENCODING PHENYLPROPANOID BIOSYNTHETIC ENZYMES IN SOYBEAN ROOTS INOCULATED WITH B. japonicum, REQUIRES NOD GENE INDUCTION AND OCCURS INDEPENDENT OF ANY KNOWN HOST FUNCTIONS.

Elizabeth Estabrook, Carol Potenza, A. Inez Feder and Champa Sengupta-Gopalan. New Mexico State University, Mol. Biol. Program, Dept. of Agron. & Hort., Las Cruces, NM 88003 USA

ABSTRACT

To address the possible roles of phenylpropanoid compounds in events occurring in roots following inoculation with the compatible symbiont, we have monitored expression of the gene members encoding phenylalanine ammonia lyase (PAL), chalcone synthase (CHS) and chalcone isomerase (CHI) during nodule development in soybeans. Plant and bacterial mutants that arrest nodule development at defined stages were analyzed to correlate changes in expression of PAL, CHS and CHI genes with distinct events in nodule development. Our results suggest that induction of the 'symbiosis specific' PAL and CHS gene members occurs prior to any known host responses like root hair curling, infection thread formation and cortical cell proliferation. Furthermore, our results show a direct correlation in the level of PAL and CHS transcripts with the number of successful nodule foci suggesting that the resulting phenylpropanoid compounds may play a role in cortical cell proliferation associated with nodule development.

Introduction

Plants produce a wide variety of compounds from the phenylpropanoid pathway. Phenylpropanoid compounds have various roles in plant growth and development. A host defense response towards a pathogen includes phenylpropanoid derived phytoalexins and the first chemical signals inducing the nod genes in legume-(Brady) rhizobium interaction are also phenylpropanoid derived compounds. Among the other functions that flavonoids perform, two of the pertinent functional roles proposed for them in relation to nodule initiation is as natural auxin transport inhibitors [5] and cytokinins [1].

Three of the enzymes involved in the synthesis of phenylpropanoid compounds are phenylalanine ammonia-lyase (PAL), which catalyzes the irreversible deamination of L-phenylalanine to yield trans-cinnamic acid and NH_4^+, chalcone synthase (CHS) which catalyzes the first committed step in flavonoid biosynthesis to produce chalcone and chalcone isomerase (CHI) which catalyzes the stereospecific conversion of chalcone to their corresponding flavanones. Both PAL and CHS are members of multigene families and the members are differentially regulated utilizing different induction signals [4].

Initial studies from our laboratory showed a dramatic increase in the level of expression of specific members of CHS and PAL genes in roots of soybean following infection with the compatible symbiont [3]. The results suggest a possible role of flavonoid/isoflavonoids in nodule initiation. In an attempt to determine the functional significance and regulatory mechanism underlying enhanced expression of PAL and CHS genes in the early stages of nodule development, we have extended our study to different mutant soybean-B. japonicum associations that are defective in some early steps of nodule development. We have also included CHI in our analysis since it catalyzes the second committed step in flavonoid biosynthesis.

E. W. Nester and D. P. S. Verma (eds.),
Advances in Molecular Genetics of Plant-Microbe Interactions, 385–389.

Materials and Methods

Table 1 describes the different plant mutants used in this study. Three day old seedlings were inoculated with either YM (uninoculated) or with B. japonicum (inoculated) and the roots harvested at different days following inoculation (DPI). Growing conditions were as described previously [7]. Total RNA was extracted by the LiCl precipitation method and poly(A)-RNA was isolated by poly(u)-sepharose chromatography. Genomic DNA was isolated by the procedure of Richter et. al. [7]. Southern blot analysis and northern blot analysis was done according to standard procedures. In vitro translation of the hybrid selected mRNA was done in the wheat germ system as described earlier [7].

Table 1. Description of Plants

Name	Description	RHC[a]	IT[b]	CCP[c]	Aborted CCP
cv Bragg	Wild-type	++	++	++	++
nts 382	Supernodulating	++	++	++	+[d]
nod 49	Nonnodulating	–[e]	–	+	+
nod 139	Nonnodulating	–	–	–	NA[f]

[a]Root hair curling
[b]Infection thread formation
[c]Cortical cell proliferation
[d]Below wild-type levels
[e]No occurrence
[f]Not applicable

Results

GENE SPECIFIC PROBES OF PAL AND CHS IDENTIFY A SUBSET OF THE TOTAL GENE MEMBERS MAKING UP THE GENE FAMILIES

Soybean cDNAs for PAL and CHS were isolated from a soybean nodule cDNA library, sequenced and compared to published sequences of the corresponding genes. Fragments corresponding to the conserved regions (5' PAL and 5' CHS) and 3' untranslated nonconserved regions (3' PAL and 3' CHS) were used as probes in Southern and northern analysis. The CHI probe was from french bean.

Restricted soybean DNA when analyzed by Southern blot analysis using the different probes showed that while the 5' PAL and 5' CHS conserved probes hybridized to multiple bands, the 3' probes hybridized to only two of the multiple bands in each case. The bean CHI fragment hybridized to 1 or 2 restriction fragments suggesting that CHI is encoded by a single gene or a closely linked gene family (data not shown).

THE GENE SPECIFIC PROBES IDENTIFY GENES WHOSE EXPRESSION IS ENHANCED BY SYMBIOTIC DEVELOPMENT

Poly(A)-RNA isolated from roots of cv. Bragg at different times following inoculation with B. japonicum U110, were subjected to northern analysis with the 5' and 3' PAL and CHS probes and with the bean CHI probe. As seen in Fig. 1A, PAL and CHS gene expression (as monitored by conserved and gene specific probes) was enhanced in the infected roots when compared to the uninfected roots at the same developmental stage. Expression of CHI gene(s), however, was not affected as a result of infection with the WT B. japonicum strain. As seen in Fig. 1B, the 3' PAL besides hybridizing to the 2.5 kb PAL transcript, also hybridized to a 1.7 kb transcript, whose identity has not yet been determined. The pattern of expression of this unidentified gene was similar to that of PAL except at 0 DPI.

A

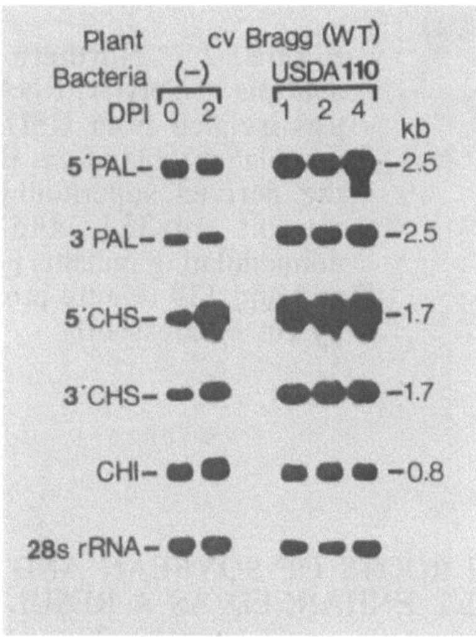

B

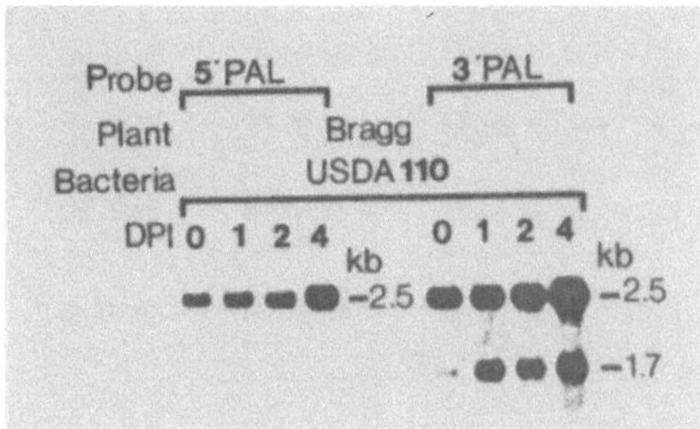

Figure 1A. Northern blot analysis of total poly(A)-RNA (2.0 μg) isolated from roots of wild type cv Bragg at different times following inoculation (DPI = Days post inoculation) with USDA 110. This blot was probed with a 28S rRNA gene to normalize RNA loads. 1B. Northern blot analysis of total poly(A)-RNA from cv. Bragg roots at different days post inoculation with U110 and probed with the 5' PAL and 3' PAL probes.

ENHANCEMENT IN THE EXPRESSION OF PAL AND CHS GENES IN SOYBEAN ROOTS FOLLOWING INOCULATION WITH B. japonicum, OCCURS INDEPENDENT OF ALL THE HOST-RELATED EVENTS

To understand the functional significance of the enhanced expression of PAL and CHS genes in the roots following inoculation with the compatible symbiont, soybean mutants defective in some early stage of nodule development, were analyzed for PAL, CHS and CHI gene expression. As seen in Fig. 2, the supernodulating mutant showed higher level of PAL and CHS gene expression between 2 and 4 DPI when compared to WT. The PAL transcript level was higher at 4DPI in nod 49 and nod 139 roots, while the CHS transcript level appeared similar to that in the WT cultivar. The level of CHI transcript in the roots of the supernodulating and the nonnodulating mutant nod 49 appeared identical to that in the WT cv. Bragg. However, nod 139 appeared to show increased expression of CHI at 2 and 4 DPI. Analysis of roots of nts 382 inoculated with R. fredii U193, showed the same enhanced expression of PAL and CHS genes as seen with roots inoculated with B. japonicum U110 (data not shown). R. fredii U193 show limited to no infection thread formation but allows normal level of cortical cell proliferation.

These results suggest that increased expression of CHS and PAL genes occurs independent of root hair curling, infection thread formation or cortical cell proliferation.

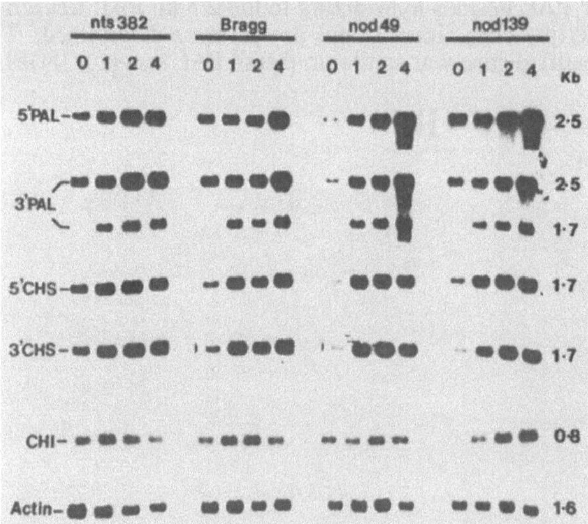

Figure 2. Northern blot analysis of poly(A)-RNA (2.5 μg) isolated from USDA110 inoculated roots of cv. Bragg, the derived supernodulating mutant nts382 and the nonnodulating mutants <u>nod</u> 49 and <u>nod</u> 139 (kindly provided by Dr. P. Gresshoff).

THREE CHS GENES ARE EXPRESSED IN ROOTS OF SOYBEAN AND THE EXPRESSION OF THESE GENES IS HIGHLY ENHANCED AS A RESULT OF INFECTION WITH <u>B. japonicum</u>

Analysis of CHS hybrid select translation products by 2D gel electrophoresis showed that three distinct translatable CHS mRNAs were selected from both uninfected and infected roots (Figure 3B). However, based on the relative level of translation products, it would appear that the amount of CHS mRNAs selected from infected roots was several fold higher than from uninfected roots.

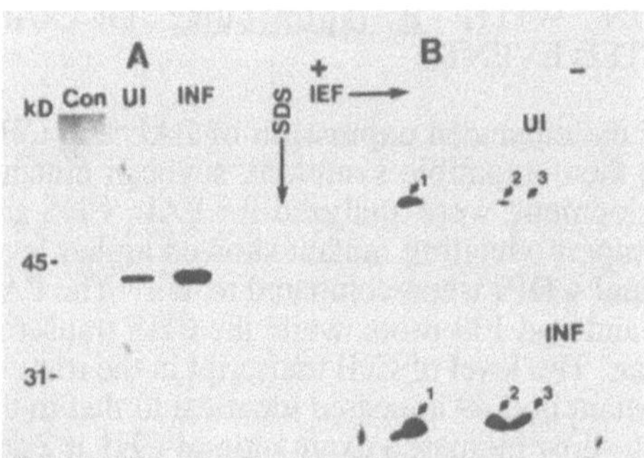

Figure 3. Analysis of CHS hybrid select translation products. Immobilized CHS DNA was used to hybrid select RNA from total poly(A)-RNA from uninoculated (UI) and inoculated nts382 roots (4 days post inoculation) (IN) under identical conditions. The selected RNA was translated <u>in vitro</u> in the wheat gern system and the translation products subjected to ID SDS PAGE (panel A) and 2D PAGE (panel B) followed by autoradiography. No RNA control is represented as CON.

Discussion

We have presented data that would suggest that some factor from the symbiont enhances the expression of CHS and PAL genes in the roots of soybean. Furthermore, our results also suggest that induction of these genes is independent of all the early host-related events in nodule development. Increased expression of the PAL and CHS genes probably leads to increased synthesis of flavonoid compounds. An increase in flavonoid synthesis and secretion as a result of infection with the symbiont has been reported for V. sativa and for white clover [6]. Increase in phenylpropanoid compounds probably is to accommodate for increased expression of the NOD genes or to function as growth regulators. The fact that CHI gene expression in soybean roots is not affected as a result inoculation with B. japonicum, would suggest that the biochemical conversion mediated by CHI may not be the limiting step in the pathway.

Based on the analysis of CHS and PAL gene expression in the inoculated roots of the different nodulation mutants, we propose a model that is an extension of the model proposed by Caetano-Annolles and Gresshoff [2], for the regulation of nodule formation. The NOD gene inducing flavonoids are secreted by the roots which in turn induce the NOD genes. The NOD factor synthesized by the bacteria, directly or indirectly induces the roots to produce a factor X. Factor X causes an increase in the expression of PAL and CHS genes which are responsible for increased synthesis of phenylpropanoid compounds. The phenylpropanoid compounds act as growth regulators and initiate cortical cell proliferation. A result of cortical cell proliferation is the production of factor Q which is translocated to the shoots where the shoot derived inhibitor (SDI) is produced. The SDI is translocated to the roots and acts as a repressor for PAL and CHS gene expression. Since SDI is not produced in the supernodulating mutant, PAL and CHS gene expression is not repressed, and more phenylpropanoid compounds are formed to continue meristematic activity at the nodule foci.

Work is in progress to isolate the 'symbiosis enhanced' CHS and PAL gene members and determine how they differ from pathogen induced gene members.

References

1. Binn, A.N., Chen, R.H., Wood, H.N., and Lynn, DG. (1987). Cell division promoting activity of naturally occurring dehydrodiconiferyl glucosides: Do cell wall components control cell division. Proc. Natl. Acad. Sci., USA 84, 980-984.
2. Caetano-Anolles, G., and Gresshoff, P.M. (1990). Early induction of feedback regulatory responses governing nodulation in soybean. Plant Sci. 71, 69-81.
3. Estabrook, E.M., and Sengupta-Gopalan, C. (1991). Differential expression of phenylalanine ammonia-lyase and chalcone synthase during soybean nodule development. Plant Cell 3, 299-308.
4. Hahlbrock, K., and Scheel, D. (1989). Physiology and molecular biology of phenylpropanoid metabolism. Annu. Rev. Plant Physiol. Plant Mol. Biol. 40, 347-369.
5. Jacobs, M., and Rubery, P.H. (1988). Naturally occurring auxin transport regulators. Sci. 241, 346-349.
6. Recourt, K., Schripsema, J., Kijne, J.W., Van Brussel, A.A.N., and Lugtenberg, B.J.J. (1991). Inoculation of *Vicia sativa* subsp. *nigra* roots with *Rhizobium leguminosarum* biovar *viciae* results in release of *nod* gene activating flavanones and chalcones. Plant Mol. Biol. 16, 841-852.
7. Richter, H.R., Sandal, N.N., Marcker, K.A., and Sengupta-Gopalan, C. (1991). Characterization and genomic organization of a highly expressed late nodule gene subfamily in soybeans. Mol. Gen. Genet. 229, 445-452.

Section 7 / Bacteria-Plant Interactions: Plant Response

ANALYSIS OF THE *ARABIDOPSIS* DEFENSE RESPONSE TO *PSEUDOMONAS* PATHOGENS.

FREDERICK M. AUSUBEL, JANE GLAZEBROOK, JEAN GREENBERG, MICHAEL MINDRINOS, GUO-LIANG YU
Department of Genetics, Harvard Medical School
and
Department of Molecular Biology
Massachusetts General Hospital
Boston, MA 02114 USA

ABSTRACT. We have studied the response of *Arabidopsis thaliana* to the bacterial pathogen *Pseudomonas syringae* pv. *maculicola* (*Psm*) strain ES4326. Several previously unknown *Arabidopsis* defense-related genes were identified including ones encoding a glutathione-S-transferase, a superoxide dismutase, a lipoxygenase, and two calmodulin-like proteins. Interestingly, mRNA corresponding to each of these genes displayed markedly different patterns of accumulation during the defense response to *Psm* ES4326. We have isolated three categories of *Arabidopsis* mutants that show an aberrant defense response to *Psm* ES4326. Three mutants were isolated that do not mount a hypersensitive response (HR) when infiltrated with *Psm* ES4326/*avrRpt2* but are still able to display an HR in response to other *avr* genes. At least two of these mutants are allelic and map to chromosome IV. To facilitate the identification of additional *Arabidopsis* mutants that do not mount an HR in response to an *avr* gene, we developed a new method that involves vacuum infiltration of seedlings growing in petri plates. We also isolated three mutants that synthesize decreased levels of camalexin, an indole-based *Arabidopsis* phytoalexin. Two of the three camalexin mutants are significantly more permissive for the growth of *Psm* ES4326 than wild-type plants. Finally, five *Arabidopsis* mutants were isolated that display accelerated disease symptoms in response to *Psm* ES4326. These latter mutants, which were given the name *acd* for accelerated cell death, were assigned to two complementation groups.

Introduction

Plant defense responses have not yet been subjected to an in-depth genetic analysis involving the generation and analysis of mutant plants defective in particular responses. To fill in this void, we and others have recently turned to the easily manipulated model crucifer, *Arabidopsis thaliana* (Meyerowitz, 1989; Redei, 1975), and have developed a new model pathogenesis system that involves the infection of *Arabidopsis* with phytopathogenic bacteria and fungi (Ausubel, *et al.*, 1991; Davis and Ausubel, 1989; Bent *et al.*, 1991; Dangl *et al.*, 1991; Dangl *et al.*, 1992; Daniels *et al.*, 1991; Davis *et al.*, 1989, 1991; Debener *et al.*, 1991; Dong *et al.*, 1991; Keith et al., 1991; Koch and Slusarenko, 1990a, 1990b; Li and Simon, 1990; Schott *et al.*, 1990; Simpson and Johnson, 1990; Tsuji and Somerville, 1988; Tsuji *et al.*, 1990; Tsuji *et al.* 1992; Whalen *et al.*, 1991). *Arabidopsis* offers several advantages compared to other plants that have been used previously to study plant-pathogen interactions. Because of its small stature, fast generation time, copious production of tiny seeds, and small genome (Redei 1975;

E. W. Nester and D. P. S. Verma (eds.),
Advances in Molecular Genetics of Plant-Microbe Interactions, 393–403.

Koornneef *et al.*, 1983; Meyerowitz, 1987, 1989; Koornneef, 1987), it should be possible to genetically identify genes involved in the defense response and then clone these genes using techniques such as chromosome walking (Chang *et al.*, 1988; Nam *et al.*, 1989), gene tagging (Feldman, 1991), or genomic subtraction (Sun *et al.*, 1992).

Pseudomonas pathogens of *Arabidopsis* and cloning an *Arabidopsis*-specific *avr* gene

Despite the advantages of *Arabidopsis* as a model laboratory plant, at the time we initiated our studies, there were no published reports on *Arabidopsis* bacterial pathogens. We therefore screened *Pseudomonas* strains that are known pathogens of crop plants related to *Arabidopsis* (family Crucifereae which includes radishes, mustard, cabbages, turnips and rape). In cooperation with Brian Staskawicz's lab at the University of California at Berkeley, two highly virulent strains, *P. syringae* pv. *maculicola* (*Psm*) ES4326 and *P. syringae* pv. *tomato* (*Pst*) DC3000, and two avirulent strains, *Pst* MM1065, and *P. cichorii* 83-1, were chosen for further analysis (Davis *et al.*, 1991; Dong *et al.*, 1991; Whalen *et al.*, 1991). *Psm* ES4326 and *Pst* DC3000 multiply 10^4 fold when approximately 10^3 to 10^4 cells are infiltrated into *Arabidopsis* leaves. Water-soaked lesions appear over the course of two days. In contrast, *Pst* MM1065 and *P. cichorii* 83-1 multiply at most 10 fold in *Arabidopsis* leaves. *Pst* MM1065 elicits the appearance of a mildly chlorotic dry lesion after 48 hours and *P. cichorii* 83-1 elicits a typical dry brown hypersensitive response (HR) lesion within 16 hours.

One specific genetic strategy that we have pursued with these *Pseudomonas* strains is to clone an *avr* gene from the avirulent strain *Pst* MM1065 so that it could be transferred to virulent strains for use in eliciting an HR. We reasoned that this strategy would increase the chances of isolating *Arabidopsis* mutants that fail to respond to *avr*-generated signals because it avoids the problem that an observed response is a consequence of the activation of parallel signal transduction pathways by independent *avr* genes. Using standard techniques (Staskawicz *et al.*, 1984), we cloned an *avr* gene, *avrRpt2*, from the avirulent strain *Pst* MM1065 (Dong *et al.*, 1991). *Psm* ES4326/*avrRpt2* elicits a strong HR-like response within 16 hours and multiplies 50 fold less than *Psm* ES4326 in *Arabidopsis* leaves (Dong *et al.*, 1991). Coincidentally, Brian Staskawicz's lab at UC Berkeley also cloned *avrRpt2* (from *Pst* JL1065, the parent of MM1065; Whalen *et al.*, 1991). In an analogous set of experiments, Debener *et al.* (1991) cloned a putative *avr* gene, *avrRpm1*, from *Psm* strain "M2".

Cloning *Arabidopsis* defense-related genes

We have used RNA blot analysis and nuclear run-on analysis of mRNA initiation to monitor the induction of individual *Arabidopsis* defense-related genes during the defense response. To obtain the DNA probes for these experiments, we cloned several *Arabidopsis* genes whose counterparts in other plants are pathogen-induced. We also screened a cDNA library constructed with RNA isolated following infiltration of *Arabidopsis* leaves with *Pst* MM1065 for clones corresponding to pathogen-induced genes. The *Arabidopsis* genes cloned using heterologous genes as hybridization probes included the single *Arabidopsis* gene encoding chalcone synthase (*CHS*) (Feinbaum and Ausubel, 1988), one of several *Arabidopsis* genes encoding phenylalanine ammonia lyase (*PAL1*) (Davis *et al.*, 1991), and three physically adjacent *Arabidopsis* genes encoding β-1,3-glucanases (*BGL1, BGL2, BGL3*) (Dong *et al.*, 1991). Ohl *et al.* (1990) have also cloned the *Arabidopsis PAL1* gene.

The *Arabidopsis* defense-related genes identified in the *Arabidopsis* cDNA library included one of at least three genes encoding glutathione-S-transferase (*GST1*, *GST2*, *GST3*), two different genes called *PIG2* and *PIG18* (pathogen-induced-gene) encoding calmodulin-like proteins, and a gene called *PIG27* that encodes a protein of unknown function (G.-L. Yu and F. Ausubel, unpublished results). The observation that *GST* gene(s) are induced following pathogen attack is interesting in light of experiments carried out recently in other laboratories, indicating that an important feature of the plant defense response may be a membrane-generated oxidative burst (Apostal *et al.*, 1989; Doke, 1983a, 1983b, 1985; Doke and Chai, 1985; Doke and Ohashi, 1988). A likely role for GST in an oxidative burst is the detoxification in plant membranes of hydroxyalkenals, toxic byproducts of lipid peroxidation (Alin, 1985; Babior, 1984; Chai and Doke, 1987; Croft *et al.*, 1990; Dudler *et al.*, 1991; Johnson and Kitagawa, 1985; Kato and Misawa, 1976; Keppler and Novacky, 1987; Morgenstern and de Pierre, 1983; Morgenstern *et al.*, 1989; Prohaska, 1980; Rogers *et al.*, 1988; Thompson *et al.*, 1987) A pathogen-induced *GST* gene has also been identified in wheat (Dudler *et al.*, 1991).

In addition to the *Arabidopsis* defense-related genes described above, we have also monitored *Pseudomonas*-mediated induction of several *Arabidopsis* genes cloned in other laboratories, including *DHS1, DHS2, SOD1,* and *LOX1*. *DHS1* and *DHS2* encode 3-deoxy-D-arabino-heptulosonate 7-phosphate (DAHP) synthase which catalyzes the first committed step in aromatic amino acid biosynthesis (Keith *et al.*, 1991). We reasoned that increased usage of phenylalanine in the phenylproponoid pathway during a defense response might lead to depletion of the phenylalanine pools unless additional DAHP synthase were synthesized. *SOD1* encodes one of several superoxide dismutases (Hindges and Slusarenko, 1991), which, like glutathione-S-transferase, is probably involved in protecting plant cells from the deleterious effects of an oxidative burst. *LOX1* encodes one of at least two *Arabidopsis* lipoxygenases (T.K. Peterman, personal communication). The physiological function of lipoxygenase is not clear. It may generate fatty acid hydroperoxides and free radicals that are toxic to microorganisms, it may be involved in the degeneration of plant membranes as part of the HR, or it may generate signal molecules like jasmonic acid (Siedow, 1991).

The results of a large number of RNA blot experiments are summarized in Table 1. The most striking conclusion arising from this series of experiments was that different *Arabidopsis* defense-related genes display markedly different patterns of mRNA accumulation. In previously published work we showed that virulent strains such as *Psm* ES4326 strongly elicited the gradual accumulation of *BGL1*, *BGL2*, and *BGL3* mRNAs but had only a modest effect on *PAL1* and *DHS1* mRNA accumulation (Dong *et al.*, 1991; Davis *et al.*, 1991; Keith *et al.*, 1991). In contrast, avirulent strains such as *Pst* MM1065 or *P. cichorii* strain 83-1 strongly elicited a rapid and transient accumulation of *PAL1* and *DHS1* mRNA but had only a minimal effect on *BGL1* and *BGL3* mRNA accumulation. *BGL2* mRNA is gradually induced at a moderate level by the avirulent strains. Importantly, *Psm* ES4326/*avrRpt2* elicited the accumulation of *PAL1* and *DHS1* mRNA similarly to *Pst* MM1065 (Dong *et al.*, 1991; Davis *et al.*, 1991; Keith *et al.*, 1991). In contrast to *PAL1, BGL1, BGL2, BGL3,* and *DHS1, CHS* and *DHS2* mRNA levels did not increase following infection with virulent or avirulent strains (Dong *et al.*, 1991; Keith *et al.*, 1991). These results suggested that the *Arabidopsis PAL1* and *DHS1* genes respond to a signal generated by *avrRpt2* whereas the *BGL* genes respond to a signal that is not pathotype specific (Dong *et al.*, 1991).

In unpublished RNA blot experiments, the use of a *GST1*-specific probe showed that *GST1* mRNA levels transiently increased following infection with both virulent and avirulent strains, reaching a peak at about 3 hours after infection. However, at 6-12 hours post infection, leaves infiltrated with avirulent strains (*Pst* MM1065 and *Psm*

ES4326/*avrRpt2*) had considerably higher levels of *GST1* mRNA than leaves infiltrated with the virulent strain (*Psm* ES4326). *LOX1* mRNA also accumulated more following infiltration of *Psm* ES4326/*avrRpt2* than *Psm* -ES4326. However, mRNA corresponding to *SOD1* accumulated to the same extent following infiltration with either *Psm* ES4326 or *Psm* ES4326/*avrRpt2*. We also carried out a series of nuclear run-on experiments which showed that at least the *PAL1, GST1, PIG2, PIG18,* and *PIG27* genes are regulated at the level of transcription initiation (data not shown).

Table 1. Accumulation of mRNA Corresponding to Putative *Arabidopsis* Defense-Related Genes in ecotype Columbia.

Gene	Induction by *Psm* ES4326	Induction by *Psm* ES4326/*avrRpt2*	Fold Maximum Induction
PAL1	+	++	5-10
DHS1	+	++	5-10
GST1	++	+++	30-40
LOX	+	++	5-10
PIG18	+	++	5-8
PIG27	+	+	3-5
SOD1	++	++	~5
PIG2	+	+	5-10
BGL3	+++	+/-	5-10
CHS	-	-	-

Isolation of *Arabidopsis* mutants that do not mount an HR in response to *avrRpt2*

At the present time, we have isolated three *Arabidopsis* mutants that do not give an HR when infiltrated with *Pseudomonas* strains carrying *avrRpt2*. The gene(s) defined by these mutants are called *rpt* for resistance to an virulence gene isolated from *Pseudomonas syringae* pv. *tomato*. Two of the *rpt* mutants were identified phenotypically on the basis that in wild-type plants, the HR elicited by *Psm* ES4326/*avrRpt2* appears within 16-24 hours whereas disease symptoms elicited by *Psm* ES4326 do not appear until 36 to 48 hours. In the case of one of these mutants, approximately 3000 five or six week old *Arabidopsis* M2 ecotype Columbia plants generated by EMS mutagenesis were hand infiltrated with *Psm* ES4326/*avrRpt2* and a single mutant, *rpt-2* was identified that displayed disease symptoms instead of an HR.

Hand infiltration of plants involves the use of a syringe without a needle to force a small inoculum of a bacterial suspension through the stomatal openings on the underside of a leaf. To simplify the process of identifying *rpt* mutants, a procedure was developed to mass-infiltrate mature *Arabidopsis* plants. Densely sown seeds were germinated in soil in small flats and grown through a nylon mesh. When the plants were five to six weeks old, the flats were inverted, the plants were partially submerged in a tray containing a culture of *Psm* ES4326/*avrRpt2*, and the plants were mass-infiltrated by drawing a vacuum for one minute in a vacuum desiccator. Plants inoculated this way developed an HR within 24 hours or disease symptoms within 48 hours. Using this procedure, approximately 40,000 *Arabidopsis* M2 ecotype Nossen plants generated by gamma ray irradiation were infiltrated and at least one mutant, *rpt-1,* was identified that reproducibly displayed disease symptoms instead of an HR in the M3 and M4 generations when inoculated with *Psm* ES4326/*avrRpt2*.

The third *Arabidopsis rpt* mutant, *rpt-3*, was isolated using a novel method that involved vacuum infiltration of seedlings growing on agar in petri dishes. This new method relied on the observation that the bean pathogen *P. syringae* pv. *phaseolicola* strain NPS3121 does not elicit disease symptoms or a defense response in mature *Arabidopsis* leaves. However, *Psp* NPS3121/*avrRpt2* elicits a strong HR. Interestingly, we found that if ten-day-old *Arabidopsis* seedlings growing on petri plates were vacuum infiltrated with *Psp* NPS3121 or *Psp* NPS3121/*avrRpt2*, about 90% of the plants infiltrated with *Psp* NPS3121 survived whereas about 90%-95% of the plants infiltrated with *Psp* NPS3121/*avrRpt2* died. We interpreted this result as follows: Vacuum infiltration of an entire small *Arabidopsis* seedling with *Psp* NPS3121/*avrRpt2* elicits a systemic HR which usually kills the seedling. In contrast, seedlings infiltrated with *Psp* NPS3121 survive because *Psp* NPS3121 is such a weak *Arabidopsis* pathogen. Importantly, in a reconstruction experiment, *rpt-2* seedlings mostly survived when infiltrated with *Psp* NPS3121/*avrRpt2*. Because about 200 plants can be screened simultaneously per plate, this method effectively increases the number of plants that can be easily screened by a factor of 10-20.

To test the seedling infiltration method, approximately 4,000 15-day-old M2 seedlings from an EMS-mutagenized population were vacuum infiltrated with 8×10^7 cfu/ml *Psp* NPS3121/*avrRpt2*. Approximately 200 seedlings (5%) survived 4 days after infiltration. The survivors were transplanted to soil, grown to maturity, and tested for their ability to give an HR by hand infiltration with *Psm* ES4326/*avrRpt2*. Among these 200, one plant gave disease symptoms instead of an HR. This mutant plant, *rpt-3*, was retested in the M3 and M4 generations which confirmed that it did not give an HR when infiltrated with either *Psm* ES4326/*avrRpt2* or *Psp* NPS3121/*avrRpt2*.

All three *rpt* mutants are specific for *avrRpt2* since all three mutants display an HR when infiltrated with *P. cichorii* 83-1 or *Psm* ES4326/*avrRpm1*. On the other hand, *rpt-1* has one feature which distinguishes it from *rpt-2* and *rpt-3*. Following infiltration of *Psm* ES4326/*avrRpt2*, the kinetics of *PAL1* and *GST1*,mRNA accumulation in *rpt-2* and *rpt-3* were the same as the kinetics following infiltration of wild-type with *Psm* ES4326. However, in contrast to ecotype Columbia, no difference in *PAL1* and *GST1* mRNA accumulation could be observed in wild-type Nossen (the parent ecotype of *rpt-1*) or in *rpt-1* infiltrated with *Psm* ES4326/*avrRpt2* or *Psm* ES4326.

When the three *rpt* mutants were crossed to their wild-type parents, in all three cases the heterozygous F1 progeny displayed a defense response that was intermediate between those displayed by the wild-type and homozygous mutant plants. In these F1 plants, an HR response could be elicited by *Psm* ES4326/*avrRpt2*; however, the HR appeared later than in the wild-type and required a higher initial inoculum. In the F2 generation, the response to *Psm* ES4326/*avrRpt2* segregated approximately as 1 strong HR: 2 intermediate HR: 1 susceptible response. .Complementation analysis showed that the two Columbia mutants, *rpt-2* and *rpt-3*, are allelic. RFLP analysis mapped *rpt-2* to chromosome 4 between the markers at600 (Chang *et al.*, 1988) and 3088 (Nam *et al.*, 1989). Kunkel *et al.* (this volume) have also isolated an *Arabidopsis* Columbia mutant that does not respond to *avrRpt2*. This latter mutant appears to have very similar phenotypes to *rpt-2* and *rpt-3* and also maps to the same region of chromosome IV. We have shown by complementation analysis that *rpt-2* is allelic to the mutant described by Kunkel *et al.*

When the F1 heterozygote generated by crossing *rpt-1* and *rpt-2* was inoculated with *Psm* ES4326/*avrRpt2*, a response intermediate between an HR and susceptibility was observed. In the F2 generation, no wild-type HR symptoms were observed although the expected frequency was at least 1 in 16 (homozygous wild type at each of the two mutant loci). The interpretation of these results is complicated because *rpt-1* and *rpt-2* were isolated in different ecotypes. Thus, at this point, we are not sure whether *rpt-1* and *rpt-2* are allelic or not. Additional crosses are being carried out and *rpt-1* is being mapped to resolve this ambiguity.

Arabidopsis phytoalexin mutants

Tsuji *et al.* (1992) have recently determined the structure of an *Arabidopsis* phytoalexin called "camalexin" which, like other Brassica phytoalexins, consists of an indole ring bearing a sulfur-containing substituent (Takasugi *et al.* 1987; Devys *et al.* 1988; Takasugi *et al.* 1988; Devys *et al.* 1990; Monde *et al.* 1990; Browne *et al.* 1991). We have begun to investigate the role of camalexin in the interaction between *Arabidopsis* and the *P. syringae* strains that we are studying.

We observed that the maximum levels of camalexin accumulation occur approximately 40 hours post-inoculation, consistent with the results of Tsuji et al. (1992). We found that the virulent strain, *Psm* ES4326, induced high levels of camalexin accumulation in ecotype Columbia (Col-0), while strains *Pst* MM1065 and *Psp* NPS3121, which do not grow in Col-0, induce very little camalexin accumulation. Strains *Psm* ES4326/*avrRpt2* and *Psm* ES4326/*avrRpm1*, which, as a consequence of the cloned avirulence genes they carry, are avirulent on Col-0 and elicit a hypersensitive response, were also tested. Interestingly, these strains and *Psm* ES4326 all elicited similar levels of camalexin accumulation. Thus, the regulation of camalexin accumulation seems to be independent of the defense response(s) to strains carrying avirulence genes.

To investigate the regulation of camalexin biosynthesis and its role in this plant-pathogen interaction, EMS-generated *Arabidopsis* M2 plants were screened for those that failed to accumulate camalexin following infection with *Psm* ES4326. The M2 plants were infiltrated with *Psm* ES4326 and infected leaves were excised 40 hours after inoculation. Camalexin was extracted and separated from other compounds by thin-layer chromatography (TLC). The presence of camalexin was judged by examination of the TLC plates under long-wave ultraviolet light, looking for a characteristic blue-purple fluorescence. 7,058 plants were screened in this manner and three camalexin-deficient mutants were identified. These have been temporarily designated 169, 331, and 454.

The accumulation of camalexin over time following infection with *Psm* ES4326 in the three mutant lines was measured and compared to that of wild-type Col-0. Lines 169, 331, and 454 accumulated camalexin to roughly 30%, 10%, and 0% of wild-type levels, respectively. These reduced camalexin levels were also evident in smaller or non-existent zones of growth inhibition in a TLC-fungal growth bioassay.

We found that when Col-0 was infiltrated with *Psm* ES4326 at very low doses, the bacteria grew to a lower maximum density than they did when the initial inoculum was higher. For example, when the starting inoculum was 10^3 bacteria/leaf disk, the maximum density was 5×10^6 bacteria/leaf disk, but when the starting inoculum was 10^2 bacteria/leaf disk, the maximum density was 3×10^5 bacteria/leaf disk. This observation suggests that the plant has a defense against *Psm* ES4326 which is effective in limiting the growth of the pathogen provided that the initial inoculum is sufficiently low.

To determine whether failure of the host plant to synthesize camalexin affects this phenomenon, the growth rates of *Psm* ES4326 in Col-0 and in the three mutant lines were compared. The growth of *Psm* ES4326 was comparable in line 454 and in Col-0. In contrast, in lines 169 and 331, the bacteria reached a final density 1.5 log units higher than in Col-0. This maximum density was reached faster in line 169 than in line 331. Interestingly, line 169, which had the weakest camalexin defect, had the most severe *Psm* ES4326 growth phenotype, while line 454, which made no camalexin, was not significantly altered in *Psm* ES4326 growth.

One model that explains this observation is the following: Camalexin accumulation is required for the limitation of *Psm* ES4326 growth seen in Col-0. Growth is still limited in line 454, which makes no camalexin, because a biosynthetic intermediate in the camalexin pathway, which is also toxic to *Psm* 4326, accumulates in these plants. Lines 169 and 331 have leaky mutations affecting other steps in the camalexin biosynthetic pathway, such that the toxic intermediate does not accumulate and camalexin levels are reduced, leading to increased bacterial growth.

Another possibility is that camalexin accumulation has no effect on *Psm* ES4326 growth *in planta*. This is the reason that growth of *Psm* ES4326 in line 454 and in Col-0 is similar. Lines 331 and 169 do not have lesions in the camalexin biosynthetic pathway, but rather in regulatory elements. These regulatory elements are pleiotropic, affecting camalexin synthesis as well as expression of another factor(s) that is required for limiting *Psm* ES4326 growth. At this early stage in the analysis of the camalexin-deficient mutants, many other models are also possible.

The genetic analysis of the camalexin-deficient mutants is in progress. Each mutant has been backcrossed to Col-0. The resulting F1 progeny were tested for camalexin accumulation. In all three cases, the F1 hybrids accumulated as much camalexin as Col-0, suggesting that all of the mutations are recessive. In the case of mutant 169, a small number of F2 progeny have also been tested. 73 plants accumulated as much camalexin as Col-0, and 23 plants accumulated reduced levels. This is consistent with the hypothesis that the phenotype is a consequence of a mutation in a single nuclear gene.

Arabidopsis accelerated cell death (*acd*) mutants

Plant mutants with accelerated symptom development in response to pathogens might help to identify plant components important for predisposing a plant to resisting disease. We reasoned that such a phenotype might be due to the misregulation of a pathogen-inducible gene(s) and/or protein(s). We have isolated and characterized two complementation groups of *Arabidopsis* mutants by screening for accelerated symptom development after inoculation with *Psm* ES4326. We have named these mutants *acd1* and *acd2* for accelerated cell death. When *acd* leaves are exposed to a virulent pathogen, they show symptoms that differ from those seen in the wild type in at least three ways: First, there is no evidence of a progressive loss of chlorophyll (chlorosis); instead leaves rapidly become necrotic and turn brown. Second, the mutants exhibit more membrane damage in infected leaves than wild-type. Third, the spread of symptoms is faster and more extensive in the mutants than the wild type. One trivial explanation for why both disease and HR symptoms appear more rapidly in *acd* mutants is that pathogens grow better in these plants. However, there is no difference between the initial growth rate of virulent or avirulent bacteria in *acd* mutants and in wild-type plants. The frequency of obtaining the *acd* mutants was 1/1300 plants in an *Arabidopsis* ecotype Col M2 population mutagenized with EMS. Genetic analysis has indicated that the mutants are all recessive and segregate in the expected 3:1 ratio in the F2 generation.

In addition to displaying accelerated disease symptoms when infiltrated with a pathogen, *acd* mutants develop what appear to be spontaneous lesions. Interestingly, these "spontaneous" lesions are actually extensively colonized with resident soil bacteria and fungi. Bacterial strains found in these lesions include *Xanthomonas maltophilia*,

Klebsiella planticola, and *Pseudomonas putida,* species which are never found colonizing wild-type plants. We also found that the mutants show relaxed growth inhibition with the non-host pathogen *Psp* NPS3121. It is likely that lesions form first which are then infected by "opportunistic" pathogens because plants grown aseptically still form lesions. Plants exhibiting the *acd* phenotype of spontaneous necrotic lesions have been seen in mutants of maize (called *les* for **les**ion mutants) (e.g., Hoisington *et al.*, 1982; Walbot, 1991). Most of these mutants contain dominant mutations in contrast to five *acd* mutations that we have studied which are all recessive. These maize mutants have not been characterized with respect to their sensitivity to bacterial pathogens or ethylene.

Acknowledgments

Unpublished work was supported by a grant from Hoechst AG to Massachusetts General Hospital. J. Glazebrook and J. Greenberg are recipients of National Science Foundation postdoctoral fellowships.

Literature Cited

Alin, P., Danielson, U. H., & Mannervik, B. (1985). 4-hydroxylalkenals are substrates for glutathione transferase. FEBS Lett., 179, 267-270.

Apostol, I., Heinstein, P. F., & Low, P. S. (1989). Rapid stimulation of an oxidative burst during elicitation of cultured plant cells: Role in defense and signal transduction. Plant Physiol., 90, 109-116.

Ausubel, F.M., Davis, K.R., Schott, E.J., Dong, X & Mindrinos, M. (1991) Identification of signal transduction pathways leading to the expression of *Arabidopsis thaliana* defense genes. In H. Hauke & D. P. S. Verma (Eds.), Advances in Molecular Genetics of Plant-Microbe Interactions, Current Plant Science and Biotechnology in Agriculture (pp. 357-364). Dordrecht: Kluwer Academic.

Babior, B. (1984). The respiratory burst of phagocytes. J. Clin. Invest., 73, 599-601.

Bent, A., Carland, F., Dahlbeck, D., Innes, R., Kearney, B., Ronald, P., Roy, M., Salmeron, J., Whalen, M., & Staskawicz, B. (1991). Gene-for-gene relationships specifying disease resistance in plant-bacterial interactions. In H. Hauke & D. P. S. Verma (Eds.), Advances in Molecular Genetics of Plant-Microbe Interactions, Current Plant Science and Biotechnology in Agriculture (pp. 32-36). Dordrecht: Kluwer Academic.

Browne, L. M., Conn, K. L., Ayer, W. A., & Tewari, J. P. (1991). The camalexins: new phytoalexins produced in the leaves of *Camelina sativa* (Cruciferae). Tetrahedron, 47, 3909-3914.

Chai, H.B. & Doke, N. (1987) Superoxide anion generation: A response of potato leaves to infection with *Phytophthora infestans*. Phytopathol. 77, 645-649.

Chang, C., Bowman, J. L., DeJohn, A. W., Lander, E. S., & Meyerowitz, E. M. (1988). Restriction fragment length polymorphism linkage map for *Arabidopsis thaliana*. Proc. Natl. Acad. Sci. USA, 85, 6856-6860.

Croft, K.P.C., Voisey, C.R. & Slusarenko, A.J. (1990). Mechanism of hypersensitive cell collapse: Correlation of increased lipoxygenase activity with membrane damage in leaves of *Phaseolus vulgaris* (L.) inoculated with an avirulent race of *Pseudomonas syringae* pv. *phaseolicola*. Physiol. Mol. Plant Pathol., 36:49-62.

Dangl, J. L., Lehnackers, H., Kiedrowski, S., Debener, T., Rupprecht, C., Arnold, M., & Somssich, I. (1991). Interactions between *Arabidopsis thaliana* and the phytopathogenic *Pseudomonas* pathovars: a model for the genetics of disease resistance. In H. Hauke & D. P. S. Verma (Eds.), Advances in Molecular Genetics of

Plant-Microbe Interactions, Current Plant Science and Biotechnology in Agriculture, Vol. 1 (pp. 78-83). Dordrecht: Kluwer Academic.

Dangl, J.L., Holub, E.B., Debener, T., Lehnackers, H., Ritter, C. & Crute, I.R. (1992) Genetic definition of loci involved in *Arabidopsis*-pathogen interactions. In C. Koncz, J. Schell & N.-H. Chua (Eds.), Methods in *Arabidopsis* Research, Singapore:World Scientific Publishing, in press.

Daniels, M. J., Fan, M. J., Barber, C. E., Clarke, B. R., & Parker, J. E. (1991). Interaction between *Arabidopsis thaliana* and *Xanthomonas campestris*. In H. Hauke & D. P. S. Verma (Eds.), Advances in Molecular Genetics of Plant-Microbe Interactions, Current Plant Science and Biotechnology in Agriculture, Vol. 1 (pp. 84-89). Dordrecht: Kluwer Academic.

Davis, K. R., & Ausubel, F. M. (1989). Characterization of elicitor-induced defense responses in suspension-cultured cells of *Arabidopsis*. Molec. Plant Microbe Interact., 2, 363-368.

Davis, K. R., Schott, E., Dong, X., & Ausubel, F. M. (1989). *Arabidopsis thaliana* as a model system for studying plant-pathogen interactions. In B. J. J. Lugtenberg (Ed.), Signal Molecules in Plants and Plant-Microbe Interactions (pp. 99-106). Berlin: Springer-Verlag.

Davis, K.R., Schott, E. & Ausubel, F.M. (1991). Virulence of selected phytopathogenic pseudomonads in *Arabidopsis thaliana*. Molec. Plant Microbe Interact., 4, 477-488.

Debener, T., Lehnackers, H. Arnold, M. & Dangl, J.L. (1991). Identification and molecular mapping of a single *Arabidopsis thaliana* locus determining resistance to a phytopathogenic *Pseudomonas syringae* isolate. Plant J. 1, 289-302.

Devys, M., Barbier, M., Kollmann, A., Rouxel, T., & Bousquet, J.-F. (1990). Cyclobrassinin sulfoxide, a sulfur-containing phytoalexin from *Brassica juncea*. Phytochemistry, 29, 1087-1088.

Devys, M., Barbier, M., Loiselet, I., Rouxel, T., Sarniguet, A., Kollman, A., & Bousquet, J.-F. (1988). Brassilexin, a novel sulfur-containing phytoalexin from *Brassica juncea L.* (Cruciferae). Tetrahedron Letters, 29, 6447-6448.

Doke, N. (1983a). Generation of superoxide anion by potato tuber protoplasts during the hypersensitive response to hyphal wall components of *Phytophthora infestans* and specific inhibition of the reaction by suppressors of hypersensitivity. Physiol. Plant Pathol., 23, 359-367.

Doke, N. (1983b). Involvement of superoxide anion generation in the hypersensitive response of potato tuber tissues to infection with an incompatible race of *Phytophthora infestans* and to the hyphal wall components. Physiol. Plant Pathol., 23, 345-357.

Doke, N. (1985). NADPH-dependent O_2-generation in membrane fractions isolated from wounded potato tubers inoculated with *Phytophthora infestans*. Physiol. Plant Pathol., 27, 311-322.

Doke, N., & Chai, H. B. (1985). Activation of superoxide generation and enhancement of resistance against compatible races of *Phytophthora infestans* in potato plants treated with digitonin. Physiol. Plant Pathol., 27, 323-334.

Doke, N., & Ohashi, Y. (1988). Involvement of an O_2-generating system in the induction of necrotic lesions on tobacco leaves infected with tobacco mosaic virus. Physiol. Molec. Plant Pathol., 32, 163-175.

Dong, X., Mindrinos, M., Davis, K.R. & Ausubel, F.M. (1991). Induction of *Arabidopsis thaliana* defense genes by virulent and avirulent *Pseudomonas syringae* strains and by a cloned avirulence gene. Plant Cell, 3, 61-72.

Dudler, R., Hertig, C., Rebmann, G., Bull, J. & Mauch, F. (1991) A pathogen-induced wheat gene encodes a protein homologous to glutathione-S-transferases. Molec. Plant Microbe Interact. 4, 14-18.

Feinbaum, R. L., & Ausubel, F. M. (1988). Transcriptional regulation of the *Arabidopsis thaliana* chalcone synthase gene. Mol. Cell. Biol., 8, 1985-1992.

Feldman, K. A. (1991). T-DNA insertion mutagenesis in *Arabidopsis*: Mutational spectrum. Plant J, 1, 71-82.

Hindges, R. & Slusarenko, A. (1991) cDNA and derived amino acid sequence of a cytosolic Cu,Zn superoxide dismutase from *Arabidopsis thaliana* (L.) Heyhn. Plant Mol. Biol., 17, 1-3.

Hoisington, D. A., Neuffer, M. G., & Walbot, V. (1982). Disease lesion mimics in maize. Developmental Biology, 93, 381-388.

Johnson, R. B. J., & Kitagawa, S. (1985). Molecular basis for the enhanced respiratory burst of activated microphages. Fed. Proc., 44, 2927-2932.

Kato, S. & Misawa, T. (1976). Lipid peroxidation during the appearance of hypersensitive reaction in cowpea leaves infected with cucumber mosaic virus. Ann. Phytopathol. Soc. Jpn. 42, 472-480.

Keith, B., Dong, X., Ausubel, F.M. & Fink, G.R. (1991). Differential induction of 3-deoxy-D-heptulosonate 7-phosphate synthase gene in *Arabidopsis thaliana* by wounding and pathogen attack. Proc. Natl. Acad. Sci. USA 88, 8821-8825.

Keppler, L.D. & Novacky, A. (1987). The initiation of membrane lipid peroxidation during bacteria-induced hypersensitive reaction. Physiol. Mol. Plant Pathol. 30, 233-246.

Koch, E., & Slusarenko, A. J. (1990a). *Arabidopsis* is susceptible to infection by a downy mildew fungus. Plant Cell, 2, 437-445.

Koch, E., & Slusarenko, A. J. (1990b). Fungal pathogens of *Arabidopsis thaliana* (L.). Heynh. Bot. Helv., 100, 257-269.

Koornneef, M. (1987). Linkage map to *Arabidopsis thaliana* In S.J. O'Brien (Ed.), Genetic Maps 1987: A Compilation of Linkage and Restriction Maps of Genetically Studied Organisms, pp. 742-745, New York: Cold Spring Harbor Laboratory Press.

Koornneef, M., van Eden, J., Hanhart, C. J., & de Jongh, A. M. M. (1983a). Genetic fine-structure of the GA-1 locus in the higher plant *Arabidopsis thaliana* (L.). Heynh. Genet. Res. Camb., 41, 57-68.

Koornneef, M., van Eden, J., Hanhart, C. J., Stam, P., Braaksma, F. J., & Feenstra, W. J. (1983b). Linkage map of *Arabidopsis thaliana*. J. Hered., 74, 265-272.

Li, X.-H., & Simon, A. E. (1990). Symptom intensification on cruciferous hosts by the virulent satellite RNA of turnip crinkle virus. Phytopathology, 80, 238-242.

Meyerowitz, E. M. (1987). *Arabidopsis thaliana*. Ann. Rev. Genet., 21, 93-111.

Meyerowitz, E. M. (1989). *Arabidopsis*, a useful weed. Cell, 56, 263-269.

Monde, K., Sasaki, K., Shirata, A., & Takasugi, M. (1990). 4-methoxybrassinin, a sulfur-containing phytoalexin from *Brassica oleracea*. Phytochemistry, 29, 1499-1500.

Morgenstern, R., & de Pierre, J. W. (1983). Microsomal glutathione transferase: Purification in unactivated form and further characterization of the activation process, substrate specificity and amino acid composition. Eur. J. Biochem., 134, 591-597.

Morgenstern, R., Lundquist, G., Jornvall, H., & De Pierre, J. W. (1989). Activation of rat liver microsomal glutathione transferase by limited proteolysis. Biochem. J., 260, 577-582.

Nam, H.-G., Giraudat, J., den Boer, B., Moonan, F., Loos, W. D. B., Hauge, B. M., & Goodman, H. M. (1989). Restriction fragment length polymorphism linkage map of *Arabidopsis thaliana*. Plant Cell, 1, 699-705.

Ohl, S., Hedrick, S., Chory, J., & Lamb, C. J. (1990). Function properties of a phenylalanine ammonia-lyase promoter from *Arabidopsis*. Plant Cell, 2, 837-848.
Prohaska, J. R. (1980). The glutathione peroxidase activity of glutathione S-transferases. Biochim. Biophys. Acta, 611, 87-98.
Redei, G. P. (1975). *Arabidopsis* as a genetic tool. Ann. Rev. Genet., 9, 111-127.
Rogers, K.R., Albert, F. & Anderson, A.J. (1988). Lipid peroxidation is a consequence of elicitor activity. Plant Physiol. 86, 547-553.
Schott, E. J., Davis, K. R., Dong, X., Mindrinos, M., Guevara, P., & Ausubel, F. M. (1990). *Pseudomonas syringae* infection of *Arabidopsis thaliana* as a model system for studying plant-bacterial interactions. In S. Silver, A. M. Chakrabarty, B. Iglewski, & S. Kaplan (Eds.), *Pseudomonas*: Biotransformation, Pathogenesis, and Evolving Biotechnology (pp. 82-90). Washington: Amer. Soc. Microbiology.
Siedow, J.N. (1991). Plant lipoxygenase: structure and function. Ann. Rev. Plant Physiol. Plant Mol. Biol. 42, 145-188.
Simpson, R. B., & Johnson, L. J. (1990). *Arabidopsis thaliana* as a host for *Xanthomonas campestris* pv. *campestris*. MPMI, 3, 233-237.
Staskawicz, B. J., Dahlbeck, D., & Keen, N. T. (1984). Cloned avirulence gene of *Pseudomonas syringae* pv. *glycinea* determines race-specific incompatibility on *Glycine max* (L.) *Merr*. Proc. Natl. Acad. Sci. USA, 81, 6024-6028.
Sun, T.-p., Goodman, H.M. & Ausubel, F.M. (1992). Cloning the *Arabidopsis GA1* locus by genomic subtraction. Plant Cell 4, 119-128.
Takasugi, M., Monde, K., Katsui, N., & Shirata, A. (1987). Spirobrassinin, a novel sulfur-containing phytoalexin from the Daikon *Raphanus sativus* L. var. *hortensis* (Cruciferae). Chem. Letters, 1631-1632.
Takasugi, M., Monde, K., Katsui, N., & Shirata, A. (1988). Novel sulfur-containing phytoalexins from the Chinese cabbage *Brassica campestris* L. spp. *pekinensis* (Cruciferae). Bull. Chem. Soc. Japan, 61, 285-289.
Thompson, J.E., Legge, R.L. & Barber, R.F. (1987). The role of free radicals in senescence and sounding. New Phytol. 105, 317-344.
Tsuji, J., Jackson, E. P., Gage, D. A., Hammerschmidt, R., & Somerville, S. S. (1992). Phytoalexin accumulation in *Arabidopsis thaliana* during the hypersensitive reaction to *Pseudomonas syringae* pv.*syringae*. Plant Physiol., 98, 1304-1309.
Tsuji, J., & Somerville, S. C. (1988). *Xanthomonas campestris* pv. *campestris*-induced chlorosis in *Arabidopsis*. *Arabidopsis* Inf. Serv., 26, 1-8.
Tsuji, J., Somerville, S. C., & Hammerschmidt, R. (1990). Identification of a gene in *Arabidopsis thaliana* that controls resistance to *Xanthomonas campestris* pv. *campestris*. Physiol. Mol. Plant Pathol., 37, 1-8.
Walbot, V. (1991). Maize mutants for the 21st century. Plant Cell, 3, 851-856.
Whalen, M. C., Innes, R. W., Bent, A. F., & Staskawicz, B. J. (1991). Identification of *Pseudomonas syringae* pathogens of *Arabidopsis* and a bacterial locus determining avirulence on both *Arabidopsis* and soybean. Plant Cell, 3, 49-59.

Genetic Approaches to an Understanding of Specific Resistance Responses of *Arabidopsis thaliana* against phytopathogenic Pseudomonads

Jeff Dangl, Thomas Debener, Maren Gerwin, Siegrid Kiedrowski, Claudia Ritter, Abdelhafid Bendahmane, Hiltrud Liedgens and Jürgen Lewald

Max-Delbrück Laboratory
Carl-von-Linné Weg 10
D-5000-Köln-30
Germany

Introduction

Many interactions between plants and microbes begin with specific recognition. The nature of this recognition, and the interpretation of subsequent signal transduction by both plant and microbe have profound impact on the outcome of the interaction.Plants have evolved effective mechanisms to recognize pathogenic microbes and halt their biotrophic or necrotrophic growth in the plant. Active plant defense mechanisms obviously force the selection of microbe variants which can evade the plant´s recognition capabilities. This evolutionary tug of war has led to a complex set of both plant and microbe genes, whose interactions lead to a successful plant resistance reaction. As well as a potentially large array of cognitive gene functions, a number of subsequent signal transduction steps may be necessary to generate a completely effective resistant phenotype.

The number and function of plant genes necessary for a resistance response is unknown. Genetic analyses in many systems over the last 50 years have demonstrated that recognititon functions are provided by dominant alleles of genes in the plant (Resistance, or *R*-genes) which interact, either directly or indirectly, with either the direct or indirect product of a single pathogen gene (avirulence, or *avr* genes) (Crute, 1985; Ellingboe, 1981, 1982, 1984; Flor, 1955 1971; Keen, 1982, 1990; Keen and Staskawicz, 1988). The "gene-for-gene" hypothesis is a genetic explanation for interactions between plants and all classes of pathogens: fungal, bacterial, viral, and insect. Yet despite intense research, we remain anive regarding the molecular structure of *R*-genes, and the function of their products. Moreover, although a great deal is known about the multitude of genes activated subsequent to triggering of the defense response, virtually nothing is known regarding the number or nature of downstream transduction steps which are truly necessary for a resistance reaction.

The goal of projects in our group is to use a model plant species, *Arabidopsis thaliana*, to identify genetically the plant loci necessary for a resistance reaction against phytopathogenic bacteria and fungus. One powerful application of *A. thaliana* is in the dissection of the processes by which plants recognize and respond to microbial pathogens

E. W. Nester and D. P. S. Verma (eds.),
Advances in Molecular Genetics of Plant-Microbe Interactions, 405–415.

(reviewed by Dangl, 1992; Davis, 1992). Several groups have established screening systems to define differential responses (resistance and susceptibility) of *A. thaliana* land races (ecotypes) to various bacterial and fungal pathogens, focussing on isolates of *Pseudomonas syringae* and *Peronospora parasitica*, and several chapters in this book detail their progress.

Our work revolves around five themes. First, identification and isolation of recognition function genes (*R* or Resistance genes). Second, use of mutation analysis to identify loci which give rise to either loss of recognition, or constitutive induction of a resistance-like phenotype. Depending on the nature of the selection scheme, mutants generated in this way will either map to predefined *R*-gene loci, or to loci involved in either specific or general interpretation of subsequent signals. Third, we also use a generalizable genetic approach to assess whether a particular activated defense response is strictly tied to *R*-gene mediated recognititon, and to critically assay the role that gene´s activation in the outcome of a plant-pathogen interaction.. Fourth, we also have isolated five potentially T-DNA tagged mutants exhibiting constitutive phenotypes reminiscent of "lesion mimic" mutants which exist in many plant species, notably maize, rice, tomato, and barley. These have potentially constitutive or "hair-trigger" activation of some step in one or more pathway(s) leading to fomation of an HR-like lesion. Fifth, we are identifying the bacterial pathogen genes which are causal to triggering of a specific plant defense response. These fall into two categories: classically defined *avr* genes, and, using Tn*phoA* as a mutagen, genes encoding membrane localized or secreted products necessary for delivery of a specific *avr* function. Progress in some of these areas is detailed below. Literature citations are by no means exhaustive, and tend to reflect work that has been done in our group.

Identification and molecular mapping of an *Arabidopsis* resistance gene against a *Pseudomonas syringae* pv. *maculicola* isolate, and characterization of the corresponding *avr* gene

We and others have shown that *A. thaliana* is a host for phytopathogenic isolates of *Pseudomonas syringae* pathovar *maculicola* (Psm), normally pathogenic on Brassicas and tomato (Dangl et al., 1991; Debener et al., 1991; see also Ausubel et al., 1991; Bent et al., 1991; Davis et al., 1991; Dong et al., 1991; Whalen et al., 1991). All groups found that there is variability in this interaction that is dependent on both host and pathogen genotype. Thus, some Psm isolates are pathogenic on all tested *A. thaliana* genotypes, some on none, and some on only a subset. Based on this series of "differential responses", we began to ask whether or not single loci in both plant and pathogen controlled the generation of a hypersensitive resistance response (HR) as predicted by the gene-for-gene hypothesis. The *A. thaliana* ecotypes Col-0 and Oy-0, among others, are resistant to Psm isolate m2, and resistance is manifested, in one type of assay, as a rapidly forming HR. Ecotype Nd-0 is, among others, susceptible to Psm isolate m2. Resistance to Psm m2 segregated as a single locus in crosses Col-0 x Nd-0 and Oy-0 x Nd-0 in analysis of F2 progeny and F3 progeny derived from selfed F2 individuals. We used RFLP mapping, a technology highly developed in the *A. thaliana* research community, to localize this resistance gene, which we named *RPM1*, to an interval of around 6 map units near the top of chromosome three.

Concurrent with this plant genetic analysis, we cloned the avirulence gene from Psm isolate m2 predicted to trigger the HR in combination with the *RPM1* product. A mobilizable cosmid library of high molecular weight DNA from Psm isolate m2 was constructed in a broad host range vector. Single clones were transferred from *E. coli* to a

Psm isolate, m4, which is virulent on all tested *A. thaliana* ecotypes, including Col-0, Nd-0, and Oy-0. After screening 220 transconjugant clones, we identified three clones containing overlapping insert DNA from Psm isolate m2 which now rendered isolate m4 avirulent on Col-0 and Oy-0, but not on ecotype Nd-0. This plant genotype dependent conversion of a virulent isolate to avirulence is a cornerstone of the gene-for-gene hypothesis. We used three different assays, formation of an HR, *in planta* bacterial growth, and diminution of symptoms, to solidly identify the *avrRpm1* gene (Debener et al., 1991).

We have subsequently defined *avrRpm1* to a functionally active 2.5kb fragment, using a subcloning scheme depicted in figure 1. Tn3spice mutagenesis further localized *avrRpm1* to a 1.0kb fragment (figure 2). In collaboration with Marjorie Humphrey and Alan Vivian at Bristol Polytechnic University in Bristol, the *avrRpm1* and the highly related *avrPpiA1* from *P. s.* pv. *pisi* were both sequenced (Dangl et al., submitted). In keeping with Alan Vivian's suggested uniform nomenclature of *avr* genes, we also refer to *avrRpm1* as *avrPmaA1*. The single ORF defined by insertion mutagenesis and sequencing of both genes predicts a 28kd M_r translation product, which is extremely hydrophilic, has no homology to any protein in available databases or other *avr* gene products, and has no signal or leader peptide. Not surprisisngly, such crypticism is typical avr genes analysed to date. Perhaps of interest is the lack of a potential *hrp*-box regulatory sequence upstream from the ORF.

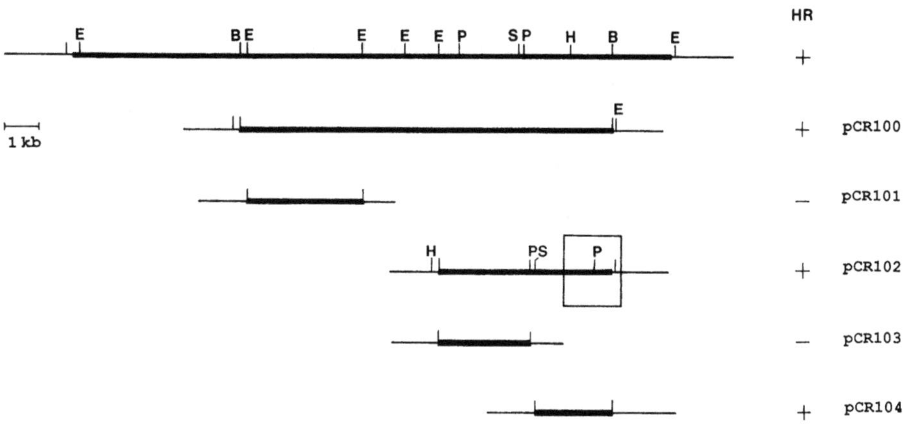

Figure 1. Identification of a small, *avrRpm1* active DNA fragment. Subclones in pLAFR5 were conjugated into Psm m4 and tested for ability to trigger HR in *A. thaliana* ecotype Col-0. Boxed segment is enlarged in figure 2.

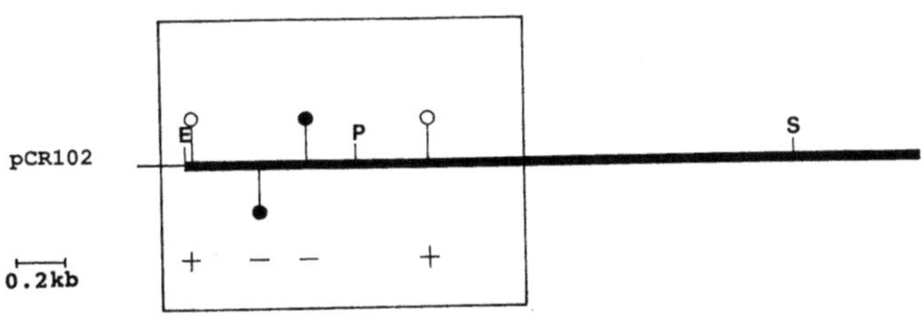

Figure 2. Localization of *avrRpm1* activity via Tn3spice mutagenesis. Solid circles represent HR^- insertions.

We further demonstrated that the cloned *avrRpm1* gene was, in fact, responsible for *RPM1*-mediated recognition of the Psm isolate m2 by *A. thaliana* ecotype Col-0. We had used Psm isolate m2 to define *RPM1* in the segregating population, and had identified F3 families which were either homozygous resistant or homozygous susceptible to isolate m2. We screened 9 individuals from 12 F3 families of each class with the cloned *avrRpm1* gene in the Psm isolate m4 background. Every plant from each family gave a reaction to the cloned *avrRpm1* containing strain identical to that of Psm isolate m2 itself, thus proving that isolate m2 carries one avr gene, *avrRpm1*, which interacts with the product of the *RPM1* locus to trigger resistance. This was the first demonstration of a gene-for-gene interaction in *A. thaliana* (Debener, et al., 1991).

We isolated several YAC (yeast artificial chromosome) clones containing large *A. thaliana* DNA inserts which hybridize to the RFLP markers genetically closest to *RPM1*. One of them, with a 270kb insert, was isolated for us by Joe Ecker and must contain *RPM1*. This exciting conclusion is based on three pieces of data, summarized in figure 3. First, the RFLP marker 583 is telomeric to *RPM1* and hybridizes to the 270kb YAC clone, while the RFLP marker centromeric to *RPM1*, 17341, hybridizes to two different 160kb YAC clones. Second, an end specific probe from the 270kb YAC hybridizes to both of the 160kb YAC clones, and end probes from one of them cross-hybridizes to the 270kb clone. Thus, the genetic interval known to contain *RPM1* is physically overlapped by this set of YAC clones. Finally, the end probe from the centromeric 160kb YAC which detects the overlap, was used as an RFLP probe. It is genetically closer to *RPM1* than the RFLP probe used to isolate it, thus proving that our chromosome walk is proceeding in the correct direction. *RPM1*, then, is contained on less than 270kb of *A. thaliana* DNA, encompasing an astounding 5 map units! The *RPM1* region is, therefore, highly recombinogenic, a trait often associated with disease resistance loci in crop plants (Pryor, 1987; Bennetzen et al., 1991; Hulbert and Bennetzen, 1991).

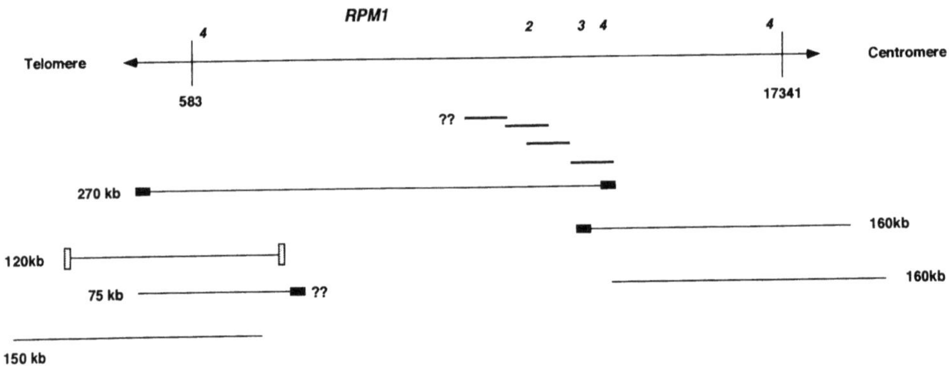

Figure 3. Current status of the *RPM1* chromosome walk. A schematic of chromosome 3, with the two RFLP markers originally used to define the genetic interval containing *RPM1*, and the number of recombinants detected by various probes (aligned below the respective numbers) are shown on top. YAC clones are shown in thin lines, with sizes indicated, cosmids in thick lines. Informative YAC ends are solid, non-informative stippled. The exact nature of YAC overlaps on the telomeric end of this contig should be regarded as preliminary.

We have recently adopted several parallel approaches to finding the *RPM1* gene on the 270kb YAC. A disadvantage of our first set of YACS was that neither end probe from an important 120kb YAC detected an RFLP (stippled ends in figure 3). Two new YACS were isolated using its centromeric end as probe. One of the new YACs, of 75kb, is also detected by the telomeric end of the 270kb YAC, as shown in figure 3. Its ends will now be used as RFLP probes, and to probe a collection of cosmids isolated via hybridization with the 270kb YAC. Of particular interest is the end shown with question marks in the figure. In parallel we have screened a cosmid library of Col-0 DNA in a T-DNA vector, provided by Neil Olszewski. We have walked several steps in from the centromeric YAC using these clones, losing another recombinant and covering a distance of at least 60kb (thick lines in figure 3). Finally, we are also screening various cDNA libraries with the 270kb YAC. Preliminary data suggests a small number of cDNA clones are detected. They will also be used to help further narrow the location of *RPM1*. Progress in this part of the project was greatly aided by the prepublication availability of *A. thaliana* YAC libraries and the international *A. thaliana* physical mapping network.

Another outstanding question is, how many resistance specificities are there in *A. thaliana* which detect *avr* specificities of phytopathogenic Pseudomonads? We are taking two approaches to this. We first returned to our initial series of differential responses (Debener et al., 1991). Based solely on differential ecotype specificities, we can conclude that new *avr-R* interaction specificities are contained in this data set. We are concentrating on two, which are summarized in figure 4, and detailed via growth curve analysis in figure 5.

Arabidopsis genotypes:

	Col-0	Mt-0	Nd-0
bacteria:			
m2 (*avrPmaA1*)	I	C	C
m6 (*avrPmaA2+ ??*)	I	I*	C
m8	I	I^	C
m4	C	C	C
transconjugants:			
m2/*avrPmaA2*	I	C	C
m2/*avrPmaB*	I	I*	C

Figure 4. New resistance specificities defined through *A. thaliana* ecotype differential reactions. I: incompatible interaction, HR at 10-15 hours p.i.; I^{*}: light HR at around 36 hours p.i.; $I^{\wedge}$: strong HR at around 20 hours p.i..

The HR generated by isolates Psm m6 and Psm m8 on *A. thaliana* ecotype Mt-0 already indicates that they define resistance specificities different from that determined by *avrRpm1*, since it is not recognized by Mt-0. Morever, we believe that the two distinct HR^{+} phenotypes triggered by these two strains signify distinct *avr* gene functions. It is interesting to note that Psm m6 carries an allele of *avrRpm1*, cloned and sequenced in Alan Vivian´s group and called *avrPmaA2* (Dangl et al., submitted). We wondered if this allele was responsible for the HR on Mt-0 and tested this following conjugation of a plasmid containing it, pAV500, into Psm m2. The clearly negative result, and corresponding growth curves shown in figure 5, strongly suggest that another *avr* gene is present Psm m6. It is designated *avrPmaB* in figure 4, and we have candidate clones which satisfy the requirements of this gene. Besides *avrRpm1*, two other cloned *avr* genes detecting *R*-gene specificities in *A. thaliana* are available (Whalen et al., 1991; Brian Staskawicz, personal communication). Genomic DNA blot hybridization experiments that neither nor either Psm m6 nor Psm m8 contain sequences hybridizing to the other two *avr* genes. We are currently screening further transconjugants from genomic libraries of both Psm m6 and Psm m8 to identify these *avr* genes.

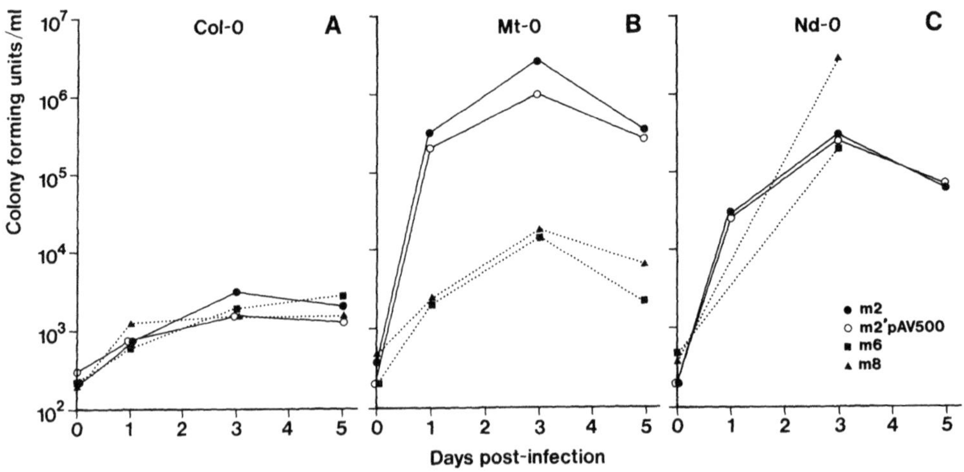

Figure 5. *In planta* growth of new Psm strains in various *A. thaliana* ecotypes. pAV500 carries the *avrPmaA2* allele from Psm m6. Inoculum was 10^5 cfu/ml. Growth curves were done as described in Debener et al. (1991). These data points are mean values from three independent experiments each.

The second approach to find new resistance specificities was to test whether known *avr* genes, detecting resistance gene specificities in crop plant species, could be recognized by *A. thaliana*. In collaboration with Alan Vivian and John Mansfield and their co-workers (Bristol Polytechnic, and Wye College, UK) we tested the *avrPpiA1* or *avrPpiB2* from *P.s.* pv. *pisi* races 2 and 3, respectively (Taylor et al., 1989; Vivian et al., 1989; Bavage et al., 1991), and *avrPph3* or *avrPph2* from *P. s.* pv. *phaseolicola* races 3 and 4, respectively (Hitchin et al, 1989; Jenner et al., 1991) by conjugating these cloned *avr* genes into our virulent Psm isolate, m4. Transconjugants were tested on our "differential series" of *A. thaliana* ecotypes and, to our amazement, the *P. s.* pv. *pisi avrPpiA1* genes generated HR on exactly the same set of ecotypes as the *avrRpm1* gene described above. We subsequently showed that *avrRpm1* and the *P. syringae* pv. *pisi*A2 gene are nearly identical, with only 10 amino acid exchanges from 220 total in the putative protein product. Moreover, it was recently shown that the *P. s.* pv. *pisi avrPpiA1* gene determined cultivar specific resistance and bean (Fillingham et al., 1992). Based on this, we showed that the *avrRpm1* gene also conditions resistance on the apropriate pea and bean cultivars, as well as on soybean (N.T. Keen, personal communication). These data strongly suggested, but did not prove, that recognition of *avrPpiA1* was mediated via *RPM1*. We therefore tested *avrPpiA1* recognition by families of F3 *A. thaliana* plants derived from selfed F2 individuals from the Col-0 x Nd-0 cross used previously to RFLP map *RPM1*. Thirty F2 progeny had been genotyped as either homozygous resistant (*RPM1/RPM1*, n=15) or homozygous susceptible (*rpm1/rpm1*, n=15) based on HR phenotype following infiltration of F3 progeny with Psm m2. We infiltrated nine individuals from each F3 family with Psm m4 carrying the *avrPpiA1* gene, and scored for development of HR 10-15 hours later. Each member of all *RPM1/RPM1* families developed HR, and no individuals from *rpm1/rpm1* families developed HR ($p=10^{-9}$; Allard, 1956). It is still theoretically possible that a second gene very tightly linked to *RPM1* is recognizing the *avrPpiA1* signal, and we are investigating this further. At a minimum, these data show that *A. thaliana* contains genes of potential agronomic interest; at best, the *RPM1* locus contains a gene or genes conditioning resistance to several phytopathogenic *Pseudomonas syringae* isolates.

Isolation and characterization of *A. thaliana* mutants with an altered resistance response

We have three current activities in this area. First, we isolated and are characterizing an EMS mutant of ecotype Col-0 which now exhibits chlorosis after inoculation with a *P.s.* pv. *tabaci* isolate that makes no lesion on any tested wild type *A. thaliana* ecotype, even at very high inoculum density. The mutant becomes very chlorotic, and supports around 30-50 fold higher levels of bacterial growth when re-challenged with the selecting bacteria. F1 individuals from an outcross to ecotype Nd-0 shows an intermediate phenotype, which subsequently segregates as a single dominant locus in F2. We are analysing this mutant further and are beginning RFLP mapping

With the cloned *avrRpm1* gene characterised, we are also screening for *A. thaliana* mutants unable to recognize its signal. These mutants should either be in the *RPM1* gene, and their analysis will further understanding of the critical recognition event mediated by its product, or they will be in loci necessary to transduce the RPM1 signal. Several technical criteria must be met before beginning this screen. First, in order to do thorough mutagenesis, we must screen tens of thousands of mutant seedlings. Therefore, we use vaccum infiltration of hundreds of sterile mutant seedlings per petri dish, devised in Fred Ausubel´s group, and search for susceptible individuals no longer able to recognize *avrRpm1*. Psm isolate m4, our normal virulent conjugation test strain for delivery of the *avrRpm1* signal, is inappropriate for this assay, since it will kill seedlings. We therefore constructed a useful bacterial strain by conjugating *avrRpm1* into a *P.s.* pv *phaseolicola* strain which is non-pathogenic and generates no macroscopic reaction on *A. thaliana* . Thus, loss of recognition mutants will continue to grow, while all seedlings which still recognize *avrRpm1* will undergo a "whole seedling HR" and die. We prepared large Col-0 and Oy-0 mutant populations using X-rays and fast neutrons, as well as available EMS and T-DNA insertion mutant populations, in order to isolate potential insertion/deletion mutants as well as point mutants. The M2 generation seed has been screened for embryo defects, and is known to have an appropriate mutation frequency. So far, around 30,000 M2 seedlings have been screened, and several interesting candidates selfed for re-testing.

Assessing the necessity of induced gene expression during specific resistance reactions: The *ELI3* example

Imre Somssich (MPI Köln) and colleagues cloned 19 different parsley cDNAs, encoding predicted proteins of known and unknown function, which were transcriptionally activated in cultured cells by fungal elicitor (Somssich et al., 1989). We are collaborating with him to analyze pathogen activated expression of recently cloned *A. thaliana* homologs after infiltration of *A. thaliana* leaves with various Psm isolates. Of the many clones tested, only the *ELI3* homolog is of immediate interest (Kiedrowski et al., submitted). Its mRNA accumulates rapidly (3-5 hours p.i.) to extremely high levels in incompatible interactions involving Psm isolate m2, while the accumulation is slower, and the amount of mRNA lower, in compatible interactions involving m2. Interestingly, the virulent Psm isolate m4 also induces *ELI3* mRNA to high levels, but much later in the interaction. A Psm m4 transconjugant harboring *avrRpm1* generates the rapid *ELI3* induction kinetic idiosyncratic of the Psm m2 interaction with Col-0, suggested, but by no means proved, that gene-for-gene recognition mediates early *ELI3* activation. We have now shown convincingly that the early activation of *ELI3* in m2 incompatible interactions is stricly governed by the presence of a dominant allele at *RPM1*. Again, we used the homozygous F3 families from the Col-0 x Nd-0 cross described above to prove this point. At the appropriate time point after infiltration with Psm isolate m2, chosen to clearly illustrate the differential induction of *ELI3* gene activity observed in Col-0 and Nd-0 parents, leaves from nine individuals from each F3 family were pooled and RNA isolated. In RNA blot experiments, all 15 *RPM1/RPM1* families had very high levels of *ELI3* mRNA, as did the resistant Col-0 parent; all 15 *rpm1/rpm1* families had very low levels, as did the susceptible Nd-0 parent. These families were essentially "genotype selected" (Michelmore et al., 1991) at *RPM1*, and are expected to be freely segregating at all other loci. This random assortment virtually eliminates any possible effects of genetic background on the phenotype being analysed. Therefore, the evidence that *ELI3* activation requires *RPM1* function is overwhelming ($p = 10^{-9}$). This is the first clear example that an induced plant defense phenomenon is regulated through a specific *R*-gene, and this type of "functional co-segregation" analysis is applicable to any induced plant defense response. Two highly related full length cDNA clones from *A. thaliana* have been sequenced, and only two genes are present in *A. thaliana*. Other than the expected high homology to the parsley *ELI3* gene, no other similarities were found in the various databases. Thus, the function of this conserved plant defense gene remains unknown. We are now constructing anti-sense and sense constructs in order to make phenocopy mutants of *ELI3* and ask whether the expected mutants have an altered phenotype after infiltration with Psm isolate m2 and other incompatible pathogens.

Identification of bacterial membrane proteins necessary for delivery of *avr* gene function

We are interested in probing the bacterial membrane and periplasm for proteins involved in the HR. This project is based on the observation that, although many bacterial *avr* genes have been cloned, none encodes an obviously membrane bound product. This prompts the question of how the avr signal reaches the plant. It has also been recently shown that *hrp* cluster encoded proteins can be members of two-component regulatory systems. One component of these systems is always a membrane bound "sensor". We reported the utilization of the specialized transposon mutagen Tn*phoA* to generate the desired mutants. The preliminary studies, however, were performed in a *P. cichorri* strain which is of limited value (Dangl et al., 1991). Therefore, we are now applying this method to the *P.s.* pv *phaseolicola* strain described above. It makes no reaction on any tested *A. thaliana* ecotype, but is capable of delivering *avr* gene function from any of the three genes recognized by *A. thaliana*. Over 92,000 Tn*phoA* insertions were generated in the *P.s.* pv *phaseolicola* strain harboring *avrRpm1*. Of these, around 1500 were PhoA$^+$ as screened on plates. To date, over 500 of these mutants have been tested for attenuation of *avrRpm1* dependent HR on Col-0. Three mutants are no longer able to deliver *avrRpm1* function in from 4 to 7 independent repetitions. At last 5 other candidates are being re-tested. Several necessary experiments are currently being done. First, although it is clear that Tn*phoA* did not physically disrupt the avr gene, we must still reisolate cosmid from each mutant to prove that the *avrRpm1* gene is still functional. Second, appropriate phenotypic and DNA blot experiments are being performed to ask whether the PhoA$^+$HR$^-$ phenotypes are due to *hrp* mutations. We enriched against this possibility by isolating the PhoA$^+$ mutants on rich media, thought to repress *hrp* genes. Finally, we will cure the *avrRpm1* containing cosmid from these mutants, introduce either of the other two *avr* genes recognized by *A. thaliana*, and ask whether loss of *avr* signal delivery is specific to *avrRpm1*, or a general phenomenon. Molecular characteization of these mutants will lead to furthered understanding of mechanisms controlling *avr* gene signal delivery.

Summary

A great deal of progress has been made, by several groups, in the exploitation of *A. thaliana* as a model in the molecular analysis of plant-pathogen interactions since the last meeting of this august Society only two years ago (Dangl, 1992). This trend will only continue as further *R*.gene loci are defined and mapped, and as loss of recognition and altered recognition mutants are isolated using a variety of strategies discussed by others in this volume. We hope to gain a basic understanding of *R* gene structure and function through molecular isolation of *RPM1*, and its potential homologs in a variety of crop species. Moreover, via utilization of either natural or induced variation at *RPM1*, we hope to dissect signals reaching the plant cell nucleus, and effector pathways involved in establishing and maintaining the resistant state. Finally, we hope to integrate this knowledge with definition of specific *avr* genes, and the genes encoding passage of *avr* signals through the bacterial membrane. This presentation has outlined our current progress in these areas.

References

Allard, R.W. (1956) Formulas and tables to facilitate the calculation of recombination values in heredity. *Hilgardia,* **24**, 235-278.

Ausubel, F.M., Davis, K.R., Schott, E.J., Dong, X. and Mindrinos, M. (1991) Identification of signal transduction pathways leading to the expression of *Arabidopsis thaliana* defense genes. In *Advances in Molecular Genetics of Plant-Microbe Interactions, Current Plant Science and Biotechnology in Agriculture,* Volume 1 (eds.) H. Hennecke and D.P.S. Verma, Kluwer Academic Publishers, Dordrecht pp. 357-364.

Bavage, A.D., Vivian, A., Atherton, G.T., Taylor, J.D. and Malik, A.N. (1991) Molecular genetics of *Pseudomonas syringae* pathovars: Plasmid involvement in cultivar-specific incompatibility. *J. Gen. Microbiol.* **137,** 2231-2239.

Bennetzen, J.L., Hulbert, S.H., Lyons, P.C. (1991) Genetic fine structure analysis of a maize disease resistance gene. In: "Molecular Strategies of Pathogens and Host Plants", S.P. Patil, S. Ouchi, D. Mills, C. Vance (eds.), Springer Verlag, New York, Berlin pp. 177-188.

Bent, A., Carland, F., Dahlbeck, D., Innes, R., Kearney, B., Ronald, P., Roy, M., Salmeron, J., Whalen, M. and Staskawicz, B. (1991) Gene-for-gene relationships specifying disease resistance in plant-bacterial interactions. In *Advances in Molecular Genetics of Plant-Microbe Interactions, Current Plant Science and Biotechnology in Agriculture,* Volume 1 (eds.) H. Hennecke and D.P.S. Verma, Kluwer Academic Publishers, Dordrecht pp. 32-36.

Crute, I.R. (1985) The genetic bases of relationships between microbial parasite and their hosts. In: "Mechanisms of Resistance to Plant Disease".R.S.S. Fraser (ed.) Martinus Nijhoff and W. Junk, Dordrecht, pp. 80-142.

Dangl, J.L, Lehnackers, H., Kiedrowski, S., Debener, T., Rupprecht C., Arnold, M. and Somssich, I. (1991) Interactions between *Arabidopsis thaliana* and phytopthogenic *Pseudomonas* pathovars: a model for the genetics of disease resistance. In *Advances in Molecular Genetics of Plant-Microbe Interactions, Current Plant Science and Biotechnology in Agriculture,* Volume 1 (eds.) H. Hennecke and D.P.S. Verma, Kluwer Academic Publishers, Dordrecht pp. 78-83.

Dangl, J.L. (1992) Applications of *Arabidopsis thaliana* to outstanding issues in plant pathology. *Int. Rev. Cytol.* **(in press)**.

Dangl, J.L., Ritter, C., Humphrey, M.J, Wood, J.R., Mur, L.A.J., Goss, S., Mansfield, J.W. and Vivian, A. (1992) Functional homologs of the *Arabidopsis RPM1* disease resistance gene in bean and pea. **(submitted).**

Davis, K.R., Schott, E. and Ausubel, F.M. (1991) Virulence of selected phytopathogenic Pseudomonads in *Arabidopsis thaliana. Mol. Plant-Microbe Interact.* **4**, 477-488.

Davis, K.R. (ed.) (1992) *Arabidopsis as a model for plant-pathogen interactions.* APS Press, St. Paul, Mn. (in press).

Debener, T., H Lehnackers, H., Arnold, M. and Dangl, J.L. (1991) Identification and molecular mapping of a single *Arabidopsis* locus conferring resistance against a phytopathogenic *Pseudomonas* isolate. *The Plant Journal* **1,** 289-302

Dong, X. Mindrinos, M., Davis, K.R. and Ausubel, F.M. (1991) Induction of *Arabidopsis* defense genes by virulent and avirulent *Pseudomonas syringae* strains and by a cloned avirulence gene. *Plant Cell* **3**, 61-72.

Ellingboe, A.H. (1981) Changing concepts in host-pathogen genetics. *Ann. Rev. Phytopathol.* **19**, 125-143.

Ellingboe, A.H. (1982) Genetical aspects of active defense. In *Active Defense Mechanisms in Plants* (Wood, R.K.S., ed.) New York: Plenum Press, pp. 179-192.

Ellingboe, A.H. (1984) Genetics of host-parasite relations: an essay. *Adv. Plant Pathol.* **2**, 131-151.

Fillingham, A.J., Wood, J., Bevan, J.R., Crute, I.R., Mansfield, J.W., Taylor, J.D. and Vivian, A. (1992) Avirulence genes from *Psudomonas syringae* pathovars *phaseolicola* and *pisi* confer specificity towards both host and non-host species. *Physiol. Mol. Plant Pathol.* **40,** 1-15.

Flor, A.H. (1955) Host-parasite interactions in flax rust - its genetics and other implications. *Phytopathol.* **45**, 680-685.

Flor, H. (1971) Current status of gene-for-gene concept. *Ann. Rev. Phytopathol.* **9**, 275-296.

Hitchin, F.E., Jenner, C.E., Harper, S., Mansfield, J.W., Barber, C.E. and Daniels, M.J. (1989) Determinnant of cultivar specific avirulence cloned from *Pseudomonas syringae* pv. *phaseolicola* race 3. *Physiol. Mol. Plant Pathol.* **34,** 309-322.

Hulbert, S.H., Bennetzen, J.L. (1991) Recombination at the *Rp1* locus of maize. *Mol. Gen. Genet.* 226, 377-382.

Jenner, C., Hitchin, E., Mansfield, J., Walters, K., Betteridge, P., Teverson, D. and Taylor, J. (1991) Gene-for-gene interactions between *Pseudomonas syringae* pv. *phaseolicola* and *Phaseolus. Mol. Plant-Microbe Interact.* **4,** 553-562.

Keen, N.T. (1982) Specific recognition in gene-for-gene host-parasite systems. *Adv. Plant Pathol.* **2,** 35-82.

Keen, N.T. (1990) Gene-for-gene complementarity in plant-pathogen interactions. *Ann. Rev. Genet.* **24**, 447-463.

Keen, N.T. and Staskawicz, B. (1988) Host range determinants in plant pathogens and symbionts. *Ann. Rev. Microbiol.* **42**, 421-440.

Kiedrowski, S., Kawalleck, P., Hahlbrock, K., Somssich, I.E. and Dangl, J.L. (1992) Rapid activation of a novel plant defense gene is strictly dependent on the *Arabidopsis RPM1* disease resistance locus. **(submitted)**

Michelmore, R.W., Paran, I., and Kesseli, R. V. (1991) Identification of markers linked to disease-resistance genes by bulked segregant analysis: A rapid method to detect markers in specific genome regions by using segregating populations. *Proc. Natl. Acad. Sci., USA* **88,** 94828-9832.

Pryor, A. (1987) The origin and structure of fungal disease resistance genes in plants. *Trends Genet.* 3, 157-161.

Somssich, I.E., Bollman, J., Hahlbrock,K., Kombrink, E. and Schulz, W. (1989) Differential early activation of defense-related genes in elicitor-treated parsley cells. *Plant Mol. Biol.* **12,** 227-234.

Taylor, J.D., Bevan, J.R., Crute, I.R. and Reader, S.L. (1989) Genetic relationship between races of *Pseudomonas syringae* pv. *pisi* and pea (*Pisum sativum*). *Plant Pathology* **38,** 364-375.

Vivian, A., Atherton, G., Bevan, J., Crute, I.R., Mur, L. and Taylor, J. (1989) Isolation and characterization of cloned DNA conferring specific avirulence in *Pseudomonas syringae* pv. *pisi* to pea (*Pisum sativum*) cultivars which possess the resistance allele R2. *Physiol. Mol. Plant Pathol.* **34,** 335-344.

Whalen, M.C., Innes, R.W., Bent, A.F. and Staskawicz, B.J. (1991) Identification of *Pseudomonas syringae* pathogens of *Arabidopsis* and a bacterial locus determing avirulence on both *Arabidopsis* and soybean. *Plant Cell,* **3**, 49-59.

IDENTIFICATION OF AN *ARABIDOPSIS* LOCUS THAT GOVERNS AVIRULENCE GENE-SPECIFIC DISEASE RESISTANCE.

B. N. KUNKEL, A. F. BENT, D. DAHLBECK, R. W. INNES[1] AND B. J. STASKAWICZ.
Department of Plant Pathology
University of California, Berkeley
Berkeley, CA 94720, USA.

[1]Present address:
Department of Biology
Indiana University
Bloomington, IN 47405

ABSTRACT. We are using a molecular genetic approach to identify and characterize plant genes that control resistance in *Arabidopsis thaliana* to the bacterial pathogen *Pseudomonas syringae* pv. tomato (*Pst*). A screen for *Arabidopsis* mutants that are altered in their ability to express resistance to *Pst* carrying the bacterial avirulence gene *avrRpt2* has been initiated. Four fully susceptible and one partially susceptible mutants have been isolated to date and are now being further characterized. Most progress has been made in the characterization of the fully susceptible mutant D203. Mutant D203 is altered specifically in its ability to recognize bacteria expressing *avrRpt2*, as it retains resistance to bacteria carrying other avirulence genes. Susceptibility in mutant D203 is due to a defect at a single locus mapping to chromosome 4. Identification of a resistance locus in *Arabidopsis* with specificity for a single bacterial avirulence gene suggests that this locus, designated *RPT2*, may control the specific recognition of bacteria expressing *avrRpt2*. In a second approach we are taking advantage of the natural variation that exists among wild isolates (ecotypes) to study the genetic basis of disease resistance in *Arabidopsis*. Two ecotypes, Wii-0 and Po-1, that are susceptible to *Pst* strains expressing *avrRpt2* have been identified. Genetic analysis of Wii-0 indicates that this ecotype lacks a functional allele of *RPT2*, demonstrating that there is natural variation at the *RPT2* locus. Genetic analysis of ecotype Po-1 suggests that resistance to *Pst* strains expressing *avrRpt2* may be controlled by more than one locus.

1. Introduction

Race-specific disease resistance in plants appears to be triggered by specific recognition of the pathogen. Genetic analysis of plant pathogens and their hosts suggests that in many cases pathogen recognition is determined by single dominant or semidominant resistance genes in the host with specificity for single, dominant avirulence (*avr*) genes in the pathogen. Little is understood about these "gene-for-gene" interactions at the molecular level, however.

Arabidopsis thaliana has recently been established as a model host in which to study the molecular genetic basis of plant disease resistance to the bacterial pathogen *Pseudomonas syringae* (Whalen et al, 1991; Dong et al, 1991; Debener et al, 1991). *P. syringae* strains that are pathogenic on *Arabidopsis* have been identified and several bacterial genes, including *avrRpt2*, *avrB* and *avrRpm1*, that control avirulence in a host genotype-specific manner have been isolated. When introduced into *Pst* any of these avirulence genes converts a normally

E. W. Nester and D. P. S. Verma (eds.),
Advances in Molecular Genetics of Plant-Microbe Interactions, 417–421.

virulent strain, such as *Pst* strain DC3000, into an avirulent one no longer capable of causing disease on *Arabidopsis* ecotype Colombia, (Col-0, see Table 1).

We are using two different genetic approaches to identify and characterize plant genes that control disease resistance in *Arabidopsis*. The first approach involves mutational analysis of resistant ecotypes. Specifically, we have initiated a screen for mutants of ecotype Col-0 that are altered in their ability to express disease resistance to *Pst* strain DC3000 carrying the avirulence gene *avrRpt2*. Our second approach takes advantage of the natural variation that exists among *Arabidopsis* ecotypes to study the genetic basis of disease resistance. Using these two approaches we expect to uncover mutations or alterations both in the plant resistance gene corresponding to *avrRpt2* and in genes required for expression of resistance subsequent to pathogen recognition.

2. Methods: Development of an efficient inoculation procedure

To facilitate the isolation of susceptible mutants, we have taken advantage of the surfactant Silwet L-77 (Union Carbide) to develop an efficient procedure for screening large numbers of *Arabidopsis* plants for susceptibility to *Pst* strain DC3000 expressing *avrRpt2* (Whalen et al, 1991). L-77 is a silicon-based co-polymer that reduces the surface tension of aqueous solutions sufficiently to allow bacterial solutions to spread evenly over the leaf surface. To carry out the screen, *Arabidopsis* seeds are planted in 3-1/2" pots covered with fiberglass screen and grown for 5 weeks in Conviron growth chambers under an 8 hour photoperiod at 24° C. Five week old plants are dipped into a bacterial suspension of 2-3 x 10^8 c.f.u./ml containing 0.02% L-77 and then placed under plastic domes for 24 hours. Three to four days after inoculation the plants are scored for disease symptoms, which consist of small, individual water-soaked lesions surrounded by a halo of chlorosis.

3. Identification of Col-0 mutants with altered resistance to *Pst* DC3000 expressing *avrRpt2*.

6, 000 M2 plants derived from diepoxybutane (DEB)-mutagenized Colombia (Col-0) seed have been screened to date using the inoculation procedure described above. Five mutants with altered resistance to *Pst* strain DC3000 expressing *avrRpt2* have been isolated. The mutants fall into two classes: a fully susceptible class (4 mutants) and a partially susceptible class (1 mutant). Most progress has been made in the characterization of the fully susceptible mutant, D203.

4. Characterization of susceptible mutant D203

MUTANT D203 IS FULLY SUSCEPTIBLE TO *PST* DC3000(*AVRRPT2*)

Mutant D203 exhibits severe disease symptoms when inoculated with *Pst* strain DC3000 expressing *avrRpt2* (Table 1). To determine whether this fully susceptible phenotype is correlated with the unrestricted growth of *Pst* DC3000 (*avrRpt2*), we monitored bacterial growth in the mutant. As is shown in Figure 1, growth of *Pst* DC3000 (*avrRpt2*) is unrestricted in mutant D203, reaching a final concentration of 10^7 to 10^8 cfu/cm^2 of leaf tissue. Growth of *Pst* DC3000 (*avrRpt2*) in wild-type Col-0 is restricted and reaches a final leaf concentration of only 10^4 cfu/cm^2. The susceptible phenotype of mutant D203 is also associated with the inability of the mutant to exhibit a visible hypersensitive response (HR) when inoculated with high levels of *Pst* expressing *avrRpt2*.

Table 1. Inoculation phenotypes of Col-0 resistance mutant D203 and the susceptible ecotypes Wii-0 and Po-1.

Pst strain	*Arabidopsis* lines Col-0(wt)	D203	Wii-0	Po-1
DC3000	S	S	S	S
DC3000 (*avrRpt2*)	R	S	S	S
DC3000 (*avrB*)	R	R	R	S
DC3000 (*avrRpm1*)	R	R	R	S

S: susceptible. R: resistant

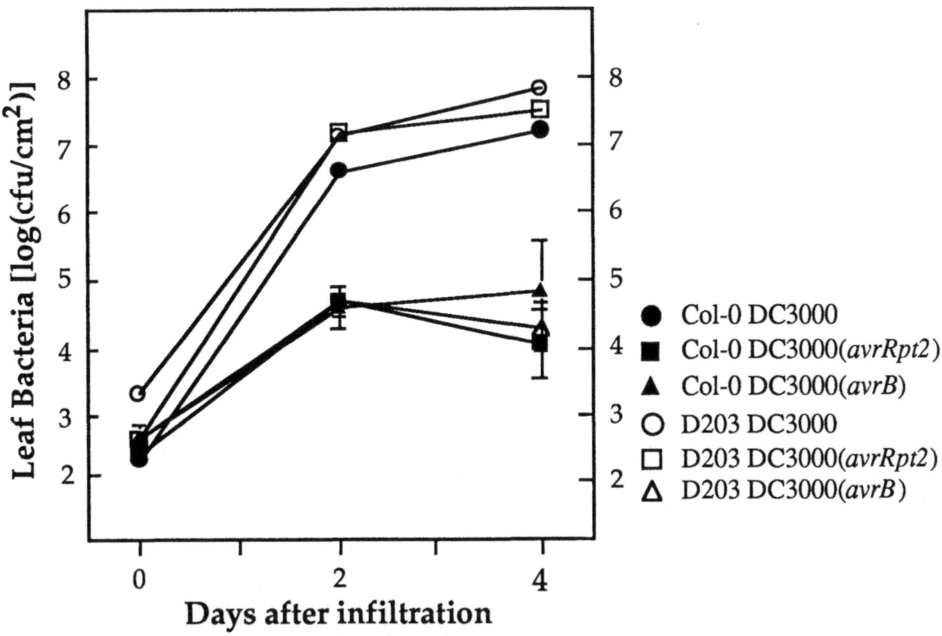

Figure 2. Growth of *Pst* DC3000 strains in Col-0 mutant D203. In planta population sizes of *Pst* strains were monitored by grinding leaf samples and plating on selective media. Plants were inoculated by vacuum infilatration of 1 x 10^5 cfu/ml bacterial suspension. Data points represent mean of three replicates +/- standard error. The avirulence gene clones were carried on vector pVSP61.

SUSCEPTIBILITY IN THE D203 MUTANT IS SPECIFIC TO *PST* STRAINS EXPRESSING *AVRRPT2*

Susceptibility in the D203 mutant is specific to *Pst* strains expressing *avrRpt2*, as the mutant retains resistance to *Pst* strains expressing *avrB* or *avrRpml* (Table 1). Resistance to *Pst* DC3000 (*avrB*) is also reflected in the restricted growth of this strain in mutant tissue (Figure 1). These results suggest that the D203 mutant is altered primarily in its ability to specifically recognize *Pst* strains expressing *avrRpt2*.

SUSCEPTIBILITY IN MUTANT D203 IS DUE TO A DEFECT AT A SINGLE LOCUS

In order to determine the genetic basis of susceptibility in the mutant we crossed D203 to wild type Col-0. The F1 progeny from this cross were resistant to Pst DC3000(*avrRpt2*) when inoculated by the L-77 dipping method described above. When assayed for the ability to express the HR after inoculation with high levels of strains expressing *avrRpt2*, the F1 progeny exhibited a delayed HR, visible only 36 to 48 hours after inoculation in contrast to the usual 16 to 24 hours observed for wild type Col-0. Thus the F1 plants exhibit an intermediate resistance phenotype, suggesting that the D203 mutation is semidominant.

The F2 progeny from the D203 x Col-0 cross segregated in a ratio of 3 resistant to 1 susceptible when scored for resistance to *Pst* DC3000(*avrRpt2*) using the L-77 inoculation procedure (Kunkel et al, manuscript in preparation). F2 progeny analysis from a reciprocal cross (Col-0 x D203) also segregated in a ratio of 3 : 1. These results indicate that susceptibility in the D203 mutant is due to a defect at a single locus.

The identification of a single gene that controls resistance to *Pst* strains expressing *avrRpt2* indicates that resistance in this *Arabidopsis/Pseudomonas syringae* interaction is governed by a gene for gene interaction. Thus we can think of the locus defined by the D203 mutation as a classical resistance locus with specificity for the *avrRpt2* avirulence gene. We are now referring to this resistance locus as *RPT2* (resistance to *P. syringae* pv. tomato)

5. The *RPT2* locus maps to chromosome 4.

We are carrying out RFLP linkage analysis to map the *RPT2* locus, utilizing progeny from a cross of D203 (Col-0 background) to the resistant ecotype No-0. We have mapped the *RPT2* locus to chromosome 4 between the RFLP markers At557 and At214. Our current data indicate that the *RPT2* locus maps within 1 cM of marker At600. We are now in the process of identifying additional closely linked markers for use in map-based cloning strategies.

6. Characterization of the additional susceptible mutants.

The additional mutants isolated in the screen for mutants with altered resistance to *Pst* strain DC3000 expressing *avrRpt2* fall into two classes: a fully susceptible class and a partially susceptible class. The three additional fully susceptible mutants are phenotypically identical to mutant D203. The one partially susceptible mutant exhibits an intermediate resistance phenotype, which is characterized by mild disease symptoms when inoculated with *Pst* DC3000 (*avrRpt2*), a delayed HR when inoculated with high levels of strains expressing *avrRpt2*, and intermediate levels of growth of *Pst* DC3000 (*avrRpt2*) in the mutant tissue. The most interesting point concerning all five of our susceptible mutants is that they are susceptible specifically to *Pst* strains expressing *avrRpt2*. We are now in the process of carrying out complementation tests among these mutants and D203 to determine the number of loci we have identified by mutation.

7. Genetic analysis of susceptible ecotypes Wii-0 and Po-1.

We are also taking advantage of the natural variation that exists among wild isolates (ecotypes) to study the genetic basis of disease resistance in *Arabidopsis*. Two ecotypes, Wii-0 and Po-1, that are susceptible to *Pst* strain DC3000 expressing *avrRpt2* have been identified (Table 1). To examine the genetic basis of susceptibility in these lines, both Wii-0 and Po-1 have been crossed to the resistant ecotype Col-0.

SUSCEPTIBLE ECOTYPE WII-0 DOES NOT CARRY A FUNCTIONAL ALLELE OF *RPT2*

Like mutant D203, Wii-0 is susceptible specifically to *Pst* strains expressing *avrRpt2* (Table 1). This susceptibility is due to an alteration at a single locus, as resistance to *Pst* DC3000 (*avrRpt2*) in F2 progeny from a cross of Col-0 to Wii-0 segregates as a single mendelian trait.

To determine whether Wii-0 carries a functional allele of *RPT2*, we crossed Wii-0 and mutant D203. None of the F1 and F2 progeny from this cross exhibited an HR when inoculated with high levels of strains expressing *avrRpt2* (Kunkel et al, manuscript in preparation). The results of this allelism test indicate that Wii-0 is lacking a functional allele of *RPT2*. These findings demonstrate that there is genetic variation at the *RPT2* locus among wild isolates of *Arabidopsis*.

GENETIC ANALYSIS OF ECOTYPE PO-1 SUGGESTS THAT RESISTANCE MAY BE CONTROLLED BY MORE THAN ONE LOCUS.

Genetic analysis of the susceptible ecotype Po-1 has revealed a somewhat surprising result. Resistance to *Pst* DC3000(*avrRpt2*) in the F2 progeny of a cross of Po-1 x Col-0 did not segregate in a ratio of 3 resistant to 1 susceptible, indicating that resistance in this cross is governed by more than one locus. In fact, the segregation ratio observed in the progeny from this cross was consistent with a ratio of 9 resistant to 7 susceptible plants, suggesting that resistance in Col-0 is controlled by two dominant loci, both of which are required and both of which are inactive or missing in Po-1. Further genetic analysis of Po-1 x Col-0 progeny has revealed that one of the genes segregating in this cross is required for specific resistance to *Pst* expressing *avrRpt2*. The role of the second gene has not yet been determined. One possibility is that the second gene may be required for the expression of disease resistance subsequent to pathogen recognition. The observation that ecotype Po-1 is susceptible to several different *Pst* strains expressing several different avirulence genes (Table 1) supports this hypothesis.

8. Conclusions

Using two separate but complementary genetic approaches we have identified at least two loci in *Arabidopsis* that control resistance to *P. syringae* pv. tomato. One of these loci, the *RPT2* locus, which is defined by the susceptible mutant D203, appears to govern the specific recognition of bacteria expressing *avrRpt2*. We are now in the process of carrying out genetic analysis of resistance to Pst strains that carry other avirulence genes, such as avrB. In the process, we will learn whether it is possible to identify additional genes that control pathogen recognition, as well as genes that are required for the expression of disease resistance subsequent to the recognition event.

9. Acknowledgements

B.N.K is a DOE -Energy Biosciences Postdoctoral Research Fellow of the Life Sciences Research Foundation. A.F.B. and R.W.I. were supported by National Science Foundation postdoctoral fellowships awarded in 1989 and 1987, respectively. This work was funded by a Department of Energy grant to B. J. S.

10. References

Debener, T., Lehnackers, H., Arnold, M. and Dangl, F. L. (1991). 'Identification and molecular mapping of a single *Arabidopsis thaliana* locus determining resistance to a phytopathogenic *Pseudomonas syringae* isolate', The Plant Journal 1, 289-302.

Dong, X., Mindrinos, M., Davis, K. R. and Ausubel, F. M. (1991). 'Induction of *Arabidopsis* defense genes by virulent and avirulent *Pseudomonas syringae* strains and by a cloned avirulence gene. Plant Cell, 3, 61-72.

Whalen, M. C., Innes, R. W., Bent, A. F., and Staskawicz, B. J. (1991). 'Identification of *Pseudomonas syringae* pathogens of *Arabidopsis* and a bacterial locus determining avirulence on both *Arabidopsis* and soybean. Plant Cell, 3, 49-59.

PLANT AND BACTERIAL GENES INVOLVED IN INTERACTIONS BETWEEN *XANTHOMONAS* AND CRUCIFERS

M.J. DANIELS, C.E. BARBER, J.M. DOW, B. HAN, S.A. LIDDLE,
M.A. NEWMAN, J.E. PARKER, S.D. SOBY AND T.G.J. WILSON
The Sainsbury Laboratory
John Innes Centre
Norwich Research Park
Norwich NR4 7UH
UK

ABSTRACT. Several classes of genes required for full pathogenicity of *Xanthomonas campestris* pathovar *campestris* have been characterised. These include 1. structural genes encoding extracellular plant degrading enzymes, 2. genes involved in synthesis of extracellular polysaccharide, 3. a multigene system encoding a secretion system for extracellular enzymes, 4. complex regulatory systems controlling enzyme and EPS synthesis, 5. *hrp* genes required for pathogenicity and for ability to induce a hypersensitive response on resistant non-host plants, 6. an avirulence gene which interacts with *Arabidopsis thaliana*, and 7. genes whose function is unknown. *Brassica* plants respond to challenge with *Xanthomonas* by differential expression of many genes, including genes encoding hydrolases such as ß-1,3 glucanase and chitinases. These show characteristic patterns of expression depending on the genotype of the challenging bacterium. The effect on glucanase induction produced by using certain classes of bacterial mutant to challenge the plants is described.

1. Introduction

Xanthomonas campestris pathovar *campestris* (hereafter *X.c. campestris*) is the causal agent of black rot of crucifers (Williams, 1980). The bacterium is seed-borne, and plants derived from contaminated seed may become infected. The primary portal of entry into the plant is the hydathode, from which vascular system becomes colonised. Typical lesions have the appearance of wedge-shaped chlorotic and necrotic areas advancing along veins from the leaf margins. In later stages the bacteria may break out from the veins and colonise adjacent tissues.

The interaction of *X.c. campestris* with plants has been studied in some detail, both from the bacterial and the plant point of view. The system constitutes an excellent model pathosystem for molecular genetic analysis for several reasons. The pathogen can be readily manipulated genetically and several potential pathogenicity or virulence factors have been identified from physiological studies. As a result many classes of gene required for pathogenicity have been identified. Moreover as the host range of *X.c.campestris* includes *Arabidopsis thaliana*, a number of powerful strategies are available which exploit the genetic potential of this plant to investigate the host's role in the interaction. In this paper we summarise some information on pathogenicity genes of *X.c. campestris*, emphasising the multiple "balancing" regulatory systems which control

E. W. Nester and D. P. S. Verma (eds.),
Advances in Molecular Genetics of Plant-Microbe Interactions, 423–433.

synthesis of enzymes and extracellular polysaccharide (EPS). The response of plants to infection is described in terms of one facet of the process, namely ß-1,3 glucanase induction. In particular the value of using bacterial mutants as probes to dissect the plant response is discussed.

2. The Pathogen

Genes required for pathogenicity of *X.c. campestris* have been cloned using several strategies, for example (i) complementation of mutants with cloned DNA, (ii) synthesis of an easily-detected protein directed by cloned DNA in a suitable bacterial host, (iii) detection of genes showing enhanced expression when the bacteria are in plants, (iv) hybridisation to related genes (or parts thereof) already cloned. The role of cloned genes in pathogenicity has usually been determined by localised mutagenesis of the chromosomal copy of the gene, using the cloned DNA as a vehicle for mutagenesis. Pathogenicity is tested by inoculation into plants, preferably using different inoculation methods, and observing symptom development and bacterial growth *in planta*.

Here we describe some classes of gene found to be important in pathogenicity. These genes may be considered relatively specific for pathogenicity because mutations have little effect on bacterial growth *in vitro*. Needless to say, mutations in genes encoding enzymes of intermediary metabolism or proteins involved in other basic physiological functions ("housekeeping genes") may have adverse effects on growth and will probably also reduce virulence. Such genes would not normally be considered to be "pathogenicity genes". However it is unrealistic to separate the two aspects of bacterial life since plant tissues constitute the major habitat of *X.c. campestris*, and it can be argued that the bacterial metabolic systems will have evolved to allow optimal exploitation of this environment. Nevertheless, the concept that there exists a set of pathogenicity genes has been important operationally.

2.1. EXTRACELLULAR ENZYMES

X.c. campestris produces protease, endoglucanase, polygalacturonate lyase, lipase and amylase activities. Since these enzymes have the capacity to degrade plant cell components, they are obvious candidates as pathogenicity factors. Structural genes for all these enzymes have been cloned (Tang et al., 1987; Gough et al., 1988; Dow et al., 1989; C.E. Barber, M.K. Sawczyc and M.J. Daniels, unpublished).

2.1.1. *Structural Genes.* Three proteases are produced, one being a serine metalloprotease and the others metalloproteases (Dow et al., 1990). The sequence of the first has been determined (Liu et al, 1990). The three enzymes differ in their proteolytic cleavage specificity, as judged by the array of peptides produced by digestion of a model substrate, ß-casein. All strains of *X.c. campestris* examined produce apparently the same proteases. *X.c. campestris* mutants lacking protease produce similar symptoms to the wild type after injection into seedlings or into mature leaf panels, but growth of bacteria in plant tissues is lower than the wild type. However, if the bacteria are applied to small wounds at leaf margins (vein endings), simulating natural hydathode infection, a marked reduction in virulence is apparent (Tang et al., 1987; Dow et al., 1990). These results suggest that protease is an essential pathogenicity factor at early stages of the disease process, but once infection is well advanced the enzyme is less significant.

Endoglucanase ("cellulase") is the major extracellular protein produced by *X.c. campestris*, and has a structure typical of other prokaryotic endoglucanases (Gough et al., 1990). Surprisingly, mutants lacking endoglucanase show little reduction in virulence, even when measured by stringent tests (Gough et al., 1988). The role of the enzyme is not understood; one possibility is that it contributes to bacterial nutrition during the saprophytic phase of the life cycle. *X.c. campestris* may be unable to degrade cellulose significantly in living plants, but in dead plant residues the polymer may be less refractory, particularly when subject to attack by a larger, more diverse microflora.

The structural gene for one polygalacturonate lyase has been cloned and the chromosomal gene disrupted (Dow et al., 1989). Little effect on virulence can be detected because the loss of the enzyme is masked by the presence of other isozymes (Dow et al., 1987). Amylase and lipase genes have not been studied in detail.

2.1.2.*Secretion of extracellular enzymes*. A non-pathogenic mutant of *X.c. campestris* was found to be pleiotropically defective in extracellular enzyme production, and was used to clone a DNA fragment which restored concomitantly enzyme production and pathogenicity. The important genes were found to lie in a cluster of *ca.* 11kb. Mutations in this region did not affect the synthesis of the set of extracellular enzymes, but the enzymes were not secreted and accumulated in the bacterial periplasm (Dow et al. 1987). Thus the mutants were phenotypically enzyme-defective and were unable to give disease symptoms in plants. Most of the secretion genes (designated *xps*) have been sequenced and the system shows a striking resemblance in gene organisation and protein sequence to protein secretion systems of other Gram-negative bacteria (Dums et al., 1991; Hu et al., 1992; Pugsley et al., 1990).

2.2. EXTRACELLULAR POLYSACCHARIDE

X.c. campestris produces the extracellular polysaccharide (EPS) called xanthan gum, which is widely used industrially as a thickening agent. The biosynthesis of xanthan has been studied both biochemically (Ielpi et al., 1981) and genetically (Thorne et al., 1987; Harding et al., 1987; Barrère et al., 1986; Hötte et al., 1990). The genes form a large cluster, and the functions of the protein products have been deduced (Vanderslice, 1989). The effect on pathogenicity of mutations which reduce EPS synthesis is reminiscent of the protease situation described above; EPS^- strains show severely reduced virulence if inoculated at leaf margins, but appear normal if introduced into the middle of leaves (unpublished observations). Many plant pathogenic bacteria produce EPS which is usually found to be necessary for pathogenicity, but the biochemical basis of EPS action is not known (Coplin and Cook 1990).

2.3. REGULATORY GENES

Extracellular enzymes and EPS may be regarded as primary pathogenicity factors. The synthesis of these substances is regulated in a complex manner. At the simplest level, synthesis of some of the factors is regulated individually. For example, protease synthesis is induced by peptides or proteins and repressed by amino acids, and polygalacturonate lyase synthesis is induced by polygalacturonate. However there are in addition at least three independent systems which coordinately control expression of all the structural genes.

2.3.1. *Positive Regulatory Genes.* A group of regulatory genes were discovered by complementation of a non-pathogenic mutant which was defective in protease and polygalacturonate lyase production (Daniels et al., 1984). Analysis of the cloned DNA by transposon mutagenesis revealed a cluster of at least seven genes, mutation of any of which resulted in substantial reduction in synthesis of all the extracellular enzymes and EPS. The genes have been designated *rpf*, for *r*egulation of *p*athogenicity *f*actors (Tang et al., 1991). Sequencing indicated that one of the genes, *rpfC*, is a member of the class of two-component regulatory genes which are widespread in prokaryotes and usually modulate gene expression in response to environmental changes (Stock et al., 1989). Another member of the cluster, *rpfG*, may also belong to this class of regulator (Y.N. Liu, J-L. Tang and M.J. Daniels, unpublished). *rpfC* is unusual in that both the sensor and the regulator modules are encoded by the same gene, whereas in most cases two separate genes are found. Only a small number of such fused two-component genes are known; interestingly one of these is *lemA* which is involved in pathogenicity of *Pseudomonas syringae* pv *syringae* (Hrabak and Willis, 1992). The predicted size of the fused RpfC protein is 74 kDa, and a product of this size is produced following expression of the cloned *rpfC* gene in *Escherichia coli* cells or extracts. It is not known whether the gene product functions in *X.c. campestris* as a fused sensor-regulator unit or whether it is post-translationally cleaved in a linker region which joins the two modules. In the latter case the cleavage reaction to give the functional system would represent a further potential level of regulation. The RpfC protein appears to be able to undergo autophosphorylation when incubated with ATP, suggesting that the signal transduction mechanism is similar to that of other two-component systems (Stock et al., 1989). Mutations in the seven genes *rpfA-G* have the same phenotype in terms of enzyme and EPS synthesis. It is not known why there are so many genes in the system, or how they interact, or what environmental signal they detect.

Osbourn et al. 1990(a) exploited the relatedness of two-component regulatory systems to clone further genes of this class from *X.c. campestris*, using oligonucleotide probes corresponding to conserved domains. Of 12 independent genes found, most did not seem to affect pathogenicity following mutation. One did seem to be involved in EPS production, but the reduction in EPS level in the mutants was insufficient to have a noticeable effect on pathogenicity.

2.3.2. *Negative Regulatory Gene.* The positive regulatory genes behave as activators of gene expression. A gene acting negatively on the same target genes (enzymes and EPS biosynthesis) was found by Tang et al. (1990). Mutation of the chromosomal copy of this gene results in over-production of enzymes and (probably) EPS. A protein has been detected in extracts of *X.c. campestris* which binds to a conserved sequence upstream of the protease and cellulase structural genes. This level of this binding protein in extracts depends on the copy number of the negative regulatory gene (called *rpfN*) and it is not found in *rpfN* mutants in which the gene has been disrupted. It is tempting to believe that the RpfN product is the DNA binding protein, or at least a component of the regulatory pathway (B. Han, S.D. Soby & M.J.Daniels, unpublished). Transposon mutagenesis, sequencing and expression studies indicate that the RpfN protein has a mass of 46 kDa and may be σ^{54} dependent.

2.3.3. *The clp Gene.* An *X.c. campestris* gene called *clp* which encodes a protein similar to the *E. coli* catabolite activator protein (CAP) was described by de Crécy-Lagard et al (1990). However the *X.c. campestris* gene product does not seem to function in the same way as CAP. Mutations in the *clp* gene affect synthesis of extracellular enzymes and EPS, and reduce virulence.

Endoglucanase and polygalacturonate lyase synthesis is reduced in the *clp* mutant, but levels of amylase and protease are unchanged. Less EPS is produced and the polymer has different properties.

Mutations in the regulatory genes affect pathogenicity to susceptible plants. However the mutants are still fully competent to induce a hypersensitive response (HR) in non-hosts such as pepper. It is not known why *X.c. campestris* has such a complex regulatory network, or how the several systems interact. It is likely that the genes act to "fine-tune" enzyme and EPS synthesis as the physicochemical properties of the bacterial environment (which includes infected plant tissues) change. This might enable the bacteria to exploit the resources of the host efficiently. It is clear that analysis of the regulatory systems is important for gaining an understanding of the disease process. Table 1 summarises the effect of mutations in the regulatory genes and also the *hrp* genes (see next section).

Table 1. Pleiotropic effects of certain genes in *X.c. campestris*.
Mutations either do (Yes) or do not (No) affect the phenotypic characters indicated.

	Gene:			
	rpfA-G,N	Other two-component	*clp*	*hrp*
Character:				
Protease	Yes	No	No	No
Cellulase	Yes	No	Yes	No
Polygalacturonate lyase	Yes	No	Yes	No
Amylase	Yes	No	No	No
EPS production	Yes	Yes	Yes	No
Pathogenicity	Yes	No	Yes	Yes
HR induction	No	No	No	Yes

2.4. *hrp* GENES

hrp mutants defective in both pathogenicity and in the ability to incite HR have been isolated in *Erwinia, Pseudomonas* and *Xanthomonas* (Willis et al., 1991). In all cases the mutations fall in clusters of genes covering at least 20kb. Boucher et al. (1987) found that the *P. solanacearum hrp* cluster hybridised with DNA from all *X. campestris* pathovars tested, but not with the other pathogen species/genera. Arlat et al. 1991 showed that the homologous DNA of *X.c. campestris* and *X.c. vitians* indeed represented *hrp* genes, and it was possible to cross-complement mutations both between *X.c. campestris* and *X.c. vitians* and also, to some extent, between *X.c. campestris* and *P. solanacearum*. *hrp* mutants of *X.c. campestris* were not affected in enzyme or EPS production, and hence the pathogenicity defect is not caused by inability to produce these factors. It is not known whether the *hrp* gene described by Kamoun and Kado (1990) falls within the same class. Expression of *hrp* genes measured by gene fusions is regulated by the physiological state of the bacteria. Nutritional shift experiments in which *X.c. campestris* was first grown in minimal medium and then transferred into the same medium with or without supplementary nutrients showed that all supplements give lower expression levels of the *hrp* genes, irrespective of their effect on the growth of the bacteria. The kinetics of induction suggest that the onset of starvation promotes *hrp* expression.

2.5. AVIRULENCE GENES

Avirulence genes of pathogens interact genetically with resistance genes of plants to trigger responses leading to an incompatible host-pathogen interaction. Only one avirulence gene has been isolated from crucifer-infecting *X. campestris*. This is discussed below (section 3.1).

2.6. OTHER GENES

In addition to the classes of gene described above, there are various examples of characterised pathogenicity genes for which no biochemical function is known. Osbourn et al. 1990(b) described such a gene from *X.c. campestris*. All attributes which have been studied appear to be normal in strains mutant in this gene, with the exception of symptom production.

The question arises of how many genes are required for pathogenicity to plants. The number of genes in *X.c. campestris* represented by the six classes described above probably approaches 60. Arguments based on relative frequency of mutations affecting pathogenicity in several bacteria suggest a figure of 10^2 (Daniels 1988). However, the "pathogenicity genes" considered here are concerned mainly with symptom production in infected plants. The total disease cycle involves many more bacterial attributes, and we must expect the number of documented pathogenicity genes (*sensu latu*) to grow considerably.

3. The Role of the Host Plant

Plants should not be regarded as passive victims of pathogen attack. Infection is followed by intense metabolic and biosynthetic activity. A common conceptual model postulates that the initial encounter between the plant and the pathogen involves a recognition event determined by the allelic state of matching plant resistance genes and pathogen avirulence genes (Lamb et al, 1989). If both the matching alleles are in their dominant functional forms a chain of events is then initiated leading to an incompatible (no disease) interaction. Alternative states lead to compatible interactions (disease). Two approaches have been used to probe plant resistance and response to *X.c. campestris*. One of these is genetical, based on the use of *A. thaliana* as a host, and the other is primarily biochemical, for which *Brassica* is a more convenient host. *X.c. campestris* appears to behave similarly in the two plants, so it is likely that findings with one approach can be used to illuminate interpretations of the other.

3.1. GENETICAL STUDIES

The biochemical strategy described below (section 3.2) is mainly useful for detecting plant gene products which show major changes in level following infection. Constitutively-expressed genes would be overlooked. To overcome this problem a genetic approach has been used. Little is known about the genetics of resistance of *Brassica* to *X.c.campestris* (Dickson and Hunter, 1987). The paucity of sources of resistance is a major problem for plant breeders because black rot is the most serious *Brassica* disease worldwide. The host range of *X.c. campestris* encompasses all cultivated and many wild crucifers, and recently several laboratories have developed the use of *A. thaliana* as an experimental host for the pathogen (Simpson and Johnson, 1990; Tsuji et al., 1991; Daniels et al., 1991). The evidence that the system can be regarded as a realistic model disease is (1) the symptoms incited by *X.c. campestris* are similar to those observed in *Brassica* inoculated by the same methods (*via* wounds or leaf infiltration) (ii) the bacteria grow and spread within the plant tissue, whereas *X. campestris* pathovars which do not infect crucifers do not, and (iii) mutants of *X.c. campestris* with defects in pathogenicity genes show reduced virulence to *A. thaliana* (Daniels et al., 1991).

The interaction of many wild type *X.c. campestris* strains with many *A. thaliana* accessions (ecotypes) has been studied in attempts to find natural variation in resistance/virulence. While most strain-accession combinations result in the same compatible interaction, a small number of combinations suggest the existence of resistance and avirulence genes. Tsuji et al. (1991) studied F2 progeny of a cross between two accessions, Col-O (resistant) and Pr-O (susceptible), and obtained evidence for a single dominant resistance gene. Daniels et al. (1991) cloned an avirulence gene from an *X. campestris* strain (1067, probably pathovar *raphani*) which rendered otherwise virulent bacterial strains avirulent on most *A. thaliana* accessions. However the avirulence gene had no effect on the pathogenicity to *Brassica* lines, and the *A. thaliana* accession Kas-1 is also susceptible to bacteria carrying the gene. Thus the material exists for a genetic analysis of resistance/susceptibility to near isogenic bacteria differing only in a single, sequenced avirulence gene.

Transposon mutagenesis and sequencing indicate that the avirulence gene (*avrXca*) encodes a protein of 66 kDa (J.E. Parker, C.E. Barber, M.J. Fan and M.J. Daniels, submitted for publication). Comparison of the sequence with data base entries provides no clues about the biochemical function of the protein. The upstream region of the gene includes a perfect "harp" box (Fellay et al. 1991). However the expression of *avrXca* is not reduced in *hrp* mutants of *X.c. campestris*, including a strain from which the whole *hrp* cluster has been deleted (S.A. Liddle and M.J. Daniels, unpublished). Furthermore expression is not dependent on the physiological state of the bacteria. The N terminus of the AvrXca protein may have a signal peptide, suggesting that the protein is located in the cell envelope or is secreted into the medium. However further experiments are required to test this possibility. So far no other characterised avirulence gene products from bacteria have been shown to be extracellular.

It is interesting to note that the resistance induced by this avirulence-resistance gene combination is of the "null" type i.e. the incompatible interaction produces no visible effect, although bacterial multiplication is severely reduced. Most avirulence genes so far studied in bacteria manifest their activity by inducing HR in resistant plants. In *A. thaliana* some *Pseudomonas* strains (Dong et al., 1991; Whalen et al., 1991) and at least one *Xanthomonas* strain (J.E. Parker, C.E. Barber and M.J. Daniels, unpublished) produce a typical rapid HR. Thus *A. thaliana* provides examples of different types of resistance mechanisms to bacteria.

One of the advantages of *A. thaliana* is the ease with which mutants can be isolated (Estelle and Somerville, 1986). By direct screening of M2 progeny plants from chemically-mutagenised seed, it has been found that mutants in many pathways can be isolated at a frequency of 10^{-3} to 10^{-4}. Since inoculation of *A. thaliana* with *X.c. campestris* is not difficult, it should be possible to screen enough plants to detect mutants altered in some disease phenotype. Mutational analysis of pathways has been exceptionally valuable in understanding many aspects of microbial physiology, including phytopathogenicity. The use of an analogous approach to understanding plant resistance and response to infection may be expected to be similarly informative, particularly since map-based strategies can be used to clone genes from *A. thaliana* defined only by phenotypic differences.

3.2. BIOCHEMICAL STUDIES

It has been known for some time that plant gene expression is required for resistance to pathogens to be manifested (Collinge & Slusarenko, 1987). Turnip leaves were infiltrated with suspensions of either *X.c. campestris* (virulent) or *X.c. vitians*, a lettuce pathogen which induces a hypersensitive response. Disease symptoms caused by *X.c. campestris* are not apparent until 48 hr after inoculation, but thereafter they intensify and spread through the leaf tissue. The HR is visible as localised collapse of infiltrated tissue *ca.* 12-18 hr after inoculation. (Collinge et al. 1987). Between 10 and 20 plant genes showed major changes in expression level during HR induction (ie up to 12 hr after inoculation). Qualitatively similar, but kinetically different changes were observed during the compatible interaction; accumulation was not observed until 48 hr post-inoculation (Collinge et al., 1987).

Subsequent studies have concentrated on a subset of the induced gene products, namely β1,3 glucanase and chitinase. Gene expression can be followed by RNA accumulation and the mature proteins can be detected immunologically and assayed enzymatically. There are two major isoforms of chitinase in turnip leaves. Although total chitinase activity increases to a similar extent in both compatible and incompatible interactions, there are interesting differences in the patterns of induction. In unchallenged leaves isoform 1 (CHL 1) predominates, and this isoform increases markedly during HR induction in tissues adjacent to the lesion. In contrast challenge of leaves with *X.c. campestris*, *E. coli* or killed *X.c. vitians* causes an increase only in CHL 2 level (Conrads-Strauch et al., 1990). The basis of this striking pattern of differential gene expression in response to closely related pathogens is currently being studied.

Turnip leaves have one glucanase species which increases in activity after inoculation with *X. campestris* pathovars. However inoculation with *E. coli* or killed *Xanthomonas* does not induce the enzyme, which accumulates in the vacuoles of the plant cells. The induction of transcription of the glucanase gene has been measured by probing northern blots of RNA from inoculated leaves with a turnip glucanase cDNA clone. In leaves challenged with virulent *X.c. campestris* glucanase mRNA begins to accumulate after 12 hr, reaching a maximum at 24 hr and subsequently declining. Interestingly, the pattern of accumulation is indistinguishable if the leaves are challenged with a *hrp* mutant of *X.c. campestris*, which gives no disease symptoms and much reduced growth in the plant tissue. Glucanase mRNA also accumulates in response to challenge with a protease-defective mutant, which gives some symptoms but reduced bacterial growth compared with the wild type. Preliminary data indicate that in this case the mRNA may begin to accumulate earlier than with the wild type bacteria.

Strains of *X.c. raphani* or *X.c. armoraciae* which give HR cause glucanase mRNA to accumulate strongly by 12 hr after inoculation (at least 12 hr earlier than in the compatible interaction). However *hrp* mutants derived from such strains produce lower and delayed induction. Although these studies are at an early stage, they suggest that the strategy of using defined bacterial mutants to trigger plant responses will illuminate the mechanisms involved in the resistance-response process.

The role of chitinase and glucanase in the response of plants to pathogens is not fully understood. The enzymes do not seem to have significant antibacterial properties, although the chitinases do have lysozyme activity (Conrads-Strauch et al, 1990). However they are useful markers of the response of plants to pathogens.

Acknowledgments

The Sainsbury Laboratory is supported by the Gatsby Charitable Foundation. Some of the work from the author's Laboratory described in this paper has been supported by the Agricultural and Food Research Council, the Commission of the European Communities, NATO and the Royal Society.

References

Arlat, M., Gough, C.L., Barber, C.E., Boucher, C. & Daniels, M.J. (1991) *Xanthomonas campestris* contains a cluster of *hrp* genes related to the larger *hrp* cluster of *Pseudomonas solanacearum*. Molecular Plant-Microbe Interactions 4: 593-601.

Barrère, G.C., Barber, C.E. & Daniels, M.J. (1986) Molecular cloning of genes involved in the production of the extracellular polysaccharide xanthan by *Xanthomonas campestris* pv. *campestris*. International Journal of Biological Macromolecules 8: 372-374.

Boucher, C.A., Van Gijsegem, F., Barberis, P.A., Arlat, M. & Zischek, C. (1987) *Pseudomonas solanacearum* genes controlling both pathogenicity on tomato and hypersensitivity on tobacco are clustered. Journal of Bacteriology 169: 5626-5632.

Collinge, D.B. & Slusarenko, A.J. (1987) Plant gene expression in response to pathogens. Plant Molecular Biology 9: 389-410.

Collinge, D.B., Milligan, D.E., Dow, J.M., Scofield, G. & Daniels, M.J. (1987) Gene expression in *Brassica campestris* showing a hypersensitive response to the incompatible pathogen *Xanthomonas campestris* pv. *vitians*. Plant Molecular Biology 8: 405-414.

Conrads-Strauch, J., Dow, J.M., Milligan, D.E., Parra, R. & Daniels, M.J. (1990) Induction of hydrolytic enzymes in *Brassica campestris* in response to pathovars of *Xanthomonas campestris*. Plant Physiology 93: 238-243.

Coplin, D.L. & Cook, D. (1990) Molecular genetics of extracellular polysaccharide biosynthesis in vascular phytopathogenic bacteria. Molecular Plant Microbe Interactions 3: 271-279.

Crécy-Lagard, V. de., Glaser, P., Lejeune, P., Sismeiro, O., Barber, C.E., Daniels, M.J. & Danchin, A. (1990) A *Xanthomonas campestris* pv. campestris protein similar to catabolite activation factor is involved in regulation of phytopathogenicity. Journal of Bacteriology 172: 5877-5883.

Daniels, M.J. (1988) Molecular genetics of host pathogen interactions. Molecular Genetics of Plant-Microbe Interactions 1988. pp 229-234. ed. Palacios, R & Verma, D.P.S. APS Press.

Daniels, M.J., Barber, C.E., Turner, P.C., Sawczyc, M.K., Byrde, R.J.W. & Fielding, A.H. (1984) Cloning of genes involved in pathogenicity of *Xanthomonas campestris* pv. *campestris* using the broad host range cosmid pLAFR1. EMBO Journal 3: 3323-3328.

Daniels, M.J., Fan, M.J., Barber, C.E., Clarke, B.R. & Parker, J.E. (1991) Interaction between *Arabidopsis thaliana* and *Xanthomonas campestris*. Advances in Molecular Genetics of Plant-Microbe Interactions 1, pp 84-89. ed. Hennecke, H. & Verma, D.P.S. Kluwer Dordrecht.

Dickson, M.D. & Hunter, J.E. (1987) Inheritance of resistance in cabbage seedlings to black rot. Horticultural Science 22: 108-109.

Dong, X., Mindrinos, M., Davis, K.R. & Ausubel, F.M. (1991) Induction of *Arabidopsis* defense genes by virulent and avirulent *Pseudomonas syringae* strains and by a cloned avirulence gene. The Plant Cell 3: 61-72.

Dow, J.M., Scofield, G., Trafford, K., Turner, P.C. & Daniels, M.J. (1987) A gene cluster in *Xanthomonas campestris* pv. *campestris* required for pathogenicity controls the excretion of polygalacturonate lyase and other enzymes. Physiological and Molecular Plant Pathology 31: 261-271.

Dow, J.M., Milligan, D.E., Jamieson, L., Barber, C.E. & Daniels, M.J. (1989) Molecular cloning of a polygalacturonate lyase gene from *Xanthomonas campestris* pv. *campestris* and role of the gene product in pathogenicity. Physiological and Molecular Plant Pathology 35: 113-120.

Dow, J.M., Clarke, B.R., Milligan, D.E., Tang, J.-L., Daniels, M.J. (1990) Extracellular proteases from *Xanthomonas campestris* pv. campestris, the black rot pathogen. Applied and Environmental Microbiology 56: 2994-2998.

Dums, F., Dow, J.M. & Daniels, M.J. (1991) Structural characterization of protein secretion genes of the bacterial phytopathogen *Xanthomonas campestris* pathovar *campestris*: relatedness to secretion systems of other gram-negative bacteria. Molecular and General Genetics 229: 357-364.

Estelle, M.A. & Somerville, C.R. (1986) The mutants of *Arabidopsis*. Trends in Genetics 2: 89-93.

Fellay, R., Rahme, L.G., Mindrinos, M.N., Frederick, R.D., Pisi, A. & Panopoulos, N.J. (1991) Genes and signals controlling the *Pseudomonas syringae* pv. *phaseolicola*-plant interaction. Advances in Molecular Genetics of Plant-Microbe Interactions 1 pp. 45-52. ed. Hennecke, H. & Verma, D.P.S. Kluwer, Dordrecht.

Gough, C.L., Dow, J.M., Barber, C.E. & Daniels, M.J. (1988) Cloning of two endoglucanase of *Xanthomonas campestris* pv. *campestris*: analysis of the role of the major endoglucanase in pathogenesis. Molecular Plant-Microbe Interactions 1: 275-281.

Gough, C.L., Dow, J.M., Keen, J., Henrissat, B. & Daniels, M.J. (1990) Nucleotide sequence of the *engXCA* gene encoding the major endoglucanase of *Xanthomonas campestris* pv. *campestris*. Gene 89: 53-59.

Harding, N.E., Cleary, J.M., Cabanas, D.K., Rosen, I.G. & Kang, K.S. (1987) Genetic and physical analyses of a cluster of genes essential for xanthan gum biosynthesis in *Xanthomonas campestris*. Journal of Bacteriology 169: 2854-2861.

Hötte, B., Ruth-Arnold, I., Pühler, A. & Simon, R. (1990) Cloning and analysis of a 35.3 kilobase DNA region involved in exopolysaccharide production in *Xanthomonas campestris* pv. *campestris*. Journal of Bacteriology 172: 2804-2807.

Hrabak, E.M. & Willis, D.K. (1992). The *lemA* gene required for pathogenicity of *Pseudomonas syringae* pv. syringae on bean is a member of a family of two-component regulators. Journal of Bacteriology 174: 3011-3020.

Hu, N.T., Hung, M.N., Chiou, S.J., Tang, F., Chiang, D.C., Huang, H.Y. & Wu, C.Y. (1992) Cloning and characterisation of a gene required for the secretion of extracellular enzymes across the outer membrane by *Xanthomonas campestris* pathovar campestris. Journal of Bacteriology 174: 2679-2687.

Ielpi, L., Couso, R. & Dankert, M. (1981). Lipid linked intermediates in the biosynthesis of xanthan gum. FEBS Letters 130: 253-256.

Kamoun, S., & Kado, C.I. (1990) A plant inducible gene of *Xanthomonas campestris* pv. campestris encodes an exocellular component required for growth in the host and hypersensitivity on non hosts. Journal of Bacteriology 172: 5165-5172.

Lamb, C., Lawton, M., Dron, M. & Dixon, R. (1989) Signals and transduction mechanisms for activation of plant defenses against microbial attack. Cell 56: 215-224.

Liu, Y.N., Tang, J.L., Clarke, B.R., Dow, J.M. & Daniels, M.J. (1990) A multipurpose broad host range cloning vector and its use to characterise an extracellular protease gene of *Xanthomonas campestris* pv. *campestris*. Molecular and General Genetics 220: 433-440.

Osbourn, A.E., Clarke, B.R., Stevens, B.J.H. & Daniels, M.J. (1990a) Use of oligonucleotide probes to identify members of two-component regulatory systems in *Xanthomonas campestris* pathovar *campestris*. Molecular and General Genetics 222: 145-151.

Osbourn, A.E., Clarke, B.R. & Daniels, M.J. (1990b) Identification and DNA sequence of a pathogenicity gene of *Xanthomonas campestris* pv. *campestris*. Molecular Plant-Microbe Interactions 3: 280-285.

Pugsley, A.P., D'Enfert, C., Reyss, I. & Kornacker, M.G. (1990) Genetics of extracellular protein secretion by gram negative bacteria. Annual Review of Genetics 24: 67-90.

Simpson, R.B. & Johnson, L.J. (1990) *Arabidopsis thaliana* as a host for *Xanthomonas campestris* pv. *campestris*. Molecular Plant-Microbe Interactions 3: 233-237.

Stock, J.B., Ninfa, A.J. & Stock, A.M. (1989) Protein phosphorylation and regulation of adaptive responses in bacteria. Microbiological Reviews 53: 450-490.

Tang, J.L., Gough, C.L., Barber, C.E., Dow, J.M. & Daniels, M.J. (1987) Molecular cloning of protease gene(s) from *Xanthomonas campestris* pv. *campestris*: Expression in *Escherichia coli* and role in pathogenicity. Molecular and General Genetics 210: 443-448.

Tang, J.L., Gough, C.L. & Daniels, M.J. (1990) Cloning of genes involved in negative regulation of production of extracellular enzymes and polysaccharide of *Xanthomonas campestris* pv. *campestris*. Molecular and General Genetics 222: 157-160.

Tang, J.L., Liu, Y.N., Barber, C.E., Dow, J.M., Wootton, J.C. & Daniels, M.J. (1991) Genetic and molecular analysis of a cluster of *rpf* genes involved in positive regulation of synthesis of extracellular enzymes and polysaccharide in *Xanthomonas campestris* pv. *campestris*. Molecular and General Genetics 226: 409-417.

Thorne, L., Tansy, L. & Pollock, T.J. (1987) Clustering of mutations blocking synthesis of xanthan gum by *Xanthomonas campestris*. Journal of Bacteriology 169: 3593-3600.

Tsuji, J., Somerville, S.C. & Hammerschmidt, R. (1991) Identification of a gene in *Arabidopsis thaliana* that controls resistance to *Xanthomonas campestris* pv. *campestris*. Physiological and Molecular Plant Pathology 38: 57-65.

Vanderslice, R.W., Doherty, D.H., Capage, M.A., Betlach, M.R., Hassler, R.A., Henderson, N.M., Ryan-Graniero, J. & Tecklenburg, M. (1989) Genetic engineering of polysaccharide in *Xanthomonas campestris*. In: Recent Developments in Industrial Polysaccharides: Biomedical and Biotechnological Advances. pp. 145-156. ed. Crescenzi, V., Dea, I.C.M. & Stivola, S.S. Gordon & Breach Science Publishers, New York.

Whalen, M.C., Innes, R.W., Bent, A.F. & Staskawicz, B.J. (1991) Identification of *Pseudomonas syringae* pathogens of *Arabidopsis* and a bacterial locus determining avirulence on both *Arabidopsis* and soybean. The Plant Cell 3: 49-59.

Williams, P.H. (1980) Black rot, a continuing threat to world crucifers. Plant Disease 64: 736-742.

Willis, D.K., Rich, J.J. & Hrabak, E.M. (1991) *hrp* genes of phytopathogenic bacteria. Molecular Plant-Microbe Interactions 4: 132-138.

Section 8 / Fungal Plant Interactions: Plant Response

THE IDENTIFICATION AND MAPPING OF LOCI IN *ARABIDOPSIS THALIANA* FOR RECOGNITION OF THE FUNGAL PATHOGENS: *PERONOSPORA PARASITICA* (DOWNY MILDEW) AND *ALBUGO CANDIDA* (WHITE BLISTER)

IAN R. CRUTE and ERIC B. HOLUB
Horticulture Research International
East Malling
Kent
ME19 6BJ
UK

MAHMUT TOR, EDEMAR BROSE and JAMES L. BEYNON
Wye College
University of London
Ashford
Kent
TN25 5AH
UK

ABSTRACT. Forms of the specific crucifer pathogens: *Peronospora parasitica* (*Pp*) (downy mildew) and *Albugo candida* (*Ac*) (white blister) occur naturally on *Arabidopsis thaliana* (*At*) in the UK. Accessions of *At* have been shown to vary for isolate specific response to both pathogens. The existence of at least eight specific recognition (resistance) alleles are required to explain the pattern of isolate x accession responses observed. All field-collected isolates of *Pp* characterised so far have proved to express different specific virulence phenotypes. A fourteen parent diallel cross has been completed and, for a sub-set of eight parental accessions, the segregation of response to two isolates of *Pp* at F_2 has been studied. In two crosses involving three parents (*Col-gl*, *Nd0* and *Oy0*) segregation at F_3 and cosegregation studies with RFLPs has enabled the identity of five *RPp* alleles to be confirmed and the chromosome location of three to be established. Each *RPp* allele conditions a different and characteristic response phenotype. *RPp1* (from *Nd0*) is located on chromosome 3 between the locus identified by probe M249 and the morphological marker *glabrous-1*; this allele conditions a response characterised by lack of asexual sporulation and the occurrence of large, spreading, necrotic "pits". *RPp2* and *RPp4* (from *Col-gl*) are located together on chromosome 4 between the loci identified by probes M326 and M600 and probably on either side of the locus identified by probe M557. *RPp2* conditions a response characterised by lack of asexual sporulation and necrotic "flecking" while *RPp4* also conditions a response characterised by "flecking" but this is accompanied by light asexual sporulation that is delayed in its appearance in comparison with plant lines lacking an active *RPp* allele. The chromosome location of *RPp3* (from *Oy0*) and *RPp7* (from *Col-gl*) have yet to be determined. An allele for specific recognition of *Ac* from the field accession *Kes37* (*RAc1*) conditioning a response phenotype characterised by lack of asexual sporulation has been mapped to chromosome 1 below the locus identified by probe M215. Several different response phenotypes to *Ac* have been observed among *At* accessions and on this basis the existence of further *RAc* alleles is

E. W. Nester and D. P. S. Verma (eds.),
Advances in Molecular Genetics of Plant-Microbe Interactions, 437–444.

suspected; genetic analyses are in progress. These studies are the prelude to efforts to clone one of more specific recognition alleles employing a map-based strategy.

1. INTRODUCTION

Plant genes that facilitate the genotype specific recognition of pathogens (specific resistance genes) were identified and began to be utilised by plant breeders soon after the turn of the century (Biffen, 1905). This was not long after the rediscovery of Mendel's seminal studies on heredity. Since this time, hundreds of such genes have been described in many plant species and breeders continue to utilise them with differing degrees of effectiveness in programmes of crop improvement. Despite all the time that has elapsed and an enormous amount of research effort, the way in which genotype specific recognition of pathogens is effected remains obscure. Here is one of the great unanswered questions in plant biology. Three aspects of genotype specific recognition of pathogens by plants are well established and likely to be of functional importance. These are briefly reemphasised.

The capacity of plants to discriminate among genotypes of potential pathogens is vast. Where n is the number of loci involved, at least 2^n host genotypes express a unique pathogen genotype recognition capability; this number will be larger when multiple allelic forms occur at a single locus. In wheat, for example, approximately 97 different alleles (ie n= c.97) have been described that condition genotype specific recognition of three rust fungi (*Puccinia striiformis*, *P. graminis* and *P. recondita*) and a powdery mildew fungus (*Erysiphe graminis*). This enormous capacity of plants to discriminate pathogen genotypes is also evident in natural plant pathosytems and is not a consequence of domestication and plant breeding (Clark *et al.*, 1987, 1990).

Characteristic plant response phenotypes are often associated with the operation of different specific recognition alleles. There are degrees of compatibility/incompatibility and specific recognition is not always reflected in an all or nothing response (Crute, 1985). In unselected natural pathosystems there is some evidence that alleles conditioning responses characterised by incomplete resistance are more common than those where the response phenotype is observed as complete resistance. It is highly probable that in crop species alleles conditioning complete resistance have been specifically sought and selected during the plant breeding process.

Where detailed genetic investigations have been conducted, it has been demonstrated that loci involved in genotype specific recognition of pathogens tend to be organised in distributed clusters through the plant genome. Allelic variants at a single locus expressing different specific recognition characteristics are also not infrequent (Crute, 1985; Islam and Shepherd, 1991).

Increasingly detailed biochemical and histological descriptions of the phenotypic expression of the action of specific recognition alleles are available but beyond these studies only speculation is possible about the way recognition is effected. A full molecular understanding of genotype specific recognition of pathogens by plants is likely to be preceded by the isolation and sequencing of several genes involved in the recognition of several pathogens.

The goals of our research programme are:

i) to provide extensive genetic definition of the interactions between *Arabidopsis thaliana* (*At*) and two biotrophic fungal pathogens: *Albugo candida* (*Ac*) (white blister) and *Peronospora parasitica* (*Pp*) (downy mildew);

ii) to utilise the unique combination of biology and technology presented by *At* to isolate alleles for genotype specific recognition of these fungi and understand how they exert their effect;

iii) to utilise alleles isolated from *At* to explore genotype specific recognition of the same fungi in *Brassica* species.

The advantages of *At* for a study of this type are numerous. Apart from being a host to several different types of microbial pathogen (Davis, *et al*., Koch and Slusarenko, 1990a,b; Simpson and Johnson, 1990; Tsuji *et al*., 1990; Whalen *et al*., 1991), the species has a small genome with little repeat sequence (c. 140Mbp), a haploid chromosome number of 5, produces prolific quantities of seed by inbreeding, has a rapid life cycle and a small stature. Additionally, genetic transformation is routine, several partly integrated genetic maps including molecular and morphological markers have been produced and a growing, cooperative research community has developed an ethos of resource sharing and information exchange.

2. CHARACTERISATION OF HOST X PATHOGEN INTERACTIONS

In the springs of 1990 and 1991, thirty-two populations of *At* were examined in Kent, UK. Of these populations, 9% had plants infected with both fungi, 19% had plants infected with *Pp* alone, and 19% with *Ac* alone. Neither fungus was found in the remaining 53% of populations.

Variation in host response to each fungus was investigated using two isolates of *Pp* and one isolate of *Ac* from Kent. The responses of twenty-two accessions of *At* (originating from ten European countries, USA, Japan, Libya, Kashmir, and Cape Verde Islands) were examined. The methods employed in this work and subsequently have been described fully elsewhere (Dangl *et al*., 1992). It was possible to classify these accessions into four phenotypic groups according to the presence (S) or absence (N) of asexual sporulation following inoculation with the two *Pp* isolates. Both isolates sporulated on 18% of the accessions (S/S), and neither sporulated on 55% (N/N). The remaining accessions responded differently to the two isolates (18% N/S; 9% S/N). Three phenotypic groups of reaction (S/S, S/N, N/S) were also found among progeny of *At* plants collected from a local population at East Malling. The *Ac* isolate sporulated asexually on all twenty-two ecotypes but did not sporulate on *At* accessions collected from four other locations in the U.K.

Further important differentiation of phenotypic responses of *At* to *Pp* has been possible. The different response phenotypes now recognised are as follows: full susceptibility with sporophores profuse and visible 3 days after inoculation (dai); sporulation sparse to moderate and delayed in appearance by at least 48 h compared to the fully susceptible phenotype; no sporophores produced but flecking lesions of necrotic host cells are visible to the naked eye 7 dai; and no sporophores produced but necrotic "spots" or "pits" are visible to the naked eye 3 dai. "Spots" and "pits" are thought to represent different host responses. Both responses are observed as discrete epidermal lesions c. 1mm in diameter 3 dai. Both types of necrotic lesion usually have a chlorotic halo, but, compared to "spots", "pits" comprise more pronounced cellular collapse forming a depression in the cotyledon surface. Both responses can be further characterised 7 dai as either determinate or expanding. This information is summarised in Table 1.

Thirteen accessions of *At* have been identified on which sporulation either does not occur or is delayed following inoculation with an isolate of *Ac* collected from East Malling. Nine of these putatively resistant accessions were collected from locations in the

UK and four from Germany. In addition to full susceptibility characterised by the formation of large profusely sporulating pustules, four different response phenotypes of *At* to *Ac* that can be discriminated: no pustules and no macroscopically visible necrosis; necrotic flecking without pustules; discrete chlorotic patches without pustules; and delayed production of numerous small pustules.

TABLE 1. Categorisation of response phenotypes to *Peronospora parasitica*

Timing of sporulation:	Early Delayed	
Intensity of sporulation:	Heavy Low to moderate Rare None	
Lesion type:	Flecking Pitting Chlorotic spot	deTerminate eXpanding

3. GENETICS OF HOST RESPONSE TO *PERONOSPORA PARASITICA*

A half-diallel cross involving fourteen *At* accessions as parents has been completed to provide segregating populations to investigate the inheritance of response to *Pp*. These accessions include examples of commonly studied laboratory lines as well as accessions from locations where the isolates were collected. Nearly all F2 populations from crosses made between the parents: *Wein*, *Nd0*, *Oy0*, *Col0*, *La-er*, *RLD*, *Tsu0*, and *Kes37* have now been tested against two isolates, CALA1 and EMOY2. At least five specific recognition alleles are needed to explain the differential interactions between these accessions and the two isolates and to accommodate the segregation data obtained. The identity of three such *RPp* alleles (*RPp1*, *RPp2* and *RPp4*) has been confirmed by cosegregation studies with molecular markers (see below). Confirmation that *RPp3* and *RPp8* are indeed alleles at single loci is still to be obtained. Three additional recognition alleles (*RPp5*, *RPp6* and *RPp7*) explain responses to other isolates and the identity of two of these has been confirmed in other laboratories (*RPp5*, J. Parker and J. Jones, personal communication; *RPp6*, A. Slusarenko and B. Mauch-Mani, personal communication). A summary of our working model to explain specific recognition between accessions of *At* and isolates of *Pp* is presented in Table 2. It is highly probable that continued analysis of the half-diallel cross and subsequent F3 progeny tests will reveal new genes.

4. GENETICS OF HOST RESPONSE TO *ALBUGO CANDIDA*

The phenotypic response to *Ac* in accession *Kes37* characterised by necrotic flecking but the complete absence of asexual sporulation has been shown to be due to a dominant allele at a single locus (*RAc1*) and this has been confirmed by cosegregation studies with molecular markers among progeny of the cross *Wein* x *Kes37*. There is evidence

TABLE 2. Gene-for-gene model (June 1992) to explain specific recognition between *Arabidopsis* accessions and isolates of *Peronospora parasitica*

At Accession	*RPp* locus								*Pp* isolate EMOY2	CALA2	HIKS1	EMWA1	NOCO2	WELA1
									1	*	1	*	*	?
									*	2	*	*	*	*
									*	3	*	*	?	*
Avr locus									4	*	*	4	*	*
									*	*	*	*	5	*
									*	*	*	*	*	6
									*	*	7	*	*	*
									8	*	*	?	*	*
Nd0	1	*	*	*	*	*	*	*	NPX[a]	EH	NPX	DH	EH	N
Oy0	*	2	3	*	?	?	*	*	EH	NF	EH	EH	NF	N
Col0	*	2	*	4	*	6	7	*	DLF	RF	NF	DLF	EH	NF
La-er	*	*	*	4	5	*	?	8	NF	EH	NF	NF	NF	EH
Tsu0	*	*	*	*	*	*	*	*	EL	EL	DL	EL	E	E
RLD	*	?	*	?	*	?	*	?	NF	NF	NF	NF	EH	NF
Ws0	?	?	?	*	?	?	*	*	NPT	NCT	NPT	DH	NCT	?

[a] Response phenotypes are as described in Table 1.

that a second allele is present in *Kes37* conditioning a phenotype expressed only in mature plants and characterised by the appearance of many small pustules, as distinct from the fewer large pustules associated with full suceptibility. Confirmation that this phenotype is conditioned by a further single *RAc* allele is being sought.

5. MAPPING OF SPECIFIC RECOGNITION ALLELES

5.1 *Albugo candida*

Probes previously shown to identify RFLPs between other *At* accessions have also been shown to identify polymorphisms between the accessions *Kes37* and *Wein*. Twenty-five homozygous fully susceptible and 18 homozygous fully resistant F_3 families were selected with which to map *RAc1*. This allele has now been firmly located on chromosome 1 approximately 13cM below the locus identified by probe M215 (Chang *et al.*, 1988) and in a region of numerous closely linked RFLPs.

5.2 *Peronospora parasitica*

We have initially concentrated on the mapping of three *RPp* alleles which can be distinguished phenotypically using two isolates of *Pp* and segregate in the single cross: *Col-gl* x *NdO* (Table 2). This same cross has been used previously to develop a genetic map of *At* (Chang *et al.*, 1988) and to determine the location of a gene for specific recognition of the bacterial pathogen *Pseudomonas syringae* pv. *maculicola* (Debener *et al.*, 1991).

In response to CALA1, progeny in the F_2 generation segregate for full susceptibility (the parental response of *NdO*) and the flecking response phenotype of *Col-gl*. In response to EMOY2, progeny in the F_2 generation segregate for the delayed sporulation response of *Col-gl* and the "pitting" phenotype of *NdO*. By progeny testing the same F_3 families from this cross in separate inoculations with each isolate, the existence of three different loci was confirmed at which were alleles conditioning the response phenotypes: "pitting", "flecking" and delayed sporulation. Cosegregation studies, primarily concentrating on selected homozygous susceptible F_3 families (in excess of 60), have demonstrated that *RPp1* is located on chromosome 3 approximately 23cM below *gl1* and 4cM above the locus identified by probe M249. *RPp2* and *RPp4* are linked and approximately 28cM apart on either side of the locus identified by probe M557. *RPp2* is approximately 22cM above the locus identified by probe M600 while *RPp4* is approximately 40cM below the locus identified by probe M326.

In collaboration with Dr. J. Dangl (Cologne, Germany) we are also mapping *RPp3* using the cross *NdO* x *OyO* (Table 2). This investigation will confirm whether the second allele in *OyO* resulting in incompatibility with isolate CALA2 is indeed *RPp2* as we postulate.

In other laboratories, the cross *Col0* x *La-er* is being used to locate allele *RPp5* from La-er (incompatible response to isolate NOCO2) (J. Parker and J. Jones, personal communication) and allele *RPp6* from *Col0* (incompatible response to isolate WELA1) (A. Slusarenko and B. Mauch-Mani, personal communication) (Table 2).

6. THE FUTURE

With our current resources we are continuing to refine the genetic definition of both pathosystems and have begun to focus on the fine-scale mapping of *RPp2* and *RAc1* with the view to imminently commencing a chromosome walk to *RPp2*. We are also actively seeking more diagnostic isolates to allow us to identify new recognition alleles segregating in the same crosses with which we are already working; the first success of this approach was the identification of *RPp7* in *Col-gl* (Table 2).

If further resources become available to us, we intend to tackle the isolation of more than a single allele using the same map-based strategy but, in addition, we will embark on a programme to isolate mutants of *RPp1*. This allele has the advantage of a highly characteristic and readily observed phenotype from which it will be possible to observe any deviation. This could provide an aid to cloning if the mutations are deletions and could also provide access to other genes involved in the expression of this phenotype if indeed such genes do exist.

7. CONCLUDING REMARKS

The two pathosytems we have begun to explore in *At* appear to have the same three significant characteristics that were alluded to in the Introduction. There seem to be a lot of alleles for genotype specific recognition of *Ac* and *Pp* in *At*; these alleles are associated with different and characteristic response phenotypes with differing degrees of compatibility/incompatibility and there is evidence for the distribution of recognition loci through the genome as well as for linkage. We look forward to exciting times ahead.

8. ACKNOWLEDGEMENTS

We are grateful to Professor Paul Williams and Dr Bill Barlow who together first alerted us to the natural occurrence of *Pp* and *Ac* on *At* in Kent. The generosity of Professor Elliot Meyerowitz in the provision of probes used to identify RFLPs is gratefully acknowledged. Our work is funded by the Agricultural and Food Research Council to whom we record our thanks.

9. REFERENCES

Biffen, R.H. (1905) 'Mendel's laws of inheritance and wheat breeding', *Journal of Agricultural Science* **1**, 4-48.

Chang, C., Bowman, J.L., DeJohn, A.W., Lander, E.S. and Meyerowitz, E.M. (1988) 'Restriction fragment length polymorphism linkage map for *Arabidopsis thaliana*', *Proceedings of the National Academy of Sciences, USA* **85**, 6856-6860.

Clark, D.D., Bevan, J.R. and Crute, I.R. (1987) 'Genetic interactions between wild plants and their parasites', in P.R. Day and G.J. Jellis (eds.), *Genetics and Plant Pathogenesis*, Blackwell Scientific Publications, Oxford, pp. 195-206.

Clarke, D.D., Campbell, F.S. and Bevan, J.S. (1990) 'Genetic interactions between *Senecio vulgaris* and the powdery mildew fungus *Erysiphe fischeri*', in J.J. Burdon and S.R. Leather (eds.), *Pests, Pathogens and Plant Communities*, Blackwell Scientific publications, Oxford, pp. 189-201.

Crute, I.R. (1985) 'The genetic bases of relationships between microbial parasites and their hosts', in R.S.S. Fraser (ed.), *Mechanisms of Resistance to Plant Disease*, Martinus Nijhoff/Dr W. Junk Publishers, Dordrecht, pp. 80-142.

Davis, K.R., Schott, E. and Ausubel, F.M. (1991) 'Virulence of selected phytopathogenic Pseudomonads in *Arabidopsis thaliana*', *Molecular Plant Microbe Interactions* **4**, 477-488.

Debener, T., Lehnackers, H., Arnold, M. and Dangl J.L. (1991) 'Identification and molecular mapping of a single *Arabidopsis thaliana* locus determining resistance to a phytopathogenic *Pseudomonas syringae* isolate', *The Plant Journal* **1**, 289-302.

Dangl, J.L., Holub, E., Debener, T., Lehnackers, H., Ritter, C. and Crute, I.R. (1992) 'Genetic definition of loci involved in *Arabidopsis*-pathogen interactions', in C. Konez, N.H. Chua and J. Schell (eds.), *Methods in Arabidopsis Research*, World Publishing Company, Singapore, (In Press).

Islam, M.R. and Shepherd, K.W. (1991) 'Present status of genetics of rust resistance in flax', *Euphytica* **55**, 255-267.

Koch, E. and Slusarenko, A.J. (1990a) 'Fungal pathogens of *Arabidopsis thaliana* (L.). Heyhn', *Botanica Helvetica* **100**, 257-269.

Koch, E. and Slusarenko, A.J. (1990b) '*Arabidopsis* is susceptible to infection by a downy mildew fungus', *The Plant Cell* **2**, 437-445.

Simpson, R.B. and Johnson, L.J. (1990) '*Arabidopsis thaliana* as a host for *Xanthomonas campestris* pv. *campestris*', *Molecular Plant-Microbe Interactions* **3**, 233-237.

Tsuji, J., Somerville, S.C. and Hammerschmidt, R. (1990) 'Identification of a gene in *Arabidopsis thaliana* that controls resistance to *Xanthomonas campestris* pv. *campestris*', *Physiological and Molecular Plant Pathology*. **37**, 1-8.

Wahlen, M.C., Innes, R.W., Bent, A.F. and Staskawicz, B.J. (1991) 'Identification of *Pseudomonas syringae* pathogens of *Arabidopsis* and a bacterial locus determining avirulence on both *Arabidopsis* and soybean', *The Plant Cell* **3**, 49-59.

SCREENING METHODS FOR *ARABIDOPSIS* MUTANTS AFFECTED IN THE SIGNAL TRANSDUCTION PATHWAYS LEADING TO DEFENSE RESPONSES

Ruud A. de Maagd, Robin K. Cameron, *Richard A. Dixon, and Christopher J. Lamb.
Plant Biology Laboratory
Salk Institute for Biological Studies
10010 North Torrey Pines Road,
La Jolla, California 92037

Plant Biology Division
*Noble Foundation
P.O. Box 2180
Ardmore, Oklahoma 73402

ABSTRACT. In this paper we describe the development of screening methods for *Arabidopsis* mutants affected in defense responses. Mutants putatively affected in the oxidative burst are being isolated by screening for decreased reactivity of seedlings with nitro blue tetrazolium. Mutants affected in the activation of a rice basic chitinase promoter in transgenic *Arabidopsis* plants are being selected by allyl alcohol treatment of seedlings after mutagenesis, using the expression of alcohol dehydrogenase controlled by the chitinase promoter. We also describe systemic acquired resistance (SAR) in *Arabidopsis*, induced by inoculation with an avirulent pathogen. Plants inoculated in this way become systemically protected against a secondary challenge with a virulent strain, showing a HR-like reaction to this strain, which normally would cause disease symptoms. Moreover inoculation with the avirulent strain also results in protection against other bacterial pathogens. A screening method for isolation of mutants affected in SAR is presented.

1. General Introduction.

Plants react to wounding or attack by pathogens by activating an array of defense responses [3]. The signal transduction pathways leading to these responses, are not well characterized. While advances have been made in biochemical and molecular biological studies of these pathways, a genetic approach to this problem may form a valuable addition. The first goal in such an approach is the isolation of mutants affected in one or more components of the defense response. Such mutants would not only identify the components of the signal transduction pathways, but could also help to determine the role and relative importance of specific components in the overall defense reaction. Finally, with the development of more efficient techniques for cloning of plant genes, such mutants will be useful in the cloning of genes involved in the signal transduction pathways.

E. W. Nester and D. P. S. Verma (eds.),
Advances in Molecular Genetics of Plant-Microbe Interactions, 445–449.

2. The Oxidative Burst.

2.1. INTRODUCTION.

One of the first reactions that is observed after plant cells come in contact with an elicitor is the increased production of active oxygen species at the cell surface. These active oxygen species are hydrogen peroxide (H_2O_2), superoxide (O_2^-), perhydroxyl (HO_2), and hydroxyl radicals ($OH^\cdot$) (reviewed in [4]). This reaction to elicitors or phytopathogens has been described for various systems such as cell cultures, potato tuber tissue and leaf discs [4]. One or more plasma membrane localized NAD(P)H oxidoreductase may be responsible for the production of superoxide. The exact role of the oxdative burst has not yet been established. Possible functions are: 1. Killing or growth inhibition of the pathogens. 2. Oxidative cross-linking of cell wall proteins, possibly providing a physical barrier to pathogen invasion. Thus, it has been shown in soybean cell cultures that elicitation results in a very rapid, hydrogen peroxide-dependent insolubilization of two groups of cell wall proteins [2]. Moreover, in soybean leaves infiltrated with *P. syringae* pv. *glycinea* the same cross-linking occurs in an incompatible interaction (HR), but not in a compatible interaction (L. Brisson, unpublished results). 3. Hydrogen peroxide may act as a second messenger in the signal transduction pathway leading to other defense responses, such as activation of genes of the phenylpropanoid pathway [1]. Given the possible importance of the oxidative burst, the isolation and characterization of mutants affected in this rapid response may help to clarify its role and importance as well as the identification of components involved in it. For this reason we have developed a screening protocol to obtain *Arabidopsis* mutants which are putatively affected in the production of superoxide.

2.2. RESULTS AND CONCLUSIONS.

Nitro blue tetrazolium has been used widely as an indicator for superoxide production. Reaction of superoxide with NBT, which is soluble and yellow, results in the formation of an insoluble, purple diformazan, which can easily be distinquished at the surfaces where the reaction product precipitates. When comparing 10-day-old *Arabidopsis thaliana* Col-O seedlings grown in liquid MS-sucrose medium, on MS-sucrose agar, or in soil, only cotyledons of liquid grown seedlings showed staining with NBT (0.05% NBT in 10 mM sodium phosphate, pH7.2). Staining became visible after 15 min and was complete in about 60 min. Preincubation of seedlings with *Phytophthora megasperma* cell wall elicitor (50 μg/ul), or treatment with pectolyase (0.5 mg/ml, 10 min) increased the rate of staining considerably. These results indicated that NBT might be used to demonstrate inducible production of superoxide and hence be used in a screen for mutants putatively affected in the oxidative burst.

To test the feasibility of a screen as described above, 10.000 M2-seedlings of EMS-mutagenized *Arabidopsis* Col-0 were tested for NBT-staining following pretreatment with pectolyase. Six mutants survived the selection and produced seed. Their progeny were retested, with and without elicitation. The progeny of 5 mutants showed the same decrease in reactivity, indicating that the phenotype observed in this test is hereditary and that the screening results are reproducible. Three of the mutants showed decreased activity in the absence of elicitor but were still inducible to some extent. The other two showed reduction mainly in the induced staining. The phenotypes of these mutants are now being characterized in more detail, i.e. superoxide production will be assayed and the effects of the mutations on the defense response will be studied. If this approach yields interesting mutants, a large scale screen will then be conducted.

3. **Localized induction of defense genes.**

3.1. Introduction.

Wounding or infection induces transcription of a large number of defense genes, such as those encoding the biosynthetic enzymes of the phenylpropanoid pathway (e.g. PAL, 4CL) and lytic enzymes (e.g. glucanases, chitinases). We have developed a method for selecting *Arabidopsis* mutants affected in the activation of a transgenic rice basic chitinase promoter [7], in order to obtain mutants in the signal transduction pathways leading to the activation of defense genes. The promoter of interest, pRCH10 is a promoter of a rice basic chitinase in which upstream sequences up to 160 base pairs 5' of the transcription start site have been deleted. Whereas the full promoter both in rice as well as in transgenic tobacco shows considerable developmental expression in different plant parts, the deletion in pRCH10 has abolished most of the developmental expression, while retaining inducibility both by wounding as well as by wounding plus elicitor (Qun Zhu, unpublished results).

3.2. Results and Conclusions.

We have constructed a pBI101-based transformation vector that contains the RCH10 promoter fused to the selectable gene Alcohol dehydrogenase (ADH) from *Arabidopsis*, as well as a copy of the promoter fused to the indicator gene GUS. This vector was used to transform *Arabidopsis* line ROO2, a Be-0 ADH-null mutant. We are currently characterizing the transgenic lines and establishing procedures for optimal induction of the transgenic promoter, using the ß-glucuronidase-fusion. Preliminary results indicate that the expression pattern and inducibility in *Arabidopsis* are as described for transgenic tobacco. T2 seeds, homozygous for the transgenic construct will be mutagenized and resulting plants will be selfed. M2 progeny will subsequently be screened for ADH-activity. The selection of mutants will be based on their ability to survive allyl alcohol treatment. In plants in which the RCH10 promoter is activated normally, the ADH produced will convert allyl alcohol into the toxic acrolein, killing the plant. Mutants in which the promoter is no longer induced will not produce ADH and therefore will survive the treatment. By assaying inducible GUS-activity in selected mutants, cis-mutations in the promoter or the ADH gene can be ruled out as the cause of allyl alcohol resistance.

Using two already available plant lines, we have developed procedures for allyl alcohol treatment of *Arabidopsis* seedlings. Both plant lines contain constructs similar to that described above for pRCH10, with either the *Arabidopsis* PAL promoter (PAG1-1) or the *Arabidopsis* CABII-promoter (POCA108). Both promoters are highly active in cotyledons and primary leaves. Although the PAL-promoter is also an inducible defense promoter, the developmental expression precludes detectable increases in expression by elicitation, at least in young plants. Ten-day-old liquid grown seedlings in 24-well tissue culture plates (4-8 plants per well) were treated with MS-sucrose medium containing 0.5 to 10 mM allyl alcohol for 30 minutes and subsequently washed 4 times with growth medium. Effects of allyl alcohol treatment were visible after 2 days and are complete after 4-5 days. Sensitive plants were bleached during this time period, particularly in the cotyledons. With sublethal concentrations, cotyledons were completely bleached, but primary leaves which were just developing were not affected and allowed the plant to survive. Both POCA108 as well as PAG 1-1 were killed at 2 - 4 mM allyl alcohol, whereas the control R002 survived concentrations higher than 10 mM. MS-sucrose agar seedlings (10 days old) were exposed to allyl alcohol vapor in 1 liter beakers closed with parafilm, for 30 minutes. Whereas the control plants survive 8 µl of allyl alcohol per beaker, PAG 1-1 and POCA108 are killed at 1-2 µl. These results show that allyl alcohol indeed specifically kills plants containing active promoter-ADH-fusions. Obviously these procedures will have to be adjusted for each plant/promoter-system.

4. Systemic acquired resistance in *Arabidopsis*.

4.1. Introduction.

It has been demonstrated for several plant/pathogen systems that a primary infection with a necrosis inducing pathogen can render the host plant resistant to subsequent infections by other pathogens. This phenomenon, Systemic Acquired Resistance (SAR), requires the formation and transport of a signal from the original site of infection to the rest of the plant. Recently salicylic acid was identified as a putative signal molecule which can induce SAR and the systemic activation of genes that are correlated with SAR [5]. These observations, together with the versatility of *Arabidopsis* as a model in genetic studies, prompted us to study the occurrence of SAR in *Arabidopsis* following inoculation with an avirulent bacterial pathogen.

4.2. Results and conclusions.

We have used the bacterial pathogen *Pseudomonas syringae* pv. *tomato* strain DC3000 (virulent) and DC3000 containing the cloned avirulence gene *avrRrpt2* (avirulent), kindly provided by Dr. B. Staskawicz, in our SAR studies. Inoculation of leaves of 6 - 8 week old plants with avirulent bacteria (10-20 µl, 10^7 cfu/ml) resulted in the formation of dry hypersensitive lesions within 24 hrs. Inoculation with virulent bacteria resulted in the formation of a wet lesion surrounded by chlorotic regions after 2 days. A lower inoculum (10^5 cfu/ml) of avirulent bacteria produced no symptoms, whereas with virulent bacteria disease symptoms occured after 4-5 days.

When two separate leaves were inoculated with avirulent bacteria (10^7 cfu/ml) followed two days later by inoculation of other leaves with virulent bacteria (10^5 cfu/ml), no disease symptoms were observed. However, at higher inoculum concentrations of the virulent strain (10^7/ml), plants were observed to have HR-like symptoms (Table 1). Consistent with these findings, growth of virulent bacteria was inhibited when compared to that in control virulent inoculations (no "immunization" with avirulent bacteria) (Table 1). Moreover in transgenic plants containing PAL-GUS and CHS(bean)-GUS promoter fusions both promoters were induced to the level normally seen after inoculation with avirulent bacteria (results not shown). Mock inoculations or inoculations with virulent bacteria did not result in protection against a subsequent challenge with virulent bacteria. Protection by inoculation with avirulent bacteria was observed to occur for challenges up to 10 days after the initial inoculation, but may continue considerably longer. Because immunization of one or two leaves resulted in protection from virulent challenges in the untreated leaves, this bacterially induced acquired resistance is also systemic in nature.

One of the defining characteristics of earlier observed examples of SAR was cross-protection against infection by many different virulent pathogens. We have therefore tested whether inoculation with avirulent *P. syringae* pv. *tomato* strain DC3000(*avrRpt2*) confers protection against a second challenge with two other *Arabidopsis* pathogens, *P. syringae* pv. *maculicola* strain ES4326 and *P. syringae* pv. *tomato* strain 3455 [6]. In both cases we observed protection

A. CONTROL INFECTIONS							B. VIRULENT CHALLENGE OF "IMMUNIZED"PLANTS				
	BACTERIAL GROWTH (cfu/leaf disc) & SYMPTOMS							BACTERIAL GROWTH (cfu/leaf disc)& SYMPTOMS			
DAYS AFTER INFECTION	AVIRULENT INFECTION 10^7 cfu/ml [a]		VIRULENT INFECTION 10^5 cfu/ml [b]		VIRULENT INFECTION 10^7 cfu/ml [b]		DAYS AFTER VIRULENT CHALLENGE	CHALLENGE TITRE 10^5 cfu/ml [c]		CHALLENGE TITRE 10^7 cfu/ml [c]	
0	+/-[d]		+/-		+/-		0	+/-		+/-	
1	+	HR	+	NS	+++	NS	1	+	NS	+++	NS
3	++	"	++	NS	++++	SD	3	++	"	+++	HR-L
5	++	"	+++	SD	++++	D	5	+++	"	+++	"
8	+++	"	++++	D	++++	D	8	+++	"	+++	"
10	+++	"	+++	"	++++	D	10	+++	"	+++	"

TABLE 1 - a - Inoculation with *P. syringae* pv *tomato* (DC3000 [*avrRpt2*]) , b - Virulent inoculation (DC3000) , c - Virulent challenge (DC3000) 2 days after Avirulent Immunization (10^7 cfu/ml), d - [+/- (10^2), + (10^3), ++ (10^4) , +++ (10^5), ++++ (10^6) cfu/leaf disc], NS - no symptoms, HR - hypersensitive response, HR-L - HR-like lesion, SD - start of disease, D - disease

against these virulent strains. These results confirm that the observed protection is indeed caused by systemic acquired resistance. In future experiments we will extend these experiments to fungal pathogens and also study the systemic induction of defense genes in more detail.

We are currently developing a procedure for isolating *Arabidopsis* mutants affected in SAR, based on the plant/bacterium-system described above. Whereas it has been described recently that SAR in *Arabidopsis* may also be induced by the (putative) systemic signal salicylic acid or its analogue 2,6-dichloroisonicotinic acid (INA) [5], use of actual pathogens for induction of SAR may lead to the isolation of a broader range of mutants, including those affected in the production or the transmission of the systemic signal(s).

Acknowledgments. We thank Peter Doerner for helpful discussions. R. d. M. was supported by a fellowship of the Dutch Organization for Scientific Research. R. C. is a Noble Foundation Fellow. C. J. L. was supported by research grants from the Samuel Roberts Noble Foundation, the Rockefeller Foundation and the National Science Foundation (DCB 91.04551)

Literature cited

1. Apostol, I., P.F. Heinstein, and P.S. Low. (1989). Rapid stimulation of an oxidative burst during elicitation of cultured plant cells. Plant Physiol. 90, 109-116.
2. Bradley, D.J., P. Kjellbom, and C.J. Lamb. (1992) Elicitor- and wound-induced oxidative cross-linking of a proline-rich plant cell wall protein: a novel, rapid defense response. Cell 70, in press.
3. Lamb, C.J., M.A. Lawton, M. Dron, and R.A. Dixon.(1989) Signals and transduction mechanisms for activation of plant defenses against microbial attack. Cell 56, 215-224.
4. Sutherland, M.W. (1991) The generation of oxygen radicals during host plant responses to infection. Physiol. Mol Plant Pathol. 39, 79-93.
5. Uknes, S., B. Mauch-Mani, M. Moyer, S. Potter, S. Williams, S. Dincher, D. Chandler, A. Slusarenko, E. Ward, and J. Ryals. (1992) Acquired resistance in *Arabidopsis*. Plant Cell 4, 645-656.
6. Whalen, M.C., R.W. Innes, A.F. Bent, and B.J. Staskawicz. (1991) Identification of *Pseudomonas syringae* pathogens of *Arabidopsis* and a bacterial locus determining avirulence on both *Arabidopsis* and soybean. Plant Cell 3, 49-59.
7. Zhu, Q., and C.J. Lamb. (1991) Isolation and characterization of a rice gene encoding a basic chitinase. Mol. Gen. Genet. 226, 289-296.

TOWARDS POSITIONAL CLONING OF THE *Pto* BACTERIAL RESISTANCE LOCUS FROM TOMATO

G. Martin, C. de Vicente, M. Ganal, L. Miller, and S. Tanksley.
Department of Plant Breeding & Biometry
Cornell University
Ithaca, NY 14853 USA

ABSTRACT. Resistance to *Pseudomonas syringae* pv. *tomato* (*Pst*) in tomato is conferred by a single dominant locus, *Pto*. We are attempting to isolate the *Pto* locus using a multi-step positional cloning strategy. First, using near-isogenic lines (NILs) we identified RFLP and RAPD markers that are tightly linked to the locus (4). Second, we developed and screened a large F_2 population (251 plants) that segregated for *Pst* resistance. This population was used to create a high resolution (0.2 cM) linkage map that allowed the precise placement of *Pto* in relation to molecular markers (6). One marker, TG538, was found to cosegregate with *Pto*. Third, we screened a tomato YAC library (5) with TG538 and identified five YACs ranging from 400 to 600 kb that carry sequences derived from the *Pto* region. Mapping the ends of the cloned tomato DNA from the smallest YAC (PTY538-1) showed that this YAC encompasses the *Pto* locus. Fourth, PTY538-1 was used to identify cDNA clones from a library made from *Pst*-induced leaf tissue. Two cDNA clones CD106 and CD127-1, were isolated by this method. Linkage mapping of these clones showed that CD106 maps 0.4 cM from the *Pto* locus, whereas CD127 cosegregates with *Pto*. Once we have identified a full-length clone of CD127 we will introduce it (and the corresponding genomic sequence) into tomato plants in an effort to transform a *Pst*-susceptible line to resistance. The eventual isolation of *Pto* should open up many avenues of research into the molecular mechanism of plant disease resistance.

1. Introduction

The tomato - *Pseudomonas syringae* pv. *tomato (Pst)* interaction is an attractive experimental system for several reasons. First, *Pst* resistance in tomato is conferred by a single locus (*Pto*) and the resistance allele displays dominant gene action (7, 8). Secondly, tomato lends itself to genetic and molecular studies: it's easily hybridized, has a relatively small genome, and is routinely transformed. Thirdly, the pathogen is amenable to molecular analysis and an avirulence gene corresponding to *Pto* has recently been cloned (*avr*Pto; 9). Finally, the unusual observation that an organophosphate insecticide, Fenthion, elicits necrotic lesions similar to a hypersensitive response (HR) only on tomato lines carrying the *Pto* gene offers an easy screening method for the gene and eventually may provide insight into its function (2, 3). Because of its commercial importance and potential for elucidating the gene-for-gene hypothesis we have targeted the *Pto* locus for positional cloning.

E. W. Nester and D. P. S. Verma (eds.),
Advances in Molecular Genetics of Plant-Microbe Interactions, 451–455.

Positional cloning of a gene proceeds in several steps. First, molecular markers are identified that are tightly linked to the target gene and these markers are then ordered on a high resolution linkage map. Secondly, the physical distance is determined between the closest markers and the target gene using pulsed field gel electrophoresis techniques. This step reveals whether the markers are physically close enough to the gene to make positional cloning feasible. Thirdly, a chromosome walk is undertaken by using various genomic libraries constructed in lambda or yeast artificial chromosome (YAC) libraries (5). The final step requires a comparison of potential transcript differences in resistant and susceptible lines or, in the case of plants, complementation of the recessive phenotype by transformation.

2. Materials and Methods

The F_2 population was derived from a cross between Rio Grande-PtoR and Rio Grande, two near-isogenic lines (NILs) differing for *Pst* susceptibility (6). F_2 plants were checked for sensitivity to Fenthion by spraying with a solution of 0.15% Fenthion/0.05% Silwet L-77 dispersed in sterile distilled water (Silwet L-77 source: Union Carbide, Southbury, CT, [11]; Fenthion source: Mobay Corp., Kansas City, MO). Symptoms (small necrotic lesions, 1-2 mm) were recorded 3-4 days after treatment. F_3 progeny from F_2 plants having crossovers near the *Pto* locus were screened for their reaction to *Pseudomonas syringae* pv. *tomato* (strain PT11, kindly provided by L. Walling) in the greenhouse as described (Martin et al. 1991) except that instead of using cotton swabs the plants were dipped in a solution of 10^6 colony-forming units per ml *Pst* strain PT11/ 0.05% Silwet L-77/ 10 mM $MgCl_2$ dispersed in sterile distilled water (6).

3. Results and Discussion

3.1. IDENTIFICATION OF MARKERS

Three approaches were used to identify a total of 17 informative markers linked to the *Pto* locus (6). First, the global mapping effort in the lab to place 1000 markers in the tomato genome identified many markers in the general *Pto* region and these were surveyed on the near-isogenic lines (NILs) to identify informative clones (6, 10). Secondly, surveys of the resistant and susceptible NILs were probed with pools of 5 random clones (600 total clones) to identify polymorphic probes ("multiprobing"; 12). Finally, RAPD analysis using 150 primers of arbitrary sequence (each amplifying about 4 bands) was used to identify additional markers (4). Overall, marker representation was: 3 cDNAs, 4 RAPDs, 10 random RFLP markers. The number of informative markers identified from each approach outlined above was: global mapping (9 markers); multiprobing (4); RAPD analysis (4).

3.2. CONSTRUCTION OF A HIGH RESOLUTION LINKAGE MAP

Since *Pto* displays dominant gene action it is necessary to test the progeny of plants resistant to *Pst* with potential recombination events in the *Pto* region to determine the allelic state at the *Pto* locus. In order to avoid progeny testing of a large number of plants we choose to identify and analyze only those plants that were homozygous recessive (*pto/pto*). To accomplish this we relied on the unusual observation made by French plant breeders that an organophosphate insecticide, Fenthion, elicits small necrotic lesions on tomato plants carrying the dominant *Pto* allele (2, 3). It is unknown whether this reaction is a pleiotropic effect of the *Pto* locus or the result of a tightly linked gene. Whatever the case, no plant showing recombination between insensitivity to Fenthion and susceptibility to *Pst* has been identified in populations of over 650 plants making this a useful screen for identifying homozygous susceptible plants (G. Martin, unpublished data).

Approximately 1200 F_2 plants were treated with Fenthion and only those healthy plants (251 total) showing insensitivity (no necrotic lesions) were used in mapping (6). The selected plants were placed in the field and analyzed with flanking markers CD31 and TG619 to detect recombinants in the *Pto* region. A total of 85 such plants were identified and these were then analyzed with the remaining 15 markers. The 17 markers mapped to 9 loci and span a region of almost 20 cM (Figure 1). Over one-half of the map distance in this region can be accounted for by the distance between TG504 and TG538 (12 cM; Figure 1). In contrast to the TG504-TG538 interval, elsewhere 12 of the markers were found to cluster in a 0.6 cM region. It is not yet known if this heterogeneity of marker density corresponds to the physical sizes of these intervals. The linkage analysis revealed that one marker, TG538, cosegregated with the *Pto* locus. Considering the size of the population and the corresponding standard error TG538 lies less than 0.6 cM from *Pto* (95% confidence interval). Since no crossover event was detected between TG538 and *Pto* it is unknown whether the locus is contained in the TG475-TG538 interval or in the TG538-TG504 interval (Figure 1).

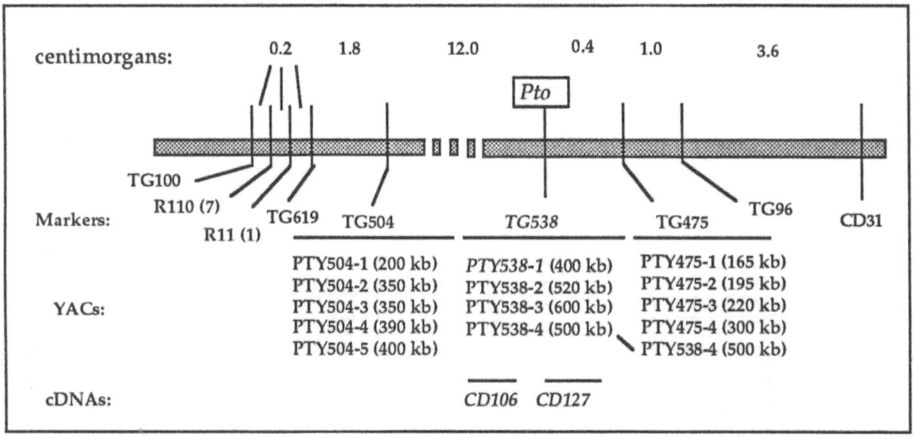

Figure 1: Linkage map of a 20 cM region of tomato chromosome 5 developed from the F_2 population segregating for *Pto*. Also shown are the YAC clones identified with TG504, TG538 and TG475 and the two cDNAs identified with PTY538-1. Number of additional markers at some loci are given in parentheses. Map is not drawn to scale.

3.3. ESTIMATION OF PHYSICAL DISTANCE

Because our goal was to use the linkage map to positionally clone the *Pto* locus we used pulsed field gel electrophoresis to estimate the maximum physical distance encompassed by the intervals on each side of TG538 (6). A total of 8 rare-cutting restriction enzymes were surveyed (*BssHI*, *NarI*, *NruI*, *MluI*, *SacII*, *SalI*, *SfiI* and *SmaI*) and those five enzymes which gave fragments between 100 and 900 kb when probed with TG538 were followed up by probing with TG475 and TG504. The experiments revealed that TG538 and TG475 detected 9 identical restriction fragments ranging in size from 340 kb to more than 800 kb. In only two enzyme digests, *Mlu*I and *Sfi*I, were unique fragments identified differentiating these two markers. Although the smallest fragment hybridizing to both TG538 and TG475 was 340 kb (with *Nru*I), this fragment only occurred in the *Pst* susceptible line, Rio Grande. Since insertions or deletions could exist in this region that differ between the resistant and susceptible lines we were primarily interested in the smallest common fragment that existed in Rio Grande-PtoR. The analysis showed that two fragments of 435 and 450 kb were in common between TG475 and TG538 (*Sal*I and *Sfi*I digests) in Rio Grande-PtoR. Thus these two markers are located no further apart than this distance on the chromosome. The minimum distance between them can not be determined from these data. If the large genetic distance between TG538 and TG504 is even moderately reflected in the physical distance we felt it would be unlikely to detect common restriction fragments between them. In fact, the survey with TG504 of 8 rare-cutting enzymes found no fragments in common with TG538.

3.4. IDENTIFICATION OF YAC CLONES AND MAPPING OF YAC END PROBES

TG475, TG504, and TG538 were used to screen the tomato YAC library (5) and clones were identified which hybridized to each of these markers (Figure 1). Interestingly, one YAC was found to hybridize to both TG475 and TG538 indicating that this clone contains sequences spanning the interval between these markers. The size of this YAC (500 kb) agrees well with our estimate of physical distance for this interval. End probes were isolated from PTY538-1 using inverse PCR and the ends were placed on the linkage map (Figure 1). PTY538-1(Left arm) placed 1.8 cM from *Pto* in the TG504-TG538 interval while PTY538-1 (Right arm) cosegregated with *Pto*. In order to determine if this YAC contained *Pto* a search was undertaken to find a plant carrying a crossover event between TG538 and PTY538-1(Right arm). A *Pst*-resistant plant(*Pto*/-; TG538 R/R) was identified that was homozygous at the alternate allele for PTY538-1(Right arm; S/S) and all other markers distal to this end-clone. These data indicate that PTY538-1 contains the *Pto* locus.

3.5. SCREENING A LEAF TISSUE cDNA LIBRARY WITH PTY538-1

Two strategies are planned for identifying the *Pto* locus in PTY538-1. First we will subclone large segments of the YAC into *Agrobacterium* vectors in an attempt to transform *Pst*-susceptible plants to resistance. A second more directed strategy is to use the YAC to identify cDNA clones derived from the *Pto* region. Towards this second goal we have constructed a cDNA library from Rio Grande-PtoR leaf tissue that was treated with both Fenthion and *Pst*. This library was probed with radiolabeled PTY538-1 and over 100 cDNA clones have been identified. From the first 30 cDNAs analyzed we identified two (CD106 and CD127) that detected RFLPs when probed on the NILs and which were confirmed to be homologous to sequences contained on PTY538-1.

3.6. MAPPING CD106 and CD127

When placed on the high resolution linkage map CD106 was found to map 0.4 cM from *Pto* in the TG504-TG538 interval. CD127 however cosegregated with TG538 (and *Pto*) and is therefore a candidate for the *Pto* gene (Figure 1).

4. Future Research

We are currently determining if CD127 is a full-length clone. Once a full-length clone is identified we will introduce it via *Agrobacterium* into a *Pst*-susceptible tomato line in an effort to complement it for resistance. In addition, we will identify the corresponding genomic sequence to use in similar experiments. Analyses of potential transcript differences between Rio Grande-PtoR and Rio Grande are also underway.

The eventual isolation of the *Pto* locus should benefit studies of the gene-for-gene hypothesis first postulated by Flor over 50 years ago (1) and may provide insights into the potential for genetic engineering of plants for disease resistance.

5. Acknowledgments

This work was supported, in part, by grants from the National Research Initiative Competitive Grants Program, USDA nos. 91-37300-6418 and 91-37300-6565. GBM was supported by a 1989 NSF Postdoctoral Research Fellowship in Plant Biology

6. References

1. **Flor AH** (1947) Host-parasite interactions in flax-rust - its genetics and other implications. Phytopath 45:680-685.
2. **Laterrot H** (1985) Susceptibility of *Pto* plants to Lebaycid insecticide: a tool for plant breeders? Tomato Genet. Coop Rep 35:6.
3. **Laterrot H, and Moretti A** (1989) Linkage between *Pto* and susceptibility to Fenthion. Tomato Genet. Coop Rep 39:21-22.
4. **Martin GB, Williams JGK, Tanksley SD** (1991) Rapid identification of markers linked to a *Pseudomonas* resistance gene in tomato using random primers and near-isogenic lines. Proc. Natl. Acad. Sci. USA **88,** 2336-2340.
5. **Martin GB, Ganal MW, Tanksley SD** (1992) Construction of a yeast artificial chromosome library of tomato and identification of cloned segments linked to two disease resistance loci. Mol Gen Genet 233:25-32.
6. **Martin GB, de Vicente MC, Tanksley SD** (1992) High resolution linkage analysis and physical characterization of a segment of tomato chromosome 5 containing the *Pto* bacterial resistance locus. (Submitted).
7. **Pitblado RE, MacNeill BH** (1983) Genetic basis of resistance to *Pseudomonas syringae* pv. *tomato* in field tomatoes. Can. J. Plant Path. 5:251-255.
8. **Pitblado RE, MacNeill BH, Kerr EA** (1984) Chromosomal identity and linkage relationships of *Pto,* a gene for resistance to *Pseudomonas syringae* pv. *tomato* in tomato. Can. J. Plant Path. 6:48-53.
9. **Ronald PC, Salmeron JM, Carland FM, Staskawicz BJ** (1992) The cloned avirulence gene *avrPto* induces disease resistance in tomato cultivars containing the *Pto* resistance gene. J Bacteriol 174:1604-1611.
10. **Tanksley SD, Ganal MW, Prince JP, de Vicente MC, Bonierbale MW, Broun P, Fulton TM, Giovanonni JJ, Grandillo S, Martin GB, Messeguer R, Miller JC, Miller L, Paterson AH, Pineda O, Roder MS, Wing RA, Wu W, Young ND** (1992) High density molecular maps of the tomato and potato genomes. Genetics (in press).
11. **Whalen MC, Innes RW, Bent AF, Staskawicz BJ** (1991) Identification of *Pseudomonas syringae* pathogens of *Arabidopsis* and a bacterial locus determining avirulence on both *Arabidopsis* and soybean. Plant Cell 3:49-59.
12. **Young ND, Zamir D, Ganal MW, Tanksley** (1988) Use of isogenic lines and simultaneous probing to identify DNA markers linked to the *Tm2-a* gene in tomato. Genetics 120:579-585.

STUDIES ON THE MECHANISM BY WHICH TOMATO *Cf* (*Cladosporium fulvum*) RESISTANCE GENES ACTIVATE PLANT DEFENCE

Kim Hammond - Kosack*, Richard Oliver+, Kate Harrison*,Tom Ashfield* and Jonathan Jones*. The Sainsbury Laboratory*, John Innes Centre, Colney Lane, Norwich, NR4 7UH, UK. The University of East Anglia+, Norwich ,NR4 7TJ, UK.

ABSTRACT. Tomato *Cf* genes 2,4,5 and 9 control phenotypically distinct incompatible reactions when challenged by pathogen or race-specific elicitors. Each *Cf* gene is semi-dominant in action; the homozygous state is more effective than the heterozygous. Two approaches have been taken to understand the mechanism by which plant defence is specifically activated by *Cf* resistance genes. Our first strategy is genetic and has involved the generation (by EMS treatment) and the characterisation of tomato mutants of the *Cf* -9 genotype which no longer express resistance to *C. fulvum* . Nine EMS M2 families containing disease sensitive individuals have been identified. Three of the fully susceptible mutants and one partially susceptible mutant map to the *Cf* -9 locus while 2 partially susceptible mutants appear to be at other loci. The mutants were identified on an in planta fungal biomass assay using a transformed *C. fulvum* race which constitutively expresses GUS activity. In an attempt to improve our selection for further aberrant *Cf* - 9 defence responses we aim to generate Cf 0 plants constitutively expressing the fungal avirulence peptide AVR9. We have already produced transgenic tobacco expressing highly active AVR9 peptide; intercellular fluids retrieved from leaves induce a grey necrotic response specifically on Cf 9 but not Cf 0 leaves within 24h of injection. In our second approach to investigate plant defence we have determined the order and delineated the relationship between various physiological and molecular events specifically induced in Cf 9 by both pathogen and race specific elicitor (IF) challenges. Early (<12h) *Cf* -9 -specific responses to IF include an oxidative burst, ethylene production, loss of membrane integrity and salicylic acid synthesis. The EMS mutants' responses are partially characterised. Ultimately, we wish to determine the effects of distinct mutations on the action of the other *Cf* genes and to map, clone and identify the function of genes whose products participate in the activation of plant defence.

1 Introduction

The molecular basis for the specificity frequently exhibited in host-pathogen interactions remains poorly understood, and may prove critical to developing stable genetic resistance to pathogens of crop plants. Several detailed genetic studies on the inheritance of virulence in the pathogen indicate that single host resistance genes are matched by single virulence genes in the pathogen. Specificity is likely to be determined by the product of a single host gene coding for resistance interacting with the product of a single avirulence gene in the pathogen (1). In the absence of pathogen recognition by the plant, defence genes are not turned on, and the pathogen is not contained. The nature of the two primary gene products and the set of molecular mechanisms initiated subsequent to positive recognition which selectively activate defence genes are unknown. Components of plant defences are activated in a number of different situations including compatible and incompatible pathogen attack, mechanical wounding , autonecrosis and treatment with abiotic compounds (2). It seems likely that plants can utilise an array of cues for defence gene activation, so the processes operating in each situation could be distinct. It is also possible that common mediators are involved in the induction of an array of defence responses.

E. W. Nester and D. P. S. Verma (eds.),
Advances in Molecular Genetics of Plant-Microbe Interactions, 457–461.

We study the interaction between tomato *Lycopersicon esculentum* and the biotrophic fungus *Cladosporium fulvum (Fulvia fulva)*, the causative agent of leaf mould disease. It is interpreted as an example of a "gene for gene" type of interaction (3). This interaction is of particular interest because different dominant *Cf* genes confer resistance against distinct physiological races of the pathogen and the resistance genes *Cf*-2, *Cf*-4, *Cf*-5 and *Cf*-9 are available as near isogenic lines of the cultivar Moneymaker (Cf 0). Also the products of the dominant fungal avirulence (*avr*) genes can be easily isolated in intercellular washing fluid (IF) obtained from infected leaves supporting a compatible interaction (4). These IFs are used to synchronously trigger the *Cf* genes into action in the absence of the pathogen by injecting IF into the airspaces of healthy leaves or cotyledons. Our studies have revealed that each *Cf* genes' action conditions a different incompatible phenotype. The restriction of hyphal ingress into the mesophyll is decreased in the gene order *Cf*-2,9,5,4. With the *Cf* -2 gene, hyphal ingress is restricted to the substomatal cavity and is accompanied by a brown reaction of a few mesophyll cells in contact with the hyphae. In *Cf*-9, *Cf*-5 and *Cf*-4 interactions, hyphae ingress 1-3, 1-5 and 6-8 mesophyll cell lengths respectively from the point of penetration. In the latter three interactions, a brown reaction of contacted host cells is rarely observed. The *Cf* -gene specific responses to IF injection are macroscopic necrosis or chlorosis. The speed of appearance of these symptoms and their overall severity decrease in the gene order *Cf* -9,2,5,4 and each is macroscopically distinct. *Cf* -9 conditions a grey necrotic phenotype within 24h of injection, *Cf*-2 a chlorotic / necrotic phenotype by day 4-5, *Cf*-5 a strong chlorosis by day 5 and *Cf*-4 a weak chlorosis by day 5-7. These responses are restricted to the injected area. Interestingly, all the *Cf* genes so far examined, namely 2,3,4,5,9 and 11 are **semi-dominant** in action both to fungus and IF, ie. the homozygous state is more effective than the heterozygous. Semi-dominance was quantified by IF dilution experiments. Each *Cf* gene in a homozygous state could respond to an IF titre 2 fold lower than when the gene was present in a heterozygous state (unpublished data). Collectively, these observations indicate that each *Cf* gene conditions a distinct host defence response to the same pathogen. The *Cf* genes are known to have been introgressed from various wild *Lycopersicon* species eg. *Cf*-2 and *Cf*-9 from *L. pimpinellifolium*, *Cf*-5 from *L. esculentum var. cerasiforme* and *Cf*- 4 from *L.hirsutum* (5,6).

The tomato plant is attractive for study because many of the defence genes have already been isolated and characterised from its genome or from the related *Solanaceous* species, tobacco, potato and petunia. Tomato is also readily transformable, its reproductive cycle is relatively short (16 weeks) and although naturally self-pollinating it can be easily crossed to yield useful numbers of progeny. It has good genetic and restriction fragment length polymorphism (RFLP) maps (5,7), and has been shown to be a plant in which the maize transposon *Ac* is active (8).

2 Strategies To Isolate Genes Involved In Plant Defence

2.1 MUTANTS

We have made and are genetically characterising tomato mutants with reduced resistance to *C. fulvum* invasion. Theoretically, these may have mutations at the resistance locus, at a locus involved in the signal transduction pathway or in an essential defence locus. Homozygous seed of the *Cf* -9 genotype was mutagenised with EMS (60mM for 16h at 24° C). From 568 EMS M2 families, 9 were identified as containing fully (4 families) or partially (5 families) disease sensitive individuals. In a separate *Cf*-9 genotype stability test (plants used Cf 9 x Cf 0, F1) no susceptibles were obtained from screening 15000 progeny. From each M1 plant 25 progeny seeds were screened in case the fruit was chimeric for mutant sectors or linked to gametic or zygotic lethals thus lowering the frequency of the mutant allele's transmission Our screen involved inoculations with a transformed race of *C. fulvum* which constitutively expresses high levels of ß-glucuronidase (9) activity. To screen many families efficiently MUG (4-methylumbelliferyl ß-D-glucuronide) assays were performed on pooled samples of cotyledons, one obtained from each seedling in the family. The MUG assays were done in vivo by infiltrating the assay buffer into the tissue pieces and quantifying MU (methylumbelliferone) in the surrounding buffer after overnight incubation at 37°C. In lines which showed a pooled MUG activity significantly greater than the value obtained for the control interaction with Cf 9, the second cotyledon was screened on an individual basis. Interestingly, none of the fully susceptible individuals identified now respond to IF whereas the partially susceptible individuals still do. Mutants have been tested for complementation. Mutants at the *Cf*-9 locus when crossed to a Cf 0 plant (no resistance gene) give a susceptible F1 progeny. A resistant F1 progeny indicates a mutation at another locus. Three of the fully susceptible mutants and one partially susceptible mutant map to the *Cf*-9 locus while 2 partially susceptible mutants appear by this criterion to be at another locus. By selfing the mutant x Cf 0 (F1) plants and identifying wild-type Cf 9 plants amongst the progeny we will verify the presence of a locus affecting the action of *Cf*-9. Linkage data between this locus and the *Cf*-9 locus will also be obtained from this experiment. The three fully susceptible mutants which map to *Cf*-9 do not complement each other. Cytological and fungal biomass determinations have revealed a susceptible phenotype of the Cf 9 mutants which is distinct from that of Cf 0. Mycelial density is lower and few conidiophores with conidia are produced on the mutants. The mutated *Cf*-9 locus gives weak biological activity. Interestingly the phenotype described is similar to Cf 1 (10). The *Cf*-1 locus is mapped to chromosome 1 in the vicinity of *Cf*-9 and *Cf*-4 (7). The two partially susceptible mutants have distinct phenotypes, both permit 3-5 times the amount of hyphal ingress, but in one this is accompanied by excess retention of various histochemical stains by surrounding host cells. A crossing programme will ascertain if the mutations also affect the action of the other *Cf* genes. All distinct loci will be mapped by crossing the mutant to *L. pennellii* and then looking for linkage between the mutant phenotype and an RFLP marker in the F2.

2.2 THE AVR9 PROBE

The *avr*9 gene of *C.fulvum* encodes for a 28 amino acid peptide (3). When purified AVR9 peptide is injected into healthy *Cf*9 and *Cf*0 leaves a grey necrotic response is observed specifically on *Cf* 9 within 24h (4). Our second strategy to isolate abberant defence response mutants has involved developing a direct visual screen to assess *Cf*-9 gene action. We are attempting to create transgenic Cf 0 plants which constitutively produce AVR9 peptide targeted to the plant's extracellular airspace regions. We anticipate that when Cf 0 plants homozygous for the T-DNA and expressing high levels of biologically active AVR9 peptide are crossed to Cf 9 the F1 plants will exhibit a grey necrotic phenotype and probably die. If the *Cf*-9 resistance gene is expressed soon after pollination and a defence response is initiated in the developing embryo no F1 seed may be produced. If low amounts of AVR9 are produced a chlorotic phenotype would be observed. To isolate abberant defence response mutants Cf 0 plants homozygous for the AVR9 construct and expressing high levels of biologically active peptide would be used as the female parent in crosses involving mutagenised Cf 9 pollen or pollen obtained from a Cf 9 line containing an active Ac transposable element. Mutants of interest would either appear healthy or have green sectors on a grey necrotic background. With a direct selection for *Cf*-9 gene action enormous numbers of plants could be efficiently screened.

Two *avr*9 gene constructs have been synthetically generated using 5 overlapping oligonucleotide primers and sequential PCR reactions (unpublished data). They contain different signal peptide sequences to target the AVR9 peptide to outside the plant cell. The signal peptide sequence from the bacterial chitinase gene of *Serratia marcescens* and the tobacco PR1a gene were selected because they are known to target proteins extracellularly in plants (11,12). The codon preferences used to generate the sequence were based on those of tomato. Both constructs were placed between 35S promoter and nos terminator sequences and the T-DNAs introduced into tomato and tobacco. Tobacco was selected because of its faster regeneration time. Analysis of IF preparations retrieved from 10 independent tobacco transformants with the *S.marcescens* signal peptide sequence showed two had reproducible biological activity specific to Cf 9. The response was however a mild chlorosis and activity was not correlated with transcript abundance or T-DNA copy number. To confirm the presence of AVR9 peptide in the IF, preparations were also injected into Cf 9 and Cf 0 plants containing a ß-1,3 glucanase-GUS reporter gene fusion. This gene is induced in a *Cf*-dependent manner after IF or pathogen challenge (unpublished data). The fluids retrieved from the AVR9 expressing tobacco plants induce the GUS reporter gene in a *Cf*-gene dependent manner. Nine out of ten independent tobacco transformant with the PR1a signal peptide construct produces AVR9 peptide with full biological activity. A grey necrosis is observed within 24h of injecting IF into Cf 9 plants down to a titre in the range 1/32 to 1/64. Authentic IF from a compatible Cf 0 - race 0 interaction routinely has a titre of 1/64. The analysis of the tomato transformants has yet to commence.

3 How Do The *Cf* Genes Operate ?

As a prelude to the characterisation of the abberant defence response mutants we have explored whether known events specific to incompatibility operate in the *C. fulvum* / tomato pathosystem. We have concentrated on exploring the early responses to IF as in theory this is a synchronous challenge to the *Cf*-gene product. We have developed an in vivo cotyledon assay on 14 day old seedlings. IF is injected 2h into a 16h photoperiod where the light intensity is supplied at 550 $\mu mol\ m^2\ s^{-1}$. The earliest *Cf*-9 -dependent physiological response identified to date was an oxidative burst in the mesophyll layer commencing within 3h of IF injection. Active oxygen radicals are detected by injecting the dye nitroblue tetrazolium into the cotyledon mesophyll layer and observing the formation of an insoluble blue formazan product (13). The oxidative burst continues until about 15h and then diminishes. A cotyledon epinastic response and ethylene production are evident by 9-10h and appear coincident with the onset of loss of membrane integrity. The latter is determined by the ability of cells to retain the vital stain sodium fluorescein and to be plasmolysed by a 2.0M sucrose solution (14). From 12h onwards significant increases in salicylic acid levels are detected and reach a peak of 1.5µg/g fresh weight by 24h (Raskin and Silverman, unpublished data). By 24 h most mesophyll cells have lost membrane integrity and extensive electrolye leakage and alkalinization of the tissues' extracellular regions can be detected. Grey necrosis in cotyledons first appears around 28-30h. In continuous dark essentially similar physiological events occur except that the onset of the oxidative burst does not commence until 6h, SA induction, electrolyte leakage and alkalinization are delayed and their final magnitude reduced. No macroscopic necrotic phenotype is observed in the dark. The kinetics of loss in cell viability and cotyledon epinasty in light and dark are identical. Another reported *Cf*-9 dependent early response to IF is increased lipoxygenase activity (15). Although we have not characterised this response in our cotyledon assay system, again no differences were reported in the induced levels of lipoxygenase activity in light and dark. Our microscopic observations show individual mesophyll cells are the first to undergo detectable physiological alterations and that these act as the foci from which the response increases. Although most of the events reported are probably well downstream of *avr* gene product perception, they are *Cf*-gene specific.

By analysis of the mutants' responses to IF and fungal challenge we can ascertain if a single *Cf* specific event is absent and if its absence is linked to the loss or delay of a second event. By this approach we will determine the relatedness of responses and their potential importance to plant defence. Preliminary characterisation of two of the partially susceptible EMS mutants indicates that responses to IF are reduced; the dilution titre at which macroscopic symptoms are observed were 2-4 fold lower. Also the epinastic response of one of the mutants appears delayed. We intend to characterise the phenotypes of each distinct mutant locus and that of the double mutants when generated. This should provide the basis of a genetic dissection of the tomato plant defence response to *C. fulvum*.

4 Some Concluding Thoughts

Plant breeders have recognised for a long while that non-R-gene components influence the outcome of plant-pathgogen interactions (1). To dissect resistance genetically we may need to generate many independent mutants to obtain the required number of distinct mutations. Initial efforts should concentrate on performing saturation mutagenesis using various agents, to isolate multiple mutations in more than one genetic locus. However with our current pool of mutants which affect the Cf 9 incompatibility phenotype we can genetically test whether the action of the *Cf*-2,4 or 5 genes require these loci or not. Also we can determine if these mutations affect the tomato plant's resistance to other pathogens, or reduces a susceptible plant's responsiveness to resistance-inducing chemicals such as 2,6-dichloroisonicotinic acid, INA (16). The availability of a race 9 of *C. fulvum* permits this latter type of analysis. Finally without the quantitative assay for in planta fungal biomass levels none of the partially susceptible mutants would have been recovered.This may not be the situation with other pathogens. In the barley-powdery mildew pathosystem Jørgenson et al (17) have also identified a mutation at a separate locus that affects the action of the resistance gene *Ml-a12*. This mutant was isolated by macroscopic observation of disease symptoms. These interesting loci will themselves become the targets of gene isolation experiments.

5. Acknowledgements

This work has been supported by the Gatsby Foundation. The experiments with the transformed *C. fulvum* GUS+ race were done under MAFF licence No. 1185B / 30 (101). We wish to thank Ilya Raskin and Paul Silverman for quantifying salicylic acid levels in samples.

6. References

1. Fraser, F.S.S. (1985) Mechanisms of resistance to plant disease, M.Nijhoff/ Dr W. Junk Publishers, Dordrecht.
2. Lamb,.C.J., Lawton, M.A. Dron, M. and Dixon,R.A. (1989) Signals and transduction mechanisms for activation of plant defenses against microbial attack. Cell 56, 215-224.
3. Van den Ackerveken, G.F.J.M., Van Kan, J.A.L. and DeWit, P.J.G.M. (1992) The Plant Journal 2, 359-366.
4. DeWit, P.J.G.M. and Spikeman,G. (1982) Evidence for the occurence of race and cultivar-specific elicitors of necrosis in intercellular fluids of compatible interactions of *Cladosporium fulvum* (syn. *Fulvia fulva*).Physiol. Plant Pathol. 21, 1-11.
5. Stevens, M.A. and Rick, C.M. (1986) 'Genetics and plant breeding' in J.G. Atherton, J.Rudich (eds), The tomato crop, 35-109, Chapman and Hall, London.
6. Dickinson,M., Jones,D.A. and Jones, J.D.G. (1992) Molecular evidence for close linkage between the *Cf-2*/*Cf-5* and *Mi* loci in tomato. Mol. Plant Mic Int. (submitted).
7. Tanksley, S.D. and Mutschler,M.A. (1990) 'Linkage map of the tomato (*Lycopersicon esculentum*) (2N=24)' in O'Brian S.J. (eds) Genetic maps; locus maps of complex genomes. Cold Spring Harbor, Laboratory Press, Cold Spring Harbor, New York,6.3-6.15.
8. Belzile,F. and Yoder,J.I. (1992) Patterns of somatic transposition in a high copy *Ac* tomato line. The Plant Journal 2, 173-179.
9. Roberts,I.N., Oliver,R.P., Punt, P.J., and van den Hondel, C.A.M.J.J. (1989) Expression of the *Escherichia coli* ß-glucuronidase gene in industrial and phytopathogenic filamentous fungi. Curr. Genet.15, 177-180.
10. Lazarovits,G. and Higgins,V.J., (1976) Histological comparison of *Cladosporium fulvum* race 1 on immune,resistant and susceptible tomato varieties. Can. J. Bot. 54, 224-234.
11. Jones,J.D.G., Grady,K.L., Suslow,T.V., and Bedbrook, J.R. (1986) Isolation and characterisation of genes encoding two chitinase enzymes from *Serratia marcescens*. The EMBO Journal 5, 467-473.
12. Cornelissen,B.J.C., Horowitz,J. van Kan,J.A.L., Goldberg, R.B. and Bol,J.F. (1987) Structure of tobacco genes encoding pathogenesis-related proteins from the PR-1 group. Nuc. Acids Research 15,6797-6811.
13. Doke,N. (1983) Involvement of superoxide anion generation in the hypersensitive response of potato tuber tissues to infection with an incompatible race of *Phytophthora infestans* and to hyphal wall components. Physiol. Plant Pathol. 23, 345-357.
14. Holliday,M.J., Keen, N.T. and Long, M. (1981) Cell death patterns and accumulation of fluorescent material in the hypersensitive response of soybean leaves to *Pseudomonas syringae* pv. *glycinea*. Physiol. Plant Pathol. 18, 279-287.
15. Peevers, T.L. and Higgins, V.J. (1989) Electrolyte leakage, lipoxygenase, and lipid peroxidation induced in tomato leaf tissue by specific and nonspecific elicitors from *Cladosporium fulvum*. Plant Physiol. 90, 867-875.
16. Ward,E.R., Unkes,S., Williams,S.C. Dincher,S.S., Wiederhold,D.IL., Alexander, D.A., Metraux, J.P. and Ryals, J.A. (1991) Coordinate gene activity in response to agents that induce systemic acquired resistance.The Plant Cell 3, 1085-1094.
17. Jørgensen,J.H. (1987) Genetic analysis of barley mutants with modifications of powdery mildew resistance gene *Ml*-a12. genomes 30, 129-132.

MOLECULAR BIOLOGY AND BIOCHEMISTRY OF *Hm1*, A MAIZE GENE FOR FUNGAL RESISTANCE

R.B. MEELEY and J.D. WALTON
DOE-Plant Research Lab
Michigan State University
East Lansing, MI 48824 USA

ABSTRACT. The *Hm1* (or *Hm*) gene of *Zea mays* L. conditions resistance to the fungus *Cochliobolus carbonum* Nelson race 1. This allele confers insensitivity to HC-toxin, a fungal-derived compound required for successful pathogenesis. *Hm*-mediated resistance in maize is genetically associated with the activity of HC-toxin reductase, an enzyme that inactivates the toxin by specific reduction of a critical carbonyl group. HC-toxin reductase activity is increased substantially in response to pathogen inoculation of resistant seedlings. In addition, HC-toxin reductase activity is detectable in crude extracts from several other monocot, but not dicot, species.

1. Introduction

Complementary genic interactions form the foundation for most molecular models in the study of plant/pathogen interactions. In many cases, plant disease resistance genes have been defined in context with dominant pathogen characteristics for avirulence. However, several economically significant and biologically interesting interactions involve gene-for-gene relationships centered around specific traits for pathogenicity. Among these cases are interactions involving fungi that produce host-selective toxins. There are several well-developed models where phytotoxic molecules represent the end products of pathogenicity genes. Host resistance is contingent upon specific "recognition" of the pathogenicity gene product. A molecular genetic and biochemical definition of "recognition" of host-selective toxins is sought to advance to our understanding of the role of toxins in the specificity of host-pathogen interactions.

Only HC-toxin-producing isolates of *C. carbonum* are pathogenic against *hm/hm* maize (Scheffer *et al.*, 1967). Production of HC-toxin presumably confers a pathogenic advantage to a spore that has initiated infection on a susceptible genotype, but how this occurs remains unknown. Pathogenicity in this case has been defined by a biochemical and molecular genetic description of HC-toxin production (Panaccione *et al.*, 1992; Walton, 1991). HC-toxin is a cyclic tetrapeptide that contains Aeo (2-amino-9,10-epoxy-8-oxodecanoate) an unusual keto-epoxy amino acid essential for biological activity (Walton and Earle, 1983; Kim *et al.* 1987).

E. W. Nester and D. P. S. Verma (eds.),
Advances in Molecular Genetics of Plant-Microbe Interactions, 463–467.

Metabolism at the keto-epoxide group of Aeo is central to our hypothesis that *Hm*-mediated "recognition" of HC-toxin by maize involves detoxification of HC-toxin. An enzymatic basis for HC-toxin inactivation was described in maize (Meeley and Walton, 1991). The enzyme HC-toxin reductase (HCTR) utilizes NADPH to reduce the 8-keto group of the Aeo side chain (Figure 1). The presence of HCTR activity is the biochemical phenotype of the *Hm* allele: susceptible inbreds, susceptible progeny from a segregating population, and transposon-induced susceptible mutants of maize are significantly impaired in the ability to detoxify HC-toxin (Meeley *et al.*, 1992).

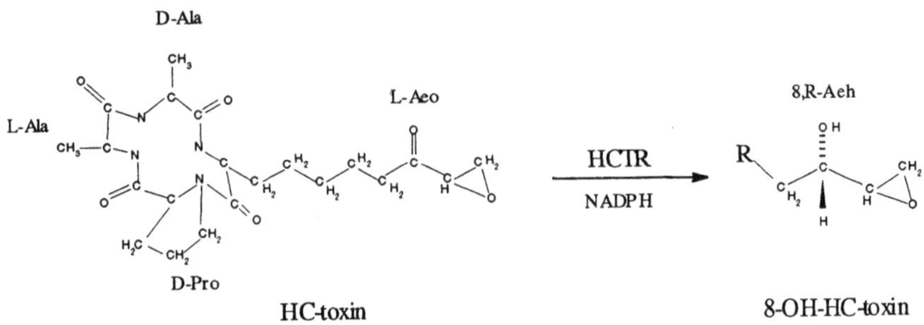

Figure 1. HC-toxin (cyclo-[D-Pro-L-Ala-D-Ala-L-Aeo]) is inactivated by the enzyme HC-toxin reductase (HCTR) by NADPH-dependent reduction of the 8-keto group of Aeo. The NMR signal for the new proton at C-8 is exclusive to δ = 3.43ppm, consistent with stereospecific formation of the 8,R-isomer described by Kim *et al.* (1987).

The simplest scenario concerning specific resistance of maize to *C. carbonum* race 1 is that the resistance allele (*Hm*) encodes the gene for the enzyme HC-toxin reductase (HCTR). Sequence data from the cloned *Hm* allele is forthcoming and supports the notion that *Hm* is the structural gene for HCTR (G. Johal and S. Briggs, Pioneer Hi-Bred International, pers. comm.)

2. HCTR Activity Is Induced By Fungal Inoculation.

Evidence of fungal induction of HCTR activity is shown in Figure 2. Twenty-four hours post-inoculation of resistant seedlings, the level of HCTR activity is increased approximately 2.5-fold over untreated controls. This observation is biologically interesting because it reinforces a defense-related role for the *Hm* gene. When pondering the origin and specificity of *Hm*-mediated resistance, one must consider the possibility that HCTR activity is a fortuitous function conferred by an enzyme involved in some other aspect of cellular metabolism. Evidence of fungal induction however, suggests that the purpose of this gene in maize is intimately related to defense; quite possibly for specific defense against HC-toxin or other fungal cyclic tetrapeptides with similar structures (see Walton, 1990). It is now necessary to test the specificity of induction. What is the effect on HCTR activity by other fungi or pathogens non-pathogenic to maize? Is *C. carbonum* race 2 as effective an inducer of HCTR activity, or is this induction a direct response to HC-toxin?

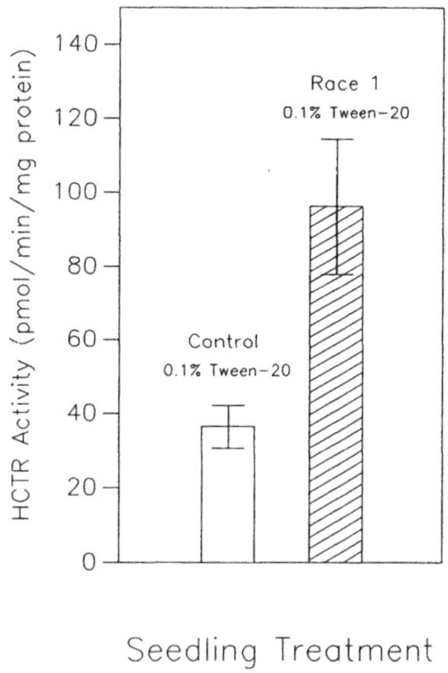

Figure 2. Induction of HCTR activity by inoculation of resistant seedlings with *C. carbonum* race 1. Etiolated resistant (*Hm/-*) seedlings were grown for 5d in the dark and treated with a heavy race 1 spore suspension in 0.1% Tween-20. Control plants were treated with Tween-water alone. 5g of seedling tissue was extracted in maize extraction buffer (Meeley and Walton, 1991) and assayed for HC-toxin reductase activity. The results represent the average of duplicate samples in replicate experiments.

3. HCTR Activity Is Present In Other Monocots, But Not Dicots.

Figure 3 illustrates the results of a preliminary survey of HCTR activity in seedling extracts. To date, we have detected the ability to metabolize HC-toxin only in monocot species. None of the dicot hosts tested possess this capability *in vitro*. Spinach leaves were found to convert HC-toxin when it was delivered through the transpiration stream (not shown), similar to the phenomenon described for susceptible maize leaves (Meeley and Walton, 1991). As we have discussed, HC-toxin is converted in the transpiration stream of susceptible leaves at a rate equal to that of resistant leaves (Meeley and Walton, 1991), but the nature of this conversion has not been characterized. Based on HCTR activity differences *in vitro*, and *in vivo* when the transpiration stream is avoided, we conclude that HC-toxin conversion during transpiration is unrelated to HCTR activity. In fact, we do not know if HC-toxin conversion in the transpiration stream is enzymatic.

4. HC-toxin metabolism and *Hm2*?

HCTR activity is induced by fungal inoculation and appears peculiar to monocot species. We are interested in exploring the co-evolutionary questions underlying these observations. How do the genes for HCTR compare among monocot species? In addition, we perhaps have an opportunity to investigate a duplicated function within maize itself. A separate locus, called *Hm2*, confers developmentally regulated resistance to *C. carbonum* race 1. We have some preliminary evidence that *Hm2*-mediated resistance may involve a similar mechanism of HC-toxin inactivation (see Figure 4). What is needed is a reliable method to assay HCTR activity in extracts of mature tissues, where *Hm2* is fully expressed.

HC-toxin metabolism influenced by *Hm2* may be relevant to some of our earlier findings. In previous data showing low levels of HCTR activity in transposon-induced mutants of *Hm* (Meeley *et al.*, 1992), our interpretation was that these low levels were due to positional effects of the inserted elements, or somatic excision of the element. However, most maize lines that are naturally resistant to *C. carbonum* race 1 due to *Hm/-* also have the genotype *Hm2/-* (Nelson and Ullstrup, 1964). It is possible that the *Hm* mutant lines that contained low-levels of HCTR activity are as such because they were present in an *Hm2/-* background. In turn, this may support a mechanism of HC-toxin metabolism controlled by *Hm2*. Genetic manipulations to resolve *Hm* mutant alleles from *Hm2* are in progress (G.S. Johal, pers. comm.)

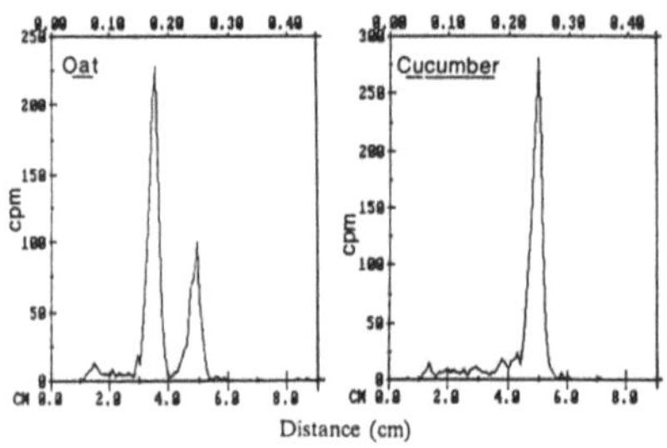

Plant	% HC-toxin converted
cucumber	n.d.*
soybean	n.d.
barley	75.4
sorghum	17.2
wheat	81.6
oat	67.2
pea	n.d
Arabidopsis	n.d.

*n.d. - none detected

Figure 3. Evidence of HCTR activity in species other than maize. The HCTR assays were conducted as described (Meeley *et al.*, 1992) on 150mg of etiolated tissue from each of the species listed. HCTR activity is expressed as the percentage of native HC-toxin converted to the 8-OH-derivative in each sample. Representative traces of quantitative TLC scans are shown for oats *vs.* cucumber. Native HC-toxin is represented by the peak at 5 cm, the metabolite, 8-OH-HC-toxin, occurs at 3.2 cm.

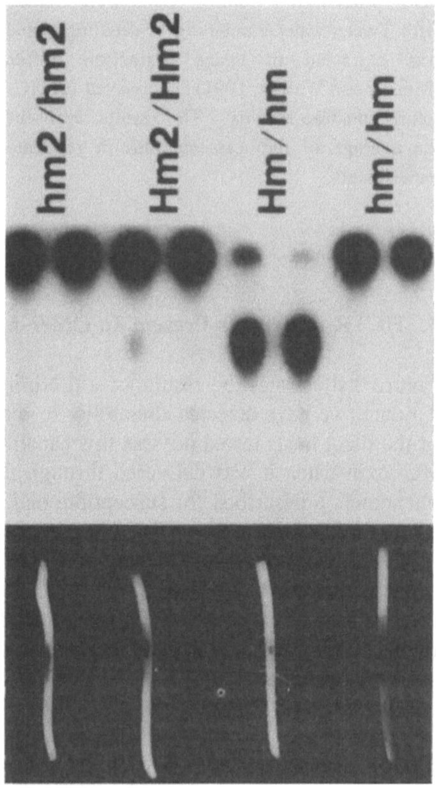

Figure 4. Evidence of HC-toxin metabolism associated with *Hm2*. These assays were performed as described (Meeley *et al.*, 1992). Resistance to *C. carbonum* race 1 by *Hm2* is expressed poorly at the seedling stage (Nelson and Ullstrup, 1964). The disease phenotype is indicative of this, but a low level of HCTR activity is evident in comparison with *hm2/hm2* plants.

6. Outlook

The interaction between *C. carbonum* race 1 and maize is well characterized at the genetic, biochemical, and molecular genetic levels. In this case, a biochemical definition of pathogen "recognition" involves inactivation of its phytotoxin. HC-toxin is active against resistant maize when the toxin concentration is raised 100-fold. We propose that HCTR activity accounts for this differential. One might predict that if *C. carbonum* race 1 could produce 100-fold more HC-toxin, the HCTR mechanism could become saturated, and perhaps resistant maize or even other crops would become infected. Our description of specificity in this disease is significant, but meaningful questions remain about how HC-toxin functions during pathogenesis.

5. References

Kim, S.-D., Knoche, H.W., and Dunkle, L.D. (1987) 'Essentiality of the ketone function for toxicity of the host-selective toxin produced by *Helminthosporium carbonum*', Physiol. Mol. Plant Path. 30, 433-440.

Meeley, R.B., Johal, G.S., Briggs, S.P., and Walton, J.D. (1992) 'A biochemical phenotype for a disease resistance gene of maize', Plant Cell 4, 71-77.

Meeley, R.B., and Walton, J.D. (1991) 'Enzymatic detoxification of HC-toxin, the host-selective cyclic peptide from *Cochliobolus carbonum*', Plant Physiol. 97, 1080-1086.

Nelson, O.E., and Ullstrup, A.J. (1964) 'Resistance to leaf spot in maize; genetic control of resistance to race 1 of *Helminthosporium carbonum* Ull.', J. Heredity 55, 195-199.

Panaccione, D.G., Scott-Craig, J.S., Pocard, J.-A., and Walton, J.D. (1992) 'A cyclic peptide synthetase gene required for pathogenicity of the fungus *Cochliobolus carbonum* on maize', Proc. Natl. Acad. Sci. USA 89, (in press).

Scheffer, R.P., Nelson, O.E., and Ullstrup A.J. (1967). 'Inheritance of toxin production and pathogenicity in *Cochliobolus carbonum* and *Cochliobolus victoriae*', Phytopathology 57, 1288-1291.

Walton, J.D. (1990) 'Peptide phytotoxins from plant pathogenic fungi', in H. Kleinkauf and H. von Döhren (eds.) Biochemistry of Peptide Antibiotics, de Gruyter, Berlin, New York, pp. 179-203.

Walton, J.D. (1991) 'Genetics and biochemistry of toxin synthesis in *Cochliobolus* (*Helminthosporium*)', in S.A. Leong and R.M. Berka (eds.), Molecular Industrial Mycology, Systems and Applications for Filamentous Fungi, Marcel Dekker, New York, pp. 225-249.

Walton, J.D., and Earle, E.D. (1983) 'The epoxide in HC-toxin is required for activity against susceptible maize', Physiol. Plant Path. 22, 371-376.

TRANSPOSON TAGGING OF A RUST RESISTANCE GENE IN MAIZE

A. PRYOR
CSIRO Plant Industry
P.O. Box 1600
Canberra ACT
Australia 2601

ABSTRACT. Transposon tagging provides a successful method for cloning genes of unknown function. The distal tip of the short arm of chromosome 10 of maize contains at least 18 *Rp* genes specifying resistance against the rust pathogen *Puccinia sorghi*. Previous studies have shown that the *Rp1* gene is unstable giving susceptible progeny due to an unequal crossing over mechanism. The *Rp5* gene located 2 map units distal to *Rp1*, appears to be stable. The paper describes the recovery and characterization of transposon tagged mutations at both the *Rp5* and *Rp1* loci using the *Ac/Ds* and *Mu* transposon systems in maize. These mutants are being used to clone genes specifying rust resistance.

1. Introduction

Transposon tagging in maize has proved a successful means of isolating genes for which there is no information of the gene product. A gene, *Hc1*, specifying resistance in maize to *Cochliobolus carbonum* has been recently cloned using this approach (Meeley et al., 1992). Major gene resistance to the rust, *Puccinia sorghi*, in maize is specified by a number of *Rp* genes some 18 of which map in a tandem array at the tip of the short arm of chromosome 10. These resistance genes show different responses to different races of rust.

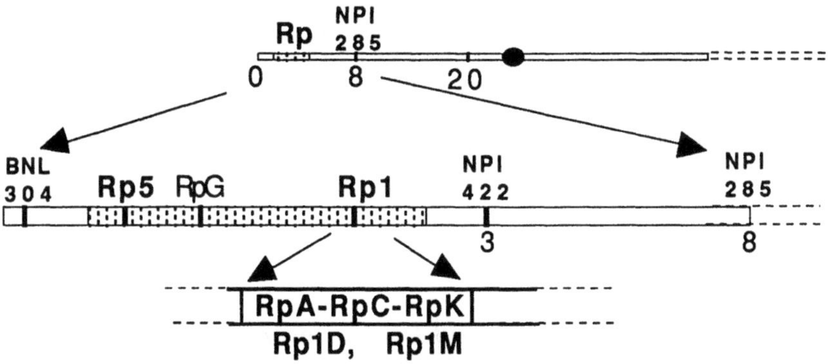

Figure 1. The *Rp* gene complex on chromosome 10 in maize showing in expanded detail the resistance region and some of the *Rp* genes which map to the tip of the short arm. The

E. W. Nester and D. P. S. Verma (eds.),
Advances in Molecular Genetics of Plant-Microbe Interactions, 469–475.

genetic distances are based on Saxena and Hooker, (1968),Hulbert and Bennetzen (1991) and Pryor (unpubl.).The genetic symbols are; *NPI285,NPI422,BNL304* - RFLP markers; *Rp*- resistance to *P.sorghi*;.

Recovery of tagged mutants at the *Rp1* locus has been difficult because of a high background of unequal crossing over events that occur at this locus (Pryor,1987, Bennetzen et al.,1988). The phenotypes of unequal cross-overs and tagged mutants cannot be distinguished because they are both susceptible. Data presented in this paper suggests that this background event is not a problem at the *Rp5* resistance gene and describes the isolation of susceptible mutants which have arisen in crosses involving the *Ac/Ds* and *Mu* transposons in maize. In addition an unstable mutant of the *Rp1* locus is described.

2. Methods

The rust races, R1 and R2, were derived from single pustule isolates and maintained in the glasshouse on appropriate differentials. R1 is avirulent on *Rp1D* and *Rp5* and virulent on *Rp1M*. R2 is avirulent on *Rp1M* and *Rp5* but is virulent on *Rp1D*. Maize lines carrying the different *Rp* genes were obtained from Hooker (see Hooker, 1985). The *Pvv* (variegated pericarp) line used as a source of active *Ac* was obtained from Kermicle and the *Mu* stock was a bronze mutable (*bz-muM1*) line from Walbot. The various transposon lines and the *Rp* lines were inter-crossed and selfed. Genotypes homozygous for the target *Rp* gene and carrying the transposon were recovered for use in a test cross outlined in Fig.2.

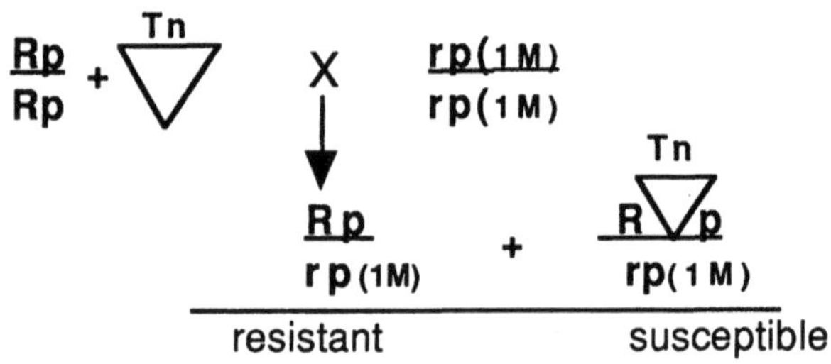

Figure 2. Outline of the general cross used to recover transposon tagged mutants of an *Rp* gene. Plants homozygous for the *Rp* gene (either *Rp1D* or *Rp5*) and carrying the **Tn** (transposon *Ac* or *Mu*) were crossed by a pollen parent carrying a gene *Rp1M* which is susceptible to rust R1 but resistant to R2. Test cross progeny were screened with R1 to recover susceptible mutants. These can be recovered from F2 progeny by screening with the R2 rust which distinguishes the *Rp1M* gene derived from the pollen parent.

3. Results

3.1. TAGGING *Rp1D*

3.1.1 *Recovery of Rp1D*-5 Tagged with Ds.* Most maize lines contain more than 50 *Ds* elements which transpose only in the presence of an active *Ac* element. A line homozygous for the *Rp1D* gene and carrying a single copy of *Ac* at the *Pvv* locus was test crossed by a pollen parent carrying the *Rp1M* gene. Progeny were screened with rust race R1 and 30 seedlings with altered resistance (called *Rp1D*-1* --> *Rp1D*-30*) were observed from a total of 171,028 (Table 2). 28 of the seedlings were fully susceptible. The two exceptions were *Rp1D*-5* which was intermediate in resistance and *Rp1D*-21* which was a high necrotic (see Pryor,1987). Rust race R2, which recognises the *Rp1M* pollen parent chromosome, was used to screen selfed progeny to allow recovery of homozygous mutant seedlings. All but 2 of the mutants were recovered and in all cases except one, the progeny maintained the same altered resistance i.e. susceptible, intermediate or necrotic. The exception was *Rp1D*-13* which had reverted to the parental resistance phenotype.

Mutants that are tagged with *Ds* should be unstable in the presence of *Ac* and revert to the parental resistance. Eighteen of the 28 recovered mutants were tested for reversion in the presence or absence of *Ac*.and no mutant reverted to resistance. Only *Rp1D*-5* was unstable in the presence of *Ac* mutating from the intermediate resistance to fully susceptible at a frequency of 0.24% (Table 1). *Rp1D*-5* was stable in the absence of *Ac*. Thus genetically *Rp1D*-5* is tagged with *Ds*. None of the other mutants tested are unstable and are most probably due to unequal crossing over which has been shown occur at the *Rp1D* gene at a frequency of about 0.015% (Pryor,1987). In a sample of 170,000, 25 unequal cross overs were expected.

TABLE 1. Stability of Rp1D* mutants in the presence or absence of Ac.

	Number of seedlings scored			
	Ac present		*Ac* Absent	
Mutant	Susceptible	Resistant	Susceptible	Resistant
Rp1D-1*	2406	0	117	0
-2	2973	0	783	0
-3	1477	0	838	0
-4	2953	0	2095	0
-5	8839	21*	6693	0
-7	1953	0	-	-
-8	914	0	-	-
-9	2115	0	-	-
-12	2881	0	528	0
-14	525	0	1065	0
-16	3669	0	-	-
-17	-	-	1666	0
-20	2503	0	980	0
-21	2000	0	700	0
-22	1838	0	689	0
-23	1389	0	-	-
-24	3135	0	1109	0
-27	2435	0	354	0
-28	2509	0	-	-

* the 21 seedlings were fully susceptible and were derived from the parental *Rp1D*-5* which was intermediate in resistance.

3.2. TAGGING *Rp5*

3.2.1. *Recovery of Ac/Ds tagged Rp5 mutants*. Because of the significant background frequency of unequal crossing over events at *Rp1*, it was replaced with the *Rp5* gene as a target for transposon mutagenesis. Test cross seedlings from an analogous cross to that outlined for *Rp1D* were screened with rust race R1 and from 120,092 seedlings, four susceptibles were observed (Table 2).

TABLE 2. The number of susceptible progeny observed in test crosses designed to recover transposon tagged mutants at *Rp1D* or *Rp5* . The general crossing procedure is outlined in Figure 2.

Target Gene	Tn	# seedlings	# susceptible	# tagged	Identity
Rp1D	*Ac/Ds*	171,078	30	1	*Rp1D*-5*
Rp5	*Ac/Ds*	120,092	4	4	*Rp5*-2. Rp5*-6* *Rp5*-9,Rp5*-10*
Rp5	*Mu*	85,005	4	4	*Rp5*-mu1,Rp5*-mu2* *Rp5*-mu3,Rp5*-mu4*

The four mutants were recovered and three have been tested for stability in the presence and absence of *Ac* (Table 3). *Rp5*-2* and *Rp5*-10*, in the presence of *Ac*, reverted to resistance which was indistinguishable from the parental *Rp5* resistance (type 1-2). A resistant seedling was recovered from the *Rp5*-6* mutant in a cross in which *Ac* was apparently absent . In all these crosses the presence of *Ac* was determined by *Pvv* which determines a visible phenotype of variegated pericarp caused by frequent transpositions of the active *Ac* at the pericarp color locus. It is possible that in the *Rp5*-6* stock which lacked the *Pvv* phenotype, possessed a transposed *Ac* else where in the genome. The explanation for this instability remains to be determined. However in the case of *Rp5*-2* and *Rp5*-10* the data are consistent with the tagging by a *Ds* transposon. In the case of *Rp5*-2* , further support for this conclusion is provided by studies on the stability of the revertants.

TABLE 3. Stability of Rp5* mutants in the presence or absence of Ac.

	Number of seedlings scored			
	Ac present		Ac Absent	
Mutant	Susceptible	Resistant	Susceptible	Resistant
Rp5 -2*	2	1730	0	793
-6	0	5921	1	6705
-9	-	-	-	3204
-10	1	1608	0	2646

3.2.2. *Second cycle tagging using revertants of Rp5*-2*. *Ac* tends to transpose preferentially to closely linked sites (Dooner and Belachew, 1989). The corollary is that transposon mutants of a gene will occur at much higher frequencies when the transposon is located close to the target gene. If reversion of *Rp5*-2* was a result of *Ds* excision, then in either one or both of the recovered revertants, *Ds* may be at a nearby location. Second cycle tagging using the revertants should generate a higher frequency of insertional mutants but only in the presence of an active *Ac*. This prediction was met for both revertants. From 27,887 test cross seedlings 17 susceptibles were observed but only when *Ac* was present (Table 4). Five of the susceptibles were good plants, nine were sterile and three died before flowering. At present there is no explanation for the defective plants but chromosome breakage events are known to be associated with *Ds* transposition (Dooner and Belachew,1991).

TABLE 4. Second cycle tagging in the presence or absence of *Ac* using *Rp5-2*Revertants.

	Number of seedlings scored			
	Ac present		*Ac* Absent	
Mutant	Susceptible	Resistant	Susceptible	Resistant
Rp5*-2RA	4	5,141	0	3757
Rp5*-2RB	13	22,746	0	7513

For the *Rp5*-2* mutant, the data fit a model of tagging by the *Ds* transposon. The initial mutant was recovered as a rare event (1/50,000) presumably due to transposition of the *Ds* from an unlinked site some where in the genome. *Rp5*-2* is unstable only in the presence of Ac and reverts to parental resistance at a frequency of about 1/1000. This is in general agreement the observed frequency of *Ds* excision events. Both revertants can now undergo second round tagging but again only produce susceptibles in the presence of *Ac* but now at a much higher frequency (1/1600). This is consistent with transposition of Ds from a closely linked site.

***Ac* dependent events**

Rp5 (Resistant) —1/50,000→ **Rp5*-2** (Suscptible) —1/1,000→ **Rp5*-2R** (Resistant) —1/1600→ **Rp5**** (Suscptible)

3.2.3. *Recovery of Mu tagged Rp5 mutants.* The *Rp5* gene is located about 2 map units distal to the *Rp1* gene complex which is a group of closely linked genes including *Rp1M*. A recombinant chromosome with *Rp5* and *Rp1M* in coupling was used as the target to recover *Mu* tagged mutants of *Rp5*.(Fig. 3). Plants homozygous for the cis linked *Rp5-Rp1M* and carrying an active *Mu* (followed by the mutable *bz-muM1*) were test crossed by a susceptible pollen parent. Four seedlings susceptible to race R1 were observed from 85,005 progeny. All four proved to be resistant to R2 indicating the continued presence of the *Rp1M* gene and showing that only the *Rp5* specificity had been lost. The closely linked *Rp1M* resistance gene can be used to recover the the *Rp5*-mu* mutants.

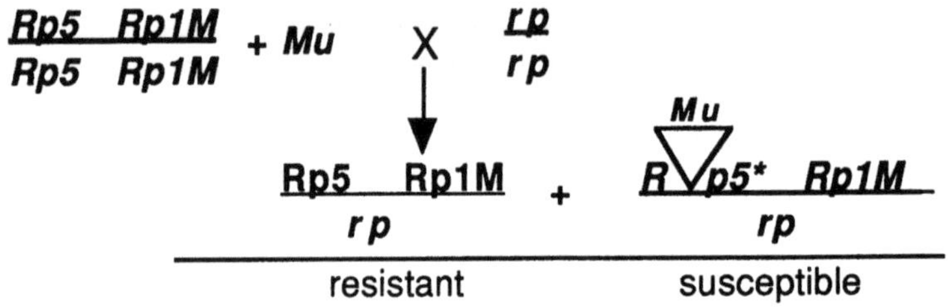

Figure 3. Outline of the cross used to screen for *Mu* tagged mutants of the *Rp5* resistance gene. The recombinant *Rp5 Rp1M* chromosome was crossed into a *Mu* active background which was ascertained by the presence of the mutable *bz-muM1* reporter gene.

4. Conclusions

The final proof of tagging depends on a demonstration of the actual insertion of transposon DNA into the resistance gene. Since using these tagged mutants to isolate the resistance gene was the purpose of these experiments, this is a somewhat circular argument. In the case of presumptive *Ac/Ds* tagged mutants, the phenotype of 'instability of the mutant in the presence of an active *Ac*', provides a degree of confidence in the tagging event. For mutants derived from the *Mu* material this corroborating evidence may or may not be available. The four mutants reported have not yet been tested for stability. Some *Mu* tagged mutants, such as the reporter gene, *bz-muM1*, used in this study, are unstable and have a variegated phenotype. Other *Mu* derived mutants are stable and cannot be distinguished from non-tagging events such as deletions which would also be selected in the screen for susceptible seedlings. By using the cis linked *Rp5 Rp1M* chromosome as a target for *Rp5* tagging, the retention of the *Rp1M* resistance in the susceptible seedlings can at least limit the size of any potential deletion . Unlike the high frequency unequal crossing over that is observed at the *Rp1* locus (Pryor,1987, Bennetzen et al.,1988), the *Rp5* gene appears to be stable. Only four susceptible seedlings were observed from over 100,000 progeny from the *Ac/Ds* cross. The three that have been tested all show instability and two only when *Ac* is present. Since a similar frequency of susceptibles was recovered from the *Mu* experiment, it is likely that these too represent tagging events. These mutants are the subject of further work aimed at cloning a rust resistance gene.

5. References:

Bennetzen,J.L., Qin,M-M., Ingels,S., and Ellingboe,A.H. (1988). Allele-specific and *Mutator*-associated instability at the *Rp1* disease-resistance locus of maize. *Nature* (London) 332, 369-370.

Dooner, H.K. and Belachew, A. (1989). Transposition pattern of the Maize Element Ac from the bz-m2(Ac) Allele. *Genetics* 122, 447-457.

Dooner, H.K. and Belachew, A. (1991). Chromosome breakage by pairs of closely linked transposable elements of the *Ac-Ds* family in maize *Genetics* 129, 855-862.

Hooker, A.L. (1985) Corn and Sorghum Rusts. In *The Cereal Rusts. Volume 2.*

Diseases, distribution, epidemiology, and control. Eds. Roelfs, A.P. and Bushnell, W.R. Orlando, Academic,pp 207-233.

Hulbert,S.H., and Bennetzen J.L. (1991). Recombination at the *Rp1* locus of maize. *Molecular and General Genetics* 226, 377-382

Meeley, R.B., Johal, G.S., Briggs, S.P. and Walton, J.D. (1992). A Biochemical Phenotype for a Disease Resistance Gene of Maize. *The Plant Cell*, 4, 71-77.

Pryor,T. (1987). The origin and structure of fungal disease resistance genes in plants. *Trends in Genetics* 3, 157-161.

Saxena, K.M.S., and Hooker, A.L. (1968). On the structure of a gene for disease resistance in maize. *Proceedings of the National Academy of Science (USA)* 611, 1300-1305.

GLUCAN ELICITOR-BINDING PROTEINS AND SIGNAL TRANSDUCTION IN THE ACTIVATION OF PLANT DEFENCE

J. EBEL, E.G. COSIO, M. FEGER, T. FREY, U. KISSEL,
S. REINOLD, and T. WALDMÜLLER
Biologisches Institut II
Universität Freiburg
Schänzlestr. 1
D-7800 Freiburg
Federal Republic of Germany

ABSTRACT. Soybean (*Glycine max* L.) tissues respond to infection with the fungus *Phytophthora megasperma* f. sp. *glycinea*, the pathogen causing stem and root rot in this plant, by the rapid, cultivar-specific activation of a phytoalexin defence response. The phytoalexin response is also expressed in cultured soybean cells following treatment with an elicitor derived from the cell walls of the fungus. The best characterized elicitors for soybean are the branched (1→3)- and (1→6)-linked ß-glucans from the fungus. The glucans are naturally released during the early stages of germination of the fungal cysts in a host-independent manner. Surface-localized glucan-binding proteins exist in soybean cells which display high affinity and specificity for the fungal ß-glucans, including an elicitor-active hepta-ß-glucoside fragment derived from the polysaccharide, suggesting that elicitor perception involves a receptor-mediated process on the host cell membrane. The main component of the ß-glucan-binding sites is a 70-kDa protein in SDS/PAGE, as identified by photoaffinity labelling and glucan-affinity chromatography of detergent-solubilized binding proteins. The elicitor-mediated transmembrane signalling process which leads to the activation of the phytoalexin response very likely involves Ca^{2+}. Purified *P. megasperma* ß-glucans and synthetic hepta-ß-glucoside showed differential elicitor activity when tested in different experimental systems. In contrast to bioassays utilizing cotyledons, cell cultures responded only weakly to the presence of these carbohydrates. A strong and synergistic response, however, was observed when the cells were simultaneously treated with glucan elicitors and compound K-252a, a known inhibitor of mammalian protein kinases.

1. Introduction

The biochemical mechanisms of cultivar resistance and non-host resistance of plants against pathogens are similar and include a wide range of inducible defence responses. Many of these responses involve gene activation, e. g. the phytoalexin response. It is postulated that inducible plant defence mechanisms share a number of features which include the perception of pathogen-derived signal(s) and the intracellular signal transduction leading to the initiation of the defence response [1]. Inducible plant defences can be activated, not only upon challenge of plant tissues by microbes, but also upon exposure to elicitors. Elicitors are now widely used in selected experimental systems to study the above-mentioned steps of the inducible defence reactions.

E. W. Nester and D. P. S. Verma (eds.),
Advances in Molecular Genetics of Plant-Microbe Interactions, 477–484.

A plant host-pathogen system in which the phytoalexin defence response, elicitor recognition and elicitor-mediated signal transduction have been studied in some detail is that of soybean (*Glycine max* L.) and the root and stem-rot fungus *Phytophthora megasperma* f. sp. *glycinea* [1, 2]. Soybean cultivars are differentially resistant to several *P. megasperma* races and both pterocarpanoid phytoalexin accumulation and expression of enzymes of phytoalexin biosynthesis reflect the known differences in the physiological plant-fungus interaction. Among the enzymes which are strongly expressed in the incompatible plant-fungus and in the plant cell culture-elicitor interaction are phenylalanine ammonia-lyase, 4-coumarate:CoA ligase, and chalcone synthase [3; Uhlmann and Ebel, unpublished results].

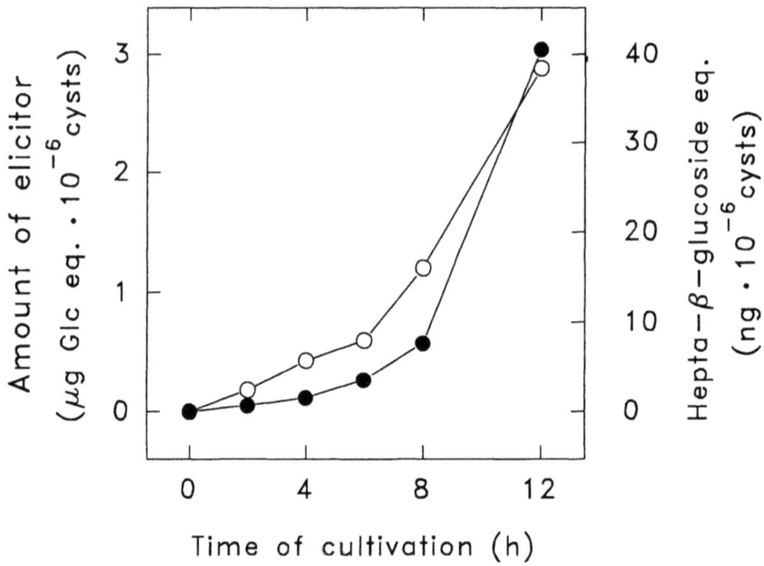

Figure 1. Time course of release of elicitor-active ß-glucans from germinating cysts of *Phytophthora megasperma* f. sp. *glycinea*. Elicitor activity (○) of the culture fluids was analyzed with a bioassay using soybean cotyledons and quantified using a dose-response curve for an elicitor fraction that was released 24 h after the onset of germination. The amounts of ß-glucans (●) released were determined in a competitive ligand-displacement assay with a soybean microsomal fraction and a radiolabelled hepta-ß-glucoside conjugate as ligand. A dose-response curve for the synthetic heptaglucoside as ligand was used for quantification.

2. Glucan Elicitor Release and Recognition

One type of phytoalexin elicitor active in soybean was characterized as (1→3)- and (1→6)-linked ß-glucans [4] which are structural polysaccharides of the mycelial cell walls of *P. megasperma* and other oomycetes. An in vitro culture system was used to show an early and rapid release of ß-glucan elicitors from germinating cysts of *P. megasperma* (Fig. 1) [5]. Linkage-composition analysis of the glucan elicitors showed that they are primarily

TABLE 1. Characteristics of ß-glucan-binding sites of soybean

Ligand specificity:	(1→ 3, 1→ 6)-ß-D-glucans
Glucan-binding affinity fungal ß-glucan fractions: hepta-ß-glucoside:	 app. K_d = 10 - 40 nM app. K_d = 3 nM
Number of binding sites in microsomal fractions:	 B_{max} = 1 - 2 pmol/mg protein
Optimum of binding:	pH 6 - 8
App. size of binding proteins:	70, 100, 170 kDa
Localization:	plasma membrane
Distribution:	roots, hypocotyls, cotyledons, cell cultures

(1→ 3)-ß-linked with (1→ 6)-ß-branches, a composition similar to that of glucans obtained by heat release from mature mycelium [6, 7] but different to that of elicitors obtained by acid hydrolysis [8] or from spontaneous autohydrolytic release by senescent cultures [9]. The time course of release of the glucans from germinating cysts (Fig. 1) as measured in a competitive ligand-displacement assay [10] with a soybean microsomal fraction and radiolabelled hepta-ß-glucoside (see below) was similar to that of the release of elicitor activity (Fig. 1) when assayed in a cotyledon bioassay [9, 11]. The release of ß-glucans, which is independent of the presence of host-produced glucanases, indicates that the complex processes taking place in the apical region during mycelial growth involve both glucan synthesis and degradation [for review see 12] and may be a significant source of elicitors during the earliest stages of cyst germination. While the time course of ß-glucan release and defence activation coincide in the incompatible plant-fungus interaction, there is a significant delay for the induction of the phytoalexin response in the compatible interaction [3, 5]. Since the *P. megasperma* ß-glucan elicitors are released from the virulent race 3 of the fungus at a rate similar to that reported here for the avirulent race 1 (data not shown), compounds other than the glucans might be exchanged in the early plant-fungus communication and might be necessarily involved in the establishment of resistance and susceptibility.

A model proposed for non-self recognition of pathogens by plants involves the interaction of pathogen or plant-derived elicitors with receptor-like target sites on the plant plasma membrane [4]. We have characterized on soybean membranes specific binding sites for the *P. megasperma* ß-glucan elicitor [8, 10, 13]. The results have been confirmed recently by Cheong and Hahn [14]. The binding of radioactively labelled conjugates of the (1→ 3, 1→ 6)-ß-glucan and of a synthetic hepta-ß-glucoside derived from it is of high affinity, saturable, and ß-glucan-specific (Table 1). Affinity measurements with the hepta-ß-glucoside as ligand gave apparent K_d values of about 3 nM. These experiments also showed that the abundance of the binding sites in a total soybean membrane fraction is low, not exceeding 1 to 2 pmol/mg membrane protein, the highest binding activity being present in the plasma membrane fraction. A close direct correlation existed between the binding affinities of the binding sites for a series of fungal and synthetic oligoglucosides and the phytoalexin elicitor activities of the ligands [10, 14].

TABLE 2. Effects of Ca^{2+} antagonists on the elicitor-mediated enhancement of chalcone synthase activity and on glucose 6-phosphate dehydrogenase activity in soybean cell cultures. Values for half-maximal inhibitor concentration (IC_{50}) were obtained from dose-response curves for the inhibition of elicitor-stimulated chalcone synthase activity and the inhibition of glucose 6-phosphate dehydrogenase activity that is not affected by elicitor.

Substance	Chalcone synthase	Glucose 6-phosphate dehydrogenase
	IC_{50} (μM)	
Nifedipine	100	> 250
Nitrendipine	20	> 500
Verapamil	200	300
D 888	> 250	> 250
(−)-Bepridil	> 250	> 250
Diltiazem	> 500	> 500
Prenylamine	60	70
Fendiline	200	250
Pimozide	20	40
Flunarizine	70	> 100
La^{3+}	200	> 500

Glucan-binding proteins were identified by photoaffinity labelling of detergent-solubilized proteins from soybean root membranes using a photoreactive, radiolabelled conjugate of the hepta-ß-glucoside [15]. The major component of the solubilized ß-glucan-binding sites appears to be a protein with apparent molecular mass of 70,000, possessing ligand-affinity characteristics very similar to those of the membrane-associated sites. Further characterization of the ß-glucan-binding proteins was achieved by employing affinity purification on a glucan-affinity matrix after detergent solubilization of the membrane proteins [15, 16]. The purified fraction contained a 70-kDa protein which was radiolabelled by the photoaffinity ligand along with two additional proteins of 100 and 170 kDa labelled to a lesser extent. The 70-kDa protein represented also the major protein as visualized by staining after SDS/PAGE. These results provide a clear identification of membrane proteins with high affinity for fungal signal molecules of carbohydrate nature capable of triggering plant defence. Unambiguous proof that the glucan-binding proteins are functional receptors, i. e. are involved in elicitor signal perception and transduction, requires detailed studies on their characteristics including the reconstitution of a ligand-response system with the isolated binding proteins.

3. Signal Transduction and Activation of Defence

The ß-glucan elicitor from *P. megasperma* activates defence-related genes in soybean such as those involved in phytoalexin formation [1-3]. The activation of the cellular response requires the transduction of the elicitor signal from the site of primary perception, occurring very likely at the plasma membrane, to the nucleus. The stimulation of elicitor responses in several plant cell systems including soybean depend on the presence of Ca^{2+} in the culture medium [1]. In addition, in several plant cell-elicitor systems rapid permeability changes of the plasma membrane to ions including Ca^{2+}, H^+, K^+, Cl^- have been suggested to be involved in signal transduction [1].

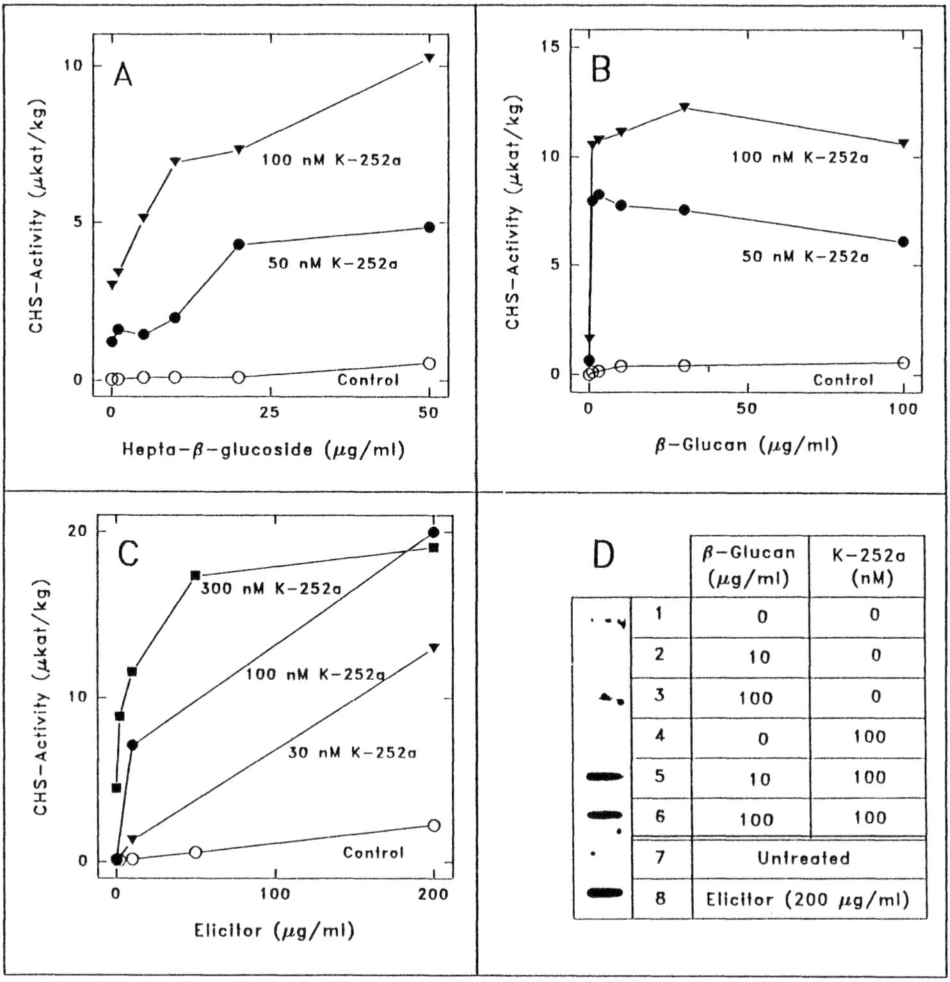

Figure 2. Enhancement by elicitor of chalcone synthase activity (A-C) and mRNA amount (D) in the absence or presence of various concentrations of K-252a. Chalcone synthase activity was measured after treatment of soybean cell cultures for 12 h with either hepta-ß-glucoside (A), a ß-glucan fraction purified by ion-exchange chromatography on DEAE cellulose and chromatography on Concanavalin A Sepharose (B), or a crude elicitor preparation (C). Chalcone synthase mRNA amount (D) was determined by slot-blot hybridization using a bean chalcone synthase cDNA [3] after treatment of the cells for 5 h with the purified ß-glucan fraction (1-6) or a crude elicitor preparation (7,8).

The permeability changes could be caused by temporary increases in the activity of ion channels [17]. Experiments with soybean cell cultures using pharmacological drugs referred to as calcium channel antagonists indicated that some of the compounds have strong inhibitory effects on the elicitor-mediated stimulation of chalcone synthase, an enzyme involved in phytoalexin formation in soybean (Table 2). These compounds include nifedipine and nitrendipine. Other compounds such as verapamil, fendiline, pimozide, and prenylamine not only affected the induction of chalcone synthase but also the viability of the cells as estimated from their effects on the activity of an enzyme of primary metabolism, glucose 6-phosphate dehydrogenase (Table 2). A third group of compounds showed only little effect in the concentration range tested. Furthermore, La^{3+} not only inhibited the elicitor-mediated stimulation of chalcone synthase activity (Table 2) but also the elicitor-enhanced uptake of $^{45}Ca^{2+}$ by the soybean cells (U. Kissel and J. Ebel, unpublished results). The results indicate that some of the channel blockers might be suitable for biochemical studies on the identification of putative calcium channels [18] in the plasma membrane of soybean cells and on their role in mediating the transduction of the elicitor signal(s).

Elicitor-mediated changes in protein phosphorylation patterns in vivo in different cell cultures including soybean [19-21] indicated that protein kinases appear to be involved in signal transduction. This view is supported by the finding [21, 22] that the protein kinase inhibitors, K-252a and staurosporine, not only suppressed elicitor responses in tomato cells, but also abolished elicitor-stimulated changes in protein phosphorylation in these cells. Unexpectedly, when applied simultaneously with different ß-glucan elicitor fractions to soybean cells, K-252a markedly enhanced the induction of chalcone synthase activity as compared to the ß-glucans given alone (Fig. 2). The synergistic effect was especially pronounced between the hepta-ß-glucoside or a partially purified ß-glucan fraction and K-252a. Concentrations of each of the carbohydrates that by themselves were too low to elicit significant levels of chalcone synthase activity were effective elicitors when added together with K-252a. The experiments (Fig. 2) also demonstrated a synergism of the effectors on the level of chalcone synthase mRNA that was largely similar to that observed for chalcone synthase activity. The results support the notion that as yet unknown components of the transduction chain mediating the activation of elicitor-responsive genes might be regulated by phosphorylation-dephosphorylation.

4. Acknowledgements

This work was supported by the Deutsche Forschungsgemeinschaft (SFB 206), the Bundesminister für Forschung und Technologie, and Fonds der Chemischen Industrie.

5. References

1. Ebel, J. and Scheel, D. (1992) Elicitor recognition and signal transduction, in T. Boller and F. Meins, Jr. (eds.), Plant Gene Research, Vol. 8, Springer, Wien, in press.
2. Ebel, J. and Grisebach, H. (1988) Defense strategies of soybean against the fungus *Phytophthora megasperma* f.sp. *glycinea*: a molecular analysis, Trends Biochem. Sci. 13, 23-27.
3. Habereder, H., Schröder, G. and Ebel, J. (1989) Rapid induction of phenylalanine ammonia-lyase and chalcone synthase mRNAs during fungus infection of soybean (*Glycine max* L.) roots or elicitor treatment of soybean cell cultures at the onset of phytoalexin synthesis, Planta 177, 58-65.

4. Darvill, A.G. and Albersheim, P. (1984) Phytoalexins and their elicitors - A defense against microbial infection in plants, Ann. Rev. Plant Physiol. 35, 243-275.
5. Waldmüller, T., Cosio, E.G., Grisebach, H. and Ebel, J. (1992) Release of highly elicitor-active glucans by germinating zoospores of *Phytophthora megasperma* f.sp. *glycinea*, Planta, in press.
6. Ayers, A.R., Ebel, J., Valent, B. and Albersheim, P. (1976) Host-pathogen interactions X. Fractionation and biological activity of an elicitor isolated from mycelial walls of *Phytophthora megasperma* var. *sojae*, Plant Physiol. 57, 760-765.
7. Ayers, A.R., Valent, B., Ebel, J. and Albersheim, P. (1976) Host-pathogen interactions XI. Composition and structure of wall-released elicitor fractions, Plant Physiol. 57, 766-774.
8. Schmidt, W.E. and Ebel, J. (1987) Specific binding of a fungal glucan phytoalexin elicitor to membrane fractions from soybean *Glycine max*, Proc. Natl. Acad. Sci. USA 84, 4117-4121.
9. Ayers, A.R., Ebel, J., Finelli, F., Berger, N. and Albersheim, P. (1976) Host-pathogen interactions IX. Quantitative assays of elicitor activity and characterization of the elicitor present in the extracellular medium of cultures of *Phytophthora magasperma* var. *sojae*. Plant Physiol. 57, 751-759.
10. Cosio, E.G., Frey, T., Verduyn, R., van Boom, J. and Ebel, J. (1990) High-affinity binding of a synthetic heptaglucoside and fungal glucan phytoalexin elicitors to soybean membranes, FEBS Letters 271, 223-226.
11. Hahn, M.G., Darvill, A.G. and Albersheim, P. (1981) Host-pathogen interactions XIX. The endogenous elicitor, a fragment of a plant cell wall polysaccharide that elicits phytoalexin accumulation in soybeans, Plant Physiol. 68, 1161-1169.
12. Gooday, G.W. and Gow, N.A.R. (1990) Enzymology of tip growth in fungi, in I.B. Heath (ed.), Tip Growth in Plant and Fungal Cells, Academic Press, San Diego, pp. 31-58.
13. Cosio, E.G., Pöpperl, H., Schmidt, W.E. and Ebel, J. (1988) High-affinity binding of fungal ß-glucan fragments to soybean (*Glycine max* L.) microsomal fractions and protoplasts, Eur. J. Biochem. 175, 309-315.
14. Cheong, J.-J. and Hahn, M.G. (1991) A specific, high-affinity binding site for the hepta-ß-glucoside elicitor exists in soybean membranes, The Plant Cell 3, 137-147.
15. Cosio, E.G., Frey, T. and Ebel, J. (1992) Identification of a high-affinity binding protein for a hepta-ß-glucoside phytoalexin elicitor in soybean, Eur. J. Biochem. 204, 1115-1123.
16. Frey, T., Cosio, E.G. and Ebel, J. (1992) Affinity purification and characterization of a binding protein for a hepta-ß-glucoside phytoalexin elicitor in soybean, Phytochemistry, submitted for publication.
17. Scheel, D., Colling, C., Hedrich, R., Kawalleck, P., Parker, J.E., Sacks, W.R., Somssich, I.E. and Hahlbrock, K. (1991) Signals in plant defense gene activation, in H. Hennecke and D.P.S. Verma (eds.), Advances in molecular genetics of plant-microbe interactions, Vol. 1, Kluwer Academic Publishers, Dordrecht, pp. 373-380.
18. Schroeder, J.I. and Thuleau, P. (1991) Ca^{2+} channels in higher plant cells, The Plant Cell 3, 555-559.
19. Grab, D., Feger, M. and Ebel, J. (1989) An endogenous factor from soybean (*Glycine max* L.) cell cultures activates phosphorylation of a protein which is dephosphorylated in vivo in elicitor-challenged cells, Planta 179, 340-348.

20. Dietrich, A., Mayer, J.E. and Hahlbrock, K. (1990) Fungal elicitor triggers rapid, transient, and specific protein phosphorylation in parsley cell suspension cultures, J. Biol. Chem. 265, 6360-6368.
21. Felix, G., Grosskopf, D.G., Regenass, M. and Boller, T. (1991) Rapid changes of protein phosphorylation are involved in transduction of the elicitor signal in plant cells, Proc. Natl. Acad. Sci. USA 88, 8831-8834.
22. Grosskopf, D.G., Felix, G. and Boller, T. (1990) K-252a inhibits the response of tomato cells to fungal elicitors in vivo and their microsomal protein kinase in vitro, FEBS Letters 275, 177-180.

ELICITOR RECOGNITION AND INTRACELLULAR SIGNAL TRANSDUCTION IN PLANT DEFENSE

WENDY R. SACKS, PATRIK FERREIRA, KLAUS HAHLBROCK, THORSTEN JABS, THORSTEN NÜRNBERGER, ANNETTE RENELT AND DIERK SCHEEL.
Max-Planck-Institut fuer Zuechtungsforschung
Department of Biochemistry
Carl-von-Linne-Weg 10
D-5000 Koeln 30
Germany

ABSTRACT. Treatment of cultured parsley cells or protoplasts with a purified extracellular glycoprotein from *Phytophthora megasperma* f.sp *glycinea* induces the transcription of the same set of defense-related genes as are activated in parsley leaves upon infection. Elicitor activity was shown to reside in a specific portion of the protein moiety. Partial amino acid sequences were used to generate two oligonucleotides which served as primers for PCR using fungal DNA as template. Several cDNA clones were isolated by screening a fungal cDNA library with the PCR product. Neither the sequenced tryptic peptides nor the protein predicted from one nearly full-length, sequenced cDNA showed significant homology to sequences in data bases. Southern blot experiments indicated that the elicitor is encoded by a small gene family in *P. megasperma* f.sp. *glycinea*. Binding of the glycoprotein elicitor to target sites on the parsley plasma membrane appears to be the initial event in defense gene activation. The subsequent intracellular transduction of the elicitor signal was shown to involve rapid and transient influxes of Ca^{2+} and H^{+} as well as effluxes of K^{+} and Cl^{-}. Blocking of all ion fluxes together or Ca^{2+} influx alone by ion channel blockers also inhibited phytoalexin synthesis, indicating the participation of ion channels in elicitor signalling. Ca^{2+} and K^{+} ionophores alone or in combination did not induce phytoalexin accumulation. The polyene antibiotic, amphotericin B, however, which triggered similar ion fluxes in cultured parsley cells as did the elicitor from *P. megasperma* f.sp. *glycinea*, stimulated the production of phytoalexins and activated the complete set of defense-related genes in the absence of elicitor, demonstrating that all four ion fluxes are necessary and sufficient. Increased cytosolic Ca^{2+} levels were shown to result from elicitor-stimulated Ca^{2+} influx. The involvement of calmodulin was suggested by the inhibitory effect of calmodulin antagonists on phytoalexin accumulation as well as by changes in the pattern of calmodulin-binding proteins in elicitor-treated parsley cells. Additional

E. W. Nester and D. P. S. Verma (eds.),
Advances in Molecular Genetics of Plant-Microbe Interactions, 485–495.

components of this signal transduction chain appear to be protein kinases and/or protein phosphatases and inositol phosphates, whereas cAMP and GTP-binding proteins were not found to be involved.

1. Introduction

Most interactions between plants and potential pathogens are incompatible, since the plant is not a host for the pathogen. Relatively few true host/pathogen combinations have evolved, which are either compatible or incompatible depending on the genotypes of the interacting plant and pathogen. In defining the different types of interactions, Peter Day (1974) wrote: "We may conveniently distinguish here between two kinds of resistance. One is *nonhost resistance*, such as that shown by wheat to the potato late blight organism *Phytophthora infestans* or by potato to the stem rust organism *Puccinia graminis tritici*. The other, *host resistance*, is the result of genetic modifications of the host which render it resistant to pathogens that would otherwise grow on it. Although nonhost resistance may be more complicated than host resistance, both could in fact be due to the same mechanisms. However, a simple test of this idea will have to wait on our developing the skills needed to cross wheat and potatoes." Although we are still not able to carry out this simple test, increasing evidence suggests that the molecular basis of both types of resistance are identical.

A number of race/cultivar-specific protein elicitors, many of which have been shown to be encoded by avirulence genes, as well as nonspecific elicitors, such as proteins, glycoproteins, carbohydrates or fatty acid derivatives, released from common components of pathogen cell walls, have been isolated from plant-pathogenic fungi (Scheel and Parker, 1990; Ebel and Scheel, 1992). Upon treatment of the appropriate plant system with such compounds, a typical defense response is elicited, which closely mimics at least part of the multicomponent reaction triggered in the intact plant by infection with the corresponding pathogen (Ebel and Scheel, 1992). Such systems have been used to investigate mode of elicitor action and signal transduction. Preliminary results indicate the presence of elicitor binding sites on the plant plasma membrane that initiate specific signalling processes upon elicitor binding, leading ultimately to the activation of defense responses (Scheel and Parker, 1990; Ebel and Scheel, 1992).

We have purified a glycoprotein elicitor from culture filtrate of the fungal soybean pathogen, *Phytophthora megasperma* f.sp. *glycinea*, which in cultured parsley cells and protoplasts activates most of the defense responses that are also part of the nonhost resistance reaction of parsley leaves to infection with spores of this fungus (Parker *et al.*, 1991). This elicitor appears to bind to specific binding sites on parsley plasma membranes (Renelt *et al.*, 1992), thereby initiating signal transduction chains involving ion fluxes through the plasma membrane (Scheel *et al.*, 1991) and phosphorylation/dephosphorylation of proteins (Dietrich *et al.*, 1990). We have now extended our studies on this system by cloning cDNAs encoding the fungal elicitor and using permeabilized parsley protoplasts to detect additional components of signal transduction.

2. Elicitor characterization

A 42-kDa glycoprotein was one of several elicitors of phytoalexin formation in parsley present in the culture filtrate of *Phytophthora megasperma* f.sp. *glycinea* (Parker *et al.*, 1991). Although the elicitor activity was found to reside in the protein portion of the purified molecule, it was also found to be heat-resistant. This suggests that the activity is a property of amino acid sequence rather than native structure. We therefore treated the pure glycoprotein with different proteases, estimated the size of the resulting fragments by SDS-PAGE and determined their elicitor activity by measuring phytoalexin accumulation in the media of parsley protoplasts 24 h after treatment with these fragments (Table 1). All but one of the proteases tested either digested the protein into fragments and destroyed its elicitor activity or left both intact. Endoprotease Glu-C, however, when applied using conditions allowing cleavage only C-terminal of glutamate, cut the

TABLE 1. Sensitivity of the glycoprotein elicitor from *Phytophthora megasperma* f.sp. *glycinea* to treatment with different proteases.

Treatment	Cleavage[1]	Elicitor activity[2]
None	-	100
Carboxypeptidase A	-	92
α-Chymotrypsin	+	1
Elastase	+	6
Endoprotease Glu-C	+	71
Pepsin	+	4
Pronase E	+	0
Thrombin	-	112
Trypsin	+	3

[1]The degree of protein digestion was determined by SDS-PAGE, '+' indicating digestion and '-' no cleavage.

[2]Elicitor activity was determined by measuring the accumulation of phytoalexins in the media of parsley protoplasts 24 h after addition of autoclaved, protease-treated elicitor. The activity is given as % of the amount of phytoalexins produced by protoplasts in response to treatment with equivalent amounts of intact elicitor.

protein into fragments of different sizes, most of which passed through a 30-kDa cut-off filter. This low molecular weight fraction retained high levels of elicitor activity. The size of these elcitor-active fragments was estimated by gel filtration to be approximately 3 and 7 kDa, respectively. We now intend to purify and sequence the smallest elicitor fragment retaining activity and use this fragment for binding studies.

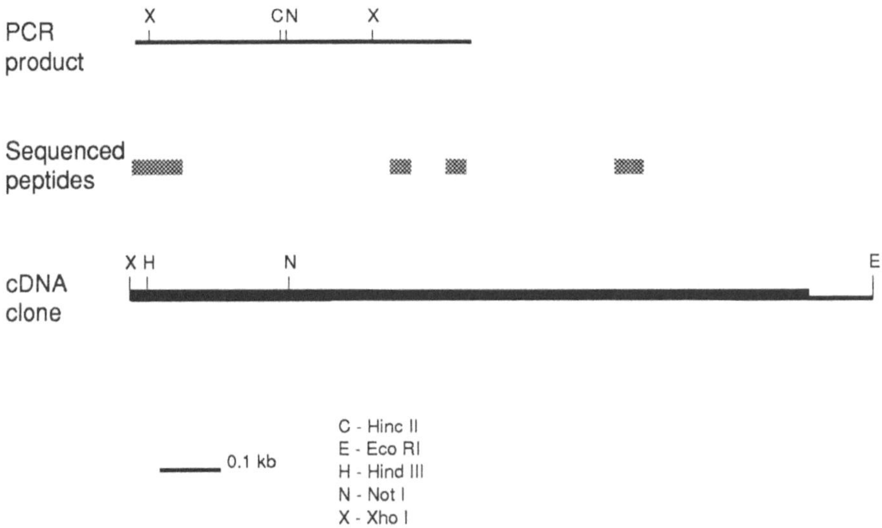

FIGURE 1. Schematic representation of sequenced regions of the elicitor glycoprotein from *Phytophthora megasperma* f.sp. *glycinea* indicating their location with respect to the PCR product and the largest cDNA clone.

The *N*-terminus as well as internal tryptic peptides of the glycoprotein elicitor were sequenced and used to design and synthesize degenerate oligonucleotides, two of which were employed as primers in PCR with fungal DNA as template. The PCR product generated by this approach was 571 bp in size and included sequences corresponding to the *N*-terminus of the mature protein and two internal tryptic peptides (Fig. 1). As a probe in Northern experiments with fungal RNA, it detected a message of approximately 2 kb in size. Several cDNA clones were isolated by screening a fungal cDNA library with the PCR product. These clones represented three classes of cDNAs, none of which were completely identical to the PCR product. At the nucleotide level, the homologies between the individual cDNAs and the PCR product ranged between 80 and 90%, suggesting the presence of at least three to four homologous isoforms of the elicitor glycoprotein in *Phytophthora megasperma* f.sp. *glycinea*. Southern blot experiments were also indicative of a small gene family.

The largest cDNA clone, which began with sequences corresponding to the *N*-terminus of the mature protein and extended through the poly-A tail, lacked a signal peptide and a 5' untranslated region (Fig. 1). The polypeptide predicted from this sequenced cDNA, which comprised the entire mature protein, did not show any significant homology to sequences in data bases. Since we were also unable to detect endopoly-galacturonase activity with the pure native glycoprotein, its function is still unclear. The *N*-terminal half is predominantly hydrophilic while the C-terminal half is hydrophobic and the protein contains one *N*-glycosylation site, eight cystein residues and a cluster of four tyrosines close to the C-terminus. This latter observation might be interesting with respect to the complete loss of elicitor activity observed upon direct iodination of the protein (Renelt *et al.*, 1992). The cDNA shown in Figure 1 has now been cloned into an appropriate expression system for elicitor production and structure/activity studies.

3. Signal transduction

Treatment of cultured parsley cells or freshly isolated protoplasts with crude elicitor preparations from *Phytophthora megasperma* f.sp. *glycinea* or with the pure glycoprotein elicitor results in transient induction of transcription of the same set of defense-related genes that are also activated in parsley leaves upon inoculation with spores

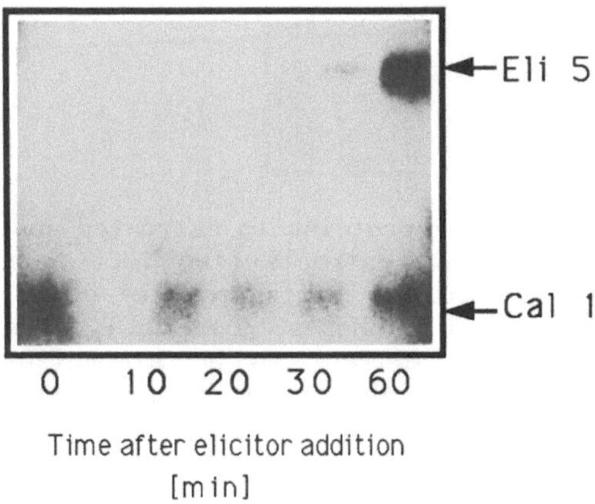

FIGURE 2. Calmodulin mRNA in elicitor-treated parsley cells. Poly-A$^+$ RNA (2 μg/lane) isolated from elicitor-treated parsley cells at the times indicated was separated on agarose-formaldehyde gels, transferred to nylon filters and hybridized to random primed, labelled calmodulin (Braam and Davis, 1990) and ELI 5 cDNAs (Somssich *et al*, 1988).

from the fungus (Dangl *et al.*, 1987; Somssich *et al.*, 1988; Schmelzer *et al.*, 1989; Parker *et al.*, 1991). The first events, observed within a few minutes upon addition of elicitor, are transient increases in ion fluxes through the plant plasma membrane, probably caused by transient changes in the activity of ion channels (Scheel *et al.*, 1991). Influxes of Ca^{2+} and H^+ and effluxes of K^+ and Cl^- were found to be both necessary and sufficient for elicitor-induced activation of defense-related genes (Scheel *et al.*, 1991). From a number of Ca^{2+} channel inhibitors tested, only those reported to be specific inhibitors of slow T-type channels, such as cinnarizine and flunarizine, inhibited both elicitor-stimulated Ca^{2+} fluxes ($IC_{50} \sim 50\ \mu M$) and phytoalexin synthesis ($IC_{50} \sim 20 \mu M$), indicating that specific types of Ca^{2+} channels are regulated by this elicitor. The Ca^{2+} influxes result in increases of cytosolic Ca^{2+} levels from approximately 150 nM to 350 nM within 30 min upon addition of elicitor.

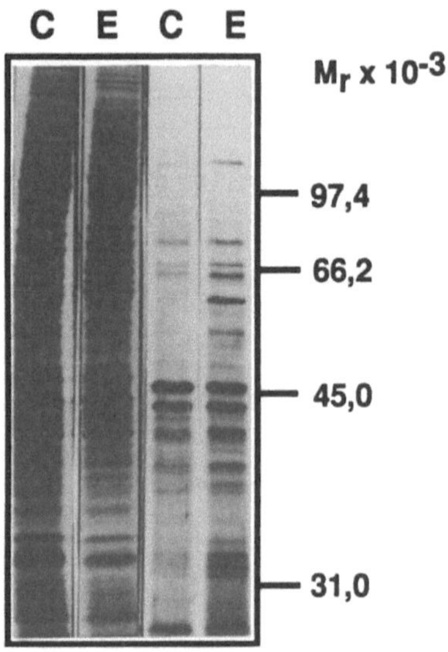

FIGURE 3. Calmodulin-binding proteins in untreated and elicitor-treated parsley cells. Crude protein extracts (two left lanes) were prepared from untreated (C) and elicitor-treated parsley cells (E) 30 min after addition of elicitor in 20 mM Hepes, 2 mM $CaCl_2$, 1 mM $MgCl_2$, pH 7.4. After application of this extract to a calmodulin agarose (Sigma, Deisenhofen, Germany) column followed by extensive washing, calmodulin-binding proteins were eluted with 20 mM Hepes, 0.5 M NaCl, 10 mM EGTA, pH 7.4, and separated on 13% SDS gels (two right lanes).

A 45 min treatment of parsley protoplasts with calmodulin antagonists W-7 and trifluoperazin at 100 - 500 μM prior to elicitor addition resulted in 50 - 100% inhibition of phytoalexin accumulation with a 20 - 50% reduction in viability, respectively, suggesting an involvement of calmodulin in signal transmission. In contrast to other plant signal transduction chains involving calmodulin (Braam and Davis, 1990; Braam, 1992), no increases in calmodulin mRNA levels were observed in cultured parsley cells within 8 h after elicitor addition. In Figure 2, Northern analysis is shown for the first 60 min following treatment, in comparison to the elicitor-responsive gene, *eli* 5 (Somssich *et al.*, 1989).

Significant differences were observed between the patterns of calmodulin-binding proteins extracted from elicitor-treated and untreated parsley cells (Fig. 3.). Although the structure of these proteins and their possible function within this specific signal transduction chain are unknown and must be elucidated, it is possible, for example, that they represent protein kinases or protein phosphatases involved in transmitting the elicitor signal downstream of calmodulin. The requirement for the presence of Ca^{2+} in the culture medium of parsley cells for elicitor-stimulated *in vivo* phosphorylation (Dietrich *et al.*, 1990) may be an indication of such a possibility. However, the protein kinase inhibitors, staurosporine and K 252 a, which efficiently block *in vivo* and *in vitro* phosphorylation as well as elicitor responsiveness in cultured tomato cells (Großkopf *et al.*, 1990; Felix *et al.*, 1991), had no effect on the elicitor-induced phytoalexin accumulation in parsley, which, on the other hand, was completely abolished by the phosphatase inhibitor, okadaic acid, suggesting that protein dephosphorylation rather than phosphorylation might be crucial in this signalling process (Renelt *et al.*, 1992).

Although the pattern of *in vivo* phosphorylated proteins of parsley cells transiently changed upon elicitor treatment (Dietrich *et al.*, 1990), we were unable to detect any differences in *in vitro* protein phosphorylation patterns between extracts from untreated or elicitor-treated cells (Renelt *et al.*, 1992). However, when protein extracts were prepared from untreated cells and crude cell wall elicitor from *Phytophthora megasperma* f.sp. *glycinea* was added together with [γ-^{32}P]ATP, one protein of approximately 55 kDa was transiently phosphorylated (Fig. 4). This phosphorylation was not stimulated by the purified glycoprotein elicitor from the same fungus, which is an efficient elicitor of phytoalexin formation in parsley, but by pronase-treated crude cell wall elicitor, which does not stimulate phytoalexin production. Therefore, the two responses are triggered by different components of the crude cell wall elicitor and this protein phosphorylation is not involved in signal transduction leading to activation of defense-related genes and phytoalexin accumulation. Since the crude cell wall elicitor is known to contain additional elicitors of other types of defense responses, this very rapidly and transiently occurring phosphorylation might belong to another signal transduction pathway of an early reaction, such as an oxidative burst.

Indirect evidence suggests the involvement of inositol phosphate metabolism in elicitor signalling in parsley. Neomycin, which interferes with phospholipase C activity, strongly inhibits elicitor-stimulated phytoalexin accumulation at levels where it does not severely reduce cell viability (Renelt *et al.*, 1992). In order to investigate whether inositol phosphate metabolites affect this induction process or initiate it in the absence of elicitor, we developed a protoplast permeabilization system based on electroporation (Renelt *et al.*, 1992). This technique allowed us to load parsley protoplasts with low molecular-weight effector molecules without significantly reducing cell viability and elicitor responsiveness. While most

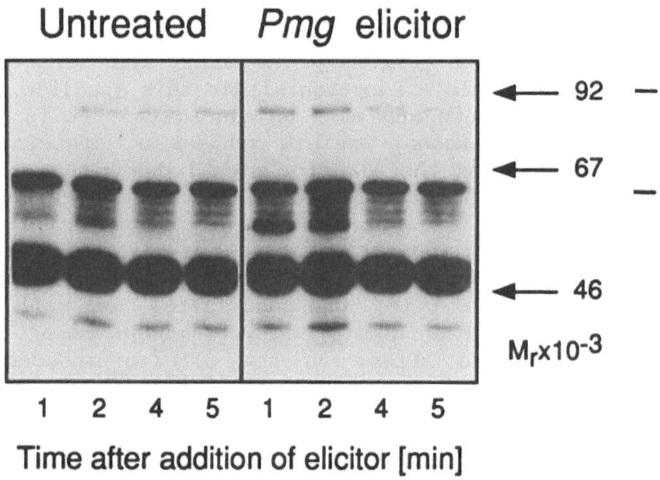

FIGURE 4. Elicitor-stimulated *in vitro* phosphorylation of parsley proteins. Proteins were extracted from parsley cells with 40 mM Hepes/Tris, pH 7.4, containing 10% glycerol, 10 mM $MgCl_2$, 2 mM $MnSO_4$, 0.5 mM phenylmethylsulfonyl fluoride and 15 mM 2-mercaptoethanol. After precipitation of cell debris, aliquots of the supernatant were combined with equal amounts of 40 mM Hepes/Tris, pH 7.4, containing 4 mM $CaCl_2$, 20 mM NaF and 10 mM 4-nitrophenyl phosphate. Phosphorylation was initiated by addition of crude cell wall elicitor from *Phytophthora megasperma* f.sp. *glycinea* or water and [γ-^{32}P]ATP. After incubation at 30°C for the periods indicated, the reactions were terminated by boiling the samples in SDS-PAGE sample buffer and the proteins were separated on 10% SDS gels.

inositol derivatives did not affect phytoalexin accumulation, inositol 1,4,5-trisphosphate and its metabolically stable analogue, inositol 1,4,5-trisphosphorothioate, stimulated the elicitor response by a factor of more than two when applied at concentrations of 500 µM and 40 nM, respectively (Renelt *et al.*, 1992). Although these results suggest that inositol phosphates are involved in this specific signalling chain, these effects might also be caused by release of Ca^{2+} from internal stores, which may then sensitize the cells by increasing cytosolic Ca^{2+} levels prior to elicitor treatment.

The protoplast permeabilization technique was used further to determine whether heterotrimeric GTP-binding proteins play a role in this signal transduction chain. None of the effector molecules tested (GTP-γ-S, GDP-β-S, guanylyl-imidodiphosphate, $[AlF_4]^-$) caused a stimu-

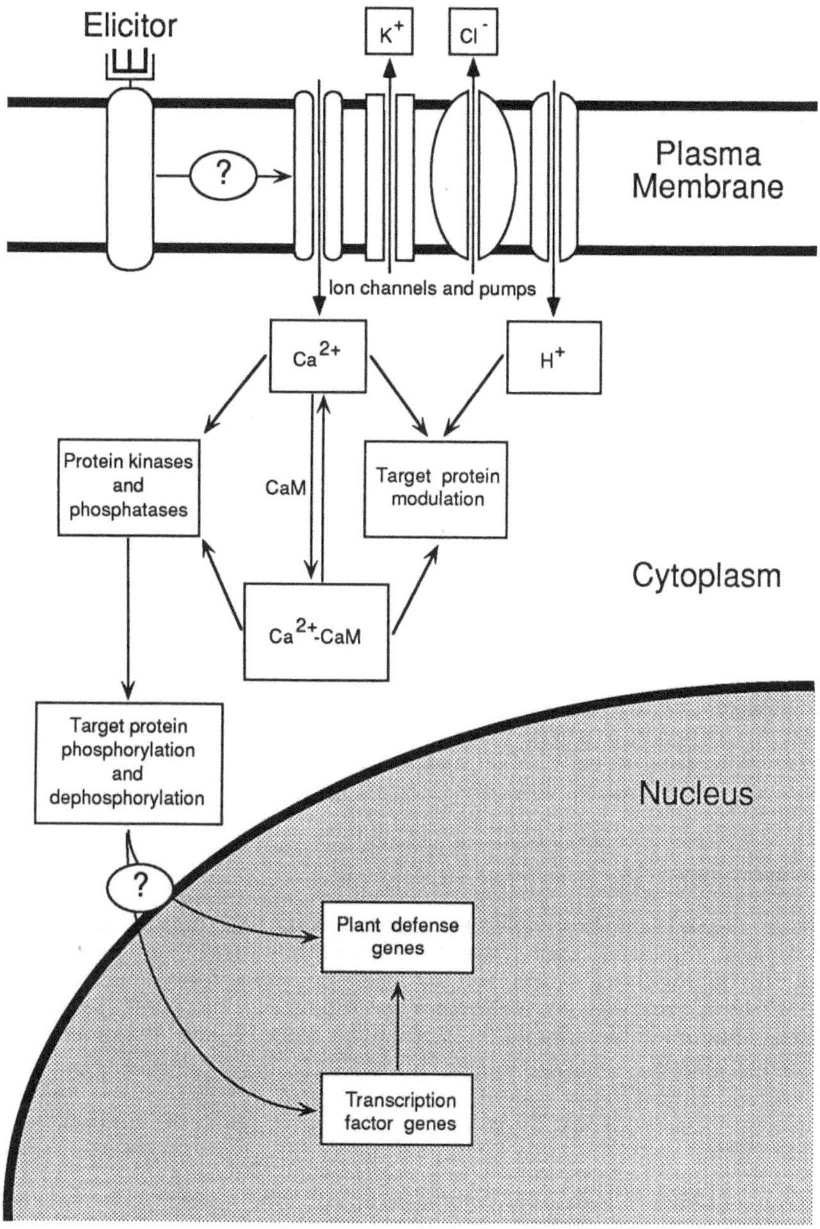

FIGURE 5. Hypothetical model of elicitor signal transduction in parsley.

lation of defense reactions or affected the elicitor response (Renelt *et al.*, 1992). Since the G-protein activator, mastoparan, was also ineffective and the pattern of G-proteins cross-linked *in vivo* to azido-[α-^{32}P]GTP was not changed by elicitor treatment, we concluded that these otherwise important components of transmembrane signalling are not involved in this particular signalling process.

Cyclic AMP appears to be present in parsley cells at low levels, but its concentration was not significantly changed upon elicitor treatment. In addition, neither dibutyryl nor caged cAMP loaded into parsley protoplasts stimulated phytoalexin accumulation in the absence of elicitor or affected the elicitor response (Renelt *et al.*, 1992). These results indicate that cAMP is not involved in elicitor signal transduction in parsley.

4. Conclusions

Although it is not yet possible to link individual signalling components into a complete network, a hypothetical scheme is presented in Figure 5, which attempts to combine and summarize the individual elements. Preliminary results from binding studies suggest that parsley plasma membranes contain a specific binding site for a glycoprotein elicitor from *Phytophthora megasperma* f.sp. *glycinea*, which by unknown mechanisms regulates ion channels of the plasma membrane. Changes in the activity of these channels lead to decreases in cytosolic levels of K^+ and Cl^- and to increases in the amounts of cytosolic Ca^{2+} and H^+. These changes could, possibly in combination with inositol 1,4,5-trisphosphate, trigger protein phosphorylation or dephosphorylation. These proteins may then directly or indirectly be involved in regulation of the expression of plant defense-related genes.

5. Acknowledgements

We thank Magdalena Jung for excellent technical assistance and Drs. Janet Braam, Houston, and Imre E. Somssich, Köln, for providing cDNA clones. This work was supported by the Deutsche Forschungsgemeinschaft (Sche 235/3-1) and by doctoral and postdoctoral fellowships from Bayer Chemical Compony, Leverkusen, to Annette Renelt and Thorsten Nürnberger, respectively.

6. References

Braam, J. (1992) 'Regulated expression of the calmodulin-related *TCH* genes in cultured *Arabidopsis* cells: Induction by calcium and heat shock', Proc. Natl. Acad. Sci. USA 89, 3213-3216.

Braam, J. and Davis, R.W. (1990) 'Rain-, wind-, and touch-induced expression of calmodulin and calmodulin-related genes in *Arabidopsis*', Cell 60, 357-364.

Dangl, J.L., Hauffe, K.D., Lipphardt, S., Hahlbrock, K. and Scheel, D.

(1987) 'Parsley protoplasts retain differential responsiveness to u.v. light and fungal elicitor', EMBO J. 6, 2551-2556.
Day, P.R. (1974) Genetics of Host-Parasite Interactions, W.H. Freeman and Company, San Francisco.
Dietrich, A., Mayer, J.E. and Hahlbrock, K. (1990) 'Fungal elicitor triggers rapid, transient, and specific protein phosphorylation in parsley cell suspension cultures', J. Biol. Chem. 265, 6360-6368.
Ebel, J. and Scheel, D. (1992) 'Elicitor recognition and signal transduction', in T. Boller and F. Meins (eds.), Plant Gene Research. Genes Involved in Plant Defense, Vol. 8, Springer-Verlag, Wien, in press.
Felix, G., Großkopf, D.G., Regenaß, M. and Boller, T. (1991) 'Rapid changes of protein phosphorylation are involved in transduction of the elicitor signal in plant cells', Proc. Natl. Acad. Sci. USA 88, 8831-8834.
Großkopf, D.G., Felix, G. and Boller, T. (1990) 'K-252a inhibits the response of tomato cells to fungal elicitors in vivo and their microsomal protein kinase in vitro', FEBS Lett. 275, 177-180.
Parker, J.E., Schulte, W., Hahlbrock, K. and Scheel, D. (1991) 'An extracellular glycoprotein from *Phytophthora megasperma* f.sp. *glycinea* elicits phytoalexin synthesis in cultured parsley cells and protoplasts', Mol. Plant-Microbe Interact. 4, 19-27.
Renelt, A., Colling, C., Hahlbrock, K., Nürnberger, T., Parker, J.E., Sacks, W.R. and Scheel, D. (1992) 'Studies on elicitor recognition and signal transduction in plant defence', J. Exp. Bot. in press.
Scheel, D. and Parker, J.E. (1990) 'Elicitor recognition and signal transduction in plant defense gene activation', Z. Naturforsch. 45c, 569-575.
Scheel, D., Colling, C., Hedrich, R., Kawalleck, P., Parker, J.E., Sacks, W.R., Somssich, I.E. and Hahlbrock, K. (1991) 'Signals in plant defense gene activation', in H. Hennecke and D.P.S. Verma (eds.), Advances in Molecular Genetics of Plant-Microbe Interactions, Vol. 1, Kluwer Academic Publishers, Dordrecht, pp. 373-380.
Schmelzer, E., Krüger-Lebus, S. and Hahlbrock, K. (1989) 'Temporal and spatial patterns of gene expression around sites of attempted fungal infection in parsley leaves', Plant Cell 1, 993-1001.
Somssich, I.E., Schmelzer, E., Kawalleck, P. and Hahlbrock, K. (1988) 'Gene structure and in situ transcript localization of pathogenesis-related protein 1 in parsley', Mol. Gen. Genet. 213, 93-98.

TRANSCRIPTIONAL REGULATION OF PHYTOALEXIN BIOSYNTHETIC GENES.

Richard A. Dixon[1], Madan K. Bhattacharyya[1], Maria J. Harrison[1], Ouriel Faktor[2], Christopher J. Lamb[2], Gary J. Loake[1,2], Weiting Ni[1], Abraham Oommen[1], Nancy Paiva[1], Bruce Stermer[1] and Lloyd M. Yu[1,2].

[1]Samuel Roberts Noble Foundation, Plant Biology Division, P. O. Box 2180, Ardmore, Oklahoma 73402.
[2]Salk Institute for Biological Studies, Plant Biology Laboratory, 10010 N. Torrey Pines Road, La Jolla, California 92037.

Abstract

In legumes, isoflavonoid derivatives function as antimicrobial phytoalexins, whereas phytoalexins of solanaceous species are of terpenoid origin. The phenylpropanoid and isoprenoid pathways leading to these phytoalexins are involved in the synthesis of a wide range of secondary metabolites with important functions in plant growth, development and responses to the environment. Elicitation of phytoalexin biosynthesis involves transcriptional activation of the genes encoding enzymes of **general** phenylpropanoid/terpenoid biosynthesis, and of the genes for the **specific** branch pathways leading to antimicrobial compounds. In order to understand the molecular controls determining the developmental and environmental regulation of the general and specific enzymes of phytoalexin synthesis, we are studying the promoter regions of three elicitor inducible genes, chalcone synthase (*chs*, isoflavonoid pathway, general), isoflavone reductase (*ifr*, isoflavonoid pathway, specific) and 3-hydroxy-3-methylglutaryl CoA reductase (*hmgr*, terpenoid pathway, general). We describe *cis*-elements and *trans*-factors involved in the expression of these genes in relation to their tissue specific expression and response to biotic stress. Two elements, the G-box and H-box, located within 50 bp of the TATA box, are important for regulation of expression of *chs* and probably *hmgr*, but are not present in the alfalfa *ifr* promoter.

E. W. Nester and D. P. S. Verma (eds.),
Advances in Molecular Genetics of Plant-Microbe Interactions, 497–509.

Biosynthesis of Isoflavonoid and Terpenoid Phytoalexins.

Phytoalexins are defined as low molecular weight antimicrobial compounds that are synthesized by and accumulate in plants in response to infection. A large body of indirect evidence (reviewed in VanEtten *et al.*, 1989) suggests that phytoalexins are important factors in disease resistance. Phytoalexins are products of plant secondary metabolism; in legumes, they are derived from phenylalanine via the flavonoid branch of phenylpropanoid metabolism, whereas in solanaceous species they are terpenoid products of the isoprenoid pathway. Thus, infection activates different biosynthetic pathways in different plant species.

The enzymatic steps in the formation of the isoflavonoid-derived pterocarpan phytoalexins of the Leguminosae are now well understood (Fig. 1A, reviewed in Dixon *et al.*, 1992). Less is understood of the biosynthesis of the sesquiterpene phytoalexins of solanaceous species such as potato or tobacco (Fig. 1B). In both cases, phytoalexin biosynthesis can be divided into three groups of reactions: Group 1, the initial, general pathway which links primary metabolism to the formation of all members of the particular secondary metabolite group; Group 2, specific branch pathways of secondary metabolism directed toward the formation of a particular phytoalexin or class of phytoalexins; Group 3, pathways of primary metabolism which provide substrates for the secondary metabolic pathways. These groups are indicated for the reactions leading to isoflavonoid biosynthesis in Fig. 1A. Group 2 may be subdivided if branch pathways diverge sequentially; thus the enzymes in group 2a are specific for both flavonoid and isoflavonoid biosynthesis.

In many cases studied, the activities of all or most of the enzymes in the three groups of reactions increase at the onset of phytoalexin accumulation in response to the inducing stimuli of fungal infection or exposure of cells to abiotic elicitors or to elicitor molecules from fungal pathogens (biotic elicitors). In elicitor-treated alfalfa cell suspension cultures, accumulation of the pterocarpan phytoalexin medicarpin is preceeded by increases in the enzymatic activities and, where demonstrated, transcript levels of PAL, CA4H, 4CL (group 1); CHS and CHI (group 2a); IFS, IOMT, IFOH, IFR and PTS (group 2b); ACC (group 3) (Dixon *et al.*, 1992; B. Shorrosh and R.A. Dixon, unpublished results). During the biosynthesis of terpenoid phytoalexins, increases in activities and transcripts have been demonstrated for the group 1 enzyme HMGR and the group 2 enzymes sesquiterpene cyclase in tobacco (Vogeli and Chappell, 1990) and casbene synthase in castor bean (Lois and West, 1990). How the increases in these various biosynthetic enzymes are induced and coordinated is a major question in molecular plant pathology.

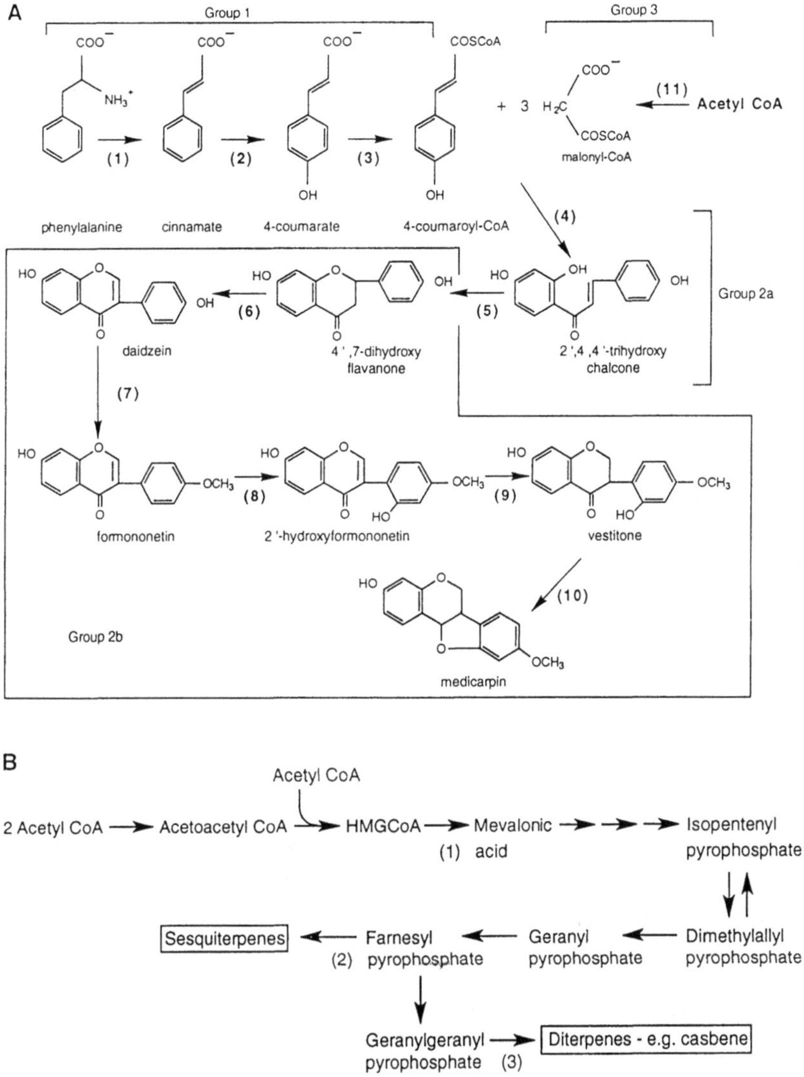

Figure 1. **A.** Biosynthesis of isoflavonoid phytoalexins in a leguminous species (alfalfa). The enzymes are: (1) L-phenylalanine ammonia-lyase (PAL); (2) cinnamic acid 4-hydroxylase (CA4H); (3) 4-coumarate: CoA ligase (4CL); (4) chalcone synthase (CHS); (5) chalcone isomerase (CHI); (6) isoflavone synthase (IFS); (7) isoflavone O-methyltransferase (IOMT); (8) isoflavone 2'-hydroxylase (IFOH); (9) isoflavone reductase (IFR); (10) pterocarpan synthase (PTS); (11) acetyl CoA carboxylase (ACC). **B.** Simplified scheme for the biosynthesis of terpenoid phytoalexins in solanaceous species. The enzymes are: (1) 3-hydroxy 3-methylglutaryl CoA reductase (HMGR); (2) sesquiterpene cyclase, (3) casbene synthase.

Transcriptional Activation of Phytoalexin Biosynthetic Genes.

Elicitor- and infection-induced increases in enzymatic activities generally result from increased transcription of the gene(s) encoding these enzymes. In legumes, nuclear transcript run on experiments have confirmed transcriptional activation underlying increases in the phytoalexin biosynthetic enzymes PAL, CHS and IFR (Lawton and Lamb, 1987; A. Oommen, R.A. Dixon and N.L. Paiva, unpublished results). A similar conclusion was reached for sesquiterpene cyclase based on thiouridine-labelling experiments (Vogeli and Chappell, 1990). Transcriptional activation also underlies the elicitor-induced appearance of other plant defense products including hydrolytic enzymes (Roby *et al.*, 1991), hydroxyproline rich glycoproteins (HRGPs) (Lawton and Lamb, 1987) and enzymes of lignin biosynthesis such caffeic acid O-methyltransferase (COMT) (W. Ni and R.A. Dixon, unpublished results). The transcriptional activation of *pal* and *chs* genes is very rapid, occurring within a few minutes of exposure to elicitors (Lawton and Lamb, 1987). This suggests that the signal transduction pathway from elicitor to defense response gene promoters is relatively direct and probably in place in the unelicited state. The rates of transcription of some defense response genes are closely coordinated. Thus, transcription rates of *pal* and *comt* in alfalfa nuclei increase and decrease with very similar kinetics (W. Ni and R.A. Dixon, unpublished results). Significant differences do, however, exist in the kinetics of enzyme activity increase and decline for these and other phenylpropanoid pathway enzymes. These changes can reflect differential rates of recruitment of transcripts into polysomes, mRNA stability, and enzyme activation and inactivation. Transcription of *hrgp* genes is delayed compared to that of *pal* and *chs* in elicited and infected bean cells (Lawton and Lamb, 1987), suggesting that different signal transduction pathways may operate for different functional classes of defense response genes.

Functional Analysis of *Cis*-Elements for Developmental and Environmental Regulation of Phytoalexin Biosynthetic Genes.

Enzymes of general secondary product biosynthesis are involved in the formation of a wide variety of functionally different end products, several of which are regulated by internal developmental cues in addition to environmental stimuli such as infection or light. For example, the flavonoid derivatives of the leguminosae function as phytoalexins, UV-protectants, inducers of rhizobial nodulation genes and potential regulators of auxin transport. These diverse functions necessitate expression of flavonoid biosynthetic genes in specific plant tissues at specific times during development. Environmental induction may be superimposed upon these developmental expression patterns. Several laboratories have begun to dissect the elements responsible for the activation of plant defense gene promoters by developmental and environmental signals. Conclusions have been based primarily on analysis of the behavior of promoter-reporter gene constructs using stable transformation in heterologous or homologous plant systems, or by transient expression studies in electroporated protoplasts.

CHS is encoded by a family of approximately eight genes in bean (*Phaseolus vulgaris* L.) (Ryder *et al.* 1987). A 1.4 kb fragment of the bean *chs8* promoter confers strong expression on a ß-glucuronidase (GUS) reporter gene in the root apical meristem and inner epidermal cells of petals of transgenic tobacco, with only weak expression in other floral organs, mature leaves and stems (Schmid *et al.* , 1990). This promoter is, however, activated in the leaves of transgenic tobacco in response to UV-light, mercuric chloride (an abiotic elicitor), fungal elicitor or infection with *Pseudomonas syringae* (Doerner *et al.*, 1990; Stermer *et al.*, 1990). A 490 bp fragment of the bean *chs15* promoter is likewise activated by UV-light and mercuric chloride, but not *Pseudomonas* infection, in stably transformed tobacco (Stermer *et al.*, 1990). The *chs15* promoter is activated by fungal elicitor and glutathione in electroporated bean, soybean and alfalfa protoplasts (Dron *et al.*, 1988; Choudhary *et al..*, 1990; Harrison *et al.*, 1991a), and in stably transformed bean and alfalfa cell suspension cultures (T. Fahrendorf, C.J. Franklin and R.A.Dixon, unpublished results). These observations form the basis of deletional and mutational analysis for identification of *cis*-elements mediating tissue specific and pathogen/elicitor-induced expression.

An increasing body of evidence suggests that intermediates of the phenylpropanoid pathway act as regulators of the flux through that pathway by effects on gene transcription (Mavandad *et al.* ,1990). In this respect, expression of the *chs15* promoter in electroporated alfalfa protoplsts is strongly down regulated by *trans*-cinnamic acid, the first intermediate in the phenylpropanoid pathway, but stimulated up to 5-fold by the next intermediate, 4-coumaric acid (Loake *et al.*, 1991). This raises the question of whether internal pathway regulation utilizes the same *cis*-elements and *trans*-factors as are invovled in developmental and environmental expression.

5'-Deletion analysis implicates elements of the *chs15* promoter between positions -326 and -173 in the quantitative control of expression, in both transgenic tobacco and electroporated soybean or alfalfa protoplasts. This region acts overall as a silencer or enhancer depending upon the source of protoplasts used for elicitation (Dron *et al.*, 1988; Harrison *et al.*, 1991a), and 5'-deletions through this region progressively reduce expression in floral tissue (J. Kooter, O. Faktor, R.A. Dixon and C.J. Lamb, unpublished results). 5'-Deletion to -72 strongly reduces basal expression, and prevents elicitor-inducibility in electroporated soybean protoplasts (Dron *et al.*, 1988). Likewise, sequences to -84 are sufficient to confer expression in the pigmented regions of petals and in root tissue, but these expression patterns are lost on deletion to -74 (J. Kooter, O. Faktor, R.A. Dixon and C.J. Lamb, unpublished results). A chimeric construct consisting of a multimer of the *chs15* promoter sequence from -80 to -42 fused to a minimal 35S promoter exhibits similar patterns of tissue specific expression to the -326 or -84 *chs15* promoters in transgenic tobacco (O. Faktor, R.A. Dixon and C.J. Lamb, unpublished results). Taken together, these observations indicate that the *chs15* promoter is very compact, comprising TATA proximal elements confering environmental, developmental, and internal pathway regulation, and 5'-elements effecting quantitative expression levels.

GRA and *in vitro* DNase I footprinting have defined more precisely putative *cis*-elements in the *chs15* promoter (Fig. 2A). The silencer/enhancer region contains three regions (boxes I, II, III) which are strongly footprinted *in vitro* with bean nuclear extracts (Lawton *et al.*, 1991). Each box contains a one base variant of the GT-1 (Green *et al.*, 1988) core consensus sequence GGTTAA, and appears to be recognized by a single sequence-specific binding factor, SBF-1 (Harrison *et al.*, 1991). The motif CCTACC(N_7)CT(N_4)A occurs three times in the *chs15* promoter, centered on positions -144, -130 (reverse orientation) and -53. This motif, which we have termed the H-box, appears to bind more than one sequence-specific binding factor (see below). Base substitutions in the 3'-most H-box (H-III) in the context of the -174 promoter strongly inhibit floral specific expression and activation of the promoter by tobacco mosaic virus infection in transgenic tobacco, but do not affect wound induction (O. Faktor, R.A. Dixon and C.J. Lamb, unpublished results). Mutation of this H-box, and to a lesser extent H-box I, also abolishes the stimulation of promoter activity by 4-coumaric acid (Loake *et al.*, 1992). Commencing 13 nucleotides 5' of H-box III is a canonical G-box element (core consenses CACGTG) (Giuliano *et al.*, 1988) followed by a GATA motif. A linker scanning mutation of this G-box, in the context of the -174 promoter, strongly reduces floral expression and abolishes responsiveness of the promoter to 4-coumaric acid (O. Faktor, R.A. Dixon and C.J. Lamb, unpublished results; Loake *et al.*, 1992). Plant G-box sequences are recognized by members of a family of basic leucine zipper (bZip) transcription factors (Oeda *et al.*, 1991; Williams *et al.*, 1992; Schindler *et al.*, 1992). The close spatial organization of G-box and H-box in the *chs15* promoter, and the apparent requirement for both elements for floral expression and responsiveness to 4-coumaric acid, suggests the possibility of interactions between G-box- and H-box-specific binding factors. Stimulation of the -174 *chs15* promoter by 4-coumaric acid is prevented by co-electroporation *in trans* of a multimerized H-box element, but not by an element in which the first two C residues of the H-box are replaced by G residues (Loake *et al.*, 1992). This provides indirect evidence for the presence of an H-box binding factor involved in transducing the response to 4-CA.

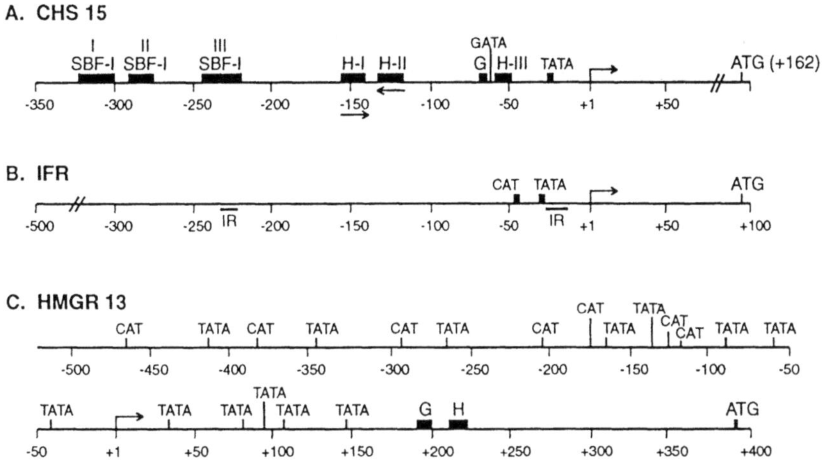

Figure 2. Diagram showing promoter regions of the bean chalcone synthase *chs15* gene (**A**), the alfalfa isoflavone reductase gene (**B**) and the potato HMG CoA reductase *hmgr13* gene (**C**). Numbers indicate base pairs from the transcription start site (+1). Translation initiation codons are marked (ATG). See text for description of SBF-1, H-box (H), G-box (G) and GATA motifs. CAT and TATA indicate occurrences of the canonical CAT and TATA box sequences; in the case of the *hmgr13* promoter, current data suggest that the functional TATA box is at +47. IR = inverted repeat.

We have recently isolated an approximately 2 kb promoter region from the single alfalfa *ifr* gene (A. Oommen, R.A. Dixon and N. Paiva, unpublished results). Although resident *ifr* transcripts exhibit similar tissue specific expression to *chs*, and are likewise inducible by fungal elicitor in suspension cultures (Paiva *et al.* 1991) and by fungal infection in leaves (N. Paiva and R.A. Dixon, unpublished results), the alfalfa *ifr* promoter contains none of the elements (SBF-1, G-box, H-box) identified as functionally important in the bean *chs15* promoter (Fig. 2B). Preliminary results indicate that *ifr* promoter-GUS fusions are expressed in both roots and stems of transgenic tobacco. Stem expression is primarily localized to vascular tissue, and appears most intense in regions where aerial roots are emerging. Work is now in progress to analyze *ifr* promoter expression patterns in transgenic alfalfa.

HMGCoA reductase in potato is encoded by a family of at least seven genes (M.K. Bhattacharyya, N. Paiva, R.A. Dixon and B. Stermer, unpublished results). Analysis of promoter-GUS fusions in transgenic tobacco reveals that the *hmgr13* promoter is activated by arachidonic acid, but not by TMV infection. Arachidonic acid is believed to be the physiologically active elicitor produced by the potato blight fungus *Phytophthora infestans* (Bostock *et al.*, 1981). The promoter establishes a pattern of developmental expression characterized by weak activity in petals, stems, roots, sepals, ovaries and styles, and very strong expression in pollen (M.K. Bhattacharyya, N. Paiva, R.A. Dixon and B. Stermer, unpublished results). A G-box (TACACGTGTC) and an H-box derivative (ACTACC(N_7)CT) occur within 14 nucleotides of one another, but in this promoter they are 3' of the transcription start site (Fig. 2C). 3'-Deletions which remove these two elements greatly reduce expression in all tissues (M.K. Bhattacharyya, N. Paiva, R.A. Dixon and B. Stermer, unpublished results). Further analysis will be necessary to determine whether the G-box and/or H-box, or other 3' sequences, are the important elements for tissue specific and elicitor-induced expression of the *hmgr13* gene.

Sequence-Specific Binding Proteins that Interact with the Bean *chs15* Promoter.

In vitro DNase I footprinting and competition GRA indicate sequence-specific binding of bean and alfalfa nuclear factor(s) to the GT-1-like motifs (boxes I-III) in the upstream silencer region of the bean *chs15* promoter (Lawton *et al.*, 1991; Harrison *et al.*, 1991a). Using an oligonucleotide corresponding to the most 3' of the boxes (box III) as GRA probe, a protein of subunit Mr 95 kDa specifically recognizing this sequence was purified from bean nuclear extracts (Harrison *et al.*, 1991b) and termed SBF-1 (Silencer Box Factor-1). Purified SBF-1 can bind to boxes I, II and III in the intact *chs15* promoter. Its binding, in South Western blots, is competed by oligonucleotides containing the box III sequence, but not by mutant oligonucleotides in which the GG dinucleotide of the GT-1 core sequence is mutated to CC. Although SBF-1 retains sequence specific binding after gel electrophoresis and blotting to Immobilon membranes, we have been unable to isolate SBF-1 cDNA clones by screeening expression library plaque lifts with the box III sequence. This suggests that SBF-1 is either toxic to *E. coli*, or requires some post-translational modification to activate binding. Dephosphorylation of partially purified SBF-1 destroys DNA binding activity (Harrison *et al.*, 1991b) and this can be restored by incubation with the regulatory subunit of bovine heart cAMP-dependent protein kinase (Fig. 3A). By analogy with yeast heat-shock transcription factors (Sorger *et al.*, 1987), the lower mobility of the re-phosphorylated form(s) of SBF-1 suggests that native

SBF-1, as isolated from bean nuclei, is phosphorylated at multiple sites, and that not all these sites are re-phosphorylated *in vitro*. We have been unable to re-activate dephosphorylated SBF-1 with spermine-stimulated (Li and Roux, 1992) or Ca^{++}-dependent (Li *et al.*, 1992) protein kinases from pea nuclei (M.J. Harrison, H. Li, S. Roux and R.A. Dixon, unpublished results) implying some specificity for the kinase(s) acting on SBF-1 *in vivo*. Alkaline phosphatase treatment of preformed SBF-1/DNA complexes results in loss of the retarded complex (as assayed by GRA, Fig. 3B), suggesting that phosphorylation of SBF-1 is required to maintain stable binding. This provides an additional regulatory mechanism by which complexes could be destabilized and SBF-1 removed from the promoter. SBF-1 activity in extracts from control and elicitor-treated bean cell suspension cultures is quantitatively similar and associated with complexes of the same mobility in gel retardation assays, consistent with the apparent lack of involvement of boxes I-III in elicitor responsiveness of the *chs15* promoter. The physiological relevance of potential phosphorylation control remains to be determined.

In order to isolate factors binding to the H-box, we designed an oligonucleotide containing the conserved regions of the consensus CCTACC(N_7)CT(N_4)A but lacking homology in its N_7 and N_4 regions to the sequences found in the three H-boxes in the *chs15* promoter. We avoided introducing known *cis*-element motifs into this "generic" H-box probe. A purification protocol involving ion-exchange and DNA afffinity chromatography resulted in the isolation from bean whole cell extracts of two protein factors, KAP-1 and KAP-2, which specifically recognize the H-box consensus sequence (L. Yu, C.J. Lamb and R.A. Dixon, unpublished results). These factors are distinct as they possess different subunit molecular weights (97 kDa for KAP-1 and 76 and 56 kDa for KAP-2), are differentially sensitive to trypsin when bound to the H-box oligonucleotide and produce retarded complexes of different mobilities in gel shift assays (Figure 3C). The large subunit molecular weight of KAP-1 suggests that it is unlikely to be a classical bZip protein. In contrast to the situation with SBF-1, de-phosphorylation of KAP-1 or KAP-2 does not significantly reduce DNA-binding activity, but increases the gel mobility of the DNA-protein complexes. Phosphorylation/de-phosphorylation could regulate transcriptional activation or interactions with other transcription factors.

KAP-1 and KAP-2 are present in equal amounts in whole cell extracts from elicited (glutathione-treated) and unelicited bean cell cultures. However, analysis of extracts from isolated nuclei reveals a significantly increased concentration of the two factors in response to glutathione treatment. This suggests either that elicitation results in movement of KAP-1 and KAP-2 to the nucleus, in a manner analogous to steroid hormone receptor action, or that a pre-existing nuclear pool of the factors is somehow activated in response to

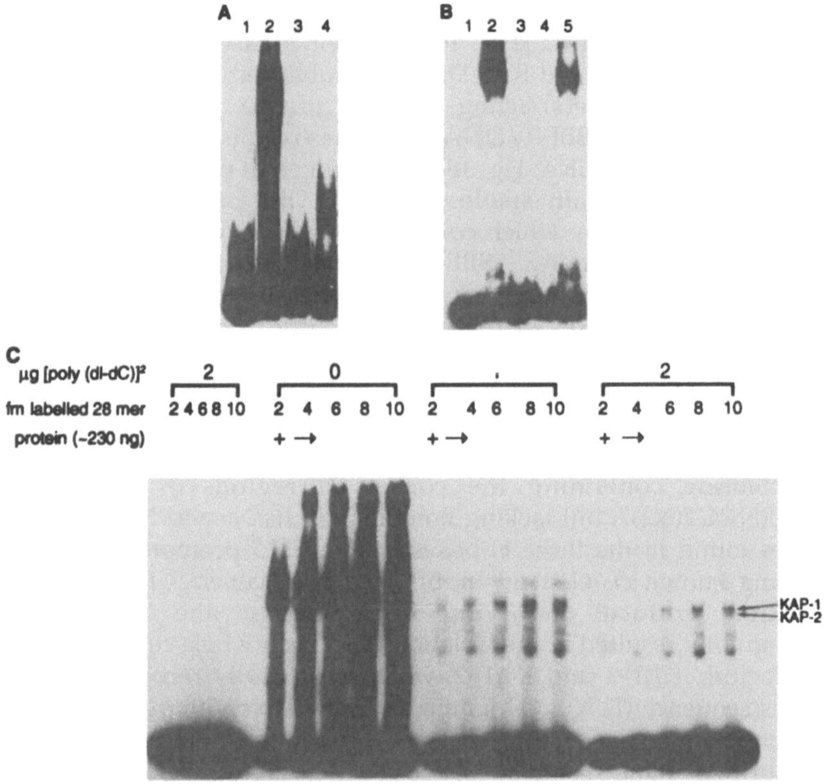

Figure 3. (**A, B**). Gel retardation analysis of SBF-1. (**A**) A *chs15* promoter fragment (-326 to -140) (Lane 1) was incubated with purified SBF-1 (Lane 2), SBF-1 pretreated with immobilized alkaline phosphatase (Lane 3), or de-phosphorylated SBF-1 incubated with bovine cAMP-dependent protein kinase (Lane 4). (**B**) *chs15* Box III tetramer (Lane 1) was incubated with untreated (Lane 2) or de-phosphorylated (Lane 3) bean nuclear extract. Preformed SBF-1/Box III complex was incubated with immobilized alkaline phosphatase (Lane 4) or buffer alone (Lane 5). (**C**) Gel retardation analysis of H-box binding factors. Partially purified H-box binding activity (anion and cation exchange and heparin agarose) was incubated with increasing amounts of labeled H-box ds oligonucleotide and non-specific DNA prior to analysis. Specifically bound KAP-1 and KAP-2 are marked with arrows.

elicitation. Either model is consistent with a role for KAP-1 and KAP-2 in mediating rapid transcriptional activation of the *chs* gene in response to elicitor. Work is now in progress to clone the KAP-1 and KAP-2 genes and to study the factors further in a plant *in vitro* transcription system. In this way we hope to elucidate the events which couple environmentally and developmentally triggered signal transduction pathways to the activation of a key plant defense response gene.

References

Choudhary, A.D., Lamb, C.J. and Dixon, R.A. (1990) 'Stress responses in alfalfa (*Medicago sativa* L.) VI. Differential responsiveness of chalcone synthase induction to fungal elicitor or glutathione in electroporated protoplasts', Plant Physiol. 94, 1802-1807.

Dixon, R.A., Choudhary, A.D., Dalkin, K., Edwards, R., Fahrendorf, T., Gowri, G., Harrison, M.J., Lamb, C.J., Loake, G.J., Maxwell, C.A., Orr, J. and Paiva, N.L. (1992) 'Molecular biology of stress-induced phenylpropanoid biosynthesis in alfalfa', Rec. Adv. Phytochem., in press.

Doerner, P.W., Stermer, B.A., Schmid, J., Dixon, R.A. and Lamb, C.J. (1990) 'Plant defense gene promoter-reporter gene fusions in transgenic plants: tools for identification of novel inducers', Bio/Technology 8, 845-848.

Dron, M., Clouse, S.D., Dixon, R.A., Lawton, M.A. and Lamb, C.J. (1988) 'Glutathione and fungal elicitor regulation of a plant defense gene promoter in electroporated protoplasts', Proc. Natl. Acad. Sci. USA 85, 6738-6742.

Giuliano, G., Pichersky, E., Malik, V.S., Timko, M.P., Scolnik, P.A. and Cashmore, A.R. (1988) 'An evolutionarily conserved protein binding sequence upstream of a plant light-regulated gene', Proc. Natl. Acad. Sci. USA, 85, 7089-7093.

Green, P.J., Yong, M.-M., Cuozzo, M., Kano-Murakami, Y., Silverstein, P. and Chua, N.-H. (1988) 'Binding site requirements for pea nuclear protein factor GT-1 correlate with sequences required for light-dependent transcriptional activation of the *rbcS-3A* gene', EMBO J. 7, 4035-4044.

Harrison, M.J., Choudhary, A.D., Dubery, I., Lamb, C.J. and Dixon, R.A. (1991a) '*Cis*-elements and *trans*-acting factors for the quantitative expression of a bean chalcone synthase gene promoter in electroporated alfalfa protoplasts', Plant Mol. Biol. 16, 877-890.

Harrison, M.J., Lawton, M.A., Lamb, C.J. and Dixon, R.A. (1991b) 'Characterization of a nuclear protein which binds to three elements within the silencer region of a bean chalcone synthase gene promoter', Proc. Natl. Acad. Sci. USA, 88, 2515-2519.

Lawton, M.A. and Lamb, C.J. (1987) 'Transcriptional activation of plant defense genes by fungal elicitor, wounding and infection', Mol. Cell Biol. 7, 335-341.

Lawton, M.A., Dean, S.J., Dron, M., Kooter, J., Kragh, K., Harrison, M.J., Yu, L., Tanguay, L., Dixon, R.A. and Lamb, C.J. (1991) 'Silencer region of a chalcone synthase promoter contains multiple binding sites for a factor, SBF-1, closely related to GT-1', Plant Mol. Biol. 16, 235-249.

Li, H., Dauwalder, M. and Roux, S.J. (1991) 'Partial purification and characterization of a Ca^{2+}-dependent protein kinase from pea nuclei', Plant Physiol. 96, 720-727.

Li, H. and Roux, S.J. (1992) 'Purification and characterization of a caesin kinase II-type protein kinase from pea nuclei. Plant Physiol., in press.

Loake, G., Choudhary, A.D., Harrison, M.J., Mavandad, M., Lamb, C.J. and Dixon, R.A. (1991) 'Phenylpropanoid pathway intermediates regulate transient expression of a chalcone synthase gene promoter in electroporated protoplasts', Plant Cell 3, 829-480.

Loake, G.J., Faktor, O., Lamb, C.J. and Dixon, R.A. (1992) 'Combination of H-box (CCTACC(N_7)CT) and G-box (CACGTG) *cis*-elements are necessary for feedforward stimulation of a chalcone synthase promoter by the phenylpropanoid pathway intermediate *p*-coumaric acid', Proc. Natl. Acad. Sci. USA, in press.

Lois, A.F. and West, C.A. (1990) 'Regulation of expression of the casbene synthetase gene during elicitation of castor bean seedlings with pectic fragments', Arch. Biochem. Biophys. 276, 270-277.

Mavandad, M., Edwards, R., Liang, X., Lamb, C.J. and Dixon, R.A. (1990) 'Effects of *trans*-cinnamic acid on expression of the bean phenylalanine ammonia-lyase gene family', Plant Physiol. 94, 671-680.

Oeda, K., Salinas, J. and Chua, N.-H. (1991) 'A tobacco bZip transcription activator (TAF-1) binds to a G-box like motif conserved in plant genes', EMBO J. 10, 1793-1802.

Paiva, N.L., Edwards, R., Sun, Y., Hrazdina, G. and Dixon, R.A. (1991) 'Stress responses in alfalfa (*Medicago sativa* L.) XI. Molecular cloning and expression of alfalfa isoflavone reductase, a key enzyme of isoflavonoid phytoalexin biosynthesis', Plant Mol. Biol. 17, 653-667.

Roby, D., Broglie, K.,Gaynor, J. and Broglie, R. (1991) 'Regulation of a chitinase gene promoter by ethylene and elicitors in bean protoplasts', Plant Physiol. 97, 433-439.

Ryder, T.B., Hedrick, S.A., Bell, J.N., Liang, X., Clouse, S.D. and Lamb, C.J. (1987) 'Organization and differential activation of a gene family encoding the plant defense enzyme chalcone synthase in *Phaseolus vulgaris*', Mol. Gen. Genet. 210, 219-233.

Schindler, U., Menkens, A.E., Beckmann, H., Ecker, J.E. and Cashmore, A.R. (1992) 'Heterodimerization between light-regulated and ubiquitously expressed *Arabidopsis* GBF bZIP proteins', EMBO J. 11, 1261-1273.

Schmid, J., Doerner, P.W., Clouse, S.D., Dixon, R.A. and Lamb, C.J. (1990) 'Developmental and environmental regulation of a bean chalcone synthase promoter in transgenic tobacco', Plant Cell 2, 619-631.

Sorger, P.K., Lewis, M.J. and Pelham, H.R.B. (1987) 'Heat shock factor is regulated differently in yeast and HeLa cells', Nature 329, 81-84.

Stermer, B.A., Schmid, J., Lamb, C.J. and Dixon, R.A. (1990) 'Infection and stress activation of bean chalcone synthase promoters in transgenic tobacco', Molecular Plant Microbe Interact. 3, 381-388.

VanEtten, H.D., Matthews, D.E. and Matthews, P.S. (1989) 'Phytoalexin detoxification: Importance for pathogenicity and practical implications', Ann. Rev. Phytopathol. 27, 143-164.

Vogeli, U. and Chappell, J. (1990) 'Regulation of a sesquiterpene cyclase in cellulase-treated tobacco cell suspension cultures', Plant Physiol. 94, 1860-1866.

Williams, M.E., Foster, R. and Chua, N.-H. (1992) 'Sequences flanking the hexameric G-box core CACGTG affect the specificity of protein binding', Plant Cell 4, 485-496.

MUTAGENESIS OF A RACE-SPECIFIC RUST RESISTANCE GENE IN *ANTIRRHINUM MAJUS* USING A TRANSPOSON-TAGGING PROTOCOL

H.J. NEWBURY, E.A.B. AITKEN and J.A. CALLOW
School of Biological Sciences
The University of Birmingham
PO Box 363, Birmingham B15 2TT
United Kingdom

ABSTRACT. A transposon-tagging strategy has been employed in order to allow the isolation of a race-specific rust resistance gene from *Antirrhinum*. Six mutant (susceptible) plants have been produced. Evidence that the R gene has been tagged is: i) repeats of the tagging protocol in which transposons are not active have yielded no susceptible mutants, ii) on selfing, the progeny of all six plants include some revertants to resistance, and iii) for the two mutants so far studied, the mutations map at, or very close to, the R gene. Molecular analyses, using cloned *Antirrhinum* sequences, are now under way. This work is being published in a fuller form in *The Plant Journal* vol. 2(5).

Introduction

A major objective in molecular plant pathology is the isolation and characterisation of genes conferring race-specific resistance to plant diseases. Such genes ('major', R genes), commonly utilised by plant breeders, are currently known only for their phenotypic effects. Because of the lack of information concerning the structure of either R genes or their protein products, and the fact that R genes are commonly held to be constitutively expressed, it is usually impossible to apply 'conventional' techniques for gene isolation. Transposon-tagging offers an opportunity to isolate an R gene by first identifying mutants with an altered race-specific resistance phenotype caused by insertional inactivation of the target gene, and then isolating the tagged R gene by using the transposon as a hybridisation probe (Bennetzen, 1984; Ellis *et al.*, 1988; Newbury, 1992).

The rust disease of *Antirrhinum majus* caused by *Puccinia antirrhini* provides an excellent model system for this approach. A series of transposons have now been cloned from *A. majus* and these have been used to generate mutations at a range of genetic loci (e.g. Luo *et al.*, 1991). Temperature has been shown to influence the rate of movement of some *Antirrhinum* transposons (Luo *et al.*, 1991; Harrison and Carpenter, 1973) with Tam3, for example, moving 1000-fold more frequently at 15° than at 25°. We have previously reported the identification of two races of *Antirrhinum* rust and resistance to race α was shown to be inherited in a manner consistent with the existence of a single, dominant R gene (Aitken *et al.*, 1989).

E. W. Nester and D. P. S. Verma (eds.),
Advances in Molecular Genetics of Plant-Microbe Interactions, 511–515.

Here, we report the results of experiments in which plants of a line of *Antirrhinum majus* normally resistant to race α, have mutated to susceptibility following a transposon-tagging protocol. We provide several lines of evidence that the mutations are due to the insertional inactivation of the R gene by a transposon.

Materials and Methods

The high transposition character was incorporated into a number of homozygous resistant (RR) commercial varieties (see Aitken *et al.*, 1992) by crossing these with an experimental high transposition genotype (Line 75, susceptible to both races α and β) donated by the *Antirrhinum* group at the John Innes Institute, Norwich. F1 heterozygotes (Rr) were selfed and RR(Hitrans) individuals selected. RR(Hitrans) plants were held at 15° to increase the likely frequency of transpositional events and were then used in crosses with genotypes homozygous for the recessive form of the R gene (r r). Progeny were then screened for disease reaction to race α (see Aitken *et al.*, 1989). Throughout the study, 'differential' lines were used to distinguish between rust race α and β.

Results

THE TAGGING PROTOCOL

As shown in Figure 1, RR(Hitrans) plants) were held at 15° to encourage transposition, allowed to flower, and pollen used to fertilise r r genotypes. The resulting progeny should all have been resistant (Rr) except where the R gene had been inactivated. Of the 15 susceptible plants recovered from the screening programme, three were stunted (one was later confirmed to be haploid) and the

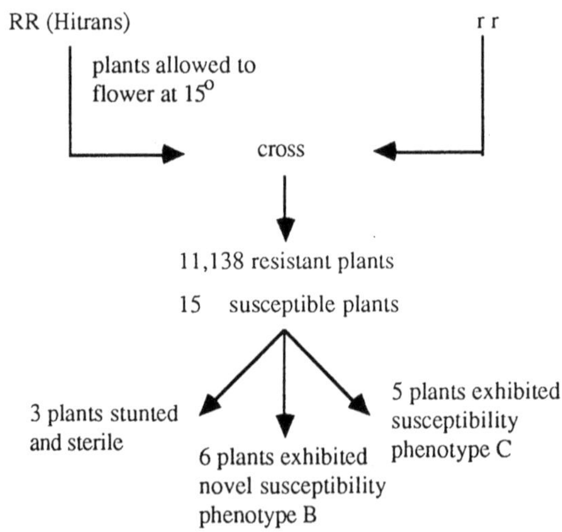

Figure 1

other eleven exhibited symptoms clearly different from the usual susceptible response, which we have designated susceptibility phenotype A. In the novel phenotype B, a larger number of smaller pustules were observed with obvious chlorotic halos. In phenotype C, only a few pustules are observed, often associated with mid-veins or leaf edges (see Aitken *et al.*, 1992).

EVIDENCE THAT THE MUTANTS ARE CAUSED BY TRANSPOSON INACTIVATION OF THE R GENE

1. *Background mutation rate at the R locus.* In order to test whether the mutations to susceptibility could be caused by an instability at the R locus unassociated with transposon movement, two 'control crosses' were made and the progeny (expected to be Rr) were screened for susceptible mutants. In the first cross, lines that were homozygous resistant (RR), but which do not exhibit measurable transposition rates, were used. No susceptible plants were observed in a screen of 6,243 progeny. In the second cross, the same genotypes were employed as in the main tagging programme, but the RR(Hitrans) parent was held at 25° during pollen formation to inhibit transposon movement. No susceptible plants were observed in a screen of 2,873 progeny.

2. *Reversion to resistance.* Reversion to an active state is a classic though not invariable characteristic of transposon-tagged genes; it results from the excision of the transposon from the locus, frequently leaving the gene in a functional state. We have tested for reversion and find that, when selfed, a proportion of the progeny of all six susceptibility phenotype B plants exhibit resistance. This is the only situation in which we have obtained resistant progeny upon selfing susceptibles. This result does not allow us to distinguish between germinal and somatic reversion, but the results of a later experiment (Figure 2) indicate that the bulk of reversion is somatic: i.e. it occurs during the growth of vegetative tissues. Vegetative clones of the mutant H1 were held at two temperatures during flowering (i.e. during gamete formation), and following selfing, the progeny were also held at two temperatures. The higher frequency of transposon movement at 15° correlates with the higher rate of reversion to resistance when the seedlings are held at this temperature.

Figure 2

			Inoculated with rust in glasshouse			
Mutant clone	Temp. at which clone grown and seed collected	Temp. at which seedlings grown	Number resistant	Number Suscep. Phen. C	Number Suscep. Phen. B	Number Suscep. Phen. A
H1 G	15°	15°	19	8	1	10
		25°	1	3	20	9
H1 EA	25°	15°	19	5	3	11
		25°	1	1	25	12

Note: After growth at 15° or 25°, the seedlings were placed together in a glasshouse for inoculation. A number of controls have shown that the pre-inoculation temperature at which the plants were grown does not affect the pathological interactions with between race α and our 'differential' lines; also, all of our plants, including mutants and their revertants, emained susceptible to race β.

3. *Mapping mutations to the R locus*. Both our own studies and previous work (Sampson, 1960) have shown that the race α-specific resistance gene is closely linked to the *eosina* locus which is involved in anthocyanin production. One of our mutants, H1, is homozygous recessive at the *eosina* locus, whilst another, T1, is heterozygous. Crossing these mutants has shown that i) the mutations to susceptibility in H1 and T1 are allelic, and ii) the mutations map as close to *eosina* as the R gene does.

Discussion

Evidence that the susceptible plants produced in the experimental protocol actually possess a transposon-tagged R gene come from three lines of investigation: when the transposition rate was reduced, no 'background' mutation to susceptibility was detected: the progeny of susceptible mutants exhibited temperature-dependent reversion to resistance: two of the mutants map at the R locus.

The susceptibility phenotype exhibited by the six prime candidates from the tagging protocol is clearly different to that previously encountered and has been classified as phenotype B. If a numerical scoring scheme were to be introduced for the *Antirrhinum*/rust race α interactions, the six mutants would be rated as less susceptible than genotypes exhibiting susceptibility phenotype A. The production of mutants with apparently lower levels of susceptibility has been reported by other workers. Chemical treatments have been used to produce a range of mutants for powdery mildew resistance in barley where mutations conferring infection types between 0 (complete resistance as conferred by *Ml-a12*) and (3-)4 (close to, but not complete, compatibility; Torp and Jorgensen, 1986; Jorgensen, 1987) were observed. Chemical mutagenesis of the wheat rust resistance genes *Lr20* and *Sr15* also resulted in a range of race-specific infection types (McIntosh, 1977).

Plants exhibiting phenotype B can clearly give rise to progeny exhibiting phenotype C. We suggest that the interaction observed is determined by the extent of reversion to resistance during leaf development: i.e. by the proportion of leaf cells that are resistant because the R gene has lost the transposon. Thus, leaves in which all, or a high proportion, of the cells possess an active R gene (i.e. free of the transposon) would exhibit a resistant response. Leaves in which a lower proportion of the cells possess an active R gene would exhibit an intermediate phenotype which we have categorised as susceptibility phenotype C. Leaves in which all, or possibly most, of the cells possess the transposon-mutagenised R gene would exhibit susceptibility phenotype B.

Molecular analyses of the mutants and their derivatives will focus upon finding a correlation between the occurrence of a particular transposon-bearing DNA restriction fragment and the pathology phenotype: i.e. in Southern analysis, the presence of a particular band (a fragment containing the tagged R gene) should correlate with the development of susceptibility phenotype B on challenge with race α. In order to facilitate these correlations, a range of plant material has been developed. This includes populations in which the mutation is segregating, and revertants for comparison with original mutants. The analyses may prove complex since there are nine cloned *Antirrhinum* transposons, each of which can exist in high numbers within the genome. It may therefore take some time to identify a DNA fragment bearing the tagged R gene.

Acknowledgements

This work was supported by the Agricultural and Food Research Council. We thank E.S. Coen and R. Carpenter for the experimental *A. majus* line and for their helpful advice, and M. Griffiths for valuable technical assistance with the plant material.

References

Aitken, E.A.B., Newbury, H.J and Callow, J.A. (1989) Races of rust (*Puccinia antirrhini*) of *Antirhinum majus* and the inheritance of host reistance. *Plant Pathology,* 38, 169-175.

Aitken, E.A.B., Callow, J.A. and H.J. Newbury (1992) Mutagenesis of a race-specific rust resistance gene in *Antirrhinum majus* using a transposon-tagging protocol. *The Plant Jourhal* (In Press).

Bennetzen, J.L. (1984) Genetic engineering for improved crop resistance. In *Applications of Genetic Engineering to Crop Improvement* , (Collins, G.B. and J.G. Petolino, J.G., eds). Dordrecht: Martinus Nijhof/Dr. W. Junk, pp. 491-525.

Ellis, J.G., Lawrence, G.J., Peacock W.J. and Pryor, A.J. (1988) Approaches to cloning plant genes conferring resistance to fungal pathogens. *Annual Review of Phytopathology*, 26, 245-263.

Harrison, B.J. and Carpenter, R. (1973) A comparison of the instabilities at the *nivea* and *pallida* loci in *Antirrhinum majus. Heredity*, 31, 309-323.

Jorgensen, J.H. (1987) Genetic analysis of barley mutants with modifications of powdery mildew resistance gene *Ml-a12. Genome*, 30, 129- 132.

Luo, D., Coen, E.S., Doyle, S. and Carpenter, R. (1991) Pigmentation mutants produced by transposon mutagenesis in *Antirrhinum majus. The Plant Journal*, 1, 59-69.

McIntosh, R.A. (1977) Nature of induced mutations affecting disease reaction in wheat. In '*Induced mutations against plant diseases*' International Atomic Energy Agency, Vienna. pp. 551-564.

Newbury, H.J. (1992) Fungal resistance: the isolation of a plant R gene by transposon-tagging. In '*Gene manipulation of pest and pathogen resistance in plants*', (Gatehouse, A.M.R., Hilder, V.A. and Boulter, D., eds). Oxford: CAB International, In Press.

Sampson, D.R. (1960) Linkage of a rust-resistance gene with two flower-colour genes in *Antirrhinum majus. Canadian Journal of Genetics and Cytology*, 2, 216-219.

Torp, J. and Jorgensen, J.H. (1986) Modification of barley powdery mildew resistance gene Ml-a12 by induced mutation. *Canadian Journal of Genetics and Cytology*, 28, 725-731.

MOLECULAR MARKERS AND GENOME ANALYSIS IN THE MANIPULATION OF LETTUCE DOWNY MILDEW

R. W. MICHELMORE, P. A. ANDERSON, H. WITSENBOER, R. V. KESSELI[1], I. PARAN[2], D. M. FRANCIS, & O. OCHOA.
Department of Vegetable Crops
University of California
Davis, CA 95616, USA
Current addresses:
[1] *Department of Biology,*
University of Massachusetts, Harbor Campus,
Boston, MA 02125-3393
[2] *Department of Field Crops,*
The Hebrew University of Jerusalem,
Rehovot, P.O. Box 12, 76100, Israel.

ABSTRACT. We are developing approaches for map-based cloning of disease resistance genes. Specificity between lettuce and its fungal pathogen, *Bremia lactucae,* is conditioned by an unambiguous gene-for-gene interaction. Classical genetics of host and pathogen has defined 13 resistance genes *(Dm)* and matching avirulence genes. The resistance genes are clustered in four regions. The current genetic map comprises 319 markers. We are now focusing on saturating the regions containing resistance genes with PCR-based markers using near-isogenic lines, bulked segregant analysis, and deletion mutants. We are also mapping many genes for disease resistance in lettuce to test whether they are all clustered. Random amplified polymorphic DNA (RAPD) markers are being converted to sequence characterized amplified region (SCAR) markers to provide reliable markers flanking each resistance locus. Recombinants selected from large populations are being used to order closely linked markers in the vicinity of resistance genes. We are developing long-range restriction maps around the *Dm* genes and analyzing recombinants for changes in the map to orientate the physical map relative to the genetic map and to determine the relationship between genetic and physical distance in the region. We are currently preparing various libraries for chromosome walking and will start isolating overlapping clones when a physically close marker has been confirmed.

1. Introduction

The interaction between lettuce (*Lactuca sativa* L.) and its fungal pathogen, *Bremia lactucae* Regel, is determined by a clear gene-for-gene interaction. The precise genetic definition makes this disease an attractive target for molecular dissection. One of the goals of our laboratory is to clone and characterize matching genes for resistance and avirulence from host and pathogen respectively. We are pursuing two strategies, map-based cloning and transposon tagging, that require few assumptions about the nature and pattern of expression of resistance genes. When possible, we are utilizing non-brute force approaches for the analysis of specific genomic regions that may be applicable to many species. In

E. W. Nester and D. P. S. Verma (eds.),
Advances in Molecular Genetics of Plant-Microbe Interactions, 517–523.

particular, we emphasize genetic analyses as prerequisites to molecular studies. This chapter discusses our progress towards map-based cloning of *Dm* genes.

2. Lettuce and Downy Mildew

Lettuce (*Lactuca sativa* L., Compositae) is a diploid (2n = 18) inbreeding species. Most cultivars are highly inbred and, therefore, exhibit extensive genetic homozygosity. Crosses can be made readily, although as occasional selfing cannot be completely excluded the generation of large backcross and testcross populations is difficult. Most genetic studies utilize F_2 and F_3 populations for analysis. Two to five generations are possible per year depending on the genotype. Each plant produces large numbers of seeds (1,500 to 5,000) allowing rare recombinants and mutants to be detected. The haploid genome size is approximately 2 pg (Galbraith and Michelmore, unpublished).

Bremia lactucae (Peronosporaceae, Oomycotina) is an obligately biotrophic fungal pathogen which can readily be cultured in the laboratory on lettuce seedlings (Michelmore *et al.*, 1988). Incompatible and compatible interactions can be clearly distinguished one week after inoculation of week-old seedlings. Therefore, many plants can be easily and accurately evaluated under controlled laboratory conditions. In a compatible interaction, *B. lactucae* penetrates an epidermal cell and then grows intercellularly through the host; there is minimal disturbance of the host until sporulation starts about a week after infection. In an incompatible interaction, the penetrated epidermal cell undergoes rapid cell death and there is little further fungal development.

3. Classical Genetics

Lettuce downy mildew is now one of the genetically best-characterized plant pathogen interactions. Parallel studies on host and pathogen have identified 13 dominant genes for resistance (*Dm*) in *L. sativa* (Crute & Johnson 1976; Johnson *et al.*, 1978; Hulbert & Michelmore, 1985; Farrara *et al.*, 1987) which are individually matched by dominant avirulence genes (*Avr*) in *B. lactucae* in a precise gene-for-gene interaction (Ilott *et al.*, 1989). Each *Dm* gene usually confers complete resistance when the corresponding *Avr* gene is expressed in the challenging fungal isolate. The 13 characterized *Dm* genes are clustered in four genetic linkage groups; the largest cluster contains seven *Dm* genes and *Ra*, a gene for resistance to root aphid (Farrara *et al.*, 1987). Several further resistances have been identified but not characterized genetically (Norwood *et al.*, 1981; Farrara & Michelmore, 1987). We and others have investigated possible complexities of the gene-for-gene interaction. In every case, however, the basic gene-for-gene interaction was shown to be effective, although occasionally further loci influenced the *Dm-Avr* interaction (Ilott *et al.*, 1989). These studies have provided a detailed understanding of the classical genetics of lettuce downy mildew and large numbers of genetically defined lines of *L. sativa* and *B. lactucae* for further analysis at the molecular level.

4. The Genetic Map of Lettuce

A detailed genetic map of lettuce is being developed from the analysis of a single F_2 population from cv. Calmar x cv. Kordaat. The basic mapping population comprises 66 F_2 plants and F_3 families derived from them. DNA was analyzed from F_2 plants and their F_3 derivatives. Classical markers exhibiting dominance, such as resistance genes, were scored in F_3 families to determine the progenitor F_2 genotype precisely. An intra-specific cross was utilized so that the resultant genetic map would have direct utility to intraspecific analyzes and breeding programs. Cvs. Calmar and Kordaat were chosen as they represent the two most important cultivated plant types of lettuce, crisphead and butterhead respectively, and they maximize the amount of intra-specific variation. This cross also segregated for six resistance genes (five *Dm* genes and *Tu*). The intra-specific nature of the cross slowed the development of the map but has yielded markers of greater utility than if an interspecific cross had been used.

The genetic map of lettuce currently comprises 319 loci, including 152 RFLP, 130 RAPD, 7 isozyme, 19 disease resistance, and 11 morphological markers (Kesseli *et al.*, 1992). Thirteen major and four minor linkage groups have been identified. Nine loci remain unlinked. The map has yet to coalesce into nine linkage groups representing the nine chromosomes. We are attempting to complete the map by several methods. We are continuing to map random RAPD loci to increase the number of loci. We will analyze the segregation of loci at the end of linkage groups on the expanded progeny; this will reduce the standard error on the recombination frequency and therefore allow us to detect linkage over greater genetic distances. We are targeting markers to the end of linkage groups using bulked segregant analysis (Michelmore *et al.*, 1991). We are also analyzing additional crosses, particularly inter-specific crosses to increase the probability of detecting polymorphic loci. Finally, we will use telomere specific probes to help define the ends of linkage groups.

5. Saturation of Genomic Regions Containing *Dm* Genes

The total number of markers of a genetic map is less important for map-based cloning than the numbers of markers close to the target gene. Techniques are now available to identify many markers in any region of the genome. The quickest markers to identify are RAPDs. There are currently over 1,500 arbitrary 10-mer oligonucleotide primers available. In lettuce, we obtain an average of over eight bands (loci) per primer. Therefore, we can survey over 12,000 loci over a period of several weeks. As our basic mapping population exhibits an average of approximately 10% polymorphism, these primers can detect over 1,200 loci scattered through the genome with an average of 0.6 markers per cM.

In order to increase the number of markers linked to *Dm* genes, we have utilized either near-isogenic lines (NILs), when available, or bulks of individuals from segregating populations. NILs differing for resistance to downy mildew were screened for differences using either pools of random cDNA clones or RAPD primers (Paran *et al.*, 1991). Over 500 cDNA clones were screened to identify four RFLP loci that were subsequently shown to be linked to the *Dm* genes. Two hundred and twelve primers identified 10 RAPD loci linked to *Dm* genes. One of the cDNA clones detected in this study detected a large, polymorphic multigene family. Different members of this family were linked to different clusters of resistance genes. Sequence analysis revealed similarity to the glycolytic enzyme, triose phosphate isomerase (TPI). Amplification of the TPI-related sequences in the genome was due to duplication of genomic sequences rather than some form retrotransposition as intron five had been conserved in all genomic sequences (Paran *et al.*, 1992). While it is difficult to imagine how TPI might be involved in resistance, these observations provide evidence for sequence duplication associated with the clusters of resistance genes and therefore support the hypothesis that many resistance genes are related at the sequence level.

NILs are not available for most regions of the lettuce genome; therefore, we developed bulked segregation analysis as a method for identifying markers in any region of the genome (Michelmore *et al.*, 1991). This approach was demonstrated by bulking individuals from the basic mapping population of contrasting genotypes for *Dm5/8*. Seventeen homozygous individuals were used to make each bulk. One hundred arbitrary 10-mer primers were screened and three RAPD differences were detected between the bulks. These were then shown to be linked to *Dm5/8* by conventional segregation analysis of the progenitor population. Bulk segregant analysis allows the use of only two samples to identify markers linked to the gene or region on which the bulks are based. It is therefore a rapid method for adding markers to specific regions and for placing classical markers on the map.

We have used bulk segregant analysis to saturate *Dm*-containing regions and to map genes that do not segregate in the basic mapping population. We now have numerous RAPD markers linked to each cluster of *Dm* genes; some cosegregate with a *Dm* gene in the limited population size so far analyzed. We are currently analyzing a variety of populations to map as many disease resistance genes in lettuce as possible. This will test the hypothesis that most resistance genes in lettuce are clustered. So far we have mapped a recessive gene for resistance to *Plasmopara lactucae radicis* (R. Kesseli, H. Witsenboer, M. Stangellini, and R. Michelmore, unpublished) and a new dominant gene for resistance to downy mildew (B. Maisoneuve, R. Michelmore, unpublished) to the *Dm5/8* cluster.

The use of segregating populations to identify linked markers requires polymorphism at a locus for it to be detected. Therefore, we are failing to detect many (90%, see above) of the loci in a specific region. In order to detect all potential markers amplified from a region, we are generating deletion mutants for each cluster of resistance genes. So far we have generated 5,000 M_2 families following gamma irradiation, one confirmed mutant that fails to express *Dm1* has been identified from the first 1,300 families screened. Analysis of this mutant revealed that all the markers flanking *Dm1* are still present and that no changes in 1.5 Mb fragments in regions flanking this gene were detected. Therefore, this mutant is not due to a large deletion at the *Dm1* locus; we are currently investigating other possibilities. We have also generated 4,000 M_2 families following irradiation of seeds with fast neutrons (courtesy of H. Brunner, International Atomic Energy Commission, Vienna). So far one putative mutant has been identified in the first 150 families screened simultaneously for inactivation of each of four *Dm* genes. When deletions for each resistance gene have been detected, we will screen for RAPD differences between the mutant and the wildtype progenitor. The existing RAPD primers and the use of deletion mutants will allow us to detect an average of 0.6 markers per 100 Kb. This frequency implies that laborious chromosome walks involving selection of several overlapping clones should be avoidable. The deletion mutants will also be useful in localizing *Dm* genes in clones of the region.

6. Sequence Characterized Amplified Regions

While RAPD loci can be quickly identified, they are sensitive to reaction conditions and are therefore not always reliable markers. We are developing Sequence Characterized Amplified Regions (SCARs) as reliable derivatives of RAPD markers (Paran and Michelmore, 1992). The RAPD fragment is cloned and checked by hybridization to confirm that the insert detects the correct RAPD band in a segregating population. It is important to check that the target fragment has indeed been cloned as the RAPD reaction amplifies many sequences; approximately 20% of our clones contain incorrect sequences. The insert is sequenced and primers designed for each of the ends. We use approximately 24 nt primers. PCR with high annealing temperatures ($60^{o}C$) results in the reliable amplification of single fragments. In some cases, amplification occurs from both genotypes with the products being different sizes. Therefore, this locus can be scored as a codominant marker.

7. Ordering of Loci

As a region becomes progressively more saturated with markers, it becomes increasingly difficult to determine the order of the loci. The only informative individuals are those resulting from recombinant gametes in the region. We have expanded the basic mapping population to over 300 individuals. This provides an average of 0.6 recombinants per 100 Kb. In order to identify more recombinants, we will select recombinant individuals from a population of 2,000 additional F_2 plants. This will be done by determining the genotypes for SCAR markers flanking each of the resistance genes using DNA extracted from seedlings. Only those plants with recombinant genotypes will be grown for further analysis. This population should provide an average of 4 recombinants per 100 Kb; as this is an average for the whole genome, the actual frequency may be considerably higher in the vicinity of *Dm* genes.

8. Correlation of Genetic versus Physical Distance

The relationship between genetic and physical distance varies greatly over eukaryotic genomes. It is critical to determine to relationship in the target region prior to initiating a chromosome walk. In particular, it will indicate how physically close the nearest marker is to the target gene and therefore whether the search for additional markers in the region should be continued. It will also indicate which type of genomic library is required. If a marker is within 50 Kb, it is probably unnecessary to work with YAC clones.

We are developing long-range maps around markers linked to *Dm* genes. We can routinely isolate and separate high molecular weight DNA from *L. sativa* following procedures developed for yeast (Chu *et al.*, 1986) and tomato (Ganal & Tanksley, 1989). This yields DNA of greater than 5 Mb which can be restricted by a variety of endonucleases to yield fragments that can be separated by pulsed-field gel electrophoresis (PFGE) in a contour-clamped homogeneous electric field (CHEF) gel. The same blot can be reprobed several times to determine whether linked probes hybridize to the same sized fragment. Single and multiple digests of two cultivars have provided a preliminary map around marker *CL922* which is tightly linked to *Dm15*. Analyzing recombinants for polymorphisms observed between the parents of the basic mapping population allowed us to orientate the genetic and physical maps (Francis, 1991; D. Francis P. Anderson and R. Michelmore, unpublished). The analysis of more recombinants will provide the estimates of the relationship between genetic and physical distance in *Dm* containing regions.

9. Studies in progress.

We are continuing to screen RAPD loci linked to *Dm* genes as new primers become available. More recombinants will be selected using SCAR markers flanking *Dm* genes. We are making a major effort to identify additional mutants for *Dm* expression. Hopefully, some mutants will be caused by deletions at these loci. Others may be due to lesions at other loci and will therefore be useful in dissection of the mechanisms of resistance. We are preparing multiple libraries for cloning *Dm* genes once physically close markers have been identified.

The location of sequences encoding *Dm* genes will rely on several strategies, including analysis of recombinants and mutants in the region. Ultimately, the identity of *Dm* genes will be confirmed by transformation into a susceptible genotype. Lettuce can be routinely transformed using *Agrobacterium tumefaciens* (Michelmore *et al.*, 1987). The target genes have already been transferred into the susceptible genetic background by conventional backcrossing; therefore, expression of the transgene should not be obstructed by background effects.

There are several potential complexities of *Dm* genes that could make their cloning difficult. In particular, if resistance genes are members of multigene families as their genetics suggests, cloning strategies relying solely on hybridization will be difficult. PCR-based approaches should help in this regard. SCARs represent single loci even though the amplified fragment may contain repeated sequences. SCARs and PCR will be used to screen libraries and we are developing methods for analysis of pulsed-field gels by PCR. PCR will also allow the design and use of *Dm* gene specific primers once they are cloned.

Several technical challenges remain before map-based cloning is routine. It is currently possible to saturate any region of the genome with RAPD markers. However, these may not be readily converted to hybridization-based (RFLP) markers as the amplified fragment often contains repeated sequences; therefore, methods are required for the analysis of pulsed field gels using PCR-based markers. Cloning still currently requires making and screening libraries of the whole genome. This a time consuming and laborious step. Technologies that allow cloning of selected large fragments are likely to be developed as part of the human genome project. These will greatly increase the feasibility of routine map-based cloning. Finally, methods are required for the rapid identification of genes in large fragments of genomic DNA. At present, the use of recombinants and mutants are the most powerful tools.

10. References

Chu, G.D., Vollrath, D. and Davis R.W. (1986). Separation of large DNA molecules by contour-clamped homogeneous electric fields. *Science* **234**:1582-1585.

Crute, I.R. and Johnson, A.G. (1976). The genetic relationship between races of *Bremia lactucae* and cultivars of *Lactuca sativa*. *Ann. Appl. Biol.* **83**:125-137.

Farrara, B. and Michelmore, R.W. (1987). Identification of new sources of resistance in *Lactuca* spp. *HortScience* **22**:647-649.

Farrara, B., Ilott, T.W. and Michelmore, R.W. (1987). Genetic analysis of factors for resistance to downy mildew (*Bremia lactucae*) in lettuce (*Lactuca sativa*). *Plant Pathol.* 36:499-514.

Francis, D.M. (1991). The physical organization of genes mediating specificity in the lettuce downy mildew interaction. Ph. D. Thesis, University of California, Davis.

Ganal, M.W. and Tanksley, S.D. (1989). Analysis of tomato DNA by pulsed-field gel electrophoresis. *Plant Molec. Biol. Reptr.* **7**:17-27.

Hulbert, S.H. and Michelmore, R.W. (1985). Linkage analysis of genes for resistance to downy mildew (*Bremia lactucae*) in lettuce (*Lactuca sativa*). *Theor. Appl. Genet.* **70**:520-528.

Ilott, T.W., Hulbert, S.H. and Michelmore, R.W. (1989). Genetic analysis of the gene-for-gene interaction between lettuce (*Lactuca sativa*) and *Bremia lactucae*. *Phytopathology* **79**:888-897.

Johnson, A.G., Laxton, S.A., Crute, I.R., Gordon,P.L. and Norwood J.M. (1978). Further work on the genetics of race specific resistance in lettuce (*Lactuca sativa*) to downy mildew (*Bremia lactucae*). *Ann. Appl. Biol.* 89:257-264.

Kesseli, R. V., Paran, I., Ochoa, O., Wang, W.-C., and Michelmore, R.W. (1992). Linkage map of lettuce (Lactuca sativa). In Genetic Maps, Vol. 5, S. O'Brien (ed.). Cold Spring Harbor Press, NY. In press.

Michelmore, R.W., Paran, I., and Kesseli, R.V. (1991). Identification of markers linked to disease resistance genes by bulked segregant analysis: a rapid method to detect markers in specific genomic regions using segregating populations. *Proc. Natl. Acad. Sci.* **88**:9828-9832.

Michelmore, R.W., Marsh, E., Seeley, S. and Landry, B.S. (1987). Transformation of lettuce mediated by *Agrobacterium tumefaciens*. *Plant Cell Reports* 6:439-442.

Michelmore, R.W., Hulbert, S.H., Ilott, T.W. and Farrara, B. (1988). Genetics of downy mildews. In: Genetics of Pathogenic Fungi (eds. G.S. Sidhu, R.H. Williams and D.S. Ingram). *Advs. Pl. Pathol.* 6:53-70.

Norwood, J.M., Crute, I.R. and Lebeda, A. (1981). The location and characteristics of novel sources of resistance to *Bremia lactucae* Regel (downy mildew) in wild *Lactuca* L. species. *Euphytica* **30**:659-668.

Paran, I., Kesseli, R.V. and Michelmore, R.W. (1991). Identification of RFLP and RAPD markers linked to downy mildew resistance genes in lettuce using near-isogenic lines. *Genome* **34**:1021-1027.

Paran, I., Kesseli, R.V., and Michelmore, R.W. (1992). Recent Amplification of triose phosphate isomerase related sequences in lettuce. *Genome* in press.

Paran, I. and R.W. Michelmore. 1992. The development of reliable PCR-based markers linked to downy mildew resistance genes in lettuce. *Theor. Appl. Genet.* in press.

Section 9 /Biotechnology

SYSTEMIC ACQUIRED RESISTANCE IN TOBACCO: USE OF TRANSGENIC EXPRESSION TO STUDY THE FUNCTIONS OF PATHOGENESIS-RELATED PROTEINS.

Danny Alexander*, Christopher Glascock*, Julie Pear*, Jeffrey Stinson*, Patricia Ahl-Goy‡, Manuella Gut-Rella‡, Eric Ward#, Robert M. Goodman*†, and John Ryals#.

**Calgene, Inc., 1920 Fifth St, Davis, CA 95616. †Univ of Wisconsin, 1630 Linden Dr, Madison, WI 53706. ‡CIBA-GEIGY Ltd., Basle, Switzerland. #CIBA-GEIGY Biotechnology, Research Triangle Park, NC 27709.*

ABSTRACT. Systemic Acquired Resistance (SAR) is the resistance to a variety of fungal, bacterial, and viral pathogens induced in many plant species by prior inoculation with a necrotizing pathogen. While the mechanism of resistance is unclear, there is a strong correlation in tobacco between the resistant state and the presence of the so-called "Pathogenesis-Related proteins" (PR-proteins). To study the involvement of the PR-proteins in SAR, we have engaged in a comprehensive program to clone and express constitutively in tobacco the cDNAs for all the PR-proteins. In addition to cloning the well-described PR 1-5 gene families, we have used differential cDNA screening to clone several new gene families which are induced by TMV inoculation. Homozygous transgenic seed lines expressing the various genes (or their anti-sense transcripts) are tested against a battery of pathogens for altered disease phenotypes. We have determined that transgenic plants expressing constitutively tobacco PR1 protein show significant resistance to blue mold (*Peronosopra tabacina*). The resistance, exhibited as a delay in symptom development, is observed in several independent transgenic PR1 lines, as well as in F1 crosses with a PR1 transgenic line as parent. Lines expressing the cDNA of a newly defined induced gene, denoted SAR8.2d, exhibit resistance to another oomycete pathogen, Phytophthora parasitica. As with the PR1 resistance, multiple transgenic events show the resistance, which is observed as much delayed symptom development.

1. Introduction

It has long been known that many plants can develop "immunity" to a variety of bacteria, fungi and viruses following an initial inoculation with a necrotizing pathogen (Chester 1933). This systemic acquired resistance (SAR) phenomenon has been most extensively studied in tobacco varieties which exhibit a hypersensitive response (HR) to tobacco mosaic virus (TMV) (Ross, 1961; Bol, *et al.* 1990; White and Antoniw, 1991). SAR can also be induced by chemicals such as salicylic acid (White, 1979) and 2,6-dichloroisonicotinic acid (Metraux, *et al.* 1991). Salt stress, acid damage, or wounding do not induce SAR.

The mechanism for this broad, inducible resistance is not known, but is clearly different from the well characterized "single gene" resistances, which generally work only against individual pathogen isolates and may exhibit high genetic diversity within a spcies (Keen,

E. W. Nester and D. P. S. Verma (eds.),
Advances in Molecular Genetics of Plant-Microbe Interactions, 527–533.

1990). Much attention has been paid to several multigene families of tobacco, termed the "pathogenesis-related proteins", whose coordinate induction correlates with the onset of SAR (Gianinazzi, et al. 1970; Van Loon, 1975; Linthorst, 1991). While several of these PR-proteins, inculding chitinases, glucanases, and permatins, have been shown to have antifungal activity *in vitro* (Mauch, *et al.* 1988; Vigers, *et al.* 1991; Woloshuk, *et al.* 1991), there is still no direct evidence that they play a causal role in disease resistance *in vivo*. Others, such as PR-1, have no known biochemical function and exhibit no homology to known genes in current databases. However, PR1 protein localization (Antoniw and White, 1986) and its high level of induction suggest an active role in disease resistance.

To evaluate the role of PR-protein genes in the development and maintenance of resistance, we have undertaken a program to express the various PR-cDNA types in transgenic tobacco, in sense and anti-sense orientations, under the control of a strong, constitutive promoter. A description of the research program and some early results are presented here.

2. General Approach

We have engaged in a three part program to find and isolate cDNAs for pathogenesis-related genes (GENE DISCOVERY), to engineer transgenic tobacco to express constitutively these PR-proteins (or their anti-sense transcripts) at high levels (TRANSGENIC EXPRESSION), and to test the resulting lines for altered SAR phenotypes (PHYTOPATHOLOGY TESTING).

2.1 GENE DISCOVERY

The gene discovery phase of the research program entailed two approaches, **a)** cloning cDNAs of the known PR proteins, and **b)** discovering cDNAs of new TMV-induced gene families for which no protein product was yet known. The first approach involved the isolation and sequencing of the PR-proteins of Xanthi nc tobacco, followed by synthesis of oligonucleotide probes for cDNA isolation. Full-length clones, generally representing the most highly expressed isoform of each family, were sequenced and used for subsequent expression engineering.

The second approach used differential screening of induced cDNA populations. A cDNA library was constructed using mRNA from upper, uninfected leaves of Xanthi nc plants which had been inoculated with TMV on the lower leaves 11 days earlier. This upper leaf tissue was chosen to avoid mRNAs involoved in the local necrotic response at the TMV lesions. Candidate cDNAs were evaluated for kinetics of induction using Northern analysis. Again the full-length cDNAs of predominant family members were used for expression engineering.

2.2 TRANSGENIC EXPRESSION

Full-length cDNAs were cloned between the enhanced CaMV promoter (double-35S) (Comai, *et al.* 1990) and the 3' terminator region of the Ti-plasmid *tml* gene. Both "sense" and "anti-sense" versions were created for most cDNAs. The reulting expression cassettes were then cloned into a binary vector (McBride and Summerfelt, 1990) containing a kanamycin (NPT II) plant selectable marker gene. The kan gene was adjacent to the left T-DNA border, and was expressed from a Ti-plasmid *mas* gene promoter. Transcription was toward the right T-DNA border. The expression cassettes were inserted adjacent to the right T-DNA border in either possible orientation.

Early experiments compared PR1 expression (as measured by ELISA) in transgenic plants transcribing the PR1a cassette either toward the right border in parallel with the kan gene, or toward the left border, in opposition to the kan gene. Measurements of approximately 25 plants in each group showed no significant difference in constitutive PR1 expression, whether compared on an average-plant or best-plant basis.

Typically, 25-30 Xanthi nc transgenic plants were generated from each chimeric gene construction. We designated primary transgenic plants obtained directly from tissue culture as T1 plants. T1 plants in soil were grown to approximately 30 cm tall, and leaf tissue was harvested for ELISA or Western blot analysis. Those "sense" orientation plants showing the highest constitutive protein expression were allowed to self-pollinate and set T2 seed. T2 seed were tested in a germination assay in the presence of kanamycin (150 mg/l), and those seed lots exhibiting 3:1 (r/s) segregation of the kan^r trait were advanced. Ten T2 plants were allowed to self-pollinate and set T3 seed. The resulting T3 seed was tested for kan segregation as above. Homozygous seed lots (*i.e.* those exhibiting 100% kan^r seed) were advanced to the phytopathology screening program. In almost all cases lines were identified which constitutively expressed the engineered protein at equal or higher levels compared to TMV-inoculated plants.

"Anti-sense" lines, and "sense" lines expressing genes for which antibodies to the encoded proteins were not yet available were selected only on the basis of kan segregation. All lines showing 3:1 (r/s) kan segregation were advanced to the homozygous T3 stage.

2.3 PHYTOPATHOLOGY TESTING

T3 lines showing highest expression of the engineered protein were used in initial diesease tests against a broad variety of pathogens. Any indications of altered disease phenotypes were followed up by larger tests against those pathogens showing the effect. Pathogens tested included TMV, PVY, *Pseudomonas tabaci*, *Pseudomonas syringae*, *Cercospora nicotianae*, *Phytophthora parasitica*, *Peronospora tabacina*, *Heliothis virescens*, and *Melodigyne incognita*.

Six to ten plants of each seed line, in six-inch clay pots, were treated in a random design. Disease ratings and data tabulations were performed in a double blind fashion. The two pathogens for which results are given here are *Phythpthora parasitica* (black shank) and *Peronospora tabicina* (blue mold). *P. parsitica* zoospores were applied to soil as a spore suspension in water. Disease was rated as degree of wilting on a scale of 1-5, with 5 representing total collapse. *P. tabacina* spore suspensions were sprayed on leaves in a dew chamber, and disease was rated as a percentage of leaf area infected.

3. RESULTS

3.1 OVERALL SCREENING RESULTS

Transgenic lines that have been tested to date include the "sense" orientations of PR1a, PR2c (acidic glucanase), PR3 (acidic chitinase), class I (basic) glucanase, class I (basic) chitinase, cucumber class III chitinase, PR-Q' (an acidic glucanase), SAR8.2d, class I glucanse less the vacuolar targeting peptide (-vtp), and class I chitinase (-vtp). Several other constructions representing new or related gene families are still advancing through the program.

To date we have identified two pathogenesis-related cDNAs which conferred significant disease tolerance to tobacco when expressed in a constitutive manner. They are the SAR8.2d transgenic lines, which showed resistance to *Phytophthora parasitica*, and the PR1a lines, which were resistant to *Peronospora tabacina*. These will be discussed in more detail below. Also, lines expressing PR3, class I (basic) chitinase, or cucumber chitinase gave clear indications of resistance in a damping-off assay against *Rhizoctonia solani*, which is in agreement with the results of Broglie, *et al.* (1991).

3.2 SAR8.2

The SAR8.2 cDNA family was discovered by differential screening of a cDNA library representing TMV-inoculated plants. The mRNA used for the library construction and for screening was taken from upper leaves of plants inoculated 11 days previously on the lower leaves with TMV. Upper leaves of mock-inouclated plants furnished mRNA for the differential comparison.

Northern analysis using SAR8.2 probes showed that the mRNAs were approximatly 550-600 nucleotides. They were strongly induced by TMV-inoculation, salcylic acid treatment, or 2,6-dichloroisonicotinic acid (INA) (Ward, *et al.* 1991). We have seen considerable variation in SAR8.2 levels in untreated plants, indicating that there may be other unknown factors affecting their regulation. However, even when the levels are relatively high in the untreated plants, TMV or the chemical inducers always cause additional induction.

We sequenced 25 cDNAs representing the SAR8.2 family, and identified five distinct cDNAs. The open reading frames (ORFs) of the cDNAs encoded small highly basic proteins with N-terminal signal peptides. Assuming processing of the signal peptides, four of the proteins were approximately 7.6 kDa, and the fifth was about 9.6 kDa.

The most interesting feature of the putative proteins is the presence of a cysteine-rich domain at the C-terminus. The four smaller proteins have this domain perfectly conserved. The fifth protein (SAR8.2e) is larger because it duplicates this domain in tandem at the C-terminus. The two cysteine-rich domains of SAR8.2e differ from the other four proteins by only one amino acid in each domain. Database searches using the full-length DNA or protein seuences turned up no significant matches. However, searches with only the cysteine-rich domain revealed many matches, based mainly on the conservation of cysteine spacing. All these proteins, mostly metallothioneins, Bowman-Birk proteinase inhibitors, or zinc-finger proteins, bind divalent cations. While the conservation and duplication of this domain in the gene family may indicate some functionl importance, we have no evidence for a specific biological role for SAR8.2.

Southern analysis showed the SAR8.2 gene family to be quite large, with 10-12 hybridizing bands generated from genomic DNA by enzymes which do not cut the cDNAs. Thus the five cDNAs may indicate only the minimum number of functional genes. Some or all of the additional bands may represent pseudogenes.

When the first SAR8.2d transgenic T3 line was tested in our phytopathology screen, most pathogen infections were indistinguishable from control lines, which included non-transformed Xanthi nc, Xanthi nc transformed with an empty expression cassette, and the SAR8.2 anti-sense line. However, *Phytophthora parasitica* infection was dramatically delayed in the "sense" expression plants. After eight days all the controls showed near total collapse, while the SAR8.2 plants were beginning to exhibit mild to moderate wilting. Two repeats of this expriment with the same line gave a similar result. Expanded experiments with additional lines, representing independent transgenic events, confirmed that the presence of the SAR8.2 chimeric gene was responsible for the observed resistance, although none of the other lines gave such spectacular results as the initial one. We investigated the possibility that exression of SAR8.2 might cause induction of some or all of the other SAR genes, leading to the resistant phenotype. Western blot analysis showed that none of the known PR proteins were detectable in the untreated SAR8.2 transgenic line. We do not know if the observed resistance is the result of decreased pathogen growth, or only a delay of symptom development.

3.3 PR1a

Early tests with PR1a "sense" transgenic lines gave indications of resistance to *Peronospoa tabacina*. To test the validity of this observation an expanded experiment was designed, incorporating several indpendent transgenic homozygous lines expressing PR1a, and F1 crosses of a PR1a line with lines expressing other transgenic PR-proteins (PR1a heterozygotes). Controls included all the other homozygous tran$genic PR-protein lines (non-PR1), untrasformed and untreated Xanthi nc, and untransformed Xanthi nc treated with an immunizing chemical (positive control).

All PR1a-expressing lines except one, whether homozygous or heterozygous, showed significant resistance seven days after inoculation. ELISA tests later revealed that the one non-resistant PR1a line was indeed not expressing PR1a. All the other PR1a lines were expressing high levels of the protein. Using pair-wise t-test comparisons, the group of PR1a homozygous plants were found to differ from the non-PR1 plant group with >99% confidence at day 7. The heterozygous plant group differed with a confidence of 91%. After 9 days the homozygous group retained the >99% confidence level, while the heterozygous group confidence had fallen to 82%. Northern analysis of transgenic PR1a plants revealed that they do not express significant levels of other PR-protein mRNAs.

It is interesting that in none of the transgenic plants do we see resistance as strong as in the chemically immunized plants, even though the PR1 levels of the transgenic plants often equals or exceeds levels in induced plants. This suggests that PR1a may be only one part of a more complex mechanism aimed at this pathogen by the SAR response.

Previous reports of transgenic expression of PR1 proteins in tobacco (Cutt, *et al.* 1989; Linthorst, *et al.* 1989) have concluded that PR1 is not involved in virus resistance. We have extended these observations to PVY, with similar results. However, none of the earlier studies addressed fungal pathogens, and our results are the first indication of a function for PR1 proteins.

4. CONCLUSION

Our goal is to elucidate the functions of the plant genes involved in the SAR response. We have shown evidence for the involvement of two pathogenesis-related protein genes in resistance to two important oomycete pathogens. The nature of the resistance is unknown, but this work provides the materials to ask important genetic and biochemical questions.

Our working hypothesis is that the systemic resistance against any particular pathogen is likely to be multigenic in nature, and that different sets of overlapping genes will be involved in resistance to various pathogens. One way to test this hypothesis is the use of "anti-sense" expression lines, an approach which we have not yet exploited fully, but for which we have a considerable amount of transgenic germplasm in hand. We feel that blocking one necessary gene in a complex function is more likely to lead to an altered disease phenotype than the constitutive expression of that gene, which by itself may not be sufficient to affect disease.

We are also engaged in a program to introduce two or more of the chimeric PR-cDNAs into tobacco, both by genetic crosses and by combinations of double or triple gene constructs and crosses. Plants expressing multiple "sense" or "anti-sense" genes should be useful in elucidating PR-protein function.

5. LITERATURE CITED

Antoniw, J.F., and White, R.F. (1986) 'Changes with time in the distribution of virus and PR protein around single local lesions of TMV infected tobacco', Plant Mol. Biol. 6:145-149.

Bol, J.F., Linthorst, H.J.M., and Cornelissen, B.J.C. (1990) 'Plant pathogenesis-related proteins induced by virus infection', Annu. Rev. Phytopathol. 28:113-138.

Broglie, K., Chet, I., Holliday, M, Cressman, R., Biddle, P, Knowlton, S., Mauvais, C.J.,and Broglie, R. (1991) 'Transgenic plants with enhcnced resistance to the fungal pathogen *Rhizoctonia solani*', Science 254:1194-1197.

Chester, K.S. (1933) 'The problem of acquired physiological immunity in plants (continued)', Quart. Rev. Biol. 8:275-324.

Comai, L., Moran, P., and Maslyar, D. (1990) 'Novel and useul properties of a chimeric plant promoter combining CaMV 35S and MAS elements', Plant Mol. Biol. 15:373-381.

Cutt, J.R., Harpster, M.H., Dixon, D.C., Carr, J.P., Dunsmuir, P., and Klessig, D.F. (1989) 'Disease response to tobacco mosaic virus in transgenic tobacco plants that constitutively express the pathogenesis-related PR1b gene', Virology 173:89-97.

Gianinazzi, S., Martin, C., and Vallée, J.C. (1970) 'Hypersensibilité aux virus, température et protéines soluble chez le *Nicotiana Xanthi*-nc. Apparition de nouvelles macromolécules lors de la répression de la synthèse virale', Acad. Sci. Paris, CR Ser D270:2383-2386.

Keen, N.T. (1990) 'Gene for gene complimentarity in plant-pathogen interactions', Annu. Rev. Genet. 24:447-463.

Linthorst, H.J.M., Meuwissen, R.L.J., Kauffmann, S., and Bol, J. (1989) 'Constitutive expression of pathogenesis-related proteins PR-1, GRP, and PR-S in tobacco has no effect on virus infection', Plant Cell 1:285-291.

Linthorst, H.J.M. (1991) 'Pathogenesis-related proteins of plants', Crit. Rev. in Plant Sci. 10:123-150.

Mauch, F., Mauch-Mani, B, and Boller, T. (1988) 'Antifungal hydrolases in pea tissue II. Inhibition of fungal growth by combinations of chitinase and ʃ-1,3-glucanase', Plant Physiol. 88:936-942.

McBride, K.E., and Summerfelt, K.R. (1990) 'Improved binary vectors for *Agrobacterium*-mediated plant transformation', Plant Mol. Biol. 14:269-276.

Métraux, J.P., Alh-Goy, P., Staub, T., Speich, J., Steinemann, A., Ryals, J., and Ward. E. (1991) 'Induced systemic resistance in cucumber in response to 2,6-isonicotinic acid and pathogens', in H. Hennecke and Verma, D.P.S., (eds.) Advances in Molecular Genetics of Plant-Microbe Interactions, Kluwer Academic Publishers, Dordrecht, pp.432-439.

Ross, A.F. (1961) 'Systemic acquired resistance induced by localized virus infections in plants', Virology 14:340-358.

Van Loon, L.C. (1975) 'Polyacrylamide disc electrophoresis of the soluble leaf proteins from Nicotiana tabacum var. 'Samsun' and 'Samsun NN' IV. Similarity of qualitative changes of specific proteins after infection with different viruses and their relationship to acquired resistance', Virology 67:566-575.

Vigers, A.J., Roberts, W.K., and Selitrennikoff, C.P. (1991) 'A new family of plant antifungal proteins', Mol. Plant-Microbe Interact. 4:315-323.

Ward, E.R., Uknes, S.J., Williams, S.C., Dincher, S.S., Wiederhold, D.L., Alexander, D.C., Ahl-Goy, P., Metraux, J.-P., and Ryals, J.A. (1991) 'Coordinate induction of gene expression in response to biological and chemical agents that induce systemic acquired resistance', Plant Cell 3:1085-1094.

White, R.F. (1979) 'Acetylsalicylic acid (asprin) induces resistance to tobacco mosaic virus in tobacco', Virology 99:410-412.

White, R.F., and Antoniw, J.F. (1991) 'Virus-induced resistance responses in plants', Crit. Rev. in Plant Sci. 9:443-455.

GENETIC AND BIOCHEMICAL DETERMINANTS OF PHENAZINE ANTIBIOTIC PRODUCTION IN FLUORESCENT PSEUDOMONADS THAT SUPPRESS TAKE-ALL DISEASE OF WHEAT

L. S. THOMASHOW[1], D. W. ESSAR[1], D. K. FUJIMOTO[1], L. S. PIERSON, III[2], C. THRANE, and D. M. WELLER[1].
[1]USDA-ARS, 362 Johnson Hall, Washington State University, Pullman, WA 99164-6430 USA; [2]Department of Plant Pathology, University of Arizona, Tucson, AZ 85721 USA

ABSTRACT. Biological control of take-all by *Pseudomonas fluorescens* strain 2-79 and *P. aureofaciens* 30-84 depends largely on the production of one or more phenazine compounds with broad spectrum antibiotic activity. Phenazine compounds are derived via the shikimic acid pathway, with chorismic acid as the probable branchpoint intermediate. Anthranilic acid is a phenazine precursor in *P. aeurginosa* and can accumulate in iron-deprived cultures of strain 2-79 that also exhibit reduced phenazine production. We have now shown that the tryptophan pathway is unlikely to provide anthranilate for phenazine synthesis in strain 2-79; inactivation of *trpE*, which encodes the large subunit of anthranilate synthase, caused auxotrophy for tryptophan but did not eliminate phenazine production. We also show that two loci, *phzP* in strain 2-79 and *phzA* in *P. aureofaciens* strain 30-84, are required for phenazine production. *PhzA* is linked to and divergently transcribed from the structural locus for phenazine production in strain 30-84, whereas *phzP*, previously identified in the mutant strain 2-79.B46, is not closely linked to the structural locus and has a pleiotropic phenotype. The biosynthetic locus is strongly homologous in strains 2-79 and 30-84 and is expressed in nonproducer pseudomonads, resulting in phenazine production. The reporter gene *lacZ* was fused to the biosynthetic loci from both strains and used to analyze their structure, kinetics of expression, and responses to environmental stimuli.

1. Introduction

Biological control, as broadly defined by the National Academy of Sciences (1987), is "the use of natural of modified organisms, genes, or gene products to reduce the effects of undesirable organisms (pests), and to favor desirable organisms such as crops, trees, animals and beneficial insects and microorganisms." Plant disease suppression by microbial biocontrol agents introduced on seeds or other planting material results from interactions among the introduced agent, the target pathogen and other inhabitants of the rhizosphere community, and the host plant. Antibiosis, competition and parasitism or predation are generally recognized as the mechanisms by which biocontrol agents suppress pathogens directly. Disease suppression also can be mediated indirectly, if the introduced agent induces a systemic resistance response in the host plant.

E. W. Nester and D. P. S. Verma (eds.),
Advances in Molecular Genetics of Plant-Microbe Interactions, 535–541.

Recent interest in bacterial biocontrol agents has focused on fluorescent pseudomonads. Strains of fluorescent *Pseudomonas* spp. are found in large numbers in association with plant roots; they are nutritionally versatile, have relatively rapid growth rates, and frequently produce antibiotics and siderophores that are inhibitory in vitro against fungal root pathogens. Isolates from the rhizosphere are generally amenable to genetic manipulation, although the methods required to successfully introduce and maintain DNA vary greatly among individual strains.

Our work with fluorescent pseudomonads has focused on two strains, *P. fluorescens* 2-79 and *P. aureofaciens* 30-84, that produce phenazine antibiotics. Phenazines are pigmented, nitrogen-containing heterocyclic compounds inhibitory in vitro to a broad spectrum of bacteria and fungi. The greenish-yellow antibiotic phenazine-1-carboxylic acid (PCA) is the sole phenazine accumulated in cultures of strain 2-79. Cultures of *P. aureofaciens* 30-84 also produce PCA in abundance, but further derivatize a portion of it to 2-hydroxyphenazine-1-carboxylate (orange) and 2-hydroxyphenazine (red). For both 2-79 and 30-84, these antibiotics are the major determinant of ability to control take-all, which probably is the most important root disease of wheat worldwide. Plants grown from seed treated with phenazine-nonproducing (Phz^-) Tn*5* mutants were reduced in height and had significantly more root disease than plants from seed treated with the wild-type strains. Complementation of the mutants concomitantly restored phenazine production and biocontrol activity to wild-type or near wild-type levels (Thomashow and Weller, 1988; Pierson and Thomashow, 1992). PCA has been isolated directly from the roots of wheat colonized by strains 2-79 or 30-84 (Thomashow *et al.*, 1990), and phenazine synthesis has been correlated with increased persistence of Phz^+ wild-type strains, as compared to Phz^- mutants, in soil and in the rhizosphere of wheat (Mazzola *et al.*, 1992). In contrast to the dominant role of the phenazines, fluorescent siderophores produced by strains 2-79 (Hamdan *et al.*, 1991) and 30-84 (Pierson and Thomashow, 1992) contributed little or nothing to the control of take-all.

Phenazine compounds synthesized by fluorescent pseudomonads are products of the shikimic acid pathway, with chorismate as the probable branchpoint intermediate (reviewed by Turner and Messenger, 1986). The first step in the pathway probably involves the condensation of two molecules of chorismate, with glutamine as the amide donor, by a mechanism analogous to that for the synthesis of anthranilate in the tryptophan pathway. Exogenously supplied anthranilate does not itself support phenazine synthesis, but anthranilate apparently is a precursor in the synthesis of pyocyanine, the blue phenazine compound characteristic of *P. aeruginosa*. In *P. aeruginosa*, anthranilate for phenazine production is derived via a unique anthranilate synthase encoded by *phnAB*, which is distinct from the *trpE* and *trpG* genes that encode a separate anthranilate synthase required for tryptophan biosynthesis (Essar *et al.*, 1990b). It is not at all clear, however (see below), that the biochemistry of the first step in the pathway and the nature of the first phenazine intermediate in strains of *P. fluorescens* and *P. aureofaciens* are the same as those in *P. aeruginosa*.

Despite their enormous potential as alternatives or supplements to chemical control strategies, few biological controls for soilborne plant pathogens have been successfully commercialized. This is due largely to the failure of microbial agents to provide consistent disease suppression in the field. Variable root colonization undoubtedly is one cause of inconsistent performance (Weller, 1988). The number of take-all lesions on roots was inversely related to the size of the population of strain 2-79 introduced on seeds or established on roots (Bull *et al.*, 1991).

Ineffective control has been observed even on well-colonized roots, however, suggesting that environmental factors influence the performance of introduced strains in situ. Ownley *et al.* (1990) showed that for wheat treated with Phz^+ derivatives of strain 2-79 and grown in ten different soil types, root disease severity was significantly correlated with 16 soil characteristics. Nine factors, of which cation exchange capacity, % silt, and iron were most important, were positively correlated with disease severity, whereas seven others, of which ammonium nitrogen, soil pH and zinc were the most significant, were directly related to disease control. These results may be of considerable value in predicting whether phenazine-producing strains will be effective in soils with particular characteristics and in reducing the number of failures in sites that are poorly suited to such strains.

Because environmental factors influence not only the biocontrol agent but also the pathogen and the host plant and their collective interactions, soil properties positively correlated with disease suppression are not necessarily those that are optimal for production or activity of phenazine antibiotics. Nevertheless, understanding how the environment impacts on the biochemistry and genetics of phenazine production will provide insight toward optimizing the performance of these strains in the field. In this report we describe the biosynthetic loci involved directly in the production of phenazines in strains 2-79 and 30-84 and show that the two strains differ in their expression of these loci in response to the environment. We also identify two *trans*-acting regulatory loci, one linked and the other unlinked to the structural locus, that are required for expression of the phenazine biosynthetic genes. Finally, we address the role of *trpE* in strain 2-79 as a potential source of anthranilate for phenazine synthesis.

2. Materials and Methods

Bacterial strains, genetic manipulations, and bioassays in vitro and in situ have been described previously (Essar *et al.*, 1990a, 1990b; Pierson and Thomashow, 1992; Thomashow and Weller, 1988).

3. Results and Discussion

3.1. ABSENCE OF A ROLE FOR *trpE* IN PHENAZINE SYNTHESIS

Anthranilic acid, a presumptive precursor in the synthesis of pyocyanine by *P. aeruginosa* PAO1 (Essar *et al.*, 1990b), can accumulate in the culture fluid of *phzP* mutants (see below) of strain 2-79 grown under iron limitation. However, several observations argue that *phzP* does not simply encode an enzyme with anthranilate synthase activity: *phzP* mutants were not tryptophan auxotrophs; the cloned *phzP* locus had no detectable homology by Southern hybridization at low stringency with cloned *phnAB*, *trpE* or *trpG* genes from *P. aeruginosa*; *phzP* did not rescue *trpE* mutants of *P. putida* from tryptophan auxotrophy; and *phzP* mutants are pleiotropic, exhibiting a variety of phenotypic changes (see below; L. Thomashow, unpublished). Moreover, sequences homologous to *phnAB* from *P. aeruginosa* were undetectable at low stringency in blots of total DNA from strain 2-79, suggesting that significant differences may exist between the two strains in an early step(s) of the phenazine biosynthetic pathway.

To help resolve uncertainty about the source or role of anthranilate in PCA synthesis by strain 2-79, we used a cloned *trpE* gene probe from *P. putida* to detect homologous sequences in a cosmid library of total DNA from strain 2-79 (D. W. Essar and L. S. Thomashow, unpublished results). A single *trpE* locus of approximately 2 kb was identified in ten separate cosmids; it restored *trpE* mutants of *P. putida* to prototrophy and mapped about 3 kilobases upstream of a *trpGDC* gene cluster, as is also the case for *trpE* in *P. putida* (Essar *et al.*, 1990a). A fragment homologous to *trpE* was subcloned, mutagenized with the transposons Tn*5* or Tn*3*HoHo1, and introduced into the genome of strain 2-79 by marker-exchange, yielding mutant derivatives auxotrophic for tryptophan. When grown in a defined medium supplemented with 20 μM D,L-tryptophan, one of these *trpE* mutants, 2-79.E1, produced about 275μg of PCA per OD_{600} of culture, as compared with 291 μg of PCA produced per OD_{600} of the unsupplemented wild-type strain (D. W. Essar and L. S. Thomashow, unpublished results). Similar amounts (275 μg and 263 μg, respectively), were produced by 2-79.E1 complemented *in trans* by pDE7, encoding the cloned *trpE* gene, and by 2-79 containing pDE7::Tn*3*HoHo1, a plasmid-borne copy of *trpE* insertionally inactivated with Tn*3*HoHo1. The rationale behind the latter is that if tryptophan synthesis in strain 2-79 is negatively regulated, as it is in *Escherichia coli* (Yanofsky and Crawford, 1987) and *Bacillus subtilis* (Kuroda *et al.*, 1988), then the presence of the mutated gene should result in full derepression of the chromosomally encoded tryptophan pathway. The fact that 2-79(pDE7::Tn*3*HoHo1) produced no more PCA than the wild type argues strongly against a role for anthranilate derived either directly, via *trpE* and *trpG*, or indirectly, via tryptophan degradation (Calhoun *et al.*, 1973), in phenazine synthesis by strain 2-79.

3.2. PHENAZINE BIOSYNTHETIC LOCI IN STRAINS 2-79 AND 30-84

In previous work, two overlapping cosmids were identified that restored phenazine production to a number of Phz⁻ Tn*5* mutants of *P. aureofaciens* 30-84, including one mutamt that produced PCA but no hydroxyphenazine derivatives. Additional evidence that these cosmids contain genes encoding a significant portion of the phenazine biosynthetic pathway was obtained when a 9.2-kb *Eco*RI fragment present in both cosmids was expressed from heterologous promoters in *Escherichia coli*, resulting in synthesis of the three phenazines normally produced by strain 30-84. Sequences required for phenazine production were further localized by Tn*5* mutagenesis and deletion analysis to a segment of approximately 2.8 kb (Pierson and Thomashow, 1992). More recently, this region has been shown to contain at least two genes designated *phzB*, encoding a 55-kD protein involved in PCA production and *phzC*, encoding a 19-kD protein involved in production of 2-hydroxyphenazine-1-carboxylic acid (L. S. Pierson, unpublished results).

Cosmid clones containing phenazine biosynthetic genes from *P. fluorescens* 2-79 were identified initially by their ability to complement Phz⁻ mutants, and subsequently, by strong hybridization with subcloned portions of the biosynthetic locus from strain 30-84. Analysis of the 12-kb clone pPHZ108A by Tn*3*HoHo1 mutagenesis revealed a 5-kb segment containing one, or possibly two, adjacent transcriptional units bordered at the 5' end by an adjoining, divergently transcribed segment of less than 1 kb. Several additional insertions up to 3 kb upstream of the transcribed region inactivated the ability of pPHZ108A to complement Phz⁻ mutants, whereas others had no effect (C. Thrane and L. S. Thomashow, unpublished results). Regulatory sequences have been identified 5' to *phzB* and *phzC* in strain 30-84 (see below), and we expect that this also will be the case in strain 2-79.

The locus present on pPHZ108A was sufficient to transfer phenazine biosynthetic capability to other strains of fluorescent pseudomonads that normally do not produce phenazines. If not already inhibitory to *Gaeumannomyces graminis* var. *tritici* (causal agent of take-all), such strains exhibited increased inhibition in vitro. Plasmid-borne copies of the biosynthetic locus in strain 2-79 also significantly increased the amount of phenazine produced in vitro. However, these overproducing strains, or others into which the biosynthetic pathway has been introduced, did not provide significantly improved disease control on wheat.

3.3. EXPRESSION OF PHENAZINE BIOSYNTHETIC LOCI

The reporter strain 30-84Z contains a genomic *phzB::lacZ* translational fusion and expresses β-galactosidase in place of PCA production (L. S. Pierson and L. S. Thomashow, unpublished results). A similar *lacZ* fusion derivative was generated in strain 2-79 by recombining a Tn*3*HoHo1 insertion in the phenazine biosynthetic region of pPHZ108A (see above) into the 2-79 genome, yielding 2-79.2A40 (C. Thrane and L. S. Thomashow, unpublished results). Studies are in progress to determine how the nutritional environment affects phenazine gene expression and whether the strains differ in their response to growth conditions.

For strains 2-79 and 30-84, both total PCA accumulation and β-galactosidase expression varied greatly as a function of nitrogen source. Thus, strain 2-79 produced only 14μg of PCA per OD_{600} of cells, and 2-79.2A40 had only 60 Miller units of β-galactosidase activity in an ammonium-based defined medium, as compared to over 500 μg of PCA per OD_{600} and 2,000 Miller units in nitrate-based medium (D. K. Fujimoto and L. S. Thomashow, unpublished results). Strain 2-79.2A40 consistently had more β-galactosidase activity that did strain 30-84Z in media containing urea, nitrate or ammonium. Whether these differences reflect a positional effect of the *lacZ* insertions in the two strains, or an actual difference in promoter strength, is still unclear; nevertheless, the ratio of β-galactosidase activity in strain 2-79.2A40 to that in 30-84Z was similar for all three nitrogen sources.

The ratio of micrograms of PCA produced by strain 2-79 to β-galactosidase units expressed by 2-79.2A40 was constant for all three nitrogen sources, at about 0.25 μg per Miller unit of enzyme activity. In each medium, gene expression was a function of biomass, with growth much reduced in the ammonium-based medium as compared to urea- or nitrate-based media. The nitrogen sources most favorable for growth also supported the greatest phenazine accumulation. If increased phenazine production results in better biocontrol (which we expect to be the case, at least up to a point), then results such as these may provide insight into how the environment should be managed to maximize the potential for expression of biocontrol activity. The relationship between phenazine gene expression and biological control is not simple, however. Thus, despite the fact that ammonium nitrogen was strongly correlated with biological control in situ (see above), it was the least supportive of the nitrogen sources tested from the standpoint of phenazine production. In the rhizosphere, ammonium nitrogen apparently has a greater effect on microbial ecology or the interactions among the biocontrol agent, the target pathogen and the host plant than it does on phenazine gene expression *per se*.

Strain 2-79 had a faster initial doubling time than did strain 30-84 when grown with glucose and urea as carbon and nitrogen sources, respectively. The kinetics of phenazine gene expression also differed significantly between the strains: an increase in β-galactosidase was detected in strain 2-79.2A40 almost immediately, whereas activity in 30-84Z declined initially and did not increase again until after 20 hours, when the growth rate was slowing and glucose had presumably become depleted. Such differences in response may help to explain why one strain consistently provides better disease suppression at some sites, and the other, at different sites.

3.4. REGULATION OF PHENAZINE GENE EXPRESSION

Two different genes that control phenazine biosynthesis have been partially characterized, one from strain 2-79 and the other, from 30-84. Mutants in *phzP*, identified originally as Phz^- in strains 2-79.B46 and 2-79.782 (Thomashow and Weller, 1988), are pleiotropic: they differ subtly in growth habit and, in contrast to the wild type, do not accumulate anthranilate when grown under iron deficiency (Thomashow and Pierson, 1991). The latter observation implicates anthranilate as a phenazine precursor in strain 2-79 (PCA accumulation also is reduced when iron is limited), although the source of anthranilate remains unknown (see above).

Tn*3*HoHo1 mutagenesis of *phzP* on the plasmid pPHZ49-61 indicated that the cloned locus spans 2.6 kb and is expressed as a single transcriptional unit. Expression of a chromosomal *phzP::lacZ* gene fusion increased in parallel with growth to a maximum of about 20 Miller units of β-galactosidase activity, plateaued at the onset of the stationary phase, and then remained stable or gradually declined.

A functional copy of *phzP* was required for expression of the phenazine biosynthetic locus. β-galactosidase activity in strain 2-79.2A40 was reduced from more than 1700 units to four units or less upon introduction of a kanamycin resistance cassette into the chromosomal *phzP* gene. Introduction of *phzP in trans* on pLAFR3 restored expression of the reporter in the biosynthetic locus, but only to about 50% of the wild-type level. Moreover, the presence of plasmid-borne copies of *phzP in trans* to the wild-type chromosomal *phzP* gene did not further stimulate expression of the the Phz::Lac reporter.

Hybridization studies indicate that *phzP* is unlinked to the major biosynthetic locus and that it is broadly distributed among other pseudomonads, including some that do not produce phenazines. In its pleiotropic phenotype, broad distribution and kinetics of expression, *phzP* resembles the *gacA* gene from *P. fluorescens* strain CHA0, which encodes a response regulator postulated to function globally in the control of secondary metabolism (Laville *et al.*, 1992).

A second activator of phenazine gene expression, *phzA*, has recently been identified in the 11.2 kb *Eco*RI fragment 5' to *phzB* and *phzC* in *P. aureofaciens* 30-84 (L. S. Pierson, unpublished results). In contrast to what was observed with *phzP*, introduction of the 11.2 kb fragment on pLAFR3 caused a six-fold increase in expression of a *phzB::lacZ* fusion in the chromosome. Subcloning, deletion analysis and mutagenesis with the transposon Tn*5*Lac further localized *phzA* to within 1.7 kb of the other phenazine biosynthetic genes and showed that it is divergently transcribed from *phzB* and *phzC*. Mutants containing a chromosomal *phzA::lacZ* fusion were Phz$^-$, but overproduced phenazines when complemented by *phzA in trans*. Complementation also resulted in 3.5-fold enhancement of *phzA::lacZ* expression, indicating that *phzA* is autoregulatory. A putative binding site for the PhzA gene product was identified within a 300 bp fragment 5' to *phzA*.

4. Acknowledgements

We thank Lauri Herman, Jim O'Hearn and Daryl Wilks for their help with some experiments. This work was supported in part by funds from the United States Department of Agriculture, National Research Initiative Competitive Grants Program.

5. Literature Cited

Bull, C. T., Weller, D. M., and Thomashow, L. S. 1991. Relationship between root colonization and suppression of *Gaeumannomyces graminis* var. *tritici* by *Pseudomonas fluorescens* strain 2-79. Phytopathology 81:954-959.

Calhoun, D. H., Pierson, D. L., and Jensen, R. A. 1973. The regulation of tryptophan synthesis in *Pseudomonas aeruginosa*. Mol. Gen. Genet. 121:117-132.

Essar, D. W., Eberly, L., and Crawford, I. P. 1990a. Evolutionary differences in chromosomal locations of four early genes of the tryptophan pathway in fluorescent pseudomonads: DNA sequences and characterization of *Pseudomonas putida trpE* and *trpGDC*. J. Bacteriol. 172:867-883.

Essar, D. W., Eberly, L., Hadero, A., and Crawford, I. P. 1990b. Identification and characterization of genes for a second anthranilate synthase in *Pseudomonas aeruginosa*: interchangeability of the two anthranilate synthases and evolutionary implications. J. Bacteriol. 172:853-866.

Hamdan, H., Weller, D. M., and Thomashow, L. S. 1991. Relative importance of fluorescent siderophores and other factors in biological control of *Gaeumannomyces graminis* var. *tritici* by *Pseudomonas fluorescens* strains 2-79 and M4-80R. Appl. Environ. Microbiol. 57:3270-3277.

Kuroda, M. I., Henner, D., and Yanofsky, C. 1988. *cis*-acting sites in the transcript of the *Bacillus subtilis trp* operon regulate expression of the operon. J. Bacteriol. 170:3080-3088.

Laville, J., Voisard, C., Keel, C., Maurhofer, M., Défago, G., and Haas, D. 1992. Global control in *Pseudomonas fluorescens* mediating antibiotic synthesis and suppression of black root rot of tobacco. Proc. Natl. Acad. Sci. USA 89:1562-1566.

Mazzola, M., Cook, R. J., Thomashow, L. S., Weller, D. M., and Pierson III, L. S. 1992. Contribution of phenazine antibiotic biosynthesis to the ecological competence of fluorescent pseudomonads in soil habitats. Appl. Environ. Microbiol. 58(8): in press.

National Academy of Sciences. 1987. Report of the Research Briefing Panel on Biological Control in Managed Ecosystems. National Academy Press, Washington, DC.

Ownley, B. H., Weller, D. M., and Alldredge, R. J. 1990. Influence of soil edaphic factors on suppression of take-all by *Pseudomonas fluorescens* 2-79. (Abstr.) Phytopathology 80:995.

Pierson III, L. S., and Thomashow, L. S. 1992. Cloning and heterologous expression of the phenazine biosynthetic locus from *Pseudomonas aureofaciens* 30-84. Mol. Plant-Microbe Interact. 5:330-339.

Thomashow, L. S., and Pierson III, L. S. 1991. Genetic aspects of phenazine antibiotic production by fluorescent pseudomonads that suppress take-all disease of wheat. Pages 433-449 in: Advances in Molecular Genetics of Plant-Microbe Interactions, Vol. 1. H. Hennecke and D. P. S. Verma, eds. Kluwer Academic Publishers, Dordrecht.

Thomashow, L. S., and Weller, D. M. 1988. Role of a phenazine antibiotic from *Pseudomonas fluorescens* in biological control of *Gaeumannomyces graminis* var. *tritici*. J. Bacteriol. 170:3499-3508.

Thomashow, L. S., Weller, D. M., Bonsall, R. F., and Pierson III, L. S. 1990. Production of the antibiotic phenazine-1-carboxylic acid by fluorescent *Pseudomonas* species in the rhizosphere of wheat. Appl. Environ. Microbiol. 56:908-912.

Turner, J. M., and Messenger, A. J. 1986. Occurrence, biochemistry and physiology of phenazine pigment production. Adv. Microbial Physiol. 27:211-275.

Weller, D.M. 1988. Biological control of soilborne plant pathogens in the rhizosphere with bacteria. Annu. Rev. Phytopathol. 26:379-407.

Yanofsky, C., and Crawford, I. P. 1987. The tryptophan operon. Pages 1453-1472 in: *Escherichia coli* and *Salmonella typhimurium*: Cellular and Molecular Biology, Vol. 2. F. C. Neidhardt, J. L. Ingraham, B. Magasanik, K. B. Low, M. Schaechter, and H. E. Umbarger, eds. American Society for Microbiology, Washington, DC.

A BIOLOGICAL SENSOR FOR IRON THAT IS AVAILABLE TO *PSEUDOMONAS FLUORESCENS* INHABITING THE PLANT RHIZOSPHERE

J. E. LOPER and M. D. HENKELS
U.S. Department of Agriculture
Agricultural Research Service
3420 N.W. Orchard Ave.
Corvallis, OR 97330

S. E. LINDOW
Department of Plant Pathology
University of California
Berkeley, CA 94720

Abstract

The availability of iron to a strain of *Pseudomonas fluorescens* inhabiting the rhizosphere of bean was evaluated using a novel biological iron sensor. The iron sensor, termed *pvd-ina*Z, was constructed by cloning an iron-regulated promoter, isolated from a gene involved in pyoverdine (*pvd*) (fluorescent siderophore) production of *P. syringae*, upstream of an ice-nucleation-activity gene (*ina*Z) that lacked its native promoter. Cells of *P. fluorescens* containing the *pvd-ina*Z fusion expressed ice nucleation activity only when grown in iron-limited media, in which the iron-regulated *pvd* promoter was active. Cells of *P. fluorescens* that contained *pvd-ina*Z expressed less ice nucleation activity in the rhizosphere of bean plants grown in soil amended with FeEDTA than in the rhizosphere of plants grown in unamended soil. Thus, the ice nucleation activity expressed by cells containing *pvd-ina*Z was regulated by the levels of iron available to *P. fluorescens* in the plant rhizosphere as well as in culture medium. Whereas conventional techniques are not useful for detection of concentrations of iron available to bacteria occupying microhabitats in the plant rhizosphere, the ice nucleation gene fusion system provided a sensitive, convenient, and inexpensive sensor of iron availability to *P. fluorescens* inhabiting these sites.

Introduction

The activities, metabolite production, and gene expression of plant-associated microorganisms are dependent on environmental factors, including the chemical composition of habitats that such microbes occupy. Iron profoundly influences gene expression of prokaryotes and is particularly meaningful to the activity and metabolism of bacteria (18). Iron also mediates certain interactions between microorganisms in the rhizosphere of agricultural plants (9). Although the iron content of an environmental sample, such as from soil or plant tissue, can be accurately assessed by chemical analysis, traditional methods can not assess the levels of iron that are available to microbes occupying microhabitats on plant surfaces. In this report, we describe a reporter gene fusion system that functions as a "biological sensor" for iron available to *Pseudomonas*

E. W. Nester and D. P. S. Verma (eds.),
Advances in Molecular Genetics of Plant-Microbe Interactions, 543–547.

fluorescens occupying the plant rhizosphere.

Iron is an essential element for living organisms by virtue of its two valences that act as cofactors in various oxidative reductive enzymatic reactions. Iron is abundant on the earth's crust; yet in aerobic environments at neutral pH, it exists as insoluble iron oxides, which are largely unavailable biologically. As a result, most organisms have systems for the specific chelation and regulated transport of iron into the cell. With some exceptions, microorganisms use siderophores and corresponding membrane receptors for iron acquisition. Siderophores are low-molecular-weight, Fe(III)-specific ligands that are produced by organisms as iron-scavenging agents when available forms of iron are limited.

Fluorescent pseudomonads produce fluorescent pigments, termed pyoverdines or pseudobactins, that function as siderophores, characterized by their synthesis only under iron-limiting conditions, specificity and high affinity for the ferric ion, and role in transport of Fe^{3+} into the bacterial cell (2). At least 12 genes located in four to five gene clusters are required for pyoverdine biosynthesis by *Pseudomonas* spp. (11,12,14). Several iron-regulated promoters have been localized within these gene clusters by constructing transcriptional fusions with reporter genes (5,13,15). A promoter that is regulated specifically by iron and is transcribed in a number of strains of *Pseudomonas* spp. was identified within a cloned genomic region required for pyoverdine production by *P. syringae* (Loper, unpublished).

The iron-regulated promoter from *P. syringae* was fused to an ice-nucleation-activity gene (*inaZ*) devoid of its native promoter. An ice nucleation gene is an extremely sensitive reporter of transcriptional activity, because theoretically, a single cell with an active ice nucleus can be detected, both on plant surfaces and in culture, by the freezing of the aqueous sample in which it is suspended. Further, the ice nucleation activity of a bacterium is related quantitatively to the amount of InaZ protein present in the outer membrane of the cell (6). Therefore, the ice nucleation reporter gene system provides a quantitative assessment of bacterial gene expression in culture, in plant tissue, and on aerial plant surfaces (6). In this report, we extend the use of the ice nucleation reporter gene system to the evaluation of transcriptional activity of bacteria inhabiting the plant rhizosphere.

Methods

CONSTRUCTION OF FUSION PLASMIDS

Plasmid pVSP61, which is maintained stably in *Pseudomonas* spp., was obtained from William Tucker (DNA Plant Technologies, Oakland, CA, USA). Constructs in pVSP61 were: i) *pvd-inaZ*, the promoterless *inaZ* gene fused to an iron-regulated promoter of a region involved in pyoverdine production of *P. syringae* (11), and ii) *iceC*, an ice nucleation gene (16) driven off its indigenous, iron-constitutive promoter.

ICE NUCLEATION ACTIVITY OF *P. FLUORESCENS* IN CULTURE

The rhizosphere bacterium, *P. fluorescens* Pf-5 (3), was obtained from Charles Howell (USDA, ARS, College Station, TX, USA). Strain Pf-5 is not ice-nucleation active unless a transcribed gene encoding the outer-membrane InaZ protein is introduced into the cell. Fusions described above were introduced into strain Pf-5 by conjugation. Transconjugants were grown at

25 C with shaking in a minimal salts medium (SM) containing glycerol (10 g/L) and glutamine (1 g/L) (10). Ice nucleation activity of cultured cells was quantified by the droplet freezing assay (7). Ice nucleation activity was normalized for the number of bacterial cells, as determined by dilution plating or turbidity measurements.

ICE NUCLEATION ACTIVITY OF *P. FLUORESCENS* IN THE RHIZOSPHERE

Roots of bean seedlings (cv. Bush Blue Lake) were dipped in aqueous bacterial suspensions (ca. 10^6 cfu/ml) and planted in pots containing Warden sandy-silt loam (-0.3 bar, pH=7.2, 14 mg/kg DTPA-extractable Fe). Iron availability of the soil was altered by adding iron chelators. The chelator EDTA, which has an Fe^{3+}-chelate stability constant of 25.00, is useful for enhancing the iron availability to soil organisms (4,17). Iron availability was reduced by adding HBED [N,N'-di-(2-hydroxybenzoyl)-ethylenediamine-N,N'diacetic acid], which has an Fe^{3+}-chelate stability constant of 39.68 (1). Plants were harvested after five days growth at 25 C. Rhizosphere bacterial populations of individual root systems were quantified by published methods (8). Ice nucleation activity of rhizosphere bacteria was determined by the droplet freezing assay (7). Inocula were grown in a culture medium containing 10^{-4} M $FeCl_3$. Because cells of *P. fluorescens* containing *pvd-inaZ* expressed negligible levels of ice nucleation activity when grown in media containing 10^{-4} M $FeCl_3$, the activity expressed by these cells after 5 days growth in the bean rhizosphere reflected *in situ* expression of the iron-regulated promoter.

Results

ICE NUCLEATION ACTIVITY IN CULTURE

The ice nucleation activity of *P. fluorescens* strain Pf-5 containing the *pvd-inaZ* gene fusion was regulated dramatically by iron. *P. fluorescens* containing *pvd-inaZ* expressed ca. 1 ice nucleus/cell when grown in an iron-deplete minimal medium, and only ca. 10^{-5} nuclei/cell when the medium was supplemented with 10^{-4} M $FeCl_3$. Ice nucleation activity of cells containing *iceC*, in which the ice nucleation gene was transcribed from its native iron-constitutive promoter, was not affected by [$FeCl_3$]. Ice nucleation activity of cells harboring *pvd-inaZ* was substantially greater than that of cells harboring *iceC* in media with less than $10^{-5.5}$ M $FeCl_3$. In contrast, at [$FeCl_3$] greater than 10^{-5} M, ice nucleation activity of cells harboring *pvd-inaZ* was substantially less than that of cells harboring *iceC*. Thus, comparison of ice nucleation activities of cells harboring *pvd-inaZ* and those harboring *iceC* provided an assessment of iron availability to *P. fluorescens*.

ICE NUCLEATION ACTIVITY IN THE RHIZOSPHERE

Cells of *P. fluorescens* that contained *iceC* served as internal controls to document that the environmental conditions encountered by cells in the rhizosphere were conducive for the expression of ice nucleation activity. Ice nucleation activity of cells containing *iceC* was similar in culture and in the rhizosphere, indicating that root exudates and soil particles did not interfere with estimates of ice nucleation activity. The availability of Fe^{3+} sensed by *P. fluorescens* in the rhizosphere was assessed by comparing the ice nucleation activity of cells containing *pvd-inaZ* to that of cells containing *iceC*. In the bean rhizosphere, cells of Pf-5 containing either *pvd-inaZ* or *iceC* expressed ice nucleation activities of ca. 10^{-2} to 10^{-3} nuclei/cell. Only background ice nucleation activity (ca. 10^{-7} nuclei/cell) was expressed by cells containing the promoterless *inaZ*. *P. fluorescens* cells containing *pvd-inaZ* were much less active (ca. 10^{-6} nuclei/cell) in soil amended with 100 μg/g FeEDTA than in unamended soil. In contrast, cells containing *pvd-inaZ* were more active (ca. 10^{-1} nuclei/cell) in soil amended with 100 μg/g HBED than in unamended soil. Neither chelator influenced the ice nucleation activity expressed by cells containing *iceC* in the rhizosphere. Thus, iron availability of the soil influenced transcriptional activity of the iron-regulated promoter in the rhizosphere.

Discussion

An iron-regulated promoter of gene(s) involved in pyoverdine production was expressed by *P. fluorescens* in the rhizosphere of bean. The level of *in situ* expression was intermediate between that in an iron-deplete medium (10^{-7} M $FeCl_3$) and that in an iron-replete medium (10^{-4} M $FeCl_3$). The expression of the iron-regulated promoter by *P. fluorescens* in the rhizosphere was manipulated by the addition of the iron chelates, FeEDTA or HBED, to the soil. Thus, the *pvd* promoter was regulated by levels of iron that were available to *P. fluorescens* in the rhizosphere as well as in culture. The *pvd-inaZ* fusion can be utilized as a sensor of concentrations of biologically-available iron in natural habitats such as the rhizosphere. The *pvd-inaZ* gene fusion also provides a tool for future evaluation of edaphic factors influencing pyoverdine production on plant surfaces.

References

1. Chaney, R. L. 1988. Plants can utilize iron from Fe-N,N'-di-(2-hydroxybenzoyl)-ethylenediamine-N,N'-diacetic acid, a ferric chelate with 10^6 greater formation constant than Fe-EDDHA. J. Plant Nutr. 11:1033-1050.
2. Demange, P. Wendenbaum, S., Bateman, A., Dell, A., and Abdallah, M. A. 1987. Bacterial siderophores: Structure and physicochemical properties of pyoverdins and related compounds. Pages 167-187 in: Iron Transport in Microbes, Plants and Animals, G. Winkelman, D. van der Helm, and J. B. Neilands, eds. VCH Publishers, Weinheim. FRG.
3. Howell, C. R., and Stipanovic, R. D. 1979. Control of *Rhizoctonia solani* on cotton seedlings with *Pseudomonas fluorescens* and with an antibiotic produced by the bacterium. Phytopathology 69: 480-482.
4. Kloepper, J. W., Leong, J., Teintze, M. and Schroth, M. N. 1980. *Pseudomonas* siderophores: A mechanism explaining disease-suppressive soils. Curr. Microbiol. 4:317-320.
5. Leong, J. Bitter, W., Koster, M., Venturi, V., and Weisbeek, P. J. 1991. Molecular analysis of iron transport in plant growth-promoting *Pseudomonas putida* WCS358. Biol. Metals 4:36-40.
6. Lindgren, P. B., Frederick, R., Govindarajan, A. G., Panopoulos, N. J., Staskawicz, B. J., and Lindow, S.E. 1989. An ice nucleation reporter gene system: identification of

inducible pathogenicity genes in *Pseudomonas syringae* pv. *phaseolicola*. EMBO J. 8:1291-1301.

7. Lindow, S. E. 1990. Bacterial ice nucleation measurements. Pages 428-434 in: Methods in Phytobacteriology, D. Sands, Z. Klement, and K. Rudolf (eds.) Akademia Kiado, Budapest.
8. Loper, J. E. 1988. Role of fluorescent siderophore production in biological control of *Pythium ultimum* by a *Pseudomonas fluorescens* strain. Phytopathology 78: 166-172.
9. Loper, J. E., and Buyer, J. S. 1991. Siderophores in microbial interactions on plant surfaces. Molecular Plant-Microbe Interactions 4:5-13.
10. Loper, J. E., and Lindow, S. E. 1987. Lack of evidence for *in situ* fluorescent pigment production by *Pseudomonas syringae* pv. *syringae* on bean leaf surfaces. Phytopathology 77:1449-1454.
11. Loper, J. E., Orser, C. S., Panopoulos, N. J., and Schroth, M. N. 1984. Genetic analysis of fluorescent pigment production in *Pseudomonas syringae* pv. *syringae*. J. Gen. Microbiol. 130: 1507-1515.
12. Marugg, J.D., van Spanje, M. Hoekstra, W. P. M., Schippers, B., and Weisbeek, P. J. 1985. Isolation and analysis of genes involved in siderophore biosynthesis in plant-growth-stimulating *Pseudomonas putida* W3S358. J. Bacteriol. 164:563-570.
13. Marugg, J. D., Nielander, H. B., Horrevoets, A. J. G., van Megen, I., van Genderen, I., and Weisbeek, P. J. 1988. Genetic organization and transcriptional analysis of a major gene cluster involved in siderophore biosynthesis in *Pseudomonas putida* WCS358. J. Bacteriol. 170:1812-1819.
14. O'Sullivan, D. J., Morris, J., and O'Gara. F. 1990. Identification of an additional ferric-siderophore uptake gene clustered with receptor, biosynthesis, and *fur*-like regulatory genes in fluorescent *Pseudomonas* sp. strain M114. Appl. Environ. Microbiol. 56:2056-2064.
15. O'Sullivan, D. J., and O'Gara, F. 1991. Regulation of iron assimilation: nucleotide sequence analysis of an iron-regulated promoter from a fluorescent pseudomonad. Mol. Gen. Genet. 228:1-8.
16. Orser, C. Staskawicz, B. J., Panopoulos, N. J., Dahlbeck, D., and Lindow, S. E. 1985. Cloning and expression of bacterial ice nucleation genes in *Escherichia coli*. J. Bacteriol. 164:359-366.
17. Scher, F. M., and Baker, R. 1982. Effect of *Pseudomonas putida* and a synthetic iron chelator on induction of soil suppressiveness to fusarium wilt pathogens. Phytopathology 72:1567-1573.
18. Weinberg, E. D. 1990. Roles of trace metals in transcriptional control of microbial secondary metabolism. Biol. Metals 2:191-196.

CHARACTERISTICS AND APPLICATIONS OF ANTIBODIES PRODUCED IN PLANTS

Andrew Hiatt, Julian K-C. Ma., Department of Cell Biology, The Scripps Research Institute, La Jolla, CA 92037

Abstract. The majority of antibodies, such as those used in diagnostic immunoassays or for *in vivo* clinical applications, are produced in animals and isolated from serum or ascites fluid, or are produced in batch culture. With the advent of recombinant DNA technology, it became possible to clone the genes for the antibody proteins and express them in other organisms. This process was most easily achieved in cells of vertebrate organisms, because the specialized cellular machinery for processing of the antibodies already existed in such cells. Part of the difficulty with the synthesis of antibodies in recombinant organisms is that the production of a normal, functional antibody requires the synthesis of two proteins and therefore the expression of two genes. Both of these genes must therefore be transferred to the recombinant organism in order to achieve antibody expression in the foreign host. This heterologous expression also requires the orderly assembly of the protein tetramer. The IgG molecule is composed of two pairs of subunits of different molecular weight and different amino acid sequence. Each IgG molecule has two identical large molecular weight (heavy chain) subunits linked together by disulfide bonds. Each heavy chain is also disulfide bonded to one of two identical lower molecular weight (light chain) subunits. In addition, there are extensive intramolecular disulfide bonds which are necessary for antigen binding and other immunoglobulin mediated functions.

The fidelity of such assembly is ensured in cells which normally synthesize immunoglobulins because of the recognition of processing and signal sequences in the immunoglobulin genes and because of specialized cellular machinery to accomodate these functions. Expression in heterologous systems which do not possess these characteristics is somewhat more difficult and still poorly understood.

A variety of bioactive compounds have been stably introduced into plants using genetic engineering techniques (1-12). These compounds, by and large, are derived from mammalian, bacterial, or viral genes which encode well characterized protein products. In some instances other bioactive plant products have been introduced into a new plant environment to obtain a useful trait. The initial studies of antibodies in plants focussed on the IgG class of antibodies, although as discussed later, we are now also investigating the production of multimeric forms of immunoglobulin. In this paper, we shall review our findings on the characteristics of antibodies derived from plants and discuss the potential applications of these antibodies.

CHARACTERISATION OF ANTIBODY EXPRESSED IN TOBACCO.

Expression strategy An IgG_1 antibody (6D4) was chosen for expression in tobacco. This antibody recognizes a synthetic phosphonate ester, P3, and can catalyze the hydrolysis of certain carboxylic esters (13). Gamma and kappa chain cDNAs derived

E. W. Nester and D. P. S. Verma (eds.),
Advances in Molecular Genetics of Plant-Microbe Interactions, 549–560.

from the 6D4 hybridoma were cloned separately into the plant expression vector, pMON530 (14). Transformation of tobacco was mediated by co-cultivation of leaf segments with *Agrobacterium tumefaciens* containing the γ or κ constructs (14). The levels of expression of γ and κ chain transformed plants are shown in Table 1A. A comparison between plants transformed with vectors that included a leader sequence (γL and κL) and vectors without leader (γNL and κL) is shown. In both cases, γ chain expression was higher than κ chain expression. However, the presence of leader sequence significantly increased the accumulation of both γ and κ chains. Transgenic plants containing γ or κ chains were then crossed to produce progeny expressing both chains. The accumulation levels of γ -γL(κL) and κ −κL(γL) chains in the F_1 plants are also shown in Table 1A. Again, the presence of leader sequence was important for the levels of expression. However, in these plants, γ and κ chains were expressed at much higher levels and in equivalent amounts - 3330ng/mg and 3700ng/mg respectively.

The pattern of antibody chain expression in the F_1 progeny is shown in Table 1B. In a cross between γ and κ chain plants that had leader sequences (κL x γL), 11/18 of the offspring expressed γ and κ chains together. Furthermore, assembly was observed in the majority of these plants. A proportion expressed only one chain and 4 plants expressed neither. If no leader sequence was present (κNL x γNL), no assembly was seen in the plants that expressed both γ and κ chains. That the F_1 plants contained assembled functional antibody, was determined by the following criteria: 1) Western blots of plant extracts under reducing conditions contained equimolar amounts of γ and κ chains which migrated at 50 kD and 25 kD respectively. Under non-reducing conditions both γ and κ bands migrated at about 160 kD. 2) ELISA assays in which plant extracts were added to microtiter plates coated with goat anti-mouse γ chain, then detected with goat anti-mouse κ chain-HRPO indicated the presence of equimolar amounts of γ chain bound to κ chain with no detectable free γ or κ chains. 3) ELISA assays with a P3-BSA conjugate as antigen, gave similar results: the affinity of γ–κ complexes for P3 was identical to that of the hybridoma-derived antibody. The specificity for P3 was indicated by inhibition of P3-BSA binding by free P3 in which half-maximal inhibition was about 10 mM for plant derived or hybridoma derived antibody.

A surprisingly high level of accumulation of functional antibody was observed. In the case of 6D4, greater than 1% of total extractable protein was found to be functional antibody. Other antibodies, which have subsequently been expressed in tobacco using the same strategy have resulted in similar levels of accumulation.

Functional activity of the purified plant antibody. Further analysis of the plant antibody was performed after purification on Sephacryl-FPLC and Protein-A sepharose. A crude homogenate of tobacco leaf in Tris buffered saline (pH 8) with 1 mM PMSF was centrifuged to remove insolubles and concentrated by filtration. Sephacryl-FPLC fractions were assayed by ELISA, pooled and adsorbed to Protein-A Sepharose using buffers that enhance binding of IgG1. The eluted antibody was evaluated by Coomassie blue staining and Western blotting and was found to be virtually pure. The quantitative retention of the antibody on Protein A Sepharose indicates that the interdomain conformation of Fc recognized by Protein A (between CH2 and CH3) is intact. The purified antibody eluted from the Protein A column was dialyzed against the appropriate buffers and used for analysis of catalysis, glycosylation, and amino acid sequence.

Catalytic activity was measured by incubation of the antibody with substrate in the presence and absence of an inhibitor to derive K_M, K_I, V_{max} and k_{cat} values (13,15). The time course of the reaction was measured spectrophotometrically. The results showed that for each parameter, plant derived and ascites derived antibodies differed by less than an order of magnitude (Table 2). These measurements are within the observed range for different batches of the ascites produced antibody (Janda, unpublished data).

Glycosylation. Glycosylation of the antibody heavy chain has been investigated by lectin binding analysis (16,17), comparing plant derived antibodies with those of murine ascites origin. Purified antibody was first Western blotted to nitrocellulose; various biotinylated lectins were then incubated with the blots and lectin binding was visualized with streptavidin-alkaline phosphatase and bromo-chloro-indoyl-phosphate. In some cases, the purified antibody was incubated with endoglycosidase H prior to blotting. The results showed that the plant-derived γ chain was bound by Concanavalin A (specific for mannose and glucose), whereas the ascites γ chain was recognized by Concanavalin A as well as the lectins from *Ricinus communis* (specific for terminal galactose and N-acetylgalactosamine) and wheat germ agglutinin (N-acetylglucosamine dimers, terminal sialic acid); (18). The lectins from *Datura stromonium* (N-acetyl glucosamine oligomers, N-acetyl lactosamine) and *Phaseolus vulgaris* (galactose b1,4 N-acetyl glucosamine b1,2 mannose) did not bind to either plant or ascites-derived γ chain.

Elution of the lectin from the blots using α-methylglucoside was used to compare the relative affinity of Con A binding to the plant and ascites heavy chains (19). The results showed that the two antibodies are indistinguishable by Con A affinity as well as by the quantity of Con A bound per mg of γ chain. Blots in which the antibodies were first digested with endoglycosidase H (18,20) displayed no reduction in Con A binding under conditions where Con A binding to ovalbumin (containing a high mannose type carbohydrate) was diminished. Since endoglycosidase H resistance is characteristic of complex carbohydrates processed in the Golgi apparatus, these results indicate that the transgenic antibody is processed in a similar fashion to complex mammalian glycoproteins.

N-linked glycosylation of proteins in plants is similar to the glycosylation process in mammals (21,22). A core high mannose oligosaccharide is attached to the asparagines contained within the canonical Asn-X-Ser/Thr sequence. This occurs in the endoplasmic reticulum and can be modified in the Golgi apparatus where α-mannosidase removes some mannose residues and terminal sugars are attached. In mammals, the predominant terminal residue is N-acetyl neuraminic acid (NANA); this carbohydrate has not been identified in plants. Terminal residues in plants have been found to consist of xylose, fucose, N-acetylglucosamine, mannose, or galactose (22). In other respects, such as the size and extent of branching, plant glycans are very similar to mammalian glycans. Although we have not attempted a structural characterization of the oligosaccarides attached to the antibody, the comparative analysis of plant-derived and tobacco-derived glycans by lectin binding has revealed both differences and similarities. Both glycans are resistant to endoglycosidase H under conditions of carbohydrate digestion of the control glycoprotein, ovalbumin, which contains a high mannose carbohydrate structure. This demonstrates processing of the high mannose carbohydrate to the complex type in the Golgi. In addition, both glycans have approximately the same affinity for Con A since they were not distinguishable by competition with α-methyl mannoside. This type of assay has previously been used to distinguish a variety of plant glycans with respect to their affinity for lectin (19). Two of the lectins used in the binding assays were found to distinguish the plant-derived from the mammalian antibody. *Ricinus communis* agglutinin, which can bind terminal galactose and N-acetyl galactosamine residues, and wheat germ agglutinin, which can bind to terminal NANA, did not bind to the plant glycan. This suggests a distinct composition of terminal residues on the plant glycan and is consistent with the absence of NANA in plants.

NH2-terminal amino acid sequence. Evidence that the antibody has been proteolytically processed in the lumen of the ER is provided by the N-terminus amino acid sequence. Purified antibody was blotted onto a polyvinylidene difluoride (PVDF) membrane (23) after which the heavy or light chain bands were located on the blot by Coomassie blue staining. Pieces of PVDF membrane with bound immunoglobulin were then subjected to automated sequence analysis. Heavy chain from either plant antibody or mouse derived 6D4 was intractable to sequencing indicating a blocked N-terminus. The light chain N-terminal sequence was asp-val-val-leu for both plant and mouse antibody. This demonstrates appropriate proteolytic processing of the mouse signal sequence by the plant ER.

Antibody secretion. A series of experiments were performed to determine whether the antibody is secreted, or remains within the confines of the cell membrane or cell wall. The best antibody producing plant was propagated as a callus suspension culture. The callus proliferates as de-differentiated cells (hormonally induced from leaf cells) which grow at high density in a defined medium consisting primarily of sucrose, nitrate and the appropriate plant hormone. Functional antibody could be detected in spent suspension culture growth medium, which suggests that the majority of the antibody pool was being secreted into the medium.

To investigate secretion directly, protoplasts containing no cell walls (24) were prepared from antibody-producing leaves and the antibody was labelled by incubating the protoplasts with ^{35}S-methionine for 2 hours. At that time, radiolabel in the cell-associated immunoglobulin was compared to radiolabel in extracellular immunoglobulin. A significant fraction of newly synthesized antibody was found in the growth medium (Table 3). After a 2 hr chase with 100 mM methionine, most of the labelled antibody was extracellular indicating that secretion of the antibody had occurred.

Since plant cells are surrounded by a cell wall which restricts the passage of large molecules (25), it is possible that the secretion of relatively large protein molecules, such as immunoglobulin, would be restricted from diffusing into the extracellular medium. To evaluate the movement of secreted antibodies from the protoplast through the cell wall barrier, antibody secretion was also measured in established suspension cells which contain a primary cell wall. An identical radiolabelling protocol was employed with the suspension cells as with protoplasts.

The comparison of antibody secretion from cells with intact cell walls (callus) and protoplasts indicated that most of the newly synthesized antibody was secreted from the cells and reached the incubation medium (Table 3). A pulse chase protocol was necessary to characterize active secretion of immunoglobulin in culture, since cell death can also result in accumulation of macromolecules in the growth medium. The results showed that after a two hour chase, accumulation of the majority of radiolabeled immunoglobulin occurs in the growth medium. In the case of protoplasts, greater than 85% of the labelled antibody was secreted; from callus cells, approximately 70% of the antibody was found in the medium. These results demonstrate antibody secretion through both the plasma membrane and the cell wall of callus cells. This was somewhat surprising, since the permeability of plant cells to macromolecules is thought to be very limited. Evaluation of macromolecule migration through plant cell walls from a variety of sources has demonstrated an exclusion limit equivalent to a 20,000 dalton globular protein (25). Secretion of the antibody (M_r = 150,000) through the cell wall of a callus suspension cell may reflect the presence of a small population of large pores in the plant cell wall to allow passage of large proteins or extracellular carbohydrates (25).

Antibody expression in protoplasts. Electroporation into tobacco protoplasts of two vectors encoding immunoglobulin heavy or light chains respectively resulted in the synthesis, assembly and secretion of functional antibodies. This transient expression system has been used to evaluate a variety of DNA constructs for their ability to support antibody expression in plant cells. Some of these constructs have also been used to transform leaf discs followed by plant regeneration. A comparison was made among vectors based on differences in the type and arrangement of promoters. These included single vectors with separate promoters, tandem promoters, and heat shock inducible promoters. In addition, DNA mutagenesis was used to generate heavy and light chain coding regions in which the glutamine target for N-glycosylation was removed, the protease sensitive sites were removed, or the majority of the heavy chain constant region was deleted.

In general, we found that different antibody structures can be produced from plant cells by a number of promoter arrangements in the vector. In previous experiments (26,17) we have transformed individual plants with either heavy or light chains; assembly of functional antibody then results from the sexual cross of γ and κ expressing plants. Since it would clearly be less time consuming to express both gamma and kappa chains in the initial transformant, we first tested promoter arrangments whereby gamma and kappa cDNAs are encoded in the same vector.

A vector, pHi202 (26) was used which contains the leader and γ chain insert. In addition, a Hind III fragment containing an additional 35S promoter (14), the light chain cDNA, and the nopaline synthase gene 3' end were introduced. The two orientations of the double expression vector were then introduced into protoplasts. Only the "head-to-tail" orientation was resulted in expression of assembled antibody. The efficiency of expression (as measured by the ng of antibody produced from 10^6 cells electroporated with 20 mg DNA) was significantly higher than electroporation using two vectors.

Two types of constructs were made to measure the effects of variations in levels of transcription on the accumulation of secreted antibody. In the first construct, the 35S promoter was substituted by the HSP 70 heat shock inducible promoter from soybean (30). Protoplasts were first electroporated then exposed to 37° C at varying times after electroporation. None of the induction protocols resulted in production of antibody. To test the functionality of these constructs, plants were regenerated after leaf disc transformation using the vectors. We found that both γ and κ chains were expressed only after exposure to 37 C for one hour. Some of the progeny from a cross between γ and κ expressing plants expressed functional antibody. It is interesting to note that antibody expressing plants were only derived from a cross between plants transformed with the HSP 70 construct. No antibody was expressed if one parental plant contained the 35S promoter and the other expressed the HSP 70 promoter.

In the second construct, the heavy or light chain expression cassette containing the 35S promoter, cDNA and 3' nos region in pMON530 (14) were introduced into a different but similar vector (pKYLX71) (31) at the Hind III site. This resulted in a vector with two 35S promoters being upstream from the cDNA. The expression of secreted antibody from protoplasts electroporated with these vectors was at least 10-fold more efficient than constructs employing a single 35S promoter.

The results from analysis of regenerated plants expressing heavy and light chain constructs suggested that the native signal sequence on the mouse transcript contributes significantly to the accumulation of heavy or light chains (Table 1). This effect is presumably due to the sequestering of the chains in the endomembrane system and their subsequent secretion. In addition, we observed that levels of expression of either chain alone, was significantly lower than when both chains were expressed in the same plant. Furthermore, in all constructs, the heavy chain expressed alone accumulated to significantly higher levels than the light chain (Table 1,35).

To examine further the accumulation of heavy and light chains derived from constructs containing no leader sequences, we have introduced cDNAs with no signal sequence into the protoplasts by electroporation. Electroporated cells were washed to remove extracellular medium then lysed by freezing and thawing in 10 volumes of water. After concentrating the clarified extract, expression of the individual chains was measured by ELISA. Expression of individual chains was barely detectible and when both chains were co-expressed, there was no assembly of γ–κ complexes, nor was there any change in the expression level of the two chains. The low titer of immunoglobulin may have been responsible for the absence of assembly and could be due to a protease localized in the cytoplasm. To test this hypothesis, a Lys-Lys pair in the heavy chain constant region and a Lys-Arg pair in the light chain constant region were mutagenised to Lys-Leu pairs. These basic amino acids had previously been shown to be target sites for immunoglobulin degradation in transgenic plants. However, no significant difference in the accumulation of the modified or unmodified heavy or light chains was observed. Assembly of γ–κ complexes did not occur, suggesting that proteolysis at these sites was not a significant factor preventing the accumulation or assembly of immunoglobulin chains. Clearly, more extensive modifications are needed to overcome the high turnover of immunoglobulin in the cytosol.

Immunoglobulin assembly in plants. Plants must contain an assembly and processing apparatus which can recognize mouse immunoglobulins. The efficiency of assembly in plants was surprising since there was a large disparity in parental levels of expression of individual γ or κ chains, the heavy chain parent produced 30 fold more Ig than the light chain parent (Table 1). After crossing, one of the antibody producing progeny contained 5-fold more γ and 160-fold more κ chain than the respective parental plants. This suggests that the assembled antibody is more stable in plants than the individual heavy or light chains. It is apparent that cognate mechanisms function in plants to coordinate the assembly of oligomers, to direct N-glycosylation and to effect processing in the golgi apparatus, to recognize and to process immunoglobulins efficiently. Assembly of immunoglobulin chains in mammalian cells is thought to occur via a native component of the endoplasmic reticulum, immunoglobulin heavy chain binding protein (BiP), which is involved in the post-translational processing of heavy chains (32). BiP has been found associated with unassembled immunoglobulin heavy chains prior to assembly with light chains. Unassembled heavy chains remain associated with BiP and are not transported to the Golgi apparatus (32). In addition to a role in processing Ig heavy chains, the extent of BiP association can be inversely correlated with the efficiency of secretion of human factor VIII, human tissue plasminogen activator, and human von Willebrand Factor expressed in transgenic CHO cells (32). A BiP-like protein in plants has recently been characterized (33,34).

APPLICATIONS OF PLANT DERIVED MONOCLONAL ANTIBODIES.

The transformation of plants to produce monoclonal antibodies (MAb) involves at least two generations of plants and is a lengthy process, compared with other more conventional techniques. Although improvements of the vectors and regeneration steps will undoubtedly occur, the technology is probably better suited to well characterised antibodies, rather than projects which may require some level of screening for the correct choice of antibody. However, plants have several advantages over other expression systems.

One of the most attractive potential uses for plant antibodies is in passive immunisation. Monoclonal antibodies have been used *in vivo* for many human diseases with varying degrees of success. The prospect of harvesting monoclonal antibody on an agricultural scale would mean that therapeutic antibodies would be available extremely cheaply in almost limitless amounts. We have also been interested in the possibility of orally delivered MAb, so that if the right transgenic plants were used, purification of the antibody would be unneccessary. The efficacy of topically applied MAb to prevent oral disease has been demonstrated in the case of dental caries in both sub-human primates (35) and humans (36). Dental caries is predominantly caused by *Streptococcus mutans* which colonises the teeth of children at an early age, and later becomes established as part of the commensal flora. Topically applied MAb, that was raised against the cell surface adhesin (SA I/II) of *S. mutans* prevented establishment of the bacteria in sub-human primates, and also reduced the levels of disease. In humans, MAb against SA I/II that were applied directly to the teeth confered long term protection against *S. mutans* colonisation in adults (37). Antibodies against *S. mutans* have also been delivered topically in food and were shown to prevent colonisation and disease (38,39). Transgenic plants that produce MAb Guy's 13 that was used in the human studies, are currently being constructed.

Once plants have been transformed to produce a MAb, there are a number of advantages with respect to vaccine storage and distribution. Plant genetic material is readily stored in seeds, which are extremely stable and require little or no maintenance. Seeds have an almost unlimited shelf life in ambient conditions, unlike the considerably more stringent requirements of bacterial or mammalian cells. Immortalisation of the plant line therefore is extremely simple, furthermore, mature plants can be self fertilised to produce identical offspring, a technique that has been established in plant breeding programmes for many years. In addition, as the number of transformable plant species increases, to include several of the major food crops, such as rice, potato, cassava and peppers, the possibility arises of delivering antibody vaccines on a global scale, using plants that are indigenous to particular regions. In this way, vaccines could be administered cheaply to third world countries by utilising the existing agricultural infrastructure.

In addition to the economic advantages of using plant derived antibodies, the ease with which genetic material can be exchanged, simply by cross fertilisation, may facilitate the construction of multimeric forms of antibody. The ability of plants cells to assemble heavy and light antibody chains correctly, has been demonstrated (26). The possibility of constructing antibodies that require more than two polypeptide chains is being explored, and the secretory form of IgA (sIgA) is of particular interest. sIgA is usually present at mucous membranes, as a dimer, consisting of two monomeric IgA molecules joined by a small polypeptide, known as the J chain, and complexed with a larger polypeptide, the secretory component. Although J chain is synthesised by the plasma cells that secrete the IgA, secretory component is expressed by epithelial cells that are present in the secretory glands. Because of this, it has not been possible to generate monoclonal sIgA using standard hybridoma techniques. However, using transgenic plants, the genes that encode the four component chains of sIgA could be incorporated into one plant by simple cross breeding, although it remains to be determined, whether correct assembly of antibody would occur.

The relative importance of sIgA in protection against infection is still not clear. However, as thc majority of infectious agents are initially encountered at mucosal sites of the body, a considerable amount of effort has been directed towards inducing immune responses in mucosal secretions (for reviews, see 40,41). The IgA response in secretions to *S. mutans* was one of the first and most extensively studied models. Oral ingestion or local injection of bacteria elicited a sIgA response in rodents, which led to protection against disease (42,43). In humans, salivary sIgA was induced following oral ingestion of *S. mutans*, in some (44), but not all studies (45). Similarly inconclusive results were encountered in studies using sub-human primates (46-48).

SUMMARY. Of the variety of compounds expressed in transgenic plants, antibodies offer probably the widest range of applications. The antibodies appear to possess all of the functional characteristics of antibody derived from hybridoma cells, although further study will be required to determine the effect of the difference in heavy chain glycosylation. More work will also be directed toward the assembly, accumulation, stability and secretion of plant antibodies. The effect of the signal sequence on the expression and assembly of antibodies has been shown and further methods for optimising transgenic protein accumulation in plants almost certainly exist.

Plant antibody technology is still in its infancy. However, it offers enormous potential in "mix-and-match" antibody engineering, and the construction of multimeric immunoglobulin complexes may be feasible relatively easily, for the first time. Furthermore, as there is an enduring interest in using antibodies for therapeutic purposes, agricultural production and distribution offers a means of obtaining large quantities of antibodies at a relatively low cost.

Table 1. <u>Expression and Assembly of Immunoglobulin Gamma and Kappa Chains in Tobacco.</u>

A. Accumulation of gamma or kappa chains in transformed plants. Results are expressed as mean ± sd ng/mg total protein, estimated by ELISA assay.

γ	$\gamma(\kappa)$[a]
1412±270	3330±2000
(2400)	(12800)

κ	$\kappa(\gamma)$[a]
56±5	3700±2300
(80)	(12800)

B. Distribution and assembly in crosses. Results are expressed as the number of plants expressing γ or κ chains among the progeny of a sexual cross.

	γonly	κonly	$\gamma\kappa$	null
κ x γ	3	10	11 (95±16% assembly)	4

[a] $\gamma(\kappa)$ refers to gamma chains in a plant that also expresses kappa chains, and vice versa. Numbers in parentheses are values for plants with the highest levels of accumulation.

Table 2. Catalytic activity of the 6D4 antibody produced in tobacco.

Source	Tobacco	Ascites
K_M (M)	1.41×10^{-6}	9.8×10^{-6}
V_{max} (M sec^{-1})	0.057×10^{-8}	0.31×10^{-8}
K_I (M)	0.47×10^{-6} (competitive)	1.06×10^{-6} (competitive)
k_{cat} (sec^{-1})	0.008	0.025

Table 3. Secretion of antibodies from protoplasts and callus cells.

35 S- METHIONINE INCORPORATION INTO 6D4 ANTIBODY IN CELLS OR MEDIUM AT 2 HOURS

		INCORPORATION (%)	
CELL TYPE	METHOD	CELLS	MEDIUM
PROTOPLASTS	PROTEIN A	75	25
PROTOPLASTS	SDS-PAGE	76	24
CALLUS SUSPENSION	PROTEIN A	72	28
CALLUS SUSPENSION	SDS-PAGE	80	20

INCORPORATION INTO 6D4 AFTER 2 HOUR CHASE

PROTOPLASTS	PROTEIN A	13	87
PROTOPLASTS	SDS-PAGE	14	86
CALLUS SUSPENSION	PROTEIN A	27	73
CALLUS SUSPENSION	SDS-PAGE	32	68

REFERENCES

1 Sijmons, P.C., Dekker, B.M.M., Schrammeijer, B., Verwoerd, T.C., van den Elzen, P.J.M. and Hoekema, A. (1990) *Bio/Technology* 8, 217-221.

2 Vandekerckhove, J., Van Damme, J., Van Lijsebettens, M. Botterman, J., De Block, M., Vandewiele, M., De Clercq, A., Leemans, J., Van Montagu, M. and Krebbers, E. (1989) *Bio/Technology* 7, 929-932.

3 Chaleff, R.S. and Ray, T.B. (1984) *Science* 223,1148-1151.

4 Shaner, D.L. and Anderson, P.C. (1985) In Biotechnology in Plant Science. Relevance to Agriculture in the Eighties. Zaitlin, M., Day, P. and Hollaender, A. (eds), Academic Press, Inc., NY, p. 287.

5 Comai, L., Facciotti, D., Hiatt, W.R., Thompson, G., Rose, R.E. and Stalker, D.M. (1985) *Nature* 317, 741-744.

6 Shah, D., Horsch, R., Klee, H., Kishore, G., Winter, J., Turner, N., Hironaka, C., Sanders, P., Gasser, C., Aykent, S., Siegel, N., Rogers, S. and Fraley, R. (1986) *Science* 233, 478-481.

7 De Block, M., Botterman, J., Vandewiele, M., Dockx, J., Thoen, C., Gossele', V., Rao Movva, N., Thompson, C., Van Montagu, M. and Leemans, J. (1987) *The EMBO Journal* 6, 2513-2518.

8 Powell-Abel, P., Nelson, R.S., De, B., Hoffman, N., Rogers, S.G., Fraley, R.T. and Beachy, R.N. (1986) *Science* 232, 738-743.

9 Van Dun, C.M.P., Overduin, B., van Vloten-Doting, L. and Bol, J.F. (1988) *Virology* 164, 383-389.

10 Van Dun, C.M.P. and Bol, J.F. (1988). *Virology* 167, 649-652.

11 Lawson, C., Kaniewski, W., Haley, L., Rozman, R., Newell, C., Sanders, P. and Tumer, N.E. (1990) *Bio/Technology* 8, 127-134.

12 Hilder, V.A., Gatehouse, A.M.R., Sheerman, S.E., Barker, R.F. and Boulter, D. (1987) *Nature* 330, 160-163.

13 Rogers, S.G., Klee, H.J., Horsch, R.B. and Fraley, R.T. (1987) *Methods in Enzymology*. 153, 253-276.

14 Tramontano, A., Janda, K., and Lerner, R. (1986) *Science* 234,1566-1569.

15 Tramontano, A., K. Janda and R.A. Lerner (1986) *Proc. Natl. Acad. Sci.* 83, 6736-6740.

16 Goldstein, I.J. and C. Hayes (1978) *Adv. Carbohydr. Chem. Biochem.* 35,127-340.

17 Hein, M.B. et al. (1991) *Biotechnology Progress* 7, in press.

18 Kijimoto-Ochiai, S., Katagiri, Y.U., Hatae, T. and Okuyama, H. (1989) *Biochem. J.* 257, 43-49.

19 Faye, L. and Crispeels, M.J. (1985) *Anal. Biochem.* 149, 218-224.

20 Trimble, R.B. and Maley, F. (1984) *Anal. Biochem.* 141, 515-522.

21 Jones, R.L. and Robinson, D.G. (1989) *Tansley Review No. 17 17*, 567-588.

22 Sturm, A., Kuik, A.V., Vliegenthart, J.F.G. and Crispeels, M.J. (1987) *J. Biol. Chem.* 262, 13392-13403.

23 Matsudaira, P. (1987) *J. Biol. Chem.* 262, 10035-10038.

24 Tricoli, D.M., Hein, M.B. and Carnes, M.G. (1986) *Plant Cell Reports 5*, 334-337.

25 Carpita, N., Sabularse, D., Montezinos, D., Helmer, D.P. (1979) *Science 205*, 1144-1147.

26 Hiatt, A.C., R. Cafferkey and Bowdish, K. (1989) *Nature* 342, 76-78.

27 Huston, J.S. et al., (1988) *Proc. Natl. Acad. Sci. USA* 85, 5879.

28 Chaudhary, V.K. et al., (1990) *Proc. Natl. Acad. Sci. USA* 87, 1066.

29 Bird, R.E. et al., (1988) *Science* 24, 423.

30 Baumann G, Raschke E, Bevan M, Schoffl F. (1987) *EMBO J.* 6, 1161-1166.

31 Schardl, C.L. et al., (1987) *Gene* 61, 1-11.
32 Rothman J.E (1989) *Cell* 59, 591-601.
33 Fontes ,E.B.P., Shank B.B., Wrobel, R.L., Moose S.P., OBrian G.R., Wurtzel, E.T., Boston, R.S. (1991) *Plant Cell* 3, 483-496.
34 Boston, R.S., Fontes E.B.P., Shank B.B., Wrobel, R.L. (1991) *Plant Cell* 3, 497-505.
35 Lehner,T., Caldwell,J., Smith, R. (1985) *Infect.Immun.*, 50,796.
36 Ma J. K-C., Smith R., Lehner T. (1987) *Infect.Immun.,* 55, 1274.
37 Ma J. K-C., Hunjan M., Smith R., Lehner T. (1989) *Clin.exp.Immunol.* 77,331.
38 Michalek S.M. et al., (1987) *Infect.Immun.* 55,2341-2347.
39 Hamada S. et al., (1991) *Infect.Immun.* 59,4161-4167.
40 Bergmann K-C., Waldman R.H. (1988) *Rev.Infect.Dis.* 10,939-950.
41 Childers N.K., (1989) *Annu.Rev.Microbiol.* 43,503-536.
42 Taubman M.A., Smith D.J. (1974) *Infect.Immun.* 9,1079-1091.
43 Michalek S.M., McGhee J.R., Mestecky J., Arnold'R.R., Bozzo L. (1976) *Science* 192,1238-1240.
44 Mestecky J., McGhee J.R., Arnold R.R., Michalek S.M., Prince S.J., Babb J.L. (1978) *J.Clin.Invest.* 61,731-737.
45 Gahnberg L., Krasse B. (1983) *Infect.Immun.* 39,514-519.
46 Lehner T., Challacombe S.J., Caldwell J. (1980) *Immunol.* 41,857-864.
47 Walker J. (1981) *Infect.Imun.* 31,61-70.
48 Linzer R., Evans R.T., Emmings F.G., Genco R.J. (1981) *Infect. Immun.* 31,345-351.

APPLICATION OF MICROINJECTION TECHNIQUE FOR THE ANALYSIS OF GENE EXPRESSION DURING HOST-PARASITE INTERACTION

Y. MATSUDA, H. TOYODA, and S. OUCHI
Faculty of Agriculture
Kinki University
3327-204 Nakamachi
Nara 631
Japan

ABSTRACT. Microinjection is a precise and reliable method for the analysis of the transcriptional and translational events in a specific single cell under host-parasite interactions. In this paper, we established an efficient system for the microinjection of foreign genes into different structures of Erysiphe graminis f. sp. hordei and barley coleoptile epidermal cells. The expression of these genes was detected by an in situ hybridization with photobiotin-labeled RNA probes or with specific enzyme-linked antibody.

1. INTRODUCTION

The microinjection technique is especially useful for directly introducing foreign material into the right portion within plant cell [2] and fungal structures [1, 4], suggesting that it could be used for the analysis of gene expression in host and parasite cells by the use of in situ hybridization cytochemistry. In the present paper, we defined conditions for in situ hybridization in powdery mildew of barley by the use of non-radioactive photobiotin-labeled probes, and describe 1) the procedure for in situ hybridization of rRNA with complementary, labeled DNA and RNA probes in conidiospores, appressoria, haustoria, and secondary hyphae as a model system for analyzing gene expression at the site of cell/cell interactions, 2) an application of microinjection to introduction of a foreign gene into fungal cells for providing an experimental basis for the genetic engineering of obligate parasites, and 3) a microinjection method for the in situ cytochemistry for estimating gene expression in barley coleoptile cells.

2. MATERIALS AND METHODS

2.1. Construction of fungal genomic library and isolation of rDNA

Chromosomal DNA was extracted from conidia produced on a susceptible barley cultivar, partially digested with Sau3A I, and fractionated by

E. W. Nester and D. P. S. Verma (eds.),
Advances in Molecular Genetics of Plant-Microbe Interactions, 561–565.

density-gradient centrifugation. The DNA fragments of 15-20 kilobase pairs (kb) were collected, ligated with Bam HI-digested arms of λ EMBL3 phage vector to construct the genomic library of the fungus through plaque hybridization. The rDNA sequence of the fungus was cloned by plaque hybridization with cDNA (CF49) of 28S rRNA of F. oxysporum f. sp. lycopersici and subcloned with E. coli transformed by pUC19-ligated restriction fragments. Finally, the 2.0 kb DNA fragment (Eco RI fragment) was isolated and designated as EGR10.

2.2. In situ hybridization in fungal cells

Single-stranded RNAs were artificially synthesized by inserting the EGR10 fragment into the polylinker site of plasmid vector pSP70 (Promega Corporation, Madison, USA). Sense and antisense RNAs (EGR10-sp and -t7) were produced by bidirectionally in vitro-transcribing the EGR10 fragment from the two promoter sequences (SP6 and T7 promoters) oppositely orientated in this plasmid. The probe was conjugated with photobiotin according to the manufacturer's protocols (Photoprobe; Vector Laboratories, Burlingame, CA, USA).

Fresh conidiospores were dusted onto a glass slide smeared with egg albumin, and air-dried for immobilization. A modified method of Raikhel et al. [3] was used for in situ hybridization. Fixed spores were pricked with an autoclaved glass needle (outer diameter, 0.7 μm) using an Olympus injectoscope [4], prehybridized with salmon sperm DNA in a hybridization buffer and hybridized to photobiotin-conjugated probe (ss RNA or heat-denatured DNA). Hybridization was detected by incubating the prehybridized spores in enzyme substrates, NBT and BCIP.

2.3. Microinjection for plant cells

Microinjection into inner coleoptile epidermal cells was conducted using the Olympus injectoscope in two ways; i) a direct injection of plasmid DNAs or sense RNAs [2], and ii) a pricking introduction [6] of labeled antisense RNA probes or specific antibodies. Beta-glucuronidase (GUS) gene, coat protein (CP) gene of TMV (OM-strain), and their sense RNAs were tested for their expression in injected cells. The plasmid pBI121 contains the GUS gene linked with the cauliflower mosaic virus (CaMV) 35S promoter and nopalin synthase terminator. The CP gene isolated from cDNAs of TMV-RNA was amplified by polymerase chain reaction (PCR). The GUS gene of pBI121 was replaced by the CP gene for constructing a plasmid vector pTCP1. To obtain sense RNAs of GUS and CP genes, new plasmids (pMCP1 and pMGS1) were constructed by inserting these genes into polylinker sites of a plasmid vector pSP64-poly (A) (Promega Co., Madison, WI, USA). Sense RNA strands were then synthesized by in vitro transcribing the insert from the promoter sequence (SP6 promoter) of this vector using Riboprobe System (Promega Co.). The transcription of sense RNA was initiated in the presence of a cap analogue.

2.4. Detection of gene expression; *In situ* hybridization and *in situ* immunoassay

Antisense CP- and GUS-RNAs prepared from pSACM1 and pSAGM1, respectively, were used as probes for *in situ* hybridization. The procedures for fixation and *in situ* hybridization with photobiotin-labeled probes were similar to those used for fungal cells. Translation products was detected by pricking the gene-introduced epidermal cells in a buffer containing BSA and rabbit anti-GUS-antibody or anti-TMV-antibody. The specimen was then treated with goat anti-rabbit IgG-antibody conjugated with alkaline phosphatase and with enzyme substrates (NBT and BCIP).

3. RESULTS AND DISCUSSION

3.1. A novel method for *in situ* hybridization in fungal cells by pricking microinjection of photobiotin-labeled probes

The most direct way of examining gene expression is to detect and measure the transcription products in gene-introduced cells. For quantifying the expression of genes, it is of primary importance to clarify the conditions that ensure a specific and effective hybridization between probes and corresponding transcripts within the cells. In this study, rRNA was chosen as a hybridization target, because it is constant in a large quantity in cytoplasm, hence will make it easier to determine the conditions for *in situ* hybridization in infection structures at various stages. The rDNA was isolated from a genomic clone and subsequently subcloned through a series of plaque and colony hybridization. The DNA fragment (EGR10) was approximately 2.0 kb in length, and the Northern analysis indicated that a single RNA strand (EGR10-t7) derived from this fragment was complementary to 28S RNA of the powdery mildew fungus.

Prior to actual *in situ* hybridization, we confirmed that the positive reaction was observed only when the specific hybrid was formed between the probe and rRNA. When a complementary RNA strand (EGR10-t7) was introduced into pricked spores, more than 80 % of pricked spores were densely and uniformly stained all over their cytoplasm. On the other hand, no hybridization was detected in non-pricked spores even when they were treated with EGR10-t7 probe for extended period of time (5-8 hr), indicating that microinjection of genes and pricking introduction of complementary probes are efficient procedure for *in situ* hybridization in fungal cells.

Similar *in situ* hybridization detected GUS gene expression in appressorium, haustorium, and secondary hyphae at a similar frequency, considerably higher than those observed in conidia.

In theory, *in situ* hybridization cytochemistry could be used for the detection of specific mRNAs in these structures. In fact, the expression of β-glucuronidase (GUS) gene linked with the CaMV 35S promoter was detected in these infection structures by two different methods; i) immunocytochemical detection of the transla-tion product and ii) *in situ* hybridization of GUS-mRNA with micro-injected antisense RNA probe. The present method will further be applied to the *in situ* detection of a particular mRNA in the fungal cell system.

3.2. Expression of Foreign Genes in Barley Powdery Mildew Fungus

The plasmid pBI221 was injected into the fungal cells under the same condition mentioned above. Primordium-forming appressoria (16 hr after inoculation) were injected with pBI221 (1 µg/µl) and incubated for 6 hr to confirm haustorial maturation. Approximately 90 % of injected appressoria developed mature hasustoria, indicating that the microinjection at this stage of fungal development was appropriate for gene introduction. The GUS gene was expressed in more than 70 % of haustoria which had been developed from gene-injected appresoria, but not in haustoria from non-injected appressoria, indicating that the powdery mildew fungus can read sequences of the CaMV 35S promoter which has been known to function in the transcription of GUS gene in higher plant cells. These results substantiate the notion that the microinjection is a promising technique for the transformation of this fungus using chimeric genes carrying the 35S promoter.

3.3. Foreign gene expression in barley coleoptile epidermis; An improved system for gene transfer and in situ detection of gene expression

In this study, we established an efficient method for detecting gene expression, based on *in situ* hybridization with antisense RNA probes and *in situ* immunoassay with enzyme-conjugated antibodies. The *in situ* hybridization has been developed for identification of transcripts primarily in sectioned specimens of higher animals and plants. As described above, we were able to detect cytoplasmic rRNA or mRNA in various fungal infection structures by pricking introduction of DNA or RNA probes labeled with non-radioactive photobiotin. The method has been further improved to detect gene products in higher plant cells. In fact, the GUS gene and the TMV-CP gene introduced into barley coleoptile epidermal cells were unequivocally expressed at the level of transcription and translation as demonstrated by positive reaction with RNA probes or specific antibody.

More than 80 % of CP gene or GUS gene-injected coleoptile epidermal cells were positive in *in situ* hybridization with photobiotin-conjugated antisence CP-RNA probe or GUS-RNA probe introduced by pricking. The rates of the expression were considerably stable and higher than those determined by β-glucuronidase assay [5].

This method was extended to the detection of translation products of introduced genes by simply replacing hybridization buffer with a solution containing enzyme-linked antibody. Both the gene-injected and non-injected cells were pricked in solution containing enzyme-conjugated antibody and then incubated in the substrates of the enzyme. The positive reaction was detected only in gene-injected cells, but not in non-injected cells. Morevoer, this method allowed us to determine efficiency of translation of sense mRNAs microinjected into coleoptile cells. The sense mRNAs of CP and GUS genes were *in vitro*-translated from pMCP1 and pMGS1, respectively. The rates of positive reaction were constantly high regardless of the types of injected mRNAs and were comparable to those in cells injected with pBI121 or pTCP1.

In summary, foreign genes microinjected into barley coleoptile epidermal cells were functionally expressed regardless of the type of genes, indicating that a barley coleoptile tissues can be used as a model system for the detailed analysis of gene expression in higher plants. In fact, using the barley coleoptile and the microinjection method, we have isolated from barley chromosomal DNA a promoter sequence that specifically responds to infection. A chimeric GUS gene linked with this sequence was expressed only in pathogen-invaded coleoptile epidermal cells (unpublished data). Thus, the method described in this paper will certainly provide an experimental basis for the manipulation and analysis of gene expression in higher plant cells.

References

1. Matsuda, Y., Toyoda, H. and Ouchi, S. (1989). 'Application of microinjection to appressoria and haustoria of *Erysiphe graminis* f. sp. *hordei*', Ann. Phytopath. Soc. Japan 55, 67-68.
2. Potrykus, I. (1990). 'Gene transfer to cereals: An assessment', *Bio/Technology* 8, 535-542.
3. Raikhel, N. V., Bednarek, A. Y. and Lerner, D. R. (1989). '*In situ* hybridization in plant tissues', in S. B. Gelvin, R. A. Schilperoot and D. P. S. Verma (eds), Plant Molecular Biology Manual, Kluwer Academinc Publisher, Boston, pp. B9:1-32.
4. Toyoda, H., Matsuda, Y., Shoji, R. and Ouchi, S. (1986). 'A microinjection technique for conidia of *Erysiphe graminis* f. sp. *hordei*', Phytopathology 77, 815-818.
5. Toyoda, H., Yamaga, T., Matsuda, Y. and Ouchi, S. (1990). 'Transient expression of the β-glucuronidase gene introduced into barley coleoptile cells by microinjection', Plant Cell Reports 9, 299-302.
6. Yamamoto, F., Furusawa, M., Furusawa, I. and Obinata, M. (1982). 'The 'pricking' method. A new efficient technique for mechanically introducing foreign DNA into nuclei of culture cells', Experimental Cell Research 142, 79-84.

RESISTANCE TO *RHIZOCTONIA SOLANI* IN TRANSGENIC TOBACCO.

Pamela Dunsmuir, William Howie, Ed Newbigin, Larry Joe, Eva Penzes and Trevor Suslow.
DNA Plant Technology Corporation,
6701 San Pablo Avenue,
Oakland. CA.94608.
USA

ABSTRACT. We are attempting to develop significant resistance to fungal disease through the expression of novel activities in genetically engineered plants. Transgenic tobacco plants which express high levels of the *Serratia marcescens* chitinase gene (*chiA*) have been shown to have significant resistance to *Rhizoctonia solani* in greenhouse trials. These genetically engineered tobacco lines have been demonstrated to also have significant resistance to *R.solani* in two seasons of field trials.

Background

Chitin, a *B* 1,4-linked polymer of N-acetyl glucosamine, is a structural component of the cell walls of all fungi except the oomycetes (Monreal and Reese, 1969). Many organisms, including bacteria, fungi and higher plants, produce enzymes that hydrolyse chitin. *Serratia marcescens* QMB 1466, a Gram-negative enteric soil bacterium, secretes high levels of chitinase activity (Roberts and Cabib, 1982) and this organism has been the source of the chitinase genes which we have isolated and characterized.

E. W. Nester and D. P. S. Verma (eds.),
Advances in Molecular Genetics of Plant-Microbe Interactions, 567–571.

Several lines of evidence suggest that the production of chitinase by higher plants may be part of the natural defense mechanism against fungal pathogens. Firstly, chitinase enzymes are induced to high levels in plant tissue following fungal infection or wounding (Boller, 1988). Secondly, purified plant chitinases strongly inhibit the growth of a wide range of pathogenic fungi (Schlumbaum *et al.*, 1986, Mauch *et al.*, 1988).

We have engineered *Nicotiana tobaccum* to express high levels of the *S. marcescens* ChiA and/or ChiB protein in an effort to augment the natural defense mechanism which exists in these plants.

Summary of Research

CHARACTERIZATION OF CHIA AND CHIB PROTEINS

We have cloned and characterized two diferent chitinase genes from *S. marcescens*, *chiA* (Jones et al, 1985) and *chiB* (Harpster et al., 1989). Each of these genes has been expressed at high levels in *E.coli* cells and the resulting protein has been purified for the purposes of polyclonal antibody production as well as protein characterization. These data are summarized in Table 1.

TABLE 1. The properties of *S. Marcescens* ChiA and ChiB protein purified from recombinant *E.coli.*

	ChiA	ChiB
Molecular Size	58 kD	52 kD
Distribution (*E.coli*)	Periplasm	Intracellular
pI	5.8	5.0
pH optimum	8.0-8.5	7.5-8.0
Activity		
(GlcNac)3-MeU	Km 2.8*u*M	-
(GlcNac)4-MeU	Km 83*u*M	Km 12.6*u*M

The two proteins are distinct immunologically and also have quite different activities *in vitro*. Both *S.marcescens* chitinases are differentiated immunologically and in activity from the endogenous plant chitinases which enables direct analyses of transgenic plants.

GENETIC ENGINEERING OF PLANTS

We have prepared a set of fusions which direct the expression of the *chiA* and *chiB* coding regions in transgenic plant tissue. The description of these constructions and the molecular characterization of the transgenic plants which carry these genes have been published- (Taylor *et al.*, 1986, Jones *et al.*, 1988, Lund *et al.*, 1989, Lund and Dunsmuir, 1992).

We have found that each of these bacterial genes can be engineered to be expressed at high levels in plant cells so that in selected transformants the ChiA or ChiB protein steady state level approximates 0.1% of the soluble protein in the cell. Furthermore it has been possible though the manipulation of the signal sequence, to direct the accumulation of the ChiA protein intra- or extracellularly in transgenic tobacco plants (Lund and Dunsmuir, 1992).

FUNGAL DISEASE RESISTANCE IN TRANSGENIC PLANTS

We have prepared lines of tobacco which express high levels of the ChiA and/or the ChiB protein and we have tested whether these populations have changed properties with respect to infection by selected fungal pathogens. Initially a large numbers of independent lines was screened in greenhouse trials and those lines which exhibited significant disease resistance in the greenhouse were subsequently tested in the field.

The fungal pathogen which we have focused on is *R.solani* and the results from our greenhouse trials indicated that in general there was a correlation between the level of ChiA expression in a given transgenic tobacco line, and the degree of resistance to infection by *R.solani* . The reduction in disease severity in these trials was at best 40% and data from a standard experiment are shown in the following Table.

TABLE 2 Biological control of *Rhizoctonia solani* in transgenic tobacco-Greenhouse Trial.

Genotype ^	Disease Incid.*	Plant Weight (g)*
Control	11.7 a	5.3 a
ChiAsg-3	10.0 a	4.8 a
ChiAsg-9	6.8 b	4.7 a
ChiAs-6	6.4 b	5.0 a
ChiAns-4	6.3 b	4.9 a

^ s-secreted, ns-non secreted, g-glycosylation minus.
* Duncan's Multiple Range Test was used for mean separation (P=0.05)

Transgenic tobacco lines which consistently exhibited decreased sensitivity to *R.solani* infection in greenhouse assays were tested in the field for resistance to infection. We found that the lines which gave the highest degree of resistance in the greenhouse also exhibited the most sifnificant resistance in the field. The data from a typical field experiment are shown in the following Table.

TABLE 3. Evaluation of transformants to infection by *Rhizoctonia solani* -April planting.

Genotype ^	# lesions*	Shoot wt.(g)*
Control	5.4 a	12.1 b
ChiAs-3	4.2 b	7.9 d
ChiAs-6	4.0 b	10.8 ab
ChiAsg-3	3.7 bc	10.8 ab
ChiAsg-9	3.6 b	9.2 c
ChiAns-4	2.6 d	11.8 ab
Control		
-Benlate	2.9 cdb	8.6 cd
-Uninfected	0 e	11.2 ab

^ s-secreted, ns-non secreted, g-glycosylation minus.
* Duncan's Multiple Range Test was used for mean separation (P=0.05)

These field trials have now been performed multiple times in two seasons with no significant differences in the results. In the best lines we see a level of disease resistance which is similar to that which results after application of Benlate at the manufacturer's suggested dosage.

References.

Boller T. (1987) Hydrolttic enzymes in plant disease resistance. In Plant Microbe Interactions, Molecular and Genetic Perspectives 2:385-413 . Ed T. Kosuge and E. Nester, Macmillan, New York.

Harpster M, Townsend J, Jones J, Bedbrook J, Dunsmuir P. (1988) Relative strengths of the CaMV 35S, 1', 2', and nopaline synthase promoters in transformed tobacco, sugarbeet and oilseed rape callus tissue. Mol Gen Genet 212:182-190.

Harpster M, Dunsmuir P. (1989) Nucleotide sequence of the chitinase B gene of *Serratia marcescens* QMB1466. Nucl Acids Res 17:5395.

Jones J, Grady K, Suslow T, Bedbrook J. (1986) Isolation and characterization of genes encoding two chitinase enzymes from *Serratia marcescens* EMBO J 5:467-473.

Jones J, Dean C, Gidoni D, Gilbert D, Bond-Nutter D, Lee R, Bedbrook J, Dunsmuir P. (1988) Expression of bacterial chitinase protein in tobacco leaves using two photosynthetic gene promoters. Mol Gen Genet 212:536-542.

Lund P, Lee R, Dunsmuir P. (1989) Bacterial chitinase is modified and secreted in transgenic tobacco. Plant Physiol 91: 130-135.

Lund P, Dunsmuir P. (1992) A plant signal sequence enhances the secretion of bacterial ChiA in transgenic tobacco. Plant Mol Biol 18:47-53.

ANTIBACTERIAL RESISTANCE OF TRANSGENIC POTATO PLANTS PRODUCING T4 LYSOZYME

KLAUS DÜRING, MATTHIAS FLADUNG[1)] and HORST LÖRZ
University of Hamburg, Center for Applied Plant Molecular Biology
Ohnhorststr. 18, D-2000 Hamburg 52
[1)] Max-Planck-Institute for Plant Breeding, Dept. Salamini,
Carl-von-Linné-Weg 10, D-5000 Köln 30

ABSTRACT. Transgenic potato plants have been produced that express and secrete the foreign bacteriophage T4 lysozyme. The chimeric barley α-amylase signal peptide - T4 lysozyme gene is driven by the CaMV 35S promoter. Low level expression can be detected at mRNA and protein level. Subcellular localization analysis demonstrates secretion of the foreign protein to the intercellular spaces. Biological *in vitro* and greenhouse experiments demonstrate that tuber maceration following inoculation with *Erwinia carotovora atroseptica* in transgenic tissue is significantly reduced in comparison to control tissues. Moreover, explants from inoculated transgenic tuber pieces show sprouting and growth without development of disease symptoms. Expression and secretion of the foreign lysozyme obviously leads to a reduced susceptibility of the transgenic potato plants towards the phytopathogenic bacterium.

1. Introduction

Erwinia carotovora spp. *atroseptica* and *carotovora* are the most important bacterial potato pathogens. They cause soft rot of tubers and black leg of stem tissue in the field as well as during storage. The first one is able to induce both diseases whereas the second one is able to induce soft rot only. No chemical means are known to prevent disease development. Also, breeders do not dispose of valuable traits which may provide efficient resistance against bacterial infection. Only a few wild *Solanum* species which cannot be sexually crossed with potato cultivars have been reported to be resistant against *Erwinia carotovora* spp. (1).

Erwinia carotovora spp. can be detected in lenticels of all tubers (2). Upon wounding the bacteria may enter the tuber tissue via the intercellular spaces and, if favourable conditions are prevalent, multiply within the apoplast to reach high densities. Initially, about 10 - 100 bacteria are invading the tuber. When 10^6 - 10^7 bacteria per gram tissue have grown development of disease symptoms can be observed (3). Therefore, a genetechnological strategy has to interact with the invading bacteria as early as possible to prevent multiplication. As a consequence, the foreign antibacterial agent has to be localized in the intercellular spaces.

Lysozymes are widely occuring enzymes with a specific bacteriolytic activity directed against the bacterial murein layer. Also in several plants, lysozyme activities have been detected which mostly are correlated with a stronger chitinase activity within a bifunctional enzyme. Even in potato, lysozyme activity has been demonstrated (4). Endogenous chitinase / lysozyme enzymes are located either in the vacuole or associated with the cell wall (5). Bacteriophage T4 lysozyme is the most active one not only against gram-negative but also against gram-positive bacteria (6). A receptor for phage T4 has been detected in *Erwinia carotovora* spp. (7). By expressing and secreting T4 lysozyme in transgenic potato plants we attempted to improve resistance against the phytopathogenic bacterium.

E. W. Nester and D. P. S. Verma (eds.),
Advances in Molecular Genetics of Plant-Microbe Interactions, 573–577.

2. Results and Discussion

Secretion of T4 lysozyme in fusion to a plant signal peptide had already been achieved in transgenic tobacco plants (8, 9). The foreign protein was efficiently secreted to the intercellular space. An analogous gene construction has been used in the experiments described here. The barley α-amylase - T4 lysozyme fusion gene is directed by the cauliflower mosaic virus 35S promoter and the respective polyadenylation region. The chimeric gene has been integrated into the genome of the tetraploid potato Z2 genotype (10) via *Agrobacterium tumefaciens* mediated gene transfer. Molecular analysis of the individual transformants revealed correct integration of one copy of the T-DNA in all the analyzed plants. The different regenerants provide varying levels of specific mRNA which is most probably due to the position effect. Accordingly, also varying levels of T4 lysozyme could be detected in the transgenic tissue. Both mRNA and protein levels are as low as about 0.001% of total mRNA or total soluble protein, respectively (11).

T4 lysozyme is produced as a glycosylated protein in the transgenic potato plants. After chemical deglycosylation a signal could be obtained representing the mature, deglycosylated T4 lysozyme. The signal peptide has been cleaved off during the secretory pathway. The T4 lysozyme has been localized in the intercellular spaces by analysis of intercellular washing fluids and electron microscopic immunogold labelling. The low levels of foreign protein are unexpected because the CaMV 35S promoter usually is able to direct expression of a foreign protein at about 0.1% of total soluble protein. On the other hand, already during analysis of the tobocco plants bearing a mannopine synthase promoter - signal peptide-T4 lysozyme gene only low levels could be detected (8). This may be due to inefficient transcription and translation of the foreign procaryotic gene in the eucaryotic cell as well as to limited mRNA stability. Further investigations will be made to improve T4 lysozyme levels in transgenic plants.

Analysis of enzymatic activity of the lysozyme produced in transgenic tissue is not possible without purification of the foreign protein because the available assays lack sensitivity by almost 3 orders of magnitude in respect to protein levels in crude extracts. Generally, lytic activity of T4 lysozyme towards *Erwinia carotovora atroseptica* has been demonstrated using overlay gel assays (K.Düring, manuscript in preparation).

The foreign lysozyme in transgenic plants has been found to be associated with fibrillar structures not present in control plants (8, 9, 11). Similar structures have been reported from tissues infected with bacteria (13) or associated with endogenous chitinases and ß-1,3-glucanases (14, 15). Therefore, these structures most probably originate from plant tissue where lytic enzymes are expressed in the intercellular spaces.

Phytopathological analysis of the transgenic plants has been performed in *in vitro* and greenhouse assays. Release to the field for obtaining final results under natural conditions is planned in future. No standard assay for assessing resistance of potato tubers or plants against *Erwinia carotovora* spp. is available today. A variety of similar assays has been described in literature. We established a tuber disc maceration assay with *Erwinia carotovora atroseptica* for assessing the levels of maceration in transgenic and control plants (11). Tuber discs of about 1 cm diameter and 2 - 3 mm thickness were prepared and immediately inoculated with a defined amount of bacteria within 10 - 20 µl suspension. Incubation was performed for 2 to 12 days in plastic boxes or petri dishes under saturated humidity at 22°C. Inoculation densities ranged from 1,000 up to 500,000 bacteria. Significant results could be obtained from inoculations using up to 150,000 bacteria. Several independent transgenic and wildtype as well as transgenic control plants have been assayed.

Reduction of tissue maceration in some of the transgenic plants was significant. These plants were those with the highest levels of specific mRNA and protein. A representative of this group of plants is regenerant T 424. When inoculated with 1,000 - 3,000 bacteria, tissue maceration reached only 5% of the disc surface in comparison to 60 - 75% in the controls. Some data are shown in Fig. 1. As outlined above inocula under natural conditions are most likely to range even

a)

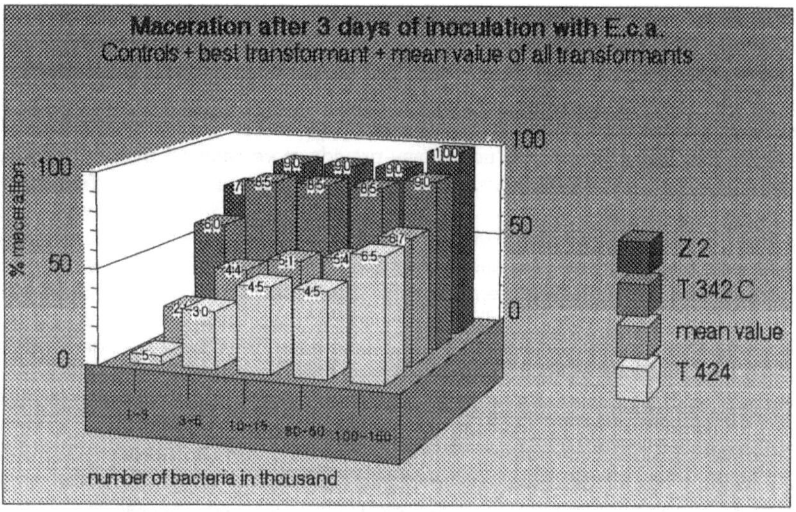

b)

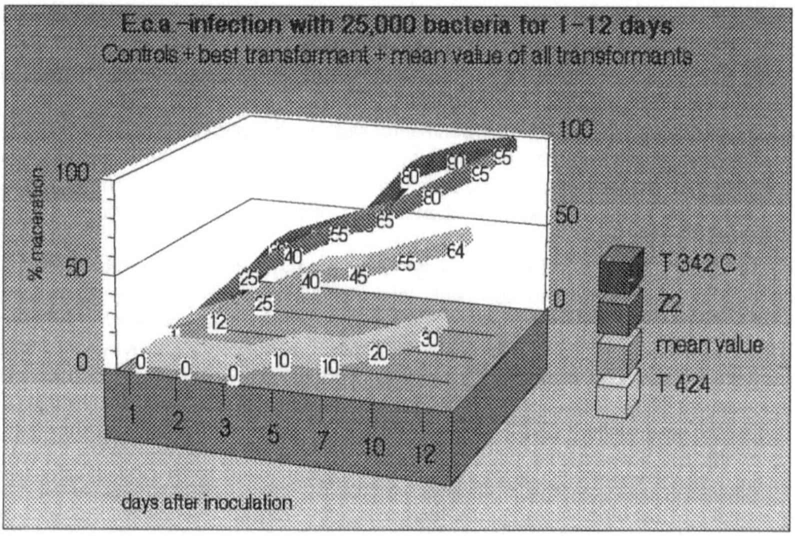

Fig. 1: Tuber disc inoculation assays as decribed in the text. *a)* Incubation of potato tuber discs with different numbers of *Erwinia carotovora atroseptica* in 10 μl of suspension for 3 days. Untransformed and transformed control plants (Z2 and T 342, resp.), the best transgenic plant (T 424) and mean value of all tested transgenic plants are displayed. Number of bacteria were: 1,000-3,000, 3,000-5,000, 10,000-15,000, 30,000-50,000 and 100,000-150,000 (x100 for bacteria per ml). Maceration is reduced by up to 85%. *b)* Incubation of potato tuber discs in plastic boxes with 25,000 *Erwinia carotovora atroseptica* in 10 μl of suspension (i.e. 2,500,000 per ml) for 12 days. The same plants were used. Maceration is displayed dependent on the length of incubation. After 12 days the reduction seen with the best plant (T 424) is still 70%. (11)

lower (about 10 - 100 bacteria per wounded site). Quantitative reproducibility from experiment to experiment is still limited whereas qualitative results were consistent throughout all experiments. Quantitative data should best be compared only within a single experiment in relation to the data produced with the control plants

Another type of assay has been performed in the greenhouse. Sprouting capacity of inoculated tuber pieces has been tested. This assay may provide more relevant data in concern of black leg disease. Spread of *Erwinia carotovora* spp. mainly occurs originating from infected seed tubers in the field. Non-emergence and tuber maceration in the soil as well as black leg development in later stages of growth can be disease symptoms. We immersed tuber pieces containing only one tuber eye in a suspension containing about 1,250,000 bacteria per ml for 2 hours and subsequently planted them into soil. Sprouting was monitored 3 - 6 weeks after planting. In a first experiment including seven independent transgenic plants and wildtype as well as transgenic control plants, each ten explants have been tested. Again, correlating with expression levels, varying numbers of sprouts have been observed. In a second experiment we concentrated on the most promising regenerants and used higher numbers of explants (39 or 78, resp.). Overall, we found significant differences between transgenic and control explants which correlate well with the results from the tuber disc assays. Sprouting in the first assay reached up to 100% for the transgenic explants (controls 0 to 20%). In the second assay growth of 37 strong sprouts from 78 T 424 explants was found whereas the wildtype control produced only 4 very little sprouts. These plantlets were obviously not healthy. Some of the T 424 sprouts have been grown in the greenhouse until tuber formation and harvest. No symptoms of disease development have been detected.

In addition, we screened some of the non-emerged transgenic and control tuber pieces and found that maceration only had occured in the controls. Sprouting of the transgenics has most probably been prevented by other environmental factors. In conclusion, data from tuber disc and greenhouse assays reveal that the highest expressing potato plants provide a significant reduction of tuber maceration in comparison to controls. Final evidence for field resistance of the T4 lysozyme producing potato plants can only be obtained during a field release experiment.

Introduction of a highly active foreign lysozyme into transgenic potato plants leads to increased resistance of tuber tissue and obviously also growing plants towards *Erwinia carotovora atroseptica*. Despite of its low level expression the foreign enzyme seems to exert a strong biological effect. This might be due solely to its own bacteriolytic activity or to strengthening the endogenous molecular protective machinery. Endogenous lysozymes localized mostly in the vacuoles can become active only after cell breakage. A wheat germ lysozyme has been localized in the cell walls (12). This might be a permanent bacteriolytic protection shield but access to the bacteria should be even faster with the T4 lysozyme which is soluble in the intercellular spaces. This is estimated to be a key feature of the genetechnological approach. Initially invading bacteria have to be lysed by the foreign lysozyme in order to prevent multiplication. Otherwise, high levels of lysozyme would be required for efficiency.

In our experiments we used 'constitutive' expression of the foreign gene in the transgenic plants. In order to keep alteration of the transgenic plant as low as possible in comparison to the respective non-transformed plant it is advisable to express the T4 lysozyme in a regulated manner. Experiments to achieve regulated expression as well as to increase expression levels are in progress. Finally, we should end up with a set of data allowing us to discriminate between necessary and sufficient expression levels and to define the optimal type of gene regulation to prevent disease development. Other pathogenic bacteria will also be included in future research. This unspecific approach may easily be transferred to other host - pathogen (bacteria) systems.

3. Acknowledgements
We thank Dr. Irmi Becker (RWTH Aachen, Dept. Prof. Dr. Fritz Kreuzaler, FRG) for cloning of the chimeric gene into her plant expression vector pS and Petra Porsch (MPI for Plant Breeding, Köln, Dept. Salamini) for transformation of the potato plants. Electron microscopy with the transgenic potato plants has been performed in cooperation with Dr. Dieter Neumann (Institute for Plant Biochemistry, Halle, FRG). This work has been funded in part by a grant from the GFP / BML (K 56/92 HS [91 HS 013]).

4. References

1 Austin,S., Lojkowska,E., Ehlenfeldt,M.K., Kelman,A. and Helgeson,J.P.; Fertile interspecific hybrids of *Solanum*: a novel source of resistance to *Erwinia*. *Phytopathol.* 78, 1216-1220 (1988)

2 Pérombelon,M.C.M.; 'The impaired host and soft rot bacteria', in: Mount,M.S. and Lacy,G.H. (eds.), Phytopathogenic procaryotes, Academic Press, New York, pp. 55 - 69 (1982)

3 Erinle,I.D.; Growth of *Erwinia carotovora* var. *atroseptica* and *Erwinia carotovora* var. *carotovora* in potato stems. *Plant Pathol.* 24, 224 - 229 (1975)

4 Kombrink,E., Hahlbrock,K., Hinze,K. and Schröder,M., in: Smith,C.J. (ed.), Biochemistry and Molecular Biology of Plant-Pathogen Interactions, Clarendon Press, Oxford, pp. 237-254 (1991)

5 Jollès,P. and Jollès,J.; What's new in lysozyme research? *Mol.Cell.Biochem.* 63, 165-189 (1984)

6 Tsugita,A.; Phage lysozyme and other lytic enzymes. In: Boyer,P.D. (ed.), The Enzymes, Volume 5, Academic Press, New York, pp. 344-411 (1971)

7 Pirhonen,M., Palva,E.T.; Occurence of bacteriophage T4 receptor in *Erwinia carotovora*. *Mol.Gen.Genet.* 214, 170-172 (1988)

8 Düring,K.; Wundinduzierbare Expression und Sekretion von T4 Lysozym und monoklonalen Antikörpern in *Nicotiana tabacum*. Ph.D. thesis, University of Cologne, FRG (1988)

9 Hippe,S., Düring,K. and Kreuzaler,F.; In situ localization of a foreign protein in transgenic plants by immunoelectron microscopy following high pressure freezing, freeze substitution and low temperature embedding. *Eur.J.Cell Biol.* 50, 230 - 234 (1989)

10 Fladung,M.; Transformation of diploid and tetraploid potato clones with rolC gene of *Agrobacterium rhizogenes* and characterization of transgenic plants. *Plant Breeding* 104, 295-304 (1990)

11 Düring,K., Porsch,P., Fladung,M. and Lörz,H.; Transgenic potato plants resistant to the phytopathogenic bacterium *Erwinia carotovora*. *Plant J.*, submitted

12 Audy,P., Benhamou,N., Trudel,J. and Asselin,A.; Immunocytochemical localization of a wheat germ lysozyme in wheat embryo and coleoptile cells and cytochemical study of its interaction with the cell wall. *Plant Physiol.* 88, 1317-1322 (1988)

13 Lyon,C.E., Lyon,G.D. and Robertson,W.M.; Observations on the structural modification of *Erwinia carotovora* subsp. *atroseptica* in rotted potato tuber tissue. *Physiol.Mol.Plant Pathol.* 34, 181-187 (1989)

14 Mauch,F. and Staehelin,L.A.; Functional implications of the subcellular localization of ethylene-induced chitinase and ß-1,3-glucanase in bean leaves. *Plant Cell* 1, 447-457 (1989)

15 Spanu,P., Boller,T., Ludwig,A., Wiemken,A., Faccio,A. and Bonfante-Fasolo,P.; Chitinase in roots of mycorrhizal *Allium porum*: regulation and localization. *Planta* 177, 447-455 (1989)

PATHOGEN-DERIVED STRATEGY TO PRODUCE TRANSGENIC PLANTS RESISTANT TO THE BACTERIAL TOXIN PHASEOLOTOXIN.

DE LA FUENTE-MARTINEZ J. M., MOSQUEDA-CANO, G., ALVAREZ-MORALES, A. AND HERRERA-ESTRELLA, L.
Departamento de Ingenieria Genética de Plantas, Centro de Investigación y Estudios Avanzados. Apartado Postal 629, 3600, Irapuato, Guanajuato, México.

ABSTRACT. Chemicals with phytotoxic activities are produced by many plant pathogenic fungi and bacteria. Toxins have been shown to be an important virulence component for most pathovars of *Pseudomonas syringae*. Here we have examined the role of phaseolotoxin in the virulence mechanism of *P. syringae* pv *phaseolicola* by producing transgenic plants that express a pathogen-derived toxin-insensitive target enzyme and showed this plants are insensitive to the toxin and less prone to be infected by this pathogen.

1. INTRODUCTION

Non host-specific toxins are generally acknowledged to be an important virulence element, and this is especially evident among the pathovars of *Pseudomonas syringae* [1]. However, considerable debate has focused on whether toxin production has a primary role in virulence since phytotoxic symptoms occur long after lesion formation. Several toxins produced by phytopathogenic bacteria have as targets enzymes involved in amino acid biosynthesis.

Phaseolotoxin is a tripeptide toxin [2,3] produced by *Pseudomonas syringae* pv. *phaseolicola*, the causal agent of the halo blight disease of common bean (*Phaseolus vulgaris* L.). Tox$^-$ isolates multiply at normal rates in inoculated leaves and cause typical lesions, but they do not cause chlorotic halos, systemic chlorosis and, more importantly, they are unable to produce systemic infections [4].

Phaseolotoxin inhibits the enzyme Ornithine Carbamoyltransferase (OCTase) [5]. This enzyme converts ornithine and carbamoyl phosphate to citrulline, a reaction involved in the biosynthesis of arginine and in the interconversion of the glutamate family of aminoacids. *In vitro* phaseolotoxin is a reversible inhibitor of OCTase, but "in planta" the toxin is cleaved by plant peptidases to produce the irreversible inhibitor Octicidin [7,8].

Phaseolotoxin is a non host-specific toxin and it causes similar chlorotic symptoms and ornithine accumulation on plants that resist infection by *P. s.* pv. *phaseolicola* [9], and

E. W. Nester and D. P. S. Verma (eds.),
Advances in Molecular Genetics of Plant-Microbe Interactions, 579–586.

it also inhibits the growth of cultivated plant cells [10] and bacteria [11]. *P. s.* pv. *phaseolicola* is able to produce two detectable OCTase activities, one sensitive and one resistant to phaseolotoxin [12,13]. The *P. s.* pv. *phaseolicola* gene encoding the phaseolotoxin-resistant OCTase (*argK*) has been cloned and sequenced [14].

Using the pathogen-derived resistance strategy [15], we aimed our research at producing phaseolotoxin-resistant transgenic plants and to investigate the role of phaseolotoxin in the virulence of *P. s.* pv. *phaseolicola.*

2. RESULTS.

2.1. PRODUCTION AND ANALYSIS OF TRANSGENIC PLANTS.

In plant cells OCTase has been shown to be located in the chloroplast [27], where citrulline biosynthesis takes place [6]. To produce transgenic plants in which the *P. s.* pv. *phaseolicola* phaseolotoxin-insensitive OCTase is located inside the chloroplast, a chimaeric gene was constructed in which the coding sequence of the *argK* gene of this bacteria is fused to the transit peptide of the small subunit of Rubisco (see figure 1). The chimaeric SSU-OCTase gene was transferred to *Agrobacterium tumefaciens* using the two step mating system.

Transgenic tobacco plants containing the SSU-OCTase construct were obtained by the leaf-disk co-cultivation method [18]. Kanamycin resistant plantlets were assayed for NPT-II activity and positive plants grown further and analyzed for the presence of intact copies of the SSU-OCTase construct by Southern blot hybridization techniques (data not shown). Three transgenic tobacco plant were chosen for further analysis.

To determine whether the SSU-OCTase construct directs the production of an active OCTase enzyme in transgenic plants, the level of OCTase activity in the leaves of control and transgenic tobacco plants was determined. The average OCTase activity of normal tobacco plants was found to be in the order of 2.5 U. In transgenic plants containing the SSU-OCTase construct OCTase activity values ranged from 2.5 to over 20 U. The activity present in three representative transgenic plants chosen for further analysis (SSO-T1, T2 and T3) is shown in figure 1. Northern blot hybridization analysis reveled the presence of a 1.1 kb transcript in trasgenic plants that is absent in the controls and that corresponds to the predicted size of the mRNA from the SSU-OCTase construct.

To determine whether the bacterial OCTase is indeed located in the chloroplasts of transgenic plants, OCTase activity was determined in isolated chloroplasts from leaves of the SSO-T3 and control plants. It was found that chloroplast from the transgenic plant have an OCTase activity a several fold higher than chloroplasts from control plants, thus confirming that the bacterial enzyme was correctly targeted to the chloroplast (see figure 1).

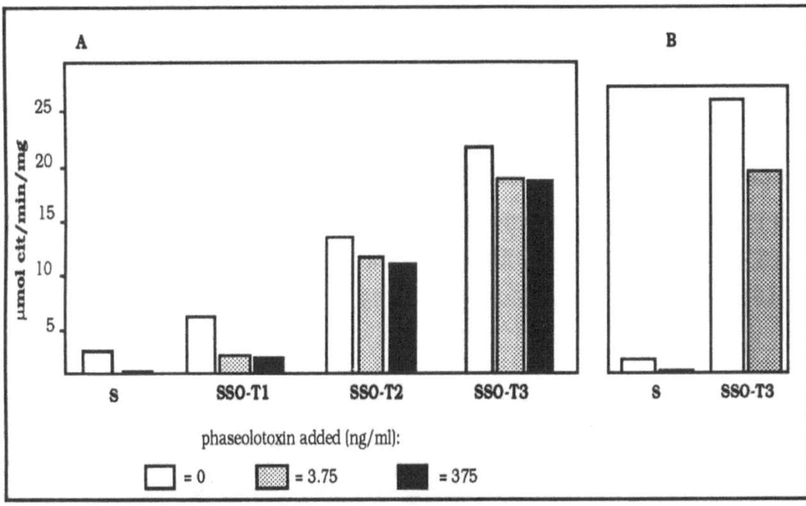

Fig. 1. Inhibition of OCTase activity by phaseolotoxin in tobacco plants.
A: OCTase activities in leaf extracts of non-transformed **S**) and transgenic tobacco plants expressing *argK* (**SSO-T1, SSO-T2 and SSO-T3**). **B**: chloroplastic OCTase activity in non-transformed (**S**) and transgenic (**SSO-T3**) tobacco plants. Enzyme assays were performed according to Ceriotti with increasing concentrations of phaseolotoxin, as indicated in the figure

2.2. EFFECT OF PHASEOLOTOXIN ON OCTASE ACTIVITY IN TRANSGENIC PLANTS.

The susceptibility to phaseolotoxin of the OCTase present in both transgenic and control plants was evaluated. It was determined that 96% of OCTase activity present in control plants is inhibited by a concentration of 37 ng/ml of phaseolotoxin and 98 % by a concentration of 375 ng/ml (figure 1). In contrast, it was found that transgenic plants containing the SSU-OCTase construction showed an OCTase inhibition of only 13 to 62 % when treated with a concentration of 37 ng/ml of phaseolotoxin. Assays using a 10-fold higher dose of phaseolotoxin produced similar results. The degree of OCTase inhibition was different for each independent transgenic plant, being lower for those having higher levels of OCTase activity. These results confirm that the bacterial OCTase is functionally expressed in transgenic plants and that the percentage of phaseolotoxin inhibited OCTase activity corresponds to the endogenous plant enzyme.

2.3. EFFECT OF PHASEOLOTOXIN AND *P. S. PHASEOLICOLA*-INFECTION ON SSU-OCTASE TRANSGENIC PLANTS.

Application of phaseolotoxin to leaves of *P. s.* pv. *phaseolicola* non-host species has been reported to result in the accumulation of ornithine, accompanied by leaf chlorosis [9]. To evaluate whether the presence of the phaseolotoxin-insensitive OCTase could have a protective effect on transgenic tobacco plants, phaseolotoxin was applied to leaves of transgenic and control plants. It was observed that control tobacco plants developed chlorotic halos when treated with 28 picomoles of toxin; These chlorotic halos were similar to those normally observed for bean plants . In contrast, no chlorotic symptoms appeared in SSU-OCTase transgenic plants when treated with the same or a 10-fold higher concentration of phaseolotoxin (see figure 1).

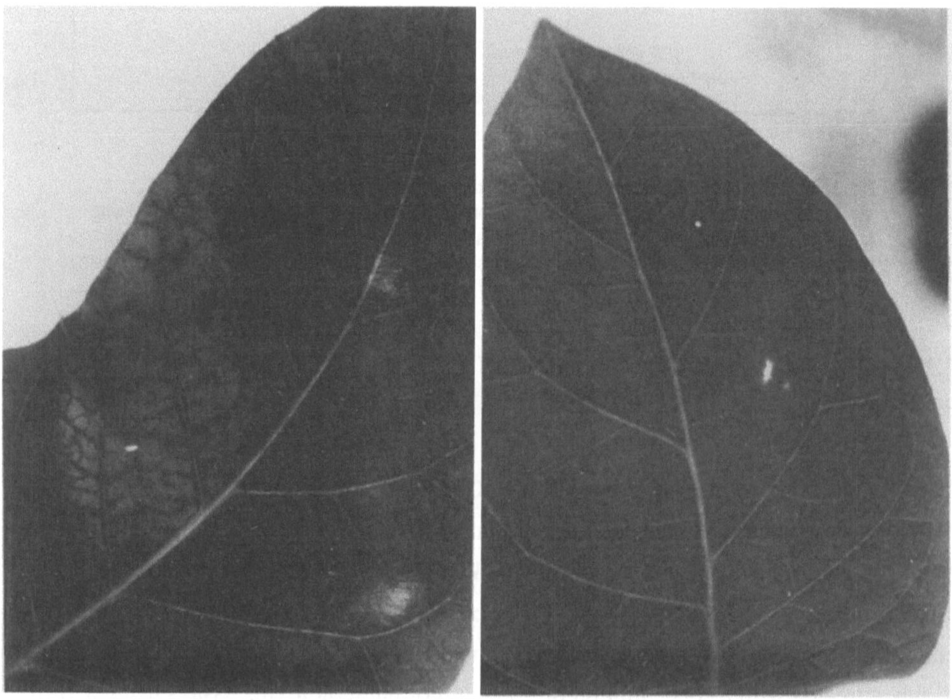

Figure1. Effect of phaseolotoxin on SSU-OCTase transgenic plants. Transgenic (right) and control tobacco plants were treated with 28 pmoles of toxin applied by puncturing young leaves with a needle. Simptoms were scored 72 h after the treatment.

To assess the role of phaseolotoxin on the virulence of *P.s.* pv. *phaseolicola*, we first evaluated different inoculation conditions to search for those in which infection of tobacco plants by this bacteria could occur. A compatible interaction between *P.s.* pv. *phaseolicola* and tobacco plants was achieved by growing inoculated plants in complete darkness immediately after infection. Using these conditions the F1 progeny of the SSO-T1 and SSO-T3 plants, including both transgenic plants expressing the bacterial OCTase and non-transformed sibling plants, were inoculated with *P.s.* pv. *phaseolicola*. Infected non-transformed tobacco plants developed chlorotic watery lesions that in some plants resulted in systemic infection. This plants did not elongate in response to darkness and evetually died (figure 3). The symptoms obtained for control plants closely resemble those observed for bean plants infected with *P.s.* pv. *phaseolicola*. In contrast, transgenic plants expressing the resistant OCTase inoculated under the same conditions showed no watery lesions or systemic infection and elongated normally in response to darkness (figure 3). Inoculated transgenic plants showed initially small necrotic lesions which later developed into larger regions of cell dead (figure 3). These results suggest that plants that do not suffer the effects of phaseolotoxin are less prone to systemic infection by *P.s.* pv. *phaseolicola* and are able to develop a hypersensitive response.

3. DISCUSSION

The characteristic symptoms of the halo blight disease, leaf chlorosis and systemic infection have been suggested to be the direct effect of phaseolotoxin. The major target of this toxin is OCTase as shown by several *in vitro* studies using different sources of the enzyme [9,11] and by the *in planta* accumulation of ornithine in chlorotic areas [28]. However, there have been several reports suggesting that phaseolotoxin may have other biochemical targets, such as chlorophyll biosynthesis, which could explain chlorosis induction [29]. In this study, it was shown that transgenic plants expressing a phaseolotoxin-insensitive OCTase do not show chlorotic symptoms when treated with high concentrations of phaseolotoxin, thus demonstrating that at least with regard to symptom development, OCTase is the only target of phaseolotoxin and that chlorosis is a consequence of the inhibition of the endogenous plant OCTase.

The importance of toxin production on the virulence of several phytopathogenic bacteria has been clearly demonstrated. The involvement of toxins as key pathogenicity elements has only been demonstrated in the case of the tabtoxin produced by *P.s.* pv *tabaci*. In this case it has been shown that transgenic tobacco plants resistant to tabtoxin developed a hypersensitive response to *P.s.* pv *tabaci*, which has normally a compatible interaction with tobacco plants [29]. In this study it was shown that when tobacco plants are kept in darkness after being infected by *Ps. pv phaseolicola* a compatible interaction is established, producing symptoms that closely resemble those observed in bean plants. Transgenic tobacco plants expressing a phaseolotoxin-insensitive OCTase showed a hypersensitive response when infected under the conditions for which control plants had a compatible interaction. These results show that phaseolotoxin or other OCTase inhibitors may play an important role in the development of a compatible interaction. Preliminar results with transgenic bean plants that express the *P.s.* pv *phaseolicola* OCTase indicate that very similar protection agains phaseolotoxin and *P.s.* pv *phaseolicola* can be obtained in this plant specie.

Figure 2. Bacterial inoculation of SSU-OCTase transgenic plants. Control (right) and transgenic (left) plants were inoculated on the 2 and 3 leaves with *P.s. phaseolicola.* Symptoms were scored 4 days after inoculation. The arrow indicates hypersensitive response in transgenic plants.

In summary our data demonstrate that OCTase is the target of phaseolotoxin, that expression of a phaseolotoxin-insensitive OCTase in transgenic plants is a valid approach to decrease the damage that *P.s.* pv *phaseolicola* produces on its host .

4. REFERENCES

1. Mitchell, R.E., 1984, The relevance of non-host-specific toxins in the expression of virulence by pathogens. Ann. Rev. Phytopathol. 22:215-245.

2. Mitchell, R.E., 1976, Isolation and structure of a chlorosis-inducing toxin of Pseudomonas phaseolicola. Phytochemistry, 15: 1941-1947.

3. Moore, R.E., Niemczura, W.P., Kwok, O.C.H. and Patil, S.S.,1984, Inhibitions of ornithine carbamoyltransferase from *Pseudomonas syringae* pv. *phaseolicola.* Revised structure of phaseolotoxin. Tetrahedron Lett. 25: 3931-3934.

4. Panopoulos, N.J., Lindgren, P.B., Peet, R.C., Thom, R.F., Hickman, M., Gies, D.R. and Willis, D.K., Molecular analysis of pathogenicity and virulence in *Pseudomonas syringae* pathovars, In Szalay, A.A. and RP. Legocki (ed), 1985, Advances in molecular genetics of the bacteria-plant interaction, Cornell University Publishers, Ithaca, NY, USA, pp. 188-192.

5. Patil, S.S., Kolattukudy, P.E. and Dimond, A.E., 1970, Inhibition of ornithine carbamoyltransferase from bean plants by the toxin of *Pseudomonas phaseolicola.* Plant Physiol. 46: 752-753.

6. Shargool, P.D., Jain, J.C. and McKay, G, 1988, Ornithine biosynthesis, and arginine biosynthesis and degradation in plant cells, Phytochemistry, 27: 1571-1574.

7. Mitchell, R.E. and Bieleski, R.L., 1977, Involvement of phaseolotoxin in halo blight of beans. Transport and conversion to functional toxin. Plant Physiol., 60: 723-729.

8. Templeton, M.D., Sullivan, R.E. and Shepherd, M.G., 1985, The inactivation of ornithine transcarbamoylase by N∂-(N'-sulpho-diaminophosphinyl)-L-ornithine, Biochem. J., 228: 347-352.

9 Ferguson, A.R. and Johnston, J.S., 1980, Phaseolotoxin: chlorosis, ornithine accumulation and inhibition of ornithine carbamoyltransferase in different plants. Physiological Ann. Pathol., 16: 269-275.

10. Hartman, C.L., Secor, G.A., Venette, J.R. and Albaugh, D.A., 1985, Response of bean calli to filtrate from Pseudomonas syringae pv. phaseolicola and correlation with whole plant disease reaction. Physiol. Mol. Plant Pathol. 28: 353-358.

11. Staskawicz, B.J. and Panopoulos, N.J., 1979, A rapid and sensitive microbiological assay for phaseolotoxin, Phytopathology, 69: 663-666.

12. Staskawicz, B.J., Panopoulos, N.J. and Hoogenraad, N.J., 1980, Phytotoxin-insensitive ornithine carbamoyltransferase of Pseudomonas syringae pv. phaseolicola. Basis for immunity to phaseolotoxin. J. Bacteriol. 142: 720-723.

13. Templeton, M.D., Sullivan, P.A. and Shepherd, M.G., 1986, Phaseolotoxin-insensitive L-ornithine transcarbamoylase from Pseudomonas syringae pv. phasseolicola. Physiol. Mol. Plant Pathol. 29: 393-403.

14. Mosqueda, G., Van den Broeck,G, Saucedo, O., Bailey, A., Alvarez-Morales, A. and Herrera-Estrella, L., 1990, Isolation and characterization of the gene from Pseudomonas syringae pv. phaseolicola encoding the phaseolotoxin-insensitive ornithine carbamoyltransferase, Mol. Gen Genet.

15. Sanford, J.C. and Johnston, S.A., 1985, The concept of parasite-derived resistance-deriving resistance genes from the parasite's own genome, J. Theor. Biol. 113: 395-405.

16. Maniatis, T., Fritsch, E.F. and Sambrook, J., 1982, Molecular cloning. A laboratory Manual. Cold Spring Harbor Laboratory, Cold Spring Harbor, NY.

17. Teeri, T.H., 1988, The study of gene regulation and protein targeting in plant cells using gene fusion techniques. Ph. D. thesis, University of Helsinki, Helsinki Finland.

18. Horsch, R.B., Fry, J.E., Hoffmann, N.L., Eichholtz, D., Rogers, S.G. and Fraley, R.T., 1985, A simple and general method for transferring genes into plants. Science. 227: 1229-1231.

19. Taylor, A.A. and Stewart, G.R., 1981, Tissue and subcellular localization of enzymes of arginine metabolism in Pisum sativum, Biochim. Biophys. Res. Comm. 101: 1281-1289.

20. Patil, S.S., Tam, L.Q. and Sakai, W.S., 1972, Mode of action of the toxin from Pseudomonas phaseolicola. I. Toxin specificity, chlorosis and ornithine accumulation. Plant Physiol. 42: 803-807.

21. Anzai, H., Yoneyama, K. and Yamaguchi, I., 1989, Transgenic tobacco resistant to a bacterial disease by the detoxification of a pathogenic toxin. Mol. Gen. Genet. 219: 492-494.

A Universally Conserved Vital Protein Revealed By Victorin Binding

D. Loschke, L. Tomaska, H. Chen, Z. Hong, B. Rolfe and D. Gabriel,* Plant-Microbe Interaction Group, RSBS, Australian National University, ACT, Australia and *Dept. of Plant Pathology, University of Florida, USA.

Abstract:

Victorin, one of the most potent and specific plant toxins known, can be derivatized with a radioactive tag without destroying its toxicity or specificity. The toxin binds via a covalent bond with specific proteins in oats. We have used victorin C labelled with either [^{125}I] or [^{35}S] to identify victorin-binding proteins with molecular weights of 98, 69, 45, 26, 21 and 15 kD and doublets at 14 and approximately 12.5 kD in both resistant and susceptible oats. We show for the first time a difference between resistant and susceptible plants *in vitro* for the binding properties of one victorin-binding protein. We have also found that all plants contain a set of proteins which specifically bind victorin and which are similar to those in oats. Surprisingly, the specific binding of victorin revealed a conserved low molecular weight protein which is found in all cells, both eukaryotic and prokaryotic and which is presumably a component of some vital physiological function.

INTRODUCTION:

Since its discovery in 1948, the fungal toxin victorin has aroused great interest because of its unusual connection to plant disease resistance genes. Victorin was discovered as a result of the sudden appearance of a new disease in oats called victoria blight, named after the parent cultivar, Victoria [1]. This disease arose after the introduction of a previously unknown susceptibility which was genetically linked to resistance against crown rust of oats (Pc-2 gene), a completely different disease. Victoria blight is caused by a toxin (named victorin) secreted by a previously unknown soil fungus, *Cochliobolus victoriae* [2,3]. Toxin susceptibility was traced to a dominant gene in oats which was named the Vb locus . Breeding studies indicated that Pc-2 and Vb were closely linked [2]. Despite an intense effort to separate rust resistance from toxin sensitivity, the linkage has never been broken [4,5,6]. The strong possibility exists that the two traits are encoded by the same gene or are members of a complex locus.

Victorin has been purified from fungal cultures as a predominant form, victorin C, and four closely related forms. All forms are chlorinated, partially cyclic pentapeptides [7,8]. Recently, Wolpert and Macko derivatized victorin C with [^{125}I] Bolton-Hunter reagent and used it to identify a 100 kD protein as a specific victorin binding protein [9]. They reported that *in vivo*, the 100 kD was labelled in the susceptible plant but not in the resistant plant. By contrast, under *in vitro* conditions, the 100 kD protein was labelled in both types of plants [9]. Subsequently, Akimitsu et al., using polyclonal antibodies raised against pure victorin C, found 100 and 45 kD victorin-binding proteins *in vivo* and 100, 65 and 45 kD victorin-binding proteins *in vitro*. They found no difference between resistant and susceptible plants[10].

In this study we have used victorin C labelled with either [^{125}I] or [^{35}S] to identify victorin-binding proteins in both resistant and susceptible oats and also in other organisms.

Toxin preparation:

Victorin C was purified from culture filtrates of *Cochliobolus victoriae* [11] and was labelled with [^{125}I]-Bolton-Hunter reagent [9] or with [^{35}S]-protein labelling reagent (a Bolton-Hunter analogue) [12]. The [^{125}I]-victorin C resulted in identical binding patterns as [^{35}S]-

E. W. Nester and D. P. S. Verma (eds.),
Advances in Molecular Genetics of Plant-Microbe Interactions, 587–591.

victorin C except that the sulfur label yielded much sharper, clearer bands, showing in several cases distinct doublets where the iodine label showed only one large blurred band. All radiographic imaging was done with a Molecular Dynamics model 400 PhosphorImager.

Specificity of radioactive victorin binding:

The specificity of labelled victorin was verified with competition experiments by comparing the labelling of oat plant extracts incubated with [^{35}S]-victorin C in the presence or absence of unlabelled victorin C. As an additional control, oat extract was incubated with the product of a mock labelling reaction in which only the labelling reagent was present (no victorin). For all cases (excepting the mock reaction), a large excess of labelled victorin was included so that all potential binding sites could be labelled. In all experiments, only reactions containing labelled victorin C with oat extract showed labelled protein bands. Presence of excess unlabelled victorin blocked binding of labelled victorin, and there was no binding of radioactivity to oat proteins with the labelling reagent alone. Examination of coomasie stained gels showed that the radioactive victorin-binding protein bands were embedded in the midst of numerous more prominent protein bands which do not bind victorin.

***in vivo* Binding studies:**

Using either [^{35}S] or [^{125}I]-victorin C, we could not reproduce the reported specific binding of [^{125}I]-victorin C to a 100 kD protein specific to susceptible oats [9]. Using the published *in vivo* procedure [9], our preparation of [^{125}I]-victorin C labelled proteins of 98, 45, 26, 21, and 14 kD in both susceptible and resistant plants.

The [^{35}S]-victorin C gave the same pattern as with [^{125}I]-victorin C except that we also saw a 69 kD protein. The experiments were repeated many times showing always that the 45 kD was labelled most strongly and that the resistant plant consistently bound more labelled victorin than did the susceptible plant for all proteins observed. In an attempt to mimic a natural interaction, we placed whole leaves cut at the base into solutions of [^{35}S]-victorin C. The label was taken up uniformly into the entire leaf by 20 minutes. However, after 2 and 4 hours of exposure, only the 45 kD protein was labelled intensely.

***in vitro* Binding studies:**

Oat extracts were prepared by grinding freshly frozen leaves of 7 to 8 day old plants in liquid nitrogen, mixing with buffer A (50mM HEPES-KOH, pH7.5, 0.3M sucrose, 20mM 2-mercaptoethanol, 10mM KCl, 2mM EDTA, 1mM phenylmethylsulfonyl flouride) (3:1 of Buffer A in mls to leaf weight in grams), straining through cheese cloth and pelleting debris by centrifugation at 20,000g for 15 min. Microsomal membranes were separated from the extract by centrifugation at 100,000g for 1 hour, washed by resuspension in 10 mls of fresh Buffer A and repelleted. Washed microsomal membrane pellets were resuspended in minimal volumes of buffer A (1:1 of estimated pellet volume to buffer volume) and compared in labelling studies with the remaining portion of the oat extract (the supernatant) from which they were originally derived, hereafter called soluble oat proteins.

Both microsomal membrane and soluble oat proteins were incubated at room temperature for two hours with a constant amount of labelled victorin C. Results were analyzed on both IEF (Pharmacia) and SDS PAGE (Gradipore) gels. When washed microsomal membranes from both resistant and susceptible oats were compared with their respective soluble oat proteins, almost all of the victorin binding activity was found in the soluble protein fraction with only a trace in the microsomal membranes.

By inspecting different exposure intensities of both [^{125}I] and [^{35}S] labelled proteins in SDS PAGE gels, 10 proteins of oats have been identified as specific victorin-binding proteins: 98 kD, 69 kD, 45 kD, 26 kD, 21 kD and 15 kD and doublets at 14 kD and 12.5 kD. When examined on IEF gels, these labelled proteins segregated into seven major bands with six to eight lesser bands depending upon the intensity of exposure. Comparison of IEF and SDS PAGE gels and SDS gels of isolated IEF radioactive bands shows that on IEF gels, the victorin-binding proteins are associated with one or more other proteins in different complexes or subcomplexes, each with different net isoelectric points.

No pronounced difference between resistant and susceptible plants was found under ordinary binding conditions except that SDS gels revealed a slightly stronger binding of the lowest molecular weight proteins in resistant plants over susceptible plants.

The victorin-binding proteins are quite stable, and the binding reactions are very fast. Labelling of all but the 98 kD protein is almost immediate. After 10 hours at room temperature, the 21 kD protein is lost, apparently broken down into a 14 kD form. The stability of the proteins are such that a single freeze-thaw cycle does not apparently damage the binding activity, and a 4 minute boiling step reduces subsequent binding activity to approximately a third of the original.

Serial competition studies:

Serial competition studies on soluble oat proteins from both resistant and susceptible plants show differences in binding properties for the low molecular weight proteins, particularly the 14 kD protein (fig.1 below). The 98 and 69 kD proteins behave essentially the same way in resistant and susceptible plants and serve as useful internal controls. However, in the case of the 14 kD protein, the [^{35}S]-victorin C signal persists to a 10 fold greater concentration level of competing unlabelled victorin C in the susceptible plant as compared to the resistant plant. This is the first reported difference in *in vitro* victorin binding between a resistant and susceptible plant. Close examination of the low molecular weight region shows that there is a protein at 15 kD, a doublet at 14 kD and a doublet at approximately 12.5 kD. These details do not appear until a higher concentration of competing unlabelled victorin is present.

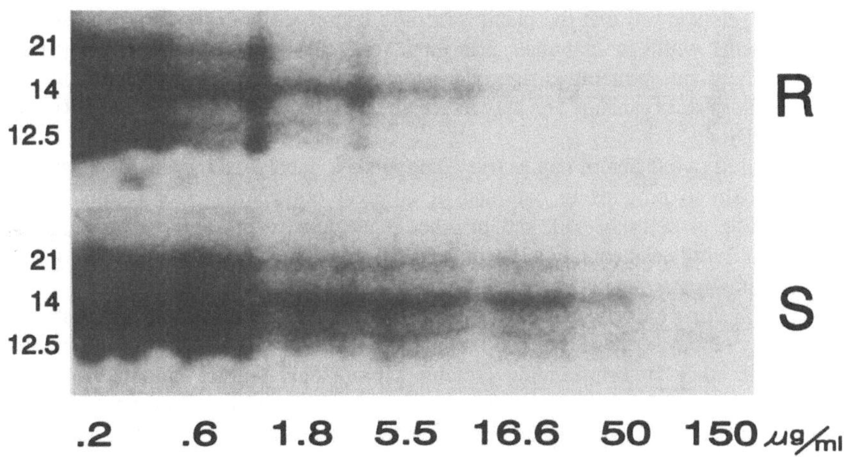

Fig. 1

Victorin-binding activity in other eukaryotes:

Soluble proteins of a number of other species were prepared exactly as described for oats. To our surprise, specific victorin-binding proteins were found in all organisms tested. In all cases binding was specific because binding of [^{35}S]-labelled victorin C could be completely blocked by an excess of unlabelled victorin C. A comparison of victorin-binding patterns in (from left to right) oat, *Arabadopsis thaliana*, tobacco, mouse liver, human fibroblasts, *Saccharomyces pombe, Saccharomyces cerevisiae* and *Escherichia coli* is shown in fig.2 below.

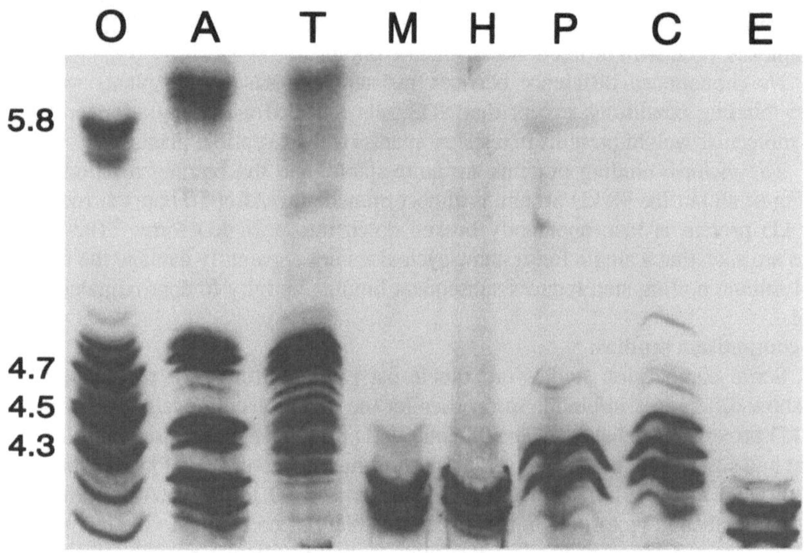

Fig. 2

The pattern on IEF gels is remarkably similar in the plants even between the monocotyledon, oats, and the two dicotyledons, *Arabidopsis* and tobacco. The pattern is clearly different in mammals and fungi, but the binding is distinct and strong. The bacterial pattern looks much like the mammalian pattern except that it is shifted to a slightly lower isoelectric point. SDS gels showed that the plants generally conserved both the large and small victorin-binding proteins while in mammals and fungi, only the smallest protein comparable to the smallest of the plant proteins is rigorously conserved. The smallest proteins bind the signal most intensely in all organisms and may therefore be present in the greatest molar concentration.

DISCUSSION:

We used two different radioactive compounds to label victorin and investigate binding of the fungal toxin to proteins in oats and in different organisms. Our results show multiple victorin-binding proteins in oats and in other plant cells. In addition, we have shown that apparently all living cells contain a low molecular weight protein which specifically and strongly binds victorin.

Our *in vitro* experiments have shown that in oats, the 21 and 14 kD proteins bind the great majority (about 90%) of the total bound label. The 98 kD protein binds much less (about 10%) and the other victorin-binding proteins capture only a trace amount in relative terms. Based upon the strength of the signals and the reproducibility of the experiments, we suspect that the 98, 21 and 14 kD proteins are functional components of the cell (as opposed to being degradation products). Even though the 45 kD protein is very weak *in vitro*, its dominance *in vivo* suggests that it too may be functional in that form. For the *in vivo* case, we consistently find that the resistant plant binds more label uniformly to all of the binding proteins than does the susceptible plant and that there is no protein which uniquely binds labelled victorin *in vivo* in either genotype.

Does the difference in the *in vitro* binding properties of the 14 kD protein between resistant and susceptible oats suggest that the 14 kD protein may be the product encoded by the Vb and its allele the vb gene? This protein appears to associate as a member of a multiprotein complex. It is equally possible that the Vb product is not a victorin-binding protein but that its mutant form (Vb) affects the quaternary structure of the complex in such a way as to affect the binding accessibility of other members of the complex which do bind victorin.

Our working hypothesis for oats is that both genotypes contain physiologically functional and equivalent structures which contain the product of the Vb or vb alleles. The function of that structure is unknown but its conservation argues that it is a vital one. In the case of resistant plants, victorin can bind to members of the complex but does not seriously impair physiological functioning. In the case of the susceptible plant, the mutated form of one member of the complex, the Vb gene product, permits normal physiological function in the absence of victorin but causes disruption of function when victorin binds [13]. It is not required that the Vb gene product bind victorin to induce this effect since its mutant form may destabilize the structure so that function is blocked when other members of the complex bind victorin.

The victorin-binding proteins and associated proteins are conserved in plants. The patterns on IEF gels are remarkably similar between oats, a monocotyledon and the two dicotyledons, *Arabidopsis* and tobacco. Since the net isoelectric point of a protein complex is affected by its charge composition, we conclude that there is general similarity in the structures revealed in plants by victorin binding.

The universal conservation across taxonomic boundaries of a protein with a specific victorin binding site is unexpected. Structural conservation of a protein is partially enforced by an indispensable role for that protein and partially by that protein's association as a member of a multiprotein complex which serves some vital function. In this case the binding site is conserved and differences between IEF and SDS PAGE gels show that the binding proteins are associated with other extracted proteins . In the one single instance provided by nature in which a mutation results in a detectable physiological response to victorin by living cells (oat cells with the Vb locus), that response is rapid cell death.

REFEFENCES

1. Meehan, F. and H.C. Murphy. 1946. A new *Helminthosporium* blight of oats. Science **104:** 413-414.
2. Litzenberger, S.C. 1949. Nature of susceptibility to *Helminthosporium victoriae* and resistance to *Puccinia coronata* in Victoria oats. *Phytopathology* 39:300-318.
3. Meehan, F. and H.C. Murphy. 1947. Differential phytoxicity of metabolic byproducts of *Helminthosporium victoriae*. Science **106:** 270-271.
4. Wheeler, H. E. and H. H. Luke. 1954. Mass screening for disease-resistant mutants in oats. *Science* 122:1229.
5. Luke, H. H. and A. T. Wallace. 1969. Sensitivity of induced mutants of an *Avena* cultivar to victorin at different temperatures. *Phytopathology* 59:1769-1770.
6. Rines, H.W. and H. H. Luke. 1985. Theor. Appl. Genet. **71:** 16-21.
7. Wolpert, T. J., V. Macko, W. Acklin, B. Jaun, J. Seibl, J. Meili and D. Arigoni. 1985. Structure of victorin C, the major host-selective toxin from *Cochliobolus victoriae*. *Experientia* 41:1524-1529.
8. Wolpert, T. J., V. Macko, W. Acklin, B. Juan and D. Arigoni. 1986. Structure of the minor host-selective toxins from *Cochliobolus victoriae*. Experientia **42:** 1296-1299.
9. Wolpert, T. J. and V. Macko. 1989. Specific binding of victorin to a 100-kDa protein from oats. *Proc. Natl. Acad. Sci. USA* 86:4092-4096.
10. Akimitsu, K., L.P. Hart, J.D. Walton and R. Hollingsworth. 1992. Covalent binding sites of victorin in oat leaf tissues detected by anti-victorin polyclonal antibodies. Plant Physiol. **98:** 121-126.
11. Mayama, S., T. Tani, T. Ueno, S. L. Midland, J. J. Sims and N. T. Keen. 1986. The purification of victorin and its phytoalexin elicitor activity in oat leaves. Physiol Mol Plant Pathol **29:** 1-18.
12. Assoian, K., P. Blix, A. Rubenstein and H. Tager. 1980. Iodotyrosylation of peptides using tertiary-butyloxycarbonyl-L-[^{125}I] iodotyrosine N-hydroxysuccinimide ester. Anal. Biochem. **103:** 70-76.
13. Gabriel, D. W., D. C. Loschke and B. G. Rolfe. 1988. Gene-for-gene recognition: the ion channel defense model. In *Molecular Genetics of Plant-Microbe Interactions*, ed. R. Palacios, D. P. S. Verma, p 3-14. St. Paul: APS Press.

Nitrogen fixation of Azorhizobium in artificially induced root para-nodules in wheat*

Ting-Wei Chen
(Soils and Fertilizer Institute,Chinese Academy of Agricultural Sciences, Beijing 100081, P.R.China)

S.Scherer
(Bacteriological Institute, Faculty of Agriculture, Technical University Munchen, D-8050 Freising,Germany)

and P.Böger
(Lehrstuhl für Physiologie und Biochemie der Pflanzen, Universitat Konstanz,D-7750 Konstanz,Germany)

Abstract

Nodule-like structures(para-nodules) can be induced in wheat roots by low concentrations of plant growth hormone (2,4-D).Growth of nodulated wheat plants was not affected by 2,4-D treatment. Infection of these para-nodules by adding oxygen-tolerant Azorhizobium caulinodans resulted in massive proliferation of bacterial cells in the intercellular spaces as well as inside the para-nodular cells. Para-nodules colonized by A.caulinodans not only reduced acetylene (about 2-4 n mol C_2H_4/plant/h), but transferred fixed nitrogen to the plant, contributing up to 16-23% of the nitrogen budget of infected wheat.

Key words:Azorhizobium.nitrogen fixation.nodules.wheat.

* *Project supported by Premier Foundation of PRC.*

E. W. Nester and D. P. S. Verma (eds.),
Advances in Molecular Genetics of Plant-Microbe Interactions, 593–606.

Ⅰ.Introduction

Considerable research has been directed towards the Rhizobium-legume symbiosis[1][2]. not only to understand the relevant molecular principles but also to establish the symbiosis in agriculturally important non-legumes. Reports published three decades ago indicated that plant hormones[3] may trigger the development of pseudo-nodular structures in non-legumes. These phenomena where reinvestigated recently[4][5][6] to promote artificial symbioses between rhizobia and non-legumes. Also. cell-wall degrading enzymes were found to induce nodule-like structures in grasses[7]. When rice was inoculated with a genetically altered Rhizobium,nodule-like structures were obsewed occasionally[8]. These artificially induced "nodular" structures, containing bacteria, may be called "para-nodules" [9]. It was first reported by Nie that pseudo-nodular structures, induced in wheat roots by 2,4-dichlorophenoxyacetic acid (2,4-D), were invaded by Rhizobium.However.no clear evidence supported that the rhizobia grew inside of the cells of the induced structure, and no nitrogen fixing activity was demonstrated. That Rhizobium indeed is invading pseudo-nodular cells of wheat roots has been shown recently[5][10][11], but no nitrogen fixing activity could be detected. To achieve nitrogen fixation in para-nodules of wheat,,we used different oxygen pressures and Azorhizobium caulinodans which was isolated from stem nodules of the tropical legume Sesbania rostrata[12][13]. This unusual species is able to fix nitrogen with high rates in liquid culture under comparatively high partial pressures of oxygen[14][15]. In this paper we report the research results on nitrogen fixation of Azorhizobium in induced root-nodules of wheat.

II.Material and methods

Organisms: Wheat (strain CA 8361. Chinese Academy of Agricultural Sciences. Beijing) and Azorhizobium caulinodans ORS 571 (strain 6465, LMG Culture Collection,Gent,Belgium) were used for the experiments. A. caulinodans ORS 571-R3 was re-isolated from an induced nodule of wheat and also for the experiments.

Induction of para-nodules in liquid culture: Seeds were surface-sterilized with Incidin (solution of formaldehyde. glyoxal. glutaraldehyde and ethanol produced by Henkel, Düsseldorf,Germany) for 15 min, followed by three washes with sterile water. After gemination (48-60 h on moist filter paper at 24℃ in the dark). seedlings were placed on a filter support in glass tubes (30mm × 200mm)containing 20ml of sterilized standard mineral medium without combined nitrogen. Plants were cultivated at 580 $\mu E\ m^{-2}s^{-1}$. When the seedlings had three leaves. 5 μM 2,4-D and Azorhizobium (5×10^{8} cels ml^{-1}) were added. Control plants received no additions or were treated with 2,4-D or with rhizobia only, 2,4-D and/or rhizobia were added to the mineral medium.

Measurement of acetylene reduction: 15-20 days after infection with rhizobia the plants were washed thoroughly,surface-sterilized for 5 min with Incidin and washed again with sterile water. The whole plant was placed into a reaction vessel (10% v/v acetylene in argon plus oxygen as indicated) and illuminated at 28℃. 400 $\mu E\ m^{-2}s^{-1}$. Acetylene reduction was followed by gas chromatography[16].

Electron microscopy: After 15-20days,para-nodules were fixed in 2.5% glutaraldehyde/25 mM potassium phosphate, pH 6.8,postfixed for 2h in 1% osmium tetroxide,dehydrated in acetone/propylene oxide series and embedded in Spurr's resin. Ultrathin sections were stained for 30 min in saturated uranylacetate/50% methanol and for 5-7 min in 0.02% lead citrate and analyzed with a Hitachi H-500 electron microscope.

^{15}N-dilution experiments. Plants were grown in sterilized 2 Kg soil/sand (1:1) culture containing a nitrogen-free standard growth medium. Irrigation was performed with sterile distilled water. At the three leaf stage, 100 ml of a 2,4-D solution (5 μM) containing Azorhizobium ORS 571 (5×10^8 cells ml^{-1}) were added. Either the original strain,A.caulinodans ORS 571 or ORS-571-R3, reisolated from infected wheat para-nodules were used. At the 5 leaf stage another 100 ml, and at the earing stage additionally 200 ml of 2,4-D solution were added. At the mature stage (approximately after 90 days) the plants were harvested, washed thoroughly, dried at 70°C for 48 h, weighed, ground and sieved. There after, 150 mg were for Kjeldahl analysis of total nitrogen or ^{15}N-analysis by mass spectroscopy. Labeled nitrogen fertilizer was applied as 33.3mg $(NH_4)_2SO_4$ per pot, including 3.35% ^{15}N. To calculate the fraction of plant nitrogen originating from air, the equations of the ^{15}N isotope dilution technique were used[17].

III.Results and discussion

In all of our experiments, induced para-nodules were visibe within 10 days after induction by 5 μM 2,4-D. The induced para-nodules have the shape of spheroid or taper and 1-2mm in diameter (Fig.1A.). In liquid culture, 20-30 para-nodules were formed per plant, while in soil culture more than hundred para-nodules developed per plant.In liquid culture, less roots were formed with 2,4-D when compared with the control. This difference disappeared in soil culture and the biomass production was not affected. Light-microscope studies revealed that the para-nodules initiated from the pericycle of the roots and contained vascular bundles. The DNA content of para-nodular tissue was approximately doubled when compared with control cells(18).This compares well with reports on polyploid nodular tissue of some Rhizobium-legume symbioses(19).

It was seen in electron-microscopical studies that rhizobia or azorhizobia first propagated in the intercellular space (Fig. 1B) and later invaded para-nodular cells (Fig. 1C). Two ways of infection of the nodules by rhizobia have been reported. In most cases[2] rhizobia infect legume roots through their root hairs, the curled tips become invaginated to form an infection thread. The thread, containing rhizobia and surrounded by a cellulose wall,grows into the root cortex both inter and intracellularly. Investigating numerous sections, we never found infection threads in wheat para-nodules(see also Bender et al.[11]).We do not know how Azorhizobium penetrates the prepara nodular tissue;in legumes, infection occasionally proceeds without infection threads. Chandler[20] reported, that infection of the legume Arachis hypogaea occurs only at the sites of emerging lateral roots, the rhizobia entering the root at the junction of the root hair and the epidermal and cortial cells. The rhizobia first distribute intercellularly via the middle lamellae and, finally, enter the cortical cells through the structurally altered cell wall. Arachis may serve as a model to elucidate the details of the infection process of auxin-induced para-nodules in wheat roots. Similar to bacteroids of nodules, the rhizobia inside the wheat para-nodular cells develop poly-β-hydroxybutyrate granules and exhibit a clear space between the cytoplasmic membrane and the cytoplasm of the host cell (Fig.ID). Occasionally a peribacteroid membrane is found.[18]

The identity of bacteria inside the para-nodules was confirmed by re-isolation and testing for Koch's postulates. In control experiments , on endophytic bacteria could be detected in the wheat plants.

In accordance with our previous findings (Chen et al.[8] [10])and those of others[11], we could not measure any acetylene reduction activity of para-nodules when plants were infected with Rhizobium or Bradyrhizobium. Since lack of nitrogen fixation could be due to the marked oxygen sensitivity of rhizobial nitrogenase and the absence of oxygen scavenging leghemoglobin in wheat nodules, an isolate of Azorhizobium was used which colonizes stem nodules of the tropical legume Sesbania rostrata. This species fixes nitrogen with high rates in pure liquid culture under relatively high oxygen concentrations[14] [16] [21]. We applied this strain for inoculation and gain positively nitrogenase activity. As shown in Tab. 1, ;acetylene reduction was observed, even under 18% oxygen. but was markedly higher under lower oxygen tensions.Under reduced oxygen pressure, nitrogenase activity was induced during the assay. As demonstrated by the control experiments (see Tab. 2-3), aceatylene reduction is mainly due to the rhizobia localized inside the para-nodules.The activity was rather low when compared to the nitrogen fixation of legumes, but was at least one order of magnitude above background.The para-nodulated wheat plants obtained as yet contain $0.4\text{-}4\times10^{8}$bacterial cells per plant, which is in the range of 0.05% of the number found per soybean plant[22]. Under 4.5% oxygen, the maximum rate of acetylene reduction of infected para-nodulated wheat plants was approximately 4 n mol plant^{-1}h^{-1},while the rates of soybeans are as high as 10-200 μ mol plant^{-1} h^{-1}. When these figures are referred to the number of rhizobial cells per plant, the rate of acetylene reduction of para-nodulated wheat plants in our artifcial incubation system (Tab.1-3) is lower but in the same order of magnitude as found for legumes.

Induction of nitrognase activity in nodules under the conditions of Tab. 1-3 may.but not necessarily does. imply that a "symbiosis" betweem host plant and Azorhizobium has been established. It might be argued that the maximum rate of acetylene reduction is induced subsequently since it is observed only under low oxygen concentrations. Furthermore, during incubation of the plant in the acetylene reduction assay Azorhizobium may have multiplied. Therefore , a transfer of fixed nitrogen to the wheat plant in situ cannot be inferred from these experiments.We tested whether rhizobia actually contribute to the nitrogen balance of para-nodulated wheat plants under more natural conditions, i.e. in soil culture. In pot experiments with ^{15}N-labeled fertilizer (Tab. 4) we measured the origin of plant nitrogen by the ^{15}N isotope dilution technique[17], using non-nitrogen-fixing control plants which were not inoculated with diazotrohs, but otherwise treated identically. When Azorhizobium ORS 571 was applied to the soil without 2,4-D, an average of 11.6% of plant nitrogen originated from air. Apparently. this is due to the nitrogen-fixing activity of rhizobia in the ecosystem which may be associated with microaerobic niches developing in well-watered soil (compare the Azospirillum -rhizosphere association as reported by Boddey and Dobereiner[23] as well as New and Kennedy[24]).This figure increased to 28%, when para-nodules were induced by 2,4-D.The strain Azorhizobium ORS 571 -R3 was re-isolated from wheat para-nodules and used to re-infect plants.This isolate contributed even more to the nitrogen balance of wheat.Tab.2 shows that an average of as high as 39.6% of the nitrogen of wheat plants is derived from air.After subtracting the control value (plants inoculated with Azorhizobium but without

induced para-nodules.Tab.4), about 16-23% of the nitrogen budget can be attributed to the artificially generated association between Azorhizobium caulinodans ORS 571-R3 and its host. It is not clesr which mechanism of carbon transfer from the host plant to Azorhizobium is in action. Conceivably, Azorhizobium merely uses carbohydrates leaking from 2,4-D penetrated para-nodular cell -membranes.

In legumes, one of the major characteristics of the Rhizobium/ legume symbiosis is its symbiont-host specificity[1] [25].In case of induced para-nodules, however,no spcificity seems to exist since not only Azorhizobium, but also different Rhizobium, Bradyrhizobium and Azospirillum species entered the para-nodules (data not shown). In our experiments, sterile growth media were used. Under natural comditions, however, not only diazotrophs will enter the roots but probably also other soil bacteria. Before any far-reaching speculations can be made, the performance of the Azorhizobium -wheat "symbiosis" established under our artificial conditions has to be investigated in more natural environments with a variety of soil bacteria present. To achieve positive results under these conditions, a genetically engineered Azorhizobium caulinodans may be useful which is able to produce significant amounts of indole -acetic acid. Since the molecular biology of Azorhizobium is currently studied (for references see De Bruijn et al.[26].Ratet et al.[27]. some tools necessary to reach this goal may soon be available.

Acknowledgements: The assistance of XIE Ying-xian, CHEN Wuan-hua, YU Dai-guan and HAN Kai in Chinese Academy of Agri. Sciences, is gratefully acknowledged. Part of the experimental work of CHEN TW was done during a leave in Konstanz and supported by Volkswagenstiftung. Hannover. Germany.

References

(1) Long, S.R., Cell, 56(1989), 203-214.

(2) Rolfe, B.G.& Gresshoff, P.M., Annu. Rev. Plant Physiol. plant Mol. BIol, 39(1988), 297-319.

(3) Arora,N.,skoog,F.& Allen,O.N.,Am.J.Bot., 46(1954),601-613.

(4) Nie, Y.F., Nature Journal (In chinese), 6(1983), 5:326.

(5) Chen,T.W. et al.,Nature Journal(In Chines),ll(1988), 3:163

(6) Tchan,Y.T.& Kennedy,I.R.,Agricultural Sciences(Mel-bourne), 2(1989),57-59.

(7) AL-Mallah, M.K. et al., J.Exp.Bot., 40(1989), 473-478.

(8) Rolfe, B.G.& Bender,G.L.,in Nitrogen fixation: Achievements and objectives(Eds.Gresshoff,P.M. et al.),Chapman and Hall, New York, London, 1990, pp.779-780.

(9) Kennedy,I.R.et al.,Trans.14th Int. Congr. Soil Sci.(Kyoto)III (1990),146.

(10) Chen,T.W.,Advances In Nitrogen Fixation For Non-leguminus Crops, China Agricultural Science and Technology Press, Beijing, 1989.

(11) Bender,G.L. et al. , in Nitrogen Fixation: Achievements and Objectives (Eds.Gresshoff, P.M. et al.), chapman and Hall, New York, london, 1990, pp. 825.

(12) Dreyfus, B.et al., Int.J. Syst. Bacteriol, 38(1989), 89-98.

(13) Tsien, H.C. et al., J.Bacteriol. 156(1983), 888-897.

(14) Dreyfus,B.L. et al.,Appl.Environ.Microbiol.,45(1983),711-713.

(15) De Bruijn, F.J., in plant- Microbe Interactions, Vol. III (Eds.Kosuge, T.and Nester E.W.), McGraw Hill, New York, 1989, pp.457-493.

(16) Weisshaar, H.& Boger,P.,Arch.Microbiol.,136(1983), 270-274.

(17) Rennie, R.J.& Rennie,D.A.,Can.J. Microbiol.,29(1983),1022-1035.

(18) Yu,D.G. et al., In: Advances In Nitrogen Fixation For Non -leguminous Crops(Ed. T.W.Chen), China Agricultural Science and Technology Press, Beijing, 1989, pp.150-157.

(19) Allen, O.N.& Allen, E.K., The leguminoceae, The University of Wisconsin Press, Wisconsin, 1981,pp.63.

(20) Chandler, M.R., J.Exp. Bot., 29(1978), 749-755.

(21) Gebhardt, C.,et al., J.Gen.Microbiol., 130(1984), 843-848.

(22) Bergersen,F.J.,Root nodules of legumes: Strure and Function, Wiley & Sons Ltd., New York, 1982, pp.33-39

(23) Boddey, R.M. & Döbereiner, J., Plant and Soil,108(1988),53-65.

(24) New, P.B.& Kennedy, I.R., Microb.Ecol., 17(1989), 299-309.

(25) Young,J.P.W.& Johnston,A.W.B.,Trends Ecol.Evol.,4(1989), 341 -349.

(26) De Bruijn,F.L. et al.,in Nitrogen Fixation: Achievements and Objectives (Eds.Gresshoff, P.M. et al.), Chapman and Hall, New York, London, 1990, pp.33-44.

(27) Ratet, P.et al., Mol.Microbiol., 3 (1989), 825-838.

Table 1

Nitrogenase activity of induced wheat para-nodules infected with Azorhizobium under different oxygen pressures※

Experimental treatment of wheat plants	Cumulative acetylene reduction (n mol C_2H_4/plant) after		
	24h	48h	72h
Control 1: plant only, no bacteria,no 2,4-D	0	0	0
Control 2: 2,4-D only, no bacteria	0	0	0
Control 3: Azorhizobium ORS 571-R3 only,no 2,4-D	0	0	0
2,4-D, Azorhizobium ORS 571,18% oxygen, V/V	9.6	16.5	19.1
2,4-D, Azorhizobium ORS 571, 9% oxygen, V/V	9.9	55.9	83.5
2,4-D, Azorhizobium ORS 571, 4.5% oxygen,V/V	20.2	102.7	167.5

※ a) Roots were surface-sterilized by Incidin for 5 min.

b) Data are means of 5 plants.

Table 2

Nitrogenase activity of induced wheat para-nodules infected with Azrhizobium(1)※

Experimental treatment of wheat plants	Cumulative acetylene reduction (n mol C_2H_4/plant) after		
	18h	42h	61h
Control 1: plant only, no bacteria no 2,4-D	0	0	0
Control 2: 2,4-D only, no bacteria	0	0	0
Control 3: Azorhizobium ORS 571-R3 only, no 2,4-D	0.24	6.97	29.97
inoculated with Azorhizobium ORS 571-R3 and treated with 2,4-D	1.78	78.25	223.43

※ a) Root withhout surface-sterilized.

b) Measurement are under 14% oxygen.

c) Data are means of 6 plants.

Table 3

Nitrogenase activity of induced wheat para-nodules infected with Azorhizobium(2)*

Experimental treatment of wheat plants	Cumulative acetylene reduction (n mol C_2H_4/plant) after		
	24h	48h	72h
Control 1: plant only, no bacteria,no 2,4-D	0	0	0
control 2: 2,4-D only, no bacteria	0	0	0
Control 3: Azorhizobium ORS 571-R3 only, no 2,4-D	0	0	7.53
Inoculated with Azorhizo bium ORS 571-R3 and treated with 2,4-D	0.14	13.27	61.85

* a) Roots were surface-sterilized by 75% alcohol for 1 min.
b) Measurement are under 14% oxygen.
c) Data are means of 6 plants.

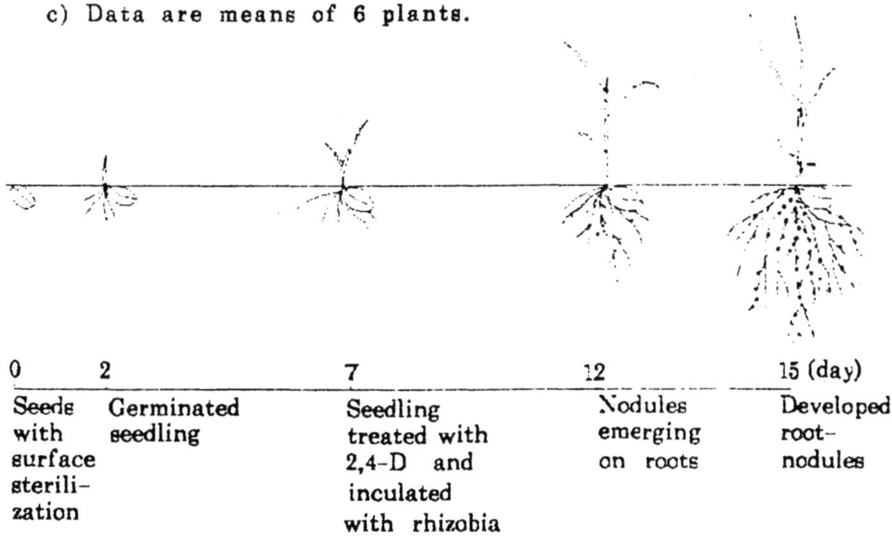

Schematic diagram of progress of induce nodulation in wheat

Table 4

Contribution of bacterial nitrogen fixation to the nitrogen balance of wheat plants with and without induced para-nodules ※

Experimental treatment	Origin of plant nitrogen fractions in percent derived from:		
	Air	fertilizer	soil
Control: no bacteria,no 2,4-D	0	1.41 ± 0.13	98.59 ± 0.13
Azorhizobium ORS 571, no 2,4-D	11.56 ± 14.52	1.25 ± 0.21	87.19 ± 14.31
Azorhizobium ORS 571, plus 2,4-D	28.00 ± 10.50	1.02 ± 0.15	70.98 ± 10.34
Azorhizobium ORS 571-R3, no 2,4-D	16.89 ± 9.76	1.17 ± 0.13	81.94 ± 9.62
Azorhizobium ORS 571-R3, plus 2,4-D	39.56 ± 9.61	0.85 ± 0.13	59.59 ± 27.66

※ Figures are expressed in percent of total plant nitrogen and represent the means(X ± SD)of 5 pots with 10 plants each.

SUBJECT INDEX

NAME INDEX

Current Plant Science and Biotechnology in Agriculture

1. H.J. Evans, P.J. Bottomley and W.E. Newton (eds.): *Nitrogen Fixation Research Progress*. Proceedings of the 6th International Symposium on Nitrogen Fixation (Corvallis, Oregon, 1985). 1985 ISBN 90-247-3255-7
2. R.H. Zimmerman, R.J. Griesbach, F.A. Hammerschlag and R.H. Lawson (eds.): *Tissue Culture as a Plant Production System for Horticultural Crops*. Proceedings of a Conference (Beltsville, Maryland, 1985). 1986 ISBN 90-247-3378-2
3. D.P.S. Verma and N. Brisson (eds.): *Molecular Genetics of Plant-microbe Interactions*. Proceedings of the 3rd International Symposium on this subject (Montréal, Québec, 1986). 1987 ISBN 90-247-3426-6
4. E.L. Civerolo, A. Collmer, R.E. Davis and A.G. Gillaspie (eds.): *Plant Pathogenic Bacteria*. Proceedings of the 6th International Conference on this subject (College Park, Maryland, 1985). 1987 ISBN 90-247-3476-2
5. R.J. Summerfield (ed.): *World Crops: Cool Season Food Legumes*. A Global Perspective of the Problems and Prospects for Crop Improvement in Pea, Lentil, Faba Bean and Chickpea. Proceedings of the International Food Legume Research Conference (Spokane, Washington, 1986). 1988 ISBN 90-247-3641-2
6. P. Gepts (ed.): *Genetic Resources of* Phaseolus *Beans*. Their Maintenance, Domestication, Evolution, and Utilization. 1988 ISBN 90-247-3685-4
7. K.J. Puite, J.J.M. Dons, H.J. Huizing, A.J. Kool, M. Koorneef and F.A. Krens (eds.): *Progress in Plant Protoplast Research*. Proceedings of the 7th International Protoplast Symposium (Wageningen, The Netherlands, 1987). 1988 ISBN 90-247-3688-9
8. R.S. Sangwan and B.S. Sangwan-Norreel (eds.): *The Impact of Biotechnology in Agriculture*. Proceedings of the International Conference 'The Meeting Point between Fundamental and Applied *in vitro* Culture Research' (Amiens, France, 1989). 1990. ISBN 0-7923-0741-0
9. H.J.J. Nijkamp, L.H.W. van der Plas and J. van Aartrijk (eds.): *Progress in Plant Cellular and Molecular Biology*. Proceedings of the 8th International Congress on Plant Tissue and Cell Culture (Amsterdam, The Netherlands, 1990). 1990 ISBN 0-7923-0873-5
10. H. Hennecke and D.P.S. Verma (eds.): *Advances in Molecular Genetics of Plant–Microbe Interactions*. Volume 1. 1991 ISBN 0-7923-1082-9
11. J. Harding, F. Singh and J.N.M. Mol (eds.): *Genetics and Breeding of Ornamental Species*. 1991 ISBN 0-7923-1094-2
12. J. Prakash and R.L.M. Pierik (eds.): *Horticulture – New Technologies and Applications*. Proceedings of the International Seminar on New Frontiers in Horticulture (Bangalore, India, 1990). 1991 ISBN 0-7923-1279-1
13. C.M. Karssen, L.C. van Loon and D. Vreugdenhil (eds.): *Progress in Plant Growth Regulation*. Proceedings of the 14th International Conference on Plant Growth Substances (Amsterdam, The Netherlands, 1991). 1992 ISBN 0-7923-1617-7
14. E.W. Nester and D.P.S. Verma (eds.): *Advances in Molecular Genetics of Plant–Microbe Interactions*. Volume 2. 1993 ISBN 0-7923-2045-X

KLUWER ACADEMIC PUBLISHERS – DORDRECHT / BOSTON / LONDON

MIX
Papier aus verantwortungsvollen Quellen
Paper from responsible sources
FSC® C105338

If you have any concerns about our products,
you can contact us on
ProductSafety@springernature.com

In case Publisher is established outside the EU,
the EU authorized representative is:
Springer Nature Customer Service Center GmbH
Europaplatz 3, 69115 Heidelberg, Germany

Printed by Libri Plureos GmbH
in Hamburg, Germany